Links between Fibrogenesis and Cancer

Links between Fibrogenesis and Cancer

Mechanistic and Therapeutic Challenges

Special Issue Editor

Esteban C. Gabazza

MDPI • Basel • Beijing • Wuhan • Barcelona • Belgrade

MDPI

Special Issue Editor
Esteban C. Gabazza
Mie University
Japan

Editorial Office
MDPI
St. Alban-Anlage 66
4052 Basel, Switzerland

This is a reprint of articles from the Special Issue published online in the open access journal *International Journal of Molecular Sciences* (ISSN 1422-0067) from 2018 to 2019 (available at: https: //www.mdpi.com/journal/ijms/special_issues/fibrogenesis).

For citation purposes, cite each article independently as indicated on the article page online and as indicated below:

LastName, A.A.; LastName, B.B.; LastName, C.C. Article Title. *Journal Name* **Year**, *Article Number*, Page Range.

ISBN 978-3-03921-706-9 (Pbk)
ISBN 978-3-03921-707-6 (PDF)

Contents

About the Special Issue Editor

Esteban C. Gabazza is a Full Professor of Medicine, Chairman of the Department of Immunology, and Director of Mie University Center for Intractable Diseases, Mie University Faculty and Graduate School of Medicine, Japan. The focus of Dr. Gabazza's research is on the mechanisms of fibrotic, malignant, and allergic diseases, diabetes mellitus, diabetic complications, and atherosclerosis, and on the role of the microbiome in the pathogenesis of chronic intractable disorders. He has received several grants from the Ministry of Education, Culture, Sports, Science, and Technology of Japan. He has been the Principal Investigator or Co-Investigator of more than 20 funded grants and has published over 300 papers in peer-reviewed journals.

International Journal of
Molecular Sciences

MDPI

Editorial

Links between Fibrogenesis and Cancer: Mechanistic and Therapeutic Challenges

Liqiang Qin [1] and Esteban C. Gabazza [2,*]

[1] Department of Nephrology, Wenzhou Medical University, Wenzhou 317000, China
[2] Department of Immunology, Mie University School of Medicine, Edobashi 2-174, Tsu City,
    Mie 514-8507, Japan
* Correspondence: gabazza@doc.medic.mie-u.ac.jp; Tel.: +81-59-231-5037; Fax: +81-59-231-5225

Received: 17 August 2019; Accepted: 3 September 2019; Published: 3 September 2019

Fibrosis is the end-stage of chronic inflammatory diseases and tissue damage resulting from a dysregulated wound-healing response [1]. Most organs can develop scarring tissue in association with different pathological states including chronic inflammation, autoimmune disorders, graft rejection, malignant tumors or unknown factors [1]. Organ fibrosis is the cause of about 45% of the worldwide total death in developed countries [2]. The high frequency of co-morbidities, the increased mortality rate, the enigma surrounding the disease pathogenesis and the lack of effective therapeutic modalities have made organ fibrosis an area of great interest and consequently the focus of many investigations in recent years [3,4]. This special issue, "Links between Fibrogenesis and Cancer: Mechanistic and Therapeutic Challenges", depicts original findings on intracellular signaling pathways playing a role in fibrogenesis and carcinogenesis associated with chronic diseases and metabolic disorders, and includes reviews describing recent discoveries on potential pathogenic factors, pathophysiological insights, and novel treatment options.

The propensity of fibrotic tissue to develop cancer is an old story, although the mechanistic pathways are still not completely understood [5]. The pro-tumorigenic microenvironment is characterized by the occurrence of persistent parenchymal cell death, compensatory regenerative overresponse, genome instability, DNA damage and gene mutations, and by an excessive local release of inflammatory cytokines, growth factors, and reactive oxygen species [5]. Liver cirrhosis caused by non-alcoholic steatohepatitis or chronic hepatitis B and C infections is an example of scarring disorder that ends up developing cancer [6–8]. As reviewed in this collection, new advances in the technology of next-generation sequencing and whole-genome sequencing have provided a huge amount of information on genetic aberrations during non-alcoholic steatohepatitis or chronic hepatitis B and C infections that trigger the activation of several intracellular oncogenic pathways [7,8]. Of these, disruption of the Hippo and the Yes-associated protein/transcriptional coactivator with PDZ-binding motif signaling pathways appears to be a critical step in the process of hepatic fibrogenesis, hepatic carcinogenesis, and resistance to therapy, angiogenesis, and proliferation of cancer-associated fibroblasts [4,9]. In established malignant tumors, the surrounding tumor microenvironment is a rich source of factors causing uncontrolled activation of intracellular pathways in cancer cells [10]. Components of the tumor microenvironment such as cancer-associated fibroblasts, macrophages, lymphocytes, endothelial cells, and the extracellular matrix may weaken the host immune response and promote the invasiveness and metastatic potential of hepatocellular carcinoma cells directly by interacting with cancer cells or indirectly by releasing an excessive amount of regulatory and suppressive factors [11,12]. The tumor microenvironment may also induce resistance to anti-cancer therapy and, as a proof-of-concept, recent studies have shown amelioration of therapeutic response to anti-cancer drugs by treatment with tumor microenvironment-modifying agents [12].

Another morbid and frequent association described here is idiopathic pulmonary fibrosis (IPF) and lung cancer [13]. Idiopathic pulmonary fibrosis is a chronic and devastating disease of unknown etiology [14]. There are similarities between IPF and cancer in terms of clinical behavior and molecular pathways underlying the pathogenesis of the disease [13,15,16]. Both diseases are characterized by poor response to therapy, short life expectancy, and frequent exposure to harmful environmental factors (e.g., cigarette smoke) [16]. Indeed, the prognosis of idiopathic pulmonary fibrosis is poorer than most types of cancer except malignant tumors of the lung or pancreas [17]. As outlined in this special issue, idiopathic pulmonary fibrosis shares common features with cancer in terms of genetic and epigenetic changes, abnormalities in cell-to-cell communication, aberrant activation of intracellular signal pathways, deregulated expression of growth factors/proteolytic enzymes, uncontrolled cell apoptosis and activation of pro-fibrotic cells [13]. This concept of close-identity between aberrant tissue scarring and cancer is not restricted to the lungs. Although at variable levels, most organs with either fibrotic, benign or malignant disease show abnormal activation of similar signaling and mechanistic pathways [15,18–20]. Fibroblast is one of these common components of cancer and fibrotic disease that plays detrimental roles by contributing to disease progression [6,10–12,21]. Cancer- and scarring tissue-associated fibroblasts may further differentiate to myofibroblasts by cytokines and growth factors such as transforming growth factorβ1 to expand the desmoplastic tissue in both pathologies and to promote cell invasiveness and resistance to therapy in malignant tumors [6,10,11]. Studies evaluating the phenotypic and behavioral alterations of cellular mediators, and biochemical and molecular abnormalities common to organ injury, fibrosis and malignant diseases may provide hints for the development of new diagnostic approaches and effective and less toxic therapeutic modalities for both diseased conditions [22–26].

Recent discoveries in science and advances in technology have revolutionized the diagnostic procedures and therapeutic approaches of many diseases leading to a significant amelioration of quality-of-life and prognosis. However, the incidence and burden of organ fibrosis and cancer are still increasing worldwide [27–29]. Much work remains to be done to achieve more timely diagnosis and to improve the therapeutic management of fibrotic and malignant diseases. The recent insights into the pathogenesis of organ fibrogenesis and malignancy and the new potential therapeutic targets described in the present special issue may be useful for identifying new biological markers and for drug discovery.

Author Contributions: Conceptualization and manuscript preparation: L.Q., E.C.G.

Funding: This work was financially supported in part by a grant from the Ministry of Education, Culture, Sports, Science, and Technology of Japan (Kakenhi No 15K09170).

Conflicts of Interest: The authors declare no conflict of interest.

References

1. Wynn, T.A. Cellular and molecular mechanisms of fibrosis. *J. Pathol.* **2008**, *214*, 199–210. [CrossRef] [PubMed]
2. Wynn, T.A.; Ramalingam, T.R. Mechanisms of fibrosis: Therapeutic translation for fibrotic disease. *Nat. Med.* **2012**, *18*, 1028–1040. [CrossRef]
3. Cano-Jimenez, E.; Hernandez Gonzalez, F.; Peloche, G.B. Comorbidities and Complications in Idiopathic Pulmonary Fibrosis. *Med. Sci.* **2018**, *6*, 71. [CrossRef] [PubMed]
4. Moon, H.; Cho, K.; Shin, S.; Kim, D.Y.; Han, K.H.; Ro, S.W. High Risk of Hepatocellular Carcinoma Development in Fibrotic Liver: Role of the Hippo-YAP/TAZ Signaling Pathway. *Int. J. Mol. Sci.* **2019**, *20*, 581. [CrossRef]
5. Ballester, B.; Milara, J.; Cortijo, J. Idiopathic Pulmonary Fibrosis and Lung Cancer: Mechanisms and Molecular Targets. *Int. J. Mol. Sci.* **2019**, *20*, 593. [CrossRef]
6. Baglieri, J.; Brenner, D.A.; Kisseleva, T. The Role of Fibrosis and Liver-Associated Fibroblasts in the Pathogenesis of Hepatocellular Carcinoma. *Int. J. Mol. Sci.* **2019**, *20*, 1723. [CrossRef] [PubMed]
7. Kanda, T.; Goto, T.; Hirotsu, Y.; Moriyama, M.; Omata, M. Molecular Mechanisms Driving Progression of Liver Cirrhosis towards Hepatocellular Carcinoma in Chronic Hepatitis B and C Infections: A Review. *Int. J. Mol. Sci.* **2019**, *20*, 1358. [CrossRef]
8. Sircana, A.; Paschetta, E.; Saba, F.; Molinaro, F.; Musso, G. Recent Insight into the Role of Fibrosis in Nonalcoholic Steatohepatitis-Related Hepatocellular Carcinoma. *Int. J. Mol. Sci.* **2019**, *20*, 1745. [CrossRef]
9. Noguchi, S.; Saito, A.; Nagase, T. YAP/TAZ Signaling as a Molecular Link between Fibrosis and Cancer. *Int. J. Mol. Sci.* **2018**, *19*, 3674. [CrossRef]

10. Hao, Y.; Baker, D.; Ten Dijke, P. TGF-beta-Mediated Epithelial-Mesenchymal Transition and Cancer Metastasis. *Int. J. Mol. Sci.* **2019**, *20*, 2767. [CrossRef]

11. Truffi, M.; Mazzucchelli, S.; Bonizzi, A.; Sorrentino, L.; Allevi, R.; Vanna, R.; Morasso, C.; Corsi, F. Nano-Strategies to Target Breast Cancer-Associated Fibroblasts: Rearranging the Tumor Microenvironment to Achieve Antitumor Efficacy. *Int. J. Mol. Sci.* **2019**, *20*, 1263. [CrossRef] [PubMed]

12. Yoshida, G.J.; Azuma, A.; Miura, Y.; Orimo, A. Activated Fibroblast Program Orchestrates Tumor Initiation and Progression; Molecular Mechanisms and the Associated Therapeutic Strategies. *Int. J. Mol. Sci.* **2019**, *20*, 2256. [CrossRef] [PubMed]

13. Kinoshita, T.; Goto, T. Molecular Mechanisms of Pulmonary Fibrogenesis and Its Progression to Lung Cancer: A Review. *Int. J. Mol. Sci.* **2019**, *20*, 1461. [CrossRef] [PubMed]

14. Barratt, S.L.; Creamer, A.; Hayton, C.; Chaudhuri, N. Idiopathic Pulmonary Fibrosis (IPF): An Overview. *J. Clin. Med.* **2018**, *7*, 201. [CrossRef] [PubMed]

15. Tzouvelekis, A.; Gomatou, G.; Bouros, E.; Trigidou, R.; Tzilas, V.; Bouros, D. Common Pathogenic Mechanisms Between Idiopathic Pulmonary Fibrosis and Lung Cancer. *Chest* **2019**, *156*, 383–391. [CrossRef]

16. Vancheri, C.; Failla, M.; Crimi, N.; Raghu, G. Idiopathic pulmonary fibrosis: A disease with similarities and links to cancer biology. *Eur. Respir. J.* **2010**, *35*, 496–504. [CrossRef] [PubMed]

17. du Bois, R.M. An earlier and more confident diagnosis of idiopathic pulmonary fibrosis. *Eur. Respir. Rev.* **2012**, *21*, 141–146. [CrossRef]

18. Barcena-Varela, M.; Colyn, L.; Fernandez-Barrena, M.G. Epigenetic Mechanisms in Hepatic Stellate Cell Activation During Liver Fibrosis and Carcinogenesis. *Int. J. Mol. Sci.* **2019**, *20*, 2507. [CrossRef]

19. Ciebiera, M.; Wlodarczyk, M.; Zgliczynska, M.; Lukaszuk, K.; Meczekalski, B.; Kobierzycki, C.; Lozinski, T.; Jakiel, G. The Role of Tumor Necrosis Factor alpha in the Biology of Uterine Fibroids and the Related Symptoms. *Int. J. Mol. Sci.* **2018**, *19*, 3869. [CrossRef]

20. Islam, M.S.; Ciavattini, A.; Petraglia, F.; Castellucci, M.; Ciarmela, P. Extracellular matrix in uterine leiomyoma pathogenesis: A potential target for future therapeutics. *Hum. Reprod. Update* **2018**, *24*, 59–85. [CrossRef]

21. Liu, T.; Zhou, L.; Li, D.; Andl, T.; Zhang, Y. Cancer-Associated Fibroblasts Build and Secure the Tumor Microenvironment. *Front. Cell Dev. Biol.* **2019**, *7*, 60. [CrossRef] [PubMed]

22. Baffour Tonto, P.; Yasuma, T.; Kobayashi, T.; D'Alessandro-Gabazza, C.N.; Toda, M.; Saiki, H.; Fujimoto, H.; Asayama, K.; Fujiwara, K.; Nishihama, K.; et al. Protein S is Protective in Acute Lung Injury by Inhibiting Cell Apoptosis. *Int. J. Mol. Sci.* **2019**, *20*, 1082. [CrossRef] [PubMed]

23. Chuang, H.M.; Ho, L.I.; Huang, M.H.; Huang, K.L.; Chiou, T.W.; Lin, S.Z.; Su, H.L.; Harn, H.J. Non-Canonical Regulation of Type I Collagen through Promoter Binding of SOX2 and Its Contribution to Ameliorating Pulmonary Fibrosis by Butylidenephthalide. *Int. J. Mol. Sci.* **2018**, *19*, 3024. [CrossRef] [PubMed]

24. On, S.; Kim, H.Y.; Kim, H.S.; Park, J.; Kang, K.W. Involvement of G-Protein-Coupled Receptor 40 in the Inhibitory Effects of Docosahexaenoic Acid on SREBP1-Mediated Lipogenic Enzyme Expression in Primary Hepatocytes. *Int. J. Mol. Sci.* **2019**, *20*, 2625. [CrossRef] [PubMed]

25. Rubisz, P.; Ciebiera, M.; Hirnle, L.; Zgliczynska, M.; Lozinski, T.; Dziegiel, P.; Kobierzycki, C. The Usefulness of Immunohistochemistry in the Differential Diagnosis of Lesions Originating from the Myometrium. *Int. J. Mol. Sci.* **2019**, *20*, 1136. [CrossRef] [PubMed]

26. Yasuma, T.; Kobayashi, T.; D'Alessandro-Gabazza, C.N.; Fujimoto, H.; Ito, K.; Nishii, Y.; Nishihama, K.; Baffour Tonto, P.; Takeshita, A.; Toda, M.; et al. Renal Injury during Long-Term Crizotinib Therapy. *Int. J. Mol. Sci.* **2018**, *19*, 2902. [CrossRef] [PubMed]

27. Hutchinson, J.; Fogarty, A.; Hubbard, R.; McKeever, T. Global incidence and mortality of idiopathic pulmonary fibrosis: A systematic review. *Eur. Respir. J.* **2015**, *46*, 795–806. [CrossRef]

28. Marengo, A.; Rosso, C.; Bugianesi, E. Liver Cancer: Connections with Obesity, Fatty Liver, and Cirrhosis. *Annu. Rev. Med.* **2016**, *67*, 103–117. [CrossRef]

29. Webster, A.C.; Nagler, E.V.; Morton, R.L.; Masson, P. Chronic Kidney Disease. *Lancet* **2017**, *389*, 1238–1252. [CrossRef]

International Journal of
Molecular Sciences

MDPI

Review

High Risk of Hepatocellular Carcinoma Development in Fibrotic Liver: Role of the Hippo-YAP/TAZ Signaling Pathway

Hyuk Moon [1,2,†], Kyungjoo Cho [1,2,†], Sunyeong Shin [1,2], Do Young Kim [1,3], Kwang-Hyub Han [1,3,*] and Simon Weonsang Ro [1,4,*]

[1] Yonsei Liver Center, Yonsei University College of Medicine, Seoul 03722, Korea; hmoon@yuhs.ac (H.M.); kyungjoo89@yuhs.ac (K.C.); SOONYOUNG94@yuhs.ac (S.S.); DYK1025@yuhs.ac (D.Y.K.)
[2] Brain Korea 21 Project for Medical Science College of Medicine, Yonsei University, Seoul 03722, Korea
[3] Department of Internal Medicine, Yonsei University College of Medicine, Seoul 03722, Korea
[4] Institute of Gastroenterology, Yonsei University College of Medicine, Seoul 03722, Korea
* Correspondence: GIHANKHYS@yuhs.ac (K.-H.H.); simonr@yuhs.ac (S.W.R.); Tel.: +82-2-2228-1949 (K.-H.H.); +82-2-2228-0811 (S.W.R.)
† These authors contributed equally to this work.

Received: 2 January 2019; Accepted: 28 January 2019; Published: 29 January 2019

Abstract: Liver cancer is the fourth leading cause of cancer-related death globally, accounting for approximately 800,000 deaths annually. Hepatocellular carcinoma (HCC) is the most common type of liver cancer, making up about 80% of cases. Liver fibrosis and its end-stage disease, cirrhosis, are major risk factors for HCC. A fibrotic liver typically shows persistent hepatocyte death and compensatory regeneration, chronic inflammation, and an increase in reactive oxygen species, which collaboratively create a tumor-promoting microenvironment via inducing genetic alterations and chromosomal instability, and activating various oncogenic molecular signaling pathways. In this article, we review recent advances in fields of liver fibrosis and carcinogenesis, and consider several molecular signaling pathways that promote hepato-carcinogenesis under the microenvironment of liver fibrosis. In particular, we pay attention to emerging roles of the Hippo-YAP/TAZ signaling pathway in stromal activation, hepatic fibrosis, and liver cancer.

Keywords: hepatocellular carcinoma; cirrhosis; regeneration; inflammation; cytokines; genetic instability; reactive oxygen species

1. Introduction

Hepatocellular carcinoma (HCC) is the most common primary liver cancer in adults, leading to an increasing number of cancer-related deaths, especially in developing economies of Asia and Africa [1]. According to the World Health Organization (WHO), about 9.5 million deaths worldwide were related to cancer in 2018, among which 800,000 deaths were due to liver cancer, making it the fourth leading cause of cancer-related death (http://gco.iarc.fr/today/fact-sheets-cancers). Various risk factors for HCC development are known, such as hepatitis B virus infection, hepatitis C virus infection, alcohol abuse, intake of aflatoxin B1 (a fungal carcinogen present in food supplies associated with mutations in a tumor suppressor gene *TP53*), and metabolic syndrome [2].

One of the most important features in liver cancer is that it is closely associated with liver fibrosis. Persistent liver damage caused by a variety of factors commonly leads to fibrosis in the liver. Hepatic fibrosis is accompanied in approximately 90% of patients with liver cancer, and the incidence rate of liver cancer within 5 years in patients with advanced liver fibrosis, or cirrhosis is as high as 5–30% [3,4]. Although a substantial increase in HCC development has been reported in cirrhotic patients as well as in animal models for hepatic fibrosis, the mechanism underlying enhanced hepato-carcinogenesis in hepatic fibrosis is not fully understood [5,6]. Several key features typically observed in fibrotic livers are suggested to create a pro-tumorigenic microenvironment, which are persistent hepatocyte death and compensatory regeneration, elevated inflammatory cytokines and growth factors, and an increase in reactive oxygen species. Recent years have seen a great advance in understanding the molecular mechanism linking liver fibrosis and cancer. Several molecular signaling pathways are found to be upregulated following liver damages and to promote hepatic fibrogenesis and liver cancer (Figure 1). Among the signaling pathways that include platelet-derived growth factor (PDGF), tumor growth factor beta (TGF-β), and sonic hedgehog (SHH) signaling pathways, we pay particular attention to the Hippo-YAP/TAZ signaling pathway in this review and introduce recent findings of new roles of YAP/TAZ signaling in hepatic fibrosis and cancer.

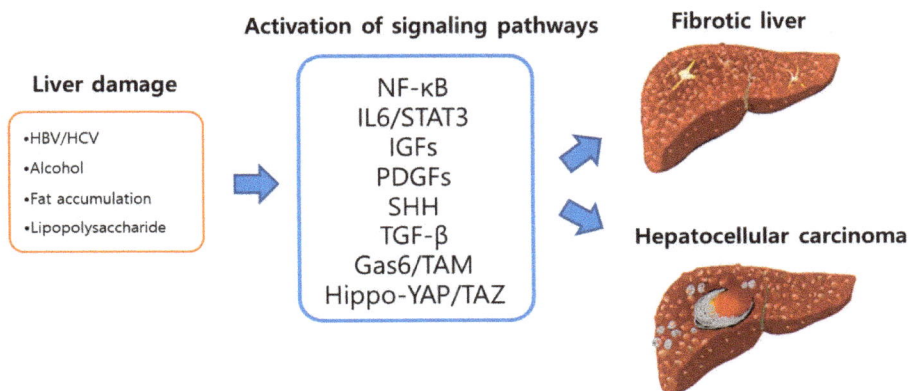

Figure 1. Schematic illustration of the mechanistic links between liver fibrosis and cancer. Persistent liver damage caused by viral infection, alcohol, fat, etc. lead to chronic inflammation and activation of various molecular signaling pathways, which contribute to both fibrogenesis and carcinogensis.

2. Liver Fibrosis

A fibrotic liver exhibits major alterations in tissue architecture and function, which results from a chronic liver damage [7–9] induced by a variety of etiological factors including hepatitis viruses, alcohol and drug abuse, autoimmune disease and hereditary disorders of metabolism [10]. Most chronic liver diseases follow a rather common pathogenic pathway. A persistent hepatic injury induces a series of pathogenic processes from mild inflammation to more severe inflammation, to fibrosis, and finally to cirrhosis. Advanced fibrosis or cirrhosis is irreversible and associated with a significant morbidity and mortality, thus it is of a importance to understand the molecular mechanism underlying liver fibrosis and to prevent or decelerate the pathological process.

In normal liver, the extracellular matrix (ECM) provides structural support of surrounding cells including various molecules for cell adhesion, and allows cells to proliferate, grow, and migrate. It also enhances hepatic function and cell differentiation, and regulates cellular behavior and tissue formation [11]. Fibrosis, or excessive deposition of extracellular matrix components in hepatic tissue, however, compromises the structure and function of the tissue as signified by decreased macromolecular transfer between sinusoids and hepatocytes. Hepatic fibrosis leads to a distorted structure of sinusoids, which not only affects hepatocytes but also non-parenchymal cells such as hepatic stellate cells and myofibroblasts [12].

Activated ECM proteins, such as type I collagen, proteoglycans and glycoproteins, and hepatic stellate cells (HSC, also known as perisinusoidal cells or Ito cells) are the major components of fibrosis in the liver [13,14]. Continuous ECM protein accumulation leads to an increase in matrix stiffness and a change in the phenotype and function of hepatocytes, endothelial cells, and HSCs. In particular, hepatic fibrosis causes the loss of hepatocyte microvilli and accumulation of lipid droplets in hepatocytes as well as a decrease in endothelial fenestration [15–17]. Distorted vascular structure and decreased endothelial fenestration reduces transport of solutes from sinusoids to hepatocytes, further contributing to functional incompetence of hepatocytes. In addition, the alterations induced by the fibrotic microenvironment enhance a stimulus for HSCs to proliferate, activate, and migrate [12]. When activated, HSCs undergo differentiation into fibrogenic myofibroblasts, this produces α-smooth muscle actin (α-SMA), collagen type I and III, fibronectin, and etc. [7,18,19].

Hepatic function is significantly compromised in fibrotic microenvironment. For example, hepatocytes cultured on dishes coated with collagen type I show a rapid change in morphology along with the loss of hepatocyte-specific functions such as albumin and cytochrome P450 expression, while hepatocytes maintained on basement membrane proteins show preservation of the functions. As well, sinusoidal endothelial cells rapidly lose their fenestrae when cultured on a substratum of collagen type I [14]. Fibrosis can also affect cellular function indirectly via up-regulating various cytokines. These include transforming growth factor β (TGF-β), platelet derived growth factor (PDGF), hepatocyte growth factor (HGF), connective tissue growth factor (CTGF), tumor necrosis factor-α (TNF-α), basic fibroblast growth factor (bFGF), and vascular endothelial growth factor (VEGF). Overall, these architectural and functional changes provoke a positive feedback loop that further amplifies fibrogenic processes, resulting in progress to liver cirrhosis and organ failure [20].

Cirrhosis is caused by prolonged liver fibrosis [21], irreversibly destroying the liver structure and impairing the capability of liver to regenerate. The cirrhotic liver contains high concentrations of several cytokines or their effectors that influence hepatocyte fates. Defenestration and capillarization of sinusoidal endothelial cells are a major contributor to hepatic dysfunction in cirrhosis. In addition, activated Kupffer cells disrupt hepatocytes and facilitate the activation of HSCs. Repeated cycles of apoptosis and regeneration of hepatocytes promote the pathogenesis of cirrhosis [22].

For homeostasis during liver injury, ECM remodeling occurs with the balance between matrix metalloproteinases (MMPs) and their inhibitors, tissue inhibitors of matrix metalloproteinases (TIMPs). While an excessive ECM is down-regulated by MMPs (MMP-1, 2, 8 and 13), progressive fibrosis is associated with high expression levels of TIMPs (TIMP-1 and TIMP-2) [12,15,23]. Several studies reported that the down-regulation of TIMPs in HSCs could be an effective therapy for liver fibrosis [24,25].

3. Genetic Instability in Fibrotic Liver

An adult liver has a remarkable regenerating potential, as demonstrated by efficient restitution of a fully functional liver mass after acute 70% partial hepatectomy in mice and humans [26,27]. Liver fibrosis is a progressive tissue change with repeated death and compensatory regeneration of hepatocytes which is induced by chronic liver damage such as infection of hepatitis virus and consumption of alcohol. Increased cell turnovers in livers with chronic inflammation or fibrosis can create a pro-tumorigenic condition by increasing genetic mutations. Furthermore, damaged livers often reveal an increased level of reactive oxygen species (ROS), which further accelerate genetic mutation in the genome of hepatocytes by creating a mutagenic genetic environment (see below). Thus, persistent cellular death and compensatory regeneration in a fibrotic liver can lead to genetic instability, which predisposes liver parenchymal cells to oncogenesis [28]. Genetic instability in hepatocytes in the fibrotic liver can be achieved in several ways.

First, regenerating hepatocytes in fibrotic livers undergo cycles of DNA replication required for cell divisions. DNA replication produces random genetic mutations such as base substitution, insertion, and deletion due to errors generated by DNA polymerase, which are imperfectly corrected by intracellular enzymes responsible for proofreading and repair. As the number of cell divisions increases during liver regeneration, mutations accumulate in the genome of hepatocytes, eventually leading to genetic alterations in cancer-related genes [29]. Tumorigenic processes can be initiated, for example, by an activating mutation in proto-oncogenes such as *RAS, SMO*, and *CTNNB1*, or a loss-of-function mutation in tumor suppressor genes such as *P53, Rb, p16^{INK4A}*, etc. [30].

Secondly, excessive cell divisions in a fibrotic liver induce telomere shortening in hepatocytes, which increases the risk of tumorigenesis via chromosomal instability. In normal progenitor cells, a telomerase RNA component (TERC) maintains genomic integrity at the chromosome terminal region via telomere elongation. Because most mature hepatocytes lack the telomere-maintaining cellular machinery, continuous cell divisions can lead to substantial shortening of the telomere [31]. Telomere shortening in normal cells can trigger DNA damages as well as chromosomal instability, which can result in neoplastic transformation of premalignant cells. Interestingly, cancer cells acquire the ability to maintain telomere during tumor progression as an increased telomerase activity is found in 90% of human cancers [32,33]. How telomere shortening, which causes genetic instability and thus promotes transformation in exhausted hepatocytes, is recovered in cancer cells is currently a topic of intensive research [34,35].

Lastly, an increased production of ROS in liver fibrosis causes toxicity to cells and tissues by generating damaged proteins, lipids, and DNA [36,37]. Typically, elevated immune responses due to chronic infection of hepatitis virus produce an excessive level of ROS, which can further damage the liver. Notably, ROS cause various DNA adducts such as 4-oxo-2-alkenals, exocyclic etheno-DNA adducts, and 8-OHdG, which lead to base modifications in DNA [38]. Therefore, continuous ROS accumulation can significantly contribute to mutation in cancer-related genes and thus tumorigenesis [39]. Further, P53 cannot efficiently activate DNA repair mechanisms in the presence of a high level of ROS, enhancing genetic instability [40]. Supporting the mutagenic and carcinogenic effects, inhibition of ROS formation by antioxidants, butylated hydroxyanisole or N-acetylcysteine suppressed HCC development [41–43].

Of note, increased intracellular ROS activate various molecular signaling pathways that are closely related to tumorigenesis [44]. Excessive levels of ROS stimulate the TGF-β and NF-κB signaling pathways, which promote cancer initiation and progression (see below). Although numerous studies have shown a strong correlation between ROS and oncogenesis, there has been a controversy regarding their intracellular actions and precise roles during tumorigenesis, which is fueled by the lack of appropriate animal models to perform ROS measurements and to study ROS-mediated tumorigenesis in vivo.

4. Increased Secretion of Growth Factors and Cytokines

The liver is an immunologically complex organ, producing plasma proteins such as TNF-α, TGF-β and albumin [45] and soluble complement components which function as the innate immune defense [46,47]. The organ also contains a population of diverse resident immune cells [48–50]. In a healthy liver, metabolism, tissue remodeling, and exposures to microbial products induce inflammation to eliminate toxic substances, damaged cells, and hepatotropic pathogens [51]. In a chronically injured liver, activation of inflammatory cells and inflammatory responses aberrantly increase [52,53], leading to pathological inflammation and disruption of tissue homeostasis. Chronic inflammation induces changes in stroma [54], establishes pro-tumorigenic microenvironment [43,55], and activates various oncogenic molecular signaling pathways [56–58].

Many pro-inflammatory cytokines, including interleukin (IL)-1, IL-6, IL-17, and TNF-α are elevated in a chronically injured liver [59,60], leading to the activation of the nuclear factor kappa B (NF-κB) and the janus kinase (JAK)/signal transducer and activator of transcription factor (STAT) signaling pathways. Increased expression of PDGF, sonic hedgehog (SHH), and TGF-β1 are frequently observed in fibrotic livers caused by various etiological factors. These cytokines and growth factors play significant roles in hepatic fibrogenesis [61,62] and tumorigenesis [63,64].

4.1. Nuclear Factor Kappa B (NF-κB)

The NF-κB transcription factor is a key regulator inducing immune and inflammatory responses [65–67]. The most potent activators of NF-kB include inflammatory cytokines such as TNF-α or IL-1, as well as Toll-like receptors (TLRs) [68,69], which can also trigger cell proliferative signals through NF-Kb [70,71]. They activate NF-κB signaling via IκB kinase (IKK)-dependent phosphorylation and degradation of the κB inhibitor (IκB) proteins [72]. IKK consists of two catalytic subunits, IKKα and IKKβ, and regulatory elements, NEMO/IKKγ, which activate IKK mainly through IKKβ [73].

NF-κB is activated by various stimuli causing liver damages such as alcohol, excessive fat accumulation, hepatitis virus, and bacterial lipopolysaccharide (LPS). These hepatotoxic stimuli activate NF-κB-mediated pro-inflammatory responses leading to the transcription of hundreds of NF-κB target genes involved in the regulation of inflammation, immune responses and cell survival [72]. NF-κB also plays major roles in hepatic fibrosis, by regulating hepatocyte injury and triggering fibrogenic responses in the liver [72,74]. For example, LPS binds to Toll-like receptor-4 (TLR4) in HSC and activates the NF-κB signaling pathway, promoting survival and activation of HSC. Activated HSCs secrete chemokines that recruit and activate Kupffer cells, which are liver resident macrophages. Activated Kupffer cells then secrete TNF-α and IL-1 as well as TGF-β (see below), further enhancing HSC activation. Activation of NF-κB signaling in HSCs leads them to secrete various inflammatory cytokines and to induce quantitative and qualitative changes in the extracellular matrix [72]. TNF-α and IL-1 are frequently up-regulated in livers of chronic inflammation, which persistently activate the NF-kB signaling pathway in hepatocytes as well as non-parenchymal cells in the liver.

NF-κB can exert a pro-tumorigenic effect via suppressing apoptosis during tumor development [75,76] through the positive regulation of anti-apoptotic factors, such as cIAPs, c-FLIP, and BclX [77]. Constitutive activation of NF-κB was frequently found in human HCC tissues compared with non-tumor tissues, and its activation was also verified in animal models of HCC [78]. Animal models with dysregulation of the NF-κB signaling pathway have shown spontaneous development of liver injury, inflammation, fibrosis and HCC, demonstrating that NF-κB acts as a mechanistic link between liver injury, inflammation, fibrosis and HCC [72,76]. Considering that patients with chronic liver inflammation and fibrosis exhibit activation of hepatic NF-κB signaling, it is of high significance and interest to investigate whether an increased incidence of HCC development in the patients can be attributed to NF-κB signaling [79].

4.2. IL6/STAT3 Signaling

IL-6 signals through a cytokine receptor complex that consists of the ligand-binding IL-6R and the signal-transducing component gp130 [80,81]. The IL6-bound hexameric signal transducing complex activates JAK tyrosine kinase, which phosphorylates and activates STAT3. Activated STAT3 dimers translocate to the nucleus and activate transcription of its target genes [82]. STAT3 inhibits the expression of mediators activating immune response against tumor cells [83] and has pro-mitogenic and anti-apoptotic effects on tumor cells [84].

Along with NF-κB signaling, IL6/STAT3 signaling is known as a major pro-inflammatory regulator in response to chronic liver damage. IL6, produced by activated Kupffer cells, is enriched in chronic liver inflammation, especially in non-alcoholic steato-hepatitis (NASH). A high STAT3 activity is also frequently observed in wounded livers, promoting survival and regeneration of hepatocytes. Upregulation of STAT3 signaling in Kupffer cells and HSC leads to subsequent pro-inflammatory and fibrogenic responses [79,83].

Expression of IL-6 and activity of STAT3 are found elevated in HCC [85]. STAT3 target genes are involved in upregulation of proliferation and downregulation of apoptosis, and have been implicated in the initiation of HCC. In hepatic inflammation, IL-6 secreted macrophages facilitated transformation of hepatocytes at early stages of hepato-carcinogenesis [86]. Thus, activated IL6/STAT3 signaling pathway in fibrotic liver due to chronic inflammation may render the liver prone to develop HCC. Of note, IL6/STAT3 can crosstalk with NF-κB signaling in inducing inflammation and liver cancer [83].

4.3. Insulin-Like Growth Factors (IGFs)

In mammals, IGFs play important roles in various cellular processes in the liver including cell growth, proliferation, and differentiation, as well as tissue repair and hepatic pathogenesis [87]. IGFs and their receptors (IGF-1R, IGF-2R) also regulate metabolic processes such as lipogenesis, and glycogen storage in the liver [88]. Although liver synthesizes IGF-1, IGF-2 and their binding proteins (IGFBPs) at a high level under the normal condition, the expression levels significantly decrease under pathogenic conditions like non-alcoholic fatty liver disease (NAFLD), cirrhosis, and HCC [88].

Experimental studies have demonstrated the roles of IGF-1 in the suppression of liver fibrosis and hepato-carcinogenesis [89]. IGF-1 reduces lipogenesis in hepatocytes and inactivates hepatic stellate cells, therefore ameliorating fibrosis with decreased serum AST and ALT levels [89]. In murine models of NASH and cirrhosis, administration of IGF-1 consistently improved steatosis, inflammation, and fibrosis with inactivation of HSC [90]. In a carbon tetrachloride (CCl_4)–treated cirrhotic model, ectopic expression of IGF-1 reduced fibrogenesis [91]. Although studies have shown that decreased levels of IGF-1 are associated with HCC development, there are lines of evidence showing tumor-promoting effects of IGF-1. IGFs can activate downstream RAS and mitogen-activated protein kinase (MAPK) signaling pathway which promotes cell proliferation and survival. In human HCC cell lines, IGF-1 has a positive effect on HCC growth and metastasis [92] and it was found that abundant IGF-1 was associated with risk of HCC development [93].

4.4. Platelet-Derived Growth Factor (PDGF)

PDGF is one of the key factors involved in hepatic fibrogenesis. Increased expression of PDGF is detected in rodent livers after liver injury and in human livers with cirrhosis [94–96]. Under physiological conditions, PDGF is mainly secreted by Kupffer cells. However, when tissue is damaged, a variety of stromal cells including fibroblasts and vascular endothelial cells can synthesize and secrete PDGF through autocrine and paracrine manners [97,98]. Members of the PDGF ligand family consist of 4 polypeptides, PDGF-A, PDGF-B, PDGF-C, and PDGF-D [99]. The function of the PDGF signaling is mediated through platelet-derived growth factor receptors (PDGFR), tyrosine kinase receptors, that are activated by PDGFs [100]. Binding of PDGFs to PDGFR triggers activation of RAS, leading to downstream activation of the Raf-1 and MAPK signaling cascade as well as phosphatidylinositide 3-kinases (PI3K) and the AKT signaling pathway [101]. Accordingly, PDGF signaling can affect a variety of cellular functions including cell growth, proliferation, and differentiation.

Campbell et al. reported that transgenic mice expressing PDGF-C developed liver fibrosis via the activation and proliferation of HCSs [102]. Of note, persistent expression of PDGF-C in the liver for more than 9 months led to the development of hepatocellular adenomas and carcinomas. Likewise, Maass et al. observed liver fibrosis in the liver of transgenic mice expressing PDGF-B [103]. Interestingly, spontaneous tumor development was rarely observed in the liver expressing PDGF-B. However, when the mice were treated with diethylnitrosamine and phenobarbital, a method for chemically induced liver carcinogenesis, the development of dysplastic lesions and their malignant transformation to HCC were significantly increased in PDGF-B transgenic mice, demonstrating a pro-tumorigenic role of PDGF in HCC development [103]. Furthermore, PDGFR-α was found to be up-regulated in early HCC when compared with dysplastic nodules [104], and an increase in PDGFR-α expression level was found in 64% cases (14/22) of HCC patients when compared with adjacent non-tumoral parenchyma [105]. Considering PDGF can promote both hepatic fibrosis and cancer, it is tempting to consider the signaling to be an important mechanistic link between the two pathological conditions in the liver.

4.5. Sonic Hedgehog (SHH)

Hedgehog (HH) signaling regulates various cellular processes including proliferation, apoptosis, migration, and differentiation [106]. The pathway plays a pivotal role in tissue morphogenesis during fetal development as well. In particular, HH signaling modulates wound-healing responses in a number of adult tissues, including the liver [107]. The pathway is activated when ligands such as SHH bind to Patched (Ptch) receptors, leading to the release of Ptch-mediated repression of G protein-coupled receptor, Smoothened (Smo). Smo can then subsequently induce the activation and nuclear accumulation of Glioblastoma (Gli) transcription factors, target genes of which are related to growth, repair, inflammation, and etc. [106]. The HH signaling pathway is found frequently activated in various types of liver injury [108].

Association of HH signaling with fibrosis has been found in fibrotic human livers and various animal models of liver fibrosis. In patients with NAFLD and NASH, a strong correlation was found between the stage of hepatic fibrosis and the degree of HH activation (determined by nuclear accumulation of Gli transcription factors) [109]. Animal models of liver fibrosis induced by various methods, such as genetic ablation of Mdr2 or Ikkβ deletion, bile duct ligation, and a methionine and choline–deficient (MCD) diet, all showed activation of SHH signaling in livers [110–112].

Synthesis of SHH is stimulated by diverse factors that can induce liver damage. SHH molecules that are released from wounded hepatocytes engage receptors on HH responsive cells, which include hepatic stellate cells, sinusoidal endothelial cells and hepatic immune cells [113]. Activation of HH signaling in the stromal cells induces various changes in the microenvironment required for liver regeneration such as growth of liver progenitor populations, tissue remodeling, angiogenesis, and hepatocyte regeneration. However, excessive and persistent activation of HH signaling sometimes overrides successful regeneration of damaged liver and contributes to pathogenesis toward liver fibrosis and cirrhosis [109,114]. A transgenic mice model ectopically expressing SHH in the liver revealed hepatic fibrosis, signifying the role of SHH signaling in fibrogenesis in the organ [5]. Secretion of SHH from hepatocytes in the model activated HH signaling in HH-responsive cells such as cholangiocytes, endothelial cells, as well as HSC. Further, activation of HH signaling in the liver induced epithelial to mesenchymal transition [108]. As HH signaling is activated in liver, expression of myofibroblast-associated genes gradually increases [108], along with the accumulation of myofibroblasts [114].

Several recent papers have demonstrated that HH signaling can significantly contribute to the initiation and promotion of hepatic cancer. Sicklick et al. reported that HH signaling was found elevated in human HCC [115], and Eichenmuller et al. showed that blocking the HH signaling pathway with an antagonist, cyclopamine suppressed cell proliferation of hepatoblastoma [116]. Moreover, SHH expression in liver promoted tumor progression by inducing the transition from hepatocellular adenoma to HCC [6]. It is noteworthy that an elevated HH signaling is found in approximately 60% of human HCC [112,115,117,118]. As HH activation is widely observed in patients with liver fibrosis induced by NASH, the signaling might contribute to a high incidence of HCC in cirrhotic livers of NASH patients.

4.6. Tumor Growth Factor Beta 1 (TGF-β1)

TGF-β signaling regulates a wide variety of cellular processes, including apoptosis of hepatocytes, activation and recruitment of inflammatory cells into injured liver, and activation of quiescent HSCs making them give rise to collagen-producing myofibroblasts [119,120]. Among the TGF-β ligands (β1, β2, and β3) TGF- β1 is linked to hepatic fibrogenesis [121]. TGF-β1 is biologically active as s 25kDa homo-dimer linked by disulfide bonds. Receptors for TGF-β1 are present on virtually all cells, suggesting that ubiquitous distribution of its target cells in tissue.

TGF-β signaling is highly involved in hepatic fibrosis, and has been known as a mater regulator of tissue fibrosis [122,123]. TGF-β signaling regulates various biological responses related to liver fibrosis including hepatic apoptosis, activation of HSC, tissue remodeling, etc. [124,125]. Ironically, TGF-β is known as a major tumor-suppressive signaling pathway that inhibits cell division and promotes apoptosis. Under certain cellular and genetic circumstances, however, TGF-β signaling can act as a tumor promoter via activating various oncogenic signaling pathways. For example, through non-canonical TGF-β signaling pathways, the signaling promotes phosphoinositide 3-kinase and various mitogen-activated protein kinase (MAP kinases) [126,127]. As well, under genetic context of p53 loss [128], YAP activation [129] and Tak1 deletion [130], TGF-β signaling is required for and/or promote tumorigenesis in the liver. Although it is still not understood how TGF-β signaling can be a tumor promoter during early stages of hepatic tumorigenesis, considering its pro-apoptotic and anti-proliferative functions, recent studies suggest that Snail, an EMT inducer and a TGF-β target, can play a pro-tumorigenic role in the liver [131–133] likely via promoting cellular proliferation.

5. Gas6/TAM Pathway in Liver Fibrosis and Cancer

Growth arrest-specific gene 6 (Gas6) product is a vitamin K-dependent protein [134] that activates a family of TAM (Tyro3, Axl, MERTK) receptors with tyrosine kinase activity [135]. TAM signaling plays a role in tissue development and homeostasis, and disposes of apoptotic cells [136]. The ligand for TAM receptors, Gas6 is overexpressed and secreted in response to both acute and chronic liver injuries [136].

In normal liver, Gas6 is mainly expressed in Kupffer cells while Axl, which is related to cell differentiation and carcinogenesis among TAM receptors, is expressed in macrophages and quiescent HSC [135]. MERTK is expressed in Kupffer cells and sinusoidal endothelial cells, but not in hepatocytes, while Tyro3 is only found in resident macrophages [137]. Upon liver injury, Gas6 is overexpressed in Kupffer cells and HSC, which promotes infiltration of monocytes into injured tissue areas [138]. Of note, serum Gas6 levels were high in patients with advanced fibrosis and cirrhosis [139,140]. In line with the findings, experimental murine models showed that increased Gas6 levels led to activation and proliferation of HSC via AKT phosphorylation and NF-κB activation, and contributed to fibrogenesis in the liver [138,140].

Numerous studies have shown that upregulation of Gas6/TAM can promote development of multiple types of cancer, including lung and gastric cancer [141]. The Gas6/TAM pathway can exert multiple pro-tumorigenic effects both on tumor cells and stromal cells [142]. Activation of Gas6/TAM signaling in cancer cells promotes their proliferation and inhibits their apoptosis while the signaling pathway can lead to a tumor-promoting microenvironment, for example, suppressing anti-tumor effects by natural killer (NK) cells. As more experimental and clinical studies are performed to reveal the hepato-carcinogenic roles of Gas6 and TAM, they are expected to uncover a mechanistic link between liver fibrosis and HCC through the Gas6/TAM signaling pathway.

6. Hippo-YAP/TAZ Signaling in Liver Fibrosis and Cancer

Hippo-YAP/TAZ signaling is activated by a variety of mechanical signals such as cell shape and ECM stiffness as well as cell–cell interactions, and transduces cell-specific transcriptional programs. Of note, the signaling is not only activated by the interaction between specific extracellular ligands and their cellular receptors, but is also regulated by modulation of cell adhesion and cell polarity [143–146]. Hippo-YAP/TAZ signaling has a major role in the regulation of cell proliferation, apoptosis, migration and differentiation, all essential for both developmental processes and homeostasis in adult organs [143]. Disruption of the Hippo signaling, or abnormal activation of Yes-associated protein (YAP; also known as YAP1) and transcriptional co-activator with PDZ-binding motif (TAZ; also known as WWTR1) leads to a number of diseases including inflammation, fibrosis and cancer [147].

The Hippo-YAP/TAZ signaling pathway consists of kinase cascades, mammalian sterile 20-like kinase 1 (MST1; also known as STK4) and MST2 (also known as STK3), large tumor suppressor kinase 1 (LATS1) and LATS2, the adaptor proteins Salvador 1 (SAV1), MOB1A and MOB1B, and YAP/TAZ [148–150]. YAP and TAZ proteins are inactivated by phosphorylation through the core Mst1/2-Lats1/2 kinase cascade. Phosphorylation of YAP/TAZ leads to cytoplasmic retention of the proteins via the interaction with 14-3-3 proteins and degradation via the ubiquitin-proteosome pathway [151]. When YAP/TAZ are dephosphorylated, they can translocate into the nucleus and activate transcription of their target genes through the interaction with the TEAD transcription factors (TEAD1–TEAD4) [152].

Hippo-YAP/TAZ signaling regulates the organ size and closely related to liver regeneration as demonstrated in animal models in which activation of YAP/TAZ promoted regeneration of the liver. Knockdown of MST1/MST2 using siRNAs or a pharmacological inhibitor targeting MST1/MST2, which led to dephosphorylation and subsequent nuclear accumulation of YAP/TAZ, augmented hepatocyte proliferation and liver regeneration after partial hepatectomy in mice [153,154]. A recent study also confirmed the role of YAP/TAZ during liver regeneration after ischemia-reperfusion (I/R). The proliferation and expansion of HSCs were prominent during liver recovery after I/R injury, in which the Hippo pathway was inactivated and YAP/TAZ was activated in HSCs. In addition, inhibition of YAP/TAZ activation by a chemical inhibitor attenuated proliferation of both hepatocytes and HSC [155]. Furthermore, YAP can function as a stress sensor, leading to the elimination of damaged hepatocytes [156]. YAP activation in damaged hepatocytes led them to migrate into the hepatic sinusoids and undergo apoptosis. In contrast, YAP activation in undamaged hepatocytes promoted cellular proliferation [156].

6.1. Hepatic Fibrosis

YAP/TAZ activation can promote hepatic fibrosis through the activation of HSCs in response to chronic liver damage [157,158]. A remarkable accumulation of nuclear YAP/TAZ was found in myofibroblasts and HSCs of human and mouse livers with fibrosis [157]. Activated YAP/TAZ upregulate ECM deposition and tissue stiffness, which facilitate fibrogenic processes [159]. In a mouse model of hepatic fibrosis induced by carbon tetrachloride (CCl_4) administration, YAP translocated from the cytoplasm into the nucleus of HSCs, and increased expression of its target genes. Notably, treatment with pharmacological inhibitors of YAP suppressed HSC activation and hepatic fibrogenesis, indicating that YAP activation is essential for liver fibrosis in mice [157]. Also, Zhang et al. found that YAP/TAZ were over-expressed in fibrotic livers of mice treated with CCl_4, and that YAP/TAZ degradation by omega-3 polyunsaturated fatty acids (ω-3 PUFAs) led to down-regulation of pro-fibrogenic genes in activated HSCs and fibrotic liver [160]. Furthermore, the level of TAZ expression in hepatocytes was elevated in a murine model of NASH, and the silencing of TAZ in the NASH model prevented or reversed inflammation, hepatocyte death, and hepatic fibrosis although there was no significant changes in the degree of steatosis. Of note, hepatocyte-targeted expression of TAZ in the NASH model promoted NASH features via the activation of the HH signaling pathway [161].

6.2. Liver Cancer

YAP/TAZ signaling is involved in multiple facets of carcinogenesis, including promotion of cellular proliferation, induction of tissue invasion of tumor cells, and maintenance of cancer stem cells (CSCs) [148,162–164]. Persistent upregulation of YAP/TAZ activity is capable of initiating tumorigenesis in the liver [162,163]. The signaling also significantly contributes to chemoresistance, metastasis, and the recurrence of cancer [162]. Increased cell survival mediated by repression of apoptosis is also a consequence of activated YAP/TAZ [165]. The connective tissue growth factor (CTGF) and extracellular matrix protein CCN1 (CYR61), which are targets of YAP/TAZ, have been reported to inhibit apoptosis in liver cells [166,167]. Additional mechanisms also contribute to repression of apoptosis by YAP/TAZ, including up-regulation of pro-survival factors such as B cell lymphoma 2 (BCL 2) family members [168].

Of note, YAP/TAZ can promote tumorigenesis via cross-talks with diverse oncogenic signaling pathways [169–172]. Increased ectopic expression of YAP in an immortalized human hepatocyte cell line confers tumorigenic potentials via AXL, a receptor tyrosine kinase, as a major downstream factor [173]. A NUAK family SNF1-like kinase 2 (NUAK2) also known as SNF1/AMP kinase-related kinase (SNARK) participates in a positive feedback loop to maximize YAP activity through promotion of actin polymerization and myosin activity. The pharmacological inactivation of NUAK2 inhibits YAP-dependent cancer cell proliferation. These results demonstrate the role of kinase NUAK2 as a mediator of YAP-driven tumorigenesis [174]. Furthermore, it was reported that YAP expression reduced cellular senescence while silencing of YAP inhibited cell proliferation and induced premature senescence [175]. In line with the pro-tumorigenic functions, activation of YAP/TAZ was found at a high frequency in liver cancer and significantly correlated with poor prognosis [176–178].

Several lines of research using genetically engineered mouse models indicate that the Hippo-YAP/TAZ signaling pathway can induce cancer initiation and progression. Liver-specific deletion of both MST1 and MST2, leading to a subsequent activation of YAP/TAZ, was found to induce liver enlargement in young adult mice due to uncontrolled cell proliferation. MST1/2 ablated livers in the mice eventually developed liver cancers exhibiting either HCC or mixed hepatocellular and cholangiocellular carcinoma (mixed HCC-CCA) [179,180]. Similarly, transgenic mice expressing YAP showed enlarged livers after 8 weeks of YAP induction, and later on exhibited a number of nodules throughout the hepatic parenchyma [181].

The significance of YAP/TAZ overexpression in liver cancer was also investigated in HCC patients. In 177 patients with HCC, YAP overexpression was detected in 62% of tumor tissues, most of which exhibited nuclear accumulation of YAP in tumor cells. In addition, overexpression of YAP in tumor cells was significantly associated with poorer differentiation and elevated levels of serum α-fetoprotein (AFP) [182].

6.3. YAP/TAZ Linking Hepatic Fibrosis and Cancer

The Hippo signaling pathway and its downstream effectors, YAP/TAZ have a strong correlation with hepatic fibrogenesis, and are critical regulators of hepatic tumorigenesis (Figure 2). The Hippo-YAP/TAZ signaling pathway also exerts significant effects on tumor microenvironment by maintaining cancer-associated fibroblasts (CAFs) and promoting neo-angiogenesis [147,157,183]. Moreover, in liver-specific conditional knockout mice, deletion of MST1/2 and SAV1 induced inflammation and elevated expression of pro-inflammatory cytokines such as IL-6 and TNF-α [179,184]. As well, it was recently reported that YAP/TAZ promoted liver inflammation and liver cancer. In hepatocytes with genetic deletion of Mst1/2, monocyte chemoattractant protein-1 (Mcp1) expression was highly up-regulated which led to massive infiltration of macrophages. In addition, macrophage ablation or Mcp1 deletion in the Mst1/2 knockout mice showed reduced hepatic inflammation and HCC development, whereas Yap elimination abolished the induction of Mcp1 expression and restored normal liver growth [185]. Another study found a strong correlation between TAZ expression in human liver tumors and secretion of pro-tumorigenic inflammatory cytokines such as IL-6 and C-X-C motif chemokine ligand 1 (Cxcl1) [186]. As hepatic fibrosis is mainly induced by chronic inflammation in the liver, YAP/TAZ might be a strong promoter for both hepatic fibrosis and liver cancer.

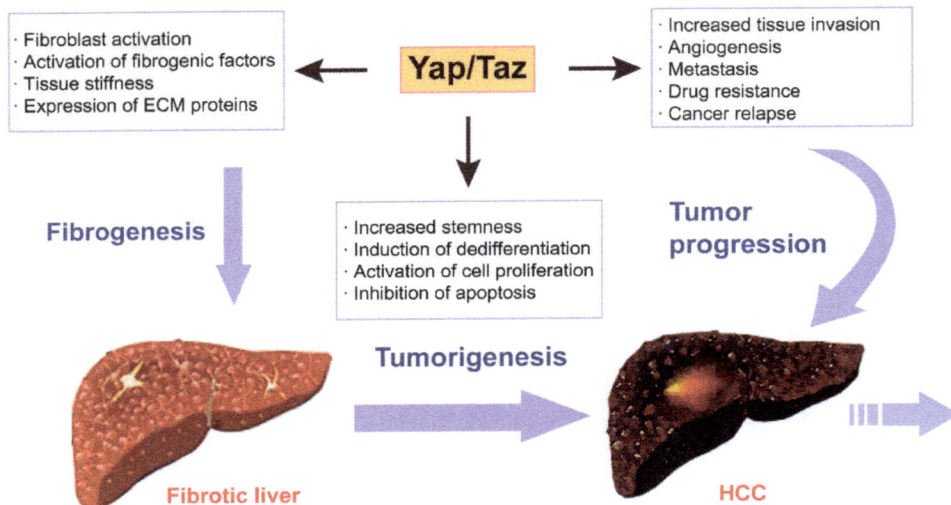

Figure 2. Schematic illustration of the roles of YAP/TAZ signaling in hepatic fibrosis and cancer.

7. Conclusions

Liver fibrosis and cirrhosis have long been regarded as major risk factors for HCC. Fibrotic livers establish a pro-tumorigenic microenvironment via an increase in genetic alterations and chromosomal instabilities, as well as activation of various oncogenic signaling pathways. Recent studies have found YAP/TAZ signaling acting as a major mechanistic link between liver fibrosis and HCC. Further study in this field is needed to better understand the pathogenic process toward liver cancer and to prevent the development of HCC in cirrhotic background, considering that there are currently no effective therapies for HCC.

Funding: This work was supported by the National Research Foundation of Korea (NRF) grant 2016R1A2B4013891 (awarded to S.W.R.), which was funded by the Korea government (MSIP).

Conflicts of Interest: The authors declare no conflict of interest.

References

1. Rawla, P.; Sunkara, T.; Muralidharan, P.; Raj, J.P. Update in global trends and aetiology of hepatocellular carcinoma. *Contemp. Oncol. (Pozn. Pol.)* **2018**, *22*, 141–150. [CrossRef]
2. Llovet, J.M.; Zucman-Rossi, J.; Pikarsky, E.; Sangro, B.; Schwartz, M.; Sherman, M.; Gores, G. Hepatocellular carcinoma. *Nat. Rev. Dis. Primers* **2016**, *2*, 16018. [CrossRef] [PubMed]
3. El-Serag, H.B. Hepatocellular carcinoma. *N. Engl. J. Med.* **2011**, *365*, 1118–1127. [CrossRef] [PubMed]
4. Fattovich, G.; Stroffolini, T.; Zagni, I.; Donato, F. Hepatocellular carcinoma in cirrhosis: Incidence and risk factors. *Gastroenterology* **2004**, *127*, S35–S50. [CrossRef] [PubMed]
5. Chung, S.I.; Moon, H.; Kim, D.Y.; Cho, K.J.; Ju, H.L.; Kim, D.Y.; Ahn, S.H.; Han, K.H.; Ro, S.W. Development of a transgenic mouse model of hepatocellular carcinoma with a liver fibrosis background. *BMC Gastroenterol.* **2016**, *16*, 13. [CrossRef] [PubMed]
6. Chung, S.I.; Moon, H.; Ju, H.L.; Cho, K.J.; Kim, D.Y.; Han, K.H.; Eun, J.W.; Nam, S.W.; Ribback, S.; Dombrowski, F.; et al. Hepatic expression of Sonic Hedgehog induces liver fibrosis and promotes hepatocarcinogenesis in a transgenic mouse model. *J. Hepatol.* **2016**, *64*, 618–627. [CrossRef] [PubMed]
7. Gressner, A.M. Hepatic fibrogenesis: The puzzle of interacting cells, fibrogenic cytokines, regulatory loops, and extracellular matrix molecules. *Zeitschrift fur Gastroenterologie* **1992**, *30* (Suppl. 1), 5–16.
8. Herbst, H.; Schuppan, D.; Milani, S. [Fibrogenesis and fibrolysis in the liver]. *Verhandlungen der Deutschen Gesellschaft fur Pathologie* **1995**, *79*, 15–27. [PubMed]

9. Han, K.H.; Yoon, K.T. New diagnostic method for liver fibrosis and cirrhosis. *Intervirology* **2008**, *51* (Suppl. 1), 11–16. [CrossRef]

10. Elsharkawy, A.M.; Oakley, F.; Mann, D.A. The role and regulation of hepatic stellate cell apoptosis in reversal of liver fibrosis. *Apoptosis Int. J. Program. Cell Death* **2005**, *10*, 927–939. [CrossRef]

11. Baiocchini, A.; Montaldo, C.; Conigliaro, A.; Grimaldi, A.; Correani, V.; Mura, F.; Ciccosanti, F.; Rotiroti, N.; Brenna, A.; Montalbano, M.; et al. Extracellular Matrix Molecular Remodeling in Human Liver Fibrosis Evolution. *PLoS ONE* **2016**, *11*, e0151736. [CrossRef] [PubMed]

12. Friedman, S.L. Mechanisms of hepatic fibrogenesis. *Gastroenterology* **2008**, *134*, 1655–1669. [CrossRef] [PubMed]

13. Benyon, R.C.; Arthur, M.J. Mechanisms of hepatic fibrosis. *J. Pediatr. Gastroenterol. Nutr.* **1998**, *27*, 75–85. [CrossRef] [PubMed]

14. Burt, A.D.C.L. Cellular and molecular aspects of hepatic fibrosis. *J. Pathol.* **1993**, *170*, 105–114. [CrossRef]

15. Elpek, G.O. Cellular and molecular mechanisms in the pathogenesis of liver fibrosis: An update. *World J. Gastroenterol.* **2014**, *20*, 7260–7276. [CrossRef] [PubMed]

16. Fraser, R.; Dobbs, B.R.; Rogers, G.W. Lipoproteins and the liver sieve: The role of the fenestrated sinusoidal endothelium in lipoprotein metabolism, atherosclerosis, and cirrhosis. *Hepatology* **1995**, *21*, 863–874.

17. Schwabe, R.F.; Maher, J.J. Lipids in liver disease: Looking beyond steatosis. *Gastroenterology* **2012**, *142*, 8–11. [CrossRef]

18. Senoo, H.; Mezaki, Y.; Fujiwara, M. The stellate cell system (vitamin A-storing cell system). *Anat. Sci. Int.* **2017**, *92*, 387–455. [CrossRef]

19. Senoo, H.; Sato, M.; Imai, K. Hepatic stellate cells—From the viewpoint of retinoid handling and function of the extracellular matrix. *Kaibogaku Zasshi J. Anat.* **1997**, *72*, 79–94.

20. Wight, T.N.; Potter-Perigo, S. The extracellular matrix: An active or passive player in fibrosis? *Am. J. Physiol. Gastrointest. Liver Physiol.* **2011**, *301*, G950–G955. [CrossRef]

21. Ellis, R.E.; Yuan, J.Y.; Horvitz, H.R. Mechanisms and functions of cell death. *Annu. Rev. Cell Biol.* **1991**, *7*, 663–698. [CrossRef] [PubMed]

22. Zhou, W.C.; Zhang, Q.B.; Qiao, L. Pathogenesis of liver cirrhosis. *World J. Gastroenterol.* **2014**, *20*, 7312–7324. [CrossRef] [PubMed]

23. Fowell, A.J.; Collins, J.E.; Duncombe, D.R.; Pickering, J.A.; Rosenberg, W.M.; Benyon, R.C. Silencing tissue inhibitors of metalloproteinases (TIMPs) with short interfering RNA reveals a role for TIMP-1 in hepatic stellate cell proliferation. *Biochem. Biophys. Res. Commun.* **2011**, *407*, 277–282. [CrossRef] [PubMed]

24. Sun, M.; Kisseleva, T. Reversibility of liver fibrosis. *Clin. Res. Hepatol. Gastroenterol.* **2015**, *39* (Suppl. 1), S60–S63. [CrossRef] [PubMed]

25. Huang, Y.; Deng, X.; Liang, J. Modulation of hepatic stellate cells and reversibility of hepatic fibrosis. *Exp. Cell Res.* **2017**, *352*, 420–426. [CrossRef] [PubMed]

26. Michalopoulos, G.K. Liver regeneration after partial hepatectomy: Critical analysis of mechanistic dilemmas. *Am. J. Pathol.* **2010**, *176*, 2–13. [CrossRef] [PubMed]

27. Oh, S.H.; Swiderska-Syn, M.; Jewell, M.L.; Premont, R.T.; Diehl, A.M. Liver regeneration requires Yap1-TGFbeta-dependent epithelial-mesenchymal transition in hepatocytes. *J. Hepatol.* **2018**, *69*, 359–367. [CrossRef] [PubMed]

28. Luedde, T.; Kaplowitz, N.; Schwabe, R.F. Cell death and cell death responses in liver disease: Mechanisms and clinical relevance. *Gastroenterology* **2014**, *147*, 765–783. [CrossRef] [PubMed]

29. Alexandrov, L.B. Understanding the origins of human cancer. *Science* **2015**, *350*, 1175–1177. [CrossRef] [PubMed]

30. Vogelstein, B.; Kinzler, K.W. Cancer genes and the pathways they control. *Nat. Med.* **2004**, *10*, 789–799. [CrossRef] [PubMed]

31. Fukutomi, M.; Enjoji, M.; Iguchi, H.; Yokota, M.; Iwamoto, H.; Nakamuta, M.; Sakai, H.; Nawata, H. Telomerase activity is repressed during differentiation along the hepatocytic and biliary epithelial lineages: Verification on immortal cell lines from the same origin. *Cell Biochem. Funct.* **2001**, *19*, 65–68. [CrossRef] [PubMed]

32. Low, K.C.; Tergaonkar, V. Telomerase: Central regulator of all of the hallmarks of cancer. *Trends Biochem. Sci.* **2013**, *38*, 426–434. [CrossRef]

33. Shay, J.W.; Zou, Y.; Hiyama, E.; Wright, W.E. Telomerase and cancer. *Hum. Mol. Genet.* **2001**, *10*, 677–685. [CrossRef] [PubMed]

34. Satyanarayana, A.; Manns, M.P.; Rudolph, K.L. Telomeres and telomerase: A dual role in hepatocarcinogenesis. *Hepatology* **2004**, *40*, 276–283. [CrossRef] [PubMed]

35. Barnard, A.; Moch, A.; Saab, S. Relationship between Telomere Maintenance and Liver Disease. *Gut Liver* **2018**, *13*, 11–15. [CrossRef]

36. Liou, G.Y.; Storz, P. Reactive oxygen species in cancer. *Free Radic. Res.* **2010**, *44*, 479–496. [CrossRef]

37. Panieri, E.; Santoro, M.M. ROS homeostasis and metabolism: A dangerous liason in cancer cells. *Cell Death Dis.* **2016**, *7*, e2253. [CrossRef]

38. Kawai, Y.; Nuka, E. Abundance of DNA adducts of 4-oxo-2-alkenals, lipid peroxidation-derived highly reactive genotoxins. *J. Clin. Biochem. Nutr.* **2018**, *62*, 3–10. [CrossRef]

39. Morry, J.; Ngamcherdtrakul, W.; Yantasee, W. Oxidative stress in cancer and fibrosis: Opportunity for therapeutic intervention with antioxidant compounds, enzymes, and nanoparticles. *Redox Biol.* **2017**, *11*, 240–253. [CrossRef]

40. Yeo, C.Q.X.; Alexander, I.; Lin, Z.; Lim, S.; Aning, O.A.; Kumar, R.; Sangthongpitag, K.; Pendharkar, V.; Ho, V.H.B.; Cheok, C.F. p53 Maintains Genomic Stability by Preventing Interference between Transcription and Replication. *Cell Rep.* **2016**, *15*, 132–146. [CrossRef]

41. Maeda, S.; Kamata, H.; Luo, J.L.; Leffert, H.; Karin, M. IKKbeta couples hepatocyte death to cytokine-driven compensatory proliferation that promotes chemical hepatocarcinogenesis. *Cell* **2005**, *121*, 977–990. [CrossRef] [PubMed]

42. Zhang, X.F.; Tan, X.; Zeng, G.; Misse, A.; Singh, S.; Kim, Y.; Klaunig, J.E.; Monga, S.P. Conditional beta-catenin loss in mice promotes chemical hepatocarcinogenesis: Role of oxidative stress and platelet-derived growth factor receptor alpha/phosphoinositide 3-kinase signaling. *Hepatology* **2010**, *52*, 954–965. [CrossRef]

43. Affo, S.; Yu, L.X.; Schwabe, R.F. The Role of Cancer-Associated Fibroblasts and Fibrosis in Liver Cancer. *Annu. Rev. Pathol.* **2017**, *12*, 153–186. [CrossRef] [PubMed]

44. Sabharwal, S.S.; Schumacker, P.T. Mitochondrial ROS in cancer: Initiators, amplifiers or an Achilles' heel? *Nat. Rev. Cancer* **2014**, *14*, 709–721. [CrossRef] [PubMed]

45. Schrodl, W.; Buchler, R.; Wendler, S.; Reinhold, P.; Muckova, P.; Reindl, J.; Rhode, H. Acute phase proteins as promising biomarkers: Perspectives and limitations for human and veterinary medicine. *Proteomics Clin. Appl.* **2016**, *10*, 1077–1092. [CrossRef] [PubMed]

46. Lubbers, R.; van Essen, M.F.; van Kooten, C.; Trouw, L.A. Production of complement components by cells of the immune system. *Clin. Exp. Immunol.* **2017**, *188*, 183–194. [CrossRef] [PubMed]

47. Berraondo, P.; Minute, L.; Ajona, D.; Corrales, L.; Melero, I.; Pio, R. Innate immune mediators in cancer: Between defense and resistance. *Immunol. Rev.* **2016**, *274*, 290–306. [CrossRef] [PubMed]

48. Crispe, I.N. The liver as a lymphoid organ. *Annu. Rev. Immunol.* **2009**, *27*, 147–163. [CrossRef] [PubMed]

49. Nemeth, E.; Baird, A.W.; O'Farrelly, C. Microanatomy of the liver immune system. *Semin. Immunopathol.* **2009**, *31*, 333–343. [CrossRef] [PubMed]

50. O'Farrelly, C.; Crispe, I.N. Prometheus through the looking glass: Reflections on the hepatic immune system. *Immunol. Today* **1999**, *20*, 394–398. [CrossRef]

51. Robinson, M.W.; Harmon, C.; O'Farrelly, C. Liver immunology and its role in inflammation and homeostasis. *Cell. Mol. Immunol.* **2016**, *13*, 267–276. [CrossRef] [PubMed]

52. Peng, W.C.; Logan, C.Y.; Fish, M.; Anbarchian, T.; Aguisanda, F.; Alvarez-Varela, A.; Wu, P.; Jin, Y.; Zhu, J.; Li, B.; et al. Inflammatory Cytokine TNFalpha Promotes the Long-Term Expansion of Primary Hepatocytes in 3D Culture. *Cell* **2018**, *175*, 1607–1619. [CrossRef] [PubMed]

53. Grunebaum, E.; Avitzur, Y. Liver-associated immune abnormalities. *Autoimmun. Rev.* **2018**, *18*, 15–20. [CrossRef] [PubMed]

54. Mazza, E.; Nava, A.; Hahnloser, D.; Jochum, W.; Bajka, M. The mechanical response of human liver and its relation to histology: An in vivo study. *Med. Image Anal.* **2007**, *11*, 663–672. [CrossRef] [PubMed]

55. Li, H.; Zhang, L. Liver regeneration microenvironment of hepatocellular carcinoma for prevention and therapy. *Oncotarget* **2017**, *8*, 1805–1813. [CrossRef] [PubMed]

56. Zhao, X.; Fu, J.; Xu, A.; Yu, L.; Zhu, J.; Dai, R.; Su, B.; Luo, T.; Li, N.; Qin, W.; et al. Gankyrin drives malignant transformation of chronic liver damage-mediated fibrosis via the Rac1/JNK pathway. *Cell Death Dis.* **2015**, *6*, e1751. [CrossRef] [PubMed]

57. Della Corte, C.M.; Viscardi, G.; Papaccio, F.; Esposito, G.; Martini, G.; Ciardiello, D.; Martinelli, E.; Ciardiello, F.; Morgillo, F. Implication of the Hedgehog pathway in hepatocellular carcinoma. *World J. Gastroenterol.* **2017**, *23*, 4330–4340. [CrossRef] [PubMed]

58. Lemberger, U.J.; Fuchs, C.D.; Karer, M.; Haas, S.; Stojakovic, T.; Schofer, C.; Marschall, H.U.; Wrba, F.; Taketo, M.M.; Egger, G.; et al. Hepatocyte specific expression of an oncogenic variant of beta-catenin results in cholestatic liver disease. *Oncotarget* **2016**, *7*, 86985–86998. [CrossRef] [PubMed]

59. Seki, E.; Schwabe, R.F. Hepatic inflammation and fibrosis: Functional links and key pathways. *Hepatology* **2015**, *61*, 1066–1079. [CrossRef]

60. Capece, D.; Fischietti, M.; Verzella, D.; Gaggiano, A.; Cicciarelli, G.; Tessitore, A.; Zazzeroni, F.; Alesse, E. The inflammatory microenvironment in hepatocellular carcinoma: A pivotal role for tumor-associated macrophages. *BioMed Res. Int.* **2013**, *2013*, 187204. [CrossRef]

61. Kodama, Y.; Kisseleva, T.; Iwaisako, K.; Miura, K.; Taura, K.; De Minicis, S.; Osterreicher, C.H.; Schnabl, B.; Seki, E.; Brenner, D.A. c-Jun N-terminal kinase-1 from hematopoietic cells mediates progression from hepatic steatosis to steatohepatitis and fibrosis in mice. *Gastroenterology* **2009**, *137*, 1467–1477. [CrossRef] [PubMed]

62. Lin, W.; Tsai, W.L.; Shao, R.X.; Wu, G.; Peng, L.F.; Barlow, L.L.; Chung, W.J.; Zhang, L.; Zhao, H.; Jang, J.Y.; et al. Hepatitis C virus regulates transforming growth factor beta1 production through the generation of reactive oxygen species in a nuclear factor kappaB-dependent manner. *Gastroenterology* **2010**, *138*, 2509–2518. [CrossRef] [PubMed]

63. Yoshida, T.; Ogata, H.; Kamio, M.; Joo, A.; Shiraishi, H.; Tokunaga, Y.; Sata, M.; Nagai, H.; Yoshimura, A. SOCS1 is a suppressor of liver fibrosis and hepatitis-induced carcinogenesis. *J. Exp. Med.* **2004**, *199*, 1701–1707. [CrossRef] [PubMed]

64. Karin, M. Nuclear factor-kappaB in cancer development and progression. *Nature* **2006**, *441*, 431–436. [CrossRef]

65. Saile, B.; Matthes, N.; El Armouche, H.; Neubauer, K.; Ramadori, G. The bcl, NFkappaB and p53/p21WAF1 systems are involved in spontaneous apoptosis and in the anti-apoptotic effect of TGF-beta or TNF-alpha on activated hepatic stellate cells. *Eur. J. Cell Biol.* **2001**, *80*, 554–561. [CrossRef]

66. Kuhnel, F.; Zender, L.; Paul, Y.; Tietze, M.K.; Trautwein, C.; Manns, M.; Kubicka, S. NFkappaB mediates apoptosis through transcriptional activation of Fas (CD95) in adenoviral hepatitis. *J. Biol. Chem.* **2000**, *275*, 6421–6427. [CrossRef]

67. Pahl, H.L. Activators and target genes of Rel/NF-kappaB transcription factors. *Oncogene* **1999**, *18*, 6853–6866. [CrossRef]

68. Karin, M.; Ben-Neriah, Y. Phosphorylation meets ubiquitination: The control of NF-[kappa]B activity. *Annu. Rev. Immunol.* **2000**, *18*, 621–663. [CrossRef]

69. West, A.P.; Koblansky, A.A.; Ghosh, S. Recognition and signaling by toll-like receptors. *Annu. Rev. Cell Dev. Biol.* **2006**, *22*, 409–437. [CrossRef]

70. Aoyama, T.; Inokuchi, S.; Brenner, D.A.; Seki, E. CX3CL1-CX3CR1 interaction prevents carbon tetrachloride-induced liver inflammation and fibrosis in mice. *Hepatology* **2010**, *52*, 1390–1400. [CrossRef]

71. Calvisi, D.F.; Pascale, R.M.; Feo, F. Dissection of signal transduction pathways as a tool for the development of targeted therapies of hepatocellular carcinoma. *Rev. Recent Clin. Trials* **2007**, *2*, 217–236. [CrossRef]

72. Luedde, T.; Schwabe, R.F. NF-kappaB in the liver—Linking injury, fibrosis and hepatocellular carcinoma. *Nat. Rev. Gastroenterol. Hepatol.* **2011**, *8*, 108–118. [CrossRef] [PubMed]

73. Schwabe, R.F.; Brenner, D.A. Mechanisms of Liver Injury. I. TNF-alpha-induced liver injury: Role of IKK, JNK, and ROS pathways. *Am. J. Physiol. Gastrointest. Liver Physiol.* **2006**, *290*, G583–G589. [CrossRef] [PubMed]

74. Ramakrishna, G.; Rastogi, A.; Trehanpati, N.; Sen, B.; Khosla, R.; Sarin, S.K. From cirrhosis to hepatocellular carcinoma: New molecular insights on inflammation and cellular senescence. *Liver Cancer* **2013**, *2*, 367–383. [CrossRef] [PubMed]

75. Pikarsky, E.; Porat, R.M.; Stein, I.; Abramovitch, R.; Amit, S.; Kasem, S.; Gutkovich-Pyest, E.; Urieli-Shoval, S.; Galun, E.; Ben-Neriah, Y. NF-kappaB functions as a tumour promoter in inflammation-associated cancer. *Nature* **2004**, *431*, 461–466. [CrossRef]

76. Luedde, T.; Beraza, N.; Kotsikoris, V.; van Loo, G.; Nenci, A.; De Vos, R.; Roskams, T.; Trautwein, C.; Pasparakis, M. Deletion of NEMO/IKKgamma in liver parenchymal cells causes steatohepatitis and hepatocellular carcinoma. *Cancer Cell* **2007**, *11*, 119–132. [CrossRef] [PubMed]

77. Kaisho, T.; Takeda, K.; Tsujimura, T.; Kawai, T.; Nomura, F.; Terada, N.; Akira, S. IkappaB kinase alpha is essential for mature B cell development and function. *J. Exp. Med.* **2001**, *193*, 417–426. [CrossRef] [PubMed]

78. Tai, D.I.; Tsai, S.L.; Chang, Y.H.; Huang, S.N.; Chen, T.C.; Chang, K.S.; Liaw, Y.F. Constitutive activation of nuclear factor kappaB in hepatocellular carcinoma. *Cancer* **2000**, *89*, 2274–2281. [CrossRef]

79. He, G.; Karin, M. NF-kappaB and STAT3—Key players in liver inflammation and cancer. *Cell Res.* **2011**, *21*, 159–168. [CrossRef] [PubMed]

80. Boulanger, M.J.; Chow, D.C.; Brevnova, E.E.; Garcia, K.C. Hexameric structure and assembly of the interleukin-6/IL-6 alpha-receptor/gp130 complex. *Science* **2003**, *300*, 2101–2104. [CrossRef]

81. Ward, L.D.; Howlett, G.J.; Discolo, G.; Yasukawa, K.; Hammacher, A.; Moritz, R.L.; Simpson, R.J. High affinity interleukin-6 receptor is a hexameric complex consisting of two molecules each of interleukin-6, interleukin-6 receptor, and gp-130. *J. Biol. Chem.* **1994**, *269*, 23286–23289. [PubMed]

82. Yu, H.; Jove, R. The STATs of cancer—New molecular targets come of age. *Nat. Rev. Cancer* **2004**, *4*, 97–105. [CrossRef] [PubMed]

83. Yu, H.; Kortylewski, M.; Pardoll, D. Crosstalk between cancer and immune cells: Role of STAT3 in the tumour microenvironment. *Nat. Rev. Immunol.* **2007**, *7*, 41–51. [CrossRef] [PubMed]

84. Catlett-Falcone, R.; Landowski, T.H.; Oshiro, M.M.; Turkson, J.; Levitzki, A.; Savino, R.; Ciliberto, G.; Moscinski, L.; Fernandez-Luna, J.L.; Nunez, G.; et al. Constitutive activation of Stat3 signaling confers resistance to apoptosis in human U266 myeloma cells. *Immunity* **1999**, *10*, 105–115. [CrossRef]

85. Svinka, J.; Mikulits, W.; Eferl, R. STAT3 in hepatocellular carcinoma: New perspectives. *Hepatic Oncol.* **2014**, *1*, 107–120. [CrossRef]

86. Schmidt-Arras, D.; Rose-John, S. IL-6 pathway in the liver: From physiopathology to therapy. *J. Hepatol.* **2016**, *64*, 1403–1415. [CrossRef]

87. Takahashi, Y. The Role of Growth Hormone and Insulin-Like Growth Factor-I in the Liver. *Int. J. Mol. Sci.* **2017**, *18*, 1447. [CrossRef]

88. Adamek, A.; Kasprzak, A. Insulin-Like Growth Factor (IGF) System in Liver Diseases. *Int. J. Mol. Sci.* **2018**, *19*, 1308. [CrossRef]

89. Nishizawa, H.; Takahashi, M.; Fukuoka, H.; Iguchi, G.; Kitazawa, R.; Takahashi, Y. GH-independent IGF-I action is essential to prevent the development of nonalcoholic steatohepatitis in a GH-deficient rat model. *Biochem. Biophys. Res. Commun.* **2012**, *423*, 295–300. [CrossRef]

90. Nishizawa, H.; Iguchi, G.; Fukuoka, H.; Takahashi, M.; Suda, K.; Bando, H.; Matsumoto, R.; Yoshida, K.; Odake, Y.; Ogawa, W.; et al. IGF-I induces senescence of hepatic stellate cells and limits fibrosis in a p53-dependent manner. *Sci. Rep.* **2016**, *6*, 34605. [CrossRef]

91. Sanz, S.; Pucilowska, J.B.; Liu, S.; Rodriguez-Ortigosa, C.M.; Lund, P.K.; Brenner, D.A.; Fuller, C.R.; Simmons, J.G.; Pardo, A.; Martinez-Chantar, M.L.; et al. Expression of insulin-like growth factor I by activated hepatic stellate cells reduces fibrogenesis and enhances regeneration after liver injury. *Gut* **2005**, *54*, 134–141. [CrossRef] [PubMed]

92. Lei, T.; Ling, X. IGF-1 promotes the growth and metastasis of hepatocellular carcinoma via the inhibition of proteasome-mediated cathepsin B degradation. *World J. Gastroenterol.* **2015**, *21*, 10137–10149. [CrossRef] [PubMed]

93. Fujiwara, N.; Friedman, S.L.; Goossens, N.; Hoshida, Y. Risk factors and prevention of hepatocellular carcinoma in the era of precision medicine. *J. Hepatol.* **2018**, *68*, 526–549. [CrossRef] [PubMed]

94. Abboud, H.E.; Grandaliano, G.; Pinzani, M.; Knauss, T.; Pierce, G.F.; Jaffer, F. Actions of platelet-derived growth factor isoforms in mesangial cells. *J. Cell. Physiol.* **1994**, *158*, 140–150. [CrossRef] [PubMed]

95. Pinzani, M.; Milani, S.; Grappone, C.; Weber, F.L., Jr.; Gentilini, P.; Abboud, H.E. Expression of platelet-derived growth factor in a model of acute liver injury. *Hepatology* **1994**, *19*, 701–707. [CrossRef] [PubMed]

96. Wong, L.; Yamasaki, G.; Johnson, R.J.; Friedman, S.L. Induction of beta-platelet-derived growth factor receptor in rat hepatic lipocytes during cellular activation in vivo and in culture. *J. Clin. Investig.* **1994**, *94*, 1563–1569. [CrossRef]

97. Friedman, S.L.; Wei, S.; Blaner, W.S. Retinol release by activated rat hepatic lipocytes: Regulation by Kupffer cell-conditioned medium and PDGF. *Am. J. Physiol.* **1993**, *264*, G947–G952. [CrossRef]

98. Ying, H.Z.; Chen, Q.; Zhang, W.Y.; Zhang, H.H.; Ma, Y.; Zhang, S.Z.; Fang, J.; Yu, C.H. PDGF signaling pathway in hepatic fibrosis pathogenesis and therapeutics (Review). *Mol. Med. Rep.* **2017**, *16*, 7879–7889. [CrossRef]

99. Li, X.; Eriksson, U. Novel PDGF family members: PDGF-C and PDGF-D. *Cytokine Growth Factor Rev.* **2003**, *14*, 91–98. [CrossRef]

100. Heldin, C.H.; Westermark, B. Mechanism of action and in vivo role of platelet-derived growth factor. *Physiol. Rev.* **1999**, *79*, 1283–1316. [CrossRef]

101. Alvarez, R.H.; Kantarjian, H.M.; Cortes, J.E. Biology of platelet-derived growth factor and its involvement in disease. *Mayo Clin. Proc.* **2006**, *81*, 1241–1257. [CrossRef] [PubMed]

102. Campbell, J.S.; Hughes, S.D.; Gilbertson, D.G.; Palmer, T.E.; Holdren, M.S.; Haran, A.C.; Odell, M.M.; Bauer, R.L.; Ren, H.P.; Haugen, H.S.; et al. Platelet-derived growth factor C induces liver fibrosis, steatosis, and hepatocellular carcinoma. *Proc. Natl. Acad. Sci. USA* **2005**, *102*, 3389–3394. [CrossRef] [PubMed]

103. Maass, T.; Thieringer, F.R.; Mann, A.; Longerich, T.; Schirmacher, P.; Strand, D.; Hansen, T.; Galle, P.R.; Teufel, A.; Kanzler, S. Liver specific overexpression of platelet-derived growth factor-B accelerates liver cancer development in chemically induced liver carcinogenesis. *Int. J. Cancer* **2011**, *128*, 1259–1268. [CrossRef] [PubMed]

104. Llovet, J.M.; Chen, Y.; Wurmbach, E.; Roayaie, S.; Fiel, M.I.; Schwartz, M.; Thung, S.N.; Khitrov, G.; Zhang, W.; Villanueva, A.; et al. A molecular signature to discriminate dysplastic nodules from early hepatocellular carcinoma in HCV cirrhosis. *Gastroenterology* **2006**, *131*, 1758–1767. [CrossRef] [PubMed]

105. Stock, P.; Monga, D.; Tan, X.; Micsenyi, A.; Loizos, N.; Monga, S.P. Platelet-derived growth factor receptor-alpha: A novel therapeutic target in human hepatocellular cancer. *Mol. Cancer Ther.* **2007**, *6*, 1932–1941. [CrossRef] [PubMed]

106. Jiang, J.; Hui, C.C. Hedgehog signaling in development and cancer. *Dev. Cell* **2008**, *15*, 801–812. [CrossRef] [PubMed]

107. Hooper, J.E.; Scott, M.P. Communicating with Hedgehogs. *Nat. Rev. Mol. Cell Biol.* **2005**, *6*, 306–317. [CrossRef]

108. Omenetti, A.; Choi, S.; Michelotti, G.; Diehl, A.M. Hedgehog signaling in the liver. *J. Hepatol.* **2011**, *54*, 366–373. [CrossRef]

109. Guy, C.D.; Suzuki, A.; Zdanowicz, M.; Abdelmalek, M.F.; Burchette, J.; Unalp, A.; Diehl, A.M. Hedgehog pathway activation parallels histologic severity of injury and fibrosis in human nonalcoholic fatty liver disease. *Hepatology* **2012**, *55*, 1711–1721. [CrossRef]

110. Omenetti, A.; Yang, L.; Li, Y.X.; McCall, S.J.; Jung, Y.; Sicklick, J.K.; Huang, J.; Choi, S.; Suzuki, A.; Diehl, A.M. Hedgehog-mediated mesenchymal-epithelial interactions modulate hepatic response to bile duct ligation. *Lab. Investig. J. Tech. Methods Pathol.* **2007**, *87*, 499–514. [CrossRef]

111. Syn, W.K.; Choi, S.S.; Liaskou, E.; Karaca, G.F.; Agboola, K.M.; Oo, Y.H.; Mi, Z.; Pereira, T.A.; Zdanowicz, M.; Malladi, P.; et al. Osteopontin is induced by hedgehog pathway activation and promotes fibrosis progression in nonalcoholic steatohepatitis. *Hepatology* **2011**, *53*, 106–115. [CrossRef] [PubMed]

112. Philips, G.M.; Chan, I.S.; Swiderska, M.; Schroder, V.T.; Guy, C.; Karaca, G.F.; Moylan, C.; Venkatraman, T.; Feuerlein, S.; Syn, W.K.; et al. Hedgehog signaling antagonist promotes regression of both liver fibrosis and hepatocellular carcinoma in a murine model of primary liver cancer. *PLoS ONE* **2011**, *6*, e23943. [CrossRef] [PubMed]

113. Chen, Y.; Choi, S.S.; Michelotti, G.A.; Chan, I.S.; Swiderska-Syn, M.; Karaca, G.F.; Xie, G.; Moylan, C.A.; Garibaldi, F.; Premont, R.; et al. Hedgehog controls hepatic stellate cell fate by regulating metabolism. *Gastroenterology* **2012**, *143*, 1319–1329. [CrossRef] [PubMed]

114. Choi, S.S.; Omenetti, A.; Syn, W.K.; Diehl, A.M. The role of Hedgehog signaling in fibrogenic liver repair. *Int. J. Biochem. Cell Biol.* **2011**, *43*, 238–244. [CrossRef] [PubMed]

115. Sicklick, J.K.; Li, Y.X.; Jayaraman, A.; Kannangai, R.; Qi, Y.; Vivekanandan, P.; Ludlow, J.W.; Owzar, K.; Chen, W.; Torbenson, M.S.; et al. Dysregulation of the Hedgehog pathway in human hepatocarcinogenesis. *Carcinogenesis* **2006**, *27*, 748–757. [CrossRef] [PubMed]

116. Eichenmuller, M.; Gruner, I.; Hagl, B.; Haberle, B.; Muller-Hocker, J.; von Schweinitz, D.; Kappler, R. Blocking the hedgehog pathway inhibits hepatoblastoma growth. *Hepatology* **2009**, *49*, 482–490. [CrossRef]

117. Patil, M.A.; Zhang, J.; Ho, C.; Cheung, S.T.; Fan, S.T.; Chen, X. Hedgehog signaling in human hepatocellular carcinoma. *Cancer Biol. Ther.* **2006**, *5*, 111–117. [CrossRef]

118. Pasca di Magliano, M.; Hebrok, M. Hedgehog signalling in cancer formation and maintenance. *Nat. Rev. Cancer* **2003**, *3*, 903–911. [CrossRef]

119. Bataller, R.; Brenner, D.A. Liver fibrosis. *J. Clin. Investig.* **2005**, *115*, 209–218. [CrossRef]

120. Schuster, N.; Krieglstein, K. Mechanisms of TGF-beta-mediated apoptosis. *Cell Tissue Res.* **2002**, *307*, 1–14. [CrossRef]

121. Hellerbrand, C.; Stefanovic, B.; Giordano, F.; Burchardt, E.R.; Brenner, D.A. The role of TGFbeta1 in initiating hepatic stellate cell activation in vivo. *J. Hepatol.* **1999**, *30*, 77–87. [CrossRef]

122. Friedman, S.L. Molecular regulation of hepatic fibrosis, an integrated cellular response to tissue injury. *J. Biol. Chem.* **2000**, *275*, 2247–2250. [CrossRef] [PubMed]

123. Seki, E.; De Minicis, S.; Osterreicher, C.H.; Kluwe, J.; Osawa, Y.; Brenner, D.A.; Schwabe, R.F. TLR4 enhances TGF-beta signaling and hepatic fibrosis. *Nat. Med.* **2007**, *13*, 1324–1332. [CrossRef]

124. Meng, X.M.; Nikolic-Paterson, D.J.; Lan, H.Y. TGF-beta: The master regulator of fibrosis. *Nat. Rev. Nephrol.* **2016**, *12*, 325–338. [CrossRef] [PubMed]

125. Stewart, A.G.; Thomas, B.; Koff, J. TGF-beta: Master regulator of inflammation and fibrosis. *Respirology* **2018**, *23*, 1096–1097. [CrossRef] [PubMed]

126. Bierie, B.; Moses, H.L. Tumour microenvironment: TGFbeta: The molecular Jekyll and Hyde of cancer. *Nat. Rev. Cancer* **2006**, *6*, 506–520. [CrossRef] [PubMed]

127. Massague, J. TGFbeta signalling in context. *Nat. Rev. Mol. Cell Biol.* **2012**, *13*, 616–630. [CrossRef] [PubMed]

128. Morris, S.M.; Baek, J.Y.; Koszarek, A.; Kanngurn, S.; Knoblaugh, S.E.; Grady, W.M. Transforming growth factor-beta signaling promotes hepatocarcinogenesis induced by p53 loss. *Hepatology* **2012**, *55*, 121–131. [CrossRef] [PubMed]

129. Nishio, M.; Sugimachi, K.; Goto, H.; Wang, J.; Morikawa, T.; Miyachi, Y.; Takano, Y.; Hikasa, H.; Itoh, T.; Suzuki, S.O.; et al. Dysregulated YAP1/TAZ and TGF-beta signaling mediate hepatocarcinogenesis in Mob1a/1b-deficient mice. *Proc. Natl. Acad. Sci. USA* **2016**, *113*, E71–E80. [CrossRef]

130. Yang, L.; Inokuchi, S.; Roh, Y.S.; Song, J.; Loomba, R.; Park, E.J.; Seki, E. Transforming growth factor-beta signaling in hepatocytes promotes hepatic fibrosis and carcinogenesis in mice with hepatocyte-specific deletion of TAK1. *Gastroenterology* **2013**, *144*, 1042–1054. [CrossRef]

131. Nakagawa, H.; Hikiba, Y.; Hirata, Y.; Font-Burgada, J.; Sakamoto, K.; Hayakawa, Y.; Taniguchi, K.; Umemura, A.; Kinoshita, H.; Sakitani, K.; et al. Loss of liver E-cadherin induces sclerosing cholangitis and promotes carcinogenesis. *Proc. Natl. Acad. Sci. USA* **2014**, *111*, 1090–1095. [CrossRef] [PubMed]

132. Moon, H.; Ju, H.L.; Chung, S.I.; Cho, K.J.; Eun, J.W.; Nam, S.W.; Han, K.H.; Calvisi, D.F.; Ro, S.W. Transforming Growth Factor-beta Promotes Liver Tumorigenesis in Mice via Up-regulation of Snail. *Gastroenterology* **2017**, *153*, 1378–1391. [CrossRef] [PubMed]

133. Shen, Q.; Eun, J.W.; Lee, K.; Kim, H.S.; Yang, H.D.; Kim, S.Y.; Lee, E.K.; Kim, T.; Kang, K.; Kim, S.; et al. Barrier to autointegration factor 1, procollagen-lysine, 2-oxoglutarate 5-dioxygenase 3, and splicing factor 3b subunit 4 as early-stage cancer decision markers and drivers of hepatocellular carcinoma. *Hepatology* **2018**, *67*, 1360–1377. [CrossRef] [PubMed]

134. Bellido-Martin, L.; de Frutos, P.G. Vitamin K-dependent actions of Gas6. *Vitam. Horm.* **2008**, *78*, 185–209. [CrossRef] [PubMed]

135. Holstein, E.; Binder, M.; Mikulits, W. Dynamics of Axl Receptor Shedding in Hepatocellular Carcinoma and Its Implication for Theranostics. *Int. J. Mol. Sci.* **2018**, *19*, 4111. [CrossRef] [PubMed]

136. Mukherjee, S.K.; Wilhelm, A.; Antoniades, C.G. TAM receptor tyrosine kinase function and the immunopathology of liver disease. *Am. J. Physiol. Gastrointest. Liver Physiol.* **2016**, *310*, G899–G905. [CrossRef] [PubMed]

137. Qi, N.; Liu, P.; Zhang, Y.; Wu, H.; Chen, Y.; Han, D. Development of a spontaneous liver disease resembling autoimmune hepatitis in mice lacking tyro3, axl and mer receptor tyrosine kinases. *PLoS ONE* **2013**, *8*, e66604. [CrossRef] [PubMed]

138. Lafdil, F.; Chobert, M.N.; Couchie, D.; Brouillet, A.; Zafrani, E.S.; Mavier, P.; Laperche, Y. Induction of Gas6 protein in CCl$_4$-induced rat liver injury and anti-apoptotic effect on hepatic stellate cells. *Hepatology* **2006**, *44*, 228–239. [CrossRef]

139. Bellan, M.; Pogliani, G.; Marconi, C.; Minisini, R.; Franzosi, L.; Alciato, F.; Magri, A.; Avanzi, G.C.; Pirisi, M.; Sainaghi, P.P. Gas6 as a putative noninvasive biomarker of hepatic fibrosis. *Biomark. Med.* **2016**, *10*, 1241–1249. [CrossRef]

140. Barcena, C.; Stefanovic, M.; Tutusaus, A.; Joannas, L.; Menendez, A.; Garcia-Ruiz, C.; Sancho-Bru, P.; Mari, M.; Caballeria, J.; Rothlin, C.V.; et al. Gas6/Axl pathway is activated in chronic liver disease and its targeting reduces fibrosis via hepatic stellate cell inactivation. *J. Hepatol.* **2015**, *63*, 670–678. [CrossRef]

141. Wu, G.; Ma, Z.; Hu, W.; Wang, D.; Gong, B.; Fan, C.; Jiang, S.; Li, T.; Gao, J.; Yang, Y. Molecular insights of Gas6/TAM in cancer development and therapy. *Cell Death Dis.* **2017**, *8*, e2700. [CrossRef] [PubMed]

142. Wu, G.; Ma, Z.; Cheng, Y.; Hu, W.; Deng, C.; Jiang, S.; Li, T.; Chen, F.; Yang, Y. Targeting Gas6/TAM in cancer cells and tumor microenvironment. *Mol. Cancer* **2018**, *17*, 20. [CrossRef] [PubMed]

143. Pan, D. The hippo signaling pathway in development and cancer. *Dev. Cell* **2010**, *19*, 491–505. [CrossRef] [PubMed]

144. Varelas, X.; Samavarchi-Tehrani, P.; Narimatsu, M.; Weiss, A.; Cockburn, K.; Larsen, B.G.; Rossant, J.; Wrana, J.L. The Crumbs complex couples cell density sensing to Hippo-dependent control of the TGF-beta-SMAD pathway. *Dev. Cell* **2010**, *19*, 831–844. [CrossRef] [PubMed]

145. Yu, J.; Zheng, Y.; Dong, J.; Klusza, S.; Deng, W.M.; Pan, D. Kibra functions as a tumor suppressor protein that regulates Hippo signaling in conjunction with Merlin and Expanded. *Dev. Cell* **2010**, *18*, 288–299. [CrossRef] [PubMed]

146. Zhang, N.; Bai, H.; David, K.K.; Dong, J.; Zheng, Y.; Cai, J.; Giovannini, M.; Liu, P.; Anders, R.A.; Pan, D. The Merlin/NF2 tumor suppressor functions through the YAP oncoprotein to regulate tissue homeostasis in mammals. *Dev. Cell* **2010**, *19*, 27–38. [CrossRef]

147. Panciera, T.; Azzolin, L.; Cordenonsi, M.; Piccolo, S. Mechanobiology of YAP and TAZ in physiology and disease. *Nat. Rev. Mol. Cell Biol.* **2017**, *18*, 758–770. [CrossRef] [PubMed]

148. Johnson, R.; Halder, G. The two faces of Hippo: Targeting the Hippo pathway for regenerative medicine and cancer treatment. *Nat. Rev. Drug Discov.* **2014**, *13*, 63–79. [CrossRef] [PubMed]

149. Meng, Z.; Moroishi, T.; Guan, K.L. Mechanisms of Hippo pathway regulation. *Genes Dev.* **2016**, *30*, 1–17. [CrossRef]

150. Goulev, Y.; Fauny, J.D.; Gonzalez-Marti, B.; Flagiello, D.; Silber, J.; Zider, A. SCALLOPED interacts with YORKIE, the nuclear effector of the hippo tumor-suppressor pathway in Drosophila. *Curr. Biol.* **2008**, *18*, 435–441. [CrossRef]

151. Zhao, B.; Wei, X.; Li, W.; Udan, R.S.; Yang, Q.; Kim, J.; Xie, J.; Ikenoue, T.; Yu, J.; Li, L.; et al. Inactivation of YAP oncoprotein by the Hippo pathway is involved in cell contact inhibition and tissue growth control. *Genes Dev.* **2007**, *21*, 2747–2761. [CrossRef] [PubMed]

152. Mo, J.S.; Yu, F.X.; Gong, R.; Brown, J.H.; Guan, K.L. Regulation of the Hippo-YAP pathway by protease-activated receptors (PARs). *Genes Dev.* **2012**, *26*, 2138–2143. [CrossRef] [PubMed]

153. Loforese, G.; Malinka, T.; Keogh, A.; Baier, F.; Simillion, C.; Montani, M.; Halazonetis, T.D.; Candinas, D.; Stroka, D. Impaired liver regeneration in aged mice can be rescued by silencing Hippo core kinases MST1 and MST2. *EMBO Mol. Med.* **2017**, *9*, 46–60. [CrossRef] [PubMed]

154. Fan, F.; He, Z.; Kong, L.L.; Chen, Q.; Yuan, Q.; Zhang, S.; Ye, J.; Liu, H.; Sun, X.; Geng, J.; et al. Pharmacological targeting of kinases MST1 and MST2 augments tissue repair and regeneration. *Sci. Transl. Med.* **2016**, *8*, 352ra108. [CrossRef] [PubMed]

155. Konishi, T.; Schuster, R.M.; Lentsch, A.B. Proliferation of hepatic stellate cells, mediated by YAP and TAZ, contributes to liver repair and regeneration after liver ischemia-reperfusion injury. *Am. J. Physiol. Gastrointest. Liver Physiol.* **2018**, *314*, G471–G482. [CrossRef] [PubMed]

156. Miyamura, N.; Hata, S.; Itoh, T.; Tanaka, M.; Nishio, M.; Itoh, M.; Ogawa, Y.; Terai, S.; Sakaida, I.; Suzuki, A.; et al. YAP determines the cell fate of injured mouse hepatocytes in vivo. *Nat. Commun.* **2017**, *8*, 16017. [CrossRef] [PubMed]

157. Mannaerts, I.; Leite, S.B.; Verhulst, S.; Claerhout, S.; Eysackers, N.; Thoen, L.F.; Hoorens, A.; Reynaert, H.; Halder, G.; van Grunsven, L.A. The Hippo pathway effector YAP controls mouse hepatic stellate cell activation. *J. Hepatol.* **2015**, *63*, 679–688. [CrossRef]

158. Caliari, S.R.; Perepelyuk, M.; Cosgrove, B.D.; Tsai, S.J.; Lee, G.Y.; Mauck, R.L.; Wells, R.G.; Burdick, J.A. Stiffening hydrogels for investigating the dynamics of hepatic stellate cell mechanotransduction during myofibroblast activation. *Sci. Rep.* **2016**, *6*, 21387. [CrossRef]

159. Herrera, J.; Henke, C.A.; Bitterman, P.B. Extracellular matrix as a driver of progressive fibrosis. *J. Clin. Investing.* **2018**, *128*, 45–53. [CrossRef]

160. Zhang, K.; Chang, Y.; Shi, Z.; Han, X.; Han, Y.; Yao, Q.; Hu, Z.; Cui, H.; Zheng, L.; Han, T.; et al. omega-3 PUFAs ameliorate liver fibrosis and inhibit hepatic stellate cells proliferation and activation by promoting YAP/TAZ degradation. *Sci. Rep.* **2016**, *6*, 30029. [CrossRef]

161. Wang, X.; Zheng, Z.; Caviglia, J.M.; Corey, K.E.; Herfel, T.M.; Cai, B.; Masia, R.; Chung, R.T.; Lefkowitch, J.H.; Schwabe, R.F.; et al. Hepatocyte TAZ/WWTR1 Promotes Inflammation and Fibrosis in Nonalcoholic Steatohepatitis. *Cell Metab.* **2016**, *24*, 848–862. [CrossRef] [PubMed]

162. Zanconato, F.; Cordenonsi, M.; Piccolo, S. YAP/TAZ at the Roots of Cancer. *Cancer Cell* **2016**, *29*, 783–803. [CrossRef] [PubMed]

163. Harvey, K.F.; Zhang, X.; Thomas, D.M. The Hippo pathway and human cancer. *Nat. Rev. Cancer* **2013**, *13*, 246–257. [CrossRef] [PubMed]

164. Moya, I.M.; Halder, G. Hippo-YAP/TAZ signalling in organ regeneration and regenerative medicine. *Nat. Rev. Mol. Cell Biol.* **2018**. [CrossRef] [PubMed]

165. Hong, L.; Li, Y.; Liu, Q.; Chen, Q.; Chen, L.; Zhou, D. The Hippo Signaling Pathway in Regenerative Medicine. *Methods Mol. Biol.* **2019**, *1893*, 353–370. [CrossRef] [PubMed]

166. Urtasun, R.; Latasa, M.U.; Demartis, M.I.; Balzani, S.; Goni, S.; Garcia-Irigoyen, O.; Elizalde, M.; Azcona, M.; Pascale, R.M.; Feo, F.; et al. Connective tissue growth factor autocriny in human hepatocellular carcinoma: Oncogenic role and regulation by epidermal growth factor receptor/yes-associated protein-mediated activation. *Hepatology* **2011**, *54*, 2149–2158. [CrossRef] [PubMed]

167. Juric, V.; Chen, C.C.; Lau, L.F. Fas-mediated apoptosis is regulated by the extracellular matrix protein CCN1 (CYR61) in vitro and in vivo. *Mol. Cell. Biol.* **2009**, *29*, 3266–3279. [CrossRef]

168. Huo, X.; Zhang, Q.; Liu, A.M.; Tang, C.; Gong, Y.; Bian, J.; Luk, J.M.; Xu, Z.; Chen, J. Overexpression of Yes-associated protein confers doxorubicin resistance in hepatocellullar carcinoma. *Oncol. Rep.* **2013**, *29*, 840–846. [CrossRef]

169. Kim, M.; Jho, E.H. Cross-talk between Wnt/beta-catenin and Hippo signaling pathways: A brief review. *BMB Rep.* **2014**, *47*, 540–545. [CrossRef]

170. Azzolin, L.; Panciera, T.; Soligo, S.; Enzo, E.; Bicciato, S.; Dupont, S.; Bresolin, S.; Frasson, C.; Basso, G.; Guzzardo, V.; et al. YAP/TAZ incorporation in the beta-catenin destruction complex orchestrates the Wnt response. *Cell* **2014**, *158*, 157–170. [CrossRef]

171. Mohseni, M.; Sun, J.; Lau, A.; Curtis, S.; Goldsmith, J.; Fox, V.L.; Wei, C.; Frazier, M.; Samson, O.; Wong, K.K.; et al. A genetic screen identifies an LKB1-MARK signalling axis controlling the Hippo-YAP pathway. *Nat. Cell Biol.* **2014**, *16*, 108–117. [CrossRef] [PubMed]

172. Zhang, W.; Nandakumar, N.; Shi, Y.; Manzano, M.; Smith, A.; Graham, G.; Gupta, S.; Vietsch, E.E.; Laughlin, S.Z.; Wadhwa, M.; et al. Downstream of mutant KRAS, the transcription regulator YAP is essential for neoplastic progression to pancreatic ductal adenocarcinoma. *Sci. Signal.* **2014**, *7*, ra42. [CrossRef] [PubMed]

173. Xu, M.Z.; Chan, S.W.; Liu, A.M.; Wong, K.F.; Fan, S.T.; Chen, J.; Poon, R.T.; Zender, L.; Lowe, S.W.; Hong, W.; et al. AXL receptor kinase is a mediator of YAP-dependent oncogenic functions in hepatocellular carcinoma. *Oncogene* **2011**, *30*, 1229–1240. [CrossRef] [PubMed]

174. Yuan, W.C.; Pepe-Mooney, B.; Galli, G.G.; Dill, M.T.; Huang, H.T.; Hao, M.; Wang, Y.; Liang, H.; Calogero, R.A.; Camargo, F.D. NUAK2 is a critical YAP target in liver cancer. *Nat. Commun.* **2018**, *9*, 4834. [CrossRef] [PubMed]

175. Xie, Q.; Chen, J.; Feng, H.; Peng, S.; Adams, U.; Bai, Y.; Huang, L.; Li, J.; Huang, J.; Meng, S.; et al. YAP/TEAD-mediated transcription controls cellular senescence. *Cancer Res.* **2013**, *73*, 3615–3624. [CrossRef]

176. Han, S.X.; Bai, E.; Jin, G.H.; He, C.C.; Guo, X.J.; Wang, L.J.; Li, M.; Ying, X.; Zhu, Q. Expression and clinical significance of YAP, TAZ, and AREG in hepatocellular carcinoma. *J. Immunol. Res.* **2014**, *2014*, 261365. [CrossRef]

177. Kim, G.J.; Kim, H.; Park, Y.N. Increased expression of Yes-associated protein 1 in hepatocellular carcinoma with stemness and combined hepatocellular-cholangiocarcinoma. *PLoS ONE* **2013**, *8*, e75449. [CrossRef] [PubMed]

178. Xiao, H.; Jiang, N.; Zhou, B.; Liu, Q.; Du, C. TAZ regulates cell proliferation and epithelial-mesenchymal transition of human hepatocellular carcinoma. *Cancer Sci.* **2015**, *106*, 151–159. [CrossRef]

179. Lu, L.; Li, Y.; Kim, S.M.; Bossuyt, W.; Liu, P.; Qiu, Q.; Wang, Y.; Halder, G.; Finegold, M.J.; Lee, J.S.; et al. Hippo signaling is a potent in vivo growth and tumor suppressor pathway in the mammalian liver. *Proc. Natl. Acad. Sci. USA* **2010**, *107*, 1437–1442. [CrossRef]

180. Song, H.; Mak, K.K.; Topol, L.; Yun, K.; Hu, J.; Garrett, L.; Chen, Y.; Park, O.; Chang, J.; Simpson, R.M.; et al. Mammalian Mst1 and Mst2 kinases play essential roles in organ size control and tumor suppression. *Proc. Natl. Acad. Sci. USA* **2010**, *107*, 1431–1436. [CrossRef]

181. Dong, J.; Feldmann, G.; Huang, J.; Wu, S.; Zhang, N.; Comerford, S.A.; Gayyed, M.F.; Anders, R.A.; Maitra, A.; Pan, D. Elucidation of a universal size-control mechanism in Drosophila and mammals. *Cell* **2007**, *130*, 1120–1133. [CrossRef] [PubMed]

182. Xu, M.Z.; Yao, T.J.; Lee, N.P.; Ng, I.O.; Chan, Y.T.; Zender, L.; Lowe, S.W.; Poon, R.T.; Luk, J.M. Yes-associated protein is an independent prognostic marker in hepatocellular carcinoma. *Cancer* **2009**, *115*, 4576–4585. [CrossRef] [PubMed]

183. Calvo, F.; Ege, N.; Grande-Garcia, A.; Hooper, S.; Jenkins, R.P.; Chaudhry, S.I.; Harrington, K.; Williamson, P.; Moeendarbary, E.; Charras, G.; et al. Mechanotransduction and YAP-dependent matrix remodelling is required for the generation and maintenance of cancer-associated fibroblasts. *Nat. Cell Biol.* **2013**, *15*, 637–646. [CrossRef] [PubMed]

184. Sun, B.; Karin, M. Inflammation and liver tumorigenesis. *Front. Med.* **2013**, *7*, 242–254. [CrossRef] [PubMed]

185. Kim, W.; Khan, S.K.; Liu, Y.; Xu, R.; Park, O.; He, Y.; Cha, B.; Gao, B.; Yang, Y. Hepatic Hippo signaling inhibits protumoural microenvironment to suppress hepatocellular carcinoma. *Gut* **2018**, *67*, 1692–1703. [CrossRef] [PubMed]

186. Hagenbeek, T.J.; Webster, J.D.; Kljavin, N.M.; Chang, M.T.; Pham, T.; Lee, H.J.; Klijn, C.; Cai, A.G.; Totpal, K.; Ravishankar, B.; et al. The Hippo pathway effector TAZ induces TEAD-dependent liver inflammation and tumors. *Sci. Signal.* **2018**, *11*, eaaj1757. [CrossRef] [PubMed]

International Journal of
Molecular Sciences

MDPI

Review

Idiopathic Pulmonary Fibrosis and Lung Cancer: Mechanisms and Molecular Targets

Beatriz Ballester [1,2,*,†], Javier Milara [2,3,4,†] and Julio Cortijo [1,2,5]

1   Department of Pharmacology, Faculty of Medicine, University of Valencia, 46010 Valencia, Spain; julio.cortijo@uv.es
2   CIBERES, Health Institute Carlos III, 28029 Valencia, Spain; xmilara@hotmail.com
3   Pharmacy Unit, University Clinic Hospital of Valencia, 46010 Valencia, Spain
4   Institute of Health Research-INCLIVA, 46010 Valencia, Spain
5   Research and teaching Unit, University General Hospital Consortium, 46014 Valencia, Spain
*   Correspondence: beaballester7@gmail.com; Tel.: +34-605148470
†   These authors contributed equally to this work.

Received: 20 December 2018; Accepted: 28 January 2019; Published: 30 January 2019

Abstract: Idiopathic pulmonary fibrosis (IPF) is the most common idiopathic interstitial pulmonary disease with a median survival of 2–4 years after diagnosis. A significant number of IPF patients have risk factors, such as a history of smoking or concomitant emphysema, both of which can predispose the patient to lung cancer (LC) (mostly non-small cell lung cancer (NSCLC)). In fact, IPF itself increases the risk of LC development by 7% to 20%. In this regard, there are multiple common genetic, molecular, and cellular processes that connect lung fibrosis with LC, such as myofibroblast/mesenchymal transition, myofibroblast activation and uncontrolled proliferation, endoplasmic reticulum stress, alterations of growth factors expression, oxidative stress, and large genetic and epigenetic variations that can predispose the patient to develop IPF and LC. The current approved IPF therapies, pirfenidone and nintedanib, are also active in LC. In fact, nintedanib is approved as a second line treatment in NSCLC, and pirfenidone has shown anti-neoplastic effects in preclinical studies. In this review, we focus on the current knowledge on the mechanisms implicated in the development of LC in patients with IPF as well as in current IPF and LC-IPF candidate therapies based on novel molecular advances.

Keywords: idiopathic pulmonary fibrosis (IPF); lung cancer (LC); non-small cell lung cancer (NSCLC)

1. Introduction

Idiopathic pulmonary fibrosis (IPF) is a form of chronic, progressive fibrosing interstitial pneumonia of unknown cause that occurs primarily in older adults. IPF is associated with the histopathologic and/or radiologic pattern of usual interstitial pneumonia (UIP) that is limited to the lungs [1]. The disease course of IPF is variable and somewhat unpredictable, nevertheless, progression to end-stage respiratory insufficiency and death after the onset of symptoms from diagnosis is 2–4 years [2]. Pulmonary and extrapulmonary comorbid conditions, such as lung cancer (LC), are commonly associated to IPF, altering the disease course and mortality. Links between pulmonary fibrosis and LC have been suggested as early as 1965 [3] and are based on the multiple common genetic, molecular, and cellular processes that connect both diseases and can predispose the patient to develop IPF and LC.

On its own, pulmonary fibrosis is a risk factor for developing lung carcinogenesis [4–7]. Moreover, elder age, male sex, history of smoking, and coexisting emphysema are also strong risk factors that contribute to developing LC in IPF patients [8–19]. The prevalence of LC in IPF patients ranges from 2.7% to 48% [8,9,13–20] (Table 1) and is significantly higher than in the general population [21].

Otherwise, the incidence of LC in IPF patients is reported to be 11.2–36 cases per 1,000 persons per year [8,20,22], which increases with each year following IPF diagnosis [15,18]. Moreover, IPF patients that are diagnosed with LC have a reduced mean survival time (1.6–1.7 years), compared to IPF patients without LC diagnosis [10,17] and Kato et al. reported 53.5%, 78.6%, and 92.9% as 1-, 3-, and 5-year all-cause mortality rates after LC diagnosis in IPF patients [8].

Table 1. Prevalence of lung cancer (LC) in idiopathic pulmonary fibrosis (IPF) patients.

Study	Number of Patients with IPF	Prevalence of LC (%)	LC-IPF Male (%)	LC-IPF Median Age	LC-IPF Smokers (%)	Reference
Nagai (1992)	99	31.3%	87.1%	70.9	87.1%	[9]
Matsusitha (1995)	20	48.2%	90%	66.4	74.3%	[13]
Park (2001)	281	22.4%	97%	66.8	88.9%	[14]
Le Jeune (2007)	1064	2.7%	ND	ND	ND	[20]
Ozawa (2009)	103	20.4%	95.2%	65.5	66.7%	[15]
Lee (2012)	1685	6.8%	94.7%	68.5	92.3%	[16]
Kreuter (2014)	265	16%	ND	ND	ND	[17]
Tomasetti (2015)	181	13%	82.6%	66.9	91.3%	[18]
Yoon (2018)	1108	2.8%	61%	65	77%	[19]
Kato (2018)	632	11.1%	94.3%	66.8	100%	[8]

ND: not determined.

2. Histological Subtypes and Parenchymal Distribution of Lung Cancer in Idiopathic Pulmonary Fibrosis

In the general population, the predominant type of LC is non-small cell lung cancer (NSCLC). Likewise, NSCLC is the predominant type of LC in LC-IPF patients. Furthermore, adenocarcinoma (ADC) is the most common subtype of histological NSCLC in the general population [23]. However, the most frequent histological subtype of LC in IPF has been controversial over the past few years (Table 2). Recently, the majority of studies have shown that squamous cell carcinoma (SQC) is the most frequent type of LC in IPF patients, while ADC is the second most frequent [8–10,14,15,17–19,24–29]. Moreover, some isolated cases of large cell carcinoma and small cell lung cancer (SCLC) have also been reported [8,14,18,27].

Table 2. Lung cancer (LC) histological subtypes in patients with idiopathic pulmonary fibrosis (IPF).

Study	Number of Patients with LC-IPF	Squamous Cell Carcinoma	Adenocarcinoma	Other Histological Subtypes	Reference
Kawai (1987)	8	12.5%	75%	12.5%	[24]
Nagai (1992)	31	45.2%	35.2%	19.6%	[9]
Park (2001)	63	35%	30%	35%	[14]
Kawasaki (2001)	53	46%	46%	8%	[25]
Aubry (2002)	24	67%	29%	4%	[10]
Ozawa (2009)	21	38%	29%	33%	[15]
Saito (2011)	28	67.9%	25%	7.1%	[26]
Lee (2014)	70	40%	30%	30%	[27]
Kreuter (2015)	42	36%	31%	33%	[17]
Tomasetti (2015)	23	39%	35%	26%	[18]
Khan (2015)	34	41%	38%	21%	[28]
Guyard (2017)	18	44%	33%	23%	[29]
Yoon (2018)	27	41%	26%	33%	[19]
Kato (2018)	70	30%	20%	50%	[8]

Similarly to fibrotic lesions, lung carcinomas are generally more frequently found in the peripheral area of the lungs in IPF patients, i.e., in the lower lobes [3,8,16,25,30–32], and are associated with honeycomb lesions, developing from honeycomb areas or in the border between honeycombing and non-fibrotic areas, and epithelial metaplasia [33–35]. In general, squamous metaplasia, but not cuboidal cell metaplasia or bronchial cell metaplasia, have been observed more frequently in LC-IPF patients than in IPF patients without lung carcinoma. Then, it is speculated that it might reflect a constitutional susceptibility of IPF patients of developing lung carcinoma [36].

3. Cell Types and Cellular Processes Involved in Lung Cancer Associated with Pulmonary Fibrosis

3.1. Cell Transformations in the Mesenchymal Phenotype

IPF is characterised by an excess of myofibroblasts that are persistently activated in fibrotic lungs [37]. The activated myofibroblasts are stellate- or spindle-shaped, and are characterised by the secretion of extracellular matrix (ECM) components (a characteristic shared with fibroblasts), such as collagen type I, and by the formation of contractile apparatus (a characteristic shared with airway smooth muscle cells), such as α-smooth muscle actin (α-SMA) microfilaments [38]. In IPF lungs, myofibroblasts have heterogeneous phenotypes [39] (Figure 1). The classic concept is that tissue injury induces the activation of resident fibroblasts to proliferate and express constituents of the ECM and α-SMA fibres [40]. One contemporary theory is that tissue injury in the presence of transforming growth factor (TGF-β) induces epithelial to mesenchymal cell transition (EMT) [41,42]. In detail, EMT is part of an unabated form of wound healing in which alveolar type II (ATII) cells [41–43] can serve as a source for the increased myofibroblast-like pool that eventually leads to organ destruction if the primary inflammatory insult that triggered the wound healing is not removed or attenuated [44]. Another contemporary theory of myofibroblast activation is that circulating fibrocytes that originate from the bone marrow are mesenchymal progenitor cells that home and extravasate into sites of tissue injury, and differentiate into the myofibroblast-like phenotype [45,46] in response to TGF-β and endothelin 1 [47–49]. Pulmonary arterial endothelial cell to mesenchymal transition (EnMT) has been suggested as another source of myofibroblasts that potentially contribute to lung fibrosis and pulmonary hypertension and is often associated with IPF, which portends a poor prognosis [50,51]. Finally, there are two emerging theories that consider pleural mesothelial cells or lung pericytes as significant sources of lung myofibroblasts in IPF [52–56].

Similarly to IPF, cancer-associated fibroblasts (CAFs) are also important players in LC because they exhibit mesenchymal-like features and have heterogeneous phenotypes (Figure 1) [57]. Lung resident fibroblasts surrounding the malignancy are thought to be the first responders to the site of insult that the tumour creates [58]. Resident epithelial cells—generally bronchiolar epithelial cells—can also undergo a partial and possibly reversible EMT during the early steps of carcinogenesis, cancer invasion, and metastasis [44,59]. Likewise, fibrocytes recruited from the peripheral circulation have also been suggested as a potential source of CAFs [58,60,61]. It has also been observed that up to 40% of CAFs arise as a consequence of EndMT in two different murine cancer models, suggesting that EndMT may play a role in tumour angiogenic sprouting into adjacent tissue [62]. In the tumour environment, vascular pericytes have also been associated with tumour vasculature. Pericytes that detach from the tumour microvasculature have been shown to undergo differentiation into stromal fibroblasts via the action of platelet-derived growth factor-BB (PDGF-BB), which significantly contributes to tumour invasion and metastasis [63]. Finally, pleural mesothelial cells are also hypothesised as a source of CAFs, where a mesothelial precursor lineage has been identified as being capable of clonally generating fibroblasts in the lungs, kidney, liver, and gut [64]. Also, it has been reported that an overexpression of mesothelin in lung ADC is induced by tobacco-related carcinogens [57].

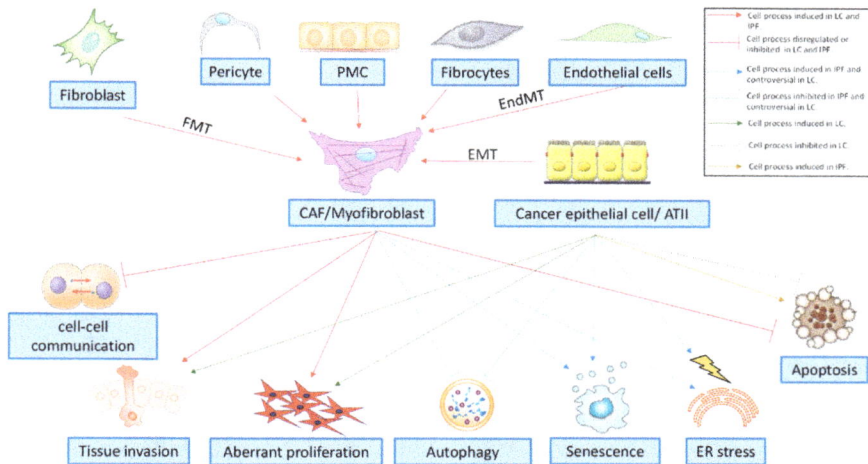

Figure 1. Cell types and cellular processes involved in lung cancer (LC) and idiopathic pulmonary fibrosis (IPF). Lung resident fibroblasts, pericytes, pleural mesothelial cells (PMC), circulating fibrocytes, vascular endothelial cells, and epithelial cells (Alveolar type II cells (ATII) in IPF and cancer epithelial cells in LC) are transformed to IPF myofibroblast or mesenchymal phenotype and cancer-associated fibroblasts (CAFs). Myofibroblasts and cancer cells are characterized by altered cell-cell communication, migration properties, and tissue invasion through basement membranes and the extracellular matrix. IPF myofibroblasts and ATII cells acquire senescent identities, but the presence of this phenotype is controversial in LC. Otherwise, endoplasmic reticulum (ER) stress is induced in IPF, while autophagy is defective in IPF. However, the function of both processes is controversial in LC. Finally, apoptosis is induced in ATII cells, but IPF myofibroblasts and cancer cells evade apoptosis.

3.2. Common Cellular Processes in Lung Cancer Associated with Pulmonary Fibrosis

3.2.1. Apoptosis and Autophagy

Growing evidence indicates a prominent role of enhanced endoplasmic reticulum (ER) stress in IPF, resulting in an unfolded protein response (UPR) [65,66]. This response mechanism activates biochemical pathways to meet the demands of protein folding. However, if that is no longer feasible, a terminal UPR directs alveolar epithelium cells towards apoptosis. By contrast, IPF myofibroblasts and cancer cells escape apoptosis [67,68]. Regarding the role of ER-stress in tumourigenesis, it is controversial [69,70], nevertheless, recent evidence shows that ER stress may attenuate senescence and promote tumorigenesis [71] (Figure 1).

Despite elevated ER stress, there is evidence that autophagy is defective in IPF [72–74] (Figure 1), which promotes lung fibroblast differentiation into myofibroblasts via excessive ECM production [72,74,75] and fibroblast resistance to apoptosis [76]. In LC, autophagy functions as a double-edged sword because it suppresses tumorigenesis in a limited number of contexts while facilitating it in most others [77]. In fact, it has been observed that autophagy can promote or inhibit apoptosis under different cellular contexts within the same tumour cell population. Therefore, therapeutic targeting of autophagy in cancer is sometimes viewed as controversial [78].

3.2.2. Cellular Proliferation

Evidence strongly supports the persistent activation of proliferative signalling pathways in IPF (Figure 1). In fact, myofibroblasts sustain their own growth through the autocrine production of TGFβ1 and partly lose their ability to produce anti-fibrotic prostaglandin E2 (PGE2) [79]. Further, they show a lack of response to the inhibitory activity of PGE2 [80], and to other antiproliferative signals [81].

The receptors for platelet-derived growth factor (PDGF), vascular endothelial growth factor (VEGF), and fibroblast growth factor (FGF) have also recently been implicated in the sustained proliferation signalling of pulmonary fibroblasts [82]. However, this persistent activation is not definitively linked with aberrant fibroblast proliferation in vivo [83–85] and the role of excessive fibroblast proliferation as a pathogenic mechanism of IPF is unclear. By contrast, aberrant proliferation of cancer cells and sustained proliferative signalling has been described as "arguably the most fundamental trait of cancer cells" [68].

3.2.3. Altered Cell-Cell Communications

Intercellular channels that are formed by connexins are essential for the synchronisation of cell proliferation and tissue repair [86]. In particular, the expression of connexin 43 (Cx43) is considered crucial in fibroblast-to-fibroblast communication. In IPF fibroblasts, Cx43 expression is strongly down-regulated, leading to a loss of proliferative control [87]. Similarly, cancer cell lines from mouse and human lung carcinoma have low levels or an absence of Cx43 expression [88], which results in reduced cell-to-cell communication (Figure 1); this may explain both the release of cells from contact-inhibition control, and the uncontrolled proliferation that characterises this disease.

3.2.4. Senescence

Analyses of cell types in the lungs of both human IPF and the bleomycin-injured mouse model have demonstrated that fibroblasts and epithelial cells acquire senescent identities [89]. Senescence appears to be a central phenotype that promotes lung fibrosis through increased production of a complex senescence-associated secretory phenotype (SASP) based on growth factors, cytokines, chemokines, and matrix metalloproteinases, as well as acquired apoptosis resistance in IPF fibroblasts [90] (Figure 1). However, therapeutic management of cell senescence is controversial in cancer. On the one hand, cell senescence could limit the replicative capacity of cells and ultimately prevent their proliferation in different stages of malignancy, while providing a protective barrier to neoplastic expansion [91]. On the other hand, it has been proposed that senescent fibroblasts may promote tumour progression, possibly by secreting an SASP based on certain matrix metalloproteases, growth factors, and cytokines [92].

3.2.5. Tissue Invasion

IPF lung fibroblasts are characterised by their ability to invade through the basement membrane and the ECM via the action of metalloproteinases [93,94]. This characteristic is also an important hallmark of cancer (Figure 1). Unlike cancers that can disseminate over long distances because they acquire further invasive mechanisms, fibrotic lung fibroblasts are restricted to local invasion [95]. The capacity of cancer cells to infiltrate the surrounding tissue is strictly related to the expression of laminin, heat shock protein 27, and fascin [96–98]. Interestingly, in IPF, it has been shown that bronchiolar basal cells surrounding the fibroblast foci express large amounts of these proteins, which induce cell motility and invasiveness of myofibroblasts [99]. Therefore, targeting these molecules may be a feasible strategy to restrain myofibroblast tissue invasion in LC-IPF patients.

3.2.6. Inflammation

The role of inflammation in IPF is controversial, although evidence shows the existence of a predominant phenotype of fibrosis-associated macrophages (FAMs) that are alternatively activated. These are an M2 phenotype of FAMs [100] that facilitate the enhanced production of FGFs [101], profibrotic cytokines [102,103], and matrix metalloproteinases [104]. Like FAMs, tumour-associated macrophages also display an M2 phenotype and support tumour growth through their ability to promote angiogenesis, activate mesenchymal cells, remodel the matrix, and suppress effector T-cell responses [105,106]. Thus, M2 macrophages could be considered key effectors in the development of LC associated with pulmonary fibrosis.

4. Principal Fibrogenic Molecules and Signal Transduction Pathways Participating in Lung Cancer Associated with Pulmonary Fibrosis

4.1. Growth Factors

TGFβ is a major profibrotic growth factor and is often chronically overexpressed in cancer and fibrosis (Table 3) [107]. TGFβ can be activated by αVβ6 integrin and, in IPF, TGFβ1 mediates fibrogenesis by antiproliferative action and apoptosis in alveolar epithelial cells (AECs) or by stimulation of fibroblast differentiation to myofibroblasts, synthesis of ECM proteins, and inhibition of ECM degradation [108,109]. TGFβ also induces the production of fibrogenic or angiogenic growth factors and is known to strongly elicit EMT and EndMT. In the early stages of cancer pathogenesis, TGFβ acts as a tumour suppressor because it inhibits the growth of many cell types and delays the appearance of primary tumors. However, after the appearance of them, TGFβ promotes tumour progression, because it can induce EMT and EndMT, and suppresses immune surveillance. Therefore, during tumour progression, TGFβ triggers the formation of spontaneous lung metastases. Finally, TGFβ is also central in the development of the tumour stroma because TGFβ1 also activates CAFs [107,110].

Tyrosine kinase receptor ligands, such as PDGF, VEGF, and FGF, are aberrantly expressed in LC and IPF (Table 3) [82]. In IPF, PDGF plays an important role in inducing the secretion of ECM components and growth factors in fibroblasts [111]. It also promotes fibroblast proliferation and recruits fibrocytes to the lung [112,113]. Furthermore, TGFβ1, FGF, and tumour necrosis factor-α exhibit PDGF-dependent profibrotic activity [114,115]. Otherwise, PDGF signalling is also important for tumour growth, angiogenesis, and lymphangiogenesis in cancer [113]. In fact, it has been shown that crenolanib (PDGF receptor inhibitor) is capable of suppressing proliferation and inducing apoptosis in a dose-dependent manner using A549 cells as a NSCLC model system. Moreover, it has been shown that crenolanib-treated A549 cells have reduced migratory activity in response to inducers of chemotaxis, and the antitumor activity of this drug has been confirmed in an NSCLC xenograft tumor model [116]. Finally, it has been observed that PDGF regulates VEGF expression in NSCLC via an autocrine mechanism [117], and is also involved in the recruitment of CAFs to the tumour mass [113]. The contribution of VEGF to IPF is not fully understood because there is still debate on the role of vascular remodelling in IPF [118]. However, in addition to the role of VEGF in tumour angiogenesis, accumulating evidence suggests that it can act directly on cancer cells to regulate growth, migration, and production of several pro-angiogenic factors [119]. FGF is also released by damaged epithelial cells and activated fibroblasts during the remodelling processes [120,121]. It was found that FGF signalling is required for fibroblast expansion within fibrotic areas [122]. FGF can also affect the proliferation, treatment sensitivity, and apoptosis of LC cells [123].

Another important fibrogenic growth factor in LC is connective tissue growth factor (CTGF). In IPF, CTGF induces fibroblast proliferation and ECM deposition [124]. By contrast, it has been observed that CTGF inhibits metastasis and invasion of human lung ADC [125], and its expression is suppressed in many NSCLCs [126] (Table 3).

Several other growth factors are involved in IPF and LC [127–129], but we have not included them in this review because they are not the focus of the new therapies currently being developed.

4.2. Lysophosphatidic Acid (LPA)

LPA, a profibrotic mediator with proinflammatory activity, is released by platelets during epithelial injury [130]. Extracellular production of LPA is catalysed by autotaxin (ATX) and further regulated by phospholipid phosphatases (PLPP). IPF patients have increased LPA levels [131] in their bronchoalveolar lavage fluid (BALF) (Table 3), and recently, it has been shown that LPA signalling mediates both the fibroblast recruitment and vascular leakage induced by lung injury in a bleomycin model of pulmonary fibrosis [131]. Moreover, it has recently been shown that the ATX/PLPP3-LPA/LPA receptor 1 (LPAR1) axis has a procarcinogenic role in lung carcinogenesis [132].

Table 3. Principal fibrogenic molecules and signal transduction pathways participating in lung cancer (LC) and idiopathic pulmonary fibrosis (IPF).

	IPF	LC
Growth Factors		
TGFβ1	Overexpressed	Overexpressed
PDGF	Overexpressed	Overexpressed
VEGF	Overexpressed	Overexpressed
FGF	Overexpressed	Overexpressed
CTGF	Overexpressed	Downregulated
Profibrotic mediators		
LPA	Overexpressed	Overexpressed
Galectin-3	Overexpressed	Overexpressed
Cytokines	Overexpressed	Overexpressed
CCL2	Overexpressed	Overexpressed
IL-13	Overexpressed	Overexpressed
Mucins		
Mucin 1	Overexpressed	Overexpressed
Mucin 4	Overexpressed	Overexpressed
Mucin 5B	Overexpressed	Overexpressed
Embryological pathways		
Wnt pathway	Overexpressed	Overexpressed
Shh pathway	Overexpressed	Overexpressed
Notch pathway	Overexpressed	Overexpressed
Proliferation-related pathways		
PI3K/AKT/mTOR pathway	Overexpressed	Overexpressed
Migration-related proteins		
Laminin	Overexpressed	Overexpressed
Fascin	Overexpressed	Overexpressed
Hsp27	Overexpressed	Overexpressed
Oxidative stress—related molecules		
NOX4	Overexpressed	Overexpressed
Nrf2	Downregulated	Downregulated
Cell-cell communication—related proteins		
Connexin 43	Downregulated	Downregulated

TGFβ1: transforming growth factor β1; PDGF: platelet derived growth factor; VEGF: vascular endothelial growth factor; FGF: fibroblast growth factor; CTGF: connective tissue growth factor; LPA: lysophosphatidic acid; CCL2: chemokine ligand 2; IL-13: interleukin 13; PI3K: phosphoinositide 3-kinase; AKT: protein kinase B; mTOR: mammalian Target of Rapamycin; Hsp27: heat shock protein 27; NOX4: NADPH oxidase 4.

4.3. Cytokines and Chemokines

Epithelial injury causes an imbalance in T-helper type 1 (Th1)/type 2 (Th2) cytokine expression, which results in a stronger Th2 response. In particular, there is evidence that interleukin 13 (IL-13) [133] plays a dominant role in the pathogenesis of fibrosis in the lungs of IPF patients (Table 3) [134]. IL-13 triggers the transformation of fibroblasts to myofibroblasts via the TGFβ-dependent and -independent pathways, while also inducing epithelial apoptosis [135–137]. Similarly to IPF, a pattern of Th2 cytokine expression has also been identified in NSCLC [138]. With respect to IL-13, a recent study observed the highest expression level of IL-13 in LC in the SQC subtype, followed by the ADC subtype [139]. Moreover, a clear association between IL-13 receptor subunit alpha-2 overexpression and poor survival in resected NSCLC patients has been shown [140]. There are other profibrotic cytokines besides IL-13 that are also associated with IPF. For example, chemokine ligand 2 (CCL2) has been reported to be present in the BALF of IPF patients at significant concentrations [141] (Table 3). In IPF, CCL2 has been shown to induce the differentiation of developing T-cells into type 2 cells [142], and to stimulate collagen synthesis and TGFβ expression in lung fibroblasts [141]. In LC, the CCL2/CCR2 (chemokine receptor type 2) axis is also important in several aspects of tumorigenesis. One of its most important roles is the generation of new vascular structures that allow tumour growth [140]. However, there is evidence of an association between CCL2 in cancer cells and better survival in NSCLC patients [143].

4.4. Reactive Oxygen Species (ROS)

ROS production by ATII cells results in oxidative stress, which induces apoptosis of epithelial cells, activates intracellular signalling pathways, and upregulates the synthesis of profibrotic cytokines that ultimately leads to tissue injury and fibrosis [144]. In this context, upregulation of NADPH oxidase 4 (NOX4) has been reported in pulmonary fibroblasts and other relevant cells in IPF. In the same way, NOX4 has been reported to be overexpressed in NSCLC (Table 3), contributing to cell proliferation and metastasis [145]. Furthermore, following ROS overproduction, alterations in DNA methylation patterns and specific histone modifications lead to aberrant gene expression, and possibly trigger the multistage process of carcinogenesis [144]. In addition, antioxidant molecules that mitigate oxidative stress, such as Nrf2, have also been reported to be dysregulated in both diseases [146,147] (Table 3). As such, they are proposed to be future targets for anti-IPF/LC treatment.

4.5. Mucins

Significant overexpression of the secreted Mucin 5B (Muc5B) protein has been found in IPF lungs (Table 3) and it is hypothesised that excess Muc5B impairs the mucosal host defence; in turn, this may interfere with alveolar repair and leads to the development of idiopathic interstitial pneumonia [148]. In this context, *MUC5B* expression has been associated with a high risk of distant metastasis in NSCLC patients and poorer prognosis in ADC patients [149,150].

We have also observed IPF overexpression of the transmembrane mucins, Muc1 [151], Muc4 [152], and Muc16 (unpublished data), which may be involved in the molecular processes that lead to the development of pulmonary fibrosis [151–153]. In addition, the extracellular region of Muc1 contains the KL-6 epitope, which is proposed to be a useful biomarker for evaluating disease activity and predicting clinical outcomes in IPF [154]. Similarly, these transmembrane mucins have previously been considered clinically relevant proteins that are aberrantly overexpressed in lung carcinogenesis [155]. In fact, Muc1 is a target in several preclinical and clinical trials for cancer treatment [156,157]. Concurrently, there is evidence that galectin 3 is a promising target for IPF [158] because it has a profibrotic action [159] that is partly mediated by binding to Muc1 [160]. Recently, the potential of galectin-3 as a therapeutic target in cancer has been highlighted since it is capable of modulating anti-tumour immunity [161].

4.6. Embryological Pathways

There is also evidence that some embryological pathways are reactivated or deregulated in fibrotic diseases (Table 3) [162]. For example, the Wnt/β-catenin pathway is overexpressed in the lung tissue of IPF [163] and LC patients [164]. This pathway regulates the expression of molecules involved in tissue invasion, such as matrilysin, laminin, and cyclin-D1, which induces the EMT process. Most importantly, this pathway is involved in biologically relevant cross talk with TGF-β [163].

The Sonic hedgehog (shh) pathway is also aberrantly activated in IPF, mainly in epithelial cells that line honeycomb cysts [165,166]. The overexpression of the shh pathway promotes increased susceptibility to epithelial cell apoptosis and increased resistance to fibroblast apoptosis [167]. This pathway is also reactivated at the early stage of oncogenesis by cancer stem cells and leads to paracrine action on other tumour cells, resulting in tumour growth, tumour spread, and EMT. In LC, reactivation of the shh pathway is involved in the development of resistance to all the main treatments of LC [168].

Finally, the Notch signalling pathway is also reactivated in AECs, induces α-SMA expression in fibroblasts, and mediates EMT in AECs [52]. In the same way, abnormal expression of the members of the Notch signalling pathway is a relatively frequent event in patients with NSCLC [169,170]. It has been demonstrated that members of the Notch signalling pathway may be potential biomarkers for predicting the progression and prognosis of patients with NSCLC. Furthermore, Notch signalling promotes the proliferation of NSCLC cells or inhibits apoptosis of NSCLC cells [171].

4.7. PI3K/AKT/mTOR Pathway

The phosphoinositide 3-kinase (PI3K)/protein kinase B (AKT)/mammalian target of rapamycin (mTOR)-dependent pathway is dysregulated in fibroproliferative diseases, like pulmonary fibrosis (Table 3) [172]. In fact, overexpression of class I isoform p110γ in lung homogenates occurs in IPF patients [173], and has been shown to activate the downstream signalling of several key profibrotic growth factors implicated in IPF, including PDGF and TGFβ1 [174,175], as well as abnormal proliferation of epithelial basal cells [173] and TGF-β-induced fibroblast proliferation and differentiation [176]. Moreover, it has been observed that the suppression of phosphatase and tensin homologue mediates matrix-mediated resistance to apoptosis [174]. Phosphatase and tensin homologue are negative regulators of PI3K that in turn activate AKT. De-regulation of the PI3K/AKT/mTOR pathway is also involved in NSCLC and has been associated with high grade tumours and advanced disease. Furthermore, abnormalities in this pathway are more common in SQC than in ADC of the lung [177].

5. Genetic and Epigenetic Alterations in Lung Cancer Associated with Pulmonary Fibrosis

5.1. Genetic Alterations

Most pulmonary fibrosis patients who have a background of familial clustering of familial interstitial pneumonia show mutations in genes that encode surfactant-associated protein C (*SFTPC*) [178,179], surfactant-associated protein A2 (*SFTPA2*) [180], telomerase components (*TERT* and *TERC*) [181,182], and genes associated with telomere biology, such as poly (a)-specific ribonuclease deadenylation nuclease (*PARN*) and regulator of telomere elongation helicase 1 (*RTEL1*) [183]. Further, several common variants in *TERT*, genomic regions near *TERC*, oligonucleotide-binding fold containing 1 (*OBFC1*), *RTEL1*, *PARN*, and toll-interacting protein (*TOLLIP*), have all been associated with a sporadic risk of developing IPF [184–186]. Nevertheless, the most important genetic risk factor for sporadic IPF is a common variant (rs35705950) in the promoter region of the *MUC5B* gene, although it is also associated with familial pulmonary fibrosis [148] (Table 4).

LC development is the result of a stepwise accumulation of multiple acquired mutations of tumour suppressor genes or candidates, and the overexpression and mutation of oncogenes (Table 4). In this context, multiple *P53* gene mutations have been found during the early stage of bronchial carcinoma [187,188]. Frequent *P53* gene alterations have also been detected in epithelial lesions from IPF patients [189] and in squamous metaplasia, distributed in the peripheral zone of the fibrotic area in patients with IPF [190]. Similarly, aberration and loss of function of the candidate tumour suppressor gene fragile histidine triad (*FHIT*) has been reported in NSCLC [191], as well as in IPF lesions [192]. Additionally, *FHIT* gene allelic loss has been seen more frequently among the metaplasias and bronchiolar epithelia samples obtained from LC-IPF patients than from IPF patients without LC [192]. Otherwise, the frequency of expression of Ras protein in ATII cells has been observed as being significantly greater in lung tissues from LC-IPF patients compared with lung tissues from IPF patients who did not have lung carcinoma. Moreover, a specific point mutation in codon 12 of the *KRAS* gene has been detected in LC-IPF patients [189]. Interestingly, this mutation has not been identified between the numerous *KRAS* mutations observed in lung carcinoma tissue [193]. However, contrasting results regarding the prevalence of *KRAS* mutations in LC-IPF patients have been reported recently [194,195]. In addition, a significantly higher prevalence of *BRAF* mutations in IPF-LC than in LC has been observed, with an equal distribution between ADC and SQC subtypes. Moreover, some of these somatic mutations have not been shown to be significant in NSCLC patients [195]. As such, it is rational to suggest that these somatic mutations in tumour suppressor genes and oncogenes, as well as oncogene overexpression, predispose IPF patients to develop LC. However, it also raises two controversial questions: Why is LC not the cause, instead of a consequence, of IPF?; and why do LC and IPF not independently and synchronously develop as a result of common pathogenetic mechanisms? In answering these questions, Hwan et al. [195] revealed a predominance of C>T somatic transitions in most of the somatic mutations detected in LC-IPF patients. By contrast, in the non-IPF SQC subtype, C>A transversions are the most frequent [196]. This suggests a potential association between APOBEC (cytidine deaminase, which converts cytosine to uracil)-related mutagenesis and the development of LC associated with pulmonary fibrosis.

Recently, germline mutations associated with familial NSCLC and predisposing to it are also being discovered [197] (Table 4). In this context, several findings suggest that some germline mutations that predispose patients to develop IPF also predispose them to developing lung carcinoma. Indeed, two heterozygous missense mutations, and a heterozygous missense mutation in *SFTPA2* [198] and *SFTPA1* [199], respectively, have been identified in LC-IPF families. All of these mutations are predicted to disrupt the structure of surfactant A protein (SP-A) and impair protein secretion [198,199], leading to protein instability and ER stress of resident ATII cells [179,200]. SP-A is produced by ATII and club cells [201], which have both been proposed as possible initiators of lung ADC [202]. Although the role of ER-stress in tumourigenesis is controversial [69,70], recent evidence showing that ER stress may attenuate senescence and promote tumorigenesis might explain the co-occurrence of LC (histological subtype ADC) and pulmonary fibrosis in families with an *SFTPA1/2* mutation [71]. Two further germline mutations in *TERT* (rs2736100) and *CDKN1A* (rs2395655), which were previously reported to confer IPF risk [203], have also been identified in several LC-IPF patients [195]. These mutations affect telomerase function and impair the cellular response to DNA damage, respectively [204]. Accordingly, they might also explain the co-occurrence of both diseases. Furthermore, the germline variant, rs2736100, has been reported to be associated with lung ADC risk [205], and other *TERT*, *TERC*, *OFBC1*, and *RTEL1* polymorphisms have also been revealed as risk factors of LC [206,207]. However, telomere functionality and its contribution to LC development is controversial. In fact, a gain at chromosomal region 5p15.33 in *TERT* is the most frequent genetic event in the early stages of NSCLC [208]. However, short telomeres in peripheral blood leukocytes have been related to an increased risk of lung carcinoma [209], probably via an increased mutation rate and the genomic instability induced by telomere dysfunction [210]. Therefore, it might be hypothesised that mutations associated with telomere biology in IPF lesions, which correlate with shortened telomeres in leukocytes and ATII cells [211], could drive the development of LC via an increased mutation rate and genomic instability.

Finally, the germline or somatic variant (rs35705950) in the *MUC5B* promoter region that consists of TT and GT genotypes (risk genotypes for IPF) has been reported to confer a survival advantage among patients with IPF [212]. However, these genotypes are associated with poorer overall survival in NSCLC patients [213]. Furthermore, significant associations between the *MUC5B* promoter polymorphism and the incidence of radiation pneumonitis in patients with NSCLC have not been identified [213]. This supports the idea that IPF underlies the development of LC and is not a consequence of it.

5.2. Epigenetic Alterations

Due to similar pathogenic mechanisms between IPF and LC, their global methylation patterns are also somewhat similar (Table 4). However, there are also differences, which may be explained partly by IPF or cancer-specific changes [214]. For example, it was found that as a consequence of promoter hypermethylation, the relative expression of the *SMAD4* gene was significantly lower in the tumours of LC-IPF patients compared to those who had LC without IPF [215]. This was a surprising finding because *SMAD4* has been identified as a tumour-suppressor gene [216], but TGFβ1 signalling is the main effector in pulmonary fibrosis. Thus, *SMAD4* over-expression would be expected in this disease. Another epigenetic alteration involved in IPF is *THY-1* promoter hypermethylation and the absence of fibroblast Thy-1 expression, which is linked to the transformation of fibroblasts into myofibroblasts [217]. Loss of this molecule has also been documented in cancer and is associated with a more invasive disease [218]. By contrast, promoter hypermethylation of the *O*-6-methylguanine

DNA methyltransferase (*MGMT*) gene is one of the early epigenetic marks in LC [219], while in IPF fibroblasts, *MGMT* is one of the most hypomethylated genes [220].

Otherwise, ~10% of miRNAs are abnormally expressed in IPF [95]. These variations are all capable of influencing EMT and inducing the regulation of apoptosis or ECM [95]. Some of these variations are also found in LC. For example, common to IPF, mir-21 is overexpressed in LC [95], which is an independent negative prognostic factor for overall survival in NSCLC patients [221]. By contrast, Let-7d expression is found to be mostly down-regulated in IPF and LC and acts as an oncogene [95,219,222].

Table 4. Mutated genes, hypermethylated genes, and non-coding RNAs with altered expression in Idiopathic pulmonary fibrosis (IPF), lung cancer (LC), and LC-IPF patients.

	IPF	LC	LC-IPF
Mutated Genes			
SFTPA1	Yes [199]	ND	Yes [199]
SFTPA2	Yes [180]	ND	Yes [198]
TERT	Yes [181,184]	Yes [206,207]	Yes [195]
TERC	Yes [181,184]	Yes [206,207]	ND
PARN	Yes [183]	ND	ND
OBFC1	Yes [184]	Yes [207]	ND
RTEL1	Yes [183]	Yes [207]	ND
TOLLIP	Yes [186]	ND	ND
MUC5B	Yes [148]	Yes [213]	ND
P53	Yes [189]	Yes [187,188]	Yes [190]
FHIT	Yes [192]	Yes [191]	Yes [192]
KRAS	ND	Yes [193]	Yes [194]
BRAF	ND	Yes [223]	Yes [195]
CDKN1A	Yes [203]	ND	Yes [195]
Hypermethylated Genes			
SMAD4	ND	ND	Yes [215]
THY-1	Yes [217]	ND *	ND
MGMT	No [220]	Yes [219]	ND
Non-coding RNAs			
Let-7d	Downregulated [95]	Downregulated [219]	ND
miR-21	Upregulated [95]	Upregulated [221]	ND

ND: Not determined. * Reported in metastatic nasopharyngeal carcinoma [218].

6. Therapeutic Management in Lung Cancer Associated with Pulmonary Fibrosis Patients

The focus of IPF treatment in previous decades has been to use anti-inflammatory/immunomodulatory drugs in combination with antioxidants. However, their therapeutic usefulness was recently questioned given the unfavourable outcome when N-acetyl L-cysteine (NAC) was used in combination with prednisolone/azathioprine [224]. Following this, NAC monotherapy results were also negative [225], although a subgroup of IPF patients with the rs3750920 (*TOLLIP*) TT genotype showed a favourable response [226]. Numerous cellular and preclinical studies hold that antioxidants protect against cancer [227,228]. However, it has been shown that NAC increases tumour progression and reduces survival in LC preclinical models [229], which contraindicates NAC treatment for LC-IPF.

In line with the antioxidant treatments tested in IPF, pirfenidone was initially considered as an antioxidant therapy since it demonstrated $O_2{}^-$ scavenging activity [230,231]. Oral NAC has been used in conjunction with pirfenidone to treat IPF, but it does not substantially alter the tolerability profile of pirfenidone and is unlikely to be beneficial in IPF [232]. Beyond its antioxidant activity, pirfenidone is a pleiotropic molecule that inhibits TGF−β, collagen synthesis, and fibroblast proliferation, and also mediates tissue repair [233–236]. It is currently approved as an IPF therapy and Miuri et al. [237] observed that the incidence of LC in IPF patients treated with pirfenidone was significantly lower than in a non-pirfenidone IPF patient group. Furthermore, recent publications have shown that pirfenidone confers anti-fibrotic effects by interfering with the shh pathway [238], which can partly explain the observed lower LC incidence in IPF patients treated with pirfenidone. It has also been observed that perioperative pirfenidone treatment reduces the incidence of postoperative acute exacerbation of IPF in LC-IPF patients [239]. Moreover, experimental data have shown that the combination of pirfenidone and cisplatin may lead to an increase of CAF and tumour cell mortality in NSCLC preclinical models [240].

Advances in the understanding of IPF pathogenesis have resulted in further preclinical and clinical trials of drugs with antiproliferative and antifibrotic effects. For instance, tyrosine kinase inhibitors (TKIs), such as imatinib, nilotinib, gefitinib, erlotinib, nintedanib, SU5918, and SU11657, are being investigated [241–247]. The important role of these inhibitors in cancer treatment was previously shown [248]. Indeed, gefitinib and erlotinib are important oral treatments for NSCLC patients with mutations that activate the epidermal growth factor receptor (EGFR). In IPF, imatinib was tested in fibrotic patients, but failed to show any benefit on survival or lung function [249]. In contrast, the VEGF, FGF, and PDGF receptor inhibitor, nintedanib, has been approved for IPF treatment. Additionally, this drug is also approved for use in combination with docetaxel as an effective second-line option for patients with advanced ADC-NSCLC who have been previously treated with one course of platinum-based therapy [250].

Another class of antifibrotic drugs are the mTOR kinase inhibitors, including everolimus, which failed as an IPF treatment [251]. However, everolimus has shown modest beneficial effects in patients with advanced NSCLC who were previously treated with chemotherapy alone, or with chemotherapy and EGFR inhibitors [252]. It is also approved as a second-line treatment in renal and breast cancer. Currently, there are efforts towards assessing the efficacy of a new mTOR kinase inhibitor (GSK-2126458) for IPF and advanced solid tumour treatment.

In addition to the previously mentioned therapeutic strategies, a broad range of IPF therapies are currently being tested in clinical trials (Table 5). Some of these therapies target molecules and mechanisms mentioned in this review, and which are hallmarks of the progression of both diseases. These include anti-IL-13 antibodies (QAX576 and Lebrikizumab), anti-CCL2 antibodies (Carlumab and CNTO-888), anti-TGFβ1 antibodies (Fresolimumab and GC1008), anti-integrin αvβ6 antibodies (BG0011 and STX-100), integrin αvβ6 antagonist drugs (GSK3008348), LPAR1 antagonist drugs (BMS-986020), ATX-inhibiting drugs (GLPG1690), angiostatic drugs (Tetrathiomolybdate), shh pathway-inhibiting drugs (vismodegib), and galectin-3-inhibiting drugs (TD139). Only two of these drugs are being clinically developed for NSCLC patients. Fresolimumab, which was tested in combination with stereotactic ablative radiotherapy in a phase I study, and tetrathiomolybdate in combination with carboplatin and pemetrexed, which is currently being tested in a phase I study. There are also preclinical studies for NSCLC that include some of these target molecules. For example, CCR2 antagonism was demonstrated to supress CCL2-mediated viability, motility, and invasion of the NSCLC cell line, A549, in vitro [253]. Likewise, galectin-3 knockdown in NSCLC cell line-derived sphere resulted in attenuation of lung carcinogenesis by inhibiting stem-like properties [254]. In the same way, genetic deletion of *ATX* and *LPAR1* was shown to attenuate lung carcinogenesis development in animal models [132]. Moreover, vismodegib is approved for the treatment of metastatic, local, or recurrent advanced basal cell carcinoma (BCC), although it has not been tested in NSCLC. Nevertheless, blockade of shh signaling synergistically has shown to increase sensitivity to EGFR-TKIs in primary and secondary resistant NSCLC cells [255].

Given the mechanistic similarities between LC and IPF diseases, and the concurrence of predominantly NSCLC and IPF, it is rational to consider the usefulness of the large number of approved NSCLC treatments for the management of pulmonary fibrosis. For example, Nivolumab is a new immunomodulatory agent that acts as a programmed death receptor-1-blocking antibody. One case study of an ADC patient with IPF showed a beneficial and sustained response in the lung, without any sign of IPF exacerbation after Nivolumab treatment [256]. This could be explained by the higher expression of programmed cell death ligand 1 reported in cancer cells from UIP-associated SQC versus non-UIP SQC patients [257]. Other examples of feasible IPF and LC-IPF treatment candidates include vanticumab, which interferes with Wnt signalling and has undergone Phase I trials for NSCLC (preclinical studies of Wnt pathway inhibition have also been performed in pulmonary fibrosis [258,259]), and Muc1-based therapeutic strategies. Indeed, four Muc1-based Phase III trials exploring cancer treatment have been completed, one of which used a Muc1 tandem repeat peptide as an immunogen (L-BLP25) in patients with stage III unresectable NSCLC after chemoradiation [260].

Table 5. Development status of drugs targeting molecules and processes involved in lung cancer (LC) and Idiopathic pulmonary fibrosis (IPF).

Therapy	IPF	LC
Anti-PDGFR, VEGFR, FGFR (nintedanib)	Approved	Approved in combination with docetaxel (second-line treatment) for ADC-NSCLC
Anti-fibrotic drug (pirfenidone)	Approved	Preclinical studies for NSCLC [240]
Anti-IL13	QAX576 (NCT00532233, NCT01266135: Phase II completed) Lebrikizumab (NCT01872689: Phase II completed)	Not studied
Anti-CCL2	Carlumab (CNTO-888) (NCT00786201: Phase II completed	Preclinical studies for NSCLC [253]
Galectin-3 inhibition	TD139 (NCT02257177: Phase I/II completed)	Preclinical studies for NSCLC [254]
Anti-TGFβ	Fresolimumab (GC1008) (NCT00125385: Phase I completed)	Fresolimumab (GC1008) (NCT02581787: Phase I/II suspended) (NSCLC patients)
Anti-αvβ6 integrin	BG0011 (STX-100) (NCT01371305: Phase II completed)	Not studied
αvβ6 antagonist	GSK3008348 (NCT02612051: Phase I completed)	Not studied
Anti-CTGF	Pamrevlumab (FG-3019) (NCT01262001: Phase II completed)	Not studied
LPAR1 antagonist	BMS-986020 (NCT01766817: Phase II completed)	Preclinical studies [132]
Autotaxin inhibition	GLPG1690 (NCT02738801: Phase II completed)	Preclinical studies [132]
Angiostatic agent	Tetrathiomolybdate (NCT00189176: Phase I/II completed)	Tetrathiomolybdate (NCT01837329: Phase I recruiting patients) (NSCLC patients)
mTOR inhibitor	GSK-2126458 (NCT01725139: Phase I completed) Sirolimus (NCT01462006: Not applicable Phase)	Not studied *
TERT gene expression induction	Nandrolone decanoate (NCT02055456: Phase I/II (unknown recruitment status))	Not studied
Shh pathway inhibitor	Vismodegib (NCT02648048: Phase Ib completed)	Preclinical studies for NSCLC [255]
Nivolumab	Not studied	Approved for NSCLC
Notch pathway inhibitor	Artesunate (preclinical studies [261])	Rovalpituzumab (approved for SCLC)
Wnt pathway inhibitor	Preclinical studies [258,259]	Vanticumab (NCT01957007: Phase I completed) (NSCLC patients)
Muc1-based therapies	Anti-KL-6 (preclinical studies [262])	Muc1 immunogen (L-BLP25 (Phase III completed [260])) (NSCLC patients)

IPF: idiopathic pulmonary fibrosis; LC: lung cancer; NSCLC: non-small cell lung cancer; SCLC: small cell lung cancer; PDGFR: platelet derived growth factor receptor; VEGFR: vascular endothelial growth factor receptor; FGFR: fibroblast growth factor receptor; IL-13: interleukin 13; CCL2: chemokine ligand 2; TGFβ: transforming growth factor β; CTGF: connective tissue growth factor; LPAR1: lysophosphatidic acid receptor type 1; mTOR: mammalian Target of Rapamycin. * (NCT02581787: Phase I/II terminated for subjects with solid advanced tumors).

Finally, rovalpituzumab treatment, although not tested for NSCLC, is currently approved for SCLC, and it could also have potential in the treatment of IPF and LC-IPF, since it interferes with the Notch signalling pathway. In fact, it has been observed that artesunate ameliorates lung fibrosis via inhibiting the Notch signaling pathway in a rat bleomycin model [261].

7. Conclusions

The course of IPF disease and its resulting mortality are altered by the frequent co-occurrence of LC. This review supports the view of LC as a consequence of a genetic predisposition in IPF patients and, common cellular processes and molecular pathways between both diseases. Currently, there is no consensus regarding the treatment of patients with both diseases. However, pirfenidone and nintedanib are two novel drugs approved for IPF that have potential for treating patients with fibrosis, possibly extending the survival time and lowering the incidence of LC. However, we are some distance from realising effective therapeutic approaches that are capable of stopping the disease process, where disease progression still occurs in most IPF patients despite treatment. Nevertheless, we now have a great deal of knowledge about cancer biology and its similarities with IPF. Therefore, it seems reasonable to investigate whether specific cancer drugs may exert beneficial anti-fibrotic effects that are effective to treat LC-IPF patients. Furthermore, clinical trials that prospectively investigate the efficacy of currently approved anti-fibrotic agents (or agents under study) as treatments for LC-IPF patients are sorely needed.

Author Contributions: Conception and design: B.B., J.M. and J.C. Data acquisition: B.B., J.M., and J.C. Analysis and interpretation: B.B., J.M. and J.C. Writing and Original Draft Preparation, B.B., J.M. and J.C.; Writing Review & Editing, B.B., J.M. and J.C.; Visualization, B.B., J.M. and J.C.; Supervision, B.B., J.M. and J.C.; Project Administration, B.B., J.M. and J.C.; Funding Acquisition, B.B., J.M. and J.C.

Funding: This work was supported by the grants SAF2017–82913-R (J.C.), FIS PI17/02158 (J.M.), CIBERES (CB06/06/0027), TRACE (TRA2009–0311; Spanish Government), and by research grants from the Regional Government Prometeo 2017/023/UV (J.C.), ACIF/2016/341 (B.B.), from "Generalitat Valenciana". Funding entities did not contribute to the study design or data collection, analysis and interpretation nor to the writing of the manuscript.

Conflicts of Interest: The authors declare no conflict of interest.

Abbreviations

IPF	Idiopathic pulmonary fibrosis
UIP	Usual interstitial pneumonia
LC	Lung cancer
LC-IPF	Lung cancer associated with idiopathic pulmonary fibrosis
NSCLC	Non-small cell-lung cancer
ADC	Adenocarcinoma
SQC	Squamous cell carcinoma
SCLC	Small cell lung cancer
ATII	Alveolar type II cells
ECM	Extracellular matrix
EMT	Epithelial to mesenchymal transition
EndMT	Endothelial to mesenchymal transition
α-SMA	α-smooth muscle actin
CAF	Cancer-associated fibroblast
ER	Endoplasmic reticulum stress
TGFβ	Transforming growth factor β
FGF	Fibroblast growth factor
VEGF	Vascular endothelial growth factor
PDGF	Platelet derived growth factor
IL-13	Interleukin-13

CCL2	Chemokine ligand 2
CCR2	Chemokine receptor 2
BALF	Bronchoalveolar lavage fluid
LPA	Lysophosphatidic acid
LPAR1	Lysophosphatidic acid receptor 1
ATX	Autotaxin
Shh	Sonic hedhehog
TKI	Tyrosine kinase inhibitor
EGFR	Epidermal growth factor

References

1. Raghu, G.; Collard, H.R.; Egan, J.J.; Martinez, F.J.; Behr, J.; Brown, K.K.; Colby, T.V.; Cordier, J.F.; Flaherty, K.R.; Lasky, J.A.; et al. An official ATS/ERS/JRS/ALAT statement: Idiopathic pulmonary fibrosis: Evidence-based guidelines for diagnosis and management. *Am. J. Respir. Crit. Care Med.* **2011**, *183*, 788–824. [CrossRef] [PubMed]

2. Ley, B.; Collard, H.R.; King, T.E., Jr. Clinical course and prediction of survival in idiopathic pulmonary fibrosis. *Am. J. Respir. Crit. Care Med.* **2011**, *183*, 431–440. [CrossRef]

3. Meyer, E.C.; Liebow, A.A. Relationship of Interstitial Pneumonia Honeycombing and Atypical Epithelial Proliferation to Cancer of the Lung. *Cancer* **1965**, *18*, 322–351. [CrossRef]

4. Karampitsakos, T.; Tzilas, V.; Tringidou, R.; Steiropoulos, P.; Aidinis, V.; Papiris, S.A.; Bouros, D.; Tzouvelekis, A. Lung cancer in patients with idiopathic pulmonary fibrosis. *Pulm. Pharmacol. Ther.* **2017**, *45*, 1–10. [CrossRef] [PubMed]

5. Hubbard, R.; Venn, A.; Lewis, S.; Britton, J. Lung cancer and cryptogenic fibrosing alveolitis. A population-based cohort study. *Am. J. Respir. Crit. Care Med.* **2000**, *161*, 5–8. [CrossRef]

6. Wells, C.; Mannino, D.M. Pulmonary fibrosis and lung cancer in the United States: Analysis of the multiple cause of death mortality data, 1979 through 1991. *South. Med J.* **1996**, *89*, 505–510. [CrossRef] [PubMed]

7. Artinian, V.; Kvale, P.A. Cancer and interstitial lung disease. *Curr. Opin. Pulm. Med.* **2004**, *10*, 425–434. [CrossRef] [PubMed]

8. Kato, E.; Takayanagi, N.; Takaku, Y.; Kagiyama, N.; Kanauchi, T.; Ishiguro, T.; Sugita, Y. Incidence and predictive factors of lung cancer in patients with idiopathic pulmonary fibrosis. *ERJ Open Res.* **2018**, *4*, 00111-2016. [CrossRef]

9. Nagai, A.; Chiyotani, A.; Nakadate, T.; Konno, K. Lung cancer in patients with idiopathic pulmonary fibrosis. *Tohoku J. Exp. Med.* **1992**, *167*, 231–237. [CrossRef] [PubMed]

10. Aubry, M.C.; Myers, J.L.; Douglas, W.W.; Tazelaar, H.D.; Washington Stephens, T.L.; Hartman, T.E.; Deschamps, C.; Pankratz, V.S. Primary pulmonary carcinoma in patients with idiopathic pulmonary fibrosis. *Mayo Clin. Proc.* **2002**, *77*, 763–770. [CrossRef] [PubMed]

11. Usui, K.; Tanai, C.; Tanaka, Y.; Noda, H.; Ishihara, T. The prevalence of pulmonary fibrosis combined with emphysema in patients with lung cancer. *Respirology* **2011**, *16*, 326–331. [CrossRef] [PubMed]

12. JafariNezhad, A.; YektaKooshali, M.H. Lung cancer in idiopathic pulmonary fibrosis: A systematic review and meta-analysis. *PLoS ONE* **2018**, *13*, e0202360. [CrossRef] [PubMed]

13. Matsushita, H.; Tanaka, S.; Saiki, Y.; Hara, M.; Nakata, K.; Tanimura, S.; Banba, J. Lung cancer associated with usual interstitial pneumonia. *Pathol. Int.* **1995**, *45*, 925–932. [CrossRef] [PubMed]

14. Park, J.; Kim, D.S.; Shim, T.S.; Lim, C.M.; Koh, Y.; Lee, S.D.; Kim, W.S.; Kim, W.D.; Lee, J.S.; Song, K.S. Lung cancer in patients with idiopathic pulmonary fibrosis. *Eur. Respir. J.* **2001**, *17*, 1216–1219. [CrossRef] [PubMed]

15. Ozawa, Y.; Suda, T.; Naito, T.; Enomoto, N.; Hashimoto, D.; Fujisawa, T.; Nakamura, Y.; Inui, N.; Nakamura, H.; Chida, K. Cumulative incidence of and predictive factors for lung cancer in IPF. *Respirology* **2009**, *14*, 723–728. [CrossRef] [PubMed]

16. Lee, K.J.; Chung, M.P.; Kim, Y.W.; Lee, J.H.; Kim, K.S.; Ryu, J.S.; Lee, H.L.; Park, S.W.; Park, C.S.; Uh, S.T.; et al. Prevalence, risk factors and survival of lung cancer in the idiopathic pulmonary fibrosis. *Thorac. Cancer* **2012**, *3*, 150–155. [CrossRef] [PubMed]

Int. J. Mol. Sci. **2019**, *20*, 593

17. Kreuter, M.; Ehlers-Tenenbaum, S.; Schaaf, M.; Oltmanns, U.; Palmowski, K.; Hoffmann, H.; Schnabel, P.A.; Heussel, C.P.; Puderbach, M.; Herth, F.J.; et al. Treatment and outcome of lung cancer in idiopathic interstitial pneumonias. *Sarcoidosis Vasc. Diffus. Lung Dis. Off. J. WASOG* **2015**, *31*, 266–274.

18. Tomassetti, S.; Gurioli, C.; Ryu, J.H.; Decker, P.A.; Ravaglia, C.; Tantalocco, P.; Buccioli, M.; Piciucchi, S.; Sverzellati, N.; Dubini, A.; et al. The impact of lung cancer on survival of idiopathic pulmonary fibrosis. *Chest* **2015**, *147*, 157–164. [CrossRef]

19. Yoon, J.H.; Nouraie, M.; Chen, X.; Zou, R.H.; Sellares, J.; Veraldi, K.L.; Chiarchiaro, J.; Lindell, K.; Wilson, D.O.; Kaminski, N.; et al. Characteristics of lung cancer among patients with idiopathic pulmonary fibrosis and interstitial lung disease—Analysis of institutional and population data. *Respir. Res.* **2018**, *19*, 195. [CrossRef]

20. Le Jeune, I.; Gribbin, J.; West, J.; Smith, C.; Cullinan, P.; Hubbard, R. The incidence of cancer in patients with idiopathic pulmonary fibrosis and sarcoidosis in the UK. *Respir. Med.* **2007**, *101*, 2534–2540. [CrossRef]

21. Ferlay, J.; Soerjomataram, I.; Dikshit, R.; Eser, S.; Mathers, C.; Rebelo, M.; Parkin, D.M.; Forman, D.; Bray, F. Cancer incidence and mortality worldwide: Sources, methods and major patterns in GLOBOCAN 2012. *Int. J. Cancer* **2015**, *136*, E359–E386. [CrossRef] [PubMed]

22. Hyldgaard, C.; Hilberg, O.; Bendstrup, E. How does comorbidity influence survival in idiopathic pulmonary fibrosis? *Respir. Med.* **2014**, *108*, 647–653. [CrossRef] [PubMed]

23. Dela Cruz, C.S.; Tanoue, L.T.; Matthay, R.A. Lung cancer: Epidemiology, etiology, and prevention. *Clin. Chest Med.* **2011**, *32*, 605–644. [CrossRef] [PubMed]

24. Kawai, T.; Yakumaru, K.; Suzuki, M.; Kageyama, K. Diffuse interstitial pulmonary fibrosis and lung cancer. *Acta Pathol. Jpn.* **1987**, *37*, 11–19. [CrossRef] [PubMed]

25. Kawasaki, H.; Nagai, K.; Yokose, T.; Yoshida, J.; Nishimura, M.; Takahashi, K.; Suzuki, K.; Kakinuma, R.; Nishiwaki, Y. Clinicopathological characteristics of surgically resected lung cancer associated with idiopathic pulmonary fibrosis. *J. Surg. Oncol.* **2001**, *76*, 53–57. [CrossRef]

26. Saito, Y.; Kawai, Y.; Takahashi, N.; Ikeya, T.; Murai, K.; Kawabata, Y.; Hoshi, E. Survival after surgery for pathologic stage IA non-small cell lung cancer associated with idiopathic pulmonary fibrosis. *Ann. Thorac. Surg.* **2011**, *92*, 1812–1817. [CrossRef] [PubMed]

27. Lee, T.; Park, J.Y.; Lee, H.Y.; Cho, Y.J.; Yoon, H.I.; Lee, J.H.; Jheon, S.; Lee, C.T.; Park, J.S. Lung cancer in patients with idiopathic pulmonary fibrosis: Clinical characteristics and impact on survival. *Respir. Med.* **2014**, *108*, 1549–1555. [CrossRef] [PubMed]

28. Khan, K.A.; Kennedy, M.P.; Moore, E.; Crush, L.; Prendeville, S.; Maher, M.M.; Burke, L.; Henry, M.T. Radiological characteristics, histological features and clinical outcomes of lung cancer patients with coexistent idiopathic pulmonary fibrosis. *Lung* **2015**, *193*, 71–77. [CrossRef]

29. Guyard, A.; Danel, C.; Theou-Anton, N.; Debray, M.P.; Gibault, L.; Mordant, P.; Castier, Y.; Crestani, B.; Zalcman, G.; Blons, H.; et al. Morphologic and molecular study of lung cancers associated with idiopathic pulmonary fibrosis and other pulmonary fibroses. *Respir. Res.* **2017**, *18*, 120. [CrossRef] [PubMed]

30. Fraire, A.E.; Greenberg, S.D. Carcinoma and diffuse interstitial fibrosis of lung. *Cancer* **1973**, *31*, 1078–1086. [CrossRef]

31. Shimizu, H. Pathological study of lung cancer associated with idiopathic interstitial pneumonia—With special reference to relationship between the primary site of lung cancer and honeycombing. *Nihon Kyobu Shikkan Gakkai Zasshi* **1985**, *23*, 873–881. [PubMed]

32. Mizushima, Y.; Kobayashi, M. Clinical characteristics of synchronous multiple lung cancer associated with idiopathic pulmonary fibrosis. A review of Japanese cases. *Chest* **1995**, *108*, 1272–1277. [CrossRef] [PubMed]

33. Calio, A.; Lever, V.; Rossi, A.; Gilioli, E.; Brunelli, M.; Dubini, A.; Tomassetti, S.; Piciucchi, S.; Nottegar, A.; Rossi, G.; et al. Increased frequency of bronchiolar histotypes in lung carcinomas associated with idiopathic pulmonary fibrosis. *Histopathology* **2017**, *71*, 725–735. [CrossRef] [PubMed]

34. Strock, S.B.; Alder, J.K.; Kass, D.J. From bad to worse: When lung cancer complicates idiopathic pulmonary fibrosis. *J. Pathol.* **2018**, *244*, 383–385. [CrossRef] [PubMed]

35. Bargagli, E.; Bonti, V.; Ferrari, K.; Rosi, E.; Bindi, A.; Bartolucci, M.; Chiara, M.; Voltolini, L. Lung Cancer in Patients with Severe Idiopathic Pulmonary Fibrosis: Critical Aspects. *In Vivo* **2017**, *31*, 773–777. [PubMed]

36. Hironaka, M.; Fukayama, M. Pulmonary fibrosis and lung carcinoma: A comparative study of metaplastic epithelia in honeycombed areas of usual interstitial pneumonia with or without lung carcinoma. *Pathol. Int.* **1999**, *49*, 1060–1066. [CrossRef] [PubMed]

39

37. Scotton, C.J.; Chambers, R.C. Molecular targets in pulmonary fibrosis: The myofibroblast in focus. *Chest* **2007**, *132*, 1311–1321. [CrossRef]

38. Singh, S.R.; Hall, I.P. Airway myofibroblasts and their relationship with airway myocytes and fibroblasts. *Proc. Am. Thorac. Soc.* **2008**, *5*, 127–132. [CrossRef]

39. Thannickal, V.J.; Toews, G.B.; White, E.S.; Lynch, J.P., 3rd; Martinez, F.J. Mechanisms of pulmonary fibrosis. *Annu. Rev. Med.* **2004**, *55*, 395–417. [CrossRef]

40. Evans, J.N.; Kelley, J.; Krill, J.; Low, R.B.; Adler, K.B. The myofibroblast in pulmonary fibrosis. *Chest* **1983**, *83*, 97S–98S. [CrossRef]

41. Kasai, H.; Allen, J.T.; Mason, R.M.; Kamimura, T.; Zhang, Z. TGF-beta1 induces human alveolar epithelial to mesenchymal cell transition (EMT). *Respir. Res.* **2005**, *6*, 56. [CrossRef] [PubMed]

42. Yao, H.W.; Xie, Q.M.; Chen, J.Q.; Deng, Y.M.; Tang, H.F. TGF-beta1 induces alveolar epithelial to mesenchymal transition in vitro. *Life Sci.* **2004**, *76*, 29–37. [CrossRef] [PubMed]

43. Willis, B.C.; Liebler, J.M.; Luby-Phelps, K.; Nicholson, A.G.; Crandall, E.D.; du Bois, R.M.; Borok, Z. Induction of epithelial-mesenchymal transition in alveolar epithelial cells by transforming growth factor-beta1: Potential role in idiopathic pulmonary fibrosis. *Am. J. Pathol.* **2005**, *166*, 1321–1332. [CrossRef]

44. Kalluri, R.; Weinberg, R.A. The basics of epithelial-mesenchymal transition. *J. Clin. Investig.* **2009**, *119*, 1420–1428. [CrossRef] [PubMed]

45. Mehrad, B.; Strieter, R.M. Fibrocytes and the pathogenesis of diffuse parenchymal lung disease. *Fibrogenesis Tissue Repair* **2012**, *5* (Suppl. 1), S22. [CrossRef]

46. Strieter, R.M. Pathogenesis and natural history of usual interstitial pneumonia: The whole story or the last chapter of a long novel. *Chest* **2005**, *128*, 526S–532S. [CrossRef] [PubMed]

47. Metz, C.N. Fibrocytes: A unique cell population implicated in wound healing. *Cell. Mol. Life Sci. CMLS* **2003**, *60*, 1342–1350. [CrossRef]

48. Phillips, R.J.; Burdick, M.D.; Hong, K.; Lutz, M.A.; Murray, L.A.; Xue, Y.Y.; Belperio, J.A.; Keane, M.P.; Strieter, R.M. Circulating fibrocytes traffic to the lungs in response to CXCL12 and mediate fibrosis. *J. Clin. Investig.* **2004**, *114*, 438–446. [CrossRef]

49. Schmidt, M.; Sun, G.; Stacey, M.A.; Mori, L.; Mattoli, S. Identification of circulating fibrocytes as precursors of bronchial myofibroblasts in asthma. *J. Immunol.* **2003**, *171*, 380–389. [CrossRef]

50. Almudever, P.; Milara, J.; De Diego, A.; Serrano-Mollar, A.; Xaubet, A.; Perez-Vizcaino, F.; Cogolludo, A.; Cortijo, J. Role of tetrahydrobiopterin in pulmonary vascular remodelling associated with pulmonary fibrosis. *Thorax* **2013**, *68*, 938–948. [CrossRef]

51. Nataraj, D.; Ernst, A.; Kalluri, R. Idiopathic pulmonary fibrosis is associated with endothelial to mesenchymal transition. *Am. J. Respir. Cell Mol. Biol.* **2010**, *43*, 129–130. [CrossRef] [PubMed]

52. Fernandez, I.E.; Eickelberg, O. New cellular and molecular mechanisms of lung injury and fibrosis in idiopathic pulmonary fibrosis. *Lancet* **2012**, *380*, 680–688. [CrossRef]

53. Hung, C.; Linn, G.; Chow, Y.H.; Kobayashi, A.; Mittelsteadt, K.; Altemeier, W.A.; Gharib, S.A.; Schnapp, L.M.; Duffield, J.S. Role of lung pericytes and resident fibroblasts in the pathogenesis of pulmonary fibrosis. *Am. J. Respir. Crit. Care Med.* **2013**, *188*, 820–830. [CrossRef] [PubMed]

54. Zolak, J.S.; Jagirdar, R.; Surolia, R.; Karki, S.; Oliva, O.; Hock, T.; Guroji, P.; Ding, Q.; Liu, R.M.; Bolisetty, S.; et al. Pleural mesothelial cell differentiation and invasion in fibrogenic lung injury. *Am. J. Pathol.* **2013**, *182*, 1239–1247. [CrossRef] [PubMed]

55. Nasreen, N.; Mohammed, K.A.; Mubarak, K.K.; Baz, M.A.; Akindipe, O.A.; Fernandez-Bussy, S.; Antony, V.B. Pleural mesothelial cell transformation into myofibroblasts and haptotactic migration in response to TGF-beta1 in vitro. *Am. J. Physiol. Lung Cell. Mol. Physiol.* **2009**, *297*, L115–L124. [CrossRef] [PubMed]

56. Mubarak, K.K.; Montes-Worboys, A.; Regev, D.; Nasreen, N.; Mohammed, K.A.; Faruqi, I.; Hensel, E.; Baz, M.A.; Akindipe, O.A.; Fernandez-Bussy, S.; et al. Parenchymal trafficking of pleural mesothelial cells in idiopathic pulmonary fibrosis. *Eur. Respir. J.* **2012**, *39*, 133–140. [CrossRef]

57. Horowitz, J.C.; Osterholzer, J.J.; Marazioti, A.; Stathopoulos, G.T. "Scar-cinoma": Viewing the fibrotic lung mesenchymal cell in the context of cancer biology. *Eur. Respir. J.* **2016**, *47*, 1842–1854. [CrossRef] [PubMed]

58. Kalluri, R. The biology and function of fibroblasts in cancer. *Nat. Rev. Cancer* **2016**, *16*, 582–598. [CrossRef]

59. Konigshoff, M. Lung cancer in pulmonary fibrosis: Tales of epithelial cell plasticity. *Respir. Int. Rev. Thorac. Dis.* **2011**, *81*, 353–358. [CrossRef]

60. Direkze, N.C.; Hodivala-Dilke, K.; Jeffery, R.; Hunt, T.; Poulsom, R.; Oukrif, D.; Alison, M.R.; Wright, N.A. Bone marrow contribution to tumor-associated myofibroblasts and fibroblasts. *Cancer Res.* **2004**, *64*, 8492–8495. [CrossRef]

61. Shiga, K.; Hara, M.; Nagasaki, T.; Sato, T.; Takahashi, H.; Takeyama, H. Cancer-Associated Fibroblasts: Their Characteristics and Their Roles in Tumor Growth. *Cancers* **2015**, *7*, 2443–2458. [CrossRef] [PubMed]

62. Zeisberg, E.M.; Potenta, S.; Xie, L.; Zeisberg, M.; Kalluri, R. Discovery of endothelial to mesenchymal transition as a source for carcinoma-associated fibroblasts. *Cancer Res.* **2007**, *67*, 10123–10128. [CrossRef]

63. Hosaka, K.; Yang, Y.; Seki, T.; Fischer, C.; Dubey, O.; Fredlund, E.; Hartman, J.; Religa, P.; Morikawa, H.; Ishii, Y.; et al. Pericyte-fibroblast transition promotes tumor growth and metastasis. *Proc. Natl. Acad. Sci. USA* **2016**, *113*, E5618–E5627. [CrossRef] [PubMed]

64. Rinkevich, Y.; Mori, T.; Sahoo, D.; Xu, P.X.; Bermingham, J.R., Jr.; Weissman, I.L. Identification and prospective isolation of a mesothelial precursor lineage giving rise to smooth muscle cells and fibroblasts for mammalian internal organs, and their vasculature. *Nat. Cell Biol.* **2012**, *14*, 1251–1260. [CrossRef] [PubMed]

65. Tanjore, H.; Blackwell, T.S.; Lawson, W.E. Emerging evidence for endoplasmic reticulum stress in the pathogenesis of idiopathic pulmonary fibrosis. *Am. J. Physiol. Lung Cell. Mol. Physiol.* **2012**, *302*, L721–L729. [CrossRef] [PubMed]

66. Wolters, P.J.; Collard, H.R.; Jones, K.D. Pathogenesis of Idiopathic Pulmonary Fibrosis. *Annu. Rev. Pathol.* **2013**. [CrossRef] [PubMed]

67. Thannickal, V.J.; Horowitz, J.C. Evolving concepts of apoptosis in idiopathic pulmonary fibrosis. *Proc. Am. Thorac. Soc.* **2006**, *3*, 350–356. [CrossRef]

68. Hanahan, D.; Weinberg, R.A. Hallmarks of cancer: The next generation. *Cell* **2011**, *144*, 646–674. [CrossRef]

69. Tsai, Y.C.; Weissman, A.M. The Unfolded Protein Response, Degradation from Endoplasmic Reticulum and Cancer. *Genes Cancer* **2010**, *1*, 764–778. [CrossRef]

70. Lin, J.H.; Walter, P.; Yen, T.S. Endoplasmic reticulum stress in disease pathogenesis. *Annu. Rev. Pathol.* **2008**, *3*, 399–425. [CrossRef]

71. Zhu, B.; Ferry, C.H.; Markell, L.K.; Blazanin, N.; Glick, A.B.; Gonzalez, F.J.; Peters, J.M. The nuclear receptor peroxisome proliferator-activated receptor-beta/delta (PPARbeta/delta) promotes oncogene-induced cellular senescence through repression of endoplasmic reticulum stress. *J. Biol. Chem.* **2014**, *289*, 20102–20119. [CrossRef] [PubMed]

72. Araya, J.; Kojima, J.; Takasaka, N.; Ito, S.; Fujii, S.; Hara, H.; Yanagisawa, H.; Kobayashi, K.; Tsurushige, C.; Kawaishi, M.; et al. Insufficient autophagy in idiopathic pulmonary fibrosis. *Am. J. Physiol. Lung Cell. Mol. Physiol.* **2013**, *304*, L56–L69. [CrossRef] [PubMed]

73. Patel, A.S.; Lin, L.; Geyer, A.; Haspel, J.A.; An, C.H.; Cao, J.; Rosas, I.O.; Morse, D. Autophagy in idiopathic pulmonary fibrosis. *PLoS ONE* **2012**, *7*, e41394. [CrossRef] [PubMed]

74. Sosulski, M.L.; Gongora, R.; Danchuk, S.; Dong, C.; Luo, F.; Sanchez, C.G. Deregulation of selective autophagy during aging and pulmonary fibrosis: The role of TGFbeta1. *Aging Cell* **2015**, *14*, 774–783. [CrossRef] [PubMed]

75. Del Principe, D.; Vona, R.; Giordani, L.; Straface, E.; Giammarioli, A.M. Defective autophagy in fibroblasts may contribute to fibrogenesis in autoimmune processes. *Curr. Pharm. Des.* **2011**, *17*, 3878–3887. [CrossRef] [PubMed]

76. Marino, G.; Niso-Santano, M.; Baehrecke, E.H.; Kroemer, G. Self-consumption: The interplay of autophagy and apoptosis. *Nat. Rev. Mol. Cell Biol.* **2014**, *15*, 81–94. [CrossRef] [PubMed]

77. White, E. The role for autophagy in cancer. *J. Clin. Investig.* **2015**, *125*, 42–46. [CrossRef] [PubMed]

78. Levy, J.M.M.; Towers, C.G.; Thorburn, A. Targeting autophagy in cancer. *Nat. Rev. Cancer* **2017**, *17*, 528–542. [CrossRef] [PubMed]

79. Vancheri, C.; Sortino, M.A.; Tomaselli, V.; Mastruzzo, C.; Condorelli, F.; Bellistri, G.; Pistorio, M.P.; Canonico, P.L.; Crimi, N. Different expression of TNF-alpha receptors and prostaglandin E(2)Production in normal and fibrotic lung fibroblasts: Potential implications for the evolution of the inflammatory process. *Am. J. Respir. Cell Mol. Biol.* **2000**, *22*, 628–634. [CrossRef] [PubMed]

80. Moore, B.B.; Ballinger, M.N.; White, E.S.; Green, M.E.; Herrygers, A.B.; Wilke, C.A.; Toews, G.B.; Peters-Golden, M. Bleomycin-induced E prostanoid receptor changes alter fibroblast responses to prostaglandin E2. *J. Immunol.* **2005**, *174*, 5644–5649. [CrossRef]

81. Nho, R.S.; Hergert, P.; Kahm, J.; Jessurun, J.; Henke, C. Pathological alteration of FoxO3a activity promotes idiopathic pulmonary fibrosis fibroblast proliferation on type i collagen matrix. *Am. J. Pathol.* **2011**, *179*, 2420–2430. [CrossRef] [PubMed]

82. Grimminger, F.; Gunther, A.; Vancheri, C. The role of tyrosine kinases in the pathogenesis of idiopathic pulmonary fibrosis. *Eur. Respir. J.* **2015**, *45*, 1426–1433. [CrossRef] [PubMed]

83. Ramos, C.; Montano, M.; Garcia-Alvarez, J.; Ruiz, V.; Uhal, B.D.; Selman, M.; Pardo, A. Fibroblasts from idiopathic pulmonary fibrosis and normal lungs differ in growth rate, apoptosis, and tissue inhibitor of metalloproteinases expression. *Am. J. Respir. Cell Mol. Biol.* **2001**, *24*, 591–598. [CrossRef] [PubMed]

84. Mio, T.; Nagai, S.; Kitaichi, M.; Kawatani, A.; Izumi, T. Proliferative characteristics of fibroblast lines derived from open lung biopsy specimens of patients with IPF (UIP). *Chest* **1992**, *102*, 832–837. [CrossRef] [PubMed]

85. Raghu, G.; Chen, Y.Y.; Rusch, V.; Rabinovitch, P.S. Differential proliferation of fibroblasts cultured from normal and fibrotic human lungs. *Am. Rev. Respir. Dis.* **1988**, *138*, 703–708. [CrossRef] [PubMed]

86. Losa, D.; Chanson, M.; Crespin, S. Connexins as therapeutic targets in lung disease. *Expert Opin. Ther. Targets* **2011**, *15*, 989–1002. [CrossRef] [PubMed]

87. Trovato-Salinaro, A.; Trovato-Salinaro, E.; Failla, M.; Mastruzzo, C.; Tomaselli, V.; Gili, E.; Crimi, N.; Condorelli, D.F.; Vancheri, C. Altered intercellular communication in lung fibroblast cultures from patients with idiopathic pulmonary fibrosis. *Respir. Res.* **2006**, *7*, 122. [CrossRef]

88. Cesen-Cummings, K.; Fernstrom, M.J.; Malkinson, A.M.; Ruch, R.J. Frequent reduction of gap junctional intercellular communication and connexin43 expression in human and mouse lung carcinoma cells. *Carcinogenesis* **1998**, *19*, 61–67. [CrossRef]

89. Schafer, M.J.; White, T.A.; Iijima, K.; Haak, A.J.; Ligresti, G.; Atkinson, E.J.; Oberg, A.L.; Birch, J.; Salmonowicz, H.; Zhu, Y.; et al. Cellular senescence mediates fibrotic pulmonary disease. *Nat. Commun.* **2017**, *8*, 14532. [CrossRef]

90. Selman, M.; Pardo, A. Revealing the pathogenic and aging-related mechanisms of the enigmatic idiopathic pulmonary fibrosis. an integral model. *Am. J. Respir. Crit. Care Med.* **2014**, *189*, 1161–1172. [CrossRef]

91. Dimri, G.P. What has senescence got to do with cancer? *Cancer Cell* **2005**, *7*, 505–512. [CrossRef] [PubMed]

92. Krtolica, A.; Parrinello, S.; Lockett, S.; Desprez, P.Y.; Campisi, J. Senescent fibroblasts promote epithelial cell growth and tumorigenesis: A link between cancer and aging. *Proc. Natl. Acad. Sci. USA* **2001**, *98*, 12072–12077. [CrossRef] [PubMed]

93. White, E.S.; Thannickal, V.J.; Carskadon, S.L.; Dickie, E.G.; Livant, D.L.; Markwart, S.; Toews, G.B.; Arenberg, D.A. Integrin alpha4beta1 regulates migration across basement membranes by lung fibroblasts: A role for phosphatase and tensin homologue deleted on chromosome 10. *Am. J. Respir. Crit. Care Med.* **2003**, *168*, 436–442. [CrossRef] [PubMed]

94. Lovgren, A.K.; Kovacs, J.J.; Xie, T.; Potts, E.N.; Li, Y.; Foster, W.M.; Liang, J.; Meltzer, E.B.; Jiang, D.; Lefkowitz, R.J.; et al. beta-arrestin deficiency protects against pulmonary fibrosis in mice and prevents fibroblast invasion of extracellular matrix. *Sci. Transl. Med.* **2011**, *3*, 74ra23. [CrossRef]

95. Vancheri, C. Common pathways in idiopathic pulmonary fibrosis and cancer. *Eur. Respir. Rev. Off. J. Eur. Respir. Soc.* **2013**, *22*, 265–272. [CrossRef] [PubMed]

96. Moriya, Y.; Niki, T.; Yamada, T.; Matsuno, Y.; Kondo, H.; Hirohashi, S. Increased expression of laminin-5 and its prognostic significance in lung adenocarcinomas of small size. An immunohistochemical analysis of 102 cases. *Cancer* **2001**, *91*, 1129–1141. [CrossRef]

97. Garrido, C.; Schmitt, E.; Cande, C.; Vahsen, N.; Parcellier, A.; Kroemer, G. HSP27 and HSP70: Potentially oncogenic apoptosis inhibitors. *Cell Cycle* **2003**, *2*, 579–584. [CrossRef]

98. Pelosi, G.; Pastorino, U.; Pasini, F.; Maissonneuve, P.; Fraggetta, F.; Iannucci, A.; Sonzogni, A.; De Manzoni, G.; Terzi, A.; Durante, E.; et al. Independent prognostic value of fascin immunoreactivity in stage I nonsmall cell lung cancer. *Br. J. Cancer* **2003**, *88*, 537–547. [CrossRef]

99. Chilosi, M.; Zamo, A.; Doglioni, C.; Reghellin, D.; Lestani, M.; Montagna, L.; Pedron, S.; Ennas, M.G.; Cancellieri, A.; Murer, B.; et al. Migratory marker expression in fibroblast foci of idiopathic pulmonary fibrosis. *Respir. Res.* **2006**, *7*, 95. [CrossRef]

100. Zhang, L.; Wang, Y.; Wu, G.; Xiong, W.; Gu, W.; Wang, C.Y. Macrophages: Friend or foe in idiopathic pulmonary fibrosis? *Respir. Res.* **2018**, *19*, 170. [CrossRef]

101. Barron, L.; Wynn, T.A. Fibrosis is regulated by Th2 and Th17 responses and by dynamic interactions between fibroblasts and macrophages. *Am. J. Physiol. Gastrointest. Liver Physiol.* **2011**, *300*, G723–G728. [CrossRef] [PubMed]

102. Ballinger, M.N.; Newstead, M.W.; Zeng, X.; Bhan, U.; Mo, X.M.; Kunkel, S.L.; Moore, B.B.; Flavell, R.; Christman, J.W.; Standiford, T.J. IRAK-M promotes alternative macrophage activation and fibroproliferation in bleomycin-induced lung injury. *J. Immunol.* **2015**, *194*, 1894–1904. [CrossRef] [PubMed]

103. Osterholzer, J.J.; Olszewski, M.A.; Murdock, B.J.; Chen, G.H.; Erb-Downward, J.R.; Subbotina, N.; Browning, K.; Lin, Y.; Morey, R.E.; Dayrit, J.K.; et al. Implicating exudate macrophages and Ly-6C(high) monocytes in CCR2-dependent lung fibrosis following gene-targeted alveolar injury. *J. Immunol.* **2013**, *190*, 3447–3457. [CrossRef]

104. Craig, V.J.; Zhang, L.; Hagood, J.S.; Owen, C.A. Matrix metalloproteinases as therapeutic targets for idiopathic pulmonary fibrosis. *Am. J. Respir. Cell Mol. Biol.* **2015**, *53*, 585–600. [CrossRef] [PubMed]

105. Murray, P.J.; Wynn, T.A. Protective and pathogenic functions of macrophage subsets. *Nat. Rev. Immunol.* **2011**, *11*, 723–737. [CrossRef] [PubMed]

106. Noy, R.; Pollard, J.W. Tumor-associated macrophages: From mechanisms to therapy. *Immunity* **2014**, *41*, 49–61. [CrossRef] [PubMed]

107. Akhurst, R.J.; Hata, A. Targeting the TGFbeta signalling pathway in disease. *Nat. Rev. Drug Discov.* **2012**, *11*, 790–811. [CrossRef] [PubMed]

108. Kelly, M.; Kolb, M.; Bonniaud, P.; Gauldie, J. Re-evaluation of fibrogenic cytokines in lung fibrosis. *Curr. Pharm. Des.* **2003**, *9*, 39–49. [CrossRef]

109. O'Riordan, T.G.; Smith, V.; Raghu, G. Development of novel agents for idiopathic pulmonary fibrosis: Progress in target selection and clinical trial design. *Chest* **2015**, *148*, 1083–1092. [CrossRef]

110. Roberts, A.B.; Wakefield, L.M. The two faces of transforming growth factor beta in carcinogenesis. *Proc. Natl. Acad. Sci. USA* **2003**, *100*, 8621–8623. [CrossRef]

111. Hetzel, M.; Bachem, M.; Anders, D.; Trischler, G.; Faehling, M. Different effects of growth factors on proliferation and matrix production of normal and fibrotic human lung fibroblasts. *Lung* **2005**, *183*, 225–237. [CrossRef] [PubMed]

112. Bonner, J.C. Regulation of PDGF and its receptors in fibrotic diseases. *Cytokine Growth Fact. Rev.* **2004**, *15*, 255–273. [CrossRef] [PubMed]

113. Noskovicova, N.; Petrek, M.; Eickelberg, O.; Heinzelmann, K. Platelet-derived growth factor signaling in the lung. From lung development and disease to clinical studies. *Am. J. Respir. Cell Mol. Biol.* **2015**, *52*, 263–284. [CrossRef] [PubMed]

114. Battegay, E.J.; Raines, E.W.; Seifert, R.A.; Bowen-Pope, D.F.; Ross, R. TGF-beta induces bimodal proliferation of connective tissue cells via complex control of an autocrine PDGF loop. *Cell* **1990**, *63*, 515–524. [CrossRef]

115. Battegay, E.J.; Raines, E.W.; Colbert, T.; Ross, R. TNF-alpha stimulation of fibroblast proliferation. Dependence on platelet-derived growth factor (PDGF) secretion and alteration of PDGF receptor expression. *J. Immunol.* **1995**, *154*, 6040–6047. [PubMed]

116. Wang, P.; Song, L.; Ge, H.; Jin, P.; Jiang, Y.; Hu, W.; Geng, N. Crenolanib, a PDGFR inhibitor, suppresses lung cancer cell proliferation and inhibits tumor growth in vivo. *Oncotargets Ther.* **2014**, *7*, 1761–1768. [CrossRef]

117. Shikada, Y.; Yonemitsu, Y.; Koga, T.; Onimaru, M.; Nakano, T.; Okano, S.; Sata, S.; Nakagawa, K.; Yoshino, I.; Maehara, Y.; et al. Platelet-derived growth factor-AA is an essential and autocrine regulator of vascular endothelial growth factor expression in non-small cell lung carcinomas. *Cancer Res.* **2005**, *65*, 7241–7248. [CrossRef] [PubMed]

118. Chaudhary, N.I.; Roth, G.J.; Hilberg, F.; Muller-Quernheim, J.; Prasse, A.; Zissel, G.; Schnapp, A.; Park, J.E. Inhibition of PDGF, VEGF and FGF signalling attenuates fibrosis. *Eur. Respir. J.* **2007**, *29*, 976–985. [CrossRef]

119. Frezzetti, D.; Gallo, M.; Maiello, M.R.; D'Alessio, A.; Esposito, C.; Chicchinelli, N.; Normanno, N.; De Luca, A. VEGF as a potential target in lung cancer. *Expert Opin. Ther. Targets* **2017**, *21*, 959–966. [CrossRef]

120. Zhang, S.; Smartt, H.; Holgate, S.T.; Roche, W.R. Growth factors secreted by bronchial epithelial cells control myofibroblast proliferation: An in vitro co-culture model of airway remodeling in asthma. *Lab. Investig. A J. Tech. Methods Pathol.* **1999**, *79*, 395–405.

121. Khalil, N.; Xu, Y.D.; O'Connor, R.; Duronio, V. Proliferation of pulmonary interstitial fibroblasts is mediated by transforming growth factor-beta1-induced release of extracellular fibroblast growth factor-2 and phosphorylation of p38 MAPK and JNK. *J. Biol. Chem.* **2005**, *280*, 43000–43009. [CrossRef] [PubMed]

122. Guzy, R.D.; Li, L.; Smith, C.; Dorry, S.J.; Koo, H.Y.; Chen, L.; Ornitz, D.M. Pulmonary fibrosis requires cell-autonomous mesenchymal fibroblast growth factor (FGF) signaling. *J. Biol. Chem.* **2017**, *292*, 10364–10378. [CrossRef] [PubMed]

123. Suzuki, T.; Yasuda, H.; Funaishi, K.; Arai, D.; Ishioka, K.; Ohgino, K.; Tani, T.; Hamamoto, J.; Ohashi, A.; Naoki, K.; et al. Multiple roles of extracellular fibroblast growth factors in lung cancer cells. *Int. J. Oncol.* **2015**, *46*, 423–429. [CrossRef] [PubMed]

124. Allen, J.T.; Knight, R.A.; Bloor, C.A.; Spiteri, M.A. Enhanced insulin-like growth factor binding protein-related protein 2 (Connective tissue growth factor) expression in patients with idiopathic pulmonary fibrosis and pulmonary sarcoidosis. *Am. J. Respir. Cell Mol. Biol.* **1999**, *21*, 693–700. [CrossRef] [PubMed]

125. Chang, C.C.; Shih, J.Y.; Jeng, Y.M.; Su, J.L.; Lin, B.Z.; Chen, S.T.; Chau, Y.P.; Yang, P.C.; Kuo, M.L. Connective tissue growth factor and its role in lung adenocarcinoma invasion and metastasis. *J. Natl. Cancer Inst.* **2004**, *96*, 364–375. [CrossRef]

126. Chien, W.; Yin, D.; Gui, D.; Mori, A.; Frank, J.M.; Said, J.; Kusuanco, D.; Marchevsky, A.; McKenna, R.; Koeffler, H.P. Suppression of cell proliferation and signaling transduction by connective tissue growth factor in non-small cell lung cancer cells. *Mol. Cancer Res. MCR* **2006**, *4*, 591–598. [CrossRef] [PubMed]

127. Allen, J.T.; Spiteri, M.A. Growth factors in idiopathic pulmonary fibrosis: Relative roles. *Respir. Res.* **2002**, *3*, 13. [CrossRef]

128. Kelly, K.; Kane, M.A.; Bunn, P.A., Jr. Growth factors in lung cancer: Possible etiologic role and clinical target. *Med. Pediatr. Oncol.* **1991**, *19*, 450–458. [CrossRef]

129. Hodkinson, P.S.; Mackinnon, A.; Sethi, T. Targeting growth factors in lung cancer. *Chest* **2008**, *133*, 1209–1216. [CrossRef]

130. Funke, M.; Zhao, Z.; Xu, Y.; Chun, J.; Tager, A.M. The lysophosphatidic acid receptor LPA1 promotes epithelial cell apoptosis after lung injury. *Am. J. Respir. Cell Mol. Biol.* **2012**, *46*, 355–364. [CrossRef]

131. Tager, A.M.; LaCamera, P.; Shea, B.S.; Campanella, G.S.; Selman, M.; Zhao, Z.; Polosukhin, V.; Wain, J.; Karimi-Shah, B.A.; Kim, N.D.; et al. The lysophosphatidic acid receptor LPA1 links pulmonary fibrosis to lung injury by mediating fibroblast recruitment and vascular leak. *Nat. Med.* **2008**, *14*, 45–54. [CrossRef] [PubMed]

132. Magkrioti, C.; Oikonomou, N.; Kaffe, E.; Mouratis, M.A.; Xylourgidis, N.; Barbayianni, I.; Megadoukas, P.; Harokopos, V.; Valavanis, C.; Chun, J.; et al. The Autotaxin-Lysophosphatidic Acid Axis Promotes Lung Carcinogenesis. *Cancer Res.* **2018**, *78*, 3634–3644. [CrossRef] [PubMed]

133. Zhu, Z.; Homer, R.J.; Wang, Z.; Chen, Q.; Geba, G.P.; Wang, J.; Zhang, Y.; Elias, J.A. Pulmonary expression of interleukin-13 causes inflammation, mucus hypersecretion, subepithelial fibrosis, physiologic abnormalities, and eotaxin production. *J. Clin. Investig.* **1999**, *103*, 779–788. [CrossRef]

134. Hancock, A.; Armstrong, L.; Gama, R.; Millar, A. Production of interleukin 13 by alveolar macrophages from normal and fibrotic lung. *Am. J. Respir. Cell Mol. Biol.* **1998**, *18*, 60–65. [CrossRef] [PubMed]

135. Ingram, J.L.; Rice, A.B.; Geisenhoffer, K.; Madtes, D.K.; Bonner, J.C. IL-13 and IL-1beta promote lung fibroblast growth through coordinated up-regulation of PDGF-AA and PDGF-Ralpha. *FASEB J. Off. Publ. Fed. Am. Soc. Exp. Biol.* **2004**, *18*, 1132–1134.

136. Kaviratne, M.; Hesse, M.; Leusink, M.; Cheever, A.W.; Davies, S.J.; McKerrow, J.H.; Wakefield, L.M.; Letterio, J.J.; Wynn, T.A. IL-13 activates a mechanism of tissue fibrosis that is completely TGF-beta independent. *J. Immunol.* **2004**, *173*, 4020–4029. [CrossRef]

137. Lee, C.G.; Kang, H.R.; Homer, R.J.; Chupp, G.; Elias, J.A. Transgenic modeling of transforming growth factor-beta(1): Role of apoptosis in fibrosis and alveolar remodeling. *Proc. Am. Thorac. Soc.* **2006**, *3*, 418–423. [CrossRef]

138. Huang, M.; Wang, J.; Lee, P.; Sharma, S.; Mao, J.T.; Meissner, H.; Uyemura, K.; Modlin, R.; Wollman, J.; Dubinett, S.M. Human non-small cell lung cancer cells express a type 2 cytokine pattern. *Cancer Res.* **1995**, *55*, 3847–3853.

139. Pastuszak-Lewandoska, D.; Domanska-Senderowska, D.; Antczak, A.; Kordiak, J.; Gorski, P.; Czarnecka, K.H.; Migdalska-Sek, M.; Nawrot, E.; Kiszalkiewicz, J.M.; Brzezianska-Lasota, E. The Expression Levels of IL-4/IL-13/STAT6 Signaling Pathway Genes and SOCS3 Could Help to Differentiate the Histopathological Subtypes of Non-Small Cell Lung Carcinoma. *Mol. Diagn. Ther.* **2018**, *22*, 621–629. [CrossRef] [PubMed]

140. Xie, M.; Wu, X.J.; Zhang, J.J.; He, C.S. IL-13 receptor alpha2 is a negative prognostic factor in human lung cancer and stimulates lung cancer growth in mice. *Oncotarget* **2015**, *6*, 32902–32913. [CrossRef]

141. Chakraborty, S.; Chopra, P.; Ambi, S.V.; Dastidar, S.G.; Ray, A. Emerging therapeutic interventions for idiopathic pulmonary fibrosis. *Expert Opin. Investig. Drugs* **2014**, *23*, 893–910. [CrossRef] [PubMed]

142. Gu, L.; Tseng, S.; Horner, R.M.; Tam, C.; Loda, M.; Rollins, B.J. Control of TH2 polarization by the chemokine monocyte chemoattractant protein-1. *Nature* **2000**, *404*, 407–411. [CrossRef] [PubMed]

143. Zhang, X.W.; Qin, X.; Qin, C.Y.; Yin, Y.L.; Chen, Y.; Zhu, H.L. Expression of monocyte chemoattractant protein-1 and CC chemokine receptor 2 in non-small cell lung cancer and its significance. *Cancer Immunol. Immunother. CII* **2013**, *62*, 563–570. [CrossRef] [PubMed]

144. Vancheri, C.; Failla, M.; Crimi, N.; Raghu, G. Idiopathic pulmonary fibrosis: A disease with similarities and links to cancer biology. *Eur. Respir. J.* **2010**, *35*, 496–504. [CrossRef] [PubMed]

145. Zhang, C.; Lan, T.; Hou, J.; Li, J.; Fang, R.; Yang, Z.; Zhang, M.; Liu, J.; Liu, B. NOX4 promotes non-small cell lung cancer cell proliferation and metastasis through positive feedback regulation of PI3K/Akt signaling. *Oncotarget* **2014**, *5*, 4392–4405. [CrossRef] [PubMed]

146. Swamy, S.M.; Rajasekaran, N.S.; Thannickal, V.J. Nuclear Factor-Erythroid-2-Related Factor 2 in Aging and Lung Fibrosis. *Am. J. Pathol.* **2016**, *186*, 1712–1723. [CrossRef] [PubMed]

147. Tong, Y.H.; Zhang, B.; Fan, Y.; Lin, N.M. Keap1-Nrf2 pathway: A promising target towards lung cancer prevention and therapeutics. *Chronic Dis. Transl. Med.* **2015**, *1*, 175–186. [CrossRef]

148. Seibold, M.A.; Wise, A.L.; Speer, M.C.; Steele, M.P.; Brown, K.K.; Loyd, J.E.; Fingerlin, T.E.; Zhang, W.; Gudmundsson, G.; Groshong, S.D.; et al. A common MUC5B promoter polymorphism and pulmonary fibrosis. *N. Engl. J. Med.* **2011**, *364*, 1503–1512. [CrossRef]

149. Yu, C.J.; Yang, P.C.; Shun, C.T.; Lee, Y.C.; Kuo, S.H.; Luh, K.T. Overexpression of MUC5 genes is associated with early post-operative metastasis in non-small-cell lung cancer. *Int. J. Cancer* **1996**, *69*, 457–465. [CrossRef]

150. Nagashio, R.; Ueda, J.; Ryuge, S.; Nakashima, H.; Jiang, S.X.; Kobayashi, M.; Yanagita, K.; Katono, K.; Satoh, Y.; Masuda, N.; et al. Diagnostic and prognostic significances of MUC5B and TTF-1 expressions in resected non-small cell lung cancer. *Sci. Rep.* **2015**, *5*, 8649. [CrossRef]

151. Ballester, B.; Milara, J.; Sanz, C.; González, S.; Guijarro, R.; Martínez, C.; Cortijo, J. Role of MUC1 in idiopathic pulmonary fibrosis. In Proceedings of the European Respiratory Society Congress, London, UK, 3–7 September 2016.

152. Ballester, B.; Milara, J.; Guijarro, R.; Morcillo, E.; Cortijo, J. Role of MUC4 in idiopathic pulmonary fibrosis. In Proceedings of the European Respiratory Society Congress, Paris, France, 15–19 September 2018.

153. Ballester, B.; Roger, I.; Contreras, S.; Montero, P.; Milara, J. Role of MUC1 in idiopathic pulmonary fibrosis: Mechanistic insights. In Proceedings of the European Respiratory Society Congress, Milan, Italy, 9–13 September 2017.

154. Ishikawa, N.; Hattori, N.; Yokoyama, A.; Kohno, N. Utility of KL-6/MUC1 in the clinical management of interstitial lung diseases. *Respir. Investig.* **2012**, *50*, 3–13. [CrossRef]

155. Bafna, S.; Kaur, S.; Batra, S.K. Membrane-bound mucins: The mechanistic basis for alterations in the growth and survival of cancer cells. *Oncogene* **2010**, *29*, 2893–2904. [CrossRef]

156. Stroopinsky, D.; Rajabi, H.; Nahas, M.; Rosenblatt, J.; Rahimian, M.; Pyzer, A.; Tagde, A.; Kharbanda, A.; Jain, S.; Kufe, T.; et al. MUC1-C drives myeloid leukaemogenesis and resistance to treatment by a survivin-mediated mechanism. *J. Cell. Mol. Med.* **2018**. [CrossRef] [PubMed]

157. Taylor-Papadimitriou, J.; Burchell, J.M.; Graham, R.; Beatson, R. Latest developments in MUC1 immunotherapy. *Biochem. Soc. Trans.* **2018**, *46*, 659–668. [CrossRef] [PubMed]

158. Lederer, D.J.; Martinez, F.J. Idiopathic Pulmonary Fibrosis. *N. Engl. J. Med.* **2018**, *378*, 1811–1823. [CrossRef] [PubMed]

159. Mackinnon, A.C.; Gibbons, M.A.; Farnworth, S.L.; Leffler, H.; Nilsson, U.J.; Delaine, T.; Simpson, A.J.; Forbes, S.J.; Hirani, N.; Gauldie, J.; et al. Regulation of transforming growth factor-beta1-driven lung fibrosis by galectin-3. *Am. J. Respir. Crit. Care Med.* **2012**, *185*, 537–546. [CrossRef]

160. Ramasamy, S.; Duraisamy, S.; Barbashov, S.; Kawano, T.; Kharbanda, S.; Kufe, D. The MUC1 and galectin-3 oncoproteins function in a microRNA-dependent regulatory loop. *Mol. Cell* **2007**, *27*, 992–1004. [CrossRef] [PubMed]

161. Farhad, M.; Rolig, A.S.; Redmond, W.L. The role of Galectin-3 in modulating tumor growth and immunosuppression within the tumor microenvironment. *Oncoimmunology* **2018**, *7*, e1434467. [CrossRef]

162. Selman, M.; Pardo, A.; Kaminski, N. Idiopathic pulmonary fibrosis: Aberrant recapitulation of developmental programs? *PLoS Med.* **2008**, *5*, e62. [CrossRef]

163. Chilosi, M.; Poletti, V.; Zamo, A.; Lestani, M.; Montagna, L.; Piccoli, P.; Pedron, S.; Bertaso, M.; Scarpa, A.; Murer, B.; et al. Aberrant Wnt/beta-catenin pathway activation in idiopathic pulmonary fibrosis. *Am. J. Pathol.* **2003**, *162*, 1495–1502. [CrossRef]

164. Stewart, D.J. Wnt signaling pathway in non-small cell lung cancer. *J. Natl. Cancer Inst.* **2014**, *106*, djt356. [CrossRef]

165. Coon, D.R.; Roberts, D.J.; Loscertales, M.; Kradin, R. Differential epithelial expression of SHH and FOXF1 in usual and nonspecific interstitial pneumonia. *Exp. Mol. Pathol.* **2006**, *80*, 119–123. [CrossRef] [PubMed]

166. Stewart, G.A.; Hoyne, G.F.; Ahmad, S.A.; Jarman, E.; Wallace, W.A.; Harrison, D.J.; Haslett, C.; Lamb, J.R.; Howie, S.E. Expression of the developmental Sonic hedgehog (Shh) signalling pathway is up-regulated in chronic lung fibrosis and the Shh receptor patched 1 is present in circulating T lymphocytes. *J. Pathol.* **2003**, *199*, 488–495. [CrossRef] [PubMed]

167. Moshai, E.F.; Wemeau-Stervinou, L.; Cigna, N.; Brayer, S.; Somme, J.M.; Crestani, B.; Mailleux, A.A. Targeting the hedgehog-glioma-associated oncogene homolog pathway inhibits bleomycin-induced lung fibrosis in mice. *Am. J. Respir. Cell Mol. Biol.* **2014**, *51*, 11–25. [CrossRef] [PubMed]

168. Giroux-Leprieur, E.; Costantini, A.; Ding, V.W.; He, B. Hedgehog Signaling in Lung Cancer: From Oncogenesis to Cancer Treatment Resistance. *Int. J. Mol. Sci.* **2018**, *19*, 2835. [CrossRef] [PubMed]

169. Ye, Y.Z.; Zhang, Z.H.; Fan, X.Y.; Xu, X.L.; Chen, M.L.; Chang, B.W.; Zhang, Y.B. Notch3 overexpression associates with poor prognosis in human non-small-cell lung cancer. *Med. Oncol.* **2013**, *30*, 595. [CrossRef]

170. Westhoff, B.; Colaluca, I.N.; D'Ario, G.; Donzelli, M.; Tosoni, D.; Volorio, S.; Pelosi, G.; Spaggiari, L.; Mazzarol, G.; Viale, G.; et al. Alterations of the Notch pathway in lung cancer. *Proc. Natl. Acad. Sci. USA* **2009**, *106*, 22293–22298. [CrossRef]

171. Zou, B.; Zhou, X.L.; Lai, S.Q.; Liu, J.C. Notch signaling and non-small cell lung cancer. *Oncol. Lett.* **2018**, *15*, 3415–3421. [CrossRef]

172. Lawrence, J.; Nho, R. The Role of the Mammalian Target of Rapamycin (mTOR) in Pulmonary Fibrosis. *Int. J. Mol. Sci.* **2018**, *19*, 778. [CrossRef]

173. Conte, E.; Gili, E.; Fruciano, M.; Korfei, M.; Fagone, E.; Iemmolo, M.; Lo Furno, D.; Giuffrida, R.; Crimi, N.; Guenther, A.; et al. PI3K p110gamma overexpression in idiopathic pulmonary fibrosis lung tissue and fibroblast cells: In vitro effects of its inhibition. *Lab. Investig. A J. Tech. Methods Pathol.* **2013**, *93*, 566–576. [CrossRef]

174. Horowitz, J.C.; Lee, D.Y.; Waghray, M.; Keshamouni, V.G.; Thomas, P.E.; Zhang, H.; Cui, Z.; Thannickal, V.J. Activation of the pro-survival phosphatidylinositol 3-kinase/AKT pathway by transforming growth factor-beta1 in mesenchymal cells is mediated by p38 MAPK-dependent induction of an autocrine growth factor. *J. Biol. Chem.* **2004**, *279*, 1359–1367. [CrossRef] [PubMed]

175. Kavanaugh, W.M.; Klippel, A.; Escobedo, J.A.; Williams, L.T. Modification of the 85-kilodalton subunit of phosphatidylinositol-3 kinase in platelet-derived growth factor-stimulated cells. *Mol. Cell. Biol.* **1992**, *12*, 3415–3424. [CrossRef] [PubMed]

176. Conte, E.; Fruciano, M.; Fagone, E.; Gili, E.; Caraci, F.; Iemmolo, M.; Crimi, N.; Vancheri, C. Inhibition of PI3K prevents the proliferation and differentiation of human lung fibroblasts into myofibroblasts: The role of class I P110 isoforms. *PLoS ONE* **2011**, *6*, e24663. [CrossRef] [PubMed]

177. Fumarola, C.; Bonelli, M.A.; Petronini, P.G.; Alfieri, R.R. Targeting PI3K/AKT/mTOR pathway in non small cell lung cancer. *Biochem. Pharmacol.* **2014**, *90*, 197–207. [CrossRef] [PubMed]

178. Mulugeta, S.; Nguyen, V.; Russo, S.J.; Muniswamy, M.; Beers, M.F. A surfactant protein C precursor protein BRICHOS domain mutation causes endoplasmic reticulum stress, proteasome dysfunction, and caspase 3 activation. *Am. J. Respir. Cell Mol. Biol.* **2005**, *32*, 521–530. [CrossRef] [PubMed]

179. Lawson, W.E.; Crossno, P.F.; Polosukhin, V.V.; Roldan, J.; Cheng, D.S.; Lane, K.B.; Blackwell, T.R.; Xu, C.; Markin, C.; Ware, L.B.; et al. Endoplasmic reticulum stress in alveolar epithelial cells is prominent in IPF: Association with altered surfactant protein processing and herpesvirus infection. *Am. J. Physiol. Lung Cell. Mol. Physiol.* **2008**, *294*, L1119–L1126. [CrossRef]

180. Van Moorsel, C.H.; Ten Klooster, L.; van Oosterhout, M.F.; de Jong, P.A.; Adams, H.; Wouter van Es, H.; Ruven, H.J.; van der Vis, J.J.; Grutters, J.C. SFTPA2 Mutations in Familial and Sporadic Idiopathic Interstitial Pneumonia. *Am. J. Respir. Crit. Care Med.* **2015**, *192*, 1249–1252. [CrossRef] [PubMed]

181. Armanios, M.Y.; Chen, J.J.; Cogan, J.D.; Alder, J.K.; Ingersoll, R.G.; Markin, C.; Lawson, W.E.; Xie, M.; Vulto, I.; Phillips, J.A., 3rd; et al. Telomerase mutations in families with idiopathic pulmonary fibrosis. *N. Engl. J. Med.* **2007**, *356*, 1317–1326. [CrossRef] [PubMed]

182. Tsakiri, K.D.; Cronkhite, J.T.; Kuan, P.J.; Xing, C.; Raghu, G.; Weissler, J.C.; Rosenblatt, R.L.; Shay, J.W.; Garcia, C.K. Adult-onset pulmonary fibrosis caused by mutations in telomerase. *Proc. Natl. Acad. Sci. USA* **2007**, *104*, 7552–7557. [CrossRef] [PubMed]

183. Stuart, B.D.; Choi, J.; Zaidi, S.; Xing, C.; Holohan, B.; Chen, R.; Choi, M.; Dharwadkar, P.; Torres, F.; Girod, C.E.; et al. Exome sequencing links mutations in PARN and RTEL1 with familial pulmonary fibrosis and telomere shortening. *Nat. Genet.* **2015**, *47*, 512–517. [CrossRef]

184. Fingerlin, T.E.; Murphy, E.; Zhang, W.; Peljto, A.L.; Brown, K.K.; Steele, M.P.; Loyd, J.E.; Cosgrove, G.P.; Lynch, D.; Groshong, S.; et al. Genome-wide association study identifies multiple susceptibility loci for pulmonary fibrosis. *Nat. Genet.* **2013**, *45*, 613–620. [CrossRef] [PubMed]

185. Petrovski, S.; Todd, J.L.; Durheim, M.T.; Wang, Q.; Chien, J.W.; Kelly, F.L.; Frankel, C.; Mebane, C.M.; Ren, Z.; Bridgers, J.; et al. An Exome Sequencing Study to Assess the Role of Rare Genetic Variation in Pulmonary Fibrosis. *Am. J. Respir. Crit. Care Med.* **2017**, *196*, 82–93. [CrossRef] [PubMed]

186. Noth, I.; Zhang, Y.; Ma, S.F.; Flores, C.; Barber, M.; Huang, Y.; Broderick, S.M.; Wade, M.S.; Hysi, P.; Scuirba, J.; et al. Genetic variants associated with idiopathic pulmonary fibrosis susceptibility and mortality: A genome-wide association study. *Lancet Respir. Med.* **2013**, *1*, 309–317. [CrossRef]

187. Bennett, W.P.; Colby, T.V.; Travis, W.D.; Borkowski, A.; Jones, R.T.; Lane, D.P.; Metcalf, R.A.; Samet, J.M.; Takeshima, Y.; Gu, J.R.; et al. p53 protein accumulates frequently in early bronchial neoplasia. *Cancer Res.* **1993**, *53*, 4817–4822. [PubMed]

188. Sozzi, G.; Miozzo, M.; Donghi, R.; Pilotti, S.; Cariani, C.T.; Pastorino, U.; Della Porta, G.; Pierotti, M.A. Deletions of 17p and p53 mutations in preneoplastic lesions of the lung. *Cancer Res.* **1992**, *52*, 6079–6082. [PubMed]

189. Takahashi, T.; Munakata, M.; Ohtsuka, Y.; Nisihara, H.; Nasuhara, Y.; Kamachi-Satoh, A.; Dosaka-Akita, H.; Homma, Y.; Kawakami, Y. Expression and alteration of ras and p53 proteins in patients with lung carcinoma accompanied by idiopathic pulmonary fibrosis. *Cancer* **2002**, *95*, 624–633. [CrossRef] [PubMed]

190. Kawasaki, H.; Ogura, T.; Yokose, T.; Nagai, K.; Nishiwaki, Y.; Esumi, H. p53 gene alteration in atypical epithelial lesions and carcinoma in patients with idiopathic pulmonary fibrosis. *Hum. Pathol.* **2001**, *32*, 1043–1049. [CrossRef] [PubMed]

191. Sozzi, G.; Veronese, M.L.; Negrini, M.; Baffa, R.; Cotticelli, M.G.; Inoue, H.; Tornielli, S.; Pilotti, S.; De Gregorio, L.; Pastorino, U.; et al. The FHIT gene 3p14.2 is abnormal in lung cancer. *Cell* **1996**, *85*, 17–26. [CrossRef]

192. Uematsu, K.; Yoshimura, A.; Gemma, A.; Mochimaru, H.; Hosoya, Y.; Kunugi, S.; Matsuda, K.; Seike, M.; Kurimoto, F.; Takenaka, K.; et al. Aberrations in the fragile histidine triad (FHIT) gene in idiopathic pulmonary fibrosis. *Cancer Res.* **2001**, *61*, 8527–8533. [PubMed]

193. Karachaliou, N.; Mayo, C.; Costa, C.; Magri, I.; Gimenez-Capitan, A.; Molina-Vila, M.A.; Rosell, R. KRAS mutations in lung cancer. *Clin. Lung Cancer* **2013**, *14*, 205–214. [CrossRef]

194. Masai, K.; Tsuta, K.; Motoi, N.; Shiraishi, K.; Furuta, K.; Suzuki, S.; Asakura, K.; Nakagawa, K.; Sakurai, H.; Watanabe, S.I.; et al. Clinicopathological, Immunohistochemical, and Genetic Features of Primary Lung Adenocarcinoma Occurring in the Setting of Usual Interstitial Pneumonia Pattern. *J. Thorac. Oncol. Off. Publ. Int. Assoc. Stud. Lung Cancer* **2016**, *11*, 2141–2149. [CrossRef] [PubMed]

195. Hwang, J.A.; Kim, D.; Chun, S.M.; Bae, S.; Song, J.S.; Kim, M.Y.; Koo, H.J.; Song, J.W.; Kim, W.S.; Lee, J.C.; et al. Genomic profiles of lung cancer associated with idiopathic pulmonary fibrosis. *J. Pathol.* **2018**, *244*, 25–35. [CrossRef]

196. Govindan, R.; Ding, L.; Griffith, M.; Subramanian, J.; Dees, N.D.; Kanchi, K.L.; Maher, C.A.; Fulton, R.; Fulton, L.; Wallis, J.; et al. Genomic landscape of non-small cell lung cancer in smokers and never-smokers. *Cell* **2012**, *150*, 1121–1134. [CrossRef] [PubMed]

197. Clamon, G.H.; Bossler, A.D.; Abu Hejleh, T.; Furqan, M. Germline mutations predisposing to non-small cell lung cancer. *Fam. Cancer* **2015**, *14*, 463–469. [CrossRef] [PubMed]

198. Wang, Y.; Kuan, P.J.; Xing, C.; Cronkhite, J.T.; Torres, F.; Rosenblatt, R.L.; DiMaio, J.M.; Kinch, L.N.; Grishin, N.V.; Garcia, C.K. Genetic defects in surfactant protein A2 are associated with pulmonary fibrosis and lung cancer. *Am. J. Hum. Genet.* **2009**, *84*, 52–59. [CrossRef] [PubMed]

199. Nathan, N.; Giraud, V.; Picard, C.; Nunes, H.; Dastot-Le Moal, F.; Copin, B.; Galeron, L.; De Ligniville, A.; Kuziner, N.; Reynaud-Gaubert, M.; et al. Germline SFTPA1 mutation in familial idiopathic interstitial pneumonia and lung cancer. *Hum. Mol. Genet.* **2016**, *25*, 1457–1467. [CrossRef] [PubMed]

200. Maitra, M.; Wang, Y.; Gerard, R.D.; Mendelson, C.R.; Garcia, C.K. Surfactant protein A2 mutations associated with pulmonary fibrosis lead to protein instability and endoplasmic reticulum stress. *J. Biol. Chem.* **2010**, *285*, 22103–22113. [CrossRef]

201. Madsen, J.; Tornoe, I.; Nielsen, O.; Koch, C.; Steinhilber, W.; Holmskov, U. Expression and localization of lung surfactant protein A in human tissues. *Am. J. Respir. Cell Mol. Biol.* **2003**, *29*, 591–597. [CrossRef]

202. Chen, Z.; Fillmore, C.M.; Hammerman, P.S.; Kim, C.F.; Wong, K.K. Non-small-cell lung cancers: A heterogeneous set of diseases. *Nat. Rev. Cancer* **2014**, *14*, 535–546. [CrossRef]

203. Kropski, J.A.; Blackwell, T.S.; Loyd, J.E. The genetic basis of idiopathic pulmonary fibrosis. *Eur. Respir. J.* **2015**, *45*, 1717–1727. [CrossRef]

204. Korthagen, N.M.; van Moorsel, C.H.; Barlo, N.P.; Kazemier, K.M.; Ruven, H.J.; Grutters, J.C. Association between variations in cell cycle genes and idiopathic pulmonary fibrosis. *PLoS ONE* **2012**, *7*, e30442. [CrossRef] [PubMed]

205. Landi, M.T.; Chatterjee, N.; Yu, K.; Goldin, L.R.; Goldstein, A.M.; Rotunno, M.; Mirabello, L.; Jacobs, K.; Wheeler, W.; Yeager, M.; et al. A genome-wide association study of lung cancer identifies a region of chromosome 5p15 associated with risk for adenocarcinoma. *Am. J. Hum. Genet.* **2009**, *85*, 679–691. [CrossRef] [PubMed]

206. Ye, G.; Tan, N.; Meng, C.; Li, J.; Jing, L.; Yan, M.; Jin, T.; Chen, F. Genetic variations in TERC and TERT genes are associated with lung cancer risk in a Chinese Han population. *Oncotarget* **2017**, *8*, 110145–110152. [CrossRef] [PubMed]

207. McKay, J.D.; Hung, R.J.; Han, Y.; Zong, X.; Carreras-Torres, R.; Christiani, D.C.; Caporaso, N.E.; Johansson, M.; Xiao, X.; Li, Y.; et al. Large-scale association analysis identifies new lung cancer susceptibility loci and heterogeneity in genetic susceptibility across histological subtypes. *Nat. Genet.* **2017**, *49*, 1126–1132. [CrossRef] [PubMed]

208. Kang, J.U.; Koo, S.H.; Kwon, K.C.; Park, J.W.; Kim, J.M. Gain at chromosomal region 5p15.33, containing TERT, is the most frequent genetic event in early stages of non-small cell lung cancer. *Cancer Genet. Cytogenet.* **2008**, *182*, 1–11. [CrossRef] [PubMed]

209. Wu, X.; Amos, C.I.; Zhu, Y.; Zhao, H.; Grossman, B.H.; Shay, J.W.; Luo, S.; Hong, W.K.; Spitz, M.R. Telomere dysfunction: A potential cancer predisposition factor. *J. Natl. Cancer Inst.* **2003**, *95*, 1211–1218. [CrossRef] [PubMed]

210. Hackett, J.A.; Feldser, D.M.; Greider, C.W. Telomere dysfunction increases mutation rate and genomic instability. *Cell* **2001**, *106*, 275–286. [CrossRef]

211. Mora, A.L.; Rojas, M.; Pardo, A.; Selman, M. Emerging therapies for idiopathic pulmonary fibrosis, a progressive age-related disease. *Nat. Rev. Drug Discov.* **2017**, *16*, 810. [CrossRef]

212. Peljto, A.L.; Zhang, Y.; Fingerlin, T.E.; Ma, S.F.; Garcia, J.G.; Richards, T.J.; Silveira, L.J.; Lindell, K.O.; Steele, M.P.; Loyd, J.E.; et al. Association between the MUC5B promoter polymorphism and survival in patients with idiopathic pulmonary fibrosis. *JAMA* **2013**, *309*, 2232–2239. [CrossRef]

213. Yang, J.; Xu, T.; Gomez, D.R.; Jeter, M.; Levy, L.B.; Song, Y.; Hahn, S.; Liao, Z.; Yuan, X. The Pulmonary Fibrosis Associated MUC5B Promoter Polymorphism Is Prognostic of the Overall Survival in Patients with Non-Small Cell Lung Cancer (NSCLC) Receiving Definitive Radiotherapy. *Transl. Oncol.* **2017**, *10*, 197–202. [CrossRef]

214. Rabinovich, E.I.; Kapetanaki, M.G.; Steinfeld, I.; Gibson, K.F.; Pandit, K.V.; Yu, G.; Yakhini, Z.; Kaminski, N. Global methylation patterns in idiopathic pulmonary fibrosis. *PLoS ONE* **2012**, *7*, e33770. [CrossRef] [PubMed]

215. Takenaka, K.; Gemma, A.; Yoshimura, A.; Hosoya, Y.; Nara, M.; Hosomi, Y.; Okano, T.; Kunugi, S.; Koizumi, K.; Fukuda, Y.; et al. Reduced transcription of the Smad4 gene during pulmonary carcinogenesis in idiopathic pulmonary fibrosis. *Mol. Med. Rep.* **2009**, *2*, 73–80. [PubMed]

216. Schutte, M.; Hruban, R.H.; Hedrick, L.; Cho, K.R.; Nadasdy, G.M.; Weinstein, C.L.; Bova, G.S.; Isaacs, W.B.; Cairns, P.; Nawroz, H.; et al. DPC4 gene in various tumor types. *Cancer Res.* **1996**, *56*, 2527–2530. [PubMed]

217. Sanders, Y.Y.; Pardo, A.; Selman, M.; Nuovo, G.J.; Tollefsbol, T.O.; Siegal, G.P.; Hagood, J.S. Thy-1 promoter hypermethylation: A novel epigenetic pathogenic mechanism in pulmonary fibrosis. *Am. J. Respir. Cell Mol. Biol.* **2008**, *39*, 610–618. [CrossRef] [PubMed]

218. Lung, H.L.; Bangarusamy, D.K.; Xie, D.; Cheung, A.K.; Cheng, Y.; Kumaran, M.K.; Miller, L.; Liu, E.T.; Guan, X.Y.; Sham, J.S.; et al. THY1 is a candidate tumour suppressor gene with decreased expression in metastatic nasopharyngeal carcinoma. *Oncogene* **2005**, *24*, 6525–6532. [CrossRef]

219. Langevin, S.M.; Kratzke, R.A.; Kelsey, K.T. Epigenetics of lung cancer. *Transl. Res. J. Lab. Clin. Med.* **2015**, *165*, 74–90. [CrossRef] [PubMed]

220. Huang, S.K.; Scruggs, A.M.; McEachin, R.C.; White, E.S.; Peters-Golden, M. Lung fibroblasts from patients with idiopathic pulmonary fibrosis exhibit genome-wide differences in DNA methylation compared to fibroblasts from nonfibrotic lung. *PLoS ONE* **2014**, *9*, e107055. [CrossRef]

221. Yang, M.; Shen, H.; Qiu, C.; Ni, Y.; Wang, L.; Dong, W.; Liao, Y.; Du, J. High expression of miR-21 and miR-155 predicts recurrence and unfavourable survival in non-small cell lung cancer. *Eur. J. Cancer* **2013**, *49*, 604–615. [CrossRef]

222. Kolenda, T.; Przybyla, W.; Teresiak, A.; Mackiewicz, A.; Lamperska, K.M. The mystery of let-7d—A small RNA with great power. *Contemp. Oncol.* **2014**, *18*, 293–301. [CrossRef]

223. Nguyen-Ngoc, T.; Bouchaab, H.; Adjei, A.A.; Peters, S. BRAF Alterations as Therapeutic Targets in Non-Small-Cell Lung Cancer. *J. Thorac. Oncol. Off. Publ. Int. Assoc. Stud. Lung Cancer* **2015**, *10*, 1396–1403. [CrossRef]

224. Idiopathic Pulmonary Fibrosis Clinical Research Network; Raghu, G.; Anstrom, K.J.; King, T.E., Jr.; Lasky, J.A.; Martinez, F.J. Prednisone, azathioprine, and N-acetylcysteine for pulmonary fibrosis. *N. Engl. J. Med.* **2012**, *366*, 1968–1977.

225. Idiopathic Pulmonary Fibrosis Clinical Research Network; Martinez, F.J.; de Andrade, J.A.; Anstrom, K.J.; King, T.E., Jr.; Raghu, G. Randomized trial of acetylcysteine in idiopathic pulmonary fibrosis. *N. Engl. J. Med.* **2014**, *370*, 2093–2101.

226. Oldham, J.M.; Ma, S.F.; Martinez, F.J.; Anstrom, K.J.; Raghu, G.; Schwartz, D.A.; Valenzi, E.; Witt, L.; Lee, C.; Vij, R.; et al. TOLLIP, MUC5B, and the Response to N-Acetylcysteine among Individuals with Idiopathic Pulmonary Fibrosis. *Am. J. Respir. Crit. Care Med.* **2015**, *192*, 1475–1482. [CrossRef] [PubMed]

227. Sablina, A.A.; Budanov, A.V.; Ilyinskaya, G.V.; Agapova, L.S.; Kravchenko, J.E.; Chumakov, P.M. The antioxidant function of the p53 tumor suppressor. *Nat. Med.* **2005**, *11*, 1306–1313. [CrossRef] [PubMed]

228. Vafa, O.; Wade, M.; Kern, S.; Beeche, M.; Pandita, T.K.; Hampton, G.M.; Wahl, G.M. c-Myc can induce DNA damage, increase reactive oxygen species, and mitigate p53 function: A mechanism for oncogene-induced genetic instability. *Mol. Cell* **2002**, *9*, 1031–1044. [CrossRef]

229. Sayin, V.I.; Ibrahim, M.X.; Larsson, E.; Nilsson, J.A.; Lindahl, P.; Bergo, M.O. Antioxidants accelerate lung cancer progression in mice. *Sci. Transl. Med.* **2014**, *6*. [CrossRef] [PubMed]

230. Mitani, Y.; Sato, K.; Muramoto, Y.; Karakawa, T.; Kitamado, M.; Iwanaga, T.; Nabeshima, T.; Maruyama, K.; Nakagawa, K.; Ishida, K.; et al. Superoxide scavenging activity of pirfenidone-iron complex. *Biochem. Biophys. Res. Commun.* **2008**, *372*, 19–23. [CrossRef] [PubMed]

231. Salazar-Montes, A.; Ruiz-Corro, L.; Lopez-Reyes, A.; Castrejon-Gomez, E.; Armendariz-Borunda, J. Potent antioxidant role of pirfenidone in experimental cirrhosis. *Eur. J. Pharmacol.* **2008**, *595*, 69–77. [CrossRef] [PubMed]

232. Behr, J.; Bendstrup, E.; Crestani, B.; Gunther, A.; Olschewski, H.; Skold, C.M.; Wells, A.; Wuyts, W.; Koschel, D.; Kreuter, M.; et al. Safety and tolerability of acetylcysteine and pirfenidone combination therapy in idiopathic pulmonary fibrosis: A randomised, double-blind, placebo-controlled, phase 2 trial. *Lancet Respir. Med.* **2016**, *4*, 445–453. [CrossRef]

233. Hisatomi, K.; Mukae, H.; Sakamoto, N.; Ishimatsu, Y.; Kakugawa, T.; Hara, S.; Fujita, H.; Nakamichi, S.; Oku, H.; Urata, Y.; et al. Pirfenidone inhibits TGF-beta1-induced over-expression of collagen type I and heat shock protein 47 in A549 cells. *BMC Pulm. Med.* **2012**, *12*, 24. [CrossRef]

234. Lin, X.; Yu, M.; Wu, K.; Yuan, H.; Zhong, H. Effects of pirfenidone on proliferation, migration, and collagen contraction of human Tenon's fibroblasts in vitro. *Investig. Ophthalmol. Vis. Sci.* **2009**, *50*, 3763–3770. [CrossRef] [PubMed]

235. Oku, H.; Shimizu, T.; Kawabata, T.; Nagira, M.; Hikita, I.; Ueyama, A.; Matsushima, S.; Torii, M.; Arimura, A. Antifibrotic action of pirfenidone and prednisolone: Different effects on pulmonary cytokines and growth factors in bleomycin-induced murine pulmonary fibrosis. *Eur. J. Pharmacol.* **2008**, *590*, 400–408. [CrossRef] [PubMed]

236. Choi, K.; Lee, K.; Ryu, S.W.; Im, M.; Kook, K.H.; Choi, C. Pirfenidone inhibits transforming growth factor-beta1-induced fibrogenesis by blocking nuclear translocation of Smads in human retinal pigment epithelial cell line ARPE-19. *Mol. Vis.* **2012**, *18*, 1010–1020. [PubMed]

237. Miura, Y.; Saito, T.; Tanaka, T.; Takoi, H.; Yatagai, Y.; Inomata, M.; Nei, T.; Saito, Y.; Gemma, A.; Azuma, A. Reduced incidence of lung cancer in patients with idiopathic pulmonary fibrosis treated with pirfenidone. *Respir. Investig.* **2018**, *56*, 72–79. [CrossRef] [PubMed]

238. Xiao, H.; Zhang, G.F.; Liao, X.P.; Li, X.J.; Zhang, J.; Lin, H.; Chen, Z.; Zhang, X. Anti-fibrotic effects of pirfenidone by interference with the hedgehog signalling pathway in patients with systemic sclerosis-associated interstitial lung disease. *Int. J. Rheum. Dis.* **2018**, *21*, 477–486. [CrossRef]

239. Iwata, T.; Yoshino, I.; Yoshida, S.; Ikeda, N.; Tsuboi, M.; Asato, Y.; Katakami, N.; Sakamoto, K.; Yamashita, Y.; Okami, J.; et al. A phase II trial evaluating the efficacy and safety of perioperative pirfenidone for prevention of acute exacerbation of idiopathic pulmonary fibrosis in lung cancer patients undergoing pulmonary resection: West Japan Oncology Group 6711 L (PEOPLE Study). *Respir. Res.* **2016**, *17*, 90. [CrossRef]

240. Mediavilla-Varela, M.; Boateng, K.; Noyes, D.; Antonia, S.J. The anti-fibrotic agent pirfenidone synergizes with cisplatin in killing tumor cells and cancer-associated fibroblasts. *BMC Cancer* **2016**, *16*, 176. [CrossRef]

241. Abdollahi, A.; Li, M.; Ping, G.; Plathow, C.; Domhan, S.; Kiessling, F.; Lee, L.B.; McMahon, G.; Grone, H.J.; Lipson, K.E.; et al. Inhibition of platelet-derived growth factor signaling attenuates pulmonary fibrosis. *J. Exp. Med.* **2005**, *201*, 925–935. [CrossRef]

242. Aono, Y.; Nishioka, Y.; Inayama, M.; Ugai, M.; Kishi, J.; Uehara, H.; Izumi, K.; Sone, S. Imatinib as a novel antifibrotic agent in bleomycin-induced pulmonary fibrosis in mice. *Am. J. Respir. Crit. Care Med.* **2005**, *171*, 1279–1285. [CrossRef]

243. Ishii, Y.; Fujimoto, S.; Fukuda, T. Gefitinib prevents bleomycin-induced lung fibrosis in mice. *Am. J. Respir. Crit. Care Med.* **2006**, *174*, 550–556. [CrossRef]

244. Li, M.; Abdollahi, A.; Grone, H.J.; Lipson, K.E.; Belka, C.; Huber, P.E. Late treatment with imatinib mesylate ameliorates radiation-induced lung fibrosis in a mouse model. *Radiat. Oncol.* **2009**, *4*, 66. [CrossRef] [PubMed]

245. Rhee, C.K.; Lee, S.H.; Yoon, H.K.; Kim, S.C.; Lee, S.Y.; Kwon, S.S.; Kim, Y.K.; Kim, K.H.; Kim, T.J.; Kim, J.W. Effect of nilotinib on bleomycin-induced acute lung injury and pulmonary fibrosis in mice. *Respir. Int. Rev. Thorac. Dis.* **2011**, *82*, 273–287. [CrossRef] [PubMed]

246. Adachi, K.; Mizoguchi, K.; Kawarada, S.; Miyoshi, A.; Suzuki, M.; Chiba, S.; Deki, T. Effects of erlotinib on lung injury induced by intratracheal administration of bleomycin (BLM) in rats. *J. Toxicol. Sci.* **2010**, *35*, 503–514. [CrossRef] [PubMed]

247. Wollin, L.; Maillet, I.; Quesniaux, V.; Holweg, A.; Ryffel, B. Antifibrotic and anti-inflammatory activity of the tyrosine kinase inhibitor nintedanib in experimental models of lung fibrosis. *J. Pharmacol. Exp. Ther.* **2014**, *349*, 209–220. [CrossRef] [PubMed]

248. Arora, A.; Scholar, E.M. Role of tyrosine kinase inhibitors in cancer therapy. *J. Pharmacol. Exp. Ther.* **2005**, *315*, 971–979. [CrossRef] [PubMed]

249. Daniels, C.E.; Lasky, J.A.; Limper, A.H.; Mieras, K.; Gabor, E.; Schroeder, D.R.; Imatinib, I.P.F.S.I. Imatinib treatment for idiopathic pulmonary fibrosis: Randomized placebo-controlled trial results. *Am. J. Respir. Crit. Care Med.* **2010**, *181*, 604–610. [CrossRef]

250. Reck, M.; Kaiser, R.; Mellemgaard, A.; Douillard, J.Y.; Orlov, S.; Krzakowski, M.; von Pawel, J.; Gottfried, M.; Bondarenko, I.; Liao, M.; et al. Docetaxel plus nintedanib versus docetaxel plus placebo in patients with previously treated non-small-cell lung cancer (LUME-Lung 1): A phase 3, double-blind, randomised controlled trial. *Lancet Oncol.* **2014**, *15*, 143–155. [CrossRef]

251. Malouf, M.A.; Hopkins, P.; Snell, G.; Glanville, A.R.; Everolimus in, I.P.F.S.I. An investigator-driven study of everolimus in surgical lung biopsy confirmed idiopathic pulmonary fibrosis. *Respirology* **2011**, *16*, 776–783. [CrossRef]

252. Soria, J.C.; Shepherd, F.A.; Douillard, J.Y.; Wolf, J.; Giaccone, G.; Crino, L.; Cappuzzo, F.; Sharma, S.; Gross, S.H.; Dimitrijevic, S.; et al. Efficacy of everolimus (RAD001) in patients with advanced NSCLC previously treated with chemotherapy alone or with chemotherapy and EGFR inhibitors. *Ann. Oncol. Off. J. Eur. Soc. Med. Oncol.* **2009**, *20*, 1674–1681. [CrossRef]

253. An, J.; Xue, Y.; Long, M.; Zhang, G.; Zhang, J.; Su, H. Targeting CCR2 with its antagonist suppresses viability, motility and invasion by downregulating MMP-9 expression in non-small cell lung cancer cells. *Oncotarget* **2017**, *8*, 39230–39240. [CrossRef]

254. Chung, L.Y.; Tang, S.J.; Wu, Y.C.; Sun, G.H.; Liu, H.Y.; Sun, K.H. Galectin-3 augments tumor initiating property and tumorigenicity of lung cancer through interaction with beta-catenin. *Oncotarget* **2015**, *6*, 4936–4952. [CrossRef] [PubMed]

255. Bai, X.Y.; Zhang, X.C.; Yang, S.Q.; An, S.J.; Chen, Z.H.; Su, J.; Xie, Z.; Gou, L.Y.; Wu, Y.L. Blockade of Hedgehog Signaling Synergistically Increases Sensitivity to Epidermal Growth Factor Receptor Tyrosine Kinase Inhibitors in Non-Small-Cell Lung Cancer Cell Lines. *PLoS ONE* **2016**, *11*, e0149370. [CrossRef] [PubMed]

256. Ide, M.; Tanaka, K.; Sunami, S.; Asoh, T.; Maeyama, T.; Tsuruta, N.; Nakanishi, Y.; Okamoto, I. Durable response to nivolumab in a lung adenocarcinoma patient with idiopathic pulmonary fibrosis. *Thorac. Cancer* **2018**, *9*, 1519–1521. [CrossRef] [PubMed]

257. Ueda, T.; Aokage, K.; Nishikawa, H.; Neri, S.; Nakamura, H.; Sugano, M.; Tane, K.; Miyoshi, T.; Kojima, M.; Fujii, S.; et al. Immunosuppressive tumor microenvironment of usual interstitial pneumonia-associated squamous cell carcinoma of the lung. *J. Cancer Res. Clin. Oncol.* **2018**, *144*, 835–844. [CrossRef] [PubMed]

258. Chen, X.; Shi, C.; Meng, X.; Zhang, K.; Li, X.; Wang, C.; Xiang, Z.; Hu, K.; Han, X. Inhibition of Wnt/beta-catenin signaling suppresses bleomycin-induced pulmonary fibrosis by attenuating the expression of TGF-beta1 and FGF-2. *Exp. Mol. Pathol.* **2016**, *101*, 22–30. [CrossRef] [PubMed]

259. Cao, H.; Wang, C.; Chen, X.; Hou, J.; Xiang, Z.; Shen, Y.; Han, X. Inhibition of Wnt/beta-catenin signaling suppresses myofibroblast differentiation of lung resident mesenchymal stem cells and pulmonary fibrosis. *Sci. Rep.* **2018**, *8*, 13644. [CrossRef] [PubMed]

260. Butts, C.; Socinski, M.A.; Mitchell, P.L.; Thatcher, N.; Havel, L.; Krzakowski, M.; Nawrocki, S.; Ciuleanu, T.E.; Bosquee, L.; Trigo, J.M.; et al. Tecemotide (L-BLP25) versus placebo after chemoradiotherapy for stage III non-small-cell lung cancer (START): A randomised, double-blind, phase 3 trial. *Lancet Oncol.* **2014**, *15*, 59–68. [CrossRef]

261. Liu, Y.; Huang, G.; Mo, B.; Wang, C. Artesunate ameliorates lung fibrosis via inhibiting the Notch signaling pathway. *Exp. Ther. Med.* **2017**, *14*, 561–566. [CrossRef]

262. Xu, L.; Yang, D.; Zhu, S.; Gu, J.; Ding, F.; Bian, W.; Rong, Z.; Shen, C. Bleomycin-induced pulmonary fibrosis is attenuated by an antibody against KL-6. *Exp. Lung Res.* **2013**, *39*, 241–248. [CrossRef]

International Journal of
Molecular Sciences

MDPI

Review

The Role of Fibrosis and Liver-Associated Fibroblasts in the Pathogenesis of Hepatocellular Carcinoma

Jacopo Baglieri [1,*], David A. Brenner [1,*] and Tatiana Kisseleva [2,*]

1   Department of Medicine, UC San Diego, La Jolla, CA 92093, USA
2   Department of Surgery, UC San Diego, La Jolla, CA 92093, USA
*   Correspondence: jbaglieri@ucsd.edu (J.B.); dbrenner@ucsd.edu (D.A.B.); tkisseleva@ucsd.edu (T.K.)

Received: 4 March 2019; Accepted: 5 April 2019; Published: 7 April 2019

Abstract: Hepatocellular carcinoma (HCC) is one of the most aggressive types of cancer and lacks effective therapeutic approaches. Most HCC develops in the setting of chronic liver injury, hepatic inflammation, and fibrosis. Hepatic stellate cells (HSCs) and cancer-associated fibroblasts (CAFs) are key players in liver fibrogenesis and hepatocarcinogenesis, respectively. CAFs, which probably derive from HSCs, activate into extracellular matrix (ECM)-producing myofibroblasts and crosstalk with cancer cells to affect tumor growth and invasion. In this review, we describe the different components which form the HCC premalignant microenvironment (PME) and the tumor microenvironment (TME), focusing on the liver fibrosis process and the biology of CAFs. We will describe the CAF-dependent mechanisms which have been suggested to promote hepatocarcinogenesis, such as the alteration of ECM, CAF-dependent production of cytokines and angiogenic factors, CAF-dependent reduction of immuno-surveillance, and CAF-dependent promotion of epithelial-mesenchymal transition (EMT). New knowledge of the fibrosis process and the role of CAFs in HCC may pave the way for new therapeutic strategies for liver cancer.

Keywords: hepatocellular carcinoma (HCC); fibrosis; cancer-associated fibroblasts (CAFs); hepatic stellate cells (HSCs); tumor microenvironment

1. Introduction

Hepatocellular carcinoma (HCC) is one of the most aggressive and fastest growing malignancies [1]. HCC is only second to lung cancer as a leading cause of cancer-related death worldwide [2]. Since, in most patients, HCC is diagnosed at a late stage, therapeutic treatments are limited and the five-year survival rate is less than 12% [3]. Most HCC cases are in southeast Asia, where the major cause of HCC is chronic hepatitis B virus (HBV) infections. In contrast, in sub-Saharan Africa, the main risk factor is exposure to aflatoxin B [4]. However, in Japan, North America, and Europe, major causes of HCC are hepatitis C virus (HCV) infections, alcoholic liver disease, and non-alcoholic fatty liver disease (NAFLD) [4]. This chronic liver injury causes liver fibrosis, which is characterized by the activation of hepatic stellate cells (HSCs) into extracellular matrix (ECM)-producing myofibroblasts [5–7]. In chronic liver injury, continuous accumulation of ECM results in the progressive substitution of the liver parenchyma by scar tissue. Regardless of the etiology of liver injury, HCC is strongly associated with liver fibrosis and cirrhosis, with about 80–90% of HCC cases having underlying fibrosis [8], and approximately one in three patients with cirrhosis will develop HCC in their lifetime [3]. However, it is still not clear whether fibrosis directly promotes HCC. Currently, there are very limited therapies for HCC treatment. Therefore, a better understanding of the role of fibrosis and myofibroblast activation in HCC development and progression may provide new therapeutic options for the treatment of HCC.

2. The Premalignant and Tumor Microenvironment in HCC

The tumor microenvironment (TME) is defined as the tumor cell population in a complex mixture of surrounding stromal cells, including fibroblasts, endothelial cells, pericytes, immune cells, and proteins like ECM elements, cytokines, chemokines, and enzymes that are secreted by both cancerous and non-cancerous cells [9]. Originally, the TME was not considered to have a role in cancer progression; however, it is now proposed that the stroma is aberrantly activated in cancer and it affects tumorigenesis. In the context of liver cancer, HCC is strongly associated with liver fibrosis and cirrhosis, suggesting that the environment in which HCC rises may influence tumorigenesis. This is different from many other tumors, where fibrosis develops as a reaction of tumor formation [10]. Therefore, it was recently proposed that the premalignant microenvironment (PME) and TME in HCC should be differentiated [10]. PME is characterized by chronic liver injury, inflammation, and fibrosis, and precedes tumor formation, whereas TME evolves in the already developed tumor.

2.1. Premalignant Microenvironment in HCC

Several mechanisms have been proposed to promote tumor formation in PME (Figure 1). First, chronic liver injury causes hepatocyte cell death. It has been demonstrated in mice that abolishing the expression of antiapoptotic proteins such as Nemo, Tak1, Mcl-1, or Bcl-xl, specifically in hepatocytes, increased hepatocyte apoptosis, fibrosis, and consequently, HCC development [11–15]. Accordingly, studies in chronic HBV and HCV patients have shown that elevated levels of ALT, which reflect hepatocyte death, positively correlate with the risk of developing HCC [16,17]. In this setting, hepatocyte death in turn triggers several other mechanisms, such as compensatory hepatocyte proliferation, liver fibrosis, inflammation, the increased generation of reactive oxygen species (ROS), and DNA damage. Hepatocyte proliferation is the consequence of injury-induced necrosis. Continuous cycles of this destructive–regenerative process are proposed to give rise to replication-related mutations in hepatocytes [18] and eventually HCC.

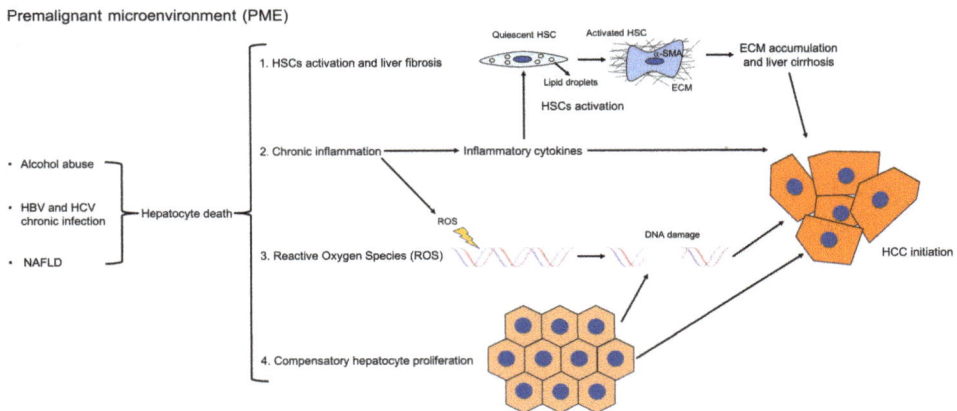

Figure 1. Mechanisms which promote HCC formation in PME. In PME, chronic liver injury causes hepatocyte death, which triggers inflammation, the activation of hepatic stellate cells into ECM-producing myofibroblasts, compensatory hepatocyte proliferation, the release of reactive oxygen species (ROS), and DNA damage. Continuous cycles of this destructive–regenerative process, which precedes tumor formation, are proposed to give rise to replication-related mutations in hepatocytes and eventually HCC. The links between the different mechanisms are indicated here.

Fibrosis is the main feature of hepatic PME [3]. Liver fibrosis starts as a protective wound healing response to acute liver damage. However, if the injury persists, fibrosis becomes chronic and dysfunctional [19]. Morphologically, liver fibrosis is characterized by the accumulation of ECM, followed by the formation of fibrous scar and subsequent cirrhosis [19,20]. HSCs are the main ECM-producing cells in the injured liver [21]. In healthy livers, quiescent HSCs localize in the space of Disse, function as pericytes, and store vitamin A. However, following continuous liver injury, HSCs activate into myofibroblasts; express alpha-smooth muscle actin (α-SMA); migrate to the site of tissue repair; and secrete ECM, chemokines, and cytokines. In the normal liver, ECM is formed by collagen type IV and VI; however, in fibrotic livers, there is a shift towards the accumulation of fibrillar collagens like type I and III, along with an increased deposition of non-collagenous glycoproteins like fibronectin, undulin, laminin, hyaluronan, elastin, and proteoglycans [20]. Moreover, the deposition of ECM is accompanied by a reduction in the activity of ECM-degrading matrix metalloproteinases (MMPs), favoring the formation of the fibrotic scar [22]. Several human studies have shown that a high fibrosis index and liver stiffness, which are indirect measurements of liver fibrosis, positively correlate with HCC risk [23–25]. Moreover, it was also demonstrated that liver fibrosis is linked to increased HCC recurrence after curative resection [26–29].

Another important feature of hepatic PME is inflammation, which, like fibrosis, is part of the protective wound healing response to acute liver damage. In the short-term, inflammation is believed to be beneficial, eliminating pathogens and favoring liver regeneration. However, chronic inflammation is detrimental and is linked to fibrosis, cirrhosis, and HCC. In fact, HSCs can be activated by several cytokines and growth factors, which are secreted by immune cells, including Kupffer cells, bone marrow-derived monocytes, Th17 cells, and innate lymphoid cells (ILC). Those inflammatory cytokines have been shown to modulate hepatic fibrogenesis in vivo and in vitro [30]. Proinflammatory mediators that have a role in HCC development include IL-1, IL-6, TNF-α, and IL-17 [31,32]. Additionally, secreted cytokines and growth factors can promote proliferative and anti-apoptotic signals in epithelial and tumor cells or induce angiogenesis, therefore favoring tumorigenesis. Interestingly, neutrophils and IL-1 promote hepatocarcinogenesis, but have a limited role in hepatic fibrosis [18,33–35]. Although IL-6 is reported to protect against liver fibrosis, it contributes to HCC development [36–39]. An additional effect of the recruitment of inflammatory cells in the PME is the production of ROS by activated macrophages, activated HSCs, and neutrophils. ROS not only promote fibrosis by facilitating HSCs activation and migration [40], but can also directly induce cancer by generating DNA damage and mutations in hepatocytes [41], or by causing the selective loss of CD4+ T lymphocytes, which mediate tumor immunosurveillance [42]. Consequently, it has been reported that antioxidants which inhibit ROS formation can effectively reduce hepatocarcinogenesis [43,44].

2.2. Tumor Microenvironment in HCC

TME in HCC consists of a dynamic network of non-tumoral stromal cells, including cancer-associated fibroblasts (CAFs), B and T cells, neutrophils, endothelial cells, and tumor-associated macrophages (TAMs) (Figure 2). Interestingly, it has recently been shown that, upon liver injury, the expression of adenine dinucleotide phosphate (NADPH) oxidase 1 (NOX1) by macrophages promotes hepatocarcinogenesis by inducing the production of inflammatory cytokines [45]. In addition to these cellular components, the TME is also characterized by profound ECM remodeling [46]. Altogether, the TME interacts bidirectionally with the tumor, generating a tumor-permissive niche. In the following paragraphs of this review, we will provide a detailed overview of CAFs and how they contribute to the development of HCC. TAMs and the other components of the TME are beyond the scope of this review.

3. Cancer-Associated Fibroblasts (CAFs)

3.1. Origin of CAFs

Fibroblasts were first described by Virchow and later Duvall in 1858 as spindle-shaped cells of the connective tissues that produce collagen. Later in 1971, Giulio Gabbiani showed that fibroblastic cells with contractile properties, called myofibroblasts, may be involved in wound healing [47]. Several studies using genetic cell fate mapping have provided strong evidence that the major precursors of α-SMA-expressing myofibroblasts in most types of experimental liver diseases are HSCs [48–50]. Therefore, CAFs most likely derive from HSCs. However, some controversies remain and, other than HSCs, the proposed sources of myofibroblasts are parenchymal cells undergoing epithelial-mesenchymal transition (EMT), bone marrow (BM)-derived cells, mesothelial cells, and portal fibroblasts (PFs).

Epithelial cells line the surfaces of the body and are located in all organs. EMT is a process in which epithelial cells lose their polarity, acquire a migratory capacity, and become myofibroblasts. Although some studies have shown that hepatocytes and cholangiocytes upregulate α-SMA and suppress epithelial markers under prolonged in vitro culturing [51,52], elegant lineage-tracing experiments have demonstrated that myofibroblasts found in experimental liver fibrosis do not originate from epithelial cells [53–55]. These results therefore suggest that myofibroblasts do not originate from EMT in fibrogenesis in vivo [56].

Two BM-derived cells which may potentially become myofibroblasts are mesenchymal stem cells (MSCs) and fibrocytes. MSCs are multipotent cells that can give rise to several cell types, including adipocytes, myocytes, chondrocytes, and osteoblasts. However, recent studies have shown that MSCs may actually have antifibrotic properties and provide a protective microenvironment in the recruited tissue [57].

In contrast, some studies have suggested that in a number of solid tumors, myofibroblasts can originate from the bone marrow [58–61]. Fibrocytes are cells with a spindle-like shape that were first described in 1994 [62]. They are characterized by co-expressing fibroblast markers (collagen type I, vimentin, and fibronectin) and hematopoietic cell markers (CD45, CD34, MHCII, CD11b, Gr-1, CD54, CD80, CD86, CCR2, CCR1, CCR7, CCR5) [63,64]. Studies have suggested that fibrocytes are recruited to the liver in response to both cholestatic and carbon tetrachloride (CCl_4)-induced liver injury, where they can differentiate into α-SMA+ myofibroblasts with a contribution range of between 3% and 50% [65–67].

Mesothelial cells form a monolayer of specialized cells which line the body's serum cavities and internal organs. They originate from the embryonic mesoderm layer and have features similar to epithelial cells. Cell fate mapping has demonstrated that during embryonic development, mesothelial cells can give rise to both PFs and HSCs [68,69]. However, it is not clear whether they can be a source of myofibroblasts in liver fibrosis. Interestingly, Asahina and coworkers have shown that mesothelial cells differentiate into both HSCs and myofibroblasts after CCl_4-induced liver injury, whereas in cholestatic liver injury, they only differentiate into HSCs, not myofibroblasts [70,71]. However, a recent study has suggested that mesothelial cells may have a role in fibrosis of the liver capsule [67].

Portal fibroblasts are a heterogenous population and reside underneath the bile duct epithelium. Since markers which can discriminate fibroblasts from other mesenchymal cells are lacking, it is challenging to identify or purify quiescent PFs. However, activated PFs were first described in cholestatic liver disease by electron microscopy, histology, and immunohistochemistry [72–74]. Cell phenotyping has demonstrated that during experimental biliary fibrosis, PFs differentiate into α-SMA-expressing myofibroblasts that produce ECM [75–77]. A study proposed that markers such as elastin, Thy1, and Ntpdase2, were specifically expressed by murine PFs, but not by HSCs [78]. The work of Iwaisako et al., using collagen promoter-driven green fluorescent protein (GFP) transgenic mice, has identified two myofibroblast populations: Vitamin A-positive HSCs and Vitamin A-negative PFs [79]. The unifying proposal is that in CCl_4-induced liver fibrosis, myofibroblasts mainly derive from HSCs, whereas in early cholestatic injury, PFs constitute the major source of myofibroblasts. However, in later cholestatic disease, HSCs again give rise to the majority of myofibroblasts. A novel signaling pathway involving the interaction of mesothelin with a MUC16-Thy1-TGFβRI complex regulates TGF-β1-mediated activation of PFs during cholestatic liver fibrosis [80].

In summary, current studies regarding the origin of hepatic myofibroblasts indicate that, depending on the type of liver injury, they mostly arise from liver-resident HSCs and to a lesser extent, from activated PFs. Mesothelial cells contribute to capsular fibrosis, whereas the contribution to liver fibrosis from BM-derived cells is quantitively small.

3.2. Markers of CAFs

In order to study and detect CAFs in the tumor, a specific marker is needed. However, a unique marker for CAFs has not been found. Several markers have been proposed to identify CAFs (Table 1); nonetheless, most of them are not unique to CAFs. For example, α-SMA is widely recognized as a robust CAFs marker [81]; however, it is also expressed by myofibroblasts [82] and its expression may vary between different CAF subtypes [83]. Another CAFs marker is the membrane-bound serine protease fibroblast activation protein α (FAPα), which is upregulated in the majority of epithelial carcinomas [84]. However, it has been shown that FAPα is also not specific to CAFs [85]. Recently, it has been demonstrated that FAPα is expressed in a certain sub-population of CAFs, but absent in others [86]. Fibroblast specific protein 1 (FSP-1) is another CAFs marker [87,88], which is also present in epithelial cells undergoing EMT [89] and in bone marrow-derived cells [90]. Additional proteins expressed in some CAFs include tenascin-C [91], periostin [92], neuron-glial antigen-2 (NG2) [93], podoplanin [94], and the novel identified marker microfibril associated protein 5 (MFAP5) [95]. In summary, the expression of CAFs markers is very heterogenous and it depends on the CAFs subpopulation being analyzed. Therefore, the discovery of CAF-specific markers will be vital to identify and therapeutically target this cell population.

Table 1. Markers of CAFs.

CAFs Markers	References
α-SMA	[81]
FAPα	[84]
FSP-1	[87,88]
Tenascin-C	[91]
Periostin	[92]
NG2	[93]
Podoplanin	[94]
MFAP5	[95]

3.3. CAFs in HCC

α-SMA-positive myofibroblasts are found in both human and murine HCC. For example, analysis by immunohistochemical technique of liver biopsy specimens from eight patients with HBV-related cirrhosis and HCC demonstrated that desmin-positive and α-SMA-positive cells were present in the perisinusoidal space and between tumor cells [96]. These results were confirmed by another study where liver specimens resected from 24 patients with HCC were analyzed by electron microscopy and immunohistochemistry. Interestingly, stromal cells strongly positive for α-SMA were found between endothelial cells and trabeculae of cancer cells [97]. Moreover, in vivo experiments demonstrated that the majority of cells producing collagens in human HCC were myofibroblasts [98]. In vitro experiments conducted in the same study then showed that HCC cell lines like HepG2, HuH17, and Hep3B, can increase ECM deposition in myofibroblasts by releasing a soluble mediator in the conditioned medium.

Multiple clinical studies have investigated the correlation between the presence of α-SMA-positive myofibroblasts and prognosis after HCC resection. For example, in 130 HCC cases, it was observed that the presence of peritumoral-activated HSCs positively correlates to poor clinical outcome after curative resection [28]. Other studies have confirmed these observations and suggested that metastasis was increased in patients expressing HSC signature genes [26,29,99,100]. Several in vitro and in vivo studies have demonstrated that HSCs can support the growth of HCC cell lines. For example, it was shown that conditioned media from human primary HSCs induced the proliferation and migration of human HCC cell lines cultured in monolayers [101], and similar results were also observed in a three-dimensional spheroid coculture system. In the same study, co-injection of HSCs and HCC cells into nude mice increased tumor growth and invasiveness, and inhibited necrosis. Another study using conditioned media from culture-activated rat HSCs and McA-RH777 rat HCC generated similar results [102]. Further in vitro studies demonstrated that activated CAFs repressed apoptosis in the Huh7 cell line by increasing the Bcl-2/BAX ratio through SDF-1/CXCR4/PI3K/AKT signaling [103]. Another work tried to discriminate the effect of human primary CAFs and primary non-tumoral fibroblasts (NTFs) on human HCC cell lines. The co-culture experiments demonstrated that CAFs up-regulated gene expressions of TGF-β1 and the fibroblast-activated protein (FAP) of HuH-7 and JHH-6, while NTF did not induce the expression of either gene [104]. Interestingly, it was also shown by co-culturing human hepatoma cells and activated human HSCs that the crosstalk between these cells is bi-directional, causing an increased expression of proinflammatory cytokines in hepatoma cells and an increased expression of VEGF and MM9 in HSCs [99]. It was demonstrated that there is a positive feedback loop between CAFs and the forkhead box Q1 (FOXQ1)/N-myc downstream-regulated gene 1 (NDRG1) axis, which drives HCC initiation [105]. Several in vivo studies have confirmed the results observed by co-culture experiments. Subcutaneous co-transplantation of an HSC cell line with MIM-R hepatocytes promoted tumor progression by inducing autocrine TGF-β signaling and nuclear β-catenin accumulation in neoplastic hepatocytes [106]. Similarly, when co-transplanted into nude mice, the HSC cell line LX2 promoted the growth of HepG2 tumors by increasing proliferation and angiogenesis and reducing HepG2 apoptosis [107]. In another study, T6 HSCs orthotopically co-injected into the livers of nude mice, together with H22 HCC cells, increased the tumorigenicity and invasiveness of the cancer cells by promoting angiogenesis [108]. Overall, evidence presented by such studies suggests that CAFs/HSCs are positive regulators of HCC. However, a recent study showed that HSCs may limit HCC progression though the orphan receptor endosialin, which may negatively regulate hepatotropic cytokines, including IGF2, RBP4, DKK1, and CCL5 [109]. This study supports the increasing recognition that HSCs not only have pro-tumorigenic functions, but may also inhibit cancer growth. For example, depleting CAFs in experimental pancreatic ductal adenocarcinoma promoted tumorigenesis [110,111].

4. CAF-Dependent Mechanisms of Hepatocarcinogenesis

Several CAF-dependent mechanisms support tumor growth in the liver (Figure 2). For example, CAFs can change the ECM stiffness and in turn affect tumorigenesis. Moreover, CAFs secrete cytokines and other factors which may promote tumor growth, tumor angiogenesis, and epithelial to mesenchymal transition (EMT). CAFs have also been shown to indirectly affect HCC by cross talking with immune cells and reducing immune surveillance.

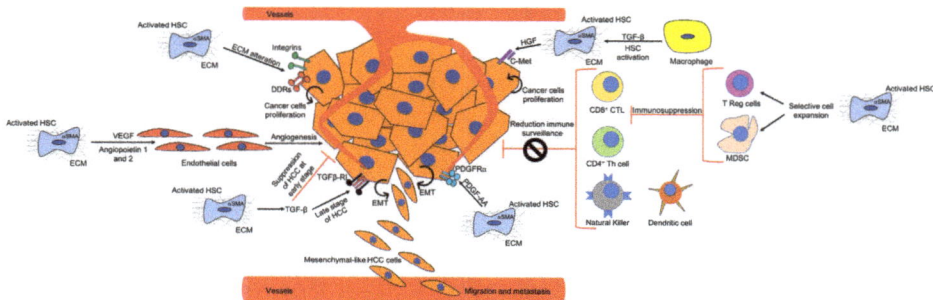

Figure 2. TME in HCC. The figure outlines the different components of the TME and the CAF-dependent mechanisms of hepatocarcinogenesis. (See text for details).

4.1. CAF-Dependent Alteration of ECM Promotes HCC

In the injured liver, activated HSCs secrete ECM proteins and there is a shift towards the accumulation of fibrillar collagens like type I and III. In this altered biomechanical environment, the ECM components can interact directly and indirectly with both cancer cells and stromal cells to change their functions [112]. For example, it was shown that laminin-5, one of the components of the ECM, secreted by primary cultures of human HSCs, stimulated cell migration in several HCC cell lines by activating the MEK/ERK pathway [113]. Moreover, the increase and reorganization of ECM created a stiff microenvironment in the liver. Interestingly, Schrader and collaborators have used "mechanically tunable" matrix-coated polyacrylamide gels to show that an increase in matrix stiffness promoted the proliferation of Huh7 and HepG2 cell lines through the PKB/Akt pathway. In contrast, a soft environment favored cellular dormancy and stem cell characteristics in HCC [114]. Another study using polyacrylamide supports of different stiffnesses suggested that HSCs are also affected by the stiff environment. In fact, primary rat HSCs required a mechanical stiff substrate to differentiate into myofibroblasts [115]. Similar results were also observed when studying CAFs in breast cancer [116]. Of importance, several studies which have measured liver stiffness by using elastography in patients with chronic liver diseases, have confirmed that stiffness correlates with the risk of HCC [24,25,117,118]. The mechanical stress caused by alteration of the ECM is transmitted to the nearby cells by integrins and discodin domain receptors (DDRs), which are responsible for mediating "outside-in" and "inside-out" signaling between ECM and the cells [119]. A study using PDGFC transgenic or Pten null mice as HCC models has shown that several collagen types and integrins were both up-regulated in tumors in these mice, suggesting a correlation in the expression of HCC-associated ECM proteins and ECM-integrins networks [120]. In accordance with these results, integrin β1 and integrin α6 were upregulated in liver biopsies of HCC patients, and the integrin expression positively correlated with pathological grade [121,122]. Integrins have been shown to promote cell proliferation by activating the MAPK and Pi3K pathways, and cell survival through antiapoptotic signaling [123]. Therefore, the altered ECM present in the HCC microenvironment may interact with integrins expressed in hepatocytes, promoting tumor proliferation, migration, and invasion. Like integrins, DDR2 expression was increased in several HCC cell lines and in 112 biopsies from HCC patients, and it was correlated with clinicopathological features of poor prognosis [124]. DDR2 was shown to facilitate HCC invasion and metastasis through activation of the ERK pathway and stabilization of the EMT marker SNAIL1, and this signaling cascade was induced by collagen type I [124]. ECM degradation by MMPs is another key process in the injured liver, which can affect tumorigenesis by releasing growth factors or generating cleavage fragments [125]. Several studies have shown that MMPs can promote tumor cell proliferation, progression, and invasion [126–129].

4.2. CAFs and Tumor Angiogenesis

Tissue hypoxia and vascular disorganization are typically observed in the injured liver. Hypoxia inhibits liver regeneration and promotes angiogenesis, fibrogenesis, and hepatocarcinogenesis [130]. Angiogenesis is the physiological process through which new blood vessels form from pre-existing vessels. Vascular endothelial growth factor (VEGF) is crucial for angiogenesis and it has been shown that it is secreted by both primary and immortalized rat hepatic stellate cells after hypoxic injury [131]. Thus, induction of VEGF may be important in the pathogenesis of liver injury and hepatocarcinogenesis. Another angiogenic factor called angiopoietin-1 was increased in a human fibrotic liver, and was expressed and secreted by activated HSCs isolated from fibrotic mice which were treated by CCl_4 or underwent bile duct ligation (BDL) surgery [132]. Another study observed that the expression of angiopoietin-1 and angiopoietin-2 was upregulated in HCC patients and correlated with tumor dedifferentiation and tumor vascularity. Moreover, the same study showed that angiopoietin-1 and angiopoietin-2 can be detected in hepatoma cells, HSCs, and smooth muscle cells [133]. Angiopoietein-2 was also found to be upregulated at both mRNA and protein levels in patients with chronic hepatitis B, suggesting that it may contribute to pathological angiogenesis and HCC progression [134]. Several studies using 3D spheroids, co-culture systems of HSCs with endothelial cells, and subcutaneous xenograft models, have shown that HSCs can promote angiogenesis by producing proangiogenic mediators [107,108,135,136].

4.3. CAF-Secreted Cytokines

In liver fibrosis, the death of hepatocytes and cholangiocytes causes the activation of HSCs directly or through several cytokines, which are secreted by immune cells. These inflammatory cytokines have been shown to modulate hepatic fibrogenesis in vivo and in vitro [137]. In turn, activated HSCs can produce cytokines that promote cancer proliferation and migration. For example, it was shown that activated HSCs secrete TGF-β, which has a bipartite role. It is a tumor suppressor at early stages of liver damage and regeneration, whereas it acts as a tumor promoter during cancer progression [138], perhaps by inducing nuclear β-catenin accumulation in neoplastic hepatocytes [106]. Studies using transgenic mice have shown that TGF-β-dependent targeting of Snail is required for the formation of liver cancers [139,140]. HSCs also produce hepatocyte growth factor (HGF), which stimulates the motility of Hep3B, HepG2, and Huh7 cells and the migration of primary HCC cells isolated from three patients. HGF promoted phosphorylation of its receptor c-Met and activation of phosphatidylinositol 3-kinase (PI3-K) [141]. A more recent study has shown, by in vitro and in vivo experiments, that HSC-secreted HGF might reduce HCC sensitization to chemotherapeutic agents by promoting epithelial-mesenchymal transition (EMT) and cancer stem cell (CSC)-like properties through the HGF/c-Met pathway [142]. Clinical studies have supported these observations, showing that the expression of HGF and its receptor c-MET was elevated in cirrhotic tissues and in 80% of HCC cases [143].

4.4. CAFs and Immune Surveillance

The immune system works as a barrier to tumor formation and progression and several studies have shown that CD8+ cytotoxic T lymphocytes (CTLs), CD4+ Th1 T cells, and natural killer (NK) and dendritic cells (DC) are critical to block tumor development [144,145]. Therefore, both the innate and adaptive immunity contribute to immune surveillance. However, cancer cells can evade the immune system either by producing immunosuppressive factors like TGF-β [146,147] or by recruiting immunosuppressive inflammatory cells such as regulatory T cells (Tregs) or myeloid derived suppressor cells (MDSC), which are able to inhibit the activity of cytotoxic lymphocyte cells [148,149]. It has been shown that populations of Tregs are increased in the tumor and peripheral blood of HCC patients [150]. MDSC in mice express CD11b and Gr-1 and they can be found in the blood, spleen, bone marrow, and tumor microenvironment [149]. In humans, MDSC are characterized by the expression of markers such as CD34, CD33, CD15, and CD16 [151]. Like Tregs, MDSC are also increased in HCC patients [152]. The mechanisms by which Tregs and MDSC limit antitumor immunity have been extensively described previously [153,154] and they are beyond the scope of this review. Using allografts, it was demonstrated for the first time that HSCs can modulate immunity in mice and inhibit T-cell responses by inducing T-cell apoptosis [155,156]. Several studies have shown that CAFs promote HCC by reducing immune surveillance. For example, a cellular transplantation model in immunocompetent mice demonstrated that HSCs prevent T-cell infiltration in tumors, creating an immunosuppressive microenvironment [157]. Immunohistochemical experiments and gene signature analysis in HCC patients have shown that activated HSCs can interact with monocytes, promoting the expression of immunosuppressive cytokines [29]. Moreover, in vivo-activated HSCs caused T-cell hyporesponsiveness, increased T-cell apoptosis, an increased number of immunosuppressive Treg cells, and T-cell mediated cytotoxicity inhibition [158]. Similarly, using the mouse hepatoma cell line H22 together with primary activated HSCs in an orthotopic liver tumor mouse model, it was demonstrated that HSCs increase the number of MDSCs in HCC [159]. More recently, another study confirmed these results and suggested that HSCs induce MDSC by the secretion of prostaglandin E2 (PGE2) [160].

Moreover, it was also suggested, by using co-culture studies, that HSCs promote the conversion of blood monocytes into MDCS in a CD44-dependent manner [161].

4.5. CAFs and EMT

EMT is a biological process in which epithelial cells lose their apicobasal polarity, thus allowing them to travel through the ECM like mesenchymal cells [162]. Cells undergoing EMT acquire increased invasiveness, enhanced production of ECM, and more resistance to apoptosis. There are three types of EMT: type 1 gives rise to primary parenchymal cells during embryogenesis; type 2 occurs during wound healing and organ fibrosis; and type 3 modifies the phenotype of cancer cells and is associated with tumor intravasation, migration, and metastasis [163,164]. Several signals, such as TGF-β, epidermal growth factor (EGF), and PDGF, produced by the tumor stroma and in particular by CAFs, may be implicated in EMT [163]. In particular, TGF-β, which signals through both Smad-dependent and -independent pathways, is considered to be the main EMT promoter in epithelial cells, including hepatocytes [165]. Thus, it was shown that TGF-β induces EMT in Ras-transformed hepatocytes [166]. Results from an in vivo HCC model where HCC cells were co-injected with myofibroblasts, and from an in vitro model with a micro-organoid HCC, suggested that the hepatic tumor-stroma crosstalk promotes tumor growth and EMT through a TGF-β and PDGF signaling axis [167].

Other studies have suggested that the TGF-β pathway may be important in the maintenance of self-renewal and pluripotent stem cells, which replicate and generate non-stem differentiated cells. It has been proposed that HCC can originate from a small subset of cancer stem cells, which are transformed from a hypothetical normal stem cell niche [168,169] or from differentiated hepatocytes [170]. Interestingly, experiments performed in rats have suggested that EMT due to chronic TGF-β stimulation produces cancer stem cells from hepatic progenitor-like cells. The same study also showed that pharmaceutical inhibition of microRNA-216a/PTEN/Akt signaling could be a novel strategy for HCC prevention [171]. Another study which employed six different human HCC cell lines has shown that tumor cells with a mesenchymal-like phenotype are refractory to sorafenib-induced cell death [172]. Therefore, these results suggest that EMT induced by TGF-β signaling derived from the tumor stroma may play an important role in supporting tumor growth, and in the generation of chemo-resistant cells, which have stem-like features in HCC.

5. Conclusions

CAFs are one of the most important components of the tumor microenvironment in HCC. Although several studies have shown different mechanisms by which those cells affect HCC growth, this area requires more study. Although the majority of studies presented in this review suggest that CAFs in HCC are positive regulators of cancer, CAFs have been shown to act both as a positive and negative regulator of tumorigenesis in different types of cancer [173]. Therefore, an open question is whether CAFs may somehow have a protective effect and prevent tumor growth in HCC as well. Further studies aiming to answer this question may be important for designing innovative therapeutic approaches. For example, CAFs may be pharmacologically targeted to secrete anti-tumor factors, which might hinder HCC growth and progression. It is also important to understand the origin of CAFs, although HSCs are considered to be the main source. Identification of CAF-specific markers will be crucial to discriminate them from HSCs in the PME and TME of HCC, and to specifically target this cell population.

Funding: This work was funded by the National Institutes of Health R01 DK099205-01A1 to T.K. and P50AA011999 to D.A.B., and by Superfund Training Core P42ES010337 to D.A.B.

Acknowledgments: The authors want to thank Karin Diggle for proofreading the manuscript.

Conflicts of Interest: The authors declare no conflict of interest. The funders had no role in the design of the study; in the collection, analyses, or interpretation of data; in the writing of the manuscript.

Abbreviations

HCC	Hepatocellular carcinoma
HSCs	Hepatic stellate cells
CAFs	Cancer-associated fibroblasts
PFs	Portal fibroblasts
ECM	Extracellular matrix
EMT	Epithelial-mesenchymal transition
PME	Premalignant microenvironment
TME	Tumor microenvironment
HBV	Hepatitis B virus
HCV	Hepatitis C virus
NAFLD	Non-alcoholic fatty liver disease
α-SMA	Alpha-smooth muscle actin
MMPs	Matrix metalloproteinases
ROS	Reactive oxygen species
TAMs	Tumor-associated macrophages
NADPH	Nicotinamide adenine dinucleotide phosphate
NOX1	NADPH oxidase 1
CCl_4	Carbon tetrachloride
BDL	Bile duct ligation
FAP	Fibroblast activated protein
NTF	Non-tumoral fibroblast
VEGF	Vascular endothelial growth factor
EGF	Epidermal growth factor
DDRs	Discodin domain receptor
PI3-K	phosphatidylinositol 3-kinase
HGF	Hepatocyte growth factor
CSC	Cancer-stem cell
FAPα	Fibroblast activation protein α
FSP-1	Fibroblast specific protein 1
NG2	Neuron-glial antigen-2
MFAP5	Microfibril associated protein 5
Treg	Regulatory T cell
MDSC	Myeloid derived suppressor cell
DC	Dendritic cell
NK	Natural killer
CTLs	Cytotoxic T lymphocytes
MSCs	Mesenchymal stem cells
GFP	Green fluorescent protein

References

1. Londoño, M.C.; Abraldes, J.G.; Altamirano, J.; Decaens, T.; Forns, X. Clinical trial watch: Reports from the AASLD Liver Meeting®, Boston, November 2014. *J. Hepatol.* **2015**, *62*, 1196–1203. [CrossRef] [PubMed]
2. Torre, L.A.; Bray, F.; Siegel, R.L.; Ferlay, J.; Lortet-Tieulent, J.; Jemal, A. Global cancer statistics, 2012. *CA Cancer J. Clin.* **2015**, *65*, 87–108. [CrossRef] [PubMed]
3. El-Serag, H.B. Hepatocellular carcinoma. *N. Engl. J. Med.* **2011**, *365*, 1118–1127. [CrossRef] [PubMed]
4. Forner, A.; Reig, M.; Bruix, J. Hepatocellular carcinoma. *Lancet* **2018**, *391*, 1301–1314. [CrossRef]
5. Friedman, S.L. Molecular regulation of hepatic fibrosis, an integrated cellular response to tissue injury. *J. Biol. Chem.* **2000**, *275*, 2247–2250. [CrossRef] [PubMed]
6. Friedman, S.L. Hepatic stellate cells: Protean, multifunctional, and enigmatic cells of the liver. *Physiol. Rev.* **2008**, *88*, 125–172. [CrossRef]

7. Lee, U.E.; Friedman, S.L. Mechanisms of hepatic fibrogenesis. *Best Pract. Res. Clin. Gastroenterol.* **2011**, *25*, 195–206. [CrossRef] [PubMed]

8. Fattovich, G.; Stroffolini, T.; Zagni, I.; Donato, F. Hepatocellular carcinoma in cirrhosis: Incidence and risk factors. *Gastroenterology* **2004**, *127*, S35–S50. [CrossRef] [PubMed]

9. Tahmasebi Birgani, M.; Carloni, V. Tumor Microenvironment, a Paradigm in Hepatocellular Carcinoma Progression and Therapy. *Int. J. Mol. Sci.* **2017**, *18*, 405. [CrossRef]

10. Affo, S.; Yu, L.X.; Schwabe, R.F. The Role of Cancer-Associated Fibroblasts and Fibrosis in Liver Cancer. *Annu. Rev. Pathol.* **2017**, *12*, 153–186. [CrossRef]

11. Luedde, T.; Beraza, N.; Kotsikoris, V.; van Loo, G.; Nenci, A.; De Vos, R.; Roskams, T.; Trautwein, C.; Pasparakis, M. Deletion of NEMO/IKKgamma in liver parenchymal cells causes steatohepatitis and hepatocellular carcinoma. *Cancer Cell* **2007**, *11*, 119–132. [CrossRef] [PubMed]

12. Inokuchi, S.; Aoyama, T.; Miura, K.; Osterreicher, C.H.; Kodama, Y.; Miyai, K.; Akira, S.; Brenner, D.A.; Seki, E. Disruption of TAK1 in hepatocytes causes hepatic injury, inflammation, fibrosis, and carcinogenesis. *Proc. Natl. Acad. Sci. USA* **2010**, *107*, 844–849. [CrossRef] [PubMed]

13. Bettermann, K.; Vucur, M.; Haybaeck, J.; Koppe, C.; Janssen, J.; Heymann, F.; Weber, A.; Weiskirchen, R.; Liedtke, C.; Gassler, N.; et al. TAK1 suppresses a NEMO-dependent but NF-kappaB-independent pathway to liver cancer. *Cancer Cell* **2010**, *17*, 481–496. [CrossRef] [PubMed]

14. Weber, A.; Boger, R.; Vick, B.; Urbanik, T.; Haybaeck, J.; Zoller, S.; Teufel, A.; Krammer, P.H.; Opferman, J.T.; Galle, P.R.; et al. Hepatocyte-specific deletion of the antiapoptotic protein myeloid cell leukemia-1 triggers proliferation and hepatocarcinogenesis in mice. *Hepatology* **2010**, *51*, 1226–1236. [CrossRef]

15. Hikita, H.; Kodama, T.; Shimizu, S.; Li, W.; Shigekawa, M.; Tanaka, S.; Hosui, A.; Miyagi, T.; Tatsumi, T.; Kanto, T.; et al. Bak deficiency inhibits liver carcinogenesis: A causal link between apoptosis and carcinogenesis. *J. Hepatol.* **2012**, *57*, 92–100. [CrossRef]

16. Chen, C.F.; Lee, W.C.; Yang, H.I.; Chang, H.C.; Jen, C.L.; Iloeje, U.H.; Su, J.; Hsiao, C.K.; Wang, L.Y.; You, S.L.; et al. Changes in serum levels of HBV DNA and alanine aminotransferase determine risk for hepatocellular carcinoma. *Gastroenterology* **2011**, *141*, 1240–1248. [CrossRef]

17. Lee, M.H.; Yang, H.I.; Lu, S.N.; Jen, C.L.; Yeh, S.H.; Liu, C.J.; Chen, P.J.; You, S.L.; Wang, L.Y.; Chen, W.J.; et al. Hepatitis C virus seromarkers and subsequent risk of hepatocellular carcinoma: Long-term predictors from a community-based cohort study. *J. Clin. Oncol.* **2010**, *28*, 4587–4593.

18. Sakurai, T.; He, G.; Matsuzawa, A.; Yu, G.Y.; Maeda, S.; Hardiman, G.; Karin, M. Hepatocyte necrosis induced by oxidative stress and IL-1 alpha release mediate carcinogen-induced compensatory proliferation and liver tumorigenesis. *Cancer Cell* **2008**, *14*, 156–165. [CrossRef] [PubMed]

19. Hernandez-Gea, V.; Friedman, S.L. Pathogenesis of liver fibrosis. *Annu. Rev. Pathol.* **2011**, *6*, 425–456. [CrossRef]

20. Bataller, R.; Brenner, D.A. Liver fibrosis. *J. Clin. Investig.* **2005**, *115*, 209–218. [CrossRef]

21. Gäbele, E.; Brenner, D.A.; Rippe, R.A. Liver fibrosis: Signals leading to the amplification of the fibrogenic hepatic stellate cell. *Front. Biosci.* **2003**, *8*, d69–d77. [PubMed]

22. Arthur, M.J. Fibrogenesis II. Metalloproteinases and their inhibitors in liver fibrosis. *Am. J. Physiol. Gastrointest. Liver Physiol.* **2000**, *279*, G245–G249. [CrossRef]

23. Suh, B.; Park, S.; Shin, D.W.; Yun, J.M.; Yang, H.K.; Yu, S.J.; Shin, C.I.; Kim, J.S.; Ahn, E.; Lee, H.; et al. High liver fibrosis index FIB-4 is highly predictive of hepatocellular carcinoma in chronic hepatitis B carriers. *Hepatology* **2015**, *61*, 1261–1268. [CrossRef] [PubMed]

24. Kim, M.N.; Kim, S.U.; Kim, B.K.; Park, J.Y.; Kim, D.Y.; Ahn, S.H.; Song, K.J.; Park, Y.N.; Han, K.H. Increased risk of hepatocellular carcinoma in chronic hepatitis B patients with transient elastography-defined subclinical cirrhosis. *Hepatology* **2015**, *61*, 1851–1859. [CrossRef] [PubMed]

25. Akima, T.; Tamano, M.; Hiraishi, H. Liver stiffness measured by transient elastography is a predictor of hepatocellular carcinoma development in viral hepatitis. *Hepatol. Res.* **2011**, *41*, 965–970. [CrossRef]

26. Zhang, D.Y.; Goossens, N.; Guo, J.; Tsai, M.C.; Chou, H.I.; Altunkaynak, C.; Sangiovanni, A.; Iavarone, M.; Colombo, M.; Kobayashi, M.; et al. A hepatic stellate cell gene expression signature associated with outcomes in hepatitis C cirrhosis and hepatocellular carcinoma after curative resection. *Gut* **2016**, *65*, 1754–1764. [CrossRef]

27. Wang, Q.; Fiel, M.I.; Blank, S.; Luan, W.; Kadri, H.; Kim, K.W.; Manizate, F.; Rosenblatt, A.G.; Labow, D.M.; Schwartz, M.E.; et al. Impact of liver fibrosis on prognosis following liver resection for hepatitis B-associated hepatocellular carcinoma. *Br. J. Cancer* **2013**, *109*, 573–581. [CrossRef]

28. Ju, M.J.; Qiu, S.J.; Fan, J.; Xiao, Y.S.; Gao, Q.; Zhou, J.; Li, Y.W.; Tang, Z.Y. Peritumoral activated hepatic stellate cells predict poor clinical outcome in hepatocellular carcinoma after curative resection. *Am. J. Clin. Pathol.* **2009**, *131*, 498–510. [CrossRef]

29. Ji, J.; Eggert, T.; Budhu, A.; Forgues, M.; Takai, A.; Dang, H.; Ye, Q.; Lee, J.S.; Kim, J.H.; Greten, T.F.; et al. Hepatic stellate cell and monocyte interaction contributes to poor prognosis in hepatocellular carcinoma. *Hepatology* **2015**, *62*, 481–495. [CrossRef]

30. Seki, E.; Schwabe, R.F. Hepatic inflammation and fibrosis: Functional links and key pathways. *Hepatology* **2015**, *61*, 1066–1079. [CrossRef]

31. Capece, D.; Fischietti, M.; Verzella, D.; Gaggiano, A.; Cicciarelli, G.; Tessitore, A.; Zazzeroni, F.; Alesse, E. The inflammatory microenvironment in hepatocellular carcinoma: A pivotal role for tumor-associated macrophages. *BioMed Res. Int.* **2013**, *2013*, 187204. [CrossRef] [PubMed]

32. Haybaeck, J.; Zeller, N.; Wolf, M.J.; Weber, A.; Wagner, U.; Kurrer, M.O.; Bremer, J.; Iezzi, G.; Graf, R.; Clavien, P.A.; et al. A lymphotoxin-driven pathway to hepatocellular carcinoma. *Cancer Cell* **2009**, *16*, 295–308. [CrossRef] [PubMed]

33. Wilson, C.L.; Jurk, D.; Fullard, N.; Banks, P.; Page, A.; Luli, S.; Elsharkawy, A.M.; Gieling, R.G.; Chakraborty, J.B.; Fox, C.; et al. NFκB1 is a suppressor of neutrophil-driven hepatocellular carcinoma. *Nat. Commun.* **2015**, *6*, 6818. [CrossRef] [PubMed]

34. Moles, A.; Murphy, L.; Wilson, C.L.; Chakraborty, J.B.; Fox, C.; Park, E.J.; Mann, J.; Oakley, F.; Howarth, R.; Brain, J.; et al. A TLR2/S100A9/CXCL-2 signaling network is necessary for neutrophil recruitment in acute and chronic liver injury in the mouse. *J. Hepatol.* **2014**, *60*, 782–791. [CrossRef]

35. Saito, J.M.; Bostick, M.K.; Campe, C.B.; Xu, J.; Maher, J.J. Infiltrating neutrophils in bile duct-ligated livers do not promote hepatic fibrosis. *Hepatol. Res.* **2003**, *25*, 180–191. [CrossRef]

36. He, G.; Dhar, D.; Nakagawa, H.; Font-Burgada, J.; Ogata, H.; Jiang, Y.; Shalapour, S.; Seki, E.; Yost, S.E.; Jepsen, K.; et al. Identification of liver cancer progenitors whose malignant progression depends on autocrine IL-6 signaling. *Cell* **2013**, *155*, 384–396. [CrossRef]

37. Naugler, W.E.; Sakurai, T.; Kim, S.; Maeda, S.; Kim, K.; Elsharkawy, A.M.; Karin, M. Gender disparity in liver cancer due to sex differences in MyD88-dependent IL-6 production. *Science* **2007**, *317*, 121–124. [CrossRef] [PubMed]

38. Kovalovich, K.; DeAngelis, R.A.; Li, W.; Furth, E.E.; Ciliberto, G.; Taub, R. Increased toxin-induced liver injury and fibrosis in interleukin-6-deficient mice. *Hepatology* **2000**, *31*, 149–159. [CrossRef]

39. Streetz, K.L.; Wüstefeld, T.; Klein, C.; Kallen, K.J.; Tronche, F.; Betz, U.A.; Schütz, G.; Manns, M.P.; Müller, W.; Trautwein, C. Lack of gp130 expression in hepatocytes promotes liver injury. *Gastroenterology* **2003**, *125*, 532–543. [CrossRef]

40. Luangmonkong, T.; Suriguga, S.; Mutsaers, H.A.M.; Groothuis, G.M.M.; Olinga, P.; Boersema, M. Targeting Oxidative Stress for the Treatment of Liver Fibrosis. *Rev. Physiol. Biochem. Pharmacol.* **2018**. [CrossRef]

41. Farazi, P.A.; DePinho, R.A. Hepatocellular carcinoma pathogenesis: From genes to environment. *Nat. Rev. Cancer* **2006**, *6*, 674–687. [CrossRef] [PubMed]

42. Ma, C.; Kesarwala, A.H.; Eggert, T.; Medina-Echeverz, J.; Kleiner, D.E.; Jin, P.; Stroncek, D.F.; Terabe, M.; Kapoor, V.; ElGindi, M.; et al. NAFLD causes selective CD4(+) T lymphocyte loss and promotes hepatocarcinogenesis. *Nature* **2016**, *531*, 253–257. [CrossRef] [PubMed]

43. Maeda, S.; Kamata, H.; Luo, J.L.; Leffert, H.; Karin, M. IKKbeta couples hepatocyte death to cytokine-driven compensatory proliferation that promotes chemical hepatocarcinogenesis. *Cell* **2005**, *121*, 977–990. [CrossRef]

44. Zhang, X.F.; Tan, X.; Zeng, G.; Misse, A.; Singh, S.; Kim, Y.; Klaunig, J.E.; Monga, S.P. Conditional beta-catenin loss in mice promotes chemical hepatocarcinogenesis: Role of oxidative stress and platelet-derived growth factor receptor alpha/phosphoinositide 3-kinase signaling. *Hepatology* **2010**, *52*, 954–965. [CrossRef]

45. Liang, S.; Ma, H.Y.; Zhong, Z.; Dhar, D.; Liu, X.; Xu, J.; Koyama, Y.; Nishio, T.; Karin, D.; Karin, G.; et al. NADPH Oxidase 1 in Liver Macrophages Promotes Inflammation and Tumor Development in Mice. *Gastroenterology* **2018**. [CrossRef]

46. Lu, P.; Weaver, V.M.; Werb, Z. The extracellular matrix: A dynamic niche in cancer progression. *J. Cell Biol.* **2012**, *196*, 395–406. [CrossRef]

47. Gabbiani, G.; Ryan, G.B.; Majne, G. Presence of modified fibroblasts in granulation tissue and their possible role in wound contraction. *Experientia* **1971**, *27*, 549–550. [CrossRef]

48. Mederacke, I.; Hsu, C.C.; Troeger, J.S.; Huebener, P.; Mu, X.; Dapito, D.H.; Pradere, J.P.; Schwabe, R.F. Fate tracing reveals hepatic stellate cells as dominant contributors to liver fibrosis independent of its aetiology. *Nat. Commun.* **2013**, *4*, 2823. [CrossRef]

49. Henderson, N.C.; Arnold, T.D.; Katamura, Y.; Giacomini, M.M.; Rodriguez, J.D.; McCarty, J.H.; Pellicoro, A.; Raschperger, E.; Betsholtz, C.; Ruminski, P.G.; et al. Targeting of αv integrin identifies a core molecular pathway that regulates fibrosis in several organs. *Nat. Med.* **2013**, *19*, 1617–1624. [CrossRef]

50. Puche, J.E.; Lee, Y.A.; Jiao, J.; Aloman, C.; Fiel, M.I.; Muñoz, U.; Kraus, T.; Lee, T.; Yee, H.F.; Friedman, S.L. A novel murine model to deplete hepatic stellate cells uncovers their role in amplifying liver damage in mice. *Hepatology* **2013**, *57*, 339–350. [CrossRef]

51. Choi, S.S.; Diehl, A.M. Epithelial-to-mesenchymal transitions in the liver. *Hepatology* **2009**, *50*, 2007–2013. [CrossRef] [PubMed]

52. Kalluri, R. EMT: When epithelial cells decide to become mesenchymal-like cells. *J. Clin. Investig.* **2009**, *119*, 1417–1419. [CrossRef]

53. Taura, K.; Miura, K.; Iwaisako, K.; Osterreicher, C.H.; Kodama, Y.; Penz-Osterreicher, M.; Brenner, D.A. Hepatocytes do not undergo epithelial-mesenchymal transition in liver fibrosis in mice. *Hepatology* **2010**, *51*, 1027–1036. [CrossRef]

54. Scholten, D.; Weiskirchen, R. Questioning the challenging role of epithelial-to-mesenchymal transition in liver injury. *Hepatology* **2011**, *53*, 1048–1051. [CrossRef]

55. Chu, A.S.; Diaz, R.; Hui, J.J.; Yanger, K.; Zong, Y.; Alpini, G.; Stanger, B.Z.; Wells, R.G. Lineage tracing demonstrates no evidence of cholangiocyte epithelial-to-mesenchymal transition in murine models of hepatic fibrosis. *Hepatology* **2011**, *53*, 1685–1695. [CrossRef] [PubMed]

56. Munker, S.; Wu, Y.L.; Ding, H.G.; Liebe, R.; Weng, H.L. Can a fibrotic liver afford epithelial-mesenchymal transition? *World J. Gastroenterol.* **2017**, *23*, 4661–4668. [CrossRef]

57. Prockop, D.J. Inflammation, fibrosis, and modulation of the process by mesenchymal stem/stromal cells. *Matrix Biol.* **2016**, *51*, 7–13. [CrossRef]

58. Worthley, D.L.; Si, Y.; Quante, M.; Churchill, M.; Mukherjee, S.; Wang, T.C. Bone marrow cells as precursors of the tumor stroma. *Exp. Cell Res.* **2013**, *319*, 1650–1656. [CrossRef]

59. Barcellos-de-Souza, P.; Gori, V.; Bambi, F.; Chiarugi, P. Tumor microenvironment: Bone marrow-mesenchymal stem cells as key players. *Biochim. Biophys. Acta* **2013**, *1836*, 321–335. [CrossRef] [PubMed]

60. Mishra, P.J.; Humeniuk, R.; Medina, D.J.; Alexe, G.; Mesirov, J.P.; Ganesan, S.; Glod, J.W.; Banerjee, D. Carcinoma-associated fibroblast-like differentiation of human mesenchymal stem cells. *Cancer Res.* **2008**, *68*, 4331–4339. [CrossRef] [PubMed]

61. Quante, M.; Tu, S.P.; Tomita, H.; Gonda, T.; Wang, S.S.; Takashi, S.; Baik, G.H.; Shibata, W.; Diprete, B.; Betz, K.S.; et al. Bone marrow-derived myofibroblasts contribute to the mesenchymal stem cell niche and promote tumor growth. *Cancer Cell* **2011**, *19*, 257–272. [CrossRef]

62. Bucala, R.; Spiegel, L.A.; Chesney, J.; Hogan, M.; Cerami, A. Circulating fibrocytes define a new leukocyte subpopulation that mediates tissue repair. *Mol. Med.* **1994**, *1*, 71–81. [CrossRef] [PubMed]

63. Abe, R.; Donnelly, S.C.; Peng, T.; Bucala, R.; Metz, C.N. Peripheral blood fibrocytes: Differentiation pathway and migration to wound sites. *J. Immunol.* **2001**, *166*, 7556–7562. [CrossRef] [PubMed]

64. Quan, T.E.; Cowper, S.; Wu, S.P.; Bockenstedt, L.K.; Bucala, R. Circulating fibrocytes: Collagen-secreting cells of the peripheral blood. *Int. J. Biochem. Cell Biol.* **2004**, *36*, 598–606. [CrossRef] [PubMed]

65. Kisseleva, T.; Uchinami, H.; Feirt, N.; Quintana-Bustamante, O.; Segovia, J.C.; Schwabe, R.F.; Brenner, D.A. Bone marrow-derived fibrocytes participate in pathogenesis of liver fibrosis. *J. Hepatol.* **2006**, *45*, 429–438. [CrossRef]

66. Scholten, D.; Reichart, D.; Paik, Y.H.; Lindert, J.; Bhattacharya, J.; Glass, C.K.; Brenner, D.A.; Kisseleva, T. Migration of fibrocytes in fibrogenic liver injury. *Am. J. Pathol.* **2011**, *179*, 189–198. [CrossRef] [PubMed]

67. Lua, I.; Li, Y.; Pappoe, L.S.; Asahina, K. Myofibroblastic Conversion and Regeneration of Mesothelial Cells in Peritoneal and Liver Fibrosis. *Am. J. Pathol.* **2015**, *185*, 3258–3273. [CrossRef]

68. Asahina, K.; Zhou, B.; Pu, W.T.; Tsukamoto, H. Septum transversum-derived mesothelium gives rise to hepatic stellate cells and perivascular mesenchymal cells in developing mouse liver. *Hepatology* **2011**, *53*, 983–995. [CrossRef]

69. Rinkevich, Y.; Mori, T.; Sahoo, D.; Xu, P.X.; Bermingham, J.R.; Weissman, I.L. Identification and prospective isolation of a mesothelial precursor lineage giving rise to smooth muscle cells and fibroblasts for mammalian internal organs, and their vasculature. *Nat. Cell Biol.* **2012**, *14*, 1251–1260. [CrossRef]

70. Li, Y.; Wang, J.; Asahina, K. Mesothelial cells give rise to hepatic stellate cells and myofibroblasts via mesothelial-mesenchymal transition in liver injury. *Proc. Natl. Acad. Sci. USA* **2013**, *110*, 2324–2329. [CrossRef]

71. Lua, I.; Li, Y.; Zagory, J.A.; Wang, K.S.; French, S.W.; Sévigny, J.; Asahina, K. Characterization of hepatic stellate cells, portal fibroblasts, and mesothelial cells in normal and fibrotic livers. *J. Hepatol.* **2016**, *64*, 1137–1146. [CrossRef] [PubMed]

72. Tang, L.; Tanaka, Y.; Marumo, F.; Sato, C. Phenotypic change in portal fibroblasts in biliary fibrosis. *Liver* **1994**, *14*, 76–82. [CrossRef]

73. Herbst, H.; Frey, A.; Heinrichs, O.; Milani, S.; Bechstein, W.O.; Neuhaus, P.; Schuppan, D. Heterogeneity of liver cells expressing procollagen types I and IV in vivo. *Histochem. Cell Biol.* **1997**, *107*, 399–409. [CrossRef]

74. Cassiman, D.; Libbrecht, L.; Desmet, V.; Denef, C.; Roskams, T. Hepatic stellate cell/myofibroblast subpopulations in fibrotic human and rat livers. *J. Hepatol.* **2002**, *36*, 200–209. [CrossRef]

75. Desmoulière, A.; Darby, I.; Costa, A.M.; Raccurt, M.; Tuchweber, B.; Sommer, P.; Gabbiani, G. Extracellular matrix deposition, lysyl oxidase expression, and myofibroblastic differentiation during the initial stages of cholestatic fibrosis in the rat. *Lab. Investig.* **1997**, *76*, 765–778.

76. Yata, Y.; Scanga, A.; Gillan, A.; Yang, L.; Reif, S.; Breindl, M.; Brenner, D.A.; Rippe, R.A. DNase I-hypersensitive sites enhance alpha1(I) collagen gene expression in hepatic stellate cells. *Hepatology* **2003**, *37*, 267–276. [CrossRef] [PubMed]

77. Dranoff, J.A.; Wells, R.G. Portal fibroblasts: Underappreciated mediators of biliary fibrosis. *Hepatology* **2010**, *51*, 1438–1444. [CrossRef]

78. Fausther, M.; Goree, J.R.; Lavoie, É.; Graham, A.L.; Sévigny, J.; Dranoff, J.A. Establishment and characterization of rat portal myofibroblast cell lines. *PLoS ONE* **2015**, *10*, e0121161. [CrossRef] [PubMed]

79. Iwaisako, K.; Jiang, C.; Zhang, M.; Cong, M.; Moore-Morris, T.J.; Park, T.J.; Liu, X.; Xu, J.; Wang, P.; Paik, Y.H.; et al. Origin of myofibroblasts in the fibrotic liver in mice. *Proc. Natl. Acad. Sci. USA* **2014**, *111*, E3297–E3305. [CrossRef] [PubMed]

80. Koyama, Y.; Wang, P.; Liang, S.; Iwaisako, K.; Liu, X.; Xu, J.; Zhang, M.; Sun, M.; Cong, M.; Karin, D.; et al. Mesothelin/mucin 16 signaling in activated portal fibroblasts regulates cholestatic liver fibrosis. *J. Clin. Investig.* **2017**, *127*, 1254–1270. [CrossRef]

81. Desmoulière, A.; Guyot, C.; Gabbiani, G. The stroma reaction myofibroblast: A key player in the control of tumor cell behavior. *Int. J. Dev. Biol.* **2004**, *48*, 509–517. [CrossRef]

82. Orimo, A.; Weinberg, R.A. Heterogeneity of stromal fibroblasts in tumors. *Cancer Biol. Ther.* **2007**, *6*, 618–619. [CrossRef] [PubMed]

83. Öhlund, D.; Handly-Santana, A.; Biffi, G.; Elyada, E.; Almeida, A.S.; Ponz-Sarvise, M.; Corbo, V.; Oni, T.E.; Hearn, S.A.; Lee, E.J.; et al. Distinct populations of inflammatory fibroblasts and myofibroblasts in pancreatic cancer. *J. Exp. Med.* **2017**, *214*, 579–596. [CrossRef]

84. Huber, M.A.; Kraut, N.; Park, J.E.; Schubert, R.D.; Rettig, W.J.; Peter, R.U.; Garin-Chesa, P. Fibroblast activation protein: Differential expression and serine protease activity in reactive stromal fibroblasts of melanocytic skin tumors. *J. Investig. Dermatol.* **2003**, *120*, 182–188. [CrossRef]

85. Roberts, E.W.; Deonarine, A.; Jones, J.O.; Denton, A.E.; Feig, C.; Lyons, S.K.; Espeli, M.; Kraman, M.; McKenna, B.; Wells, R.J.; et al. Depletion of stromal cells expressing fibroblast activation protein-α from skeletal muscle and bone marrow results in cachexia and anemia. *J. Exp. Med.* **2013**, *210*, 1137–1151. [CrossRef]

86. Li, H.; Courtois, E.T.; Sengupta, D.; Tan, Y.; Chen, K.H.; Goh, J.J.L.; Kong, S.L.; Chua, C.; Hon, L.K.; Tan, W.S.; et al. Reference component analysis of single-cell transcriptomes elucidates cellular heterogeneity in human colorectal tumors. *Nat. Genet.* **2017**, *49*, 708–718. [CrossRef]

87. Strutz, F.; Okada, H.; Lo, C.W.; Danoff, T.; Carone, R.L.; Tomaszewski, J.E.; Neilson, E.G. Identification and characterization of a fibroblast marker: FSP1. *J. Cell Biol.* **1995**, *130*, 393–405. [CrossRef] [PubMed]

88. Togo, S.; Polanska, U.M.; Horimoto, Y.; Orimo, A. Carcinoma-associated fibroblasts are a promising therapeutic target. *Cancers* **2013**, *5*, 149–169. [CrossRef]

89. Okada, H.; Danoff, T.M.; Kalluri, R.; Neilson, E.G. Early role of Fsp1 in epithelial-mesenchymal transformation. *Am. J. Physiol.* **1997**, *273*, F563–F574. [CrossRef]

90. Österreicher, C.H.; Penz-Österreicher, M.; Grivennikov, S.I.; Guma, M.; Koltsova, E.K.; Datz, C.; Sasik, R.; Hardiman, G.; Karin, M.; Brenner, D.A. Fibroblast-specific protein 1 identifies an inflammatory subpopulation of macrophages in the liver. *Proc. Natl. Acad. Sci. USA* **2011**, *108*, 308–313. [CrossRef]

91. Yoshida, T.; Akatsuka, T.; Imanaka-Yoshida, K. Tenascin-C and integrins in cancer. *Cell Adh. Migr.* **2015**, *9*, 96–104. [CrossRef]

92. Kikuchi, Y.; Kashima, T.G.; Nishiyama, T.; Shimazu, K.; Morishita, Y.; Shimazaki, M.; Kii, I.; Horie, H.; Nagai, H.; Kudo, A.; et al. Periostin is expressed in pericryptal fibroblasts and cancer-associated fibroblasts in the colon. *J. Histochem. Cytochem.* **2008**, *56*, 753–764. [CrossRef]

93. Sugimoto, H.; Mundel, T.M.; Kieran, M.W.; Kalluri, R. Identification of fibroblast heterogeneity in the tumor microenvironment. *Cancer Biol. Ther.* **2006**, *5*, 1640–1646. [CrossRef]

94. Atsumi, N.; Ishii, G.; Kojima, M.; Sanada, M.; Fujii, S.; Ochiai, A. Podoplanin, a novel marker of tumor-initiating cells in human squamous cell carcinoma A431. *Biochem Biophys Res Commun* **2008**, *373*, 36–41. [CrossRef] [PubMed]

95. Principe, S.; Mejia-Guerrero, S.; Ignatchenko, V.; Sinha, A.; Ignatchenko, A.; Shi, W.; Pereira, K.; Su, S.; Huang, S.H.; O'Sullivan, B.; et al. Proteomic Analysis of Cancer-Associated Fibroblasts Reveals a Paracrine Role for MFAP5 in Human Oral Tongue Squamous Cell Carcinoma. *J Proteome Res* **2018**, *17*, 2045–2059. [CrossRef] [PubMed]

96. Chau, K.Y.; Lily, M.A.; Wu, P.C.; Yau, W.L. Myofibroblasts in hepatitis B related cirrhosis and hepatocellular carcinoma. *J. Clin. Pathol.* **1992**, *45*, 446–448. [CrossRef]

97. Enzan, H.; Himeno, H.; Iwamura, S.; Onishi, S.; Saibara, T.; Yamamoto, Y.; Hara, H. Alpha-smooth muscle actin-positive perisinusoidal stromal cells in human hepatocellular carcinoma. *Hepatology* **1994**, *19*, 895–903.

98. Faouzi, S.; Le Bail, B.; Neaud, V.; Boussarie, L.; Saric, J.; Bioulac-Sage, P.; Balabaud, C.; Rosenbaum, J. Myofibroblasts are responsible for collagen synthesis in the stroma of human hepatocellular carcinoma: An in vivo and in vitro study. *J. Hepatol.* **1999**, *30*, 275–284. [CrossRef]

99. Coulouarn, C.; Corlu, A.; Glaise, D.; Guénon, I.; Thorgeirsson, S.S.; Clément, B. Hepatocyte-stellate cell cross-talk in the liver engenders a permissive inflammatory microenvironment that drives progression in hepatocellular carcinoma. *Cancer Res.* **2012**, *72*, 2533–2542. [CrossRef]

100. Lau, E.Y.; Lo, J.; Cheng, B.Y.; Ma, M.K.; Lee, J.M.; Ng, J.K.; Chai, S.; Lin, C.H.; Tsang, S.Y.; Ma, S.; et al. Cancer-Associated Fibroblasts Regulate Tumor-Initiating Cell Plasticity in Hepatocellular Carcinoma through c-Met/FRA1/HEY1 Signaling. *Cell Rep.* **2016**, *15*, 1175–1189. [CrossRef]

101. Amann, T.; Bataille, F.; Spruss, T.; Mühlbauer, M.; Gäbele, E.; Schölmerich, J.; Kiefer, P.; Bosserhoff, A.K.; Hellerbrand, C. Activated hepatic stellate cells promote tumorigenicity of hepatocellular carcinoma. *Cancer Sci.* **2009**, *100*, 646–653. [CrossRef] [PubMed]

102. Wang, Z.M.; Zhou, L.Y.; Liu, B.B.; Jia, Q.A.; Dong, Y.Y.; Xia, Y.H.; Ye, S.L. Rat hepatic stellate cells alter the gene expression profile and promote the growth, migration and invasion of hepatocellular carcinoma cells. *Mol. Med. Rep.* **2014**, *10*, 1725–1733. [CrossRef] [PubMed]

103. Song, T.; Dou, C.; Jia, Y.; Tu, K.; Zheng, X. TIMP-1 activated carcinoma-associated fibroblasts inhibit tumor apoptosis by activating SDF1/CXCR4 signaling in hepatocellular carcinoma. *Oncotarget* **2015**, *6*, 12061–12079. [CrossRef]

104. Sukowati, C.H.; Anfuso, B.; Crocé, L.S.; Tiribelli, C. The role of multipotent cancer associated fibroblasts in hepatocarcinogenesis. *BMC Cancer* **2015**, *15*, 188. [CrossRef]

105. Luo, Q.; Wang, C.Q.; Yang, L.Y.; Gao, X.M.; Sun, H.T.; Zhang, Y.; Zhang, K.L.; Zhu, Y.; Zheng, Y.; Sheng, Y.Y.; et al. FOXQ1/NDRG1 axis exacerbates hepatocellular carcinoma initiation via enhancing crosstalk between fibroblasts and tumor cells. *Cancer Lett.* **2018**, *417*, 21–34. [CrossRef]

106. Mikula, M.; Proell, V.; Fischer, A.N.; Mikulits, W. Activated hepatic stellate cells induce tumor progression of neoplastic hepatocytes in a TGF-beta dependent fashion. *J. Cell Physiol.* **2006**, *209*, 560–567. [CrossRef]

107. Geng, Z.M.; Li, Q.H.; Li, W.Z.; Zheng, J.B.; Shah, V. Activated human hepatic stellate cells promote growth of human hepatocellular carcinoma in a subcutaneous xenograft nude mouse model. *Cell Biochem. Biophys.* **2014**, *70*, 337–347. [CrossRef]

108. Lin, N.; Chen, Z.; Lu, Y.; Li, Y.; Hu, K.; Xu, R. Role of activated hepatic stellate cells in proliferation and metastasis of hepatocellular carcinoma. *Hepatol. Res.* **2015**, *45*, 326–336. [CrossRef]

109. Mogler, C.; König, C.; Wieland, M.; Runge, A.; Besemfelder, E.; Komljenovic, D.; Longerich, T.; Schirmacher, P.; Augustin, H.G. Hepatic stellate cells limit hepatocellular carcinoma progression through the orphan receptor endosialin. *EMBO Mol. Med.* **2017**, *9*, 741–749. [CrossRef]

110. Özdemir, B.C.; Pentcheva-Hoang, T.; Carstens, J.L.; Zheng, X.; Wu, C.C.; Simpson, T.R.; Laklai, H.; Sugimoto, H.; Kahlert, C.; Novitskiy, S.V.; et al. Depletion of carcinoma-associated fibroblasts and fibrosis induces immunosuppression and accelerates pancreas cancer with reduced survival. *Cancer Cell* **2014**, *25*, 719–734. [CrossRef]

111. Rhim, A.D.; Oberstein, P.E.; Thomas, D.H.; Mirek, E.T.; Palermo, C.F.; Sastra, S.A.; Dekleva, E.N.; Saunders, T.; Becerra, C.P.; Tattersall, I.W.; et al. Stromal elements act to restrain, rather than support, pancreatic ductal adenocarcinoma. *Cancer Cell* **2014**, *25*, 735–747. [CrossRef]

112. Carloni, V.; Luong, T.V.; Rombouts, K. Hepatic stellate cells and extracellular matrix in hepatocellular carcinoma: More complicated than ever. *Liver Int.* **2014**, *34*, 834–843. [CrossRef]

113. Santamato, A.; Fransvea, E.; Dituri, F.; Caligiuri, A.; Quaranta, M.; Niimi, T.; Pinzani, M.; Antonaci, S.; Giannelli, G. Hepatic stellate cells stimulate HCC cell migration via laminin-5 production. *Clin. Sci. (Lond.)* **2011**, *121*, 159–168. [CrossRef] [PubMed]

114. Schrader, J.; Gordon-Walker, T.T.; Aucott, R.L.; van Deemter, M.; Quaas, A.; Walsh, S.; Benten, D.; Forbes, S.J.; Wells, R.G.; Iredale, J.P. Matrix stiffness modulates proliferation, chemotherapeutic response, and dormancy in hepatocellular carcinoma cells. *Hepatology* **2011**, *53*, 1192–1205. [CrossRef] [PubMed]

115. Olsen, A.L.; Bloomer, S.A.; Chan, E.P.; Gaça, M.D.; Georges, P.C.; Sackey, B.; Uemura, M.; Janmey, P.A.; Wells, R.G. Hepatic stellate cells require a stiff environment for myofibroblastic differentiation. *Am. J. Physiol. Gastrointest. Liver Physiol.* **2011**, *301*, G110–G118. [CrossRef] [PubMed]

116. Calvo, F.; Ege, N.; Grande-Garcia, A.; Hooper, S.; Jenkins, R.P.; Chaudhry, S.I.; Harrington, K.; Williamson, P.; Moeendarbary, E.; Charras, G.; et al. Mechanotransduction and YAP-dependent matrix remodelling is required for the generation and maintenance of cancer-associated fibroblasts. *Nat. Cell. Biol.* **2013**, *15*, 637–646. [CrossRef]

117. Wang, H.M.; Hung, C.H.; Lu, S.N.; Chen, C.H.; Lee, C.M.; Hu, T.H.; Wang, J.H. Liver stiffness measurement as an alternative to fibrotic stage in risk assessment of hepatocellular carcinoma incidence for chronic hepatitis C patients. *Liver Int.* **2013**, *33*, 756–761. [CrossRef]

118. Masuzaki, R.; Tateishi, R.; Yoshida, H.; Goto, E.; Sato, T.; Ohki, T.; Imamura, J.; Goto, T.; Kanai, F.; Kato, N.; et al. Prospective risk assessment for hepatocellular carcinoma development in patients with chronic hepatitis C by transient elastography. *Hepatology* **2009**, *49*, 1954–1961. [CrossRef]

119. Provenzano, P.P.; Keely, P.J. Mechanical signaling through the cytoskeleton regulates cell proliferation by coordinated focal adhesion and Rho GTPase signaling. *J. Cell Sci.* **2011**, *124*, 1195–1205. [CrossRef]

120. Lai, K.K.; Shang, S.; Lohia, N.; Booth, G.C.; Masse, D.J.; Fausto, N.; Campbell, J.S.; Beretta, L. Extracellular matrix dynamics in hepatocarcinogenesis: A comparative proteomics study of PDGFC transgenic and Pten null mouse models. *PLoS Genet.* **2011**, *7*, e1002147. [CrossRef]

121. Begum, N.A.; Mori, M.; Matsumata, T.; Takenaka, K.; Sugimachi, K.; Barnard, G.F. Differential display and integrin alpha 6 messenger RNA overexpression in hepatocellular carcinoma. *Hepatology* **1995**, *22*, 1447–1455. [CrossRef] [PubMed]

122. Zhao, G.; Cui, J.; Qin, Q.; Zhang, J.; Liu, L.; Deng, S.; Wu, C.; Yang, M.; Li, S.; Wang, C. Mechanical stiffness of liver tissues in relation to integrin β1 expression may influence the development of hepatic cirrhosis and hepatocellular carcinoma. *J. Surg. Oncol.* **2010**, *102*, 482–489. [CrossRef]

123. Cox, D.; Brennan, M.; Moran, N. Integrins as therapeutic targets: Lessons and opportunities. *Nat. Rev. Drug Discov.* **2010**, *9*, 804–820. [CrossRef]

124. Xie, B.; Lin, W.; Ye, J.; Wang, X.; Zhang, B.; Xiong, S.; Li, H.; Tan, G. DDR2 facilitates hepatocellular carcinoma invasion and metastasis via activating ERK signaling and stabilizing SNAIL1. *J. Exp. Clin. Cancer Res.* **2015**, *34*, 101. [CrossRef]

125. Duarte, S.; Baber, J.; Fujii, T.; Coito, A.J. Matrix metalloproteinases in liver injury, repair and fibrosis. *Matrix Biol.* **2015**, *44–46*, 147–156. [CrossRef] [PubMed]

126. Hotary, K.B.; Allen, E.D.; Brooks, P.C.; Datta, N.S.; Long, M.W.; Weiss, S.J. Membrane type I matrix metalloproteinase usurps tumor growth control imposed by the three-dimensional extracellular matrix. *Cell* **2003**, *114*, 33–45. [CrossRef]

127. Hernandez-Gea, V.; Toffanin, S.; Friedman, S.L.; Llovet, J.M. Role of the microenvironment in the pathogenesis and treatment of hepatocellular carcinoma. *Gastroenterology* **2013**, *144*, 512–527. [CrossRef]

128. Okazaki, I. Novel Cancer-targeting Agents/Application Strategies Developed from MMP Science. *Anticancer Agents Med. Chem.* **2012**, *12*, 687. [CrossRef]

129. Jia, Y.L.; Shi, L.; Zhou, J.N.; Fu, C.J.; Chen, L.; Yuan, H.F.; Wang, Y.F.; Yan, X.L.; Xu, Y.C.; Zeng, Q.; et al. Epimorphin promotes human hepatocellular carcinoma invasion and metastasis through activation of focal adhesion kinase/extracellular signal-regulated kinase/matrix metalloproteinase-9 axis. *Hepatology* **2011**, *54*, 1808–1818. [CrossRef]

130. Rosmorduc, O.; Housset, C. Hypoxia: A link between fibrogenesis, angiogenesis, and carcinogenesis in liver disease. *Semin. Liver Dis.* **2010**, *30*, 258–270. [CrossRef] [PubMed]

131. Ankoma-Sey, V.; Wang, Y.; Dai, Z. Hypoxic stimulation of vascular endothelial growth factor expression in activated rat hepatic stellate cells. *Hepatology* **2000**, *31*, 141–148. [CrossRef]

132. Taura, K.; De Minicis, S.; Seki, E.; Hatano, E.; Iwaisako, K.; Osterreicher, C.H.; Kodama, Y.; Miura, K.; Ikai, I.; Uemoto, S.; et al. Hepatic stellate cells secrete angiopoietin 1 that induces angiogenesis in liver fibrosis. *Gastroenterology* **2008**, *135*, 1729–1738. [CrossRef] [PubMed]

133. Torimura, T.; Ueno, T.; Kin, M.; Harada, R.; Taniguchi, E.; Nakamura, T.; Sakata, R.; Hashimoto, O.; Sakamoto, M.; Kumashiro, R.; et al. Overexpression of angiopoietin-1 and angiopoietin-2 in hepatocellular carcinoma. *J Hepatol* **2004**, *40*, 799–807. [CrossRef] [PubMed]

134. Sanz-Cameno, P.; Martín-Vílchez, S.; Lara-Pezzi, E.; Borque, M.J.; Salmerón, J.; Muñoz de Rueda, P.; Solís, J.A.; López-Cabrera, M.; Moreno-Otero, R. Hepatitis B virus promotes angiopoietin-2 expression in liver tissue: Role of HBV x protein. *Am. J. Pathol.* **2006**, *169*, 1215–1222. [CrossRef] [PubMed]

135. Kang, N.; Yaqoob, U.; Geng, Z.; Bloch, K.; Liu, C.; Gomez, T.; Billadeau, D.; Shah, V. Focal adhesion assembly in myofibroblasts fosters a microenvironment that promotes tumor growth. *Am. J. Pathol.* **2010**, *177*, 1888–1900. [CrossRef] [PubMed]

136. Wirz, W.; Antoine, M.; Tag, C.G.; Gressner, A.M.; Korff, T.; Hellerbrand, C.; Kiefer, P. Hepatic stellate cells display a functional vascular smooth muscle cell phenotype in a three-dimensional co-culture model with endothelial cells. *Differentiation* **2008**, *76*, 784–794. [CrossRef] [PubMed]

137. Marra, F. Chemokines in liver inflammation and fibrosis. *Front. Biosci.* **2002**, *7*, d1899–d1914. [CrossRef]

138. Meindl-Beinker, N.M.; Matsuzaki, K.; Dooley, S. TGF-β signaling in onset and progression of hepatocellular carcinoma. *Dig. Dis.* **2012**, *30*, 514–523. [CrossRef] [PubMed]

139. Moon, H.; Han, K.H.; Ro, S.W. Pro-tumorigenic roles of TGF-β signaling during the early stages of liver tumorigenesis through upregulation of Snail. *BMB Rep.* **2017**, *50*, 599–600. [CrossRef]

140. Moon, H.; Ju, H.L.; Chung, S.I.; Cho, K.J.; Eun, J.W.; Nam, S.W.; Han, K.H.; Calvisi, D.F.; Ro, S.W. Transforming Growth Factor-β Promotes Liver Tumorigenesis in Mice via Up-regulation of Snail. *Gastroenterology* **2017**, *153*, 1378–1391. [CrossRef] [PubMed]

141. Nakanishi, K.; Fujimoto, J.; Ueki, T.; Kishimoto, K.; Hashimoto-Tamaoki, T.; Furuyama, J.; Itoh, T.; Sasaki, Y.; Okamoto, E. Hepatocyte growth factor promotes migration of human hepatocellular carcinoma via phosphatidylinositol 3-kinase. *Clin Exp Metastasis* **1999**, *17*, 507–514. [CrossRef]

142. Yu, G.; Jing, Y.; Kou, X.; Ye, F.; Gao, L.; Fan, Q.; Yang, Y.; Zhao, Q.; Li, R.; Wu, M.; et al. Hepatic stellate cells secreted hepatocyte growth factor contributes to the chemoresistance of hepatocellular carcinoma. *PLoS ONE* **2013**, *8*, e73312. [CrossRef] [PubMed]

143. Ljubimova, J.Y.; Petrovic, L.M.; Wilson, S.E.; Geller, S.A.; Demetriou, A.A. Expression of HGF, its receptor c-met, c-myc, and albumin in cirrhotic and neoplastic human liver tissue. *J. Histochem. Cytochem.* **1997**, *45*, 79–87. [CrossRef]

144. Teng, M.W.; Swann, J.B.; Koebel, C.M.; Schreiber, R.D.; Smyth, M.J. Immune-mediated dormancy: An equilibrium with cancer. *J. Leukoc. Biol.* **2008**, *84*, 988–993. [CrossRef]

145. Kim, R.; Emi, M.; Tanabe, K. Cancer immunoediting from immune surveillance to immune escape. *Immunology* **2007**, *121*, 1–14. [CrossRef] [PubMed]

146. Yang, L.; Pang, Y.; Moses, H.L. TGF-beta and immune cells: An important regulatory axis in the tumor microenvironment and progression. *Trends Immunol.* **2010**, *31*, 220–227. [CrossRef] [PubMed]

147. Shields, J.D.; Kourtis, I.C.; Tomei, A.A.; Roberts, J.M.; Swartz, M.A. Induction of lymphoidlike stroma and immune escape by tumors that express the chemokine CCL21. *Science* **2010**, *328*, 749–752. [CrossRef]

148. Mougiakakos, D.; Choudhury, A.; Lladser, A.; Kiessling, R.; Johansson, C.C. Regulatory T cells in cancer. *Adv. Cancer Res.* **2010**, *107*, 57–117.

149. Ostrand-Rosenberg, S. Myeloid-derived suppressor cells: More mechanisms for inhibiting antitumor immunity. *Cancer Immunol. Immunother.* **2010**, *59*, 1593–1600. [CrossRef]

150. Ormandy, L.A.; Hillemann, T.; Wedemeyer, H.; Manns, M.P.; Greten, T.F.; Korangy, F. Increased populations of regulatory T cells in peripheral blood of patients with hepatocellular carcinoma. *Cancer Res.* **2005**, *65*, 2457–2464. [CrossRef]

151. Almand, B.; Clark, J.I.; Nikitina, E.; van Beynen, J.; English, N.R.; Knight, S.C.; Carbone, D.P.; Gabrilovich, D.I. Increased production of immature myeloid cells in cancer patients: A mechanism of immunosuppression in cancer. *J. Immunol.* **2001**, *166*, 678–689. [CrossRef]

152. Hoechst, B.; Ormandy, L.A.; Ballmaier, M.; Lehner, F.; Krüger, C.; Manns, M.P.; Greten, T.F.; Korangy, F. A new population of myeloid-derived suppressor cells in hepatocellular carcinoma patients induces CD4(+)CD25(+)Foxp3(+) T cells. *Gastroenterology* **2008**, *135*, 234–243. [CrossRef]

153. Facciabene, A.; Motz, G.T.; Coukos, G. T-regulatory cells: Key players in tumor immune escape and angiogenesis. *Cancer Res.* **2012**, *72*, 2162–2171. [CrossRef] [PubMed]

154. Dilek, N.; Vuillefroy de Silly, R.; Blancho, G.; Vanhove, B. Myeloid-derived suppressor cells: Mechanisms of action and recent advances in their role in transplant tolerance. *Front. Immunol.* **2012**, *3*, 208. [CrossRef]

155. Chen, C.H.; Kuo, L.M.; Chang, Y.; Wu, W.; Goldbach, C.; Ross, M.A.; Stolz, D.B.; Chen, L.; Fung, J.J.; Lu, L.; et al. In vivo immune modulatory activity of hepatic stellate cells in mice. *Hepatology* **2006**, *44*, 1171–1181. [CrossRef] [PubMed]

156. Yu, M.C.; Chen, C.H.; Liang, X.; Wang, L.; Gandhi, C.R.; Fung, J.J.; Lu, L.; Qian, S. Inhibition of T-cell responses by hepatic stellate cells via B7-H1-mediated T-cell apoptosis in mice. *Hepatology* **2004**, *40*, 1312–1321. [CrossRef]

157. Zhao, W.; Zhang, L.; Yin, Z.; Su, W.; Ren, G.; Zhou, C.; You, J.; Fan, J.; Wang, X. Activated hepatic stellate cells promote hepatocellular carcinoma development in immunocompetent mice. *Int. J. Cancer* **2011**, *129*, 2651–2661. [CrossRef] [PubMed]

158. Zhao, W.; Su, W.; Kuang, P.; Zhang, L.; Liu, J.; Yin, Z.; Wang, X. The role of hepatic stellate cells in the regulation of T-cell function and the promotion of hepatocellular carcinoma. *Int. J. Oncol.* **2012**, *41*, 457–464. [CrossRef]

159. Zhao, W.; Zhang, L.; Xu, Y.; Zhang, Z.; Ren, G.; Tang, K.; Kuang, P.; Zhao, B.; Yin, Z.; Wang, X. Hepatic stellate cells promote tumor progression by enhancement of immunosuppressive cells in an orthotopic liver tumor mouse model. *Lab. Investig.* **2014**, *94*, 182–191. [CrossRef] [PubMed]

160. Xu, Y.; Zhao, W.; Xu, J.; Li, J.; Hong, Z.; Yin, Z.; Wang, X. Activated hepatic stellate cells promote liver cancer by induction of myeloid-derived suppressor cells through cyclooxygenase-2. *Oncotarget* **2016**, *7*, 8866–8878. [CrossRef]

161. Höchst, B.; Schildberg, F.A.; Sauerborn, P.; Gäbel, Y.A.; Gevensleben, H.; Goltz, D.; Heukamp, L.C.; Türler, A.; Ballmaier, M.; Gieseke, F.; et al. Activated human hepatic stellate cells induce myeloid derived suppressor cells from peripheral blood monocytes in a CD44-dependent fashion. *J. Hepatol.* **2013**, *59*, 528–535. [CrossRef] [PubMed]

162. Kalluri, R.; Neilson, E.G. Epithelial-mesenchymal transition and its implications for fibrosis. *J. Clin. Investig.* **2003**, *112*, 1776–1784. [CrossRef] [PubMed]

163. Kalluri, R.; Weinberg, R.A. The basics of epithelial-mesenchymal transition. *J. Clin. Investig.* **2009**, *119*, 1420–1428. [CrossRef]

164. Yoshida, K.; Murata, M.; Yamaguchi, T.; Matsuzaki, K.; Okazaki, K. Reversible Human TGF-β Signal Shifting between Tumor Suppression and Fibro-Carcinogenesis: Implications of Smad Phospho-Isoforms for Hepatic Epithelial-Mesenchymal Transitions. *J. Clin. Med.* **2016**, *5*, 7. [CrossRef]

165. Zavadil, J.; Böttinger, E.P. TGF-beta and epithelial-to-mesenchymal transitions. *Oncogene* **2005**, *24*, 5764–5774. [CrossRef]

166. Gotzmann, J.; Huber, H.; Thallinger, C.; Wolschek, M.; Jansen, B.; Schulte-Hermann, R.; Beug, H.; Mikulits, W. Hepatocytes convert to a fibroblastoid phenotype through the cooperation of TGF-beta1 and Ha-Ras: Steps towards invasiveness. *J. Cell Sci.* **2002**, *115*, 1189–1202.

167. Van Zijl, F.; Mair, M.; Csiszar, A.; Schneller, D.; Zulehner, G.; Huber, H.; Eferl, R.; Beug, H.; Dolznig, H.; Mikulits, W. Hepatic tumor-stroma crosstalk guides epithelial to mesenchymal transition at the tumor edge. *Oncogene* **2009**, *28*, 4022–4033. [CrossRef]

168. Hsia, C.C.; Evarts, R.P.; Nakatsukasa, H.; Marsden, E.R.; Thorgeirsson, S.S. Occurrence of oval-type cells in hepatitis B virus-associated human hepatocarcinogenesis. *Hepatology* **1992**, *16*, 1327–1333. [CrossRef] [PubMed]

169. Lee, J.S.; Heo, J.; Libbrecht, L.; Chu, I.S.; Kaposi-Novak, P.; Calvisi, D.F.; Mikaelyan, A.; Roberts, L.R.; Demetris, A.J.; Sun, Z.; et al. A novel prognostic subtype of human hepatocellular carcinoma derived from hepatic progenitor cells. *Nat. Med.* **2006**, *12*, 410–416. [CrossRef] [PubMed]

170. Tummala, K.S.; Brandt, M.; Teijeiro, A.; Graña, O.; Schwabe, R.F.; Perna, C.; Djouder, N. Hepatocellular Carcinomas Originate Predominantly from Hepatocytes and Benign Lesions from Hepatic Progenitor Cells. *Cell Rep.* **2017**, *19*, 584–600. [CrossRef]

171. Wu, K.; Ding, J.; Chen, C.; Sun, W.; Ning, B.F.; Wen, W.; Huang, L.; Han, T.; Yang, W.; Wang, C.; et al. Hepatic transforming growth factor beta gives rise to tumor-initiating cells and promotes liver cancer development. *Hepatology* **2012**, *56*, 2255–2267. [CrossRef] [PubMed]

172. Fernando, J.; Malfettone, A.; Cepeda, E.B.; Vilarrasa-Blasi, R.; Bertran, E.; Raimondi, G.; Fabra, À.; Alvarez-Barrientos, A.; Fernández-Salguero, P.; Fernández-Rodríguez, C.M.; et al. A mesenchymal-like phenotype and expression of CD44 predict lack of apoptotic response to sorafenib in liver tumor cells. *Int. J. Cancer* **2015**, *136*, E161–E172. [CrossRef] [PubMed]

173. Kalluri, R. The biology and function of fibroblasts in cancer. *Nat. Rev. Cancer* **2016**, *16*, 582–598. [CrossRef] [PubMed]

International Journal of
Molecular Sciences

MDPI

Review

Molecular Mechanisms Driving Progression of Liver Cirrhosis towards Hepatocellular Carcinoma in Chronic Hepatitis B and C Infections: A Review

Tatsuo Kanda [1], Taichiro Goto [2,*], Yosuke Hirotsu [3], Mitsuhiko Moriyama [1] and Masao Omata [3,4]

[1] Division of Gastroenterology and Hepatology, Department of Medicine, Nihon University School of Medicine, 30-1 Oyaguchi-kamicho, Itabashi-ku, Tokyo 173-8610, Japan; kanda2t@yahoo.co.jp (T.K.); moriyama.mitsuhiko@nihon-u.ac.jp (M.M.)
[2] Lung Cancer and Respiratory Disease Center, Yamanashi Central Hospital, 1-1-1 Fujimi, Kofu, Yamanashi 400-8506, Japan
[3] Genome Analysis Center, Yamanashi Central Hospital, Yamanashi 400-8506, Japan; hirotsu-bdyu@ych.pref.yamanashi.jp (Y.H.); m-omata0901@ych.pref.yamanashi.jp (M.O.)
[4] The University of Tokyo, 7-3-1 Hongo, Bunkyo-ku, Tokyo 113-8655, Japan
* Correspondence: taichiro@1997.jukuin.keio.ac.jp; Tel.: +81-55-253-7111

Received: 19 January 2019; Accepted: 14 March 2019; Published: 18 March 2019

Abstract: Almost all patients with hepatocellular carcinoma (HCC), a major type of primary liver cancer, also have liver cirrhosis, the severity of which hampers effective treatment for HCC despite recent progress in the efficacy of anticancer drugs for advanced stages of HCC. Here, we review recent knowledge concerning the molecular mechanisms of liver cirrhosis and its progression to HCC from genetic and epigenomic points of view. Because ~70% of patients with HCC have hepatitis B virus (HBV) and/or hepatitis C virus (HCV) infection, we focused on HBV- and HCV-associated HCC. The literature suggests that genetic and epigenetic factors, such as microRNAs, play a role in liver cirrhosis and its progression to HCC, and that HBV- and HCV-encoded proteins appear to be involved in hepatocarcinogenesis. Further studies are needed to elucidate the mechanisms, including immune checkpoints and molecular targets of kinase inhibitors, associated with liver cirrhosis and its progression to HCC.

Keywords: cirrhosis; HBV; HCV; hepatocellular carcinoma

1. Introduction

Almost all patients with hepatocellular carcinoma (HCC), a major type of primary liver cancer, have liver cirrhosis [1], the severity of which can prevent effective treatment for HCC despite recent progress in the efficacy of anticancer drugs for advanced stages of HCC [2–7]. Because most patients with liver cirrhosis are asymptomatic, it is difficult to diagnose early stages of HCC [8], and patients with hepatic symptoms and HCC are considered to have advanced-stage HCC [8,9]. These issues explain the prevalence of poor prognosis for HCC patients.

HCC is the 4th most common neoplasm and the 2nd commonest cause of cancer deaths in the world. Notably, HCC is a male-dominant disease, with the incidence of HCC ~3-fold higher in males than in females [10]. Hepatitis B virus (HBV) infection is associated with the higher HCC incidence in persons with cirrhosis, occurring in high endemic areas and in Western countries (5-year cumulative incidence, 15% and 10%, respectively) [11]. In hepatitis C virus (HCV)-related cirrhosis, the 5-year cumulative HCC risk is 30% in Japan and 17% in Western countries [11].

Histologically, liver fibrosis involves the deposition of extracellular matrix proteins, including collagen, in higher-order structures within hepatic parenchyma [12,13], with hepatic stellate cells and fibroblasts representing major producers of collagen [14]. The histological pattern of liver fibrosis is not unique, and the extent and distribution of liver fibrosis exhibits various patterns depending on different etiologies [12,15]. Excessive liver fibrosis often develops within portal tracts and extends into the hepatic parenchyma in viral hepatitis, with these activities appearing to be associated with persistent portal inflammation [12,16]. Bridging fibrosis appears following the development of periportal fibrosis and extends across lobules to connect mesenchymal structures (portal tracts and central veins) to different extents. Generally, these processes accompany intrahepatic portosystemic vascular shunting and regeneration of hepatocytes, thereby transforming from normal hepatic architecture to nodule formation and finally establishing the structure of cirrhosis [12].

In this review, we discuss the molecular mechanisms underlying the progression of liver cirrhosis to HCC. We expect that this review will help clinicians diagnose and treat patients with liver cirrhosis and/or HCC in their daily clinical practice. Notably, ~70% of HCC patients are afflicted with HBV or HCV infection [11]; therefore, we focused on the occurrence of HCC during HBV and HCV infection (Figure 1).

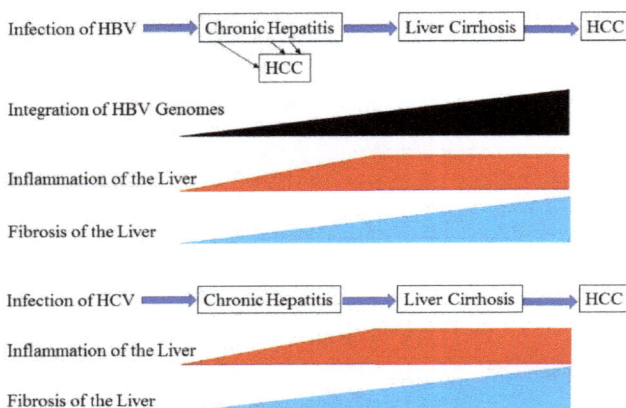

Figure 1. Occurrence of HCC in natural course of HBV and HCV infection.

2. Liver Cirrhosis and Its Progression to HCC with HBV Infection

2.1. Development of Liver Cirrhosis in Patients with Chronic Hepatitis B Infection

HBV infection causes acute and chronic hepatitis, cirrhosis, and HCC. HBV-carrier rates are higher in African and Asian countries and globally are estimated to have resulted in 786,000 deaths in 2010, the majority of which were attributed to HCC (341,400 deaths) and cirrhosis (312,400 deaths) [17]. Annual rates of development from chronic hepatitis B to liver cirrhosis ranged from 2.1 to 6.0% [18–21], and annual rates of the development of liver cirrhosis in HBV e antigen (HBeAg)-positive or anti-HBe-positive patients were 2.4% and 1.3%, respectively [18].

HBeAg-positivity and elevated HBV DNA levels are risk factors for the development of liver cirrhosis in patients with chronic hepatitis B [19]. Sumi et al. [20] reported that progression to cirrhosis is slower in HBV genotype B than that in HBV genotype C infection. Additionally, for patients with chronic hepatitis B infection, coinfection with HCV or human immunodeficiency virus (HIV) is another risk factor for the development of liver cirrhosis [21,22].

Older age (≥55 years), male gender, chronic active hepatitis, higher alanine aminotransferase (ALT) levels, history of decompensation, ferredoxin-1-associated single-nucleotide polymorphism, HLA-DQA2 rs9276370 variants, and HLA-DQB2 rs7756516 variants are also respective risk factors for the development of liver cirrhosis [18–21,23–25].

2.2. Development of HCC in Patients with HBV-related Liver Cirrhosis

Annual rates of occurrence of HCC in patients with HBV-related cirrhosis are ~2.3% [1]. Higher HBV DNA levels, HBeAg-positivity, higher HBV surface antigen (HBsAg) levels, HBV genotype C, and basal core-promoter mutations are viral risk factors for the occurrence of HCC, and older age, male gender, chronic active hepatitis, higher ALT levels, and higher α-fetoprotein levels are non-viral risk factors for the occurrence of HCC [20,23,26]. Treatment with nucleos(t)ide analogs for HBV control decreases the occurrence rates of HCC in patients with HBV-related cirrhosis [27].

3. Liver Cirrhosis and Its Progression to HCC with HCV Infection

HCV infection causes acute and chronic hepatitis, cirrhosis, and HCC. In 2015, there were 170,000 new HCV infections [28], with annual incidence rates of cirrhosis accompanying HCV infection at 1.1% [29]. Forns et al. [30] reported that chronic HCV infection displayed a high rate of progression to liver cirrhosis over a prolonged follow-up period, and that an aspartate aminotransferase (AST) value >70 IU/L was associated with cirrhosis development [odds ratio (OR): 4.22, 95% confidence interval (CI): 1.3–13.8]. Additionally, there is an association between HCV genotype 3 and steatosis, which accelerates fibrosis development over time in HCV genotype 3-infected patients [31,32]. Moreover, previous studies reported that exposure to HCV at a young age is associated with a reduced rate of fibrosis progression [33–37], and that liver fibrosis progression was mainly dependent on the age and duration of infection [36]. The evidences are growing that liver steatosis and diabetes mellitus are factors affecting progression to HCC in HCV-infected patients with compensated cirrhosis [11]. These comorbidities typically affecting aged patients might explain the increased malignant progression of the cirrhotic liver damage. Ongoing alcohol consumption and severe inflammation according to liver histology are also associated with liver fibrosis progression [35–37]. A multivariate analysis of HCV-infected patients [38] revealed that only male sex (OR: 3.17, 95% CI: 1.152–8.773) and HIV infection (OR: 6.85, 95% CI: 2.93–16.005) were associated with advanced liver fibrosis. Furthermore, HBsAg-positive HCV-coinfected patients are at high risk of developing liver disease [39], and injected-drug users [40]; patients with insulin resistance or diabetes [41]; and patients with concurrent obesity, diabetes, and steatosis [42] are also at risk of advanced liver fibrosis. Following antiviral treatment, sustained virological response (SVR) can reduce liver fibrosis progression in most patients infected with HCV [43].

Annual rates of HCC occurrence in patients with HCV-related cirrhosis are ~4.5% [1]. Caporaso et al. [44] found that mean age and male/female ratio were significantly higher in patients with HCC plus liver cirrhosis than in those with liver cirrhosis alone. Additionally, HBsAg-positive HCV-coinfected patients with liver cirrhosis are at a high risk of developing HCC [45], with elevated ALT levels [46] and hepatic steatosis [47] risk factors for the development of HCC in patients with HCV-related liver cirrhosis. SVR can reduce HCC development in most patients infected with HCV-related liver cirrhosis [48], although changes in the risk of HCC development following HCV eradication with direct-acting antivirals (DAAs) is controversial [49–51]. Further studies will be needed in DAA-era.

4. Molecular Mechanisms of Liver Cirrhosis and Its Progression to HCC

Accumulation of genetic mutations occurs during HCC progression. Recent advances in next-generation sequencing technology to augment Sanger sequencing enabled whole-genome sequencing to allow critical insights into the molecular mechanisms associated with this activity.

4.1. Driver-Gene Candidates in HCC

Totoki et al. [52] found 30 candidate driver genes [telomerase reverse transcriptase (TERT), catenin β1 (CTNNB1), tumor protein p53 (TP53), AT-rich interaction domain 2 (ARID2), axin 1 (AXIN1), TSC complex subunit 2 (TSC2), retinoblastoma protein 1 (RB1), activin A receptor type 2A (ACVR2A), bromodomain containing 7 (BRD7), cyclin dependent kinase inhibitor (CDKN)1A, menin 1 (MEN1), polypeptide N-acetylgalactosaminyltransferase 11 (GALN11), fibroblast growth factor 19 (FGF19), cyclin (CCN)D1, AT-rich interaction domain 1A (ARID1A), CDKN2A, CDKN2B, ribosomal protein S6 kinase, 90 kDa, polypeptide 3 (RPS6KA3), nuclear factor, erythroid 2 like 2 (NFE2L2), nuclear receptor corepressor 1 (NCOR1), alcohol dehydrogenase 1B, β polypeptide (ADH1B), Snf2-related CREB binding protein (CREBBP) activator protein (SRCAP), Fc receptor like 1 (FCRL1), phosphatase and tensin homolog (PTEN), heterogeneous nuclear ribonucleoprotein A2/B1 (HNRNPA2B1), cytochrome P450 family 2 subfamily E member 1 (CYP2E1), mitogen-activated protein kinase 3 (MAP2K3), tuberous sclerosis (TSC)1, transmembrane protein 99 (TMEM99), and glucose-6-phosphatase, catalytic subunit (G6PC)] and 11 core pathway modules [β-catenin, chromatin remodeling, DNA repair, estrogen-related receptor beta (ERRB), fibrinogen, mechanistic target of rapamycin kinase (mTOR), synaptic connection, TERT, p53, NOTCH, and NCOR] through the collection of data from 503 HCC genomes from different populations (Table 1).

Fujimoto et al. [53] identified 15 significantly mutated genes, including TP53, ERBB-receptor feedback inhibitor 1, Zic family member 3, CTNNB1, glucoside xylosyltransferase 1, otopetrin 1, albumin (ALB), ATM serine/threonine kinase (ATM), zinc finger protein 226, ubiquitin specific peptidase (USP)25, WW-domain-containing E3 ubiquitin protein ligase 1, immunoglobulin superfamily member 10, ARID1A, ubiquitin protein ligase E3 component n-recognin 3, and bromodomain adjacent to zinc finger domain 2B, after sequencing and analyzing the whole genomes of 27 HCCs, including 25 with HBV- or HCV-associated HCC. Additionally, whole-genome sequencing analyses of HCCs demonstrated the influence of etiological backgrounds on mutation patterns and recurrent mutations in chromatin regulators [53]. The authors found that multiple chromatin regulators, including ARID1A, ARID1B, ARID2, lysine methyltransferase 2A, and lysine methyltransferase 2C, were mutated in ~50% of the tumors, with clonal integration of the HBV genome in the TERT gene frequently detected.

Table 1. Representative driver-gene candidates in HCC.

Gene Symbol	Gene Name	Pathways	References
TERT	Telomerase reverse transcriptase	TERT	[52–56]
CTNNB1	Catenin β1	β-catenin	[52,53]
TP53	Tumor protein p53	p53–RB	[52,53]
ARID2	AT-rich interaction domain 2	Chromatin remodeling	[52]
AXIN1	Axin 1	β-catenin	[52]
TSC2	TSC complex subunit 2	PI3K–mTOR	[52]
RB1	Retinoblastoma 1	p53–RB	[52]
ACVR2A	Activin A receptor type 2A	SMAD	[52]
BRD7	Bromodomain containing 7	Chromatin remodeling	[52]
CDKN1A	Cyclin-dependent kinase inhibitor 1A	β-catenin	[52]
MEN1	Menin 1	(MEN1 syndrome)	[52]
GALN11	Polypeptide N-acetylgalactosaminyltransferase 11	NOTCH	[52]
FGF19	Fibroblast growth factor 19	β-catenin	[52]
CCND1	Cyclin D1	p53–RB	[52]
ARID1A	AT-rich interaction domain 1A	Chromatin remodeling	[52–54,57,58]
CDKN2A	Cyclin-dependent kinase inhibitor 2A	p53–RB	[52]
CDKN2B	Cyclin-dependent kinase inhibitor 2B	p53–RB	[52]
RPS6KA3	Ribosomal protein S6 kinase, 90kDa, polypeptide 3	p53–RB	[52]
NFE2L2	Nuclear factor, erythroid 2-like 2	NRF2–KEAP1	[52]
NCOR1	Nuclear receptor corepressor 1	β-catenin/chromatin remodeling	[52]
ADH1B	Alcohol dehydrogenase 1B, β polypeptide		[52]
SRCAP	Snf2-related CREB binding protein activator protein	Chromatin remodeling	[52]
FCRL1	Fc receptor like 1		[52]
PTEN	Phosphatase and tensin homolog	PI3K–mTOR	[52]
HNRNPA2B1	Heterogeneous nuclear ribonucleoprotein A2/B1	MAPK	[52]
CYP2E1	Cytochrome P450 family 2 subfamily E member 1		[52]
MAP2K3	Mitogen-activated protein kinase 3	MAPK	[52]
TSC1	Tuberous sclerosis 1	mTOR	[52]
TMEM99	Transmembrane protein 99		[52]
G6PC	Glucose-6-phosphatase, catalytic subunit	FoxO	[52]

4.2. The p53-RB Pathway

TP53 mutations have been reported in HCC in Japan and aflatoxin B1-induced HCC [59,60]. In HCC, the p53-RB pathway is altered in 72% of cases, with 68% of these a result of significantly altered genes [52]. Somatic mutations and copy number alterations in TP53, which plays a central role in apoptosis [61], are present in 31% and 37% of cases, respectively. Among the upstream genes targeted by p53, somatic mutations and copy number alterations in ATM are present in 4% and 6% of cases, respectively; those in RPS6KA3 are present in 4% and 5% of cases, respectively, and those in CDKN2A are present in 2% and 20% of cases, respectively. Among downstream genes targeted by p53, somatic mutations and copy number alterations in CDKN1A are present in 1% and 1% of cases, respectively; those in FBXW7 are present in <1% and 16% of cases, respectively, and those in CCNE1, which suppress RB1 transcription and promote cell cycle progression, are present in 0% and 2% of cases, respectively. Among upstream genes targeted by RB, somatic mutations and copy number alterations in CCND1 are present in 0% and 8% of cases, respectively. Somatic mutations and copy number alterations in RB1, which inhibit cell cycle progression, are present in 4% and 20% of cases, respectively. Additionally, RB controls the levels of p21, which is associated with hepatocarcinogenesis [62]. Overall, 72% of HCC cases harbor alterations in component genes of either the p53 or RB pathway alone or the combined p53-RB pathway (Figure 2) [52].

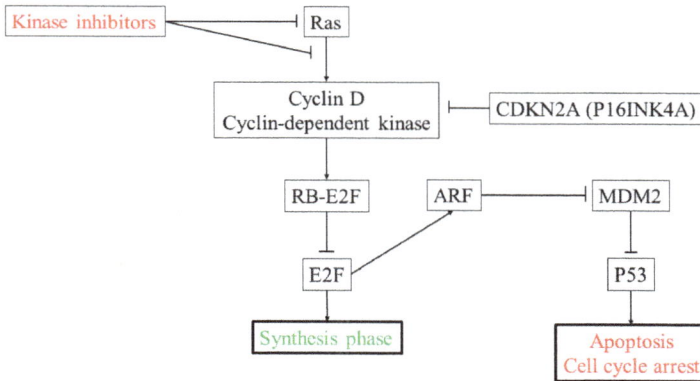

Figure 2. The p53-RB pathway in HCC.

4.3. The β-catenin Pathway (WNT Pathway)

In HCC, the β-catenin pathway is altered in 66% of cases, with 51% of these involving significantly altered genes [52]. Somatic mutations and copy number alterations in CTNNB1, which induce the expression of WNT target genes and transcription of CCND1 resulting in inhibited RB1 expression, are present in 31% and <1% of cases, respectively [52]. Among the upstream genes targeted by β-catenin, somatic mutations and copy number alterations in NCOR1 are present in 2% and 29% of cases, respectively; those in FGF19 are present in 0% and 6% of cases, respectively; those in AXIN1 are present in 6% and 15% of cases, respectively; and those in APC are present in 2% and 4% of cases, respectively [52]. Overall, 66% of HCC cases harbored WNT-pathway-related alterations.

4.4. Chromatin and Transcription Modulators

In HCC, chromatin-remodeling pathways are altered in 67% of cases, with 41% of these involving significantly altered genes [52]. Among upstream genes associated with the nucleosome-remodeling pathway, somatic mutations and copy number alterations in ARID1B are present in <1% and 15% of cases, respectively; those in ARID1A are present in 8% and 17% of cases, respectively; those in BRD7 are present in 2% and 16% of cases, respectively; those in ARID2 are present in 10% and 2% of cases, respectively; those in switch/sucrose non-fermentable (SWI/SNF)-related, matrix associated, actin-dependent regulator of chromatin (SMARC) subfamily C member 1 are present in <1% and 4% of cases, respectively; those in SMARC subfamily B member 1 are present in <1% and 6% of cases, respectively; those in SMARC subfamily A member 4 are present in 1% and 10%, respectively; those in SMARC subfamily E member 1 are present in <1% and 3% of cases, respectively; and those in SMARC subfamily A member 2 are present in 2% and 10% of cases, respectively [52]. Additionally, Li et al. reported that 18.2% of patients with HCV-associated HCC in the United States and Europe harbored ARID2-inactivation mutations [57]. Moreover, another study reported ARID1A mutation in 14 of 110 (13%) HBV-associated HCC specimens [54]. ARID1A functions as the epigenetic regulation of hepatic lipid homeostasis, and its suppression leads to nonalcoholic fatty liver disease and nonalcoholic steatohepatitis (NASH) [58].

Among upstream genes associated with the histone-modification pathway, somatic mutations and copy number alterations in NCOR1 are present in 2% and 29% of cases, respectively; those in SRCAP are present in 3% and 10% of cases, respectively; those in SET domain bifurcated histone lysine methyltransferase 1 (SETDB1) are present in 2% and <1% of cases, respectively; and those in lysine demethylase 6A (KDM6A) are present in 1% and 5% of cases, respectively [52]. Notably, the histone methyltransferase SETDB1 promotes HCC metastasis [63,64], and KDM6A is associated with the epithelial–mesenchymal transition in HCC [65].

4.5. Other Pathways

In HCC, the phosphoinositide 3-kinase (PI3K)–mTOR pathway is altered in 45% of cases, with 26% of these involving significantly altered genes [52]. Somatic mutations and copy number alterations of PI3KCA (encoding the p110α subunit of PI3K) are present in 1% and <1% of cases, respectively; those in neurofibromin 1 are present in 4% and 3% of cases, respectively; those in PTEN are present in 1% and 10% of cases, respectively; those in inositol polyphosphate-4-phosphatase, type IIB, are present in 1% and 16% of cases, respectively; those in serine/threonine kinase 11 are present in <1% and 11% of cases, respectively; those in TSC1 are present in 2% and 8% of cases, respectively; and those in TSC2 are present in 5% and 14% of cases, respectively [52]. TSC1 and TSC2 suppress mTOR expression, which promotes HCC proliferation and survival. In sorafenib-treated patients with HCC, oncogenic PI3K–mTOR-pathway alterations are associated with lower disease-control rates and decreased median progression-free survival and overall survival [66].

In HCC, the nuclear factor (erythroid-derived 2)-like 2 (NRF2)–kelch-like ECH-associated protein 1 (KEAP1) pathway is altered in 19% of cases, with 5% of these involving significantly altered genes [52]. The NRF2–KEAP1 pathway is associated with an oxidative-stress response, and persistent activation of NRF2 through the accumulation of p62 is involved in HCC development [67–69].

There are many protein-coding genes recurrently mutated at a frequency of <5% in HCC [55,70], whereas HBV integrations and frequent noncoding mutations in the TERT promoter represent prominent examples of noncoding mutations in HCC [52–55]. Nault et al. [56] identified frequent somatic mutations resulting in the activation of the TERT promoter in HCC (59%), cirrhotic preneoplastic macronodules (25%), and hepatocellular adenomas with malignant transformation in HCC (44%). Moreover, TERT-promoter mutation represents the most frequent genetic alteration in HCC arising from the cirrhotic or non-cirrhotic liver [56], resulting in reportedly enhanced TERT activity in HCC [71].

5. Molecular Mechanisms of HBV-Associated HCC

5.1. HBV Genome Integration Promotes HCC

The discovery that HBV integrates into host chromosomes calls into question whether HBV genome integration interacts with the activation of oncogenic processes in HCC [72]. Tokino et al. [72] could not detect specific HBV genome-integration sites in chromosomes. About half of HBV DNA-cell

DNA junctions are located within the so-called cohesive end region that lies between two 11-bp direct repeats (DR1 and DR2) in the HBV genome where transcription and replication of the genome are initiated. All of the integrated HBV genomes were defective in at least one site around the cohesive end region, particularly within the HBx gene [73]. Additionally, Imazeki et al. [74] detected integrated HBV DNA in all nine HBsAg-seropositive HCCs and three of 25 (12%) HBsAg-seronegative HCCs in Japan. An analysis of breakpoints within the integration region showed that 40% of breakpoints were near the 1800th nucleotide of the HBV genome, which contains an enhancer, an HBx gene, and core promoters of the HBV genome [75].

TERT is located on chromosome 5p, which is one of the most common targets for amplification in non-small cell lung cancer [76] and is reportedly directly associated with HBV genome integration [77]. Recent data from whole-genome sequencing supports this observation and confirmed the frequent observation of HBV genome integration in the TERT locus in a high clonal proportion [52–56].

HBx expression might play a role in hepatocarcinogenesis by interfering with telomerase activity during hepatocyte proliferation [78], which is supported by breakpoint analysis of the HBV genome [75]. HBx functions as a transcriptional activator and suppressor and has effects on hepatocellular apoptosis [79,80]. HBx proteins may upregulate the transcriptional activation of human telomerase transcriptase [81]. Cis-activation of human TERT mRNA by HBx gene may also be the mechanism in hepatocarcinogenesis [82].

Therefore, HBV genome integration into host genomes is a mechanism involved in HCC associated with HBV infection. Moreover, whole-genome sequencing demonstrated that the integration of HBV DNA into the host hepatocyte genome can be detected in 80% to 90% of HCC and in ~30% of non-HCC liver tissue adjacent to HCC [83], and that this integration appears prior to the occurrence of HCC [75]. Thus, it is possible that occult HBV infection may also accelerate hepatocarcinogenesis in HBsAg-negative patients to some extent [84].

5.2. Inflammation Promotes HBV-Associated HCC

In general, a high number of HBV-DNA integrations randomly distributed among chromosomes has been detected in HBV-infected liver [85]. Chronic HBV infection progresses through multiple phases, including immune tolerance, immune activation, immune control, and, in a subset of patients who achieve immune control and immune reactivation [86]. Immune-mediated liver injury is often associated with elevated ALT levels, and elevations in tumor necrosis factor-α (TNF-α) and interleukin (IL)-1β levels are often observed in the sera of HBV-infected patients [87]. Additionally, long-lasting hepatic inflammation caused by host immune responses during chronic HBV infection can promote liver fibrosis, cirrhosis and HCC progression due to accelerated hepatocyte turnover rates and the accumulation of mutations [88].

5.3. Epigenetic Mechanisms Involved in HBV-Associated HCC

Epigenetic mechanisms play a role in HBV-associated hepatocarcinogenesis. MicroRNAs (miRs) are endogenous noncoding RNAs (18–22 nucleotides in length) that posttranscriptionally regulate the expression of target genes. miRs bind to the 3'-untranslated region (UTR) of target mRNA, thereby inhibiting translation [89].

Several miRNAs involved in the Toll-like receptor (TLR) signaling pathway play a critical role in innate immunity against HBV infection [89]. For example, miR-145 and miR-148a target TLR3, miR-200b, miR-200c, miR148a, miR-455, and let-7-family members target TLR4, let-7b and miR-155 target TLR7, and miR-148a targets TLR9, and all of which are involved in HBV infection [89,90]. HBV might influence miR expression and induce inflammatory cytokine production [91]. Previous reports indicated that some hepatic miRs are involved in the pathogenesis of HBV-associated HCC [92–142], with certain serum miRs involved in HBV-associated HCC also potentially useful as biomarkers (Table 2).

Table 2. Hepatic and serum miRs involved in HBV-associated HCC.

miRs	Upregulation/Downregulation	Target Genes	References
Hepatic miRs			
miR-223	Upregulation	Stathmin 1 (STMN1)	[92]
miR-143	Upregulation	Fibronectin type III domain containing 3B (FNDC3B)	[93]
miR-602	Upregulation	Ras-association domain family member 1 (RASSF1A)	[94]
miR-224	Upregulation	N/A	[98,110,116]
miR-22	Upregulation in male HCC- adjacent tissue	Estrogen receptor alpha (ERα)	[99]
miR-96	Upregulation	N/A	[112]
miR-183	Upregulation	N/A	[112]
miR-196a	Upregulation	N/A	[112]
miR-545/374a	Upregulation	Estrogen-related receptor gamma (ESRRG)	[120]
miR-331-3p	Upregulation	Inhibitor of growth family member 5 (ING5)	[124]
miR-519a	Upregulation	Forkhead box F2 (FOXF2)	[124]
miR-106b	Upregulation	N/A	[131]
miR-1269b	Upregulation	Cell division cycle 40 homolog (CDC40)	[132]
let-7a	Downregulation	Signal transducer and activator of transcription 3 (STAT3)	[95]
miR-152	Downregulation	DNA methyltransferase 1 (DNMT1), TNFRF6B	[96,115]
miR-145	Downregulation	Cullin 5 (CUL5)	[98,125]
miR-199b	Downregulation	N/A	[98]
miR-29c	Downregulation	TNF-α-induced protein 3 (TNFAIP3)	[100]
miR-92	Downregulation	N/A	[102]
miR-338-3p	Downregulation	3' UTR region of CCND1	[104]
miR-34a	Downregulation	C-C motif chemokine ligand 22 (CCL22)	[115]
miR-101	Downregulation	DNA methyltransferase (DNMT)3A	[116]
miR-122	Downregulation	Pituitary tumor-transforming gene 1 (PTTG1) binding factor (PBF)	[117,134,139]
miR-148a	Downregulation	Hematopoietic pre-B cell leukemia transcription factor-interacting protein (HPIP)	[108]
miR-22	Downregulation	CDKN1A	[109,134]
let-7c	Downregulation	N/A	[112]
miR-138	Downregulation	N/A	[112]
miR-205	Downregulation	N/A	[113]
miR-101-3p	Downregulation	RAB GTPase 5A (RAB5A)	[114]
miR-429	Downregulation	NOTCH1	[121]
miR-216	Downregulation	Insulin-like growth factor 2 mRNA-binding protein 2 (IGF2BP2)	[122]
miR-34c	Downregulation	Transforming growth factor-β-induced factor homeobox 2 (TGIF2)	[127]
miR-18a	Downregulation	Connective tissue growth factor (CTGF)	[133]
miR-30b-5p	Downregulation	DNMT3α, USP37	[135]
miR-384	Downregulation	Pleiotrophin (PTN)	[136]
miR-125a-5p	Downregulation	erb-b2 receptor tyrosine kinase 3 (ERBB3)	[137]
miR-26a	Downregulation	N/A	[140]
miR-302c-3p	Downregulation	TNF receptor-associated factor 4 (TRAF4)	[141]
miR-1271	Downregulation	CCNA1	[142]
Serum miRs			
miR-25	Upregulation	N/A	[97]
miR-375	Upregulation	N/A	[97]
let-7f	Upregulation	N/A	[97]
miR-122	Upregulation	N/A	[99]
miR-18a	Upregulation	N/A	[103]
miR-101	Upregulation	N/A	[111]
miR-24-3p	Upregulation	N/A	[117]
miR-545/374a	Upregulation	ERR gamma (ERRG)	[120]
miR-96	Upregulation	N/A	[129]
miR-21	Downregulation	N/A	[118]
miR-222	Downregulation	N/A	[118]
miR-29	Downregulation	N/A	[119]
miR-150	Downregulation	N/A	[123]
miR-126	Downregulation	N/A	[128]
miR-125b	Downregulation	N/A	[130]
miR-143/145	Downregulation	N/A	[138]

Other noncoding RNAs, such as long noncoding RNAs (lncRNAs) and circular RNA (circRNA), are also involved in the pathogenesis of HBV-associated HCC [143–145]. Functional studies reveal that lncRNAs (100–300 kb) contribute to the onset and progression of HBV-related HCC [143], and circRNAs, which form covalently closed continuous loops by means of unique non-canonical 'head-to-tail' splicing in the absence of free 3′ or 5′ ends, might promote the development of HBV-associated HCC [144,145].

Epigenetic silencing of genes, such as methylation of promoter regions, also regulates gene expression in HBV-associated HCC and could potentially serve as a diagnostic and prognostic marker of the disease [146,147]. Additionally, HBV can cause epigenetic changes by altering the methylation state of cellular DNA, the posttranslational modification of histones, and miR expression [146,148], all of which are also critical for the pathogenesis of HBV-associated HCC.

5.4. Roles of HBV-encoded Proteins

HBV is a partially double-stranded DNA virus (genome length: 3200 bp) and a member of the Orthohepadnavirus genus and the Hepadnaviridae family [149]. The HBV genome encodes at least four proteins [HBsAg, a core protein (splice variant: HBeAg), a DNA polymerase, and the HBx protein] that are translated from mRNAs transcribed from HBV covalently closed circular DNA and/or from HBV genome sequences integrated into the host genome [150]. Viral protein translation is initiated through binding of the PreS2 domain of HBV surface antigens to protein kinase C (PKC)α/β, which triggers PKC-dependent activation of Raf-1 proto-oncogene serine/threonine kinase (RAF-1)/MAP2K signaling and transcription factors, such as activator protein-1 (AP-1) and nuclear factor kappaB (NF-kB), resulting in the increased proliferation of hepatocytes [151]. PreS2-mediated activity subsequently upregulates the expression of the transcriptional coactivator with PDZ-binding motif by repressing miRNA-338-3p, which promotes HCC proliferation and migration [152]. Moreover, a previous study showed that a truncated mutant of HBsAg increases HBV-related tumorigenesis in a mechanism potentially associated with the downregulated expression of tumor growth factor (TGF)BI associated with the TGFβ–SMAD pathway [153]. Furthermore, HBsAg enhances the IL-6–STAT3 pathway, thereby increasing the HBsAg-mediated malignant potential of HBV-associated HCC [154].

The HBV core protein enhances cytokine production [155], and HBeAg is reportedly associated with the host immune response and cytokine production [156–158], both of which play roles in HBV-associated HCC.

Numerous studies reported an association between the HBx protein and hepatocarcinogenesis [159–170]. Integrated HBV DNA harbors the conserved sequences for genes encoding the core protein, surface antigens, and HBx along with their respective promoter sequences [159,160], suggesting that HBx is important for hepatocarcinogenesis. HBx represents a transactivating factor [161], and transgenic mice expressing HBx develop HCC [162]. HBx transactivates binding sites for the transcription factors AP-1 and NF-κB [163], activates the p53-RB [164,165] and β-catenin [164–170] pathways, and is involved in chromatin remodeling [171–173] and transcriptional modulation in hepatocarcinogenesis [174].

Overexpression of the HBV polymerase due to core-gene deletion enhances HCC-cell growth by inhibiting miR-100 [175]. A previous study showed that transgenic mice expressing the reverse-transcriptase domain of HBV polymerase in their livers developed early cirrhosis with steatosis by 18 months of age, with 10% subsequently developing HCC [176].

6. Molecular Mechanisms of HCV-Associated HCC

6.1. Inflammation Promotes HCV-Associated HCC

Cytokines reflect the degree of inflammation in the liver of patients with chronic hepatitis C, with this production possibly related to HCC development [177]. Takano et al. [1] prospectively investigated the incidence of HCC in 124 cases with HCV infection, finding that HCC occurred in 13 cases that included 12 cirrhotic livers and only 1 non-cirrhotic liver and suggesting that most HCC occurs in advanced liver diseases in HCV-infected individuals [178]. Because hepatic inflammation plays a role in the development of advanced liver diseases, inflammation is an important aspect in the development of HCC associated with chronic HCV infection.

6.2. Epigenetic Mechanisms Involved in HCV-Associated HCC

Epigenetic mechanisms also play a role in HCV-associated hepatocarcinogenesis. Previous studies reported the involvement of specific hepatic and serum miRs in the pathogenesis of HCV-associated HCC [179–196], with some of the serum miRs representing potentially useful biomarkers of the disease (Table 3). Notably, the number of reports of miRs involved in HCV-associated HCC is smaller than that related to HBV-associated HCC (Table 2).

Table 3. Hepatic and serum miRs involved in HCV-associated HCC.

miRs	Upregulation/Downregulation	Target Genes	References
Hepatic miRs			
miR-193b	Upregulation	STMN1	[179]
miR-155	Upregulation	N/A	[180]
miR-122	Upregulation	Cyclin G1	[181,182,186]
miR-373	Upregulation	Wee1	[195]
miR-24	Downregulation	N/A	[183]
miR-27a	Downregulation	N/A	[183]
miR-198	Downregulation	N/A	[184]
miR-152	Downregulation	N/A	[186]
miR-181c	Downregulation	Homeobox A1	[187]
miR-431	Downregulation	N/A	[189]
miR-138	Downregulation	TERT	[192]
Serum miRs			
miR-150	Upregulation	N/A	[191]
miR-221	Upregulation	N/A	[193]
miR-101-1	Upregulation	N/A	[193]
miR-27a	Upregulation	N/A	[194]
miR-221	Downregulation	Suppressor of cytokine signaling (SOCS)1 and SOCS3	[188]
miR-16	Downregulation	N/A	[190]
miR-34a	Downregulation	Heat-shock protein 70	[194]

Previous studies reported that lncRNAs are involved in the pathogenesis of HBV-associated HCC [143–145], with functional studies revealing that these lncRNAs contribute also to the onset and progression of HCV-related HCC [195,197]. Zhang et al. [198] reported LINC01419 transcripts expressed at higher levels in early stage HCC as compared with levels observed in dysplastic tissue. Moreover, this study also reported increased and decreased levels of AK021443 and AF070632 in advanced HCC, respectively, and that LINC01419 and AK021443 regulated the expression of genes associated with cell cycle progression, whereas AF070632 was associated with cofactor binding, oxidation–reduction activity and carboxylic acid catabolic processes [198].

Upregulated lncRNAs associated with urothelial carcinoma associated-1 (lncRNA-UCA1) and WD-repeat-containing antisense to TP53 (lncRNA-Wrap53) potentially serve as novel serum biomarkers for HCC diagnosis and prognosis [199]. Additionally, the lncRNA associated with activated by TGFβ (lncRNA-ATB) is a key regulator of TGFβ signaling and is positively correlated with the development of liver cirrhosis and HCC-specific vascular invasion [200]. Moreover, lncRNA-ATB might represent a novel diagnostic biomarker and potential therapeutic target for HCV-related hepatic fibrosis [200]. A previous study showed that HCV infection upregulates the expression of miR-373 and Wee1-like protein kinase (WEE1), a pivotal player in the G2/M transition in the cell cycle (although WEE1 is a direct target of miR-373). This study also showed that miR-373 forms a complex with the LINC00657, resulting in release of their common target, WEE1, in HCV-infected cells, and the promotion of uncontrolled cell growth [195]. Another study reported that hypermethylation of promoter regions suppressed mRNA expression, which played a role in the progression of HCV-associated HCC [201].

6.3. Roles of HCV-coding Proteins

We previously reported the roles for HCV-coding proteins in hepatocarcinogenesis [178], with the HCV core proteins and NS5A reportedly playing important roles in HCC development [202–206]. HCV-infected patients showed a higher prevalence of diabetes mellitus and insulin resistance (IR) relative to those with HBV infection [207]. IR measured using a homeostasis model assessment of IR (HOMA-IR) is significantly associated with HCC development in patients with or without chronic HCV infection. Moreover, patients with NASH, which is associated with elevated HOMA-IR, have an increased risk of liver fibrosis, cirrhosis, and HCC [208,209].

A previous study showed that the HCV core protein downregulates insulin receptor substrate (IRS)1 and IRS2 by upregulating SOCS3 levels [210]. Moreover, cross-talk between the HCV core protein and molecules regulating insulin signaling might affect HCV-associated hepatocarcinogenesis [210–214]. Insulin-like growth factors initiate tyrosyl phosphorylation of IRS1 and activate multiple signaling pathways essential for liver growth and HCC [215], with the activation of IRS1-mediated signaling potentially contributing to hepatic oncogenesis.

CCAAT/enhancer-binding protein (C/EBPβ) and HCV NS5A might be essential components promoting increased gluconeogenesis associated with HCV infection [216]. Previous studies reported the involvement of HCV NS5A in enhanced gluconeogenic gene expression associated with impaired insulin signaling [217,218].

Sphere-forming hepatocytes express several cancer stem-like cell (CSC) markers, including c-Kit. Previous studies showed that the HCV core protein significantly upregulates c-Kit expression at the transcriptional level, and that HCV infection potentiates CSC generation [219,220]. Additionally, a Western diet high in cholesterol and saturated fat (HCFD) in combination with the translation of HCV NS5A stimulates TLR4–Nanog homeobox (NANOG) and leptin receptor–phosphorylated STAT3 signaling, resulting in liver tumorigenesis through an exaggerated mesenchymal phenotype involving prominent Twist1-expressing tumor-initiating stem-like cells expressing NANOG [221].

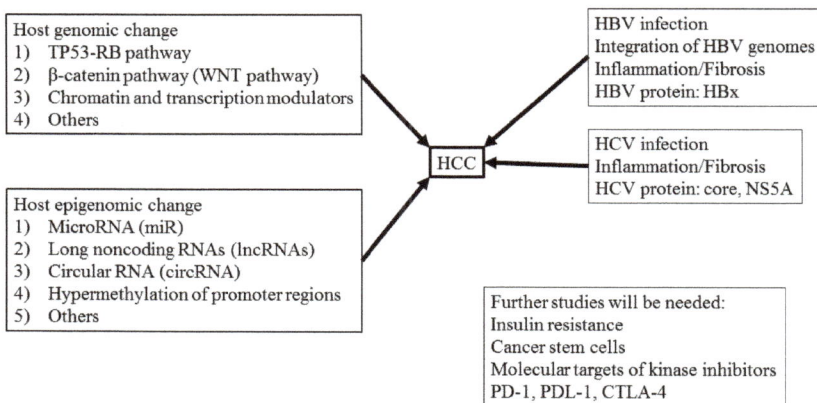

Figure 3. Molecular mechanisms of liver cirrhosis and its progression to HCC.

7. HCC-specific Molecular Mechanisms from a Therapeutic Point of View

Sorafenib, regorafenib and lenvatinib are approved therapies as oral molecular-targeting drugs for advanced stages of HCC [3–6]. Sorafenib is an oral serine/threonine kinase inhibitor targeting the extracellular-signaling-regulated kinase/MAPK pathway, vascular endothelial growth factor receptor (VEGFR), platelet-derived growth factor receptor (PDGFR), and epithelial growth factor receptor (EGFR). Additionally, it acts as a tyrosine kinase inhibitor targeting VEGFR1, VEGFR2, VEGFR3,

PDGFRβ, ret proto-oncogene (RET), and fms-related tyrosine kinase 3, resulting in the inhibition of tumor proliferation [222].

Regorafenib is an oral multiple kinase inhibitor of VEGFR1, VEGFR2, VEGFR3, TEK receptor tyrosine kinase, PDGFRβ, fibroblast growth factor receptor (FGFR), proto-oncogene receptor tyrosine kinase (KIT), RET, RAF-1, and B-RAF. Lenvatinib is an oral tyrosine kinase inhibitor targeting VEGFR1/2/3, FGFR1/2/3/4, PDGFRα, KIT, and RET to inhibit tumor angiogenesis and growth [222]. Therefore, sorafenib, regorafenib, and lenvatinib represent multiple kinase inhibitors that do not suppress a specific molecule, thereby restricting their use only in patients with advanced HCC and well-compensated liver function.

Nivolumab, pembrolizumab and tislelizumab are anti-programmed death-1 (PD-1) antibodies. The immune-checkpoint molecule PD-1 is a receptor that negatively regulates immune responses [222,223] via binding of its representative ligands PD-ligand (PD-L)1 and PD-L2. Inhibition of this pathway can eliminate tumors by recovering their immunosuppressive effects and restoring innate immune activity [222,224]. Cytotoxic T lymphocyte-associated antigen-4 (CTLA-4) is another counterreceptor for the B7 family of costimulatory molecules and a negative regulator of T cell activation. A blockade of the inhibitory effects of CTLA-4 is supposed to enhance immune responses against tumor cells [225] and also represents a type of immune checkpoint. The blockade of such immune checkpoints represents a promising therapeutic option for HCC; therefore, further studies are needed to determine other immune checkpoints, as well as potential targets of kinase inhibitors related to liver cirrhosis to HCC.

8. Conclusions

This review describes the molecular mechanisms of liver cirrhosis and its progression to HCC (Figure 3). Recent progress in next-generation sequencing has revealed several HCC-specific driver-gene candidates. Additionally, ~70% of HCC cases are caused by HBV and/or HCV infection. Although nucleos(t)ide analogs and direct-acting antivirals can control HBV and HCV replication, HCC occurrence is occasionally observed [49–51], and better biomarkers for early detection of HCC are needed. Furthermore, genetic and epigenetic factors, such as miRs, are involved in liver cirrhosis and its progression to HCC. Despite significant advances, additional studies are required to elucidate other molecular mechanisms, including immune checkpoints and molecular targets of kinase inhibitors associated with liver cirrhosis and its progression to HCC.

Author Contributions: Writing—Original Draft Preparation, T.K.; Writing—Review & Editing, T.G., Y.H., M.M.; Supervision, M.O.

Funding: This work was partly supported by JSPS KAKENHI GRANT Number 17K09404 (to T.K.).

Conflicts of Interest: The other authors declare no conflict of interest.

References

1. Takano, S.; Yokosuka, O.; Imazeki, F.; Tagawa, M.; Omata, M. Incidence of hepatocellular carcinoma in chronic hepatitis B and C: A prospective study of 251 patients. *Hepatology* **1995**, *21*, 650–655. [CrossRef] [PubMed]
2. Takayama, T.; Makuuchi, M. Segmental liver resections, present and future-caudate lobe resection for liver tumours. *Hepatogastroenterology* **1998**, *45*, 20–23.
3. Llovet, J.M.; Ricci, S.; Mazzaferro, V.; Hilgard, P.; Gane, E.; Blanc, J.F.; de Oliveira, A.C.; Santoro, A.; Raoul, J.L.; Forner, A.; et al. Sorafenib in advanced hepatocellular carcinoma. *N. Engl. J. Med.* **2008**, *359*, 378–390. [CrossRef] [PubMed]
4. Cheng, A.L.; Kang, Y.K.; Chen, Z.; Tsao, C.J.; Qin, S.; Kim, J.S.; Luo, R.; Feng, J.; Ye, S.; Yang, T.S.; et al. Efficacy and safety of sorafenib in patients in the Asia-Pacific region with advanced hepatocellular carcinoma: A phase III randomised, double-blind, placebo-controlled trial. *Lancet Oncol.* **2009**, *10*, 25–34. [CrossRef]

5. Bruix, J.; Qin, S.; Merle, P.; Granito, A.; Huang, Y.H.; Bodoky, G.; Pracht, M.; Yokosuka, O.; Rosmorduc, O.; Breder, V.; et al. Regorafenib for patients with hepatocellular carcinoma who progressed on sorafenib treatment (RESORCE): A randomised, double-blind, placebo-controlled, phase 3 trial. *Lancet* **2017**, *389*, 56–66. [CrossRef]

6. Kudo, M.; Finn, R.S.; Qin, S.; Han, K.H.; Ikeda, K.; Piscaglia, F.; Baron, A.; Park, J.W.; Han, G.; Jassem, J.; et al. Lenvatinib versus sorafenib in first-line treatment of patients with unresectable hepatocellular carcinoma: A randomised phase 3 non-inferiority trial. *Lancet* **2018**, *391*, 1163–1173. [CrossRef]

7. Schachter, J.; Ribas, A.; Long, G.V.; Arance, A.; Grob, J.J.; Mortier, L.; Daud, A.; Carlino, M.S.; McNeil, C.; Lotem, M.; et al. Pembrolizumab versus ipilimumab for advanced melanoma: Final overall survival results of a multicentre, randomised, open-label phase 3 study (KEYNOTE-006). *Lancet* **2017**, *390*, 1853–1862. [CrossRef]

8. Bruix, J.; Reig, M.; Sherman, M. Evidence-Based Diagnosis, Staging, and Treatment of Patients With Hepatocellular Carcinoma. *Gastroenterology* **2016**, *150*, 835–853. [CrossRef]

9. Obi, S.; Yoshida, H.; Toune, R.; Unuma, T.; Kanda, M.; Sato, S.; Tateishi, R.; Teratani, T.; Shiina, S.; Omata, M. Combination therapy of intraarterial 5-fluorouracil and systemic interferon-alpha for advanced hepatocellular carcinoma with portal venous invasion. *Cancer* **2006**, *106*, 1990–1997. [CrossRef] [PubMed]

10. Kanda, T.; Takahashi, K.; Nakamura, M.; Nakamoto, S.; Wu, S.; Haga, Y.; Sasaki, R.; Jiang, X.; Yokosuka, O. Androgen Receptor Could Be a Potential Therapeutic Target in Patients with Advanced Hepatocellular Carcinoma. *Cancers* **2017**, *9*, 43. [CrossRef] [PubMed]

11. Fattovich, G.; Stroffolini, T.; Zagni, I.; Donato, F. Hepatocellular carcinoma in cirrhosis: Incidence and risk factors. *Gastroenterology* **2004**, *127* (Suppl. 1), S35–S50. [CrossRef]

12. Bataller, R.; Brenner, D.A. Liver fibrosis. *J. Clin. Investig.* **2005**, *115*, 209–218. [CrossRef] [PubMed]

13. Patel, K.; Bedossa, P.; Castera, L. Diagnosis of liver fibrosis: Present and future. *Semin. Liver Dis.* **2015**, *35*, 166–183. [CrossRef]

14. Xu, L.; Hui, A.Y.; Albanis, E.; Arthur, M.J.; O'Byrne, S.M.; Blaner, W.S.; Mukherjee, P.; Friedman, S.L.; Eng, F.J. Human hepatic stellate cell lines, LX-1 and LX-2: New tools for analysis of hepatic fibrosis. *Gut* **2005**, *54*, 142–151. [CrossRef]

15. Okuda, K.; Nakashima, T.; Kojiro, M.; Kondo, Y.; Wada, K. Hepatocellular carcinoma without cirrhosis in Japanese patients. *Gastroenterology* **1989**, *97*, 140–146. [CrossRef]

16. Clouston, A.D.; Powell, E.E.; Walsh, M.J.; Richardson, M.M.; Demetris, A.J.; Jonsson, J.R. Fibrosis correlates with a ductular reaction in hepatitis C: Roles of impaired replication, progenitor cells and steatosis. *Hepatology* **2005**, *41*, 809–818. [CrossRef]

17. Lozano, R.; Naghavi, M.; Foreman, K.; Lim, S.; Shibuya, K.; Aboyans, V.; Abraham, J.; Adair, T.; Aggarwal, R.; Ahn, S.Y.; et al. Global and regional mortality from 235 causes of death for 20 age groups in 1990 and 2010: A systematic analysis for the Global Burden of Disease Study 2010. *Lancet* **2012**, *380*, 2095–2128. [CrossRef]

18. Liaw, Y.F.; Tai, D.I.; Chu, C.M.; Chen, T.J. The development of cirrhosis in patients with chronic type B hepatitis: A prospective study. *Hepatology* **1988**, *8*, 493–496. [CrossRef]

19. Fattovich, G.; Brollo, L.; Giustina, G.; Noventa, F.; Pontisso, P.; Alberti, A.; Realdi, G.; Ruol, A. Natural history and prognostic factors for chronic hepatitis type B. *Gut* **1991**, *32*, 294–298. [CrossRef] [PubMed]

20. Sumi, H.; Yokosuka, O.; Seki, N.; Arai, M.; Imazeki, F.; Kurihara, T.; Kanda, T.; Fukai, K.; Kato, M.; Saisho, H. Influence of hepatitis B virus genotypes on the progression of chronic type B liver disease. *Hepatology* **2003**, *37*, 19–26. [CrossRef]

21. Liaw, Y.F.; Chen, Y.C.; Sheen, I.S.; Chien, R.N.; Yeh, C.T.; Chu, C.M. Impact of acute hepatitis C virus superinfection in patients with chronic hepatitis B virus infection. *Gastroenterology* **2004**, *126*, 1024–1029. [CrossRef]

22. Lacombe, K.; Massari, V.; Girard, P.M.; Serfaty, L.; Gozlan, J.; Pialoux, G.; Mialhes, P.; Molina, J.M.; Lascoux-Combe, C.; Wendum, D.; et al. Major role of hepatitis B genotypes in liver fibrosis during coinfection with HIV. *AIDS* **2006**, *20*, 419–427. [CrossRef]

23. Stroffolini, T.; Esvan, R.; Biliotti, E.; Sagnelli, E.; Gaeta, G.B.; Almasio, P.L. Gender differences in chronic HBsAg carriers in Italy: Evidence for the independent role of male sex in severity of liver disease. *J. Med. Virol.* **2015**, *87*, 1899–1903. [CrossRef]

24. Al-Qahtani, A.; Khalak, H.G.; Alkuraya, F.S.; Al-hamoudi, W.; Alswat, K.; Al Balwi, M.A.; Al Abdulkareem, I.; Sanai, F.M.; Abdo, A.A. Genome-wide association study of chronic hepatitis B virus infection reveals a novel candidate risk allele on 11q22.3. *J. Med. Genet.* **2013**, *50*, 725–732. [CrossRef] [PubMed]

25. Chang, S.W.; Fann, C.S.; Su, W.H.; Wang, Y.C.; Weng, C.C.; Yu, C.J.; Hsu, C.L.; Hsieh, A.R.; Chien, R.N.; Chu, C.M.; et al. A genome-wide association study on chronic HBV infection and its clinical progression in male Han-Taiwanese. *PLoS ONE* **2014**, *9*, e99724. [CrossRef]

26. Chu, C.M.; Liaw, Y.F. Hepatitis B virus-related cirrhosis: Natural history and treatment. *Semin. Liver Dis.* **2006**, *26*, 142–152. [CrossRef]

27. Tawada, A.; Kanda, T.; Imazeki, F.; Yokosuka, O. Prevention of hepatitis B virus-associated liver diseases by antiviral therapy. *Hepatol. Int.* **2016**, *10*, 574–593. [CrossRef]

28. World Health Organization. Global Hepatitis Report. 2017. Available online: www.who.int/hepatitis/publications/global-hepatitis-report2017/en/ (accessed on 27 December 2018).

29. Yang, Y.H.; Chen, W.C.; Tsan, Y.T.; Chen, M.J.; Shih, W.T.; Tsai, Y.H.; Chen, P.C. Statin use and the risk of cirrhosis development in patients with hepatitis C virus infection. *J. Hepatol.* **2015**, *63*, 1111–1117. [CrossRef]

30. Forns, X.; Ampurdanès, S.; Sanchez-Tapias, J.M.; Guilera, M.; Sans, M.; Sánchez-Fueyo, A.; Quintó, L.; Joya, P.; Bruguera, M.; Rodés, J. Long-term follow-up of chronic hepatitis C in patients diagnosed at a tertiary-care center. *J. Hepatol.* **2001**, *35*, 265–271. [CrossRef]

31. Westin, J.; Nordlinder, H.; Lagging, M.; Norkrans, G.; Wejstål, R. Steatosis accelerates fibrosis development over time in hepatitis C virus genotype 3 infected patients. *J. Hepatol.* **2002**, *37*, 837–842. [CrossRef]

32. Rubbia-Brandt, L.; Fabris, P.; Paganin, S.; Leandro, G.; Male, P.J.; Giostra, E.; Carlotto, A.; Bozzola, L.; Smedile, A.; Negro, F. Steatosis affects chronic hepatitis C progression in a genotype specific way. *Gut* **2004**, *53*, 406–412. [CrossRef]

33. Locasciulli, A.; Testa, M.; Pontisso, P.; Benvegnù, L.; Fraschini, D.; Corbetta, A.; Noventa, F.; Masera, G.; Alberti, A. Prevalence and natural history of hepatitis C infection in patients cured of childhood leukemia. *Blood* **1997**, *90*, 4628–4633. [PubMed]

34. Kenny-Walsh, E. Clinical outcomes after hpatitis C infection from contaminated anti-D immune globulin. Irish Hepatology Research Group. *N. Engl. J. Med.* **1999**, *340*, 1228–1233. [CrossRef]

35. Poynard, T.; Bedossa, P.; Opolon, P. Natural history of liver fibrosis progression in patients with chronic hepatitis C. The OBSVIRC, METAVIR, CLINIVIR, and DOSVIRC groups. *Lancet* **1997**, *349*, 825–832. [CrossRef]

36. Poynard, T.; Ratziu, V.; Charlotte, F.; Goodman, Z.; McHutchison, J.; Albrecht, J. Rates and risk factors of liver fibrosis progression in patients with chronic hepatitis c. *J. Hepatol.* **2001**, *34*, 730–739. [CrossRef]

37. Shiffman, M.L. Natural history and risk factors for progression of hepatitis C virus disease and development of hepatocellular cancer before liver transplantation. *Liver Transpl.* **2003**, *9*, S14–S20. [CrossRef] [PubMed]

38. Fuster, D.; Planas, R.; Muga, R.; Ballesteros, A.L.; Santos, J.; Tor, J.; Sirera, G.; Guardiola, H.; Salas, A.; Cabré, E.; et al. Advanced liver fibrosis in HIV/HCV-coinfected patients on antiretroviral therapy. *AIDS Res. Hum. Retrovir.* **2004**, *20*, 1293–1297. [CrossRef] [PubMed]

39. Ribes, J.; Clèries, R.; Rubió, A.; Hernández, J.M.; Mazzara, R.; Madoz, P.; Casanovas, T.; Casanova, A.; Gallen, M.; Rodríguez, C.; et al. Cofactors associated with liver disease mortality in an HBsAg-positive Mediterranean cohort: 20 years of follow-up. *Int. J. Cancer* **2006**, *119*, 687–694. [CrossRef]

40. Wilson, L.E.; Torbenson, M.; Astemborski, J.; Faruki, H.; Spoler, C.; Rai, R.; Mehta, S.; Kirk, G.D.; Nelson, K.; Afdhal, N.; et al. Progression of liver fibrosis among injection drug users with chronic hepatitis C. *Hepatology* **2006**, *43*, 788–795. [CrossRef]

41. Petta, S.; Cammà, C.; Di Marco, V.; Alessi, N.; Cabibi, D.; Caldarella, R.; Licata, A.; Massenti, F.; Tarantino, G.; Marchesini, G.; et al. Insulin resistance and diabetes increase fibrosis in the liver of patients with genotype 1 HCV infection. *Am. J. Gastroenterol.* **2008**, *103*, 1136–1144. [CrossRef]

42. Dyal, H.K.; Aguilar, M.; Bhuket, T.; Liu, B.; Holt, E.W.; Torres, S.; Cheung, R.; Wong, R.J. Concurrent Obesity, Diabetes, and Steatosis Increase Risk of Advanced Fibrosis Among HCV Patients: A Systematic Review. *Dig. Dis. Sci.* **2015**, *60*, 2813–2824. [CrossRef] [PubMed]

43. Shiratori, Y.; Imazeki, F.; Moriyama, M.; Yano, M.; Arakawa, Y.; Yokosuka, O.; Kuroki, T.; Nishiguchi, S.; Sata, M.; Yamada, G.; et al. Histologic improvement of fibrosis in patients with hepatitis C who have sustained response to interferon therapy. *Ann. Intern. Med.* **2000**, *132*, 517–524. [CrossRef]

44. Caporaso, N.; Romano, M.; Marmo, R.; de Sio, I.; Morisco, F.; Minerva, A.; Coltorti, M. Hepatitis C virus infection is an additive risk factor for development of hepatocellular carcinoma in patients with cirrhosis. *J. Hepatol.* **1991**, *12*, 367–371. [CrossRef]

45. Benvegnù, L.; Fattovich, G.; Noventa, F.; Tremolada, F.; Chemello, L.; Cecchetto, A.; Alberti, A. Concurrent hepatitis B and C virus infection and risk of hepatocellular carcinoma in cirrhosis. A prospective study. *Cancer* **1994**, *74*, 2442–2448. [CrossRef]

46. Ishikawa, T.; Ichida, T.; Yamagiwa, S.; Sugahara, S.; Uehara, K.; Okoshi, S.; Asakura, H. High viral loads, serum alanine aminotransferase and gender are predictive factors for the development of hepatocellular carcinoma from viral compensated liver cirrhosis. *J. Gastroenterol. Hepatol.* **2001**, *16*, 1274–1281. [CrossRef]

47. Ohata, K.; Hamasaki, K.; Toriyama, K.; Matsumoto, K.; Saeki, A.; Yanagi, K.; Abiru, S.; Nakagawa, Y.; Shigeno, M.; Miyazoe, S.; et al. Hepatic steatosis is a risk factor for hepatocellular carcinoma in patients with chronic hepatitis C virus infection. *Cancer* **2003**, *97*, 3036–3043. [CrossRef] [PubMed]

48. Yoshida, H.; Shiratori, Y.; Moriyama, M.; Arakawa, Y.; Ide, T.; Sata, M.; Inoue, O.; Yano, M.; Tanaka, M.; Fujiyama, S.; et al. Interferon therapy reduces the risk for hepatocellular carcinoma: National surveillance program of cirrhotic and noncirrhotic patients with chronic hepatitis C in Japan. IHIT Study Group. Inhibition of Hepatocarcinogenesis by Interferon Therapy. *Ann. Intern. Med.* **1999**, *131*, 174–181. [CrossRef] [PubMed]

49. Alberti, A.; Piovesan, S. Increased incidence of liver cancer after successful DAA treatment of chronic hepatitis C: Fact or fiction? *Liver Int.* **2017**, *37*, 802–808. [CrossRef] [PubMed]

50. El Kassas, M.; Funk, A.L.; Salaheldin, M.; Shimakawa, Y.; Eltabbakh, M.; Jean, K.; El Tahan, A.; Sweedy, A.T.; Afify, S.; Youssef, N.F.; et al. Increased recurrence rates of hepatocellular carcinoma after DAA therapy in a hepatitis C-infected Egyptian cohort: A comparative analysis. *J. Viral Hepat.* **2018**, *25*, 623–630. [CrossRef]

51. Sasaki, R.; Kanda, T.; Kato, N.; Yokosuka, O.; Moriyama, M. Hepatitis C virus-associated hepatocellular carcinoma after sustained virologic response. *World J. Hepatol.* **2018**, *10*, 898–906. [CrossRef]

52. Totoki, Y.; Tatsuno, K.; Covington, K.R.; Ueda, H.; Creighton, C.J.; Kato, M.; Tsuji, S.; Donehower, L.A.; Slagle, B.L.; Nakamura, H.; et al. Trans-ancestry mutational landscape of hepatocellular carcinoma genomes. *Nat. Genet.* **2014**, *46*, 1267–1273. [CrossRef]

53. Fujimoto, A.; Totoki, Y.; Abe, T.; Boroevich, K.A.; Hosoda, F.; Nguyen, H.H.; Aoki, M.; Hosono, N.; Kubo, M.; Miya, F.; et al. Whole-genome sequencing of liver cancers identifies etiological influences on mutation patterns and recurrent mutations in chromatin regulators. *Nat. Genet.* **2012**, *44*, 760–764. [CrossRef] [PubMed]

54. Huang, J.; Deng, Q.; Wang, Q.; Li, K.Y.; Dai, J.H.; Li, N.; Zhu, Z.D.; Zhou, B.; Liu, X.Y.; Liu, R.F.; et al. Exome sequencing of hepatitis B virus-associated hepatocellular carcinoma. *Nat. Genet.* **2012**, *44*, 1117–1121. [CrossRef]

55. Schulze, K.; Imbeaud, S.; Letouzé, E.; Alexandrov, L.B.; Calderaro, J.; Rebouissou, S.; Couchy, G.; Meiller, C.; Shinde, J.; Soysouvanh, F.; et al. Exome sequencing of hepatocellular carcinomas identifies new mutational signatures and potential therapeutic targets. *Nat. Genet.* **2015**, *47*, 505–511. [CrossRef] [PubMed]

56. Nault, J.C.; Mallet, M.; Pilati, C.; Calderaro, J.; Bioulac-Sage, P.; Laurent, C.; Laurent, A.; Cherqui, D.; Balabaud, C.; Zucman-Rossi, J. High frequency of telomerase reverse-transcriptase promoter somatic mutations in hepatocellular carcinoma and preneoplastic lesions. *Nat. Commun.* **2013**, *4*, 2218. [CrossRef] [PubMed]

57. Li, M.; Zhao, H.; Zhang, X.; Wood, L.D.; Anders, R.A.; Choti, M.A.; Pawlik, T.M.; Daniel, H.D.; Kannangai, R.; Offerhaus, G.J.; et al. Inactivating mutations of the chromatin remodeling gene ARID2 in hepatocellular carcinoma. *Nat. Genet.* **2011**, *43*, 828–829. [CrossRef]

58. Moore, A.; Wu, L.; Chuang, J.C.; Sun, X.; Luo, X.; Gopal, P.; Li, L.; Celen, C.; Zimmer, M.; Zhu, H. Arid1a loss drives non-alcoholic steatohepatitis in mice via epigenetic dysregulation of hepatic lipogenesis and fatty acid oxidation. *Hepatology* **2018**. [CrossRef]

59. Nose, H.; Imazeki, F.; Ohto, M.; Omata, M. p53 gene mutations and 17p allelic deletions in hepatocellular carcinoma from Japan. *Cancer* **1993**, *72*, 355–3560. [CrossRef]

60. Imazeki, F.; Yokosuka, O.; Ohto, M.; Omata, M. Aflatoxin and p53 abnormality in duck hepatocellular carcinoma. *J. Gastroenterol. Hepatol.* **1995**, *10*, 646–649. [CrossRef]

61. Ray, R.B.; Meyer, K.; Ray, R. Suppression of apoptotic cell death by hepatitis C virus core protein. *Virology* **1996**, *226*, 176–182. [CrossRef]

62. Ray, R.B.; Steele, R.; Meyer, K.; Ray, R. Hepatitis C virus core protein represses p21WAF1/Cip1/Sid1 promoter activity. *Gene* **1998**, *208*, 331–336. [CrossRef]

63. Fei, Q.; Shang, K.; Zhang, J.; Chuai, S.; Kong, D.; Zhou, T.; Fu, S.; Liang, Y.; Li, C.; Chen, Z.; et al. Histone methyltransferase SETDB1 regulates liver cancer cell growth through methylation of p53. *Nat. Commun.* **2015**, *6*, 8651. [CrossRef] [PubMed]

64. Wong, C.M.; Wei, L.; Law, C.T.; Ho, D.W.; Tsang, F.H.; Au, S.L.; Sze, K.M.; Lee, J.M.; Wong, C.C.; Ng, I.O. Up-regulation of histone methyltransferase SETDB1 by multiple mechanisms in hepatocellular carcinoma promotes cancer metastasis. *Hepatology* **2016**, *63*, 474–487. [CrossRef] [PubMed]

65. Kodama, T.; Newberg, J.Y.; Kodama, M.; Rangel, R.; Yoshihara, K.; Tien, J.C.; Parsons, P.H.; Wu, H.; Finegold, M.J.; Copeland, N.G.; et al. Transposon mutagenesis identifies genes and cellular processes driving epithelial-mesenchymal transition in hepatocellular carcinoma. *Proc. Natl. Acad. Sci. USA* **2016**, *113*, E3384–E3393. [CrossRef] [PubMed]

66. Harding, J.J.; Nandakumar, S.; Armenia, J.; Khalil, D.N.; Albano, M.; Ly, M.; Shia, J.; Hechtman, J.F.; Kundra, R.; El Dika, I.; et al. Prospective Genotyping of Hepatocellular Carcinoma: Clinical Implications of Next Generation Sequencing for Matching Patients to Targeted and Immune Therapies. *Clin. Cancer Res.* **2018**. [CrossRef] [PubMed]

67. Inami, Y.; Waguri, S.; Sakamoto, A.; Kouno, T.; Nakada, K.; Hino, O.; Watanabe, S.; Ando, J.; Iwadate, M.; Yamamoto, M.; et al. Persistent activation of Nrf2 through p62 in hepatocellular carcinoma cells. *J. Cell Biol.* **2011**, *193*, 275–284. [CrossRef]

68. Zavattari, P.; Perra, A.; Menegon, S.; Kowalik, M.A.; Petrelli, A.; Angioni, M.M.; Follenzi, A.; Quagliata, L.; Ledda-Columbano, G.M.; Terracciano, L.; et al. Nrf2, but not β-catenin, mutation represents an early event in rat hepatocarcinogenesis. *Hepatology* **2015**, *62*, 851–862. [CrossRef]

69. Bartolini, D.; Dallaglio, K.; Torquato, P.; Piroddi, M.; Galli, F. Nrf2-p62 autophagy pathway and its response to oxidative stress in hepatocellular carcinoma. *Transl. Res.* **2018**, *193*, 54–71. [CrossRef]

70. Fujimoto, A.; Furuta, M.; Totoki, Y.; Tsunoda, T.; Kato, M.; Shiraishi, Y.; Tanaka, H.; Taniguchi, H.; Kawakami, Y.; Ueno, M.; et al. Whole-genome mutational landscape and characterization of noncoding and structural mutations in liver cancer. *Nat. Genet.* **2016**, *48*, 500–509. [CrossRef]

71. Kojima, H.; Yokosuka, O.; Imazeki, F.; Saisho, H.; Omata, M. Telomerase activity and telomere length in hepatocellular carcinoma and chronic liver disease. *Gastroenterology* **1997**, *112*, 493–500. [CrossRef]

72. Tokino, T.; Matsubara, K. Chromosomal sites for hepatitis B virus integration in human hepatocellular carcinoma. *J. Virol.* **1991**, *65*, 6761–6764.

73. Nagaya, T.; Nakamura, T.; Tokino, T.; Tsurimoto, T.; Imai, M.; Mayumi, T.; Kamino, K.; Yamamura, K.; Matsubara, K. The mode of hepatitis B virus DNA integration in chromosomes of human hepatocellular carcinoma. *Genes Dev.* **1987**, *1*, 773–782. [CrossRef]

74. Imazeki, F.; Omata, M.; Yokosuka, O.; Okuda, K. Integration of hepatitis B virus DNA in hepatocellular carcinoma. *Cancer* **1986**, *58*, 1055–1060. [CrossRef]

75. Wang, M.; Xi, D.; Ning, Q. Virus-induced hepatocellular carcinoma with special emphasis on HBV. *Hepatol. Int.* **2017**, *11*, 171–180. [CrossRef]

76. Meyerson, M.; Counter, C.M.; Eaton, E.N.; Ellisen, L.W.; Steiner, P.; Caddle, S.D.; Ziaugra, L.; Beijersbergen, R.L.; Davidoff, M.J.; Liu, Q.; et al. hEST2, the putative human telomerase catalytic subunit gene, is up-regulated in tumor cells and during immortalization. *Cell* **1997**, *90*, 785–795. [CrossRef]

77. Tokino, T.; Fukushige, S.; Nakamura, T.; Nagaya, T.; Murotsu, T.; Shiga, K.; Aoki, N.; Matsubara, K. Chromosomal translocation and inverted duplication associated with integrated hepatitis B virus in hepatocellular carcinomas. *J. Virol.* **1987**, *61*, 3848–3854.

78. Koike, K.; Shirakata, Y.; Yaginuma, K.; Arii, M.; Takada, S.; Nakamura, I.; Hayashi, Y.; Kawada, M.; Kobayashi, M. Oncogenic potential of hepatitis B virus. *Mol. Biol. Med.* **1989**, *6*, 151–160.

79. Kanda, T.; Yokosuka, O.; Imazeki, F.; Yamada, Y.; Imamura, T.; Fukai, K.; Nagao, K.; Saisho, H. Hepatitis B virus X protein (HBx)-induced apoptosis in HuH-7 cells: Influence of HBV genotype and basal core promoter mutations. *Scand. J. Gastroenterol.* **2004**, *39*, 478–485. [CrossRef]

80. Liu, H.; Shi, W.; Luan, F.; Xu, S.; Yang, F.; Sun, W.; Liu, J.; Ma, C. Hepatitis B virus X protein upregulates transcriptional activation of human telomerase reverse transcriptase. *Virus Genes.* **2010**, *40*, 174–182. [CrossRef]

81. Zou, S.Q.; Qu, Z.L.; Li, Z.F.; Wang, X. Hepatitis B virus X gene induces human telomerase reverse transcriptase mRNA expression in cultured normal human cholangiocytes. *World J. Gastroenterol.* **2004**, *10*, 2259–2262. [CrossRef]

82. Kojima, H.; Kaita, K.D.; Xu, Z.; Ou, J.H.; Gong, Y.; Zhang, M.; Minuk, G.Y. The absence of up-regulation of telomerase activity during regeneration after partial hepatectomy in hepatitis B virus X gene transgenic mice. *J. Hepatol.* **2003**, *39*, 262–268. [CrossRef]

83. Sung, W.K.; Zheng, H.; Li, S.; Chen, R.; Liu, X.; Li, Y.; Lee, N.P.; Lee, W.H.; Ariyaratne, P.N.; Tennakoon, C.; et al. Genome-wide survey of recurrent HBV integration in hepatocellular carcinoma. *Nat. Genet.* **2012**, *44*, 765–769. [CrossRef] [PubMed]

84. Nakano, M.; Kawaguchi, T.; Nakamoto, S.; Kawaguchi, A.; Kanda, T.; Imazeki, F.; Kuromatsu, R.; Sumie, S.; Satani, M.; Yamada, S.; et al. Effect of occult hepatitis B virus infection on the early-onset of hepatocellular carcinoma in patients with hepatitis C virus infection. *Oncol. Rep.* **2013**, *30*, 2049–2055. [CrossRef] [PubMed]

85. Mason, W.S.; Gill, U.S.; Litwin, S.; Zhou, Y.; Peri, S.; Pop, O.; Hong, M.L.; Naik, S.; Quaglia, A.; Bertoletti, A.; et al. HBV DNA Integration and Clonal Hepatocyte Expansion in Chronic Hepatitis B Patients Considered Immune Tolerant. *Gastroenterology* **2016**, *151*, 986–998. [CrossRef] [PubMed]

86. Kennedy, P.T.F.; Litwin, S.; Dolman, G.E.; Bertoletti, A.; Mason, W.S. Immune Tolerant Chronic Hepatitis B: The Unrecognized Risks. *Viruses* **2017**, *9*, 96. [CrossRef] [PubMed]

87. Wu, S.; Kanda, T.; Nakamoto, S.; Jiang, X.; Nakamura, M.; Sasaki, R.; Haga, Y.; Shirasawa, H.; Yokosuka, O. Cooperative effects of hepatitis B virus and TNF may play important roles in the activation of metabolic pathways through the activation of NF-κB. *Int. J. Mol. Med.* **2016**, *38*, 475–481. [CrossRef] [PubMed]

88. Xie, Y. Hepatitis B Virus-Associated Hepatocellular Carcinoma. *Adv. Exp. Med. Biol.* **2017**, *1018*, 11–21. [CrossRef]

89. Jiang, X.; Kanda, T.; Wu, S.; Nakamura, M.; Miyamura, T.; Nakamoto, S.; Banerjee, A.; Yokosuka, O. Regulation of microRNA by hepatitis B virus infection and their possible association with control of innate immunity. *World J. Gastroenterol.* **2014**, *20*, 7197–7206. [CrossRef]

90. Sarkar, N.; Panigrahi, R.; Pal, A.; Biswas, A.; Singh, S.P.; Kar, S.K.; Bandopadhyay, M.; Das, D.; Saha, D.; Kanda, T.; et al. Expression of microRNA-155 correlates positively with the expression of Toll-like receptor 7 and modulates hepatitis B virus via C/EBP-β in hepatocytes. *J. Viral Hepat.* **2015**, *22*, 817–827. [CrossRef] [PubMed]

91. Wang, W.; Bian, H.; Li, F.; Li, X.; Zhang, D.; Sun, S.; Song, S.; Zhu, Q.; Ren, W.; Qin, C.; et al. HBeAg induces the expression of macrophage miR-155 to accelerate liver injury via promoting production of inflammatory cytokines. *Cell. Mol. Life Sci.* **2018**, *75*, 2627–2641. [CrossRef] [PubMed]

92. Wong, Q.W.; Lung, R.W.; Law, P.T.; Lai, P.B.; Chan, K.Y.; To, K.F.; Wong, N. MicroRNA-223 is commonly repressed in hepatocellular carcinoma and potentiates expression of Stathmin1. *Gastroenterology* **2008**, *135*, 257–269. [CrossRef] [PubMed]

93. Zhang, X.; Liu, S.; Hu, T.; Liu, S.; He, Y.; Sun, S. Up-regulated microRNA-143 transcribed by nuclear factor kappa B enhances hepatocarcinoma metastasis by repressing fibronectin expression. *Hepatology* **2009**, *50*, 490–499. [CrossRef] [PubMed]

94. Yang, L.; Ma, Z.; Wang, D.; Zhao, W.; Chen, L.; Wang, G. MicroRNA-602 regulating tumor suppressive gene RASSF1A is overexpressed in hepatitis B virus-infected liver and hepatocellular carcinoma. *Cancer Biol. Ther.* **2010**, *9*, 803–808. [CrossRef] [PubMed]

95. Wang, Y.; Lu, Y.; Toh, S.T.; Sung, W.K.; Tan, P.; Chow, P.; Chung, A.Y.; Jooi, L.L.; Lee, C.G. Lethal-7 is down-regulated by the hepatitis B virus x protein and targets signal transducer and activator of transcription 3. *J. Hepatol.* **2010**, *53*, 57–66. [CrossRef]

96. Huang, J.; Wang, Y.; Guo, Y.; Sun, S. Down-regulated microRNA-152 induces aberrant DNA methylation in hepatitis B virus-related hepatocellular carcinoma by targeting DNA methyltransferase 1. *Hepatology* **2010**, *52*, 60–70. [CrossRef] [PubMed]

97. Li, L.M.; Hu, Z.B.; Zhou, Z.X.; Chen, X.; Liu, F.Y.; Zhang, J.F.; Shen, H.B.; Zhang, C.Y.; Zen, K. Serum microRNA profiles serve as novel biomarkers for HBV infection and diagnosis of HBV-positive hepatocarcinoma. *Cancer Res.* **2010**, *70*, 9798–9807. [CrossRef] [PubMed]

98. Gao, P.; Wong, C.C.; Tung, E.K.; Lee, J.M.; Wong, C.M.; Ng, I.O. Deregulation of microRNA expression occurs early and accumulates in early stages of HBV-associated multistep hepatocarcinogenesis. *J. Hepatol.* **2011**, *54*, 1177–1184. [CrossRef] [PubMed]

99. Jiang, R.; Deng, L.; Zhao, L.; Li, X.; Zhang, F.; Xia, Y.; Gao, Y.; Wang, X.; Sun, B. miR-22 promotes HBV-related hepatocellular carcinoma development in males. *Clin. Cancer Res.* **2011**, *17*, 5593–5603. [CrossRef] [PubMed]

100. Wang, C.M.; Wang, Y.; Fan, C.G.; Xu, F.F.; Sun, W.S.; Liu, Y.G.; Jia, J.H. miR-29c targets TNFAIP3, inhibits cell proliferation and induces apoptosis in hepatitis B virus-related hepatocellular carcinoma. *Biochem. Biophys. Res. Commun.* **2011**, *411*, 586–592. [CrossRef]

101. Qi, P.; Cheng, S.Q.; Wang, H.; Li, N.; Chen, Y.F.; Gao, C.F. Serum microRNAs as biomarkers for hepatocellular carcinoma in Chinese patients with chronic hepatitis B virus infection. *PLoS ONE* **2011**, *6*, e28486. [CrossRef]

102. Cardin, R.; Piciocchi, M.; Sinigaglia, A.; Lavezzo, E.; Bortolami, M.; Kotsafti, A.; Cillo, U.; Zanus, G.; Mescoli, C.; Rugge, M.; et al. Oxidative DNA damage correlates with cell immortalization and mir-92 expression in hepatocellular carcinoma. *BMC Cancer* **2012**, *12*, 177. [CrossRef]

103. Li, L.; Guo, Z.; Wang, J.; Mao, Y.; Gao, Q. Serum miR-18a: A potential marker for hepatitis B virus-related hepatocellular carcinoma screening. *Dig. Dis. Sci.* **2012**, *57*, 2910–2916. [CrossRef] [PubMed]

104. Fu, X.; Tan, D.; Hou, Z.; Hu, Z.; Liu, G. miR-338-3p is down-regulated by hepatitis B virus X and inhibits cell proliferation by targeting the 3′-UTR region of CyclinD1. *Int. J. Mol. Sci.* **2012**, *13*, 8514–8539. [CrossRef] [PubMed]

105. Yang, P.; Li, Q.J.; Feng, Y.; Zhang, Y.; Markowitz, G.J.; Ning, S.; Deng, Y.; Zhao, J.; Jiang, S.; Yuan, Y.; et al. TGF-β-miR-34a-CCL22 signaling-induced Treg cell recruitment promotes venous metastases of HBV-positive hepatocellular carcinoma. *Cancer Cell* **2012**, *22*, 291–303. [CrossRef] [PubMed]

106. Wei, X.; Xiang, T.; Ren, G.; Tan, C.; Liu, R.; Xu, X.; Wu, Z. miR-101 is down-regulated by the hepatitis B virus x protein and induces aberrant DNA methylation by targeting DNA methyltransferase 3A. *Cell. Signal.* **2013**, *25*, 439–446. [CrossRef]

107. Li, C.; Wang, Y.; Wang, S.; Wu, B.; Hao, J.; Fan, H.; Ju, Y.; Ding, Y.; Chen, L.; Chu, X.; et al. Hepatitis B virus mRNA-mediated miR-122 inhibition upregulates PTTG1-binding protein, which promotes hepatocellular carcinoma tumor growth and cell invasion. *J. Virol.* **2013**, *87*, 2193–2205. [CrossRef]

108. Xu, X.; Fan, Z.; Kang, L.; Han, J.; Jiang, C.; Zheng, X.; Zhu, Z.; Jiao, H.; Lin, J.; Jiang, K.; et al. Hepatitis B virus X protein represses miRNA-148a to enhance tumorigenesis. *J. Clin. Investig.* **2013**, *123*, 630–645. [CrossRef]

109. Shi, C.; Xu, X. MicroRNA-22 is down-regulated in hepatitis B virus-related hepatocellular carcinoma. *Biomed. Pharmacother.* **2013**, *67*, 375–380. [CrossRef]

110. Lan, S.H.; Wu, S.Y.; Zuchini, R.; Lin, X.Z.; Su, I.J.; Tsai, T.F.; Lin, Y.J.; Wu, C.T.; Liu, H.S. Autophagy suppresses tumorigenesis of hepatitis B virus-associated hepatocellular carcinoma through degradation of microRNA-224. *Hepatology* **2014**, *59*, 505–517. [CrossRef]

111. Fu, Y.; Wei, X.; Tang, C.; Li, J.; Liu, R.; Shen, A.; Wu, Z. Circulating microRNA-101 as a potential biomarker for hepatitis B virus-related hepatocellular carcinoma. *Oncol. Lett.* **2013**, *6*, 1811–1815. [CrossRef]

112. Li, J.; Shi, W.; Gao, Y.; Yang, B.; Jing, X.; Shan, S.; Wang, Y.; Du, Z. Analysis of microRNA expression profiles in human hepatitis B virus-related hepatocellular carcinoma. *Clin. Lab.* **2013**, *59*, 1009–1015. [CrossRef] [PubMed]

113. Zhang, T.; Zhang, J.; Cui, M.; Liu, F.; You, X.; Du, Y.; Gao, Y.; Zhang, S.; Lu, Z.; Ye, L.; et al. Hepatitis B virus X protein inhibits tumor suppressor miR-205 through inducing hypermethylation of miR-205 promoter to enhance carcinogenesis. *Neoplasia* **2013**, *15*, 1282–1291. [CrossRef]

114. Sheng, Y.; Li, J.; Zou, C.; Wang, S.; Cao, Y.; Zhang, J.; Huang, A.; Tang, H. Downregulation of miR-101-3p by hepatitis B virus promotes proliferation and migration of hepatocellular carcinoma cells by targeting Rab5a. *Arch. Virol.* **2014**, *159*, 2397–2410. [CrossRef] [PubMed]

115. Dang, Y.W.; Zeng, J.; He, R.Q.; Rong, M.H.; Luo, D.Z.; Chen, G. Effects of miR-152 on cell growth inhibition, motility suppression and apoptosis induction in hepatocellular carcinoma cells. *Asian Pac. J. Cancer Prev.* **2014**, *15*, 4969–4976. [CrossRef] [PubMed]

116. Lan, S.H.; Wu, S.Y.; Zuchini, R.; Lin, X.Z.; Su, I.J.; Tsai, T.F.; Lin, Y.J.; Wu, C.T.; Liu, H.S. Autophagy-preferential degradation of MIR224 participates in hepatocellular carcinoma tumorigenesis. *Autophagy* **2014**, *10*, 1687–1689. [CrossRef]

117. Meng, F.L.; Wang, W.; Jia, W.D. Diagnostic and prognostic significance of serum miR-24-3p in HBV-related hepatocellular carcinoma. *Med. Oncol.* **2014**, *31*, 177. [CrossRef]

118. Bandopadhyay, M.; Banerjee, A.; Sarkar, N.; Panigrahi, R.; Datta, S.; Pal, A.; Singh, S.P.; Biswas, A.; Chakrabarti, S.; Chakravarty, R. Tumor suppressor micro RNA miR-145 and onco micro RNAs miR-21 and miR-222 expressions are differentially modulated by hepatitis B virus X protein in malignant hepatocytes. *BMC Cancer* **2014**, *14*, 721. [CrossRef] [PubMed]

119. Xing, T.J.; Jiang, D.F.; Huang, J.X.; Xu, Z.L. Expression and clinical significance of miR-122 and miR-29 in hepatitis B virus-related liver disease. *Genet. Mol. Res.* **2014**, *13*, 7912–7918. [CrossRef]

120. Zhao, Q.; Li, T.; Qi, J.; Liu, J.; Qin, C. The miR-545/374a cluster encoded in the Ftx lncRNA is overexpressed in HBV-related hepatocellular carcinoma and promotes tumorigenesis and tumor progression. *PLoS ONE* **2014**, *9*, e109782. [CrossRef]

121. Gao, H.; Liu, C. miR-429 represses cell proliferation and induces apoptosis in HBV-related HCC. *Biomed. Pharmacother.* **2014**, *68*, 943–949. [CrossRef]

122. Liu, F.Y.; Zhou, S.J.; Deng, Y.L.; Zhang, Z.Y.; Zhang, E.L.; Wu, Z.B.; Huang, Z.Y.; Chen, X.P. MiR-216b is involved in pathogenesis and progression of hepatocellular carcinoma through HBx-miR-216b-IGF2BP2 signaling pathway. *Cell Death Dis.* **2015**, *6*, e1670. [CrossRef]

123. Yu, F.; Lu, Z.; Chen, B.; Dong, P.; Zheng, J. microRNA-150: A promising novel biomarker for hepatitis B virus-related hepatocellular carcinoma. *Diagn. Pathol.* **2015**, *10*, 129. [CrossRef] [PubMed]

124. Cao, Y.; Chen, J.; Wang, D.; Peng, H.; Tan, X.; Xiong, D.; Huang, A.; Tang, H. Upregulated in Hepatitis B virus-associated hepatocellular carcinoma cells, miR-331-3p promotes proliferation of hepatocellular carcinoma cells by targeting ING5. *Oncotarget* **2015**, *6*, 38093–38106. [CrossRef]

125. Gao, F.; Sun, X.; Wang, L.; Tang, S.; Yan, C. Downregulation of MicroRNA-145 Caused by Hepatitis B Virus X Protein Promotes Expression of CUL5 and Contributes to Pathogenesis of Hepatitis B Virus-Associated Hepatocellular Carcinoma. *Cell. Physiol. Biochem.* **2015**, *37*, 1547–1559. [CrossRef] [PubMed]

126. Shao, J.; Cao, J.; Liu, Y.; Mei, H.; Zhang, Y.; Xu, W. MicroRNA-519a promotes proliferation and inhibits apoptosis of hepatocellular carcinoma cells by targeting FOXF2. *FEBS Open Bio* **2015**, *5*, 893–899. [CrossRef] [PubMed]

127. Wang, Y.; Wang, C.M.; Jiang, Z.Z.; Yu, X.J.; Fan, C.G.; Xu, F.F.; Zhang, Q.; Li, L.I.; Li, R.F.; Sun, W.S.; et al. MicroRNA-34c targets TGFB-induced factor homeobox 2, represses cell proliferation and induces apoptosis in hepatitis B virus-related hepatocellular carcinoma. *Oncol. Lett.* **2015**, *10*, 3095–3102. [CrossRef]

128. Ghosh, A.; Ghosh, A.; Datta, S.; Dasgupta, D.; Das, S.; Ray, S.; Gupta, S.; Datta, S.; Chowdhury, A.; Chatterjee, R.; et al. Hepatic miR-126 is a potential plasma biomarker for detection of hepatitis B virus infected hepatocellular carcinoma. *Int. J. Cancer* **2016**, *138*, 2732–2744. [CrossRef] [PubMed]

129. Chen, Y.; Dong, X.; Yu, D.; Wang, X. Serum miR-96 is a promising biomarker for hepatocellular carcinoma in patients with chronic hepatitis B virus infection. *Int. J. Clin. Exp. Med.* **2015**, *8*, 18462–18468. [PubMed]

130. Chen, S.; Chen, H.; Gao, S.; Qiu, S.; Zhou, H.; Yu, M.; Tu, J. Differential expression of plasma microRNA-125b in hepatitis B virus-related liver diseases and diagnostic potential for hepatitis B virus-induced hepatocellular carcinoma. *Hepatol. Res.* **2017**, *47*, 312–320. [CrossRef] [PubMed]

131. Yen, C.S.; Su, Z.R.; Lee, Y.P.; Liu, I.T.; Yen, C.J. miR-106b promotes cancer progression in hepatitis B virus-associated hepatocellular carcinoma. *World J. Gastroenterol.* **2016**, *22*, 5183–5192. [CrossRef]

132. Kong, X.X.; Lv, Y.R.; Shao, L.P.; Nong, X.Y.; Zhang, G.L.; Zhang, Y.; Fan, H.X.; Liu, M.; Li, X.; Tang, H. HBx-induced MiR-1269b in NF-κB dependent manner upregulates cell division cycle 40 homolog (CDC40) to promote proliferation and migration in hepatoma cells. *J. Transl. Med.* **2016**, *14*, 189. [CrossRef]

133. Liu, X.; Zhang, Y.; Wang, P.; Wang, H.; Su, H.; Zhou, X.; Zhang, L. HBX Protein-Induced Downregulation of microRNA-18a is Responsible for Upregulation of Connective Tissue Growth Factor in HBV Infection-Associated Hepatocarcinoma. *Med. Sci. Monit.* **2016**, *22*, 2492–2500. [CrossRef]

134. Qiao, D.D.; Yang, J.; Lei, X.F.; Mi, G.L.; Li, S.L.; Li, K.; Xu, C.Q.; Yang, H.L. Expression of microRNA-122 and microRNA-22 in HBV-related liver cancer and the correlation with clinical features. *Eur. Rev. Med. Pharmacol. Sci.* **2017**, *21*, 742–747. [PubMed]

135. Qin, X.; Chen, J.; Wu, L.; Liu, Z. MiR-30b-5p acts as a tumor suppressor, repressing cell proliferation and cell cycle in human hepatocellular carcinoma. *Biomed. Pharmacother.* **2017**, *89*, 742–750. [CrossRef] [PubMed]

136. Bai, P.S.; Xia, N.; Sun, H.; Kong, Y. Pleiotrophin, a target of miR-384, promotes proliferation, metastasis and lipogenesis in HBV-related hepatocellular carcinoma. *J. Cell. Mol. Med.* **2017**, *21*, 3023–3043. [CrossRef] [PubMed]

137. Li, G.; Zhang, W.; Gong, L.; Huang, X. MicroRNA-125a-5p Inhibits Cell Proliferation and Induces Apoptosis in Hepatitis B Virus-Related Hepatocellular Carcinoma by Downregulation of ErbB3. *Oncol. Res.* **2017**. [CrossRef]

138. Zhao, Q.; Sun, X.; Liu, C.; Li, T.; Cui, J.; Qin, C. Expression of the microRNA-143/145 cluster is decreased in hepatitis B virus-associated hepatocellular carcinoma and may serve as a biomarker for tumorigenesis in patients with chronic hepatitis B. *Oncol. Lett.* **2018**, *15*, 6115–6122. [CrossRef] [PubMed]

139. Quoc, N.B.; Phuong, N.D.N.; Ngan, T.K.; Linh, N.T.M.; Cuong, P.H.; Chau, N.N.B. Expression of Plasma hsa-miR122 in HBV-Related Hepatocellular Carcinoma (HCC) in Vietnamese Patients. *Microrna* **2018**, *7*, 92–99. [CrossRef] [PubMed]

140. Jones, K.R.; Nabinger, S.C.; Lee, S.; Sahu, S.S.; Althouse, S.; Saxena, R.; Johnson, M.S.; Chalasani, N.; Gawrieh, S.; Kota, J. Lower expression of tumor microRNA-26a is associated with higher recurrence in patients with hepatocellular carcinoma undergoing surgical treatment. *J. Surg. Oncol.* **2018**, *118*, 431–439. [CrossRef] [PubMed]

141. Yang, L.; Guo, Y.; Liu, X.; Wang, T.; Tong, X.; Lei, K.; Wang, J.; Huang, D.; Xu, Q. The tumor suppressive miR-302c-3p inhibits migration and invasion of hepatocellular carcinoma cells by targeting TRAF4. *J. Cancer* **2018**, *9*, 2693–2701. [CrossRef]

142. Chen, Y.; Zhao, Z.X.; Huang, F.; Yuan, X.W.; Deng, L.; Tang, D. MicroRNA-1271 functions as a potential tumor suppressor in hepatitis B virus-associated hepatocellular carcinoma through the AMPK signaling pathway by binding to CCNA1. *J. Cell. Physiol.* **2018**. [CrossRef]

143. Qiu, L.; Wang, T.; Xu, X.; Wu, Y.; Tang, Q.; Chen, K. Long Non-Coding RNAs in Hepatitis B Virus-Related Hepatocellular Carcinoma: Regulation, Functions, and Underlying Mechanisms. *Int. J. Mol. Sci.* **2017**, *18*, 2505. [CrossRef]

144. Cui, S.; Qian, Z.; Chen, Y.; Li, L.; Li, P.; Ding, H. Screening of up- and downregulation of circRNAs in HBV-related hepatocellular carcinoma by microarray. *Oncol. Lett.* **2018**, *15*, 423–432. [CrossRef] [PubMed]

145. Wang, S.; Cui, S.; Zhao, W.; Qian, Z.; Liu, H.; Chen, Y.; Lv, F.; Ding, H.G. Screening and bioinformatics analysis of circular RNA expression profiles in hepatitis B-related hepatocellular carcinoma. *Cancer Biomark.* **2018**, *22*, 631–640. [CrossRef]

146. Matsuda, Y.; Ichida, T.; Genda, T.; Yamagiwa, S.; Aoyagi, Y.; Asakura, H. Loss of p16 contributes to p27 sequestration by cyclin D(1)-cyclin-dependent kinase 4 complexes and poor prognosis in hepatocellular carcinoma. *Clin. Cancer Res.* **2003**, *9*, 3389–3396. [PubMed]

147. Pezzuto, F.; Buonaguro, L.; Buonaguro, F.M.; Tornesello, M.L. The Role of Circulating Free DNA and MicroRNA in Non-Invasive Diagnosis of HBV- and HCV-Related Hepatocellular Carcinoma. *Int. J. Mol. Sci.* **2018**, *19*, 1007. [CrossRef]

148. Tian, Y.; Ou, J.H. Genetic and epigenetic alterations in hepatitis B virus-associated hepatocellular carcinoma. *Virol. Sin.* **2015**, *30*, 85–91. [CrossRef] [PubMed]

149. Wu, S.; Kanda, T.; Imazeki, F.; Arai, M.; Yonemitsu, Y.; Nakamoto, S.; Fujiwara, K.; Fukai, K.; Nomura, F.; Yokosuka, O. Hepatitis B virus e antigen downregulates cytokine production in human hepatoma cell lines. *Viral Immunol.* **2010**, *23*, 467–476. [CrossRef]

150. Hadziyannis, E.; Laras, A. Viral Biomarkers in Chronic HBeAg Negative HBV Infection. *Genes* **2018**, *9*, 469. [CrossRef] [PubMed]

151. Hildt, E.; Hofschneider, P.H. The PreS2 activators of the hepatitis B virus: Activators of tumour promoter pathways. *Recent Results Cancer Res.* **1998**, *154*, 315–329. [PubMed]

152. Liu, P.; Zhang, H.; Liang, X.; Ma, H.; Luan, F.; Wang, B.; Bai, F.; Gao, L.; Ma, C. HBV preS2 promotes the expression of TAZ via miRNA-338-3p to enhance the tumorigenesis of hepatocellular carcinoma. *Oncotarget* **2015**, *6*, 29048–29059. [CrossRef] [PubMed]

153. Wang, M.L.; Wu, D.B.; Tao, Y.C.; Chen, L.L.; Liu, C.P.; Chen, E.Q.; Tang, H. The truncated mutant HBsAg expression increases the tumorigenesis of hepatitis B virus by regulating TGF-β/Smad signaling pathway. *Virol. J.* **2018**, *15*, 61. [CrossRef] [PubMed]

154. Song, J.; Zhang, X.; Ge, Q.; Yuan, C.; Chu, L.; Liang, H.F.; Liao, Z.; Liu, Q.; Zhang, Z.; Zhang, B. CRISPR/Cas9-mediated knockout of HBsAg inhibits proliferation and tumorigenicity of HBV-positive hepatocellular carcinoma cells. *J. Cell. Biochem.* **2018**, *119*, 8419–8431. [CrossRef] [PubMed]

155. Kanda, T.; Wu, S.; Sasaki, R.; Nakamura, M.; Haga, Y.; Jiang, X.; Nakamoto, S.; Yokosuka, O. HBV Core Protein Enhances Cytokine Production. *Diseases* **2015**, *3*, 213–220. [CrossRef] [PubMed]

156. Wu, S.; Kanda, T.; Imazeki, F.; Nakamoto, S.; Tanaka, T.; Arai, M.; Roger, T.; Shirasawa, H.; Nomura, F.; Yokosuka, O. Hepatitis B virus e antigen physically associates with receptor-interacting serine/threonine protein kinase 2 and regulates IL-6 gene expression. *J. Infect. Dis.* **2012**, *206*, 415–420. [CrossRef] [PubMed]

157. Chen, M.T.; Billaud, J.N.; Sällberg, M.; Guidotti, L.G.; Chisari, F.V.; Jones, J.; Hughes, J.; Milich, D.R. A function of the hepatitis B virus precore protein is to regulate the immune response to the core antigen. *Proc. Natl. Acad. Sci. USA* **2004**, *101*, 14913–14918. [CrossRef] [PubMed]

158. Chen, M.; Sällberg, M.; Hughes, J.; Jones, J.; Guidotti, L.G.; Chisari, F.V.; Billaud, J.N.; Milich, D.R. Immune tolerance split between hepatitis B virus precore and core proteins. *J. Virol.* **2005**, *79*, 3016–3027. [CrossRef]

159. Yaginuma, K.; Kobayashi, H.; Kobayashi, M.; Morishima, T.; Matsuyama, K.; Koike, K. Multiple integration site of hepatitis B virus DNA in hepatocellular carcinoma and chronic active hepatitis tissues from children. *J. Virol.* **1987**, *61*, 1808–1813.

160. Zhou, Y.Z.; Butel, J.S.; Li, P.J.; Finegold, M.J.; Melnick, J.L. Integrated state of subgenomic fragments of hepatitis B virus DNA in hepatocellular carcinoma from mainland China. *J. Natl. Cancer Inst.* **1987**, *79*, 223–231.

161. Kim, C.M.; Koike, K.; Saito, I.; Miyamura, T.; Jay, G. HBx gene of hepatitis B virus induces liver cancer in transgenic mice. *Nature* **1991**, *351*, 317–320. [CrossRef]

162. Wollersheim, M.; Debelka, U.; Hofschneider, P.H. A transactivating function encoded in the hepatitis B virus X gene is conserved in the integrated state. *Oncogene* **1988**, *3*, 545–552. [PubMed]

163. Kekulé, A.S.; Lauer, U.; Weiss, L.; Luber, B.; Hofschneider, P.H. Hepatitis B virus transactivator HBx uses a tumour promoter signalling pathway. *Nature* **1993**, *361*, 742–745. [CrossRef]

164. Choi, B.H.; Choi, M.; Jeon, H.Y.; Rho, H.M. Hepatitis B viral X protein overcomes inhibition of E2F1 activity by pRb on the human Rb gene promoter. *DNA Cell Biol.* **2001**, *20*, 75–80. [CrossRef]

165. Staib, F.; Hussain, S.P.; Hofseth, L.J.; Wang, X.W.; Harris, C.C. TP53 and liver carcinogenesis. *Hum. Mutat.* **2003**, *21*, 201–216. [CrossRef] [PubMed]

166. Cha, M.Y.; Kim, C.M.; Park, Y.M.; Ryu, W.S. Hepatitis B virus X protein is essential for the activation of Wnt/beta-catenin signaling in hepatoma cells. *Hepatology* **2004**, *39*, 1683–1693. [CrossRef]

167. Longato, L.; de la Monte, S.; Kuzushita, N.; Horimoto, M.; Rogers, A.B.; Slagle, B.L.; Wands, J.R. Overexpression of insulin receptor substrate-1 and hepatitis Bx genes causes premalignant alterations in the liver. *Hepatology* **2009**, *49*, 1935–1943. [CrossRef] [PubMed]

168. Keng, V.W.; Tschida, B.R.; Bell, J.B.; Largaespada, D.A. Modeling hepatitis B virus X-induced hepatocellular carcinoma in mice with the Sleeping Beauty transposon system. *Hepatology* **2011**, *53*, 781–790. [CrossRef]

169. Wang, C.; Yang, W.; Yan, H.X.; Luo, T.; Zhang, J.; Tang, L.; Wu, F.Q.; Zhang, H.L.; Yu, L.X.; Zheng, L.Y.; et al. Hepatitis B virus X (HBx) induces tumorigenicity of hepatic progenitor cells in 3,5-diethoxycarbonyl-1,4-dihydrocollidine-treated HBx transgenic mice. *Hepatology* **2012**, *55*, 108–120. [CrossRef]

170. Von Olshausen, G.; Quasdorff, M.; Bester, R.; Arzberger, S.; Ko, C.; van de Klundert, M.; Zhang, K.; Odenthal, M.; Ringelhan, M.; Niessen, C.M.; et al. Hepatitis B virus promotes β-catenin-signalling and disassembly of adherens junctions in a Src kinase dependent fashion. *Oncotarget* **2018**, *9*, 33947–33960. [CrossRef]

171. Singh, A.K.; Swarnalatha, M.; Kumar, V. c-ETS1 facilitates G1/S-phase transition by up-regulating cyclin E and CDK2 genes and cooperates with hepatitis B virus X protein for their deregulation. *J. Biol. Chem.* **2011**, *286*, 21961–21970. [CrossRef]

172. Luo, L.; Chen, S.; Gong, Q.; Luo, N.; Lei, Y.; Guo, J.; He, S. Hepatitis B virus X protein modulates remodelling of minichromosomes related to hepatitis B virus replication in HepG2 cells. *Int. J. Mol. Med.* **2013**, *31*, 197–204. [CrossRef] [PubMed]

173. Saeed, U.; Kim, J.; Piracha, Z.Z.; Kwon, H.; Jung, J.; Chwae, Y.J.; Park, S.; Shin, H.J.; Kim, K. Parvulin 14 and parvulin 17 bind to HBx and cccDNA and upregulate HBV replication from cccDNA to virion in an HBx-dependent manner. *J. Virol.* **2019**, *93*, e01840-18. [CrossRef] [PubMed]

174. Swarnalatha, M.; Singh, A.K.; Kumar, V. Promoter occupancy of MLL1 histone methyltransferase seems to specify the proliferative and apoptotic functions of E2F1 in a tumour microenvironment. *J. Cell Sci.* **2013**, *126*, 4636–4646. [CrossRef] [PubMed]

175. Huang, Y.H.; Tseng, Y.H.; Lin, W.R.; Hung, G.; Chen, T.C.; Wang, T.H.; Lee, W.C.; Yeh, C.T. HBV polymerase overexpression due to large core gene deletion enhances hepatoma cell growth by binding inhibition of microRNA-100. *Oncotarget* **2016**, *7*, 9448–9461. [CrossRef] [PubMed]

176. Chung, H.J.; Chen, X.; Yu, Y.; Lee, H.K.; Song, C.H.; Choe, H.; Lee, S.; Kim, H.J.; Hong, S.T. A critical role of hepatitis B virus polymerase in cirrhosis, hepatocellular carcinoma, and steatosis. *FEBS Open Bio* **2017**, *8*, 130–145. [CrossRef]

177. Kakumu, S.; Okumura, A.; Ishikawa, T.; Yano, M.; Enomoto, A.; Nishimura, H.; Yoshioka, K.; Yoshika, Y. Serum levels of IL-10, IL-15 and soluble tumour necrosis factor-alpha (TNF-alpha) receptors in type C chronic liver disease. *Clin. Exp. Immunol.* **1997**, *109*, 458–463. [CrossRef]

178. Kanda, T.; Yokosuka, O.; Omata, M. Hepatitis C virus and hepatocellular carcinoma. *Biology* **2013**, *2*, 304–316. [CrossRef]

179. Braconi, C.; Valeri, N.; Gasparini, P.; Huang, N.; Taccioli, C.; Nuovo, G.; Suzuki, T.; Croce, C.M.; Patel, T. Hepatitis C virus proteins modulate microRNA expression and chemosensitivity in malignant hepatocytes. *Clin. Cancer Res.* **2010**, *16*, 957–966. [CrossRef]

180. Zhang, Y.; Wei, W.; Cheng, N.; Wang, K.; Li, B.; Jiang, X.; Sun, S. Hepatitis C virus-induced up-regulation of microRNA-155 promotes hepatocarcinogenesis by activating Wnt signaling. *Hepatology* **2012**, *56*, 1631–1640. [CrossRef]

181. Hsu, S.H.; Wang, B.; Kota, J.; Yu, J.; Costinean, S.; Kutay, H.; Yu, L.; Bai, S.; La Perle, K.; Chivukula, R.R.; et al. Essential metabolic, anti-inflammatory, and anti-tumorigenic functions of miR-122 in liver. *J. Clin. Investig.* **2012**, *122*, 2871–2883. [CrossRef]

182. Zhao, L.; Li, F.; Taylor, E.W. Can tobacco use promote HCV-induced miR-122 hijacking and hepatocarcinogenesis? *Med. Hypotheses* **2013**, *80*, 131–133. [CrossRef] [PubMed]

183. Salvi, A.; Abeni, E.; Portolani, N.; Barlati, S.; De Petro, G. Human hepatocellular carcinoma cell-specific miRNAs reveal the differential expression of miR-24 and miR-27a in cirrhotic/non-cirrhotic HCC. *Int. J. Oncol.* **2013**, *42*, 391–402. [CrossRef] [PubMed]

184. Elfimova, N.; Sievers, E.; Eischeid, H.; Kwiecinski, M.; Noetel, A.; Hunt, H.; Becker, D.; Frommolt, P.; Quasdorff, M.; Steffen, H.M.; et al. Control of mitogenic and motogenic pathways by miR-198, diminishing hepatoma cell growth and migration. *Biochim. Biophys. Acta* **2013**, *1833*, 1190–1198. [CrossRef] [PubMed]

185. Thomas, M.; Deiters, A. MicroRNA miR-122 as a therapeutic target for oligonucleotides and small molecules. *Curr Med Chem.* **2013**, *20*, 3629–3640. [CrossRef] [PubMed]

186. Huang, S.; Xie, Y.; Yang, P.; Chen, P.; Zhang, L. HCV core protein-induced down-regulation of microRNA-152 promoted aberrant proliferation by regulating Wnt1 in HepG2 cells. *PLoS ONE* **2014**, *9*, e81730. [CrossRef] [PubMed]

187. Mukherjee, A.; Shrivastava, S.; Bhanja Chowdhury, J.; Ray, R.; Ray, R.B. Transcriptional suppression of miR-181c by hepatitis C virus enhances homeobox A1 expression. *J. Virol.* **2014**, *88*, 7929–7940. [CrossRef]

188. Xu, G.; Yang, F.; Ding, C.L.; Wang, J.; Zhao, P.; Wang, W.; Ren, H. MiR-221 accentuates IFN's anti-HCV effect by downregulating SOCS1 and SOCS3. *Virology* **2014**, *462–463*, 343–350. [CrossRef]

189. Pan, L.; Ren, F.; Rong, M.; Dang, Y.; Luo, Y.; Luo, D.; Chen, G. Correlation between down-expression of miR-431 and clinicopathological significance in HCC tissues. *Clin. Transl. Oncol.* **2015**, *17*, 557–563. [CrossRef]

190. El-Abd, N.E.; Fawzy, N.A.; El-Sheikh, S.M.; Soliman, M.E. Circulating miRNA-122, miRNA-199a, and miRNA-16 as Biomarkers for Early Detection of Hepatocellular Carcinoma in Egyptian Patients with Chronic Hepatitis C Virus Infection. *Mol. Diagn. Ther.* **2015**, *19*, 213–220. [CrossRef]

191. Devhare, P.B.; Steele, R.; Di Bisceglie, A.M.; Kaplan, D.E.; Ray, R.B. Differential Expression of MicroRNAs in Hepatitis C Virus-Mediated Liver Disease Between African Americans and Caucasians: Implications for Racial Health Disparities. *Gene Expr.* **2017**, *17*, 89–98. [CrossRef]

192. Shiu, T.Y.; Shih, Y.L.; Feng, A.C.; Lin, H.H.; Huang, S.M.; Huang, T.Y.; Hsieh, C.B.; Chang, W.K.; Hsieh, T.Y. HCV core inhibits hepatocellular carcinoma cell replicative senescence through downregulating microRNA-138 expression. *J. Mol. Med.* **2017**, *95*, 629–639. [CrossRef]

193. Shaker, O.; Alhelf, M.; Morcos, G.; Elsharkawy, A. miRNA-101-1 and miRNA-221 expressions and their polymorphisms as biomarkers for early diagnosis of hepatocellular carcinoma. *Infect. Genet. Evol.* **2017**, *51*, 173–181. [CrossRef] [PubMed]

194. Shehata, R.H.; Abdelmoneim, S.S.; Osman, O.A.; Hasanain, A.F.; Osama, A.; Abdelmoneim, S.S.; Toraih, E.A. Deregulation of miR-34a and Its Chaperon Hsp70 in Hepatitis C virus-Induced Liver Cirrhosis and Hepatocellular Carcinoma Patients. *Asian Pac. J. Cancer Prev.* **2017**, *18*, 2395–2401. [PubMed]

195. Sur, S.; Sasaki, R.; Devhare, P.; Steele, R.; Ray, R.; Ray, R.B. Association between MicroRNA-373 and Long Noncoding RNA NORAD in Hepatitis C Virus-Infected Hepatocytes Impairs Wee1 Expression for Growth Promotion. *J. Virol.* **2018**, *92*, e01215-18. [CrossRef] [PubMed]

196. Rashad, N.M.; El-Shal, A.S.; Shalaby, S.M.; Mohamed, S.Y. Serum miRNA-27a and miRNA-18b as potential predictive biomarkers of hepatitis C virus-associated hepatocellular carcinoma. *Mol. Cell. Biochem.* **2018**, *447*, 125–136. [CrossRef]

197. Hou, W.; Bonkovsky, H.L. Non-coding RNAs in hepatitis C-induced hepatocellular carcinoma: Dysregulation and implications for early detection, diagnosis and therapy. *World J. Gastroenterol.* **2013**, *19*, 7836–7845. [CrossRef] [PubMed]

198. Zhang, H.; Zhu, C.; Zhao, Y.; Li, M.; Wu, L.; Yang, X.; Wan, X.; Wang, A.; Zhang, M.Q.; Sang, X.; et al. Long non-coding RNA expression profiles of hepatitis C virus-related dysplasia and hepatocellular carcinoma. *Oncotarget* **2015**, *6*, 43770–43778. [CrossRef] [PubMed]

199. Kamel, M.M.; Matboli, M.; Sallam, M.; Montasser, I.F.; Saad, A.S.; El-Tawdi, A.H.F. Investigation of long noncoding RNAs expression profile as potential serum biomarkers in patients with hepatocellular carcinoma. *Transl. Res.* **2016**, *168*, 134–145. [CrossRef] [PubMed]

200. Fu, N.; Niu, X.; Wang, Y.; Du, H.; Wang, B.; Du, J.; Li, Y.; Wang, R.; Zhang, Y.; Zhao, S.; et al. Role of LncRNA-activated by transforming growth factor beta in the progression of hepatitis C virus-related liver fibrosis. *Discov. Med.* **2016**, *22*, 29–42.

201. Kanda, T.; Tada, M.; Imazeki, F.; Yokosuka, O.; Nagao, K.; Saisho, H. 5-aza-2'-deoxycytidine sensitizes hepatoma and pancreatic cancer cell lines. *Oncol. Rep.* **2005**, *14*, 975–979. [CrossRef]

202. Ray, R.B.; Lagging, L.M.; Meyer, K.; Ray, R. Hepatitis C virus core protein cooperates with ras and transforms primary rat embryo fibroblasts to tumorigenic phenotype. *J. Virol.* **1996**, *70*, 4438–4443. [PubMed]

203. Moriya, K.; Fujie, H.; Shintani, Y.; Yotsuyanagi, H.; Tsutsumi, T.; Ishibashi, K.; Matsuura, Y.; Kimura, S.; Miyamura, T.; Koike, K. The core protein of hepatitis C virus induces hepatocellular carcinoma in transgenic mice. *Nat. Med.* **1998**, *4*, 1065–1067. [CrossRef]

204. Kanda, T.; Steele, R.; Ray, R.; Ray, R.B. Hepatitis C virus core protein augments androgen receptor-mediated signaling. *J. Virol.* **2008**, *82*, 11066–11072. [CrossRef] [PubMed]

205. Ghosh, A.K.; Majumder, M.; Steele, R.; Meyer, K.; Ray, R.; Ray, R.B. Hepatitis C virus NS5A protein protects against TNF-alpha mediated apoptotic cell death. *Virus Res.* **2000**, *67*, 173–178. [CrossRef]

206. Majumder, M.; Ghosh, A.K.; Steele, R.; Ray, R.; Ray, R.B. Hepatitis C virus NS5A physically associates with p53 and regulates p21/waf1 gene expression in a p53-dependent manner. *J. Virol.* **2001**, *75*, 1401–1407. [CrossRef] [PubMed]

207. Imazeki, F.; Yokosuka, O.; Fukai, K.; Kanda, T.; Kojima, H.; Saisho, H. Prevalence of diabetes mellitus and insulin resistance in patients with chronic hepatitis C: Comparison with hepatitis B virus-infected and hepatitis C virus-cleared patients. *Liver Int.* **2008**, *28*, 355–362. [CrossRef] [PubMed]

208. Neuschwander-Tetri, B.A.; Clark, J.M.; Bass, N.M.; Van Natta, M.L.; Unalp-Arida, A.; Tonascia, J.; Zein, C.O.; Brunt, E.M.; Kleiner, D.E.; McCullough, A.J.; et al. NASH Clinical Research Network. Clinical, laboratory and histological associations in adults with nonalcoholic fatty liver disease. *Hepatology* **2010**, *52*, 913–924. [CrossRef] [PubMed]

209. Haga, Y.; Kanda, T.; Sasaki, R.; Nakamura, M.; Nakamoto, S.; Yokosuka, O. Nonalcoholic fatty liver disease and hepatic cirrhosis: Comparison with viral hepatitis-associated steatosis. *World J. Gastroenterol.* **2015**, *21*, 12989–12995. [CrossRef] [PubMed]

210. Kawaguchi, T.; Yoshida, T.; Harada, M.; Hisamoto, T.; Nagao, Y.; Ide, T.; Taniguchi, E.; Kumemura, H.; Hanada, S.; Maeyama, M.; et al. Hepatitis C virus down-regulates insulin receptor substrates 1 and 2 through up-regulation of suppressor of cytokine signaling 3. *Am. J. Pathol.* **2004**, *165*, 1499–1508. [CrossRef]

211. Miyamoto, H.; Moriishi, K.; Moriya, K.; Murata, S.; Tanaka, K.; Suzuki, T.; Miyamura, T.; Koike, K.; Matsuura, Y. Involvement of the PA28gamma-dependent pathway in insulin resistance induced by hepatitis C virus core protein. *J. Virol.* **2007**, *81*, 1727–1735. [CrossRef]

212. Banerjee, S.; Saito, K.; Ait-Goughoulte, M.; Meyer, K.; Ray, R.B.; Ray, R. Hepatitis C virus core protein upregulates serine phosphorylation of insulin receptor substrate-1 and impairs the downstream akt/protein kinase B signaling pathway for insulin resistance. *J. Virol.* **2008**, *82*, 2606–2612. [CrossRef] [PubMed]

213. Banerjee, A.; Meyer, K.; Mazumdar, B.; Ray, R.B.; Ray, R. Hepatitis C virus differentially modulates activation of forkhead transcription factors and insulin-induced metabolic gene expression. *J. Virol.* **2010**, *84*, 5936–5946. [CrossRef] [PubMed]

214. Bose, S.K.; Shrivastava, S.; Meyer, K.; Ray, R.B.; Ray, R. Hepatitis C virus activates the mTOR/S6K1 signaling pathway in inhibiting IRS-1 function for insulin resistance. *J. Virol.* **2012**, *86*, 6315–6322. [CrossRef] [PubMed]

215. Tanaka, S.; Wands, J.R. Insulin receptor substrate 1 overexpression in human hepatocellular carcinoma cells prevents transforming growth factor beta1-induced apoptosis. *Cancer Res.* **1996**, *56*, 3391–3394. [PubMed]

216. Qadri, I.; Choudhury, M.; Rahman, S.M.; Knotts, T.A.; Janssen, R.C.; Schaack, J.; Iwahashi, M.; Puljak, L.; Simon, F.R.; Kilic, G.; et al. Increased phosphoenolpyruvate carboxykinase gene expression and steatosis during hepatitis C virus subgenome replication: Role of nonstructural component 5A and CCAAT/enhancer-binding protein β. *J. Biol. Chem.* **2012**, *287*, 37340–37351. [CrossRef]

217. Parvaiz, F.; Manzoor, S.; Iqbal, J.; McRae, S.; Javed, F.; Ahmed, Q.L.; Waris, G. Hepatitis C virus nonstructural protein 5A favors upregulation of gluconeogenic and lipogenic gene expression leading towards insulin resistance: A metabolic syndrome. *Arch. Virol.* **2014**, *159*, 1017–1025. [CrossRef] [PubMed]

218. Parvaiz, F.; Manzoor, S.; Iqbal, J.; Sarkar-Dutta, M.; Imran, M.; Waris, G. Hepatitis C virus NS5A promotes insulin resistance through IRS-1 serine phosphorylation and increased gluconeogenesis. *World J. Gastroenterol.* **2015**, *21*, 12361–12369. [CrossRef]

219. Kwon, Y.C.; Bose, S.K.; Steele, R.; Meyer, K.; Di Bisceglie, A.M.; Ray, R.B.; Ray, R. Promotion of Cancer Stem-Like Cell Properties in Hepatitis C Virus-Infected Hepatocytes. *J. Virol.* **2015**, *89*, 11549–11556. [CrossRef]

220. Kwon, Y.C.; Sasaki, R.; Meyer, K.; Ray, R. Hepatitis C Virus Core Protein Modulates Endoglin (CD105) Signaling Pathway for Liver Pathogenesis. *J. Virol.* **2017**, *91*, e01235-17. [CrossRef]

221. Uthaya Kumar, D.B.; Chen, C.L.; Liu, J.C.; Feldman, D.E.; Sher, L.S.; French, S.; DiNorcia, J.; French, S.W.; Naini, B.V.; Junrungsee, S.; et al. TLR4 Signaling via NANOG Cooperates with STAT3 to Activate Twist1 and Promote Formation of Tumor-Initiating Stem-Like Cells in Livers of Mice. *Gastroenterology* **2016**, *150*, 707–719. [CrossRef]

222. Kudo, M. Systemic Therapy for Hepatocellular Carcinoma: Latest Advances. *Cancers* **2018**, *10*, 412. [CrossRef] [PubMed]

223. Okazaki, T.; Honjo, T. PD-1 and PD-1 ligands: From discovery to clinical application. *Int. Immunol.* **2007**, *19*, 813–824. [CrossRef] [PubMed]

224. Iwai, Y.; Ishida, M.; Tanaka, Y.; Okazaki, T.; Honjo, T.; Minato, N. Involvement of PD-L1 on tumor cells in the escape from host immune system and tumor immunotherapy by PD-L1 blockade. *Proc. Natl. Acad. Sci. USA* **2002**, *99*, 12293–12297. [CrossRef] [PubMed]

225. Leach, D.R.; Krummel, M.F.; Allison, J.P. Enhancement of antitumor immunity by CTLA-4 blockade. *Science* **1996**, *271*, 1734–1736. [CrossRef] [PubMed]

International Journal of
Molecular Sciences

MDPI

Review

Recent Insight into the Role of Fibrosis in Nonalcoholic Steatohepatitis-Related Hepatocellular Carcinoma

Antonio Sircana [1], Elena Paschetta [2], Francesca Saba [3], Federica Molinaro [2] and Giovanni Musso [2,*]

[1] Department of Cardiology, Azienda Ospedaliero Universitaria, 07100 Sassari, Italy; ant.sircana@gmail.com
[2] HUMANITAS Gradenigo, University of Turin, 10132 Turin, Italy; elena.paschetta@alice.it (E.P.); fede.molinaro@gmail.com (F.M.)
[3] Department of Medical Sciences, Cittàdella Salute, University of Turin, 10126 Turin, Italy; francescasaba85@yahoo.it
* Correspondence: giovanni_musso@yahoo.it

Received: 1 February 2019; Accepted: 23 March 2019; Published: 9 April 2019

Abstract: Hepatocellular carcinoma (HCC) is one of the most widespread tumors in the world and its prognosis is poor because of lack of effective treatments. Epidemiological studies show that non-alcoholic steatohepatitis (NASH) and advanced fibrosis represent a relevant risk factors to the HCC development. However little is known of pathophysiological mechanisms linking liver fibrogenesis to HCC in NASH. Recent advances in scientific research allowed to discover some mechanisms that may represent potential therapeutic targets. These include the integrin signaling, hepatic stellate cells (HSCs) activation, Hedgehog signaling and alteration of immune system. In the near future, knowledge of fibrosis-dependent carcinogenic mechanisms, will help optimize antifibrotic therapies as an approach to prevent and treat HCC in patients with NASH and advanced fibrosis.

Keywords: hepatocellular carcinoma; non-alcoholic steatohepatitis; fibrosis; hepatic stellate cells; extracellular matrix; carcinogenesis

1. Introduction

As declared by the WHO (World Health Organization) Global Hepatitis Report [1], hepatocellular carcinoma (HCC) is the fifth most common cancer in the world and the second most common cause of death related to cancer in the last years. The main risk factors are chronic HBV and HCV infection, NAFLD/NASH, alcoholic hepatitis, and any disease leading to cirrhosis.

Non-alcoholic fatty liver disease (NAFLD), with global prevalence of 25% [2], is the most common cause of chronic liver disease. Its clinical spectrum ranges from simple hepatic steatosis (in the absence of secondary causes) to non-alcoholic steatohepatitis (NASH), a more aggressive form with inflammation, hepatocyte injury and varying degrees of fibrosis [3,4].

NASH prevalence has increased exponentially over the years and probably in the next decades it will be the leading cause of liver transplantation in the western industrialized countries. Epidemiological data suggest that HCC attributed to viral infection is in decline, while cases of cryptogenic or NASH-related HCC are significantly increased [5]. Across studies, the risk of HCC is not uniform among patients with NASH, and it ranges between 2.4% and 12.8%. If on the one hand it has been shown that patients with NASH can develop HCC in the absence of fibrosis or cirrhosis [6,7], on the other, it is shown that fibrosis plays a crucial role in causing HCC and its presence is correlated with poor prognosis [8,9]. A metanalisis of 17 cohort studies found a greater risk of HCC in NASH-cirrhosis cohorts than in NAFLD or NASH without fibrosis/cirrhosis [10]. HCC development is considered as result of different environmental risk factors that engage distinct genetic, epigenetic, and chromosomal alterations. It originates from chronic liver injury through a complex multistep process that involves several pathogenic mechanisms that contribute to carcinogenesis [11]. In humans, it is very complicated to identify the different molecular mechanisms underlying the pathogenesis of NASH and to understand how this evolves towards HCC. To overcome this problem, reproducible and representative preclinical models that are susceptible to genetic and functional analysis are used. Many NASH mouse models have been described, but some of them do not closely reflect human disease, thus hindering the efforts to definitively connect the various pathophysiological processes [12].

In this review we will discuss the latest findings on the pathophysiological mechanisms linking liver fibrogenesis to HCC in NASH and their potential therapeutic targets.

2. Pathophysiological Mechanisms of NAFLD Progression

NASH differs from simple steatosis by the presence of hepatocyte death, inflammation and various degrees of fibrosis. Cell death and inflammation play a leading role in disease progression, through HSCs' activation and the subsequent fibrosis. Lately, scientists put forward the idea ("the multiple parallel hits hypothesis") that progression from simple steatosis to NASH is the result of several disorders acting in parallel, including genetic predisposition, altered lipid metabolism, lipotoxicity, oxidative and endoplasmic reticulum stress, mitochondrial dysfunction, abnormal production of cytokines and adipokines, gut dysbiosis, and translocation of gut-derived LPS [13]. On this theory, hepatic inflammation constitute the "primum movens" of fibrosis progression in NASH. Genetic susceptibility and poor eating habits predispose to the development of insulin resistance and hepatic steatosis. In this context, lipotoxic metabolites of saturated fatty acids (SFA) can cause lipotoxicity, process that leads to cellular damage through excessive oxidative stress [14,15]. Damaged hepatocytes release DAMPs (damage-endogenous-associated molecular patterns) that activate pro-inflammatory signaling pathways via toll-like receptors (TLRs). Subsequent activation of Kupffer cells (KC) and inflammasome promote the massive release of pro-inflammatory, pro-fibrogenic cytokines and ligands. HSCs are then stimulated to produce high amount of extra-cellular matrix leading to progressive fibrosis [15]. KCs activation favors a pro-inflammatory microenvironment that triggers an adaptive immune response Th17-mediated. Moreover, chronic portal inflammatory infiltrate boosts a ductular reaction (DR) and hepatic progenitor cells (HPC) recruitment. All of these factors encourage progressive fibrosis that constitutes an imbalance between tissue injury and repair secondary to influence of vary inflammatory cells [16].

Acute inflammation constitutes a useful reaction to achieve tissues recovery by promoting regeneration. Conversely, chronic inflammation is maladaptive and provides a fertile soil to the development of liver fibrosis and HCC. Chronic injury triggers secretion of significant amounts of proinflammatory molecules including IL-1, IL-6, TNF-α, lymphotoxin-β that facilitate HCC development [17]. Activation of inflammatory signaling pathways and enhanced secretion of inflammatory molecules increase release of reactive oxygen species (ROS) by hepatocites. ROS can increase the tumor risk by mechanisms including DNA damage [11] and inhibition on immunosurveillance [18].

Although there is an established relationship between inflammation and fibrogenesis, it looks like that some inflammatory pathways selectively impact the tumorigenesis without affecting fibrosis. For instance, lymphotoxin-β and neutrophils, promote the development of HCC but have no known role in hepatic fibrogenesis [19]. In diethylnitrosamine (DEN) mice model, obesity and chronic inflammation enhanced production of IL-6 and TNF and promoted HCC development through activation of the oncogenic transcription factor STAT3 [20]. In an elegant study, Grohmann et al. demonstrated how obesity-associated hepatic oxidative stress can independently contribute to the pathogenesis of NASH, fibrosis, and HCC via STAT-1 and STAT-3 signaling [21]. On the contrary, some inflammatory cells such as macrophages may promote both fibrosis and HCC [17]. Activation of TLR4 by gut-derived lipopolysaccharide (LPS) promotes both as well [22]. In the light of the above, it may be concluded that inflammation promotes hepatocarcinogenesis through fibrosis-dependent and -independent pathways.

3. Fibrosis-Dependent Hepatocarcinogenesis

Fibrogenesis is a multi-cellular response that occurs whenever there is hepatic damage with hepatocellular death. Acute liver injury triggers the inflammatory and fibrogenic cascade with activation of the HSCs that constitute the main source of extracellular matrix (ECM) rich in collagen I and III [23]. The aim is to restore the architecture and function of the organ after serious damage. In fact, inflammatory signals promote hepatic regeneration, inflammatory cells provide for removal of cellular debris, while fibrosis enable the mechanical stability [24].

However, wound healing responses grow into harmful ones when the underlying trigger cannot be removed and the hepatocellular death becomes chronic, causing chronic inflammation and the development of progressive liver fibrosis which distorts the hepatic and vascular architecture [25].

Several stimuli are directly hepatocarcinogenic and the inflammation–fibrosis–cirrhosis–HCC paradigm does not provide a causal link between fibrosis and HCC. In fact, fibrosis is just one component that can be difficult to separate from other carcinogenic insults.

Because most of the data is associative rather than causal, it could be legitimate to suppose that fibrosis could only be a spectator in the process of carcinogenesis. But in the last few years, researchers have sought to create representative models that reflect the response of the human liver to injury in order to provide a mechanistic link between these two phenomena.

Potential mechanisms of fibrosis-dependent hepatocarcinogenesis include enhanced integrin signaling by ECM; paracrine crosstalk between HSCs, hepatocytes and the ECM; augmented stromal stiffness; hypoxia; imbalance between matrix metalloproteinases and tissue inhibitor of metalloproteases; excessive activation of Hedgehog pathway signaling; autophagy, hepatic progenitor cells recruitment and dysregulation of the immune system (Figure 1).

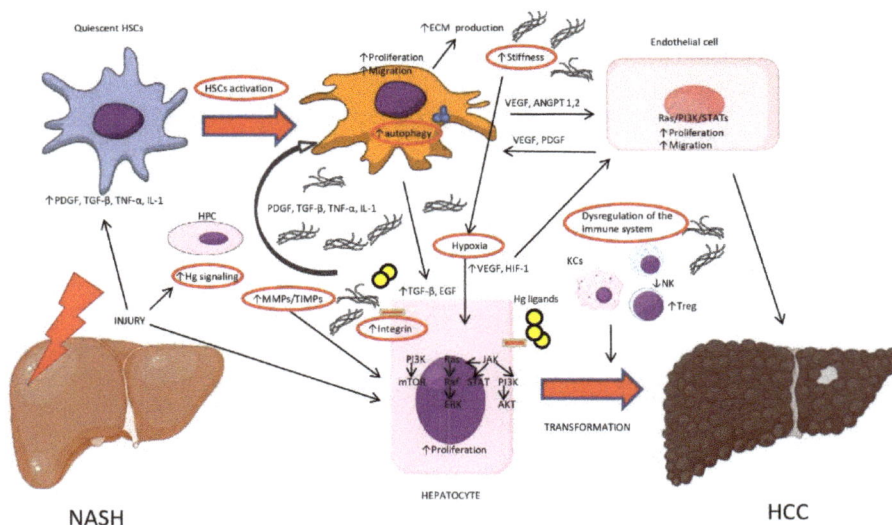

Figure 1. Fibrosis-dependent mechanisms of hepatocarcinogenesis in non-alcoholic steatohepatitis (NASH). Abbreviations: Hh, Hedgehog; PDGF, platelet derived growth factor; TGF-β, transforming growth factors-β; TNF-α, tumor necrosis factor-α; IL-1, interleukin-1; HSCs, hepatic stellate cells; ECM, extracellular matrix; PI3K, phosphatidylinositol 3-kinase; EGF, epidermal growth factor; MMPs, matrix metalloproteinases; TIMPs, tissue inhibitor of metalloproteases; VEGF, vascular endothelial growth factor; HIF-1, hypoxia-inducible factor 1; ANGPT, angiopoietin; HPCs, hepatic progenitor cells; KCs, Kupffer Cells.

3.1. Extracellular Matrix, Hepatic Stellate Cells, and Integrins

In NASH, chronic hepatocellular damage and inflammation cause the activation of regenerative pathways and the proliferation of fibrogenic cells creating a micro-environment favorable to cellular survival and proliferation [26]. HSCs are 'quiescent' liver vitamin A-storing cells located in the perisinusoidal space. During the process of liver injury, the progressive release of pro-inflammatory molecules (platelet derived growth factor (PDGF), transforming growth factors-β (TGF-β), tumor necrosis factor-α (TNF-α), interleukin (IL)-1 and several chemokine) fosters HSCs activation and successive differentiation into contractile and fibrogenic myofibroblasts, which are characterized by upregulation of mesenchymal markers (α-smooth muscle actin (α-SMA, ACTA2), desmin (DES), and collagen α1(I)) [27]. Activated HSCs produce large amounts of fibrillar type I and III collagens, but also fibronectins, laminins, and fibrinogen, creating tight and highly crosslinked collagen bundles. An excessive production of ECM and reduction of ECM turnover is the typical characteristic of liver fibrosis. Changes in the composition and structure of the ECM have been shown to provoke cellular responses through the integrin family of transmembrane receptors, which play a crucial role in the activation of TGF-β and fibrogenic response [28,29]. Integrins modulate proliferation, differentiation and survival through the activation of Hedgehog signaling [30] and other intracellular specifics pathways, including protein kinase C, phosphatidylinositol 3-kinase (PI3K) and mitogen-activated protein kinases (MAPK) [31]. Integrins may also be involved in signal transduction during angiogenesis by stimulating the intracellular signaling molecules. The integrin expression pattern are different and up-regulated in primary and metastatic HCC tissues compared to normal hepatocytes. In human HCC cell lines, integrins α1β1 and α2β1 inhibition reduces migration induced by profibrotic growth factors including TGF-β1, epidermal growth factor (EGF) and fibroblast growth factor (FGF) [32]. However there are disagreements about the role of these receptors in the onset and progression of HCC [33]. Recently, Zheng et al. showed that collagen I promotes HCC cells proliferation by regulating the β1/FAK integrin pathway in murine models of NAFLD/NASH [34].

ECM proteins interact even with DDR proteins (discoidin domain receptor). Of these, DDR2 is a tyrosine kinase of the type I collagen that promotes epithelial-mesenchymal transition (EMT), an important mechanism that favors the malignant transformation of epithelial cells [35]. In animal model of pulmonary fibrosis, treatment with an antisense oligonucleotides (ASO) or with DDR2 inhibitors, prevents myofibroblast activation and angiogenesis [36].

3.2. Matrix Stiffness

In liver fibrosis, excessive collagen deposition increases the stiffness of the ECM, which in turn promotes the further increase in collagen secretion by HSCs [37]. A meta-analysis of 17 prospective cohort studies including 7058 patients found that increasing liver stiffness (measured using transient elastography) was associated with higher risk of HCC development [38].

Higher matrix stiffness, caused by copious matrix protein deposition and crosslinking, plays an important role in cell growth, proliferation, motility and tumor metastasis in several tissues [39,40].

Experimental data suggest that matrix stiffness might modulate HCC cells proliferation, invasion, angiogenesis through several pathways including extracellular signal-regulated kinase (ERK), protein kinase B (PKB/Akt), signal transducer and activator of transcription 3 (STAT3), integrin β1/GSK-3β/β-catenin [41] β1-integrin/focal adhesion kinase (FAK) [42], and β1-integrin/phosphatidylinositol-3- kinase (PI3K)/Akt signaling pathways [43]. Recent data indicate that ECM stiffness could promote tumorigenesis through activation of Hippo-YAP/TAZ signaling pathway, a critical regulator of cell proliferation, differentiation and apoptosis [19].

Recently, You et al., showed that matrix stiffness could take an active part in the process of stemness regulation through integrin β1/Akt/mTOR/SOX2 signaling pathway [44].

3.3. Matrix Metalloproteinases and Tissue Inhibitor of Metalloproteases

Matrix metalloproteinases (MMPs), (calcium-dependent enzymes secreted mainly by HSCs but also by macrophages) and their endogenous inhibitors, tissue inhibitor of metalloproteases (TIMPs) regulate the ECM turnover [45]. In animal models, an imbalance between MMPs and TIMPs concentrations is associated not only with an alteration of ECM homeostasis, but also with alterations of various biological activities that increase the risk of developing rheumatoid arthritis, atherosclerosis, nephritis, fibrosis, cirrhosis and cancer [46]. In fact, MMPs degrade the stroma, can remove the extracellular receptors on the cell surface and are also involved in the turnover of non-matrix substrates: MMPs liberate growth factors, ligands and citokines from the ECM. In addition, they are involved in the metabolism of macromolecules (proteins, lipids, carbohydrates) and regulation of growth and cellular differentiation [47].

Animal and human data suggest that MMPs and TIMPs are implicated in phenomena such as angiogenesis, proliferation of HSCs and progression of hepatocytes from dysplasia to HCC [48]. Indeed, TIMP-1 inhibits the tumor apoptosis via SDF-1/PI3K/AKT signaling [49]. MMP-1 production enhances proliferation, invasion and fibrosis in NASH. Moreover, serum MMP-1 levels reflect disease activity and may be used as a potential biomarker for monitoring the progression of disease [50]. MMP1 and MMP7 promote re-epithelialization and smooth muscle cells de-differentiation via PAR-1; MMP8 activates HSCs; MMP3 promotes the HGF-induced invasion of human HCC; MMP-19 is involved in TGF-β signaling; MMP-25 attenuates alpha-1 proteinase inhibitor facilitating the migration. In hepatoma cells, expression of MT1-MMP, MMP-2, and MMP-9 facilitates stromal invasion [45]. These data indicate that the imbalance between MMPs expression and their endogenous inhibitors could play a key role in HCC development. Researches investigating potential drug therapies targeting MMPs and TIMPs in HCC are currently ongoing.

3.4. Hypoxia

Abundant ECM deposition and subversion of normal architecture during the fibrogenic process creates altered blood flow that reduces metabolic exchange of oxygen in the liver parenchyma [51]. Hypoxia leads to activaction of specific signalling pathways [52] (PI3K-Akt and MAPK pathways) and upregulation of angiogenic factors including vascular endothelial growth factor (VEGF) and [53] hypoxia-inducible factor 1 (HIF-1) [54]. These facilitates tumor growth, angiogenesis, EMT and metastasis [55]. High levels of HIF-1 and VEGF correlate with the aggressiveness of HCC and with worst prognosis [56].

Tuftelin1 (TUFT1) is an acidic protein expressed in several tissues and exerts multifunctional roles [57]. New study in human with HCC found that hypoxia enhances TUFT1 expression through HIF-1α/miR-671-5p/TUFT1/AKT signaling pathway [58]. TUFT1 furthered HCC cell growth and metastasis in vitro and in vivo by activating the Ca2/PI3K/AKT pathway. In this study TUFT1 knockdown minimized the promoting effects of hypoxia on tumor growth and metastasis, suggesting that TUFT1 may represent a new potential therapeutic target for HCC treatment.

3.5. Hedgehog Signaling

During embryogenesis, the Hedgehog (Hh) pathway represents an essential signaling mechanism that modulates many aspects of cell differentiation and tissue development [59]. In healthy adult liver, there is a low expression of Hh ligand and high expression of Hh ligand antagonist, so there is almost no Hh activity [60]. Liver injury, in contrast, is associated with elevated Hh signaling that stimulates liver regeneration [61]. Several studies show that in HCC there is an excessive activation of Hh signaling and this promotes proliferation, migration and invasion of HCC cells [62]. A recent study claims there is interplay between Sonic Hedgehog (Shh) and TGF-β1 in hepatic inflammatory reactions. Secreted Shh may involve activation of TGF-β1 and subsequent activation of HSCs, which together promote the progression of human NASH [63]. In Mdr2⁻/⁻ mice with chronic liver fibrosis, Philips et al.

studied the role of the Hh pathway in hepatocarcinogenesis [64]. In this work, Hh pathway activation promotes both liver fibrosis and hepatocarcinogenesis while inhibition of Hh signaling reverses both processes. Authors suggest that the carcinogenic effect of Hh could be mediated by augmented myofibroblast activation and fibrosis [64].

3.6. Hepatic Progenitor Cells

Hepatic progenitor cells (HPCs) are resident stem cells located at the level of Canals of Hering and when activated promote tissue turnover and liver regeneration [65]. Chronic hepatocyte injury is associated with HPCs activation and enhancement of several pathways identified in liver cancer, including Hg, canonical Wnt signaling and Notch [66]. Proliferation and differentiation of HPCs, depend on the up-regulation of these pathways [67]. Studies in NASH patients claimed that HPCs activation and the expansion of ductular reaction (DR) were independently correlated with progressive fibrosis both in adult and children [68,69]. During hepatic necrosis, proliferating HPCs augment their expression of profibrogenic factors while DR cells produce PDGF, TGF-β, Sonic Hg and activate HSCs [65]. These findings support the idea that HPCs activation could contribute to the initiation and sustain HCC.

3.7. Autophagy

Autophagy is an evolutionarily conserved cellular process for lysosomal degradation of damaged cell components. Cellular organelles, lipid droplets, large protein aggregates are sequestered and degradated, while amino acids generated are used for producing energy [70]. Autophagy is a complex dynamic process regulated by several signaling pathways involved in cellular proliferation and apoptosis such as PI3K/Akt/mTOR and AMPK pathway [71]. In NASH, activated HSCs develop autophagic activity as a mechanism of lipid droplet degradation from which obtain energy support for their activation. In fact, treatment with an autophagy inhibitor prevents HSCs activation in vitro while reduces lipid droplet degradation [72]. This suggests that autophagy could be a target for the treatment of NASH and fibrosis. In rats, curcumin treatment leads to protection against toxin-induced HCC through induction of autophagic pathway and inhibition of apoptosis [73].

However the role of autophagy on HCC development is controversial. A study in autophagy-deficient mice with mosaic deletion of Atg5, showed the 'double-edged sword' of autophagy, since it is important for suppression of tumorigenesis in the liver but at the same time it promotes tumor progression because of accumulation of harmful protein aggregated [74]. Conversely, Sun et al. found that autophagy-deficient Kupffer cells promote liver inflammation, fibrosis, and HCC by enhancing mitochondrial ROS- NF-kB-IL1α/β pathways [75]. These data recommend that targeting autophagy for the treatment of NASH-fibrosis and HCC should require cell-specific autophagy inhibitors.

3.8. Dysregulation of the Immune System

The innate and adaptive immune system are essential for the identification and suppression of transformed cells. In NASH, liver injury stimulates the activation of several types of immune cells [76].

Kupffer cells (KCs) could play a important role in initiation and progression of inflammation and fibrosis. Liver biopsies of patients with NAFLD found a higher expansion of KCs than in controls. This phenomenon precedes the recruitment of other inflammatory cells, in fact macrophage infiltration occurs in more advanced inflammatory stages of the disease [77]. In the early phase of NAFLD, pathogen-associated molecular patterns (PAMPs), changes in gut microbiota, increase intestinal traslocation of bacteria and toxins can activate KCs [78], which secrete TGF-β, TNF-α, pro-inflammatory cytokines such as CCL2, ROS and activate inflammasomes [76]. Activation of NLRP3 fosters secretion of IL-1β, IL-18, and IL-6 which in turn promote disease progression and HCC development [79]. Furthermore, damaged hepatocytes are be able to activate KCs through cell stress pathways such as c-Jun N-terminal kinase (JNK) and release of DAMPs factors which promote inflammation via TNF-α, nuclear factor (NF)-κB and TLR signaling activation [78,80]. In addition, in condition of hypoxia KCs can be activated by HIF-1α [81]. Thus far, the role of KC in HCC is not fully clarified and is being investigated.

The role of dendritic cells (DC) in NASH is not clear. Existing experimental data are inconsistent perhaps because differences in the NASH models utilized or the diets administered in the studies [82].

The role of neutrophils is not also entirely clarified. However increase in liver neutrophils has been reported in human with NASH [83], and the degree of infiltration is correlated with severity of disease [84]. Neutrophils could contribute to NASH progression and carcinogenesis via Myeloperoxidase (MPO)-related mechanisms [84].

Natural killer (NK) cells are significantly elevated in NASH livers, compared to normal healthy control. Their activation during injury may be due to the higher levels of several cytokines including IL-12, IL-18, and IFN-γ [85]. In the early phases of disease, NK cells act against fibrosis development through IFNγ and by inducing apoptosis of HSCs via TRAIL and FasL, while in the late stages, NK cell function is compromised leading to further increases the ECM deposition promoting HCC development [86]. NK cells play a critical role in the immune surveillance of liver tumors [87]. Several mechanisms has been proposed to explain the decrease in the NK cell functions that are associated with advance fibrosis and HCC. These include TGFβ-mediated inhibition, phagoctyosis of NK cells by HSCs, inability to make target-cell contact and the dysregulation of activating ligands [88]. However, further studies are required to clarify the role of NK cells in the carcinogenesis process during fibrosis.

NKT cells increase in human NASH with advanced fibrosis but their role in hepatocarcinogenesis is still unclear. They can secrete IL-4, IFN-γ, and TNF-α [89] as well as promote steatosis through signaling via the lymphotoxin-like inducible protein LIGHT, activate HSCs, enhance Hh signaling then fibrosis [76]. The results of experimental data in HCC are in contrast. Indeed, NKT cells may both play an anti-tumor role and promote tumor tolerance [76].

In a choline-deficient high-fat diet mice model, Wolf et al., described intrahepatic activation of CD8⁺ T cells, NKT cells, and inflammatory cytokines, similarly to NASH patients [83]. CD8⁺ T cells and NKT cells synergistically induced steatosis, NASH and HCC development. NKT cells promote steatosis via lymphotoxin (LT)-like inducible protein LIGHT, while CD8(+) T cells cause liver damage in a LTβR-independent manner. CD8+ cytotoxic T lymphocytes kill their target cells not only through their two mayor cytotoxic mechanisms (perforin/granzyme-mediated, and Fas ligand (FasL)-mediated) [90], but also by secretion of IFN-γ and TNF-α [91]. However, depletion of CD8+ T lymphocytes in different experimental mice models produced varying results on onset and progression of HCC [92,93].

Previous studies reported that CD4+ T lymphocytes are capable to inhibit HCC initiation and favor tumor regression through the expression of chemokines [94]. Study in mouse models of NAFLD and HCC, found that dysregulation of lipid metabolism induces a ROS-mediated selective loss of intrahepatic CD4+ T but not of CD8+ T cells, leading to tumorigenesis [18]. An analysis of 547 patients with HCC showed that progressive loss of CD4⁺ cytotoxic T cells was significantly correlated with an advanced stage of disease and poor prognosis [95].

Regulatory T cells (Treg) recruitment may impair the function of CD8⁺ T cells and promote cancer progression. In a DEN-induced HCC mouse model, TGF-β promoted Treg cell differentiation, and this was identified as a major inhibitory mechanism of CD8+ T cells [96]. An increased recruitment of Treg cells correlated with poor prognosis [97].

In humans, NASH with advanced fibrosis is associated with high circulating IgA⁺ cells levels that build up in fibrotic liver [98]. These cells can interfere with activation of cytotoxic CD8⁺ T lymphocytes through programmed death ligand 1 (PD-L1) and interleukin-10, and promote HCC development [99]. In mice, PD-L1 blockade induces cytotoxic T-lymphocyte-mediated regression of established HCC [93].

Although the mechanisms are not yet well-known, we can state that immune system may play a 'dual and opposite role' in the development and progression of HCC. However, extensive studies are needed.

3.9. Crosstalk between NASH and Hepatocellular Carcinoma

HSCs, fibroblasts, immune cells, endothelial and mesenchymal stem cells as well as cytokines, growth factors and ECM constitute the liver tumor microenvironment. The results of numerous studies provide evidence that the cross-talk between tumor cells and their surrounding microenvironment is essential for cell growth, proliferation, EMT and metastasis [100] (Table 1).

Activated HSCs secrete several molecules including PDGF-B and PDGF-C, TGF-α, TGF-β, EGF, VEGF, angiopoietin-1 and -2, hepatocyte growth factor (HGF), stromal-derived factor-1alpha (SDF-1), Wnt ligands, interleukin-6 (IL-6) and epimorphin (EPM) [101]. These mediators are important pro-angiogenic, proliferative, and regenerative citokines that create a favorable microenvironment facilitating tumor initiation and progression [56]. In fibrotic liver, these molecules are passively sequestered by the ECM favoring the bidirectional interaction between endothelial, stromal cells and hepatocytes in an autocrine/paracrine manner [102]. Therefore, HSCs can be the target of the molecules produced by themselves. Activated HSCs show increased expression of different receptors for soluble cytokines, including PDGF that is the most potent proliferative cytokine [103]. Generally, expression of PDGF receptors by HSCs is low but drastically increases during inflammation, NAFLD and NASH [104]. Induction of β-PDGF receptors leads to activation of the Fas-MAPK pathway and release of intracellular calcium ions that activate PKC family members then activation of a more contractile and fibrogenic phenotype of HSC [105]. In mice model, hepatic over-expression of PDGF-C induces changes in gene expression, inflammation, progressive fibrosis, neoangiogenesis, and dysplasia. In patients with NASH, Wright et al., found a important correlation between PDGF-CC levels in liver and severity of disease [106]. Thus, PDGF-C could be more crucial in modulating the microenvironment to promote HCC development in a paracrine manner than in promoting direct carcinogenic effects on hepatocytes. In vitro, EPM promotes HCC cells invasion and metastasis by activating MMP-9 expression through the FAK-ERK pathway [107].

New data suggest that HSCs could promote HCC progression through the production of IL-1β, via a mechanism that seems to be dependent on PKR activation [108].

Cancer-associated fibroblasts (CAFs) are the most abundant cell type of the tumor stroma and are similar in morphology and molecular expression profiles to the myofibroblasts (HSCs) that are activated during the wound repair process [109]. However origin of the CAFs in HCC is obscure. Indeed, they can originate from HSCs, migrated bone-marrow stem cell and EMT [110]. Thus far, there is little evidence that HSCs and CAFs drive the malignant transformation of hepatocytes, but it is established that these stromal cells create a microenvironment that supports the growth of dysplastic hepatocytes and HCC [19]. Studies in vitro suggest that HGF could be a mediator of tumor–stromal interactions through which CAFs regulate the proliferation and invasion of HCC cells [111].

MMP-2 together with MMP-9 has a fundamental role on degrading type IV collagen that is the most abundant component of ECM. Feng et al., found that co-cultures of vascular endothelial and HCC HepG2 cells increased expression of MMP-2 and MMP-9 which enhanced the invasion ability of the HepG2 cells [112].

Angiogenesis provides a source of oxygen, nutrients and is indispensable for tumor growth and metastasis [113]. Both in phase of tumor development and progression, several molecular pathways are involved in the induction of angiogenesis and in the preservation of metastasis supporting vascular networks. VEGF represents a fundamental element that controls most of mechanisms of tumor-induced angiogenesis, because stimulates vascular sprouting, tip cells formation, sprout elongation and lumen formation [114].

In this context, tumour cells may activate Treg cells and promote immune tolerance by inhibiting the anti-tumorigenic effects of NK and CD8$^+$ T cells [115].

Table 1. Main mechanisms involved in NASH progression and hepatocellular carcinoma (HCC) development.

Factor	Mechanism	Biological Effects
Inflammation	↑ PDGF, TGF-β, TNF-α, IL-1 and chemokines	HSCs and KCs activation
Activated HSCs	• ↑ Type I and III collagens, fibronectins, laminins, fibrinogen secretion • ↑ TGF-β$_1$, TGF-α, PDGF, FGF, EGF, VEGF, ANGPT 1-2, HGF, SDF-1, Wnt ligands, IL-6, EPM	• ↓ MMPs↓ and ECM turnover • ↑ Stiffness • ↑ Hypoxia • ↓ NK cell functions • ↑ Integrin signaling • ↑ HSCs activation • ↑ Angiogenesis
↑ Stiffness	• ↑ Collagen secretion • Activation of ERK, PKB/Akt, STAT3, integrin β1/GSK-3β/β-catenin; β1-integrin/FAK and β1-integrin/PI3K/Akt signaling pathways • Activation of Hippo-YAP/TAZ signaling pathway and integrin β1/Akt/mTOR/SOX2 signaling pathway	• ↑ Cell proliferation, motility and tumor metastasis • ↓ Apoptosis • Stemness dysregulation • ↓ NK cell functions
Integrins	• ↑ Hedgehog signaling, PI3K, MAPK pathways	• ↑ Proliferation, survival • ↑ Angiogenesis
MMPs/TIMPs imbalance	• ↑ Release of growth factors, ligands and citokines from the ECM • ↑ TGF-β signaling • ↓ Alpha-1 proteinase inhibitor • ↓ SDF-1/PI3K/AKT signaling	• ↓ ECM turnover • ↑ HSCs activation • Progression of hepatocytes from dysplasia to HCC • ↑ Migration • ↑ Invasion • ↓ Apoptosis
Hypoxia	• Activation of PI3K-Akt and MAPK pathways • ↑ VEGF and HIF-1 • ↑ TUFT1 →Ca2 /PI3K/AKT pathway	• KCs activation • ↑ Angiogenesis, • ↑ EMT • ↑ Tumor growth and metastasis
Hedgehog (Hg)	• ↑ TGF-β1 and HSCs activation	• NASH progression • ↑ Proliferation • ↑ Migration and invasion of HCC cells
HPCs and ductular reaction (DR)	• ↑ Hg, Wnt signaling and Notch pathways • ↑ PDGF, TGF-β, and HSCs activation	• NASH progression
Autophagy	• PI3K/Akt/mTOR and AMPK pathway • Mitochondrial ROS- NF-kB-IL1α/β pathways	• ±Cellular proliferation and apoptosis
Kupffer Cells (KCs)	• ↑ TGF-β, TNF-α, CCL2, ROS • Activation of NLRP3 →IL-1β, IL-18 and IL-6	• NASH progression • ↑ Proliferation
↓ NK cells	• ↓ IFNγ • ↓ Apoptosis of HSCs	• ↑ ECM deposition • ↓ Tumor surveillance
NKT cells	• IL-4, IFN-γ, and TNF-α • Regulate HSCs activation and Hh signaling	• Anti-tumor role or tumor tolerance
↑ Treg cells	• ↓ NK and CD8+ T cells	• ↓ Tumor surveillance

Hh, Hedgehog; PDGF, platelet derived growth factor; TGF-β, transforming growth factors-β; TNF-α, tumor necrosis factor-α; IL-1, interleukin-1; HSCs, hepatic stellate cells; ECM, extracellular matrix; PI3K, phosphatidylinositol 3-kinase; EGF, epidermal growth factor; MMPs, matrix metalloproteinases; TIMPs, tissue inhibitor of metalloproteases; VEGF, vascular endothelial growth factor; HIF-1, hypoxia-inducible factor 1; ANGPT, angiopoietin; HPCs, Hepatic progenitor cells; EPM, epimorphin.

4. Therapeutic Perspective

Although numerous drugs have been indagated, none of them have been validated in phase III.

Trials and so far, there is no medicament authorized by regulatory authorities for management of NASH. However, several molecules were shown to be effective in preclinical studies and some of these are currently being examined in humans.

TGF-β induces a complex modulation of gene expression because it can use several extracellular signals and adhesion molecules [116]. It promotes liver inflammation, activation of HCSs then fibrosis, EMT and growth of HCC [117]. TGF-β is regarded as pivotal molecule in HCC tumorigenesis since it, secreted by HSCs or by transformed hepatocytes, may inhibits NK cell functions [118], and may control the secretion of other cytokines [119]. In human HCC cell lines, treatment with TGF-β inhibitor LY2109761 stops migration and invasion by upregulating E-cadherin [120], inhibits neo-angiogenesis, reduces tumor growth and metastasis of HCC cells by inhibiting CAFs proliferation [121]. Galunisertib (LY2157299) is another promising antifibroticTGF-β inhibitor that inhibits SMAD2 phosphorylation and blocks the collagens deposition promoting their degradation [122]. Large RCT in humans are needed.

Chemokines, regulate many functions of hepatocytes, endothelial cells, HCSs, and circulating immune cells. Interactions between C-C chemokine receptors (CCR2, CCR5,CCL2,CCL5,) and their ligands, promote fibrogenesis by HSCs activation and macrophage recruitment in the liver [123]. In preclinical models of NASH, Cenicriviroc (CCR2/CCR5 antagonist) improved hepatic inflammation and fibrosis [124]. These results have been confirmed in the recent phase 2b CENTAUR study [125]. A phase III trial to evaluate the efficacy of this drug is ongoing.

Sorafenib is multikinase inhibitor of VEGFR and PDGFR approved for the treatment of HCC where increases overall survival compared to placebo [126]. In NASH rodent models, sorafenib treatment decreased inflammation, angiogenesis, HSCs activation, collagen deposition, and hepatic fibrosis [127]. It may be considered for the treatment of NASH in humans. Further studies are warranted.

Inhibition of collagen synthesis may represent a potential therapy for fibrotic liver diseases. Recent study in rats found that cationic lipid nanoparticles loaded with small interfering RNA to the procollagen α1(I) gene administration, provoked specific inhibition of type I collagen synthesis without visible side effects [128]. So far, no data on human are avaiable.

In recent years, results of phase 2-3 trials found that treatment with integrin inhibitors was ineffective for the treatment of various cancers including HCC [129]. However, it has recently been discovered that αv integrins play a key role in the fibrogenesis in the liver, skin, kidney, and lung, although many mechanisms are still unknown [130]. Many Phase 1-2 trials are underway to evaluate the efficacy of integrin inhibitors in reversing the fibrosis process (ClinicalTrials.gov).

Hh dysregulation represent a novel mechanism for hepatic fibrosis and hepatocarcinogenesis, and may considered as potential therapeutic target for patients with NASH or HCC. Recently Hh-inhibitor LDE225 was approved for the treatment of basocellular carcinoma [131]. Phase 1 study to test the safety and determine the maximum safe dose of LDE225 in patients with HCC is ongoing (NCT02151864).

In fibrotic livers, the immunosuppressive function of the fibrosis-stimulated IgA+ cells probably depends on the expression of PD-L1 and IL-10, which may promote CTL dysfunction [132]. Programmed death 1 (PD-1) inhibitors reverse CD8+ T cells dysfunction and can therefore represent a treatment option for patients with advanced HCC. Nivolumab is a PD-1 inhibitor approved for the treatment of several malignancy including melanoma, non-small cell lung cancer and renal carcinoma. In a phase 1/2 trial, Nivolumab showed a manageable safety profile and no side effects were observed in patients with advanced HCC [133]. Several phase 2/3 Trial are ongoing (ClinicalTrial.gov).

5. Conclusions and Discussion

Growing evidence suggest a mechanistic link between fibrotic microenvironment and the HCC development. However, the lack of representative animal models is hampering the efforts to understand the pathophysiological mechanisms in NASH-related HCC. NASH is a complex and extremely heterogeneous metabolic disease and despite several mouse models can mimic disease; rarely, they replicate the pathogenic sequence of human NASH-HCC. These features can explain the conflicting results among studies using different animal models.

There are 3 main categories of murine NASH models: diet models, toxins/diet-based models, and genetic/diet models [12]. For instance, a methionine/choline-deficient (MCD) diet produces histological features of NASH, but rodents treated with this diet do not develop insulin resistance [134]. A western diet (WD), induces NASH, obesity and insulin resistance but disease does not progress to advanced fibrosis [135]. Streptozotocin (STZ) and diethylnitrosamine (DEN) + HFD models, develop obesity, insulin resistance, type 2 diabetes, mild fibrosis and HCC. However an independent carcinogenic effect of toxins cannot be excluded [12]. Tsuchida et al. studied a new mice model where mice were treated with WD and weekly dosing of carbon tetrachloride (CCl4) [136]. This interesting model reproduces the progressive stages of human NAFLD, from simple steatosis, to inflammation, fibrosis, and HCC. In addition, the model replicates gene expression and immune abnormalities of human disease [136]. Nevertheless, previous reports showed that CCl4 is a potent hepatotoxin that can cause genotoxicity and oxidative DNA damage in rats [137]. Numerous genetically modified mice that are susceptible to NASH and HCC development have been described but most of them do not resemble the human progression of disease. HFD-fed MUP-uPA and DIAMOND mice develop human NASH-like disease and almost all of them progress to HCC [12]. However, in these models HCC development is significantly slower than in toxins/diet-based models. Moreover, the mutational landscape of these mice differs significantly from mouse to mouse [93].

Aim of research is overcoming this gap in order to identify a model that best replicates the several aspects of NASH-driven HCC, because studies that are performed with inappropriate models generate misleading results that delay progress in this field.

The understanding of the multiple molecular mechanisms involved in fibrogenesis make it possible to identify various therapeutic targets including cytokines, chemokines, HSCs, Hedgehog pathway signaling and other potential targets, for the purpose of reversing the fibrosis process. A better understanding of the underlying pathophysiological mechanisms could also be useful for identify non-invasive biomarkers of NASH and fibrosis because liver biopsy as the diagnostic "gold standard" it is not without risk. Moreover, imaging techniques can detect steatosis but not steatohepatitis.

The results of preclinical experiments are promising, as they have shown that it is possible to reverse the fibrogenesis process. However, it is not that simple to verify in humans. Furthermore, it is desiderable that future research establishes whether a reduction of fibrosis in patients with NASH is effective in HCC prevention [138].

Funding: This research received no external funding.

Conflicts of Interest: The authors declare no conflict of interest.

Abbreviations

NASH	Non-alcoholic steatohepatitis
NAFLD	Non-alcoholic fatty liver disease
HCC	Hepatocellular carcinoma
HSCs	Hepatic stellate cells
ECM	Extracellular matrix
PDGF	Platelet derived growth factor
TGF-β	Transforming growth factors-β
ROS	Reactive oxygen species
KCs	Kupffer cells
DC	Dendritic cells
TLR	Toll-like receptor
NF	Nuclear factor

SFA	Saturated fatty acids
DAMPs	Damage-endogenous-associated molecular patterns
PAMPs	Pathogen-associated molecular patterns
DR	Ductular reaction
HPCs	Hepatic progenitor cells
DEN	Diethylnitrosamine
LPS	Lipopolysaccharide
DDR	Discoidin domain receptor
PI3K	Phosphatidylinositol 3-kinase
MAPK	Mitogen-activated protein kinases
ERK	extracellular signal-regulated kinase
STAT	signal transducer and activator of transcription
FAK	focal adhesion kinase
JNK	Jun N-terminal kinase
VEGF	Vascular endothelial growth factor
HIF-1	Hypoxia-inducible factor 1
TUFT1	Tuftelin1
Hh	Hedgehog
HGF	Hepatocyte growth factor
EMT	Epithelial-mesenchymal transition
EGF	Epidermal growth factor
FGF	Fibroblast growth factor
EPM	Epimorphin
MMPs	Matrix metalloproteinases
TIMPs	Tissue inhibitor of metalloproteases
CAFs	Cancer-associated fibroblasts
NK	Natural killer
Treg	Regulatory T cells

References

1. World Health Organization. *Global Hepatitis Report 2017*; World Health Organization: Geneva, Switzerland, 2017.
2. Younossi, Z.M.; Koenig, A.B.; Abdelatif, D.; Fazel, Y.; Henry, L.; Wymer, M. Global epidemiology of nonalcoholic fatty liver disease-Meta-analytic assessment of prevalence, incidence, and outcomes. *Hepatology* **2016**, *64*, 73–84. [CrossRef]
3. Chalasani, N.; Younossi, Z.; Lavine, J.E.; Charlton, M.; Cusi, K.; Rinella, M.; Harrison, S.A.; Brunt, E.M.; Sanyal, A.J. The diagnosis and management of nonalcoholic fatty liver disease: Practice guidance from the American Association for the Study of Liver Diseases. *Hepatology* **2018**, *67*, 328–357. [CrossRef]
4. European Association for the Study of the Liver (EASL); European Association for the Study of Diabetes (EASD); European Association for the Study of Obesity (EASO). EASL-EASD-EASO Clinical Practice Guidelines for the management of non-alcoholic fatty liver disease. *J. Hepatol.* **2016**, *64*, 1388–1402. [CrossRef]
5. Liew, Z.-H.; Goh, G.B.; Hao, Y.; Chang, P.-E.; Tan, C.-K. Comparison of Hepatocellular Carcinoma in Patients with Cryptogenic Versus Hepatitis B Etiology: A Study of 1079 Cases Over 3 Decades. *Dig. Dis. Sci.* **2019**, *64*, 585–590. [CrossRef]
6. Paradis, V.; Zalinski, S.; Chelbi, E.; Guedj, N.; Degos, F.; Vilgrain, V.; Bedossa, P.; Belghiti, J. Hepatocellular carcinomas in patients with metabolic syndrome often develop without significant liver fibrosis: A pathological analysis. *Hepatology* **2009**, *49*, 851–859. [CrossRef]
7. Stine, J.G.; Wentworth, B.J.; Zimmet, A.; Rinella, M.E.; Loomba, R.; Caldwell, S.H.; Argo, C.K. Systematic review with meta-analysis: Risk of hepatocellular carcinoma in non-alcoholic steatohepatitis without cirrhosis compared to other liver diseases. *Aliment. Pharmacol. Ther.* **2018**, *48*, 696–703. [CrossRef]
8. Ekstedt, M.; Hagström, H.; Nasr, P.; Fredrikson, M.; Stål, P.; Kechagias, S.; Hultcrantz, R. Fibrosis stage is the strongest predictor for disease-specific mortality in NAFLD after up to 33 years of follow-up. *Hepatology* **2015**, *61*, 1547–1554. [CrossRef]

9. Llovet, J.M.; Zucman-Rossi, J.; Pikarsky, E.; Sangro, B.; Schwartz, M.; Sherman, M.; Gores, G. Hepatocellular carcinoma. *Nat. Rev. Dis. Prim.* **2016**, *2*, 16018. [CrossRef]

10. White, D.L.; Kanwal, F.; El-Serag, H.B. Association between nonalcoholic fatty liver disease and risk for hepatocellular cancer, based on systematic review. *Clin. Gastroenterol. Hepatol.* **2012**, *10*, 1342–1359. [CrossRef]

11. Farazi, P.A.; DePinho, R.A. Hepatocellular carcinoma pathogenesis: From genes to environment. *Nat. Rev. Cancer* **2006**, *6*, 674–687. [CrossRef]

12. Febbraio, M.A.; Reibe, S.; Shalapour, S.; Ooi, G.J.; Watt, M.J.; Karin, M. Preclinical Models for Studying NASH-Driven HCC: How Useful Are They? *Cell Metab.* **2019**, *29*, 18–26. [CrossRef]

13. Tilg, H.; Moschen, A.R. Evolution of inflammation in nonalcoholic fatty liver disease: The multiple parallel hits hypothesis. *Hepatology* **2010**, *52*, 1836–1846. [CrossRef]

14. De Minicis, S.; Agostinelli, L.; Rychlicki, C.; Sorice, G.P.; Saccomanno, S.; Candelaresi, C.; Giaccari, A.; Trozzi, L.; Pierantonelli, I.; Mingarelli, E.; et al. HCC development is associated to peripheral insulin resistance in a mouse model of NASH. *PLoS ONE* **2014**, *9*, e97136. [CrossRef]

15. Peverill, W.; Powell, L.W.; Skoien, R. Evolving concepts in the pathogenesis of NASH: Beyond steatosis and inflammation. *Int. J. Mol. Sci.* **2014**, *15*, 8591–8638. [CrossRef]

16. Wong, V.W.-S.; Chitturi, S.; Wong, G.L.-H.; Yu, J.; Chan, H.L.-Y.; Farrell, G.C. Pathogenesis and novel treatment options for non-alcoholic steatohepatitis. *Lancet Gastroenterol. Hepatol.* **2016**, *1*, 56–67. [CrossRef]

17. Capece, D.; Fischietti, M.; Verzella, D.; Gaggiano, A.; Cicciarelli, G.; Tessitore, A.; Zazzeroni, F.; Alesse, E. The Inflammatory Microenvironment in Hepatocellular Carcinoma: A Pivotal Role for Tumor-Associated Macrophages. *Biomed. Res. Int.* **2013**, *2013*, 187204. [CrossRef]

18. Ma, C.; Kesarwala, A.H.; Eggert, T.; Medina-Echeverz, J.; Kleiner, D.E.; Jin, P.; Stroncek, D.F.; Terabe, M.; Kapoor, V.; ElGindi, M.; et al. NAFLD causes selective CD4+ T lymphocyte loss and promotes hepatocarcinogenesis. *Nature* **2016**, *531*, 253–257. [CrossRef]

19. Affo, S.; Yu, L.-X.; Schwabe, R.F. The Role of Cancer-Associated Fibroblasts and Fibrosis in Liver Cancer. *Annu. Rev. Pathol.* **2017**, *12*, 153–186. [CrossRef]

20. Park, E.J.; Lee, J.H.; Yu, G.-Y.; He, G.; Ali, S.R.; Holzer, R.G.; Osterreicher, C.H.; Takahashi, H.; Karin, M. Dietary and genetic obesity promote liver inflammation and tumorigenesis by enhancing IL-6 and TNF expression. *Cell* **2010**, *140*, 197–208. [CrossRef]

21. Grohmann, M.; Wiede, F.; Dodd, G.T. Obesity Drives STAT-1-Dependent NASH and STAT-3-Dependent HCC. *Cell* **2018**, *175*, 1289–1306. [CrossRef]

22. Dapito, D.H.; Mencin, A.; Gwak, G.-Y. Promotion of Hepatocellular Carcinoma by the Intestinal Microbiota and TLR4. *Cancer Cell* **2012**, *21*, 504–516. [CrossRef]

23. Carloni, V.; Luong, T.V.; Rombouts, K. Hepatic stellate cells and extracellular matrix in hepatocellular carcinoma: More complicated than ever. *Liver Int.* **2014**, *34*, 834–843. [CrossRef]

24. Michalopoulos, G.K. Liver regeneration. *J. Cell. Physiol.* **2007**, *213*, 286–300. [CrossRef]

25. Seki, E.; Schwabe, R.F. Hepatic inflammation and fibrosis: Functional links and key pathways. *Hepatology* **2015**, *61*, 1066–1079. [CrossRef]

26. Stickel, F.; Hellerbrand, C. Non-alcoholic fatty liver disease as a risk factor for hepatocellular carcinoma: Mechanisms and implications. *Gut* **2010**, *59*, 1303–1307. [CrossRef]

27. Zhang, C.-Y.; Yuan, W.-G.; He, P.; Lei, J.-H.; Wang, C.-X. Liver fibrosis and hepatic stellate cells: Etiology, pathological hallmarks and therapeutic targets. *World J. Gastroenterol.* **2016**, *22*, 10512–10522. [CrossRef]

28. Leitinger, B.; Hohenester, E. Mammalian collagen receptors. *Matrix Biol.* **2007**, *26*, 146–155. [CrossRef]

29. Patsenker, E.; Stickel, F. Role of integrins in fibrosing liver diseases. *Am. J. Physiol. Gastrointest. Liver Physiol.* **2011**, *301*, G425–G434. [CrossRef]

30. Hernandez-Gea, V.; Friedman, S.L. Pathogenesis of Liver Fibrosis. *Annu. Rev. Pathol. Mech. Dis.* **2011**, *6*, 425–456. [CrossRef]

31. Howe, A.; Aplin, A.E.; Alahari, S.K.; Juliano, R.L. Integrin signaling and cell growth control. *Curr. Opin. Cell. Biol.* **1998**, *10*, 220–231. [CrossRef]

32. Yang, C.; Zeisberg, M.; Lively, J.C.; Nyberg, P.; Afdhal, N.; Kalluri, R. Integrin alpha1beta1 and alpha2beta1 are the key regulators of hepatocarcinoma cell invasion across the fibrotic matrix microenvironment. *Cancer Res.* **2003**, *63*, 8312–8317.

33. Wu, Y.; Qiao, X.; Qiao, S.; Yu, L. Targeting integrins in hepatocellular carcinoma. *Expert Opin. Ther. Targets* **2011**, *15*, 421–437. [CrossRef]

34. Zheng, X.; Liu, W.; Xiang, J.; Liu, P.; Ke, M.; Wang, B.; Wu, R.; Lv, Y. Collagen I promotes hepatocellular carcinoma cell proliferation by regulating integrin β1/FAK signaling pathway in nonalcoholic fatty liver. *Oncotarget* **2017**, *8*, 95586–95595. [CrossRef]

35. Walsh, L.A.; Nawshad, A.; Medici, D. Discoidin domain receptor 2 is a critical regulator of epithelial-mesenchymal transition. *Matrix Biol.* **2011**, *30*, 243–247. [CrossRef]

36. Zhao, H.; Bian, H.; Bu, X.; Zhang, S.; Zhang, P.; Yu, J.; Lai, X.; Li, D.; Zhu, C.; Yao, L.; et al. Targeting of Discoidin Domain Receptor 2 (DDR2) Prevents Myofibroblast Activation and Neovessel Formation During Pulmonary Fibrosis. *Mol. Ther.* **2016**, *24*, 1734–1744. [CrossRef]

37. Wells, R.G. The role of matrix stiffness in regulating cell behavior. *Hepatology* **2008**, *47*, 1394–1400. [CrossRef]

38. Singh, S.; Fujii, L.L.; Murad, M.H.; Wang, Z.; Asrani, S.K.; Ehman, R.L.; Kamath, P.S.; Talwalkar, J.A. Liver stiffness is associated with risk of decompensation, liver cancer, and death in patients with chronic liver diseases: A systematic review and meta-analysis. *Clin. Gastroenterol. Hepatol.* **2013**, *11*, 1573–1584. [CrossRef]

39. Mouw, J.K.; Yui, Y.; Damiano, L.; Bainer, R.O.; Lakins, J.N.; Acerbi, I.; Ou, G.; Wijekoon, A.C.; Levental, K.R.; Gilbert, P.M.; et al. Tissue mechanics modulate microRNA-dependent PTEN expression to regulate malignant progression. *Nat. Med.* **2014**, *20*, 360–367. [CrossRef]

40. Ulrich, T.A.; de Juan Pardo, E.M.; Kumar, S. The mechanical rigidity of the extracellular matrix regulates the structure, motility, and proliferation of glioma cells. *Cancer Res.* **2009**, *69*, 4167–4174. [CrossRef]

41. You, Y.; Zheng, Q.; Dong, Y.; Wang, Y.; Zhang, L.; Xue, T.; Xie, X.; Hu, C.; Wang, Z.; Chen, R.; et al. Higher Matrix Stiffness Upregulates Osteopontin Expression in Hepatocellular Carcinoma Cells Mediated by Integrin β1/GSK3β/β-Catenin Signaling Pathway. *PLoS ONE* **2015**, *10*, e0134243. [CrossRef]

42. Schrader, J.; Gordon-Walker, T.T.; Aucott, R.L.; van Deemter, M.; Quaas, A.; Walsh, S.; Benten, D.; Forbes, S.J.; Wells, R.G.; Iredale, J.P. Matrix stiffness modulates proliferation, chemotherapeutic response, and dormancy in hepatocellular carcinoma cells. *Hepatology* **2011**, *53*, 1192–1205. [CrossRef]

43. Dong, Y.; Xie, X.; Wang, Z.; Hu, C.; Zheng, Q.; Wang, Y.; Chen, R.; Xue, T.; Chen, J.; Gao, D.; et al. Increasing matrix stiffness upregulates vascular endothelial growth factor expression in hepatocellular carcinoma cells mediated by integrin β1. *Biochem. Biophys. Res. Commun.* **2014**, *444*, 427–432. [CrossRef]

44. You, Y.; Zheng, Q.; Dong, Y.; Xie, X.; Wang, Y.; Wu, S.; Zhang, L.; Wang, Y.; Xue, T.; Wang, Z.; et al. Matrix stiffness-mediated effects on stemness characteristics occurring in HCC cells. *Oncotarget* **2016**, *7*, 32221–32231. [CrossRef]

45. Naim, A.; Pan, Q.; Baig, M.S. Matrix Metalloproteinases (MMPs) in Liver Diseases. *J. Clin. Exp. Hepatol.* **2017**, *7*, 367–372. [CrossRef]

46. Rodríguez, D.; Morrison, C.J.; Overall, C.M. Matrix metalloproteinases: What do they not do? New substrates and biological roles identified by murine models and proteomics. *Biochim. Biophys. Acta* **2010**, *1803*, 39–54. [CrossRef]

47. Murphy, G.; Docherty, A.J.P. The Matrix Metalloproteinases and Their Inhibitors. *Am. J. Respir. Cell. Mol. Biol.* **1992**, *7*, 120–125. [CrossRef]

48. Wallace, M.C.; Friedman, S.L. Hepatic fibrosis and the microenvironment: Fertile soil for hepatocellular carcinoma development. *Gene Expr.* **2014**, *16*, 77–84. [CrossRef]

49. Song, T.; Dou, C.; Jia, Y.; Tu, K.; Zheng, X. TIMP-1 activated carcinoma-associated fibroblasts inhibit tumor apoptosis by activating SDF1/CXCR4 signaling in hepatocellular carcinoma. *Oncotarget* **2015**, *6*, 12061–12079. [CrossRef]

50. Ando, W.; Yokomori, H.; Tsutsui, N.; Yamanouchi, E.; Suzuki, Y.; Oda, M.; Inagaki, Y.; Otori, K.; Okazaki, I. Serum matrix metalloproteinase-1 level represents disease activity as opposed to fibrosis in patients with histologically proven nonalcoholic steatohepatitis. *Clin. Mol. Hepatol.* **2018**, *24*, 61–76. [CrossRef]

51. O'Rourke, J.M.; Sagar, V.M.; Shah, T. Carcinogenesis on the background of liver fibrosis: Implications for the management of hepatocellular cancer. *World J. Gastroenterol.* **2018**, *24*, 4436–4447. [CrossRef]

52. Harris, A.L. Hypoxia—A key regulatory factor in tumour growth. *Nat. Rev. Cancer* **2002**, *2*, 38–47. [CrossRef]

53. Yu, D.-C.; Chen, J.; Ding, Y.-T. Hypoxic and highly angiogenic non-tumor tissues surrounding hepatocellular carcinoma: The 'niche' of endothelial progenitor cells. *Int. J. Mol. Sci.* **2010**, *11*, 2901–2909. [CrossRef]

54. Semenza, G.L. Oxygen Sensing, Homeostasis, and Disease. *N. Engl. J. Med.* **2011**, *365*, 537–547. [CrossRef]

55. Lichtenberger, B.M.; Tan, P.K.; Niederleithner, H.; Ferrara, N.; Petzelbauer, P.; Sibilia, M. Autocrine VEGF signaling synergizes with EGFR in tumor cells to promote epithelial cancer development. *Cell* **2010**, *140*, 268–279. [CrossRef]

56. Hernandez-Gea, V.; Toffanin, S.; Friedman, S.L.; Llovet, J.M. Role of the microenvironment in the pathogenesis and treatment of hepatocellular carcinoma. *Gastroenterology* **2013**, *144*, 512–527. [CrossRef]

57. Leiser, Y.; Blumenfeld, A.; Haze, A.; Dafni, L.; Taylor, A.L.; Rosenfeld, E.; Fermon, E.; Gruenbaum-Cohen, Y.; Shay, B.; Deutsch, D. Localization, quantification, and characterization of tuftelin in soft tissues. *Anat. Rec.* **2007**, *290*, 449–454. [CrossRef]

58. Dou, C.; Zhou, Z.; Xu, Q.; Liu, Z.; Zeng, Y.; Wang, Y.; Li, Q.; Wang, L.; Yang, W.; Liu, Q.; et al. Hypoxia-induced TUFT1 promotes the growth and metastasis of hepatocellular carcinoma by activating the Ca^{2+}/PI3K/AKT pathway. *Oncogene* **2018**, *38*, 1239–1255. [CrossRef]

59. McMahon, A.P.; Ingham, P.W.; Tabin, C.J. Developmental roles and clinical significance of hedgehog signaling. *Curr. Top. Dev. Biol.* **2003**, *53*, 1–114.

60. Choi, S.S.; Omenetti, A.; Syn, W.-K.; Diehl, A.M. The role of Hedgehog signaling in fibrogenic liver repair. *Int. J. Biochem. Cell. Biol.* **2011**, *43*, 238–244. [CrossRef]

61. Ochoa, B.; Syn, W.-K.; Delgado, I.; Karaca, G.F.; Jung, Y.; Wang, J.; Zubiaga, A.M.; Fresnedo, O.; Omenetti, A.; Zdanowicz, M.; et al. Hedgehog signaling is critical for normal liver regeneration after partial hepatectomy in mice. *Hepatology* **2010**, *51*, 1712–1723. [CrossRef]

62. Zheng, X.; Zeng, W.; Gai, X.; Xu, Q.; Li, C.; Liang, Z.; Tuo, H.; Liu, Q. Role of the Hedgehog pathway in hepatocellular carcinoma (Review). *Oncol. Rep.* **2013**, *30*, 2020–2026. [CrossRef]

63. Zhou, X.; Wang, P.; Ma, Z.; Li, M.; Teng, X.; Sun, L.; Wan, G.; Li, Y.; Guo, L.; Liu, H. Novel Interplay Between Sonic Hedgehog and Transforming Growth Factor-β1 in Human Nonalcoholic Steatohepatitis. *Appl. Immunohistochem. Mol. Morphol. AIMM* **2019**. [CrossRef]

64. Philips, G.M.; Chan, I.S.; Swiderska, M.; Schroder, V.T.; Guy, C.; Karaca, G.F.; Moylan, C.; Venkatraman, T.; Feuerlein, S.; Syn, W-K.; et al. Hedgehog signaling antagonist promotes regression of both liver fibrosis and hepatocellular carcinoma in a murine model of primary liver cancer. *PLoS ONE* **2011**, *6*, e23943. [CrossRef]

65. Carpino, G.; Renzi, A.; Onori, P.; Gaudio, E. Role of Hepatic Progenitor Cells in Nonalcoholic Fatty Liver Disease Development: Cellular Cross-Talks and Molecular Networks. *Int. J. Mol. Sci.* **2013**, *14*, 20112–20130. [CrossRef]

66. Mishra, L.; Banker, T.; Murray, J.; Byers, S.; Thenappan, A.; He, A.R.; Shetty, K.; Johnson, L.; Reddy, E.P. Liver stem cells and hepatocellular carcinoma. *Hepatology* **2009**, *49*, 318–329. [CrossRef]

67. Clevers, H. The intestinal crypt, a prototype stem cell compartment. *Cell* **2013**, *154*, 274–284. [CrossRef]

68. Nobili, V.; Carpino, G.; Alisi, A.; Franchitto, A.; Alpini, G.; de Vito, R.; Onori, P.; Alvaro, D.; Gaudio, E. Hepatic progenitor cells activation, fibrosis, and adipokines production in pediatric nonalcoholic fatty liver disease. *Hepatology* **2012**, *56*, 2142–2153. [CrossRef]

69. Richardson, M.M.; Jonsson, J.R.; Powell, E.; Brunt, E.M.; Neuschwander-Tetri, B.A.; Bhathal, P.S.; Dixon, J.B.; Weltman, M.D.; Tilg, H.; Moschen, A.R.; et al. Progressive fibrosis in nonalcoholic steatohepatitis: Association with altered regeneration and a ductular reaction. *Gastroenterology* **2007**, *133*, 80–90. [CrossRef]

70. Czaja, M.J. Function of Autophagy in Nonalcoholic Fatty Liver Disease. *Dig. Dis. Sci.* **2016**, *61*, 1304–1313. [CrossRef]

71. Mao, Y.; Yu, F.; Wang, J.; Guo, C.; Fan, X. Autophagy: A new target for nonalcoholic fatty liver disease therapy. *Hepat. Med.* **2016**, *8*, 27–37. [CrossRef]

72. Thoen, L.F.R.; Guimarães, E.L.M.; Dollé, L.; Mannaerts, I.; Najimi, M.; Sokal, E.; van Grunsven, L.A. A role for autophagy during hepatic stellate cell activation. *J. Hepatol.* **2011**, *55*, 1353–1360. [CrossRef]

73. Elmansi, A.; El-Karef, A.; Shishtawy, M.; Eissa, L. Hepatoprotective Effect of Curcumin on Hepatocellular Carcinoma Through Autophagic and Apoptic Pathways. *Ann. Hepatol.* **2017**, *16*, 607–618. [CrossRef]

74. Takamura, A.; Komatsu, M.; Hara, T.; Sakamoto, A.; Kishi, C.; Waguri, S.; Eishi, Y.; Hino, O.; Tanaka, K.; Mizushima, N. Autophagy-deficient mice develop multiple liver tumors. *Genes Dev.* **2011**, *25*, 795–800. [CrossRef]

75. Sun, K.; Xu, L.; Jing, Y.; Han, Z.; Chen, X.; Cai, C.; Zhao, P.; Zhao, X.; Yang, L.; Wei, L. Autophagy-deficient Kupffer cells promote tumorigenesis by enhancing mtROS-NF-κB-IL1α/β-dependent inflammation and fibrosis during the preneoplastic stage of hepatocarcinogenesis. *Cancer Lett.* **2017**, *388*, 198–207. [CrossRef]

76. Ringelhan, M.; Pfister, D.; O'Connor, T.; Pikarsky, E.; Heikenwalder, M. The immunology of hepatocellular carcinoma. *Nat. Immunol.* **2018**, *19*, 222–232. [CrossRef]

77. Lanthier, N. Targeting Kupffer cells in non-alcoholic fatty liver disease/non-alcoholic steatohepatitis: Why and how? *World J. Hepatol.* **2015**, *7*, 2184–2188. [CrossRef]

78. Ganz, M.; Szabo, G. Immune and inflammatory pathways in NASH. *Hepatol. Int.* **2013**, *7* (Suppl. S2), 771–781. [CrossRef]

79. Kong, L.; Zhou, Y.; Bu, H.; Lv, T.; Shi, Y.; Yang, J. Deletion of interleukin-6 in monocytes/macrophages suppresses the initiation of hepatocellular carcinoma in mice. *J. Exp. Clin. Cancer Res.* **2016**, *35*, 131. [CrossRef]

80. Kubes, P.; Mehal, W.Z. Sterile Inflammation in the Liver. *Gastroenterology* **2012**, *143*, 1158–1172. [CrossRef]

81. Koh, M.Y.; Gagea, M.; Sargis, T.; Lemos, R.; Grandjean, G.; Charbono, A.; Bekiaris, V.; Sedy, J.; Kiriakova, G.; Liu, X.; et al. A new HIF-1α/RANTES-driven pathway to hepatocellular carcinoma mediated by germline haploinsufficiency of SART1/HAF in mice. *Hepatology* **2016**, *63*, 1576–1591. [CrossRef]

82. Cai, J.; Zhang, X.-J.; Li, H. The Role of Innate Immune Cells in Nonalcoholic Steatohepatitis. *Hepatology* **2019**. [CrossRef]

83. Wolf, M.J.; Adili, A.; Piotrowitz, K.; Abdullah, Z.; Boege, Y.; Stemmer, K.; Ringelhan, M.; Simonavicius, N.; Egger, M.; Wohlleber, D.; et al. Metabolic activation of intrahepatic CD8+ T cells and NKT cells causes nonalcoholic steatohepatitis and liver cancer via cross-talk with hepatocytes. *Cancer Cell* **2014**, *26*, 549–564. [CrossRef]

84. Nati, M.; Haddad, D.; Birkenfeld, A.L.; Koch, C.A.; Chavakis, T.; Chatzigeorgiou, A. The role of immune cells in metabolism-related liver inflammation and development of non-alcoholic steatohepatitis (NASH). *Rev. Endocr. Metab. Disord.* **2016**, *17*, 29–39. [CrossRef]

85. Tian, Z.; Chen, Y.; Gao, B. Natural killer cells in liver disease. *Hepatology* **2013**, *57*, 1654–1662. [CrossRef]

86. Jeong, W.-I.; Park, O.; Suh, Y.-G.; Byun, J.-S.; Park, S.-Y.; Choi, E.; Kim, J.-K.; Ko, H.; Wang, H.; Miller, A.M.; et al. Suppression of innate immunity (natural killer cell/interferon-γ) in the advanced stages of liver fibrosis in mice. *Hepatology* **2011**, *53*, 1342–1351. [CrossRef]

87. Male, V.; Stegmann, K.; Easom, N.; Maini, M. Natural Killer Cells in Liver Disease. *Semin. Liver Dis.* **2017**, *37*, 198–209. [CrossRef]

88. Albertsson, P.A.; Basse, P.H.; Hokland, M.; Goldfarb, R.H.; Nagelkerke, J.F.; Nannmark, U.; Kuppen, P.J.K. NK cells and the tumour microenvironment: Implications for NK-cell function and anti-tumour activity. *Trends Immunol.* **2003**, *24*, 603–609. [CrossRef]

89. Sachdeva, M.; Chawla, Y.K.; Arora, S.K. Immunology of hepatocellular carcinoma. *World J. Hepatol.* **2015**, *7*, 2080–2090. [CrossRef]

90. Hassin, D.; Garber, O.G.; Meiraz, A.; Schiffenbauer, Y.S.; Berke, G. Cytotoxic T lymphocyte perforin and Fas ligand working in concert even when Fas ligand lytic action is still not detectable. *Immunology* **2011**, *133*, 190–196. [CrossRef]

91. Lucifora, J.; Xia, Y.; Reisinger, F.; Zhang, K.; Stadler, D.; Cheng, X.; Sprinzl, M.F.; Koppensteiner, H.; Makowska, Z.; Volz, T.; et al. Specific and Nonhepatotoxic Degradation of Nuclear Hepatitis B Virus cccDNA. *Science* **2014**, *343*, 1221–1228. [CrossRef]

92. Endig, J.; Buitrago-Molina, L.E.; Marhenke, S.; Marhenke, S.; Reisinger, F.; Saborowski, A.; Schütt, J.; Limbourg, F.; Könecke, C.; Schreder, A.; et al. Dual Role of the Adaptive Immune System in Liver Injury and Hepatocellular Carcinoma Development. *Cancer Cell* **2016**, *30*, 308–323. [CrossRef]

93. Shalapour, S.; Lin, X.-J.; Bastian, I.N.; Brain, J.; Burt, A.D.; Aksenov, A.A.; Vrbanac, A.F.; Li, W.; Perkins, A.; Matsutani, T.; et al. Inflammation-induced IgA+ cells dismantle anti-liver cancer immunity. *Nature* **2017**, *551*, 340–345. [CrossRef]

94. Rakhra, K.; Bachireddy, P.; Zabuawala, T.; Zeiser, R.; Xu, L.; Kopelman, A.; Fan, A.C.; Yang, Q.; Braunstein, L.; Crosby, E.; et al. CD4(+) T cells contribute to the remodeling of the microenvironment required for sustained tumor regression upon oncogene inactivation. *Cancer Cell* **2010**, *18*, 485–498. [CrossRef]

95. Fu, J.; Zhang, Z.; Zhou, L.; Qi, Z.; Xing, S.; Lv, J.; Shi, J.; Fu, B.; Liu, Z.; Zhang, J.-Y.; et al. Impairment of CD4+ cytotoxic T cells predicts poor survival and high recurrence rates in patients with hepatocellular carcinoma. *Hepatology* **2013**, *58*, 139–149. [CrossRef]

96. Shen, Y.; Wei, Y.; Wang, Z.; Jing, Y.; He, H.; Yuan, J.; Li, R.; Zhao, Q.; Wei, L.; Yang, T.; et al. TGF-β regulates hepatocellular carcinoma progression by inducing Treg cell polarization. *Cell. Physiol. Biochem.* **2015**, *35*, 1623–1632. [CrossRef]

97. Fu, J.; Xu, D.; Liu, Z.; Shi, M.; Zhao, P.; Fu, B.; Zhang, Z.; Yang, H.; Zhang, H.; Zhou, C.; et al. Increased regulatory T cells correlate with CD8 T-cell impairment and poor survival in hepatocellular carcinoma patients. *Gastroenterology* **2007**, *132*, 2328–2339. [CrossRef]

98. McPherson, S.; Henderson, E.; Burt, A.D.; Day, C.P.; Anstee, Q.M. Serum immunoglobulin levels predict fibrosis in patients with non-alcoholic fatty liver disease. *J. Hepatol.* **2014**, *60*, 1055–1062. [CrossRef]

99. Shalapour, S.; Font-Burgada, J.; Di Caro, G.; Zhong, Z.; Sanchez-Lopez, E.; Dhar, D.; Willimsky, G.; Ammirante, M.; Strasner, A.; Hansel, D.E.; et al. Immunosuppressive plasma cells impede T-cell-dependent immunogenic chemotherapy. *Nature* **2015**, *521*, 94–98. [CrossRef]

100. Wang, H.; Chen, L. Tumor microenviroment and hepatocellular carcinoma metastasis. *J. Gastroenterol. Hepatol.* **2013**, *28*, 43–48. [CrossRef]

101. Kocabayoglu, P.; Friedman, S.L. Cellular basis of hepatic fibrosis and its role in inflammation and cancer. *Front. Biosci. (Schol. Ed.)* **2013**, *5*, 217–230. [CrossRef]

102. Vlodavsky, I.; Miao, H.Q.; Medalion, B.; Danagher, P.; Ron, D. Involvement of heparan sulfate and related molecules in sequestration and growth promoting activity of fibroblast growth factor. *Cancer Metastasis Rev.* **1996**, *15*, 177–186. [CrossRef]

103. Levental, K.R.; Yu, H.; Kass, L.; Lakins, J.N.; Egeblad, M.; Erler, J.T.; Fong, S.F.T.; Csiszar, K.; Giaccia, A.; Weninger, W.; et al. Matrix crosslinking forces tumor progression by enhancing integrin signaling. *Cell* **2009**, *139*, 891–906. [CrossRef]

104. Campbell, J.S.; Hughes, S.D.; Gilbertson, D.G.; Palmer, T.E.; Holdren, M.S.; Haran, A.C.; Odell, M.M.; Bauer, R.L.; Ren, H.-P.; Haugen, H.S.; et al. Platelet-derived growth factor C induces liver fibrosis, steatosis, and hepatocellular carcinoma. *Proc. Natl. Acad. Sci. USA* **2005**, *102*, 3389–3394. [CrossRef]

105. Kelly, J.D.; Haldeman, B.A.; Grant, F.J.; Murray, M.J.; Seifert, R.A.; Bowen-Pope, D.F.; Cooper, J.A.; Kazlauskas, A. Platelet-derived growth factor (PDGF) stimulates PDGF receptor subunit dimerization and intersubunit trans-phosphorylation. *J. Biol. Chem.* **1991**, *266*, 8987–8992.

106. Wright, J.H.; Johnson, M.M.; Shimizu-Albergine, M. Paracrine activation of hepatic stellate cells in platelet-derived growth factor C transgenic mice: Evidence for stromal induction of hepatocellular carcinoma. *Int. J. Cancer* **2014**, *134*, 778–788. [CrossRef]

107. Jia, Y.-L.; Shi, L.; Zhou, J.-N.; Fu, C.-J.; Chen, L.; Yuan, H.-F.; Wang, Y.-F.; Yan, X.-L.; Xu, Y.-C.; Zeng, Q.; et al. Epimorphin promotes human hepatocellular carcinoma invasion and metastasis through activation of focal adhesion kinase/extracellular signal-regulated kinase/matrix metalloproteinase-9 axis. *Hepatology* **2011**, *54*, 1808–1818. [CrossRef]

108. Imai, Y.; Yoshida, O.; Watanabe, T.; Yukimoto, A.; Koizumi, Y.; Ikeda, Y.; Tokumoto, Y.; Hirooka, M.; Abe, M.; Hiasa, Y. Stimulated hepatic stellate cell promotes progression of hepatocellular carcinoma due to protein kinase R activation. *PLoS ONE* **2019**, *14*, e0212589. [CrossRef]

109. Kubo, N.; Araki, K.; Kuwano, H.; Shirabe, K. Cancer-associated fibroblasts in hepatocellular carcinoma. *World J. Gastroenterol.* **2016**, *22*, 6841–6850. [CrossRef]

110. Cesselli, D.; Beltrami, A.P.; Poz, A.; Marzinotto, S.; Comisso, E.; Bergamin, N.; Bourkoula, E.; Pucer, A.; Puppato, E.; Toffoletto, B.; et al. Role of tumor associated fibroblasts in human liver regeneration, cirrhosis, and cancer. *Int. J. Hepatol.* **2011**, *2011*, 120925. [CrossRef]

111. Neaud, V.; Faouzi, S.; Guirouilh, J.; Le Bail, B.; Balabaud, C.; Bioulac-Sage, P.; Rosenbaum, J. Human hepatic myofibroblasts increase invasiveness of hepatocellular carcinoma cells: Evidence for a role of hepatocyte growth factor. *Hepatology* **1997**, *26*, 1458–1466. [CrossRef]

112. Feng, T.; Yu, H.; Xia, Q.; Ma, Y.; Yin, H.; Shen, Y.; Liu, X. Cross-talk mechanism between endothelial cells and hepatocellular carcinoma cells via growth factors and integrin pathway promotes tumor angiogenesis and cell migration. *Oncotarget* **2017**, *8*, 69577–69593. [CrossRef]

113. Kareva, I.; Abou-Slaybi, A.; Dodd, O.; Dashevsky, O.; Klement, G.L. Normal Wound Healing and Tumor Angiogenesis as a Game of Competitive Inhibition. *PLoS ONE* **2016**, *11*, e0166655. [CrossRef]

114. Welti, J.; Loges, S.; Dimmeler, S.; Carmeliet, P. Recent molecular discoveries in angiogenesis and antiangiogenic therapies in cancer. *J. Clin. Investig.* **2013**, *123*, 3190–3200. [CrossRef]

115. Makarova-Rusher, O.V.; Medina-Echeverz, J.; Duffy, A.G.; Greten, T.F. The yin and yang of evasion and immune activation in HCC. *J. Hepatol.* **2015**, *62*, 1420–1429. [CrossRef]

116. Derynck, R.; Zhang, Y.E. Smad-dependent and Smad-independent pathways in TGF-beta family signalling. *Nature* **2003**, *425*, 577–584. [CrossRef]

117. Fabregat, I.; Fernando, J.; Mainez, J.; Sancho, P. TGF-beta signaling in cancer treatment. *Curr. Pharm. Des.* **2014**, *20*, 2934–2947. [CrossRef]

118. Wang, Y.; Liu, T.; Tang, W.; Deng, B.; Chen, Y.; Zhu, J.; Shen, X. Hepatocellular Carcinoma Cells Induce Regulatory T Cells and Lead to Poor Prognosis via Production of Transforming Growth Factor-β1. *Cell. Physiol. Biochem.* **2016**, *38*, 306–318. [CrossRef]

119. Calon, A.; Tauriello, D.V.F.; Batlle, E. TGF-beta in CAF-mediated tumor growth and metastasis. *Semin. Cancer Biol.* **2014**, *25*, 15–22. [CrossRef]

120. Fransvea, E.; Angelotti, U.; Antonaci, S.; Giannelli, G. Blocking transforming growth factor-beta up-regulates E-cadherin and reduces migration and invasion of hepatocellular carcinoma cells. *Hepatology* **2008**, *47*, 1557–1566. [CrossRef]

121. Mazzocca, A.; Fransvea, E.; Dituri, F.; Lupo, L.; Antonaci, S.; Giannelli, G. Down-regulation of connective tissue growth factor by inhibition of transforming growth factor beta blocks the tumor-stroma cross-talk and tumor progression in hepatocellular carcinoma. *Hepatology* **2010**, *51*, 523–534. [CrossRef]

122. Luangmonkong, T.; Suriguga, S.; Bigaeva, E.; Boersema, M.; Oosterhuis, D.; de Jong, K.P.; Schuppan, D.; Mutsaers, H.A.M.; Olinga, P. Evaluating the antifibrotic potency of galunisertib in a human ex vivo model of liver fibrosis. *Br. J. Pharmacol.* **2017**, *174*, 3107–3117. [CrossRef]

123. Marra, F.; Tacke, F. Roles for chemokines in liver disease. *Gastroenterology* **2014**, *147*, 577–594. [CrossRef]

124. Puengel, T.; Krenkel, O.; Mossanen, J.; Longerich, T.; Lefebvre, E.; Trautwein, C.; Tacke, F. The Dual Ccr2/Ccr5 Antagonist Cenicriviroc Ameliorates Steatohepatitis and Fibrosis in Vivo by Inhibiting the Infiltration of Inflammatory Monocytes into Injured Liver. *J. Hepatol.* **2016**, *64*, S160. [CrossRef]

125. Friedman, S.L.; Ratziu, V.; Harrison, S.; Abdelmalek, M.F.; Aithal, G.P.; Caballeria, J.; Francque, S.; Farrell, G.; Kowdley, K.V.; Craxi, A.; et al. A randomized, placebo-controlled trial of cenicriviroc for treatment of nonalcoholic steatohepatitis with fibrosis. *Hepatology* **2018**, *67*, 1754–1767. [CrossRef]

126. Llovet, J.M.; Ricci, S.; Mazzaferro, V.; Hilgard, P.; Gane, E.; Blanc, J.-F.; de Oliveira, A.C.; Santoro, A.; Raoul, J.-L.; Forner, A.; et al. Sorafenib in advanced hepatocellular carcinoma. *N. Engl. J. Med.* **2008**, *359*, 378–390. [CrossRef]

127. Stefano, J.T.; Pereira, I.V.A.; Torres, M.M.; Bida, P.M.; Coelho, A.M.M.; Xerfan, M.P.; Cogliati, B.; Barbeiro, D.F.; Mazo, D.F.C.; Kubrusly, M.S.; et al. Sorafenib prevents liver fibrosis in a non-alcoholic steatohepatitis (NASH) rodent model. *Braz. J. Med. Biol. Res.* **2015**, *48*, 408–414. [CrossRef]

128. Jiménez Calvente, C.; Sehgal, A.; Popov, Y.; Kim, Y.O.; Zevallos, V.; Sahin, U.; Diken, M.; Schuppan, D. Specific hepatic delivery of procollagen α1(I) small interfering RNA in lipid-like nanoparticles resolves liver fibrosis. *Hepatology* **2015**, *62*, 1285–1297. [CrossRef]

129. Schuppan, D.; Ashfaq-Khan, M.; Yang, A.T.; Kim, Y.O. Liver fibrosis: Direct antifibrotic agents and targeted therapies. *Matrix Biol.* **2018**, *68–69*, 435–451. [CrossRef]

130. Henderson, N.C.; Arnold, T.D.; Katamura, Y.; Giacomini, M.M.; Rodriguez, J.D.; McCarty, J.H.; Pellicoro, A.; Raschperger, E.; Betsholtz, C.; Ruminski, P.G.; et al. Targeting of αv integrin identifies a core molecular pathway that regulates fibrosis in several organs. *Nat. Med.* **2013**, *19*, 1617–1624. [CrossRef]

131. Becker, L.R.; Aakhus, A.E.; Reich, H.C.; Lee, P.K. A Novel Alternate Dosing of Vismodegib for Treatment of Patients with Advanced Basal Cell Carcinomas. *JAMA Dermatol.* **2017**, *153*, 321–322. [CrossRef]

132. Whiteside, T.L.; Demaria, S.; Rodriguez-Ruiz, M.E.; Zarour, H.M.; Melero, I. Emerging Opportunities and Challenges in Cancer Immunotherapy. *Clin. Cancer Res.* **2016**, *22*, 1845–1855. [CrossRef]

133. El-Khoueiry, A.B.; Sangro, B.; Yau, T.; Crocenzi, T.S.; Kudo, M.; Hsu, C.; Kim, T.-Y.; Choo, S.-P.; Trojan, J.; Welling, T.H.; et al. Nivolumab in patients with advanced hepatocellular carcinoma (CheckMate 040): An open-label, non-comparative, phase 1/2 dose escalation and expansion trial. *Lancet* **2017**, *389*, 2492–2502. [CrossRef]

134. Machado, M.V.; Michelotti, G.A.; Xie, G.; de Almeida, T.P.; Boursier, J.; Bohnic, B.; Guy, C.D.; Diehl, A.M.; Diehl, A.M. Mouse Models of Diet-Induced Nonalcoholic Steatohepatitis Reproduce the Heterogeneity of the Human Disease. *PLoS ONE* **2015**, *10*, e0127991. [CrossRef]

135. Asgharpour, A.; Cazanave, S.C.; Pacana, T.; Seneshaw, M.; Vincent, R.; Banini, B.A.; Kumar, D.P.; Daita, K.; Min, H.-K.; Mirshahi, F.; et al. A diet-induced animal model of non-alcoholic fatty liver disease and hepatocellular cancer. *J. Hepatol.* **2016**, *65*, 579–588. [CrossRef]

136. Tsuchida, T.; Lee, Y.A.; Fujiwara, N.; Ybanez, M.; Allen, B.; Martins, S.; Fiel, M.I.; Goossens, N.; Chou, H.-I.; Hoshida, Y.; et al. A simple diet- and chemical-induced murine NASH model with rapid progression of steatohepatitis, fibrosis and liver cancer. *J. Hepatol.* **2018**, *69*, 385–395. [CrossRef]

137. Alkreathy, H.M.; Khan, R.A.; Khan, M.R.; Sahreen, S. CCl4 induced genotoxicity and DNA oxidative damages in rats: Hepatoprotective effect of *Sonchus arvensis*. *BMC Complement. Altern. Med.* **2014**, *14*, 452. [CrossRef]

138. Musso, G.; Cassader, M.; Paschetta, E.; Gambino, R. Thiazolidinediones and Advanced Liver Fibrosis in Nonalcoholic Steatohepatitis: A Meta-analysis. *JAMA Intern. Med.* **2017**, *177*, 633–640. [CrossRef]

International Journal of
Molecular Sciences

MDPI

Review

YAP/TAZ Signaling as a Molecular Link between Fibrosis and Cancer

Satoshi Noguchi [1],*, Akira Saito [1,2] and Takahide Nagase [1]**

[1] Department of Respiratory Medicine, Graduate School of Medicine, The University of Tokyo, 7-3-1 Hongo, Bunkyo-ku, Tokyo 113-0033, Japan; asaitou-tky@umin.ac.jp (A.S.); takahide-tky@umin.ac.jp (T.N.)
[2] Division for Health Service Promotion, The University of Tokyo, 7-3-1 Hongo, Bunkyo-ku, Tokyo 113-0033, Japan
* Correspondence: snoguchi-tky@umin.ac.jp; Tel.: +81-3-3815-5411; Fax: +81-3-3815-5954

Received: 23 October 2018; Accepted: 16 November 2018; Published: 20 November 2018

Abstract: Tissue fibrosis is a pathological condition that is associated with impaired epithelial repair and excessive deposition of extracellular matrix (ECM). Fibrotic lesions increase the risk of cancer in various tissues, but the mechanism linking fibrosis and cancer is unclear. Yes-associated protein (YAP) and the transcriptional coactivator with PDZ-binding motif (TAZ) are core components of the Hippo pathway, which have multiple biological functions in the development, homeostasis, and regeneration of tissues and organs. YAP/TAZ act as sensors of the structural and mechanical features of the cell microenvironment. Recent studies have shown aberrant YAP/TAZ activation in both fibrosis and cancer in animal models and human tissues. In fibroblasts, ECM stiffness mechanoactivates YAP/TAZ, which promote the production of profibrotic mediators and ECM proteins. This results in tissue stiffness, thus establishing a feed-forward loop of fibroblast activation and tissue fibrosis. In contrast, in epithelial cells, YAP/TAZ are activated by the disruption of cell polarity and increased ECM stiffness in fibrotic tissues, which promotes the proliferation and survival of epithelial cells. YAP/TAZ are also involved in the epithelial–mesenchymal transition (EMT), which contributes to tumor progression and cancer stemness. Importantly, the crosstalk with transforming growth factor (TGF)-β signaling and Wnt signaling is essential for the profibrotic and tumorigenic roles of YAP/TAZ. In this article, we review the latest advances in the pathobiological roles of YAP/TAZ signaling and their function as a molecular link between fibrosis and cancer.

Keywords: YAP; TAZ; Hippo pathway; fibrosis; cancer; mechanotransduction; TGF-β; Wnt

1. Introduction

Fibrosis is a pathological process that is characterized by mesenchymal cell infiltration and proliferation in the interstitial space. The fibrogenic response consists of the following phases: (1) initiation of the response driven by primary injury to the organ, (2) activation of effector cells such as fibroblasts induced by inflammatory mediators, and (3) elaboration of the extracellular matrix (ECM) [1]. Under normal conditions, the repair process is completed by provisional ECM degradation and the removal of excessive mesenchymal cells by apoptosis and phagocytosis, which leaves minimal damage and restores the normal architecture of the tissue. However, dysregulated processes can result in dynamic deposition and an insufficient resorption of ECM, which promotes progression to fibrosis and, ultimately, to end-organ failure. The transforming growth factor (TGF)-β and Wnt signaling pathways play pivotal roles in fibroblast activation and ECM deposition, and these pathways are reported to be activated in human fibrotic tissue as well as experimental models of fibrosis [2,3].

Fibrotic lesions increase the risk of cancer in various tissues, including the lung [4] and liver [5]. However, the mechanism linking fibrosis and cancer is unclear. The formation of fibrotic ECM disrupts epithelial cell polarity, and the abundance of growth factors in the profibrotic milieu stimulates the proliferation of epithelial cells, creating conditions that are favorable for cancer initiation and progression. Fibrosis and cancer have several dysregulated intercellular communication and intracellular signaling pathways in common. Indeed, cancer stroma displays fibrotic reactions to various degrees, which is termed desmoplasia, as does organ fibrosis. Of note, the TGF-β and Wnt signaling pathways are activated in cancer tissues, which are likely to contribute to cancer progression.

The Hippo pathway is an evolutionarily conserved signaling pathway that has multiple biological functions in development, homeostasis, and regeneration of tissues and organs. Yes-associated protein (YAP) and its paralog, the transcriptional coactivator with PDZ-binding motif (TAZ), which is also known as the WW-domain containing transcription regulator-1 (WWTR1), are core components of the Hippo pathway [6–9]. While YAP and TAZ have a lot of similarities in their structures, regulations, and functions, they have distinct and non-overlapping roles. YAP/TAZ signaling is involved in both fibrosis and cancer. Accumulating evidence indicates that YAP/TAZ act as sensors of mechanical forces and modulate the fibrotic response as well as the behavior of cancer cells. Moreover, YAP/TAZ function in a cooperative manner with other established signaling pathways. In particular, the crosstalk with the TGF-β and Wnt signaling pathways is closely associated with both fibrosis and cancer [10,11]. Here, we review the recent progress in the pathobiological role of YAP/TAZ signaling and its function as a molecular link between fibrosis and cancer.

2. Overview of YAP/TAZ Signaling

The core components of the Hippo pathway were initially identified by genetic screens to identify tumor suppressors in *Drosophila*, and 'Hippo' originated from the morphological phenotype of a *Drosophila* mutant [12]. Subsequent cellular and genetic studies have demonstrated that the core components of the Hippo pathway are highly conserved from *Drosophila* to mammals. The mammalian Hippo pathway includes a kinase cascade of mammalian sterile 20-like kinase 1/2 (MST1/2) and large tumor suppressor kinase 1/2 (LATS1/2). MST1/2 in complex with the regulatory protein SAV1 phosphorylate hydrophobic motifs of LATS1/2, which form a complex with the regulatory protein, MOB1 [13]. Phosphorylated and activated LATS1/2 then phosphorylate serine residues of YAP/TAZ. Upon phosphorylation by LATS1/2, YAP/TAZ interact with 14-3-3, which sequesters YAP/TAZ from nuclear translocation, leading to ubiquitination-mediated proteasomal and autolysosomal degradation [14]. The phosphorylation of YAP/TAZ results in the loss of their transcriptional coactivator function. In contrast, unphosphorylated YAP/TAZ localize to the nucleus, and act mainly through TEAD family transcription factors (TEADs) to stimulate the expression of genes—including CTGF, AXL, BIRC5, and AREG—involved in cell proliferation and the suppression of apoptosis [15]. In addition to TEADs, YAP/TAZ also interact with other transcription factors—such as Smad, Runx2, p73, and TBX5—to mediate cellular context-dependent transcriptional regulation [16]. As a negative regulator of the YAP–TEAD transcriptional complex, VGLL4 directly competes with YAP for binding to TEADs [17].

A variety of upstream signals activate or inhibit YAP/TAZ signaling. Apical-basal polarity regulates YAP/TAZ subcellular localization and activity through interactions with cell-polarity proteins (Scribble and Crumbs) or cell-junction molecules (angiomotin and α-catenin) [18]. Extracellular hormones modulate LATS1/2 kinase activity via G protein-coupled receptor (GPCR) signaling [19]. Serum-borne lysophosphatidic acid (LPA) and sphingosine-1-phosphophate (S1P) act through G12/13-coupled receptors to inhibit LATS1/2, thereby activating YAP/TAZ. Furthermore, recent evidence has shown that a variety of stress signals—such as energy stress, endoplasmic reticulum stress, oxidative stress, and hypoxia—regulate the activity of YAP/TAZ [20–23].

3. Mechanotransduction and YAP/TAZ Activity

In addition to the above-mentioned upstream signals, extracellular mechanical cues including ECM stiffness, cell attachment or detachment, and cellular tension are potent regulators of YAP/TAZ. Dupont et al. first reported the association of YAP/TAZ activity with ECM stiffness and cell spreading [24]. In cells stretched by a stiff ECM, YAP/TAZ localize predominantly to the nucleus, and their transcriptional activity is elevated. On the other hand, their localization is predominantly cytoplasmic on a soft ECM. This regulation is dependent on Rho GTPase and the tension of the actomyosin cytoskeleton. Notably, this process is independent of LATS1/2, because the depletion of LATS1/2 had a marginal effect on the regulation of YAP/TAZ activity by mechanical cues. The LATS1/2-dependent regulation of YAP/TAZ activity by stress fiber (F-actin) formation has been reported [25,26]. This finding was confirmed by the observation that the F-actin-capping/severing proteins cofilin, CapZ, and gelsolin restrict the nuclear localization of YAP [27]. Zhao et al. showed that cell detachment from ECM activates LATS1/2 by promoting cytoskeleton reorganization, which leads to YAP inactivation and apoptosis, which is a process termed anoikis [28].

The mechanisms by which cytoskeletal tension regulates YAP/TAZ are unclear, although the nucleus may play a mechanotransductive role in the regulation of YAP [29]. The focal adhesions and stress fibers that are generated on stiff substrates transduce mechanical forces to the nucleus, leading to nuclear flattening. This increases YAP nuclear import by reducing mechanical restriction in nuclear pores. In contrast, on soft substrates, mechanical forces fail to reach the nucleus, and nucleocytoplasmic shuttling of YAP through nuclear pores is balanced.

Interactions between cells and ECM are largely mediated by the proteins of the integrin family. Focal adhesions composed of integrins, focal adhesion kinase (FAK), and Src play an important role as a sensor of ECM rigidity and in the intracellular transduction of extracellular signals. FAK stimulates Rho-associated protein kinase (ROCK)-dependent actin remodeling and the formation of stress fibers. Integrin-linked kinase (ILK) inactivates NF2, which is an upstream negative regulator of YAP/TAZ signaling, by inhibiting MYPT1 via direct phosphorylation [30]. The integrin-dependent FAK–Src signaling pathway also positively regulates the activity of YAP/TAZ. Integrin-dependent adhesion leads to the activation of p21-activated kinase 1 (PAK1), which is a downstream effector of the small GTPase RAC1. PAK1 directly phosphorylate NF2, which reduces its interactions with YAP, resulting in YAP dephosphorylation and nuclear translocation [31]. Adhesion to fibronectin also enhances YAP nuclear translocation via the FAK–Src–PI3K–PDK1 pathway [32]. However, despite these significant roles of the integrin signaling, cells spreading over polylysine-coated cover glasses, where cells do not form focal adhesions, show nuclear translocation of YAP, suggesting that focal adhesion formation is not always required for YAP mechanoactivation [28]. Interestingly, YAP/TAZ directly regulate a number of proteins involved in focal adhesion and control cytoskeleton stability [33].

4. YAP/TAZ Signaling in Fibrosis

Fibrosis is a sequela of several chronic inflammatory diseases, and progressive fibrosis typically has a devastating clinical course without good therapeutic options. Injury to epithelial cells results in the release of a variety of cytokines, chemokines, and growth factors. These mediators induce fibroblast accumulation and convert fibroblasts into myofibroblasts positive for α-smooth muscle actin (α-SMA). The epithelial repair is facilitated through mutual communication between the regenerating epithelium and activated fibroblasts. In the pathogenesis of tissue fibrosis, such epithelial–mesenchymal interactions are dysregulated due to epithelial regenerative failure or aberrant activation of fibroblasts. Myofibroblasts are the effector cells that are responsible for fibrogenesis and are characterized by the production of large quantities of ECM components, such as collagen and fibronectin, the secretion of proteases, and their contractile ability. Accumulating evidence suggests that YAP/TAZ signaling is linked to the pathophysiology of fibrosis, and aberrant YAP/TAZ activation has been reported in both the epithelial compartment and fibroblasts/myofibroblasts (Table 1).

4.1. Lung Fibrosis

Idiopathic pulmonary fibrosis (IPF) is a progressive chronic interstitial lung disease with a poor prognosis that is characterized by clusters of proliferating fibroblasts termed fibroblastic foci. Remodeling of the ECM and the contraction of fibroblasts contribute to tissue tension or stiffness, which is associated with decreased vital capacity in IPF patients. Liu et al. and our group showed that YAP/TAZ are highly expressed in the nucleus of lung fibroblasts in the fibroblastic foci of the lungs of patients with IPF, indicating that YAP/TAZ in lung fibroblasts play a significant role in lung fibrosis [34,35].

In cell culture experiments, YAP/TAZ accumulated in the nucleus of lung fibroblasts grown on a pathologically stiff matrix, but not on a physiologically compliant matrix. Fibroblasts on a stiff matrix showed increased proliferation, contraction, and ECM production compared to those on a soft matrix. These fibroblast responses to matrix stiffness were ablated by YAP/TAZ knockdown, suggesting that YAP/TAZ activation contributes to such myofibroblastic features. In a follow-up study, Liu et al. showed that the overexpression of the constitutively active mutant form of TAZ in lung fibroblasts on a soft matrix increased the expression of connective tissue growth factor (CTGF) and plasminogen activator inhibitor-1 (PAI-1, also known as SERPINE1), but not that of ECM proteins such as collagens and fibronectin. However, the expression of these ECM proteins was induced by TAZ overexpression in cells cultured on a stiff matrix. These findings indicate that the full activation of TAZ requires the input of mechanical stimuli to facilitate the expression of genes related to fibrosis [36]. Gene expression profiling by RNA-sequencing revealed that TAZ-regulated genes in lung fibroblasts were associated with cell migration and motility, and partly overlapped with those regulated by TGF-β, which is a known mediator of fibrogenesis [35].

Therefore, YAP/TAZ constitute a feed-forward loop that accelerates the fibrotic process. YAP/TAZ act as sensors of ECM stiffness, and their activity is enhanced by the progression of tissue fibrosis. In their roles as effectors, YAP/TAZ, in cooperation with TGF-β signaling, stimulate the production of fibrogenic factors (CTGF and PAI-1) and ECM proteins, thereby promoting the pathogenesis of tissue fibrosis (Figure 1).

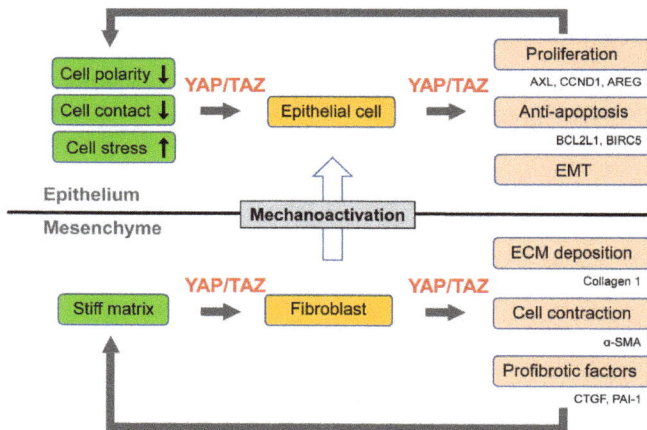

Figure 1. The activation of Yes-associated protein (YAP) and the transcriptional coactivator with PDZ-binding motif (TAZ) in epithelial cells and fibroblasts. In epithelial cells, the disruption of cell polarity, loss of cell contact, and increased cell stress signals activate YAP/TAZ, which promotes cell proliferation and the epithelial–mesenchymal transition (EMT), and inhibits apoptosis. In contrast, in fibroblasts, YAP/TAZ act as sensors of extracellular matrix (ECM) stiffness through the mechanotransduction pathway. YAP/TAZ also stimulate the production of fibrogenic factors and ECM proteins and enhance cell contraction. This process promotes tissue stiffness, thus forming a feed-forward loop of fibroblast activation and tissue fibrosis. YAP/TAZ can also be activated in epithelial cells of fibrotic tissues due to increased ECM stiffness.

Murine models of experimental pulmonary fibrosis support these profibrotic functions of YAP/TAZ. The heterozygous deletion of TAZ in mice led to resistance to lung fibrosis induced by the intratracheal administration of bleomycin [7]. Bleomycin-treated heterozygous mice had a lower tissue fibrosis score, hydroxyproline content, and lung elastance. Furthermore, the adoptive transfer of YAP/TAZ-overexpressing fibroblasts into the tail vein of wild-type mice led to elevated lung fibrogenic responses [34].

Regenerative failure of the respiratory epithelium following both acute and chronic injury is involved in lung fibrogenesis. YAP/TAZ signaling controls the proliferation and differentiation of epithelial progenitor cells in embryo and adult lungs [37]. Xu et al. performed single-cell RNA sequencing-based gene expression profiling of epithelial cells from IPF patients [38]. They identified distinct epithelial cell types with characteristics of conducting airway basal and goblet cells, and an additional subset of atypical transitional cells. Pathway analysis showed that the YAP/TAZ, TGF-β, Wnt, and PI3K signaling pathways were aberrantly activated. Immunofluorescence staining of the lung epithelial cells of IPF patients demonstrated increased nuclear YAP accumulation [39]. Interestingly, YAP was suggested to interact with mTOR/PI3K/AKT signaling to enhance the proliferation and migration, and inhibit the differentiation, of lung epithelial cells.

As such, YAP/TAZ activation in lung fibroblasts and epithelial cells differentially contributes to the pathogenesis of IPF. YAP/TAZ enhance fibrotic reactions while stimulating epithelial cell regeneration, albeit in a pathological manner (Figure 1).

4.2. Kidney Fibrosis

High TAZ expression was demonstrated in the kidneys of patients with IgA nephropathy or membranous nephropathy, both of which are pathologically characterized by renal tubulointerstitial fibrosis [40]. The YAP level is high in the renal tubular epithelium during the regeneration and fibrogenesis stages after acute kidney injury [41]. Increased TAZ nuclear accumulation in the renal tubulointerstitium was observed in three different mouse models of nephropathy (obstructive, diabetic, and toxin-induced renal injuries) [42]. Cell culture experiments involving renal fibroblasts showed that TGF-β-induced canonical Smad2/3 signaling is regulated by ECM stiffness in a manner dependent on YAP/TAZ [43].

In renal tubular epithelial cells, YAP overexpression promoted cell proliferation and YAP/TAZ activation induced by SAV1 deletion triggered epithelial–mesenchymal transition (EMT)-like phenotypic changes [40,41]. YAP/TAZ activation was also found in podocytes in a rat model of glomerular disease induced by intraperitoneal administration of puromycin aminonucleoside [44]. YAP overexpression in podocytes increased the levels of several ECM-related proteins—such as collagen 6, its receptor BCAM, and matrix metalloproteinase ADAMTS1—which promoted basement membrane thickening and stiffening.

4.3. Skin Fibrosis

YAP regulates keratinocyte proliferation and differentiation, and is essential for epidermal homeostasis and regeneration [45,46]. The nuclear accumulation of YAP/TAZ is reported in the dermis during the healing of skin wounds [47]. The closure rate of YAP/TAZ siRNA-treated skin wounds was significantly decreased in a mouse model, indicating that YAP/TAZ promote wound healing. YAP/TAZ modulated the expression of TGF-β signaling pathway components and its targets such as Smad2, p21, and Smad7.

Table 1. YAP/TAZ expression and functional relevance in organ fibrosis.

Organ	Expression in Human Tissues	Animal Model	Cell Culture (Cell Type)	Phenotype	Related Molecule	Reference
Lung	Elevated YAP/TAZ nuclear staining in lung fibroblasts of IPF patients Elevated YAP nuclear staining in lung epithelial cells of IPF patients	• Intratracheal administration of bleomycin • Adoptive transfer of fibroblasts into the tail vein	Lung fibroblast Lung epithelial cell	Cell proliferation, contraction, ECM production, migration, myofibroblast differentiation Cell proliferation, migration, loss of apical-basal polarity	CTGF, PAI-1, Collagen 1, α-SMA AXL, CTGF, AJUBA, Scribble, VANGL1	[7,34–36,39]
Kidney	Elevated TAZ nuclear staining in the renal tubular epithelium and interstitium of patients with IgA nephropathy and membranous nephropathy Elevated YAP nuclear staining in the renal tubular epithelium during the regeneration and fibrogenesis stages after acute kidney injury (AKI)	• Unilateral ureteral obstruction • Streptozotocin-induced renal injury (diabetic nephropathy) • Aristolochic acid-induced nephropathy • Ischemia/reperfusion by clamping the bilateral renal arteries (AKI–chronic kidney disease transition)	Renal fibroblast Renal tubular epithelial cell Podocyte	ECM production, myofibroblast differentiation Cell proliferation, EMT-like phenotype, ECM production ECM production	α-SMA, PAI-1, Collagen 1 TGF-β2, TGF-β receptor 2, CTGF Collagen 6, BCAM, ADAMTS1	[40–44]
Skin	Elevated YAP nuclear staining in the skin nodules of patients with Dupuytren disease Elevated YAP/TAZ nuclear staining in dermal fibroblasts of patients with SSc	• Wound generation by full-thickness punch biopsy (wound healing process) • Administration of bleomycin by osmotic minipump	Dermal fibroblast Keratinocyte	Contraction, myofibroblast differentiation, ECM production Cell proliferation, anti-apoptosis, suppressed differentiation	α-SMA, Collagen 1 CYR61	[46–49]
Liver	Elevated YAP nuclear staining in the stellate cells of human cirrhotic livers caused by infection with hepatitis C virus	• Intraperitoneal administration of carbon tetrachloride • Non-alcoholic steatohepatitis (NASH) model induced by diet rich in fructose, palmitate, and cholesterol	Hepatic stellate cell Hepatocyte	Myofibroblast differentiation, cell proliferation Secretion of profibrotic factors	α-SMA, Collagen 1, CTGF, Glutaminase, MMP2 Indian hedgehog	[50–52]

YAP/TAZ are also activated in the dermal fibroblasts of patients with fibrotic diseases. Piersma et al. showed that YAP expression levels and nuclear localization were elevated in the skin tissue of patients with Dupuytren disease, which is a fibroproliferative disorder of the hands and fingers [48]. The knockdown of YAP in dermal fibroblasts attenuated the TGF-β-mediated formation of contractile actin stress fibers and the deposition of collagen type I. In addition, YAP/TAZ proteins were localized in the nucleus of fibroblasts in skin biopsies from patients with systemic sclerosis (SSc) [49]. Notably, dimethyl fumarate exerts a potent anti-fibrotic effect in SSc dermal fibroblasts by promoting the proteosomal degradation of YAP/TAZ.

4.4. Liver Fibrosis

Hepatic stellate cells (HSCs) are the dominant source of fibrogenic myofibroblasts in patients with chronic liver diseases. The nuclear localization of YAP was found in the stellate cells of mouse fibrotic liver induced by the administration of carbon tetrachloride and in human cirrhotic livers caused by infection with hepatitis C virus [50]. In vitro, siRNA-mediated silencing of YAP or the pharmacological inhibition of YAP with verteporfin, which interferes with the interaction between YAP/TAZ and TEAD, blocked the induction of the expression of its target genes and myofibroblast differentiation from HSCs. In vivo treatment with verteporfin reduced liver fibrogenesis in mice, further demonstrating that YAP is a key regulator of HSC activation. These findings were reinforced by a recent report that the hedgehog pathway-mediated activation of YAP directs HSC differentiation to myofibroblasts [51].

Wang et al. showed that TAZ activation in hepatocytes promotes progression from steatosis to non-alcoholic steatohepatitis (NASH) [52]. TAZ expression was increased in the hepatocytes of patients with NASH. TAZ silencing in hepatocytes prevented and even reversed hepatic inflammation and fibrosis in mouse models of NASH. Mechanistically, TAZ enhanced Indian hedgehog (Ihh) secretion from hepatocytes, which upregulated the expression of profibrotic genes, including collagen 1 and TIMP metallopeptidase inhibitor 1 (TIMP1), in HSCs.

5. YAP/TAZ Signaling in Cancer Cells

YAP/TAZ are upregulated and activated in a variety of human cancers. YAP/TAZ promote tumor initiation, progression, and metastasis. Furthermore, the higher expression or activation of YAP/TAZ is correlated with a poor prognosis, as reviewed in some articles [16,53–58]. Cell culture experiments showed that YAP/TAZ promote proliferation, anti-apoptosis, anchorage-independent growth, drug resistance, and stem cell traits in a variety of cancer cell lines. Xenograft mouse models and genetically engineered mice with tissue-specific overexpression or the deletion of YAP/TAZ or Hippo signaling components demonstrated a crucial role for YAP/TAZ in the formation and growth of tumors [53]. For example, hepatocyte-specific deletion of MST1/2 in mice resulted in YAP activation, leading to significantly enlarged livers and tumor development by the age of five to six months [59]. Available evidence concerning YAP/TAZ expression and functional relevance in cancer is outlined in Table 2.

Table 2. YAP/TAZ expression and functional relevance in cancer.

Organ	Expression in Human Tissues	Animal Model	Cell Culture (Phenotype)	Related Molecule	Reference
Lung	• Elevated YAP/TAZ expression correlates with advanced TNM stage and lymph node metastases and is a predictor of worse prognosis in non-small cell lung cancer.	• TAZ knockdown impairs the growth of subcutaneous xenografts of the lung adenocarcinoma cell line A549 in mice. • YAP deletion inhibits the progression of lung adenocarcinoma in LKB1-deficient Kras^{G12D} mice.	Cell proliferation, anchorage-independent growth, resistance to EGFR-tyrosine kinase inhibitor (EGFR-TKI)	AXL, CYR61, AREG, EREG, NRG1	[60–65]
Kidney	• YAP expression is elevated in clear cell renal cell carcinoma (ccRCC) and mucinous tubular and spindle cell carcinoma. • Elevated YAP expression is a predictor of worse prognosis in ccRCC.	• YAP knockdown impairs the growth of subcutaneous xenografts of the renal cell adenocarcinoma cell line ACHN in mice.	Cell proliferation, migration, anchorage-independent growth	CYR61, c-Myc, Endothelin 1/2	[66–70]
Skin	• Elevated YAP/TAZ expression correlates with tumor thickness and lymph node metastases and is a predictor of worse prognosis in melanoma.	• YAP/TAZ knockdown inhibits lung metastasis following tail-vein injection of the melanoma cell line 1205Lu in mice. • Pharmacological inhibition of YAP/TAZ with verteporfin impairs the growth of subcutaneous xenografts of the BRAF inhibitor-resistant melanoma cell line in mice.	Cell proliferation, invasion, anchorage-independent growth, resistance to BRAF and MEK inhibitors, immune evasion	CTGF, AXL, PD-L1	[71–74]
Liver	• Elevated YAP/TAZ expression correlates with advanced TNM stage and poor tumor differentiation and is a predictor of worse prognosis in hepatocellular carcinoma (HCC). • YAP expression is elevated in cholangiocarcinoma and hepatoblastoma. • Elevated YAP expression correlates with poor response to transarterial chemoembolization in HCC.	• Liver-specific YAP overexpression leads to hepatomegaly followed by liver tumor formation in mice. • YAP/TAZ knockdown impairs the growth of subcutaneous xenografts of the hepatocellular carcinoma cell line Bel7402 in mice.	Cell proliferation, migration, invasion, EMT, resistance to irinotecan	BIRC5, c-Myc, ABCB1, ABCC1, FOXM1, Jag-1	[75–79]

Multiple mechanisms of YAP/TAZ activation have been reported in different cancer types. Several cancers harbor the amplification of YAP as part of the 11q22 amplicon [80]. TAZ-CAMTA1 or YAP-TFE3 gene fusions, which render the N-terminus of YAP/TAZ unresponsive to negative regulation by the Hippo pathway, have been reported in epithelioid hemangioendothelioma, which is a rare form of sarcoma [81,82]. A high frequency of inactivating mutation of NF2, which is a negative regulator of YAP/TAZ signaling, has been found in several types of tumor, such as meningioma, schwannoma, and malignant mesothelioma [83–85]. Except for NF2, DNA mutations in the Hippo pathway components are rare, and no activating mutation of YAP/TAZ has been demonstrated in human cancers. The epigenetic gene silencing of upstream negative regulators of YAP/TAZ such as MST1/2, LATS1/2, and RASSF1A has been reported in several cancers [86–88]. In most tumors, YAP/TAZ signaling can also be activated by a number of extrinsic signals that are enhanced in cancer tissues, which include increased mechanical forces, growth factors, inflammatory mediators, hypoxia, and altered metabolic conditions.

YAP/TAZ enhance tumor proliferation and survival by transactivating target genes associated with cell-cycle progression and anti-apoptosis. YAP physically interacts with mutant p53 protein and upregulates cyclin A (CCNA), cyclin B (CCNB), and CDK1 in cooperation with the NF-Y transcriptional factor in breast cancer cells [89]. YAP enhances the proliferation of malignant mesothelioma cells by directly upregulating cyclin D1 (CCND1) [90]. An EGFR ligand, amphiregulin (AREG), is also induced by YAP/TAZ, and stimulates cell proliferation and malignant transformation by activating EGFR signaling in a non-cell autonomous manner [60,62]. YAP and TBX5 form a complex with β-catenin and transactivate anti-apoptosis genes, including BCL2L1 and BIRC5, in various types of cancer cell [91]. Furthermore, YAP/TAZ promote cell cycle progression by inducing the expression of the proto-oncogene, c-Myc [92].

Recent genome-wide occupancy profiling studies revealed that YAP/TAZ binding sites are not restricted to gene promoters. Zanconato et al. found, by chromatin immunoprecipitation (ChIP)-sequence analysis of breast cancer cells, that most YAP/TAZ-bound cis-regulatory regions coincide with enhancer elements, which are located distant from transcription start sites [92]. Notably, the AP-1 transcription factor was present in most YAP/TAZ/TEAD-binding sites, forming a complex and synergistically activating target genes involved in cell cycle progression. Most recently, it has been shown that YAP/TAZ-bound enhancers recruit bromodomain-containing protein 4 (BRD4), which is a critical epigenetic modulator that binds to acetylated histone, boosting the expression of growth-regulating genes [93].

6. YAP/TAZ Signaling in the EMT

EMT involves the transformation of epithelial cells to mesenchymal cells. The EMT of cancer cells confers aggressive features such as invasion, resistance to apoptosis, and at advanced disease stages, chemoresistance [94]. During the EMT, epithelial cells lose their apical–basal polarity, basement membrane attachment, and cell–cell contact. In turn, they gain migratory and invasive properties associated with the mesenchymal phenotype [95]. The EMT is executed by a subset of transcription factors, including ZEB1/2, Snail/Slug, and Twist. In cancer cells, the EMT is also associated with cancer stem cell (CSC) characteristics, anti-apoptosis, and drug resistance [94]. The pathological significance of the EMT in tissue fibrosis is controversial, but the upregulation of EMT-related factors and the colocalization of epithelial and mesenchymal markers in fibrotic tissues have been demonstrated [96,97].

YAP/TAZ signaling is involved in the EMT. Lei et al. showed that the overexpression of TAZ in mammary epithelial cells induces the EMT [98]. YAP regulates multiple EMT-related genes by inducing SOX2 expression in cooperation with Oct4 in non-small lung cancer cells [99]. Moreover, YAP can rescue cell death induced by the suppression of KRAS in cancer cells [100,101]. YAP and FOS, a member of the AP-1 transcription factor family, coordinately regulate the EMT by directly binding to the promoter region of Vimentin and Slug.

Interestingly, the EMT itself can promote TAZ activation. Induction of the EMT by ectopic overexpression of Twist and Snail leads to the delocalization of Scribble, which is a scaffold protein involved in cell polarization, from the cell membrane. Next, TAZ is relieved from the Scribble polarity complex and promotes CSC traits in breast cancer cells [102]. These findings suggest that the TAZ-mediated EMT serves as a self-sustaining mechanism of TAZ activation. Increasing matrix stiffness can also induce the EMT and promote tumor invasion and metastasis in breast cancer cells, in which Twist plays an essential role as a mechanomediator [103].

ZEB1 directly binds to YAP to regulate the transcription of target genes [104]. In terms of clinical importance, the set of ZEB1/YAP target genes was a strong predictor of poor relapse-free survival, therapy resistance, and an increased risk of metastasis in breast cancer. Similarly, Snail/Slug form a complex with YAP/TAZ to regulate the self-renewal and differentiation of skeletal stem cells [105].

Therefore, YAP/TAZ in close association with ZEB1/2, Snail/Slug, and Twist govern the EMT, thereby inducing the malignant features of cancer cells. Indeed, YAP/TAZ confer CSC-related traits—such as tumor initiation, drug resistance, and metastasis—in a wide range of human cancers. The first evidence came from the work of Cordenonsi et al. [102]. This group showed that TAZ is highly expressed in the CD44^{high}/CD24^{low} subpopulation with CSC properties in primary breast tumors. TAZ was required for self-renewal and the formation of high-grade tumors. The gene expression profiling of patient-derived breast cancer stem cell lines showed that TAZ is a central mediator of metastasis and resistance to chemotherapy [106]. YAP directly upregulates SOX9 and induces CSC properties in esophageal cancer cells [107]. YAP endows urothelial cancer cells with CSC traits by directly binding to the enhancer region of SOX2 [108]. Conversely, SOX2 directly represses two negative regulators of YAP/TAZ, NF2 and WWC1, leading to activation of YAP and maintenance of CSCs in osteosarcomas [109].

7. YAP/TAZ Signaling in the Cancer Microenvironment

The interactions between cancer cells and their supporting stroma are critical determinants of cancer initiation and progression. The structural components of tumor stroma, which constitute the tumor microenvironment, include ECM, blood and lymphatic vessels, and stromal cells including endothelial cells, immune cells, and cancer-associated fibroblasts (CAFs). Fibroblasts are activated by cancer-associated soluble factors such as TGF-β and produce ECM proteins. Collagens and other ECM proteins, including fibronectins, proteoglycans, and tenascin C, are overproduced in several cancers [110]. In addition, lysyl oxidase (LOX) and LOX-like (LOXLs) enzymes mediate the process of covalent intramolecular and intermolecular crosslinking of collagen fibers and other ECM components, resulting in increased tissue stiffness [111]. This desmoplastic stiff stroma is generally considered to enhance the proliferation and motility of cancer cells [112]. Indeed, ECM stiffness is correlated with the progression and a poor prognosis of cancer [111,113]. Moreover, treatment with the ECM-degrading enzyme hyaluronidase decreased interstitial pressure and reduced the aggression of pancreatic ductal cancer, suggesting that the ECM has potential as a therapeutic target for cancer [114].

As mentioned above, the activity of YAP/TAZ is regulated by ECM stiffness. The activation of YAP/TAZ by the stiff ECM of fibrotic tissues can promote the proliferation and survival of cancer cells. Jang et al. showed that YAP and TEAD directly control the mechanical cue-dependent transcription of Skp2, which is important for cell cycle progression in breast cancer cells [115]. Agrin, an ECM component, was reported to activate YAP by suppressing the Hippo pathway, leading to the development of liver cancer [116–118]. Agrin signals matrix and cellular rigidity by activating the integrin-FAK-ILK signaling axis. This, in turn, stimulates PAK1, which inactivates NF2, a negative regulator of YAP/TAZ. Agrin also activates YAP through RhoA-dependent actin cytoskeletal rearrangements. Agrin is secreted by platelet-derived growth factor (PDGF)-stimulated HSCs and promotes hepatocarcinogenesis [119]. Glypican-3, a member of the glypican family of heparan sulfate proteoglycans, is also involved in the progression of hepatocellular carcinoma (HCC).

HN3, a conformation-specific antibody against glypican-3, inhibits the proliferation of HCC cells in vitro and in vivo by inducing cell-cycle arrest through YAP downregulation [120].

CAFs promote cancer invasion and metastasis by producing soluble factors and matrix remodeling [110]. A coculture experiment involving carcinoma cells and stromal fibroblasts revealed that CAFs promote the invasion of cancer cells by making passageways in the ECM [121]. CAFs generate a gap in the basement membrane and ECM by releasing matrix metalloproteinases and pulling ECM fibers via direct mechanical forces [122].

A stiff ECM enhances the activity of YAP/TAZ not only in cancer cells but also in stromal cells, including CAFs. Recently, Calvo et al. demonstrated that YAP/TAZ in CAFs play a crucial role in the progression of mammary tumors [123]. First, they performed a global mRNA expression profiling analysis of fibroblasts isolated from mice with breast cancer of various disease stages, and found that the gene signature of YAP/TAZ signaling is enriched in CAFs compared with in normal mammary fibroblasts (NFs). Consistently, YAP was predominantly cytoplasmic in NFs but active in the nucleus in CAFs. The siRNA-mediated silencing of YAP reduced the ability of CAFs to contract a collagen-rich matrix and promote invasion of cancer cells and angiogenesis. These findings indicate that YAP is required for the tumor-promoting functions of CAFs. Furthermore, a series of cell culture experiments revealed that the YAP-mediated activation of myosin light chain 2 is critical for generating CAFs, which subsequently promote matrix remodeling and tumor invasion. Moreover, a stiff ECM induced the formation of actin stress fibers in fibroblasts, leading to YAP activation, thus establishing a self-sustaining feed-forward loop to maintain the phenotype of CAFs (Figure 2).

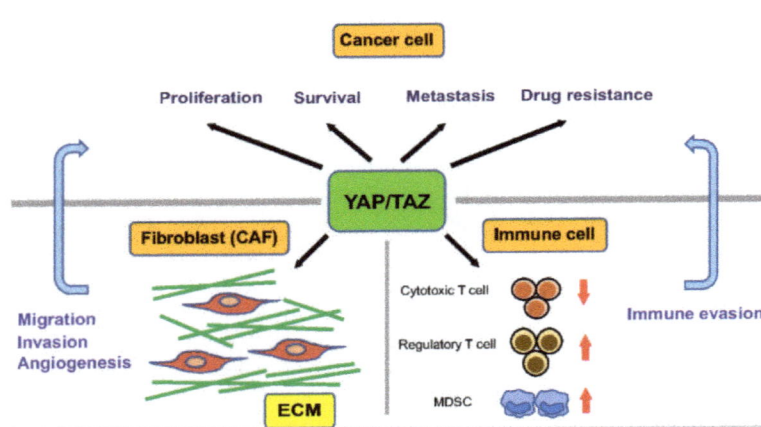

Figure 2. The function of YAP/TAZ in cancer cells and in the cancer microenvironment. YAP/TAZ enhance the proliferation, survival, metastasis, and drug resistance of cancer cells. The cancer microenvironment comprises ECM and stromal cells, such as cancer-associated fibroblasts (CAFs) and immune cells. The activation of YAP/TAZ in CAFs promotes migration and invasion of cancer cells, and angiogenesis. YAP/TAZ facilitate tumor immune evasion by suppressing cytotoxic T cells through PD-L1 expression in cancer cells, and supporting myeloid-derived suppressor cells (MDSCs) and regulatory T cells.

The myocardin-related transcription factor (MRTF)—serum response factor (SRF) pathway is also involved in the response to mechanical stress and is under the control of Rho GTPase signaling. MRTF and SRF promote the expression of dozens of cytoskeleton-related genes, including that encoding α-SMA. CAFs exhibit elevated MRTF–SRF signaling, which is required for their contractile and pro-invasive properties [124]. The MRTF–SRF and YAP–TEAD pathways can indirectly activate each other by modulating actin cytoskeletal dynamics in CAFs.

Recent studies have uncovered new roles for YAP/TAZ signaling in the host immune response in the tumor microenvironment (Figure 2). Cancers escape from host immunity by upregulating PD-L1, the interaction of which with PD-1 receptors on activated T cells reduces the proliferative capacity and effector function of cytotoxic T cells. YAP/TAZ promote immune evasion by upregulating PD-L1 in cancer cells, via directly binding to its promoter with TEADs in human cancer cells [125,126]. Coculture experiments demonstrated that the overexpression of TAZ in cancer cells is sufficient to disrupt the functions of T cells by upregulating the expression of PD-L1. TAZ-dependent upregulation of PD-L1 can be induced by activation of lactate-mediated G protein-coupled receptor 81 (GPR81) [127].

YAP/TAZ also recruit myeloid-derived suppressor cells (MDSCs) by upregulating the production of proinflammatory cytokines, including CXCL5 and TNF-α [128,129]. The hepatocyte-specific deletion of MST1/2 in mice leads to increased macrophage infiltration and HCC formation in the liver. Recently, Kim et al. showed that monocyte chemoattractant protein 1 (MCP1) expression in hepatocytes are required for tumor formation in the MST1/2 double knockout mice [130]. MCP1 recruits macrophages with both M1 and M2 characteristics, which regulate hepatocyte proliferation and survival. MCP1 is a direct transcriptional target of YAP/TAZ in hepatocytes. Furthermore, in HCC, YAP promotes differentiation of naïve T cells to regulatory T cells by directly upregulating the expression of TGF-β receptor 2 (TGFBR2) [131]. Moreover, YAP is highly expressed in regulatory T cells and bolsters their suppression of anti-tumor immunity by activating Activin signaling [132].

8. Crosstalk with the TGF-β and Wnt Pathways

YAP/TAZ signaling cooperates with the TGF-β and Wnt pathways in epithelial cells and fibroblasts, which accelerates profibrotic reactions. Such crosstalk is presumed to be also active in cancer cells and CAFs, in which it promotes tumor progression.

8.1. Crosstalk with TGF-β Signaling

TGF-β signaling is closely associated with both tissue fibrosis and cancer [133]. TGF-β drives the differentiation of quiescent fibroblasts into matrix-secreting myofibroblasts, which is a key step in tissue fibrosis. As mentioned above, CAFs promote tumor progression by enhancing ECM deposition, angiogenesis, and secretion of growth factors. Importantly, TGF-β is a strong inducer of the CAF phenotype and TGF-β signaling is activated in CAFs [134,135]. Therefore, it is conceivable that fibroblast activation and ECM remodeling in tissue fibrosis promote tumor progression.

TGF-β suppresses the proliferation of epithelial cells and acts as a tumor suppressor during the early stages of tumorigenesis. However, TGF-β is also a potent inducer of the EMT. The activation of TGF-β signaling and EMT-related transcriptional changes are also observed in epithelial cells derived from IPF lung tissues [39]. It is tempting to speculate that epithelial cells undergoing the EMT in tissue fibrosis are primed to gain malignant features following carcinogenesis by mutations in oncogenic driver genes. TGF-β binds to TGF-β receptors type I and II, which then phosphorylate the intracellular signal transducers, Smad2 and Smad3. These form a heterotrimeric complex with Smad4 and translocate into the nucleus, resulting in activation of the expression of various genes. The interaction between TAZ and Smad2/3 is critical in nucleocytoplasmic shuttling of the Smad2/3-Smad4 complex [136]. In the absence of TAZ, the Smad complex fails to accumulate in the nucleus, and TGF-β-mediated transcription is disrupted in embryonic stem cells. RASSF1A regulates the TGF-β-induced interaction between YAP and Smad2, which suppresses the invasion of cancer cells [137].

The promoters of a variety of genes harbor both Smad-binding elements (SBEs) and TEAD-binding elements [138]. In the nucleus, YAP/TAZ, TEAD, and Smad2/3 form a complex that regulates the transcription of these genes. NEGR1 and UCA1 are synergistically activated by this complex, and are necessary for maintaining the tumorigenic activity of metastatic breast cancer cells. The YAP–TEAD–Smad complex also binds to the promoter of CTGF, which enhances ECM production and proliferation by malignant mesothelioma cells [139]. A recent study has shown that TGF-β increases the TAZ protein level in fibroblasts via Smad3-independent, and p38 and MRTF-mediated mechanisms [140].

8.2. Crosstalk with Wnt Signaling

Wnt signaling also plays a key role in fibrogenesis and carcinogenesis. The activation of Wnt signaling in epithelial cells and fibroblasts is a common feature of fibrotic tissues and contributes to the proliferative and migratory activities in various organs, including the lung, kidney, heart, and skin [3,141–143]. The inhibition of Wnt signaling by a small molecule, ICG-001, prevented and even reversed bleomycin-induced lung fibrosis in mice [144]. Aberrant Wnt signaling is also involved in carcinogenesis, as genetic alterations in Wnt signaling components are frequently observed in various cancers [145].

In the absence of Wnt ligands, β-catenin is captured by the cytoplasmic destruction complex, which consists of a scaffold protein, Axin, and other components such as APC, GSK3β, and CK1. This capture results in the phosphorylation and degradation of β-catenin by β-TrCP. In the presence of Wnt ligands, the destruction complex is dissociated, leading to the stabilization and nuclear accumulation of β-catenin. In the nucleus, β-catenin interacts with the transcription factor TCF/LEF to activate the transcription of target genes.

There is complex crosstalk between Wnt and YAP/TAZ signaling, and YAP/TAZ activity is indispensable for Wnt-induced biological responses [11,146]. Azzolin et al. showed that phosphorylated β-catenin serves as a presenting factor for TAZ to β-TrCP and promotes TAZ degradation [11]. Wnt signaling stabilizes β-catenin, which leads to nuclear accumulation of TAZ and enhancement of its transcriptional activity. Gene expression profiling revealed that the regulation of a considerable portion of Wnt target genes was TAZ-dependent in both mammary epithelial cells and colorectal cancer cells. In a follow-up study, they showed that YAP/TAZ are integral components of the β-catenin destruction complex [146]. The activity of YAP/TAZ is regulated by Wnt signaling in a similar manner to β-catenin and is required for stabilization of β-catenin.

Cai et al. reported that Wnt signaling mediates the β-catenin-independent activation of YAP/TAZ [147]. APC, a core component of the β-catenin destruction complex, interacts with SAV1 and LATS1, which are upstream regulators of YAP/TAZ. The regulation of YAP/TAZ signaling by APC is essential for intestinal tumorigenesis. DVL, a scaffolding protein involved in Wnt signaling, is required for the nucleocytoplasmic shuttling of YAP [148]. Wnt can also activate YAP/TAZ via the Frizzled (a GPCR-like Wnt receptor)–Gα12/13–Rho GTPases–LATS1/2 axis, which is independent of canonical Wnt/β-catenin signaling [149]. Moreover, YAP was suggested to be a direct transcriptional target of Wnt/β-catenin signaling, and its expression is required for the growth of colorectal cancer cells [150].

Multiple layers of crosstalk between Wnt signaling and YAP/TAZ activity have been reported, and these mechanisms may be involved in the pathogenesis of tissue fibrosis and the progression of cancers.

9. Conclusions

There is abundant epidemiologic evidence that fibrosis—including pulmonary fibrosis and liver cirrhosis—increases the risk of cancer in various organs. Among the mechanisms linking these two diseases, recent studies strongly suggest that YAP/TAZ signaling plays an important role. (1) YAP/TAZ contribute to fibrosis by acting both as mechanosensors and profibrotic effectors in fibroblasts/myofibroblasts. Fibroblast activation and fibrotic changes in tissue fibrosis are similar to cancer stromal reactions and provide a tumorigenic microenvironment. (2) Increased matrix stiffness in tissue fibrosis can mechanoactivate YAP/TAZ in epithelial cells, which promotes cell proliferation and survival. Other extrinsic signals, including oxidative stress and hypoxia, also activate YAP/TAZ and promote tumorigenesis. (3) YAP/TAZ signaling cooperates with the TGF-β and Wnt pathways in epithelial cells and fibroblasts, which may exert profibrotic and tumorigenic effects.

(4) YAP/TAZ signaling, possibly in cooperation with TGF-β signaling, is capable of inducing the EMT. EMT-related alterations in epithelial cells in fibrotic tissue may promote tumorigenesis. In cancer cells, the YAP/TAZ-mediated EMT may further promote tumorigenicity and cancer stemness.

YAP/TAZ signaling has potential as a target for anti-fibrosis and anti-cancer therapies. For example, by high-throughput screening, verteporfin was identified as a small molecule inhibitor of the interaction between YAP/TAZ and TEAD, and reversed the malignant behavior of cancer cells, although many off-target effects and general toxicity diminish its therapeutic potential [151,152]. A cell-based screen of chemical reagents that induce the recruitment of YAP to the cytosol showed that dobutamine inhibits the YAP-dependent gene transcription [153]. Rho and ROCK inhibitors abolish YAP/TAZ nuclear localization and activation [24]. Several studies have suggested that both fibrosis and cancer could be treated by inhibiting Rho and ROCK [154,155]. Statins, which downregulate YAP/TAZ activity via the mevalonate metabolic pathway, delay fibrosis progression and reduce the risk of liver cancer in patients with HCV infection [156]. As YAP/TAZ directly upregulate PD-L1 in cancer cells, PD-1/PD-L1 checkpoint inhibitors, standard therapies for various types of cancer, could be effective against cancers with YAP/TAZ activation.

However, targeting YAP/TAZ signaling may cause adverse effects because YAP/TAZ play critical roles in organ homeostasis and regeneration. Therefore, molecular partners directing their tumorigenic functions, such as AP-1, can also be promising therapeutic targets. Further studies are needed to identify the group of patients with tissue fibrosis or cancer that is most likely to benefit from therapies targeting YAP/TAZ signaling.

Funding: This work was supported by KAKENHI (Grants-in-Aid for Scientific Research) from the Ministry of Education, Culture, Sports, Science, and Technology of Japan (Grant # 17K15009 to S. N., Grant # 18K08170 to A. S., and Grant # 16H02653 to T. N.).

Conflicts of Interest: The authors declare no conflict of interest.

References

1. Rockey, D.C.; Bell, P.D.; Hill, J.A. Fibrosis—A common pathway to organ injury and failure. *N. Engl. J. Med.* **2015**, *372*, 1138–1149. [CrossRef] [PubMed]
2. Saito, A.; Horie, M.; Nagase, T. TGF-beta Signaling in Lung Health and Disease. *Int. J. Mol. Sci.* **2018**, *19*, 2460. [CrossRef] [PubMed]
3. Lam, A.P.; Flozak, A.S.; Russell, S.; Wei, J.; Jain, M.; Mutlu, G.M.; Budinger, G.R.; Feghali-Bostwick, C.A.; Varga, J.; Gottardi, C.J. Nuclear beta-catenin is increased in systemic sclerosis pulmonary fibrosis and promotes lung fibroblast migration and proliferation. *Am. J. Respir. Cell Mol. Boil.* **2011**, *45*, 915–922. [CrossRef] [PubMed]
4. Naccache, J.M.; Gibiot, Q.; Monnet, I.; Antoine, M.; Wislez, M.; Chouaid, C.; Cadranel, J. Lung cancer and interstitial lung disease: A literature review. *J. Thorac. Dis.* **2018**, *10*, 3829–3844. [CrossRef] [PubMed]
5. Affo, S.; Yu, L.X.; Schwabe, R.F. The Role of Cancer-Associated Fibroblasts and Fibrosis in Liver Cancer. *Annu. Rev. Pathol.* **2017**, *12*, 153–186. [CrossRef] [PubMed]
6. Piccolo, S.; Dupont, S.; Cordenonsi, M. The biology of YAP/TAZ: Hippo signaling and beyond. *Physiol. Rev.* **2014**, *94*, 1287–1312. [CrossRef] [PubMed]
7. Mitani, A.; Nagase, T.; Fukuchi, K.; Aburatani, H.; Makita, R.; Kurihara, H. Transcriptional coactivator with PDZ-binding motif is essential for normal alveolarization in mice. *Am. J. Respir. Crit. Care Med.* **2009**, *180*, 326–338. [CrossRef] [PubMed]
8. Makita, R.; Uchijima, Y.; Nishiyama, K.; Amano, T.; Chen, Q.; Takeuchi, T.; Mitani, A.; Nagase, T.; Yatomi, Y.; Aburatani, H.; et al. Multiple renal cysts, urinary concentration defects, and pulmonary emphysematous changes in mice lacking TAZ. *Am. J. Physiol. Ren. Physiol.* **2008**, *294*, F542–F553. [CrossRef] [PubMed]
9. Morin-Kensicki, E.M.; Boone, B.N.; Howell, M.; Stonebraker, J.R.; Teed, J.; Alb, J.G.; Magnuson, T.R.; O'Neal, W.; Milgram, S.L. Defects in yolk sac vasculogenesis, chorioallantoic fusion, and embryonic axis elongation in mice with targeted disruption of Yap65. *Mol. Cell. Boil.* **2006**, *26*, 77–87. [CrossRef] [PubMed]
10. Saito, A.; Nagase, T. Hippo and TGF-beta interplay in the lung field. *Am. J. Physiol. Lung Cell. Mol. Physiol.* **2015**, *309*, L756–L767. [CrossRef] [PubMed]

11. Azzolin, L.; Zanconato, F.; Bresolin, S.; Forcato, M.; Basso, G.; Bicciato, S.; Cordenonsi, M.; Piccolo, S. Role of TAZ as mediator of Wnt signaling. *Cell* **2012**, *151*, 1443–1456. [CrossRef] [PubMed]

12. Udan, R.S.; Kango-Singh, M.; Nolo, R.; Tao, C.; Halder, G. Hippo promotes proliferation arrest and apoptosis in the Salvador/Warts pathway. *Nat. Cell Biol.* **2003**, *5*, 914–920. [CrossRef] [PubMed]

13. Chan, E.H.; Nousiainen, M.; Chalamalasetty, R.B.; Schafer, A.; Nigg, E.A.; Sillje, H.H. The Ste20-like kinase Mst2 activates the human large tumor suppressor kinase Lats1. *Oncogene* **2005**, *24*, 2076–2086. [CrossRef] [PubMed]

14. Meng, Z.; Moroishi, T.; Guan, K.L. Mechanisms of Hippo pathway regulation. *Genes Dev.* **2016**, *30*, 1–17. [CrossRef] [PubMed]

15. Ota, M.; Sasaki, H. Mammalian Tead proteins regulate cell proliferation and contact inhibition as transcriptional mediators of Hippo signaling. *Development* **2008**, *135*, 4059–4069. [CrossRef] [PubMed]

16. Pan, D. The hippo signaling pathway in development and cancer. *Dev. Cell* **2010**, *19*, 491–505. [CrossRef] [PubMed]

17. Zhang, W.; Gao, Y.; Li, P.; Shi, Z.; Guo, T.; Li, F.; Han, X.; Feng, Y.; Zheng, C.; Wang, Z.; et al. VGLL4 functions as a new tumor suppressor in lung cancer by negatively regulating the YAP-TEAD transcriptional complex. *Cell Res.* **2014**, *24*, 331–343. [CrossRef] [PubMed]

18. Kim, N.G.; Koh, E.; Chen, X.; Gumbiner, B.M. E-cadherin mediates contact inhibition of proliferation through Hippo signaling-pathway components. *Proc. Natl. Acad. Sci. USA* **2011**, *108*, 11930–11935. [CrossRef] [PubMed]

19. Yu, F.X.; Zhao, B.; Panupinthu, N.; Jewell, J.L.; Lian, I.; Wang, L.H.; Zhao, J.; Yuan, H.; Tumaneng, K.; Li, H.; et al. Regulation of the Hippo-YAP pathway by G-protein-coupled receptor signaling. *Cell* **2012**, *150*, 780–791. [CrossRef] [PubMed]

20. DeRan, M.; Yang, J.; Shen, C.H.; Peters, E.C.; Fitamant, J.; Chan, P.; Hsieh, M.; Zhu, S.; Asara, J.M.; Zheng, B.; et al. Energy stress regulates hippo-YAP signaling involving AMPK-mediated regulation of angiomotin-like 1 protein. *Cell Rep.* **2014**, *9*, 495–503. [CrossRef] [PubMed]

21. Ma, B.; Chen, Y.; Chen, L.; Cheng, H.; Mu, C.; Li, J.; Gao, R.; Zhou, C.; Cao, L.; Liu, J.; et al. Hypoxia regulates Hippo signalling through the SIAH2 ubiquitin E3 ligase. *Nat. Cell Biol.* **2015**, *17*, 95–103. [CrossRef] [PubMed]

22. Wu, H.; Wei, L.; Fan, F.; Ji, S.; Zhang, S.; Geng, J.; Hong, L.; Fan, X.; Chen, Q.; Tian, J.; et al. Integration of Hippo signalling and the unfolded protein response to restrain liver overgrowth and tumorigenesis. *Nat. Commun.* **2015**, *6*, 6239. [CrossRef] [PubMed]

23. Shao, D.; Zhai, P.; Del Re, D.P.; Sciarretta, S.; Yabuta, N.; Nojima, H.; Lim, D.S.; Pan, D.; Sadoshima, J. A functional interaction between Hippo-YAP signalling and FoxO1 mediates the oxidative stress response. *Nat. Commun.* **2014**, *5*, 3315. [CrossRef] [PubMed]

24. Dupont, S.; Morsut, L.; Aragona, M.; Enzo, E.; Giulitti, S.; Cordenonsi, M.; Zanconato, F.; Le Digabel, J.; Forcato, M.; Bicciato, S.; et al. Role of YAP/TAZ in mechanotransduction. *Nature* **2011**, *474*, 179–183. [CrossRef] [PubMed]

25. Sansores-Garcia, L.; Bossuyt, W.; Wada, K.; Yonemura, S.; Tao, C.; Sasaki, H.; Halder, G. Modulating F-actin organization induces organ growth by affecting the Hippo pathway. *EMBO J.* **2011**, *30*, 2325–2335. [CrossRef] [PubMed]

26. Wada, K.; Itoga, K.; Okano, T.; Yonemura, S.; Sasaki, H. Hippo pathway regulation by cell morphology and stress fibers. *Development* **2011**, *138*, 3907–3914. [CrossRef] [PubMed]

27. Aragona, M.; Panciera, T.; Manfrin, A.; Giulitti, S.; Michielin, F.; Elvassore, N.; Dupont, S.; Piccolo, S. A mechanical checkpoint controls multicellular growth through YAP/TAZ regulation by actin-processing factors. *Cell* **2013**, *154*, 1047–1059. [CrossRef] [PubMed]

28. Zhao, B.; Li, L.; Wang, L.; Wang, C.Y.; Yu, J.; Guan, K.L. Cell detachment activates the Hippo pathway via cytoskeleton reorganization to induce anoikis. *Genes Dev.* **2012**, *26*, 54–68. [CrossRef] [PubMed]

29. Elosegui-Artola, A.; Andreu, I.; Beedle, A.E.M.; Lezamiz, A.; Uroz, M.; Kosmalska, A.J.; Oria, R.; Kechagia, J.Z.; Rico-Lastres, P.; Le Roux, A.L.; et al. Force Triggers YAP Nuclear Entry by Regulating Transport across Nuclear Pores. *Cell* **2017**, *171*, 1397–1410. [CrossRef] [PubMed]

30. Serrano, I.; McDonald, P.C.; Lock, F.; Muller, W.J.; Dedhar, S. Inactivation of the Hippo tumour suppressor pathway by integrin-linked kinase. *Nat. Commun.* **2013**, *4*, 2976. [CrossRef] [PubMed]

31. Sabra, H.; Brunner, M.; Mandati, V.; Wehrle-Haller, B.; Lallemand, D.; Ribba, A.S.; Chevalier, G.; Guardiola, P.; Block, M.R.; Bouvard, D. Beta1 integrin-dependent Rac/group I PAK signaling mediates YAP activation of Yes-associated protein 1 (YAP1) via NF2/merlin. *J. Boil. Chem.* **2017**, *292*, 19179–19197. [CrossRef] [PubMed]

32. Kim, N.G.; Gumbiner, B.M. Adhesion to fibronectin regulates Hippo signaling via the FAK-Src-PI3K pathway. *J. Cell Boil.* **2015**, *210*, 503–515. [CrossRef] [PubMed]

33. Nardone, G.; Oliver-De La Cruz, J.; Vrbsky, J.; Martini, C.; Pribyl, J.; Skladal, P.; Pesl, M.; Caluori, G.; Pagliari, S.; Martino, F.; et al. YAP regulates cell mechanics by controlling focal adhesion assembly. *Nat. Commun.* **2017**, *8*, 15321. [CrossRef] [PubMed]

34. Liu, F.; Lagares, D.; Choi, K.M.; Stopfer, L.; Marinkovic, A.; Vrbanac, V.; Probst, C.K.; Hiemer, S.E.; Sisson, T.H.; Horowitz, J.C.; et al. Mechanosignaling through YAP and TAZ drives fibroblast activation and fibrosis. *Am. J. Physiol. Lung Cell. Mol. Physiol.* **2015**, *308*, L344–L357. [CrossRef] [PubMed]

35. Noguchi, S.; Saito, A.; Mikami, Y.; Urushiyama, H.; Horie, M.; Matsuzaki, H.; Takeshima, H.; Makita, K.; Miyashita, N.; Mitani, A.; et al. TAZ contributes to pulmonary fibrosis by activating profibrotic functions of lung fibroblasts. *Sci. Rep.* **2017**, *7*, 42595. [CrossRef] [PubMed]

36. Jorgenson, A.J.; Choi, K.M.; Sicard, D.; Smith, K.M.; Hiemer, S.E.; Varelas, X.; Tschumperlin, D.J. TAZ activation drives fibroblast spheroid growth, expression of profibrotic paracrine signals, and context-dependent ECM gene expression. *Am. J. Physiol. Cell Physiol.* **2017**, *312*, C277–C285. [CrossRef] [PubMed]

37. Lange, A.W.; Sridharan, A.; Xu, Y.; Stripp, B.R.; Perl, A.K.; Whitsett, J.A. Hippo/Yap signaling controls epithelial progenitor cell proliferation and differentiation in the embryonic and adult lung. *J. Mol. Cell Boil.* **2015**, *7*, 35–47. [CrossRef] [PubMed]

38. Xu, Y.; Mizuno, T.; Sridharan, A.; Du, Y.; Guo, M.; Tang, J.; Wikenheiser-Brokamp, K.A.; Perl, A.T.; Funari, V.A.; Gokey, J.J.; et al. Single-cell RNA sequencing identifies diverse roles of epithelial cells in idiopathic pulmonary fibrosis. *JCI Insight* **2016**, *1*, e90558. [CrossRef] [PubMed]

39. Gokey, J.J.; Sridharan, A.; Xu, Y.; Green, J.; Carraro, G.; Stripp, B.R.; Perl, A.T.; Whitsett, J.A. Active epithelial Hippo signaling in idiopathic pulmonary fibrosis. *JCI Insight* **2018**, *3*. [CrossRef] [PubMed]

40. Seo, E.; Kim, W.Y.; Hur, J.; Kim, H.; Nam, S.A.; Choi, A.; Kim, Y.M.; Park, S.H.; Chung, C.; Kim, J.; et al. The Hippo-Salvador signaling pathway regulates renal tubulointerstitial fibrosis. *Sci. Rep.* **2016**, *6*, 31931. [CrossRef] [PubMed]

41. Xu, J.; Li, P.X.; Wu, J.; Gao, Y.J.; Yin, M.X.; Lin, Y.; Yang, M.; Chen, D.P.; Sun, H.P.; Liu, Z.B.; et al. Involvement of the Hippo pathway in regeneration and fibrogenesis after ischaemic acute kidney injury: YAP is the key effector. *Clin. Sci.* **2016**, *130*, 349–363. [CrossRef] [PubMed]

42. Anorga, S.; Overstreet, J.M.; Falke, L.L.; Tang, J.; Goldschmeding, R.G.; Higgins, P.J.; Samarakoon, R. Deregulation of Hippo-TAZ pathway during renal injury confers a fibrotic maladaptive phenotype. *FASEB J. Off. Publ. Fed. Am. Soc. Exp. Boil.* **2018**, *32*, 2644–2657. [CrossRef] [PubMed]

43. Szeto, S.G.; Narimatsu, M.; Lu, M.; He, X.; Sidiqi, A.M.; Tolosa, M.F.; Chan, L.; De Freitas, K.; Bialik, J.F.; Majumder, G.; et al. YAP/TAZ Are Mechanoregulators of TGF-beta-Smad Signaling and Renal Fibrogenesis. *J. Am. Soc. Nephrol.* **2016**, *27*, 3117–3128. [CrossRef] [PubMed]

44. Rinschen, M.M.; Grahammer, F.; Hoppe, A.K.; Kohli, P.; Hagmann, H.; Kretz, O.; Bertsch, S.; Hohne, M.; Gobel, H.; Bartram, M.P.; et al. YAP-mediated mechanotransduction determines the podocyte's response to damage. *Sci. Signal.* **2017**, *10*, eaaf8165. [CrossRef] [PubMed]

45. Schlegelmilch, K.; Mohseni, M.; Kirak, O.; Pruszak, J.; Rodriguez, J.R.; Zhou, D.; Kreger, B.T.; Vasioukhin, V.; Avruch, J.; Brummelkamp, T.R.; et al. Yap1 acts downstream of alpha-catenin to control epidermal proliferation. *Cell* **2011**, *144*, 782–795. [CrossRef] [PubMed]

46. Zhang, H.; Pasolli, H.A.; Fuchs, E. Yes-associated protein (YAP) transcriptional coactivator functions in balancing growth and differentiation in skin. *Proc. Natl. Acad. Sci. USA* **2011**, *108*, 2270–2275. [CrossRef] [PubMed]

47. Lee, M.J.; Byun, M.R.; Furutani-Seiki, M.; Hong, J.H.; Jung, H.S. YAP and TAZ regulate skin wound healing. *J. Investig. Dermatol.* **2014**, *134*, 518–525. [CrossRef] [PubMed]

48. Piersma, B.; de Rond, S.; Werker, P.M.; Boo, S.; Hinz, B.; van Beuge, M.M.; Bank, R.A. YAP1 Is a Driver of Myofibroblast Differentiation in Normal and Diseased Fibroblasts. *Am. J. Pathol.* **2015**, *185*, 3326–3337. [CrossRef] [PubMed]

49. Toyama, T.; Looney, A.P.; Baker, B.M.; Stawski, L.; Haines, P.; Simms, R.; Szymaniak, A.D.; Varelas, X.; Trojanowska, M. Therapeutic Targeting of TAZ and YAP by Dimethyl Fumarate in Systemic Sclerosis Fibrosis. *J. Investig. Dermatol.* **2018**, *138*, 78–88. [CrossRef] [PubMed]

50. Mannaerts, I.; Leite, S.B.; Verhulst, S.; Claerhout, S.; Eysackers, N.; Thoen, L.F.; Hoorens, A.; Reynaert, H.; Halder, G.; van Grunsven, L.A. The Hippo pathway effector YAP controls mouse hepatic stellate cell activation. *J. Hepatol.* **2015**, *63*, 679–688. [CrossRef] [PubMed]

51. Du, K.; Hyun, J.; Premont, R.T.; Choi, S.S.; Michelotti, G.A.; Swiderska-Syn, M.; Dalton, G.D.; Thelen, E.; Rizi, B.S.; Jung, Y.; et al. Hedgehog-YAP Signaling Pathway Regulates Glutaminolysis to Control Activation of Hepatic Stellate Cells. *Gastroenterology* **2018**, *154*, 1465–1479. [CrossRef] [PubMed]

52. Wang, X.; Zheng, Z.; Caviglia, J.M.; Corey, K.E.; Herfel, T.M.; Cai, B.; Masia, R.; Chung, R.T.; Lefkowitch, J.H.; Schwabe, R.F.; et al. Hepatocyte TAZ/WWTR1 Promotes Inflammation and Fibrosis in Nonalcoholic Steatohepatitis. *Cell Metab.* **2016**, *24*, 848–862. [CrossRef] [PubMed]

53. Zanconato, F.; Cordenonsi, M.; Piccolo, S. YAP/TAZ at the Roots of Cancer. *Cancer Cell* **2016**, *29*, 783–803. [CrossRef] [PubMed]

54. Shibata, M.; Ham, K.; Hoque, M.O. A time for YAP1: Tumorigenesis, immunosuppression and targeted therapy. *Int. J. Cancer* **2018**. [CrossRef] [PubMed]

55. Low, B.C.; Pan, C.Q.; Shivashankar, G.V.; Bershadsky, A.; Sudol, M.; Sheetz, M. YAP/TAZ as mechanosensors and mechanotransducers in regulating organ size and tumor growth. *FEBS Lett.* **2014**, *588*, 2663–2670. [CrossRef] [PubMed]

56. Yu, F.X.; Zhao, B.; Guan, K.L. Hippo Pathway in Organ Size Control, Tissue Homeostasis, and Cancer. *Cell* **2015**, *163*, 811–828. [CrossRef] [PubMed]

57. Warren, J.S.A.; Xiao, Y.; Lamar, J.M. YAP/TAZ Activation as a Target for Treating Metastatic Cancer. *Cancers* **2018**, *10*, 115. [CrossRef] [PubMed]

58. Janse van Rensburg, H.J.; Yang, X. The roles of the Hippo pathway in cancer metastasis. *Cell. Signal.* **2016**, *28*, 1761–1772. [CrossRef] [PubMed]

59. Lu, L.; Li, Y.; Kim, S.M.; Bossuyt, W.; Liu, P.; Qiu, Q.; Wang, Y.; Halder, G.; Finegold, M.J.; Lee, J.S.; et al. Hippo signaling is a potent in vivo growth and tumor suppressor pathway in the mammalian liver. *Proc. Natl. Acad. Sci. USA* **2010**, *107*, 1437–1442. [CrossRef] [PubMed]

60. Noguchi, S.; Saito, A.; Horie, M.; Mikami, Y.; Suzuki, H.I.; Morishita, Y.; Ohshima, M.; Abiko, Y.; Mattsson, J.S.; Konig, H.; et al. An integrative analysis of the tumorigenic role of TAZ in human non-small cell lung cancer. *Clin. Cancer Res. Off. J. Am. Assoc. Cancer Res.* **2014**, *20*, 4660–4672. [CrossRef] [PubMed]

61. Zhao, B.; Wei, X.; Li, W.; Udan, R.S.; Yang, Q.; Kim, J.; Xie, J.; Ikenoue, T.; Yu, J.; Li, L.; et al. Inactivation of YAP oncoprotein by the Hippo pathway is involved in cell contact inhibition and tissue growth control. *Genes Dev.* **2007**, *21*, 2747–2761. [CrossRef] [PubMed]

62. Zhang, J.; Ji, J.Y.; Yu, M.; Overholtzer, M.; Smolen, G.A.; Wang, R.; Brugge, J.S.; Dyson, N.J.; Haber, D.A. YAP-dependent induction of amphiregulin identifies a non-cell-autonomous component of the Hippo pathway. *Nat. Cell Biol.* **2009**, *11*, 1444–1450. [CrossRef] [PubMed]

63. Zhang, W.; Gao, Y.; Li, F.; Tong, X.; Ren, Y.; Han, X.; Yao, S.; Long, F.; Yang, Z.; Fan, H.; et al. YAP promotes malignant progression of Lkb1-deficient lung adenocarcinoma through downstream regulation of survivin. *Cancer Res.* **2015**, *75*, 4450–4457. [CrossRef] [PubMed]

64. Hsu, Y.L.; Hung, J.Y.; Chou, S.H.; Huang, M.S.; Tsai, M.J.; Lin, Y.S.; Chiang, S.Y.; Ho, Y.W.; Wu, C.Y.; Kuo, P.L. Angiomotin decreases lung cancer progression by sequestering oncogenic YAP/TAZ and decreasing Cyr61 expression. *Oncogene* **2015**, *34*, 4056–4068. [CrossRef] [PubMed]

65. Cui, Z.L.; Han, F.F.; Peng, X.H.; Chen, X.; Luan, C.Y.; Han, R.C.; Xu, W.G.; Guo, X.J. YES-associated protein 1 promotes adenocarcinoma growth and metastasis through activation of the receptor tyrosine kinase Axl. *Int. J. Immunopathol. Pharmacol.* **2012**, *25*, 989–1001. [CrossRef] [PubMed]

66. Mehra, R.; Vats, P.; Cieslik, M.; Cao, X.; Su, F.; Shukla, S.; Udager, A.M.; Wang, R.; Pan, J.; Kasaian, K.; et al. Biallelic Alteration and Dysregulation of the Hippo Pathway in Mucinous Tubular and Spindle Cell Carcinoma of the Kidney. *Cancer Discov.* **2016**, *6*, 1258–1266. [CrossRef] [PubMed]

67. Liu, S.; Yang, Y.; Wang, W.; Pan, X. Long noncoding RNA TUG1 promotes cell proliferation and migration of renal cell carcinoma via regulation of YAP. *J. Cell. Biochem.* **2018**, *119*, 9694–9706. [CrossRef] [PubMed]

68. Schutte, U.; Bisht, S.; Heukamp, L.C.; Kebschull, M.; Florin, A.; Haarmann, J.; Hoffmann, P.; Bendas, G.; Buettner, R.; Brossart, P.; et al. Hippo signaling mediates proliferation, invasiveness, and metastatic potential of clear cell renal cell carcinoma. *Transl. Oncol.* **2014**, *7*, 309–321. [CrossRef] [PubMed]

69. Chen, K.H.; He, J.; Wang, D.L.; Cao, J.J.; Li, M.C.; Zhao, X.M.; Sheng, X.; Li, W.B.; Liu, W.J. Methylationassociated inactivation of LATS1 and its effect on demethylation or overexpression on YAP and cell biological function in human renal cell carcinoma. *Int. J. Oncol.* **2014**, *45*, 2511–2521. [CrossRef] [PubMed]

70. Rybarczyk, A.; Klacz, J.; Wronska, A.; Matuszewski, M.; Kmiec, Z.; Wierzbicki, P.M. Overexpression of the YAP1 oncogene in clear cell renal cell carcinoma is associated with poor outcome. *Oncol. Rep.* **2017**, *38*, 427–439. [CrossRef] [PubMed]

71. Fisher, M.L.; Grun, D.; Adhikary, G.; Xu, W.; Eckert, R.L. Inhibition of YAP function overcomes BRAF inhibitor resistance in melanoma cancer stem cells. *Oncotarget* **2017**, *8*, 110257–110272. [CrossRef] [PubMed]

72. Menzel, M.; Meckbach, D.; Weide, B.; Toussaint, N.C.; Schilbach, K.; Noor, S.; Eigentler, T.; Ikenberg, K.; Busch, C.; Quintanilla-Martinez, L.; et al. In melanoma, Hippo signaling is affected by copy number alterations and YAP1 overexpression impairs patient survival. *Pigment. Cell Melanoma Res.* **2014**, *27*, 671–673. [CrossRef] [PubMed]

73. Kim, M.H.; Kim, C.G.; Kim, S.K.; Shin, S.J.; Choe, E.A.; Park, S.H.; Shin, E.C.; Kim, J. YAP-Induced PD-L1 Expression Drives Immune Evasion in BRAFi-Resistant Melanoma. *Cancer Immunol. Res.* **2018**. [CrossRef] [PubMed]

74. Nallet-Staub, F.; Marsaud, V.; Li, L.; Gilbert, C.; Dodier, S.; Bataille, V.; Sudol, M.; Herlyn, M.; Mauviel, A. Pro-invasive activity of the Hippo pathway effectors YAP and TAZ in cutaneous melanoma. *J. Investig. Dermatol.* **2014**, *134*, 123–132. [CrossRef] [PubMed]

75. Dai, X.Y.; Zhuang, L.H.; Wang, D.D.; Zhou, T.Y.; Chang, L.L.; Gai, R.H.; Zhu, D.F.; Yang, B.; Zhu, H.; He, Q.J. Nuclear translocation and activation of YAP by hypoxia contributes to the chemoresistance of SN38 in hepatocellular carcinoma cells. *Oncotarget* **2016**, *7*, 6933–6947. [CrossRef] [PubMed]

76. Han, S.X.; Bai, E.; Jin, G.H.; He, C.C.; Guo, X.J.; Wang, L.J.; Li, M.; Ying, X.; Zhu, Q. Expression and clinical significance of YAP, TAZ, and AREG in hepatocellular carcinoma. *J. Immunol. Res.* **2014**, *2014*, 261365. [CrossRef] [PubMed]

77. Guo, Y.; Pan, Q.; Zhang, J.; Xu, X.; Liu, X.; Wang, Q.; Yi, R.; Xie, X.; Yao, L.; Liu, W.; et al. Functional and clinical evidence that TAZ is a candidate oncogene in hepatocellular carcinoma. *J. Cell. Biochem.* **2015**, *116*, 2465–2475. [CrossRef] [PubMed]

78. Li, H.; Wolfe, A.; Septer, S.; Edwards, G.; Zhong, X.; Abdulkarim, A.B.; Ranganathan, S.; Apte, U. Deregulation of Hippo kinase signalling in human hepatic malignancies. *Liver Int. Off. J. Int. Assoc. Study Liver* **2012**, *32*, 38–47. [CrossRef] [PubMed]

79. Xiao, W.; Wang, J.; Ou, C.; Zhang, Y.; Ma, L.; Weng, W.; Pan, Q.; Sun, F. Mutual interaction between YAP and c-Myc is critical for carcinogenesis in liver cancer. *Biochem. Biophys. Res. Commun.* **2013**, *439*, 167–172. [CrossRef] [PubMed]

80. Zender, L.; Spector, M.S.; Xue, W.; Flemming, P.; Cordon-Cardo, C.; Silke, J.; Fan, S.T.; Luk, J.M.; Wigler, M.; Hannon, G.J.; et al. Identification and validation of oncogenes in liver cancer using an integrative oncogenomic approach. *Cell* **2006**, *125*, 1253–1267. [CrossRef] [PubMed]

81. Tanas, M.R.; Sboner, A.; Oliveira, A.M.; Erickson-Johnson, M.R.; Hespelt, J.; Hanwright, P.J.; Flanagan, J.; Luo, Y.; Fenwick, K.; Natrajan, R.; et al. Identification of a disease-defining gene fusion in epithelioid hemangioendothelioma. *Sci. Transl. Med.* **2011**, *3*, 98ra82. [CrossRef] [PubMed]

82. Antonescu, C.R.; Le Loarer, F.; Mosquera, J.M.; Sboner, A.; Zhang, L.; Chen, C.L.; Chen, H.W.; Pathan, N.; Krausz, T.; Dickson, B.C.; et al. Novel YAP1-TFE3 fusion defines a distinct subset of epithelioid hemangioendothelioma. *Genes Chromosom. Cancer* **2013**, *52*, 775–784. [CrossRef] [PubMed]

83. Oh, J.E.; Ohta, T.; Satomi, K.; Foll, M.; Durand, G.; McKay, J.; Le Calvez-Kelm, F.; Mittelbronn, M.; Brokinkel, B.; Paulus, W.; et al. Alterations in the NF2/LATS1/LATS2/YAP Pathway in Schwannomas. *J. Neuropathol. Exp. Neurol.* **2015**, *74*, 952–959. [CrossRef] [PubMed]

84. Evans, D.G. Neurofibromatosis 2 [Bilateral acoustic neurofibromatosis, central neurofibromatosis, NF2, neurofibromatosis type II]. *Genet. Med. Off. J. Am. Coll. Med Genet.* **2009**, *11*, 599–610. [CrossRef] [PubMed]

85. Bueno, R.; Stawiski, E.W.; Goldstein, L.D.; Durinck, S.; De Rienzo, A.; Modrusan, Z.; Gnad, F.; Nguyen, T.T.; Jaiswal, B.S.; Chirieac, L.R.; et al. Comprehensive genomic analysis of malignant pleural mesothelioma identifies recurrent mutations, gene fusions and splicing alterations. *Nat. Genet.* **2016**, *48*, 407–416. [CrossRef] [PubMed]

86. Endoh, H.; Yatabe, Y.; Shimizu, S.; Tajima, K.; Kuwano, H.; Takahashi, T.; Mitsudomi, T. RASSF1A gene inactivation in non-small cell lung cancer and its clinical implication. *Int. J. Cancer* **2003**, *106*, 45–51. [CrossRef] [PubMed]

87. Takahashi, Y.; Miyoshi, Y.; Takahata, C.; Irahara, N.; Taguchi, T.; Tamaki, Y.; Noguchi, S. Down-regulation of LATS1 and LATS2 mRNA expression by promoter hypermethylation and its association with biologically aggressive phenotype in human breast cancers. *Clin. Cancer Res. Off. J. Am. Assoc. Cancer Res.* **2005**, *11*, 1380–1385. [CrossRef] [PubMed]

88. Seidel, C.; Schagdarsurengin, U.; Blumke, K.; Wurl, P.; Pfeifer, G.P.; Hauptmann, S.; Taubert, H.; Dammann, R. Frequent hypermethylation of MST1 and MST2 in soft tissue sarcoma. *Mol. Carcinog.* **2007**, *46*, 865–871. [CrossRef] [PubMed]

89. Di Agostino, S.; Sorrentino, G.; Ingallina, E.; Valenti, F.; Ferraiuolo, M.; Bicciato, S.; Piazza, S.; Strano, S.; Del Sal, G.; Blandino, G. YAP enhances the pro-proliferative transcriptional activity of mutant p53 proteins. *EMBO Rep.* **2016**, *17*, 188–201. [CrossRef] [PubMed]

90. Mizuno, T.; Murakami, H.; Fujii, M.; Ishiguro, F.; Tanaka, I.; Kondo, Y.; Akatsuka, S.; Toyokuni, S.; Yokoi, K.; Osada, H.; et al. YAP induces malignant mesothelioma cell proliferation by upregulating transcription of cell cycle-promoting genes. *Oncogene* **2012**, *31*, 5117–5122. [CrossRef] [PubMed]

91. Rosenbluh, J.; Nijhawan, D.; Cox, A.G.; Li, X.; Neal, J.T.; Schafer, E.J.; Zack, T.I.; Wang, X.; Tsherniak, A.; Schinzel, A.C.; et al. beta-Catenin-driven cancers require a YAP1 transcriptional complex for survival and tumorigenesis. *Cell* **2012**, *151*, 1457–1473. [CrossRef] [PubMed]

92. Zanconato, F.; Forcato, M.; Battilana, G.; Azzolin, L.; Quaranta, E.; Bodega, B.; Rosato, A.; Bicciato, S.; Cordenonsi, M.; Piccolo, S. Genome-wide association between YAP/TAZ/TEAD and AP-1 at enhancers drives oncogenic growth. *Nat. Cell Biol.* **2015**, *17*, 1218–1227. [CrossRef] [PubMed]

93. Zanconato, F.; Battilana, G.; Forcato, M.; Filippi, L.; Azzolin, L.; Manfrin, A.; Quaranta, E.; Di Biagio, D.; Sigismondo, G.; Guzzardo, V.; et al. Transcriptional addiction in cancer cells is mediated by YAP/TAZ through BRD4. *Nat. Med.* **2018**. [CrossRef] [PubMed]

94. Miyazono, K.; Katsuno, Y.; Koinuma, D.; Ehata, S.; Morikawa, M. Intracellular and extracellular TGF-beta signaling in cancer: Some recent topics. *Front. Med.* **2018**, *12*, 387–411. [CrossRef] [PubMed]

95. Noguchi, S.; Yamauchi, Y.; Takizawa, H. Novel therapeutic strategies for fibrotic lung disease: A review with a focus on epithelial-mesenchymal transition. *Recent Patents Inflamm. Allergy Drug Discov.* **2014**, *8*, 9–18. [CrossRef]

96. Rock, J.R.; Barkauskas, C.E.; Cronce, M.J.; Xue, Y.; Harris, J.R.; Liang, J.; Noble, P.W.; Hogan, B.L. Multiple stromal populations contribute to pulmonary fibrosis without evidence for epithelial to mesenchymal transition. *Proc. Natl. Acad. Sci. USA* **2011**, *108*, E1475–E1483. [CrossRef] [PubMed]

97. Willis, B.C.; Liebler, J.M.; Luby-Phelps, K.; Nicholson, A.G.; Crandall, E.D.; du Bois, R.M.; Borok, Z. Induction of epithelial-mesenchymal transition in alveolar epithelial cells by transforming growth factor-beta1: Potential role in idiopathic pulmonary fibrosis. *Am. J. Pathol.* **2005**, *166*, 1321–1332. [CrossRef]

98. Lei, Q.Y.; Zhang, H.; Zhao, B.; Zha, Z.Y.; Bai, F.; Pei, X.H.; Zhao, S.; Xiong, Y.; Guan, K.L. TAZ promotes cell proliferation and epithelial-mesenchymal transition and is inhibited by the hippo pathway. *Mol. Cell. Boil.* **2008**, *28*, 2426–2436. [CrossRef] [PubMed]

99. Bora-Singhal, N.; Nguyen, J.; Schaal, C.; Perumal, D.; Singh, S.; Coppola, D.; Chellappan, S. YAP1 Regulates OCT4 Activity and SOX2 Expression to Facilitate Self-Renewal and Vascular Mimicry of Stem-Like Cells. *Stem Cells* **2015**, *33*, 1705–1718. [CrossRef] [PubMed]

100. Shao, D.D.; Xue, W.; Krall, E.B.; Bhutkar, A.; Piccioni, F.; Wang, X.; Schinzel, A.C.; Sood, S.; Rosenbluh, J.; Kim, J.W.; et al. KRAS and YAP1 converge to regulate EMT and tumor survival. *Cell* **2014**, *158*, 171–184. [CrossRef] [PubMed]

101. Kapoor, A.; Yao, W.; Ying, H.; Hua, S.; Liewen, A.; Wang, Q.; Zhong, Y.; Wu, C.J.; Sadanandam, A.; Hu, B.; et al. Yap1 activation enables bypass of oncogenic Kras addiction in pancreatic cancer. *Cell* **2014**, *158*, 185–197. [CrossRef] [PubMed]

102. Cordenonsi, M.; Zanconato, F.; Azzolin, L.; Forcato, M.; Rosato, A.; Frasson, C.; Inui, M.; Montagner, M.; Parenti, A.R.; Poletti, A.; et al. The Hippo transducer TAZ confers cancer stem cell-related traits on breast cancer cells. *Cell* **2011**, *147*, 759–772. [CrossRef] [PubMed]

103. Wei, S.C.; Fattet, L.; Tsai, J.H.; Guo, Y.; Pai, V.H.; Majeski, H.E.; Chen, A.C.; Sah, R.L.; Taylor, S.S.; Engler, A.J.; et al. Matrix stiffness drives epithelial-mesenchymal transition and tumour metastasis through a TWIST1-G3BP2 mechanotransduction pathway. *Nat. Cell Biol.* **2015**, *17*, 678–688. [CrossRef] [PubMed]

104. Lehmann, W.; Mossmann, D.; Kleemann, J.; Mock, K.; Meisinger, C.; Brummer, T.; Herr, R.; Brabletz, S.; Stemmler, M.P.; Brabletz, T. ZEB1 turns into a transcriptional activator by interacting with YAP1 in aggressive cancer types. *Nat. Commun.* **2016**, *7*, 10498. [CrossRef] [PubMed]

105. Tang, Y.; Feinberg, T.; Keller, E.T.; Li, X.Y.; Weiss, S.J. Snail/Slug binding interactions with YAP/TAZ control skeletal stem cell self-renewal and differentiation. *Nat. Cell Biol.* **2016**, *18*, 917–929. [CrossRef] [PubMed]

106. Bartucci, M.; Dattilo, R.; Moriconi, C.; Pagliuca, A.; Mottolese, M.; Federici, G.; Benedetto, A.D.; Todaro, M.; Stassi, G.; Sperati, F.; et al. TAZ is required for metastatic activity and chemoresistance of breast cancer stem cells. *Oncogene* **2015**, *34*, 681–690. [CrossRef] [PubMed]

107. Song, S.; Ajani, J.A.; Honjo, S.; Maru, D.M.; Chen, Q.; Scott, A.W.; Heallen, T.R.; Xiao, L.; Hofstetter, W.L.; Weston, B.; et al. Hippo coactivator YAP1 upregulates SOX9 and endows esophageal cancer cells with stem-like properties. *Cancer Res.* **2014**, *74*, 4170–4182. [CrossRef] [PubMed]

108. Ooki, A.; Del Carmen Rodriguez Pena, M.; Marchionni, L.; Dinalankara, W.; Begum, A.; Hahn, N.M.; VandenBussche, C.J.; Rasheed, Z.A.; Mao, S.; Netto, G.J.; et al. YAP1 and COX2 Coordinately Regulate Urothelial Cancer Stem-like Cells. *Cancer Res.* **2018**, *78*, 168–181. [CrossRef] [PubMed]

109. Basu-Roy, U.; Bayin, N.S.; Rattanakorn, K.; Han, E.; Placantonakis, D.G.; Mansukhani, A.; Basilico, C. Sox2 antagonizes the Hippo pathway to maintain stemness in cancer cells. *Nat. Commun.* **2015**, *6*, 6411. [CrossRef] [PubMed]

110. Kalluri, R.; Zeisberg, M. Fibroblasts in cancer. *Nat. Rev. Cancer* **2006**, *6*, 392–401. [CrossRef] [PubMed]

111. Levental, K.R.; Yu, H.; Kass, L.; Lakins, J.N.; Egeblad, M.; Erler, J.T.; Fong, S.F.; Csiszar, K.; Giaccia, A.; Weninger, W.; et al. Matrix crosslinking forces tumor progression by enhancing integrin signaling. *Cell* **2009**, *139*, 891–906. [CrossRef] [PubMed]

112. Butcher, D.T.; Alliston, T.; Weaver, V.M. A tense situation: Forcing tumour progression. *Nat. Rev. Cancer* **2009**, *9*, 108–122. [CrossRef] [PubMed]

113. Lu, P.; Weaver, V.M.; Werb, Z. The extracellular matrix: A dynamic niche in cancer progression. *J. Cell Boil.* **2012**, *196*, 395–406. [CrossRef] [PubMed]

114. Provenzano, P.P.; Cuevas, C.; Chang, A.E.; Goel, V.K.; Von Hoff, D.D.; Hingorani, S.R. Enzymatic targeting of the stroma ablates physical barriers to treatment of pancreatic ductal adenocarcinoma. *Cancer Cell* **2012**, *21*, 418–429. [CrossRef] [PubMed]

115. Jang, W.; Kim, T.; Koo, J.S.; Kim, S.K.; Lim, D.S. Mechanical cue-induced YAP instructs Skp2-dependent cell cycle exit and oncogenic signaling. *EMBO J.* **2017**, *36*, 2510–2528. [CrossRef] [PubMed]

116. Chakraborty, S.; Lakshmanan, M.; Swa, H.L.; Chen, J.; Zhang, X.; Ong, Y.S.; Loo, L.S.; Akincilar, S.C.; Gunaratne, J.; Tergaonkar, V.; et al. An oncogenic role of Agrin in regulating focal adhesion integrity in hepatocellular carcinoma. *Nat. Commun.* **2015**, *6*, 6184. [CrossRef] [PubMed]

117. Chakraborty, S.; Njah, K.; Pobbati, A.V.; Lim, Y.B.; Raju, A.; Lakshmanan, M.; Tergaonkar, V.; Lim, C.T.; Hong, W. Agrin as a Mechanotransduction Signal Regulating YAP through the Hippo Pathway. *Cell Rep.* **2017**, *18*, 2464–2479. [CrossRef] [PubMed]

118. Chakraborty, S.; Hong, W. Linking Extracellular Matrix Agrin to the Hippo Pathway in Liver Cancer and Beyond. *Cancers* **2018**, *10*, 45. [CrossRef] [PubMed]

119. Lv, X.; Fang, C.; Yin, R.; Qiao, B.; Shang, R.; Wang, J.; Song, W.; He, Y.; Chen, Y. Agrin para-secreted by PDGF-activated human hepatic stellate cells promotes hepatocarcinogenesis in vitro and in vivo. *Oncotarget* **2017**, *8*, 105340–105355. [CrossRef] [PubMed]

120. Feng, M.; Gao, W.; Wang, R.; Chen, W.; Man, Y.G.; Figg, W.D.; Wang, X.W.; Dimitrov, D.S.; Ho, M. Therapeutically targeting glypican-3 via a conformation-specific single-domain antibody in hepatocellular carcinoma. *Proc. Natl. Acad. Sci. USA* **2013**, *110*, E1083–E1091. [CrossRef] [PubMed]

121. Gaggioli, C.; Hooper, S.; Hidalgo-Carcedo, C.; Grosse, R.; Marshall, J.F.; Harrington, K.; Sahai, E. Fibroblast-led collective invasion of carcinoma cells with differing roles for RhoGTPases in leading and following cells. *Nat. Cell Biol.* **2007**, *9*, 1392–1400. [CrossRef] [PubMed]

122. Glentis, A.; Oertle, P.; Mariani, P.; Chikina, A.; El Marjou, F.; Attieh, Y.; Zaccarini, F.; Lae, M.; Loew, D.; Dingli, F.; et al. Cancer-associated fibroblasts induce metalloprotease-independent cancer cell invasion of the basement membrane. *Nat. Commun.* **2017**, *8*, 924. [CrossRef] [PubMed]

123. Calvo, F.; Ege, N.; Grande-Garcia, A.; Hooper, S.; Jenkins, R.P.; Chaudhry, S.I.; Harrington, K.; Williamson, P.; Moeendarbary, E.; Charras, G.; et al. Mechanotransduction and YAP-dependent matrix remodelling is required for the generation and maintenance of cancer-associated fibroblasts. *Nat. Cell Biol.* **2013**, *15*, 637–646. [CrossRef] [PubMed]

124. Foster, C.T.; Gualdrini, F.; Treisman, R. Mutual dependence of the MRTF-SRF and YAP-TEAD pathways in cancer-associated fibroblasts is indirect and mediated by cytoskeletal dynamics. *Genes Dev.* **2017**, *31*, 2361–2375. [CrossRef] [PubMed]

125. Lee, B.S.; Park, D.I.; Lee, D.H.; Lee, J.E.; Yeo, M.K.; Park, Y.H.; Lim, D.S.; Choi, W.; Lee, D.H.; Yoo, G.; et al. Hippo effector YAP directly regulates the expression of PD-L1 transcripts in EGFR-TKI-resistant lung adenocarcinoma. *Biochem. Biophys. Res. Commun.* **2017**, *491*, 493–499. [CrossRef] [PubMed]

126. Janse van Rensburg, H.J.; Azad, T.; Ling, M.; Hao, Y.; Snetsinger, B.; Khanal, P.; Minassian, L.M.; Graham, C.H.; Rauh, M.J.; Yang, X. The Hippo Pathway Component TAZ Promotes Immune Evasion in Human Cancer through PD-L1. *Cancer Res.* **2018**, *78*, 1457–1470. [CrossRef] [PubMed]

127. Feng, J.; Yang, H.; Zhang, Y.; Wei, H.; Zhu, Z.; Zhu, B.; Yang, M.; Cao, W.; Wang, L.; Wu, Z. Tumor cell-derived lactate induces TAZ-dependent upregulation of PD-L1 through GPR81 in human lung cancer cells. *Oncogene* **2017**, *36*, 5829–5839. [CrossRef] [PubMed]

128. Wang, G.; Lu, X.; Dey, P.; Deng, P.; Wu, C.C.; Jiang, S.; Fang, Z.; Zhao, K.; Konaparthi, R.; Hua, S.; et al. Targeting YAP-Dependent MDSC Infiltration Impairs Tumor Progression. *Cancer Discov.* **2016**, *6*, 80–95. [CrossRef] [PubMed]

129. Sarkar, S.; Bristow, C.A.; Dey, P.; Rai, K.; Perets, R.; Ramirez-Cardenas, A.; Malasi, S.; Huang-Hobbs, E.; Haemmerle, M.; Wu, S.Y.; et al. PRKCI promotes immune suppression in ovarian cancer. *Genes Dev.* **2017**, *31*, 1109–1121. [CrossRef] [PubMed]

130. Kim, W.; Khan, S.K.; Liu, Y.; Xu, R.; Park, O.; He, Y.; Cha, B.; Gao, B.; Yang, Y. Hepatic Hippo signaling inhibits protumoural microenvironment to suppress hepatocellular carcinoma. *Gut* **2018**, *67*, 1692–1703. [CrossRef] [PubMed]

131. Fan, Y.; Gao, Y.; Rao, J.; Wang, K.; Zhang, F.; Zhang, C. YAP-1 Promotes Tregs Differentiation in Hepatocellular Carcinoma by Enhancing TGFBR2 Transcription. *Cell. Physiol. Biochem. Int. J. Exp. Cell. Physiol. Biochem. Pharmacol.* **2017**, *41*, 1189–1198. [CrossRef] [PubMed]

132. Ni, X.; Tao, J.; Barbi, J.; Chen, Q.; Park, B.V.; Li, Z.; Zhang, N.; Lebid, A.; Ramaswamy, A.; Wei, P.; et al. YAP Is Essential for Treg-Mediated Suppression of Antitumor Immunity. *Cancer Discov.* **2018**, *8*, 1026–1043. [CrossRef] [PubMed]

133. Caja, L.; Dituri, F.; Mancarella, S.; Caballero-Diaz, D.; Moustakas, A.; Giannelli, G.; Fabregat, I. TGF-beta and the Tissue Microenvironment: Relevance in Fibrosis and Cancer. *Int. J. Mol. Sci.* **2018**, *19*, 1294. [CrossRef] [PubMed]

134. Calon, A.; Espinet, E.; Palomo-Ponce, S.; Tauriello, D.V.; Iglesias, M.; Cespedes, M.V.; Sevillano, M.; Nadal, C.; Jung, P.; Zhang, X.H.; et al. Dependency of colorectal cancer on a TGF-beta-driven program in stromal cells for metastasis initiation. *Cancer Cell* **2012**, *22*, 571–584. [CrossRef] [PubMed]

135. Navab, R.; Strumpf, D.; Bandarchi, B.; Zhu, C.Q.; Pintilie, M.; Ramnarine, V.R.; Ibrahimov, E.; Radulovich, N.; Leung, L.; Barczyk, M.; et al. Prognostic gene-expression signature of carcinoma-associated fibroblasts in non-small cell lung cancer. *Proc. Natl. Acad. Sci. USA* **2011**, *108*, 7160–7165. [CrossRef] [PubMed]

136. Varelas, X.; Sakuma, R.; Samavarchi-Tehrani, P.; Peerani, R.; Rao, B.M.; Dembowy, J.; Yaffe, M.B.; Zandstra, P.W.; Wrana, J.L. TAZ controls Smad nucleocytoplasmic shuttling and regulates human embryonic stem-cell self-renewal. *Nat. Cell Biol.* **2008**, *10*, 837–848. [CrossRef] [PubMed]

137. Pefani, D.E.; Pankova, D.; Abraham, A.G.; Grawenda, A.M.; Vlahov, N.; Scrace, S.; O'Neill, E. TGF-beta Targets the Hippo Pathway Scaffold RASSF1A to Facilitate YAP/SMAD2 Nuclear Translocation. *Mol. Cell* **2016**, *63*, 156–166. [CrossRef] [PubMed]

138. Hiemer, S.E.; Szymaniak, A.D.; Varelas, X. The transcriptional regulators TAZ and YAP direct transforming growth factor beta-induced tumorigenic phenotypes in breast cancer cells. *J. Boil. Chem.* **2014**, *289*, 13461–13474. [CrossRef] [PubMed]

139. Fujii, M.; Toyoda, T.; Nakanishi, H.; Yatabe, Y.; Sato, A.; Matsudaira, Y.; Ito, H.; Murakami, H.; Kondo, Y.; Kondo, E.; et al. TGF-beta synergizes with defects in the Hippo pathway to stimulate human malignant mesothelioma growth. *J. Exp. Med.* **2012**, *209*, 479–494. [CrossRef] [PubMed]

140. Miranda, M.Z.; Bialik, J.F.; Speight, P.; Dan, Q.; Yeung, T.; Szaszi, K.; Pedersen, S.F.; Kapus, A. TGF-beta1 regulates the expression and transcriptional activity of TAZ protein via a Smad3-independent, myocardin-related transcription factor-mediated mechanism. *J. Boil. Chem.* **2017**, *292*, 14902–14920. [CrossRef] [PubMed]

141. He, W.; Dai, C.; Li, Y.; Zeng, G.; Monga, S.P.; Liu, Y. Wnt/beta-catenin signaling promotes renal interstitial fibrosis. *J. Am. Soc. Nephrol.* **2009**, *20*, 765–776. [CrossRef] [PubMed]

142. Duan, J.; Gherghe, C.; Liu, D.; Hamlett, E.; Srikantha, L.; Rodgers, L.; Regan, J.N.; Rojas, M.; Willis, M.; Leask, A.; et al. Wnt1/betacatenin injury response activates the epicardium and cardiac fibroblasts to promote cardiac repair. *EMBO J.* **2012**, *31*, 429–442. [CrossRef] [PubMed]

143. Beyer, C.; Schramm, A.; Akhmetshina, A.; Dees, C.; Kireva, T.; Gelse, K.; Sonnylal, S.; de Crombrugghe, B.; Taketo, M.M.; Distler, O.; et al. Beta-catenin is a central mediator of pro-fibrotic Wnt signaling in systemic sclerosis. *Ann. Rheum. Dis.* **2012**, *71*, 761–767. [CrossRef] [PubMed]

144. Henderson, W.R., Jr.; Chi, E.Y.; Ye, X.; Nguyen, C.; Tien, Y.T.; Zhou, B.; Borok, Z.; Knight, D.A.; Kahn, M. Inhibition of Wnt/beta-catenin/CREB binding protein (CBP) signaling reverses pulmonary fibrosis. *Proc. Natl. Acad. Sci. USA* **2010**, *107*, 14309–14314. [CrossRef] [PubMed]

145. Wiese, K.E.; Nusse, R.; van Amerongen, R. Wnt signalling: Conquering complexity. *Development* **2018**, *145*. [CrossRef] [PubMed]

146. Azzolin, L.; Panciera, T.; Soligo, S.; Enzo, E.; Bicciato, S.; Dupont, S.; Bresolin, S.; Frasson, C.; Basso, G.; Guzzardo, V.; et al. YAP/TAZ incorporation in the beta-catenin destruction complex orchestrates the Wnt response. *Cell* **2014**, *158*, 157–170. [CrossRef] [PubMed]

147. Cai, J.; Maitra, A.; Anders, R.A.; Taketo, M.M.; Pan, D. beta-Catenin destruction complex-independent regulation of Hippo-YAP signaling by APC in intestinal tumorigenesis. *Genes Dev.* **2015**, *29*, 1493–1506. [CrossRef] [PubMed]

148. Lee, Y.; Kim, N.H.; Cho, E.S.; Yang, J.H.; Cha, Y.H.; Kang, H.E.; Yun, J.S.; Cho, S.B.; Lee, S.H.; Paclikova, P.; et al. Dishevelled has a YAP nuclear export function in a tumor suppressor context-dependent manner. *Nat. Commun.* **2018**, *9*, 2301. [CrossRef] [PubMed]

149. Park, H.W.; Kim, Y.C.; Yu, B.; Moroishi, T.; Mo, J.S.; Plouffe, S.W.; Meng, Z.; Lin, K.C.; Yu, F.X.; Alexander, C.M.; et al. Alternative Wnt Signaling Activates YAP/TAZ. *Cell* **2015**, *162*, 780–794. [CrossRef] [PubMed]

150. Konsavage, W.M., Jr.; Kyler, S.L.; Rennoll, S.A.; Jin, G.; Yochum, G.S. Wnt/beta-catenin signaling regulates Yes-associated protein (YAP) gene expression in colorectal carcinoma cells. *J. Boil. Chem.* **2012**, *287*, 11730–11739. [CrossRef] [PubMed]

151. Stanger, B.Z. Quit your YAPing: A new target for cancer therapy. *Genes Dev.* **2012**, *26*, 1263–1267. [CrossRef] [PubMed]

152. Konstantinou, E.K.; Notomi, S.; Kosmidou, C.; Brodowska, K.; Al-Moujahed, A.; Nicolaou, F.; Tsoka, P.; Gragoudas, E.; Miller, J.W.; Young, L.H.; et al. Verteporfin-induced formation of protein cross-linked oligomers and high molecular weight complexes is mediated by light and leads to cell toxicity. *Sci. Rep.* **2017**, *7*, 46581. [CrossRef] [PubMed]

153. Bao, Y.; Nakagawa, K.; Yang, Z.; Ikeda, M.; Withanage, K.; Ishigami-Yuasa, M.; Okuno, Y.; Hata, S.; Nishina, H.; Hata, Y. A cell-based assay to screen stimulators of the Hippo pathway reveals the inhibitory effect of dobutamine on the YAP-dependent gene transcription. *J. Biochem.* **2011**, *150*, 199–208. [CrossRef] [PubMed]

154. Zhou, Y.; Huang, X.; Hecker, L.; Kurundkar, D.; Kurundkar, A.; Liu, H.; Jin, T.H.; Desai, L.; Bernard, K.; Thannickal, V.J. Inhibition of mechanosensitive signaling in myofibroblasts ameliorates experimental pulmonary fibrosis. *J. Clin. Investig.* **2013**, *123*, 1096–1108. [CrossRef] [PubMed]

155. Jansen, S.; Gosens, R.; Wieland, T.; Schmidt, M. Paving the Rho in cancer metastasis: Rho GTPases and beyond. *Pharmacol. Ther.* **2018**, *183*, 1–21. [CrossRef] [PubMed]
156. Sorrentino, G.; Ruggeri, N.; Specchia, V.; Cordenonsi, M.; Mano, M.; Dupont, S.; Manfrin, A.; Ingallina, E.; Sommaggio, R.; Piazza, S.; et al. Metabolic control of YAP and TAZ by the mevalonate pathway. *Nat. Cell Biol.* **2014**, *16*, 357–366. [CrossRef] [PubMed]

International Journal of
Molecular Sciences

MDPI

Review

TGF-β-Mediated Epithelial-Mesenchymal Transition and Cancer Metastasis

Yang Hao, David Baker and Peter ten Dijke *

Department of Cell and Chemical Biology and Oncode Institute, Leiden University Medical Center, Einthovenweg 20, 2300 RC Leiden, The Netherlands; Y.Hao@lumc.nl (Y.H.); D.A.Baker@lumc.nl (D.B.)
* Correspondence: p.ten_dijke@lumc.nl; Tel.: +31-71-526-9271

Received: 18 April 2019; Accepted: 24 May 2019; Published: 5 June 2019

Abstract: Transforming growth factor β (TGF-β) is a secreted cytokine that regulates cell proliferation, migration, and the differentiation of a plethora of different cell types. Consistent with these findings, TGF-β plays a key role in controlling embryogenic development, inflammation, and tissue repair, as well as in maintaining adult tissue homeostasis. TGF-β elicits a broad range of context-dependent cellular responses, and consequently, alterations in TGF-β signaling have been implicated in many diseases, including cancer. During the early stages of tumorigenesis, TGF-β acts as a tumor suppressor by inducing cytostasis and the apoptosis of normal and premalignant cells. However, at later stages, when cancer cells have acquired oncogenic mutations and/or have lost tumor suppressor gene function, cells are resistant to TGF-β-induced growth arrest, and TGF-β functions as a tumor promotor by stimulating tumor cells to undergo the so-called epithelial-mesenchymal transition (EMT). The latter leads to metastasis and chemotherapy resistance. TGF-β further supports cancer growth and progression by activating tumor angiogenesis and cancer-associated fibroblasts and enabling the tumor to evade inhibitory immune responses. In this review, we will consider the role of TGF-β signaling in cell cycle arrest, apoptosis, EMT and cancer cell metastasis. In particular, we will highlight recent insights into the multistep and dynamically controlled process of TGF-β-induced EMT and the functions of miRNAs and long noncoding RNAs in this process. Finally, we will discuss how these new mechanistic insights might be exploited to develop novel therapeutic interventions.

Keywords: EMT; lncRNA; metastasis; miRNA; SMAD; TGF-β; targeted therapy; tumor microenvironment

1. Introduction

Cancer treatments have been refined over a period spanning several thousand years, culminating in the primary modern approaches of surgery, chemotherapy and radiation therapy [1]. In recent decades, more targeted and personalized treatments have gained prominence. The fundamental aim of these therapies is to specifically kill tumor cells while leaving healthy tissues intact [2]. Despite demonstrable progress, these strategies frequently deliver relatively modest improvements in disease outcomes owing to acquired resistance and excessive toxicity [3]. These findings indicate the need for greater optimization of current treatments as well as the identification of alternative targets for the development of novel cancer remedies.

Cancer is a complex disease in which tumor cell heterogeneity and the reciprocal interplay between tumor cells and surrounding stromal cells and extracellular matrix are key determinants in tumor progression and therapy response. Tumorigenesis shows similarity to subverted normal embryogenic developmental processes in which communication between cells is controlled by cytokines that act in an autocrine, paracrine or juxtacrine manner. In this light, we will provide a general overview of the established roles of one such important developmental cancer signaling pathway, namely the transforming growth factor-β (TGF-β), in tumorigenesis [4]. In particular, we consider how this essential TGF-β signaling network orchestrates the epithelial to mesenchymal transition (EMT), the mechanism by which cancer cells lose polarity and separate from each other, adopt the characteristics of a mesenchymal phenotype, become motile and invade distant sites. Key TGF-β-induced effectors in this process are the transcriptional repressors of E-cadherin, e.g., SNAIL1, SNAIL2 (also termed SLUG), ZEB1/2 and TWIST. Moreover, miRNAs and long noncoding RNAs (lncRNAs) are emerging as potentially quantifiable biomarkers of cancer status and are potential targets for anti-metastatic therapies [5]. In this review, we will highlight the pivotal role of these molecules in regulating TGF-β signaling and epithelial-mesenchymal transition (EMT).

2. TGF-β and Signaling Transduction across the Plasma Membrane

TGF-β1 (hereafter termed simply TGF-β) is the prototypic member of a large family of structurally and functionally related proteins, which includes its close relatives TGF-β2 and TGF-β3 but also activins, nodal, inhibins, Mullerian-inhibiting substance (MIS), growth and differentiation factors (GDF) and bone morphogenetic proteins (BMPs) [6]. TGF-β was discovered in the early 1980s as a secreted factor that, together with TGF-β or epidermal growth factor (EGF), induced the growth of normal rat kidney (NRK) cells in soft agar. Since then, we have gained a much deeper understanding of this multifunctional cytokine [7]. The TGF-β gene encodes a pre-pro-precursor peptide of 390 amino acids. The pro-precursor peptide is proteolytically processed by furin proteases into an amino-terminal fragment and the carboxy-terminal 112 amino acids, which corresponds to the mature bioactive TGF-β [8]. The amino-terminal part is also termed the latency associated peptide (LAP) and is noncovalently attached to the mature TGF-β [9,10]. Latent TGF-β is activated by specific proteases that cleave the LAP and/or by mechanical forces generated by cell surface integrins, resulting in the release of mature, active TGF-β [11,12]. Bioactive TGF-β, which is capable of binding its cell surface receptors, is a dimeric protein linked by disulfide bonds with an apparent molecular weight of 25 kDa.

TGF-β exerts its cellular effects via cell surface TGF-β type I and type II receptors, e.g., TβRI and TβRII, respectively [13]. TβRI and TβRII are structurally related and consist of an extracellular domain characterized by the presence of cysteine residues that form disulfide bonds, a single transmembrane domain and an intracellular region harboring a conserved serine/threonine kinase domain. TGF-β initially engages the TβRII, which drives the recruitment of TβRI and the formation of a heterotetrameric complex composed of two TβRIIs and two TβRIs. Subsequently, the active TβRII kinase domain phosphorylates TβRI on specific serine and threonine residues in the glycine-serine (GS) juxtamembrane region, which leads to its activation [14]. TGF-β receptors are widely expressed in human tissues/cells and mediate various biological phenomena, such as embryonic development, tissue homeostasis, organogenesis, immune surveillance and tissue repair.

3. Intracellular SMAD and Non-SMAD Signaling Pathways

Genetic studies designed to delineate the critical components of the dauer and decapentaplegic pathway in *Caenorhabditis elegans* and *Drosophila*, respectively, led to the identification of *Small* (*Sma*) and *Mothers against dpp* (*Mad*) genes [15,16]. The mammalian homologues of the proteins encoded by these genes, termed SMADs, were found to act as intracellular transcriptional effectors of TGF-β family receptor signaling [17]. The SMAD family can be divided into receptor-regulated (R-) SMADs (in vertebrates: R-SMAD1, -2, -3, -5 and -8) that interact with and become phosphorylated by activated type I receptor kinases, the common (Co-) SMADs (in vertebrates: SMAD4) that form heteromeric complexes with activated R-SMADs and inhibitory I-SMADs (in vertebrates: I-SMAD6/7), which antagonize canonical SMAD signaling [18] (Figure 1).

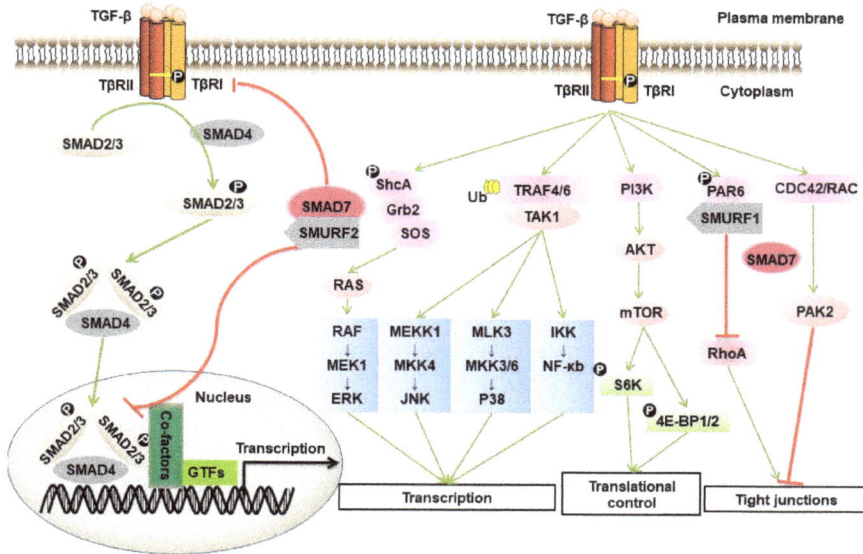

Figure 1. TGF-β/SMAD and non-SMAD signaling. TGF-β elicits its cellular responses by forming ligand-induced complex formation of TGF-β type I and type II cell surface receptors (i.e., TβRI and TβRII) that are endowed with serine/threonine kinase activity. The extracellular signal is transduced across the plasma membrane through the action of the constitutively active TβRII kinase that phosphorylates specific serine and threonine residues in the intracellular juxtamembrane GS domain of TβRI. Intracellular signaling is then initiated when the activated TβRI kinase phosphorylates or activates intracellular signal mediators. In the case of the canonical SMAD pathway, TβRI recruits and phosphorylates specific R-SMADs, e.g., SMAD2 and SMAD3, which can form heteromeric complexes with SMAD4. These transcription factor complexes then translocate into the nucleus and cooperate with other transcription regulators to regulate target gene expression. In the non-SMAD pathway, TGF-β receptors activate other pathways, including various branches of MAPK pathways, RHO-like GTPase signaling pathways and phosphatidylinositol-3-kinase (PI3K)/AKT pathways. Inhibitory signals are indicated with inhibitory red arrows; Stimulatory signals are indicated with green arrows.

The R- and Co-SMADs have two conserved domains, termed Mad homology (MH)1 and MH2 domains at the amino-terminal (MH1) and carboxy-terminal (MH2) ends of the proteins. MH1 and MH2 are separated by a flexible linker region. R-SMAD phosphorylation by type I receptors occurs on two serine residues, which comprise the SXS sequence motif, at the C-termini. SMAD3 and SMAD4, but not SMAD2, bind directly to the consensus 5′-CAGA-3′ DNA motif. Heteromeric complex formation between R-SMADs and SMAD4 is mediated by MH2 domains. Heteromeric complex formation of R-SMADs and SMAD4 exposes nuclear import signals and shields nuclear export signals, resulting in their nuclear accumulation. In the nucleus, they can act as transcription factors in concert with other DNA binding transcription factors, coactivators and repressors and chromatin remodeling factors, which enable diverse transcriptional responses depending upon the particular combination of proteins [19]. I-SMAD7, which has a carboxy terminal region with homology to R-SMADs and SMAD4 MH2 domains, antagonizes the activation of TGF-β receptor/SMAD signaling. SMAD7 achieves this via multiple mechanisms, including the recruitment of the E3 ubiquitin ligase SMURF2 to the activated receptor, thereby targeting the TGF-β receptor for proteasomal and lysosomal degradation. SMAD7 can also attenuate signaling by recruiting phosphatases to the activated TGF-β receptor, which mediate receptor inactivation by the dephosphorylation of specific amino acid residues [20].

TGF-β can also signal via non-SMAD pathways. This process can occur directly or indirectly [21]. An example of an indirect mechanism is TGF-β/SMAD-stimulated expression of growth factors, such as platelet-derived growth factor (PDGF) and epidermal growth factor (EGF), which thereafter initiate non-SMAD responses. Non-SMAD signaling can also be initiated directly downstream of the TGF-β receptor by activation of various branches of the MAP kinase (MAPK) pathway, Rho-like GTPase signaling pathways and the phosphatidylinositol-3-kinase (PI3K)/AKT pathway [22]. For example, activated TβRI can induce tyrosine phosphorylation of SHCA, which associates with GRB2, and recruits a GRB2/SOS complex, thereby triggering activation of the RAS/RAF/MAP kinase signaling cascade [23]. Activated TGF-β receptor complexes also induce K63-linked polyubiquitination of TRAF4/6, leading to the recruitment of TAK1 and triggering its activation, thus allowing TAK1 to activate JNK signaling through MKK4 or P38 MAPK signaling via MKK3/6 [18,24,25]. At tight junctions, TβRII phosphorylates PAR6 at serine residue 345. This phosphorylated PAR6 recruits SMURF1 to the activated TβRI-TβRII receptor complex. The PAR6-SMURF1 complex subsequently mediates localized ubiquitination and turnover of RHOA at cellular protrusions [26]. TGF-β can promote the interaction between CDC42 GTPase/RAC and p21-activated kinase (PAK) 2, an interaction that is dependent on SMAD7 [27]. TGF-β receptors interact with the regulatory p85 subunit of PI3K, resulting in activation of the PI3K/AKT pathway, which controls translational responses through mammalian target of rapamycin (mTOR) [28]. Activation of non-SMAD signaling occurs in a context-dependent manner, and these pathways also crosstalk with the canonical SMAD pathway.

4. Tumor Suppressive Effects of TGF-β

TGF-β can act as a potent tumor suppressor in normal and premalignant epithelial cell types (Figure 2). TGF-β triggers G1 phase cell cycle arrest by different mechanisms in different cell types [29]. For example, this molecule can activate the translation-inhibitory protein 4E-BP1 (regulator of eukaryotic translation initiation factor-4F (eIF4E)) promoter activity through SMAD4, thereby suppressing translation and cell growth and proliferation [30]. TGF-β causes late G1 cell cycle arrest by inducing the expression of cyclin-dependent kinase (CDK) inhibitors (p15INK4b, p21WAF1 and p27KIP1) to inhibit CDK-cyclin complexes [31]. In MCF10A and MDA-MB-231 cell lines, the tumor suppressor p53 is a critical SMAD partner, which promotes TGF-β-induced p21 expression to block cell cycle progression [32]. In addition, TGF-β exerts growth inhibitory effects by inhibiting the expression of CDC25a phosphatase (which is required for CDK-cyclin activation) [33] or negatively regulating Id proteins (helix-loop-helix (HLH) proteins, which are essential for inhibition of cell differentiation and growth arrest) [34] in the prostate epithelial cell line HPr-1. TGF-β directly inhibits c-MYC by binding a transcriptional repression complex containing SMAD2/3, E2F4/5, p107, and C/EBPβ to the TGF-β inhibitory element in the proximal region and thus achieving cell cycle arrest in HaCaT, COS-1, and Mv1Lu-tet-p15 cells and human leukemia MO-91 cells [35].

In addition to cell cycle arrest, TGF-β can induce cell apoptosis in the early phase of tumorigenesis [29]. However, the precise mechanisms by which TGF-β induces this effect in different cell types remain unclear. The expression of some apoptotic regulators (such as growth arrest and DNA damage (GADD) 45, Bcl-2-like protein 11 (BIM), BCL-2 interacting killer (BIK), death associated protein kinase (DAPK), FAS, and B-cell lymphoma-extra large (BCL-XL)) was shown to be regulated by the TGF-β/SMAD signaling pathway [36]. For example, BIM was found to be a key mediator of TGF-β-induced apoptosis in intestinal adenoma cells [37] and in hepatocarcinoma cells [38]. TGF-β1-associated regulatory SMAD proteins bind to the BIK (also known as NBK) promoter, which encodes a proapoptotic sensitizer protein in B cells [39]. After TGF-β1 treatment, researchers found that TGF-β induced SMAD-dependent binding between the proapoptotic effector BIM and BCL-XL in gastric carcinoma cell lines [40] and a decrease in BCL-XL expression followed by activation of the apoptosis proteins caspase-9 and caspase-3 in human hepatoma cells (HuH-7) [41]. TGF-β can activate the TAK1-p38/JNK pathway, which has been reported to lead to apoptosis in HEK 293T cells [25]. This molecule also promoted the expression of SMAD-dependent GADD45β in hepatocytes and plays an important role in cell death by mediating delayed TGF-β-induced p38 MAP kinase activation [42]. Here, the effects of TGF-β in proapoptotic signaling occur in a context-dependent manner. Additionally, the TGF-β signaling pathway can be coupled to the cell death machinery through the induction of reactive oxygen species (ROS) [43], apoptosis genes (SHIP and TIEG), modulation of epigenetic regulators (DNMTs) [44], H3K79me3 and H2BK120me1 [45], and telomere shortening through regulating human telomerase reverse transcriptase (hTERT) in the breast cancer MCF-7 cell line [46]. These effects of TGF-β in proapoptotic signaling occur in a context-dependent manner.

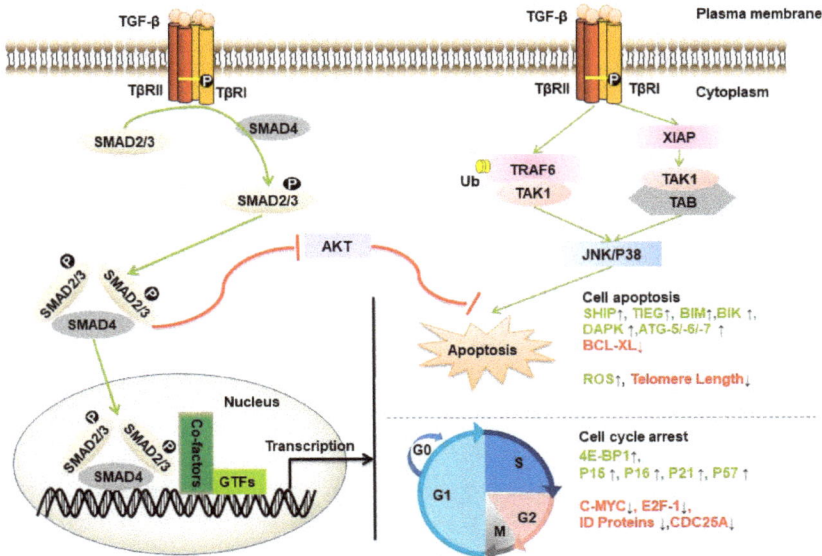

Figure 2. TGF-β-induced growth inhibition and apoptosis. The TGF-β/SMAD pathway can arrest cells in the G1 phase of the cell cycle by modulating the expression of specific genes, including induction of 4EBP1, p15INK4b, p16INK4A, p21WAF1 and p57KIP2 and repression of Id proteins, E2F, c-MYC and CDC25a genes. The TGF-β/SMAD pathway can induce cell apoptosis by inducing the expression of proapoptotic genes, such as BCL-XL, BIM, BIK, SHIP and TIEG. The TGF-β activation of TAK-1 occurs via TRAF6 and the adaptor XIAP-mediated TAB/TAK-1 complex and GADD45β. Activation of NF-κB by the TAK pathway stimulates p38/JNK phosphorylation, which has been reported to lead to apoptosis. Additionally, the TGF-β signaling pathway can be coupled to the cell death machinery through ROS, autophagy activation (ATG-5/-6/-7), induction of DAPK expression, epigenetic changes and shortening of telomere length (regulating hTERT). Inhibitory signals are indicated with inhibitory red arrows; Stimulatory signals are indicated with green arrows. SMAD-mediated transcriptional events are indicated with the black arrow.

The TGF-β signaling pathway can inhibit tumor growth by multiple other mechanisms, including activating autophagy in certain human cancer cells. For example, in human hepatocellular carcinoma cell lines, TGF-β induced the accumulation of autophagosomes and increased the expression levels of the autophagy markers Autophagy-related 5 (ATG5), Beclin1, ATG7 and death-associated protein kinase (DAPK) [47]. Moreover, siRNA-mediated silencing of autophagy genes attenuated TGF-β-mediated growth inhibition and induction of the proapoptotic genes BIM and BMF in human hepatocellular carcinoma cells [48].

Because of its central role in tumor suppression, TGF-β signaling components were found to be mutated and functionally inactivated in various cancers [49]. The first example was SMAD4, which is frequently mutated in gastrointestinal cancers [50]. Subsequently, other TGF-β signaling components, e.g., TGF-β receptors [51] and SMADs (SMAD2 and SMAD3), were found to be mutated in various cancers, including bladder, colon, breast, esophageal, stomach, brain, liver, and lung cancers [52]. Loss of tumor suppressor function and epigenome and microenvironmental changes can also affect the tumor-promoting activity of the TGF-β receptor/SMAD pathway [53]. For example, in gastrointestinal tumors, TβR1 activity was decreased due to the methylation status of the *TβR1* promoter [54]. In turn, TGF-β/SMAD can affect the epigenome of genes involved in cancer processes. TGF-β and SMAD2/3 show oncogenic activities, such as promoting glioma cell proliferation, by affecting the methylation status of the *platelet-derived growth factor-β* (*PDGF-B*) gene and autocrine PDGF-B signaling within tumor microenvironments [55]. TGF-β stimulated myofibroblast percent and invasion rate in tumor-associated fibroblasts (CAFs) that increase tumor invasion [56].

5. TGF-β-induced Tumor Promoting Effects

5.1. Cell Biology of TGF-β-induced EMT

In the late stage of cancer progression, cancer cells remain responsive to TGF-β but become resistant to its cytostatic effects. In fact, by acting directly on cancer cells, TGF-β can promote tumorigenesis by inducing the so-called epithelial to mesenchymal transition (EMT) [57]. Under normal physiological conditions, EMT plays a crucial role in the context of embryogenesis and tissue damage repair [58]. This process can be subverted and pathological, and EMT drives the development of fibrotic disease and tumorigenesis [59]. EMT is characterized by changes in the levels of three prominent biomarkers (E-cadherin, vimentin, and N-cadherin), and these changes can lead to decreased adhesion of cells, loss of polarity and tight junctions. At the same time, epithelial cells adopt the traits of a mesenchymal phenotype, notably motility and susceptibility to invasion and metastasis (Figure 3) [60]. Importantly, mesenchymal cancer cells are correlated with poor prognosis and associated with resistance to chemotherapy [61]. Interestingly, recent studies have questioned the necessity of EMT in establishing metastasis [62]. Zheng et al. reported that in genetically engineered mouse models of pancreatic adenocarcinoma development and spontaneous metastasis mouse models of breast cancer, tumor cells could metastasize without activating EMT programs, and EMT only contributed to chemoresistance [63]. Similarly, Fischer et al. also showed that EMT is not required for lung metastasis but contributes to chemoresistance in spontaneous breast-to-lung metastasis models [64]. However, these findings have been challenged by other researchers, who believed that Zheng et al. failed to completely suppress the activation of EMT, and their results can only speak to the redundancy of EMT within the transcriptional network in pancreatic carcinomas. These researchers continue to subscribe to the notion that EMT is required for metastatic dissemination in pancreatic carcinoma cells [65].

Increasing data have shown that EMT is not a single, stereotypical program but instead has a multistep process that passes through intermediate hybrid states (partial EMT state, P-EMT) during the transition from epithelial to mesenchymal cells [66]. The TGF-β-induced transition from epithelial to P-EMT is reversible, whereas the transition from P-EMT to mesenchymal cells is potentially irreversible depending on the type of cell that is involved [59]. Studies have found that there are at least seven tumor subpopulations associated with different EMT stages in skin and breast cancer tissues, which contributes to intratumor heterogeneity. These EMT subpopulations displayed differences in cellular plasticity, invasiveness, and metastatic potential, and tumor cells with an early stage of EMT were most likely to metastasize [67]. An interesting paper reported that TGF-β activates scleroderma epithelial cells to the P-EMT process in fibrotic skin [68]. Related to this, single cell RNA-seq showed that P-EMT plays an important role in head and neck cancer, and further in vitro analyses suggested that TGF-β dynamically controls the transition between P-EMT and non-P-EMT states in cells [69].

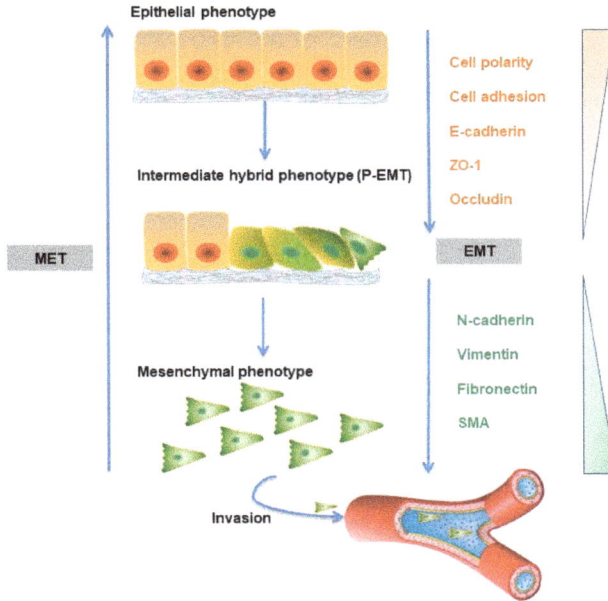

Figure 3. TGF-β mediates EMT. TGF-β is a strong promoter of EMT, which is characterized by a loss of epithelial and gain of mesenchymal markers. Polar epithelial cells remodel into highly migratory mesenchymal cells, followed by decreased adhesion of cells and loss of polarity and tight junctions. TGF-β via SMAD or non-SMAD signaling pathways can induce the expression of several EMT-TFs, such as SNAIL1, SNAIL2, ZEB1/2 and TWIST. The migratory and invasive mesenchymal-like tumor cells extravasate from primary lesions into blood or lymphatic vessels and then intravasate to distant sites via, where they form metastatic colonies upon MET.

5.2. Molecular Mechanisms in TGF-β-induced EMT

SMAD levels/activities mediate TGF-β-induced EMT by inducing the expression of E-cadherin transcriptional repressors, such as SNAIL, ZEB and TWIST, which cooperate with other transcription regulators in the nucleus [58]. Several additional lines of evidence argue for a role of SMADs in TGF-β-induced EMT, and some examples are provided below. Knockdown of SMAD4 in MDA-MB-231 breast cancer cells robustly attenuated bone metastasis in nude mice and significantly prolonged survival of the treated animals [70]. SMAD3 and SMAD4 interact to form a complex with SNAIL1 that targets the tight-junction protein (CAR) and E-cadherin during TGF-β-driven EMT in breast epithelial cells. Conversely, co-silencing of SNAIL1 and SMAD4 by siRNA inhibited repression of CAR and occludin during EMT [71]. In addition, SMAD3/SMAD4-mediated SNAIL transcription contributed to EMT during skin carcinogenesis, while SMAD2 loss significantly increased this effect [72]. Moreover, SMAD7, the transcriptional target and negative regulator of TGF-β signaling, upregulated TGF-β and inducing SMAD7 transcription prevented TGF-β-induced EMT and invasion of cancer cells [73]. Additionally, ubiquitin ligases that promote poly-ubiquitination and proteasomal degradation of SMADs affect EMT. This finding is illustrated by E3 ubiquitin ligase RNF8, which activates TWIST via K63-linked ubiquitination to promote EMT and cancer stem cell (CSC) self-renewal, resulting in enhanced metastasis and chemoresistance in breast cancer [74].

TGF-β can also induce EMT in a non-SMAD-dependent fashion, for example, by promoting cytoskeletal remodeling, which leads to activation of ERK [75]. The ERK required for cytoskeletal remodeling interacts with SHC or GRB2 to form an SHC-GRB2-ERK complex, which is a key component of TGF-β-induced tumor invasion and metastasis [76]. ERK substrates, AP-1 family members, enhance SMAD transcriptional activity to regulate gene expression and TGF-β-induced EMT [77]. However, the RHO-like GTPases, including RHOA, RAC and CDC42, are also involved in TGF-β-induced EMT. TGF-β regulates cytoskeletal changes via mediating RHO GTPase to achieve the dissolution of tight junctions among cells. TGF-β mediates the RHOA activity level and promotes the activation of LIM kinase (LIMK) by Rho-related kinase (ROCK) and phosphorylated myosin light chain (MLC) to inhibit cofilin [78]. In addition, TGF-β affects tight junctions through SMAD7-dependent CDC42-PAK1 (p21-activated kinase) and filopodia formation. The TRAF6-TAK1-JNK/P38 pathway and PI3K-AKT-mTOR signaling are also necessary non-SMAD pathways for TGF-β-mediated EMT [79,80]. Scientists have shown that PI3K/AKT signaling promotes tumor metastasis by inducing TWIST1 phosphorylation, via a crosstalk between AKT/PKB and TGF-β signaling [81]. Twist also had a significant effect on AKT signaling pathway activation by inducing expression of miR-10b in gastric cancer cells, and the miR-10b induced by TWIST increased the expression of a well-characterized pro-metastatic gene, RHOC [82]. Moreover, in an orthotopic syngeneic mouse tumor model, metastasis caused by EMT was attenuated in mice treated with the p38 inhibitor SB203580 [83]. TRAF6 knockdown inhibited the migration and invasion caused by EMT of SCCHN (squamous cell carcinoma of head and neck) cells [84].

In addition to these SMAD/non-SMAD pathways, TGF-β affects the activities of other EMT trigger signaling pathways (NOTCH, WNT, INTEGRIN, etc.) by several complexes, such as ZEB1/2, the SNAIL1-SMAD3/4 complex, the β-catenin-SMAD2 complex, the LEF1-SMAD3/4 complex and the SMAD3-AP1-1 complex [85]. As early as 20 years ago, scientists discovered that SMAD4 could form a complex with β-catenin and LEF1/TCF (lymphoid enhancer factor1), which are downstream components of the Wnt signaling cascade in vivo [86]. SMAD signaling subsequently stimulates the formation of β-catenin/LEF1 and SNAIL-LEF1 complexes, which promote EMT by inhibiting the expression of E-cadherin [87,88]. TGF-β can promote EMT associated with WNT-11 signals through the WNT-11 receptor FZD8 in prostate cancer [89]. Moreover, WNT-11 signaling mediates the nuclear entry process of TAK1 (TGF-β-activated kinase) [90]. The TGF-β and NOTCH pathways coregulate a large cohort of genes in human cancer, such as renal cell carcinoma [91]. R-SMAD activates the NOTCH ligand Jagged1 to release Notch intracellular domain (ICN) and then binds to CLS (an acronym for CBF-1/RBPJ-κ in *Homo sapiens/Mus musculus* respectively, Suppressor of Hairless in *Drosophila melanogaster*, Lag-1 in *Caenorhabditis elegans*). This ICN-CLS complex induces the binding of the transcription factor SNAIL or HEYl to the E-cadherin E-box to reduce E-cadherin expression and initiate the EMT process [92]. Moreover, SMAD signaling and MAPK/JNK signaling converge at AP1-binding promoter sites by SMAD3 and SMAD4, which cooperate with c-JUN/c-FOS [93], and the RAS-ERK MAP kinase pathways are likely to act synergistically with TGF-β and contribute to multiple aspects of the EMT, including the pro-invasive and pro-metastatic behavior of tumor cells of diverse tissue origins [94]. TGF-β increases the level of SNAIL and promotes EMT with the cooperation of oncogenic RAS [57] and the transcription factor nuclear factor κB (NF-κB) [95]. In addition, TGF-β upregulates receptors and ligands of PDGF, leading to phosphorylation of PI3K and activation of the SRC/STAT3 pathway, thereby triggering the EMT process [96].

5.3. MicroRNAs Involved in TGF-β-induced EMT

Two microRNA (a class of small noncoding RNAs approximately 22 nt in length)-dependent negative feedback loops are at the heart TGF-β-induced EMT (Figure 4). These pathways are the SNAIL1/miR-34 family/ZEB/miR-200 family feedback loop and the autocrine TGF-β/miR-200 feedback loop [97].

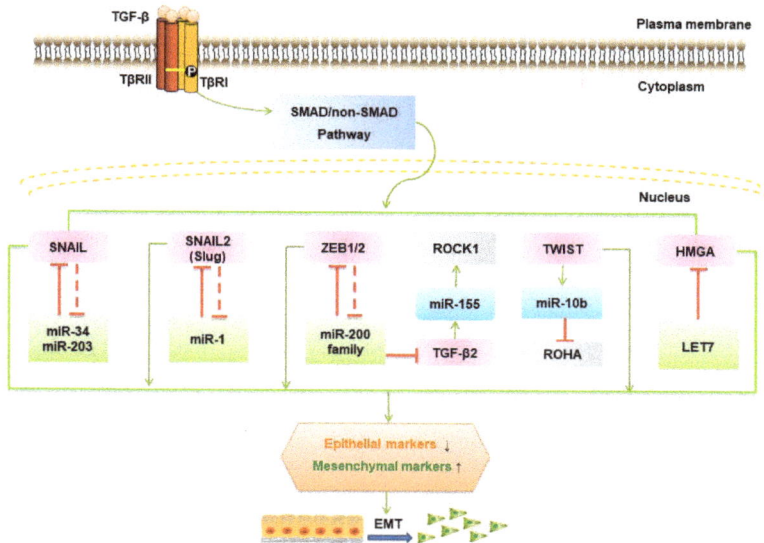

Figure 4. MicroRNAs in TGF-β-induced EMT. At the heart of TGF-β-induced EMT, there are two main double-negative feedback regulatory loops of miRNAs, e.g., the SNAIL1/miR-34 family and ZEB/miR-200 family and the autocrine TGF-β/miR-200 negative feedback loop. Specifically, TGF-β downregulates miR-200 family members, thereby increasing ZEB1 and ZEB2 mRNA levels indirectly, and ZEB binds to promoters of the miR-200 members to repress their expression, thus constituting a double-negative regulatory loop. The same situation occurs in SNAIL1 and miR-34, which are directly linked to p53 status. For the autocrine TGF-β/miR-200 system, autocrine TGF-β positively regulates the expression of SNAIL1 and then increases ZEB mRNA and protein levels, further downregulating miR-200. Inhibitory signals are indicated with inhibitory (dashed) red arrows; Stimulatory signals are indicated with green arrows.

Mechanistically, TGF-β downregulates miR-200 family members, including miR-200a/-200b/-200c/-141/-429, which augments ZEB1 and ZEB2 mRNA levels. ZEB counteracts this mechanism through binding to the promoters of the miR-200 members and thereby repressing their expression. Additionally, miR-200 family members maintain the epithelial phenotype not only by targeting ZEB1/2 but also by actively repressing genes involved in cell motility and invasion [98]. MiR-1199-5p similarly regulates ZEB1 expression [99]. A comparable mechanism governs SNAIL1/miR-34 and the control of p53 status [100]. One study showed that in colorectal cancer, Zinc Finger protein 281 (ZNF281) can be an intermediate regulator between SNAIL1 and miR-34 [101]. In addition to SNAIL and p53, miR-34b experiences epigenetic regulation (chromatin modifications and DNA methylation) by directly targeting methyltransferases and deacetylases, resulting in a positive feedback loop inducing partial demethylation and activity [102]. Silencing miR-34a promoted liver metastases of colon cancer associated with upregulation of c-MET, SNAIL, and β-catenin expression [103]. Transcriptome profiling studies have demonstrated that TGF-β signaling regulates the SMAD4/miR-34a signaling network [104]. The SNAIL1/miR-34 regulatory loop was shown to be involved in the early reversible stage of EMT (from epithelial to P-EMT), whereas the ZEB/miR-200 system is responsible for the establishment of a mesenchymal state [105]. For the autocrine TGF-β/miR-200 system, autocrine TGF-β positively regulates the expression of SNAIL1 and then increases ZEB mRNA and protein levels, further affecting miR-200 [106]. This process makes the second switch (from P-EMT to mesenchymal) irreversible, modulating the maintenance of EMT.

High mobility group protein A2 (HMGA2) has been shown to promote lung cancer progression in mouse and human cells by competing with TGF-β type III receptor for the let-7 microRNA (miRNA) family [107], while decreased let-7g levels influence the TGF-β pathway by targeting TβR1 and SMAD2 gene expression [108]. The overexpression of miR-10b induced TGF-β-driven EMT in breast cancer [109]. In contrast, silencing of miR-10b markedly suppressed the formation of lung metastases by inhibiting its target gene HOXD10 in a mouse mammary tumor model [110].

TGF-β upregulates certain miRNAs, such as miR-182, which prolong NF-κB activation by directly suppressing an NF-κB negative regulator (cylindromatosis, CYLD) [111]. Overexpression of miR-182 restrained SMAD7 expression and promoted breast tumor invasion and TGF-β-induced osteoclastogenesis and bone metastasis [73]. MiR-181a, another miRNA upregulated by TGF-β, promoted TGF-β-mediated EMT and metastasis in breast cancer [112] and via repression of SMAD7 in ovarian cancer progression [113]. TGF-β activates miR-1269 by SOX4 and thereby enhances TGF-β signaling by targeting SMAD7 and HOXD10. This positive feedback loop significantly increased the ability of colorectal cancer cells to invade and metastasize in vivo [114]. Overexpression of miR-216a/217 activated the PI3K/AKT and TGF-β pathways by targeting PTEN, and SMAD7 underlies hepatocarcinogenesis and tumor recurrence of hepatocarcinoma [115]. TGF-β induces the expression and promoter activity of miR-155 through SMAD4. This change reduces RHOA protein and disrupts tight junction formation, leading to EMT [116]. Other research has shown that miR-206 inhibits autocrine production of TGF-β as well as downstream neuropilin-1 (NRP1) and SMAD2 expression, resulting in decreased migration, invasion, and EMT in breast cancer cells [117]. Several other miRNAs, such as miR-373, miR-655, miR206, miR-155, miR-140-5p, miR-494, miR-125a/b, and miR-375, have been implicated in EMT [118–122]. However, for many of these factors, their specific functions remain to be elucidated.

5.4. LncRNAs Involved in TGF-β-induced EMT

Long non coding RNAs (LncRNAs) are a class of RNAs that do not encode proteins or have minimal coding capacity. LncRNAs are emerging as important regulators of a variety of cellular and physiologic functions, such as chromatin dynamics, gene expression, growth, differentiation, and development [123]. LncRNAs can be differentially expressed and localized within the cell. They play a role in chromatin and DNA interactions, negatively or positively affecting the stability or processing of coding mRNA, directly binding to and modulating the functions of signaling proteins, and competitively binding to and thereby controlling the function of miRNAs [124]. Aberrant expression and mutations in lncRNAs have been linked to tumorigenesis, metastasis, and tumor stage [125]. Moreover, they have been detected in the circulating blood and/or urine of cancer patients. For example, plasma levels of a novel lncRNA, p53-induced transcript (Linc-pint), were significantly lower in patients with pancreatic ductal adenocarcinoma (PDAC) than healthy controls [126]. The expression of twenty lncRNAs linked with breast cancer-associated genes (BCAGs) was detectable in human breast cancer cell lines with different expression patterns [127]. LncRNAs are novel, potential therapeutic targets and biomarkers for cancer treatment, and new functions continue to be discovered [128].

Whereas previous studies have shown the involvement of miRNAs in regulating TGF-β signaling and EMT, only a few studies have reported a prominent role of lncRNAs in these processes. Trans-acting lncRNA ELIT-1 induced by TGF-β1 forms a positive TGF-β/SMAD3 signaling feedback and promotes EMT progression by acting as a SMAD3 cofactor [129]. The miR-17~92 polycistronic miRNA cluster encoded by the lncRNA MIR17HG locus was shown to attenuate the TGF-β signaling pathway and stimulate angiogenesis and tumor growth [65,130]. In a recent study, researchers found that a lncRNA activated by TGF-β, lncRNA-ATB, induces EMT and invasion by competitively binding miR-200 family members, which promoted organ-specific metastasis by binding IL-11 mRNA. This competitive binding increased IL-11 mRNA stability, which caused autocrine induction of IL-11 and subsequent activation of STAT3 signaling [131]. These findings suggest that lncRNA-ATB, a mediator of TGF-β signaling, could predispose HCC patients to metastasis [132]. Another study showed that lncRNA-PNUTS, which is highly expressed in mesenchymal breast tumor cells, competitive binds to and neutralizes the activity of miR-205 during EMT. Moreover, elevated expression of lncRNA-PNUTS was correlated with upregulated levels of ZEB mRNAs [133]. LncRNA MEG3 can modulate the activity of TGF-β genes by binding to distal regulatory elements [134]. Another lncRNA, DNM3OS, was associated with overexpression of TWIST1 and specifically contributed to EMT in ovarian cancer [135]. Recently, a paper showed that lncRNA-MUF can directly activate WNT/β-catenin signaling and EMT by binding to ANNEXIN 2A. LncRNA-MUF can also indirectly promote EMT by competitively binding to miR-34a and upregulating SNAIL1 expression [136].

Reduced lncRNA H19 expression in hepatocarcinogenesis (HCG) tissues from patients with the epithelial TGF-β gene signature [137] but increased H19 expression promoted tumor metastasis after TGF-β treatment in Hep3B cells [138]. In a mouse model of spontaneous metastatic breast cancer, lncRNA H19 mediated EMT and MET by differentially binding to the microRNAs miR-200b/c and let-7b [139]. LncRNA H19 can also interact with SLUG and/or EZH2, which regulates E-cadherin expression [140]. Lnc-Spry1 is downregulated by TGF-β and plays a direct regulatory role in the early stage of TGF-β-induced EMT, thus affecting cell invasion and migration. This molecule also controls gene and protein expression levels through an interaction with the splicing factor U2AF65 [141]. LncRNA-KRTAP5-AS1 and lncRNA-TUBB2A control the function of CLAUDIN-4 and thereby influence EMT in gastric cancer [142]. LncRNA HOTAIR (for HOX Transcript antisense intergenic RNA) acts as a crucial player during EMT by mediating a physical interaction between SNAIL and EZH2, which form an enzymatic subunit of the polycomb repressive complex 2, the main writer of chromatin-repressive marks [143]. Together, these findings suggest that lncRNAs can be mediators of TGF-β signaling and may serve as a potential target for anti-metastatic therapies. The mechanisms by which lncRNAs regulate TGF-β signaling are largely unknown, and how they affect TGF-β pathway components in cancer metastasis remains to be discovered (Figure 5 and Table 1).

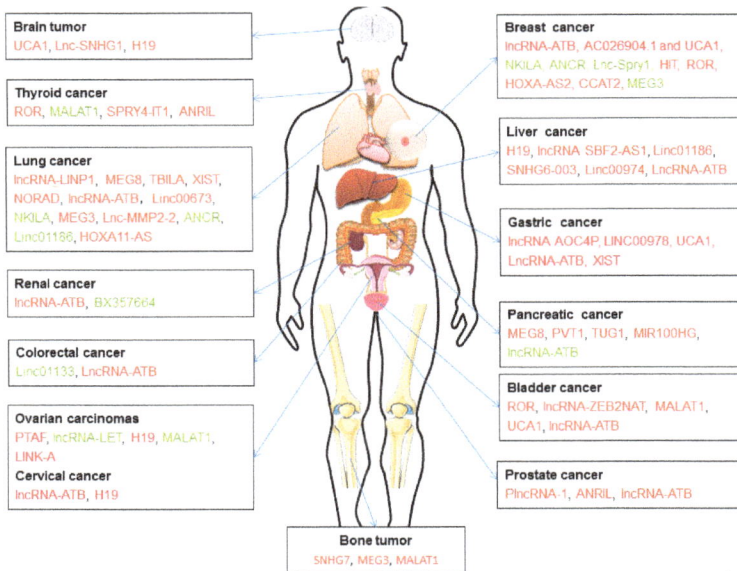

Figure 5. LncRNAs in TGF-β signaling and TGF-β-induced EMT. LncRNAs associated with TGF-β signaling and TGF-β-induced EMT in various cancer cell types. LncRNAs with high expression in tumor tissues (in red) and low expression in tumor tissues (in green).

Table 1. LncRNAs involved in TGF-β signaling, focusing on those that impact TGF-β-induced EMT. LncRNAs with high expression in tumor tissues (in red) and low expression in tumor tissues (in green).

Cancer Type	LncRNA	Function and Mechanism of Action	Example of Key Findings or Experiments
Breast cancer	lncRNA-ATB [144]	Functions as a sponge of miR-141-3p, increasing ZEB1 and ZEB2 expression.	Knockdown results in a morphological change of breast cancer cells from spindle-like to round shape and in a remarkable inhibition of cell migration and invasion.
	AC026904.1 and UCA1 [145]	Functions as an enhancer RNA (eRNA) to activate Slug gene transcription in the nucleus, whereas UCA1 exerts a competitive endogenous RNA (ceRNA) for titrating miR-1 and miR-203a to promote Slug expression at the post-transcriptional level in the cytoplasm.	Knockdown of either AC026904.1 or UCA1 prolongs survival time of the nude mice bearing D3H2LN mammary tumors, these two genes exert critical roles in TGF-β-induced EMT and promote invasion in metastatic breast cancer.
	NKILA [146]	Suppresses TGF-β-induced EMT by blocking NF-κB signaling	Overexpression reduces TGF-β-induced tumor metastasis in vivo.
	ANCR [147]	Functions as a downstream effector molecule, down-regulated by TGF-β1, and is essential for TGF-β1-induced EMT by decreasing RUNX2 expression.	Ectopic expression partly attenuates the TGF-β1-induced EMT and knockdown promotes TGF-β1-induced EMT and metastasis in breast cancer.
	Lnc-Spry1 [141]	Functions as an immediate-early regulator of EMT that is downregulated by TGF-β, affecting the expression of TGF-β-regulated gene targets; alternative splicing by U2AF65 splicing factor; isoform switching of fibroblast growth factor receptors.	Knockdown promotes a mesenchymal-like phenotype and results in increased cell migration and invasion.
	lncRNA-HIT [148]	Ectopic expression disrupts tight junction by targeting E-cadherin.	Knockdown results in decrease of cell migration, invasion, tumor growth, and metastasis.
	linc-ROR [149,150]	Functions as a sponge of miR-145 and therefore upregulate the expression of ARF6, which regulates adhesion and invasion properties of breast tumor cells through E-cadherin.	Regulates the cancer stem cell phenotype in Triple-negative Breast Cancer, which plays a critical role in drug resistance and metastasis.
	HOXA-AS2 [151]	Functions as an endogenous sponge of miR-520c-3p, and controls the expression of miR-520c-3p target genes, TβR2 and RELA, in breast cancer cells.	Knockdown inhibits the progression of breast cancer cells in vitro and in vivo.
	CCAT2 [152]	Knockdown causes cell cycle arrested in G0/G1 phase, promotes cell apoptosis and downregulates the protein expression levels of TGF-β, Smad2 and α-SMA in breast cancer cells.	Down-regulation inhibits the proliferation, invasion and migration in breast cancer cells.
	MEG3 [134,153,154]	Regulates the TGF-β pathway genes through formation of RNA-DNA triplex structures, and downregulates AKT and functions as a sponge of miR-421 to regulate E-cadherin expression.	Ectopic expression inhibits in vivo tumorigenesis and angiogenesis in a nude mouse xenograft model.
Gastric cancer	LINC00978 (known as MIR4435-2HG and AK001796) [155]	Knockdown inhibits the activation of TGF-β/SMAD signaling pathway and EMT in GC cells.	Knockdown inhibits the proliferation, migration and invasion and decreases the in vivo tumorigenicity of GC cells in mice.
	UCA1 [156]	Knockdown inhibits TGF-β1-induced-EMT process and the effect could be partly restored by TGF-β1 treatment.	A potential oncogenic factor by regulating GC cells proliferation, invasion, and metastasis under TGF-β1 induction.
	lncRNA-ATB [157–159]	Induced by TGF-β and functions as a ceRNA of miR-141-3p or miR-200s.	A novel biomarker of lncRNA, correlated with increased invasion depth, more distant metastasis and advanced tumor-node-metastasis stage.
	XIST [160]	Functions as a competing endogenous lncRNA (ceRNA) to regulate TGF-β1 by sponging miR-185 in GC.	sh-XIST inhibited GC development in vitro.

Table 1. *Cont.*

Cancer Type	LncRNA	Function and Mechanism of Action	Example of Key Findings or Experiments
Ovarian carcinomas	LncRNA-LET [161]	Regulates EMT process and the expression of TIMP2 and activates the Wnt/β-catenin and Notch signaling pathways.	Overexpression inhibits cell viability, migration and EMT process, and increases apoptosis in KGN cells
	H19 [162]	Functions by competing with miR-370-3p to regulate TGF-β-induced EMT in ovarian cancer.	Knockdown suppresses TGF-β-induced EMT, while H19 overexpression promotes TGF-β-induced EMT.
	MALAT1 [163,164]	TGF-β increases its expression by inhibiting miR-200c; MALAT1 interacts with MARCH7, which regulates TGF-β-smad2/3 pathway by interacting with TβR2, via miR-200a as a ceRNA.	Interrupts the interaction between miR-200c/MALAT1 decreases the invasive capacity of EEC cells and EMT in vitro and inhibits EEC growth and EMT-associated protein expression in vivo; This LncRNA plays an important role in TβR2-Smad2/3-MALAT1/MARCH7/ATG7 feedback loop mediated autophagy, migration and invasion in ovarian cancer.
	LINK-A [165] cervical cancer	TGF-β1 treatment has no effects on LINK-A expression, and there is no clear mechanism.	Overexpression increases expression of TGF-β1 in ovarian carcinoma cells and promotes cell migration and invasion and this effect can be attenuated by TGF-β1; Plasma levels are correlated with distant tumor metastasis but not tumor size.
	lncRNA-ATB [166]	No clear mechanism.	A promising prognostic marker that correlates with the malignant phenotypes and poor prognosis of cervical cancer
	PTAF [167]	Functions by competing with miR-25 and affecting SNAI2 to regulate the expression of many EMT-related protein-coding genes in OvCa.	A mediator of TGF-β signaling. Upregulation induces elevated SNAI2, which in turn promoted OvCa cell EMT and invasion; knockdown inhibits tumor progression and metastasis in an orthotopic mouse model of OvCa.
Bladder cancer	lncRNA-ZEB2NAT [168]	Induced by TGF-β, and can regulate EMT process by affecting ZEB2 protein level.	Knockdown reverses CAF-CM-induced EMT and invasion of cancer cells, as well as reduces the ZEB2 protein level.
	MALAT1 [169]	Induced by TGF-β and regulates EMT by negatively correlated E-cadherin, N-cadherin and fibronectin expression by zeste 12 (suz12) in vitro and in vivo.	Overexpression is significantly correlated with poor survival in patients with bladder cancer. Inhibition of malat1 or suz12 suppresses the migratory and invasive properties induced by TGF-β and inhibits tumor metastasis in animal models.
	UCA1 [170]	Induced by BMP9 through phosphorylated AKT and there are no clear mechanism.	Its BMP-9-induced expression associates with increased proliferation and migration of bladder cancer cells. The promoting effect of BMP9 is rescued after interfering with UCA1 in BMP9 overexpressed bladder cancer cells both in vitro and in vivo.
	lncRNA-ATB [171]	Is upregulated by TGF-β, acting as a molecular sponge of miR-126 and regulate the direct target of miR-126 (KRAS).	Its overexpression significantly promotes cell viability, migration, and invasion in T24 cells.
	PlncRNA-1 [172]	Regulates the cell cycle, cyclin-D1 and EMT in prostate cancer cells through the TGF-β1 pathway.	Functions as an oncogene.
	ROR [173]	Knockdown can reverse TGF-β1-induced-EMT phenotype in SGC-996 and Noz cells. However, there are no clear mechanism.	High expression is associated with poor prognosis in gallbladder cancer patients and knockdown inhibits cell proliferation, migration, and invasion.
Prostate cancer	ANRIL [174]	Regulates let-7a/TGF-β1/Smad signaling pathway.	Overexpression promotes the proliferation and migration of prostate cancer cells.
	lncRNA-ATB [175]	Upregulated by TGF-β, stimulates EMT associated with ZEB1 and ZNF217 expression levels via ERK and PI3K/AKT signaling pathways.	Overexpression promotes, and knockdown of lncRNA-ATB inhibits the growth of prostate cancer cells via regulations of cell cycle regulatory protein expression levels.
Brain tumor	lnc-SNHG1 [176]	Activates the TGFBR2/SMAD3 and RAB11A/Wnt/β-catenin pathways in pituitary tumor cells via sponging miR-302/372/373/520.	Promotes the progression of pituitary tumors, ectopic expression of lnc-SNHG1 promotes cell proliferation, migration, and invasion, as well as the EMT, by affecting the cell cycle and cell apoptosis in vitro and tumor growth in vivo.
	UCA1 [177]	Functions as a ceRNA of miR-1 and miR-203a to promote Slug expression, which underlies TGF-β-induced EMT and stemness of glioma cells.	Knockdown attenuates EMT and stemness processes and their enhancement by TGF-β.

Table 1. *Cont.*

Cancer Type	LncRNA	Function and Mechanism of Action	Example of Key Findings or Experiments
Lung cancer	lncRNA-LINP1 [178]	TGF-β1 inhibits its transcription in a SMAD4-dependent manner.	Inhibits TGF-β-induced EMT and thereby controlling cancer cell migration, invasion, and stemless in lung cancer cells.
	TBILA [179]	Induced by TGF-β and functions via cis-regulating HGAL and activating S100A7/JAB1 signaling.	Promotes non-small cell lung cancer progression in vitro and in vivo.
	XIST [180,181]	Functions as an endogenous sponge by directly binding to miR-137, negatively regulating its expression and regulating Notch gene expression.	Overexpression inhibits proliferation and TGF-β1-induced EMT in A549 and H1299 cells, regulating proliferation and TGF-β1-induced EMT in NSCLC, which could be involved in NSCLC progression.
	NORAD [182]	Affects the physical interaction of its binding partner (importin β1) with Smad3, and then inhibits the nuclear accumulation of Smad complexes in response to TGF-β.	Stimulates TGF-β signaling and regulates TGF-β-induced EMT-like phenotype in A549 cells.
	lncRNA-ATB [183]	Regulates EMT by down-regulating miR-494 in A549 cells, which in turn increases phosphorylated levels of AKT, JAK1, and STAT3.	Overexpression promotes proliferation, migration, and invasion of A549 cells. In contract, ATB silence shows the opposite influence.
	linc00673 [184]	Functions as a sponge of miR-150-5p and indirectly modulates the expression of key EMT regulator ZEB1.	Inhibition attenuates the tumorigenesis ability of A549 cells in vivo.
	NKILA [185]	Expression is regulated by TGF-β and regulates EMT process by inhibiting the phosphorylation of IκBα and NF-κB activation to attenuate Snail expression.	Inhibits migration, invasion and viability of NSCLC cells. Lower NKILA expression are correlated with lymph node metastasis and advanced TNM stage.
	MEG3 [186]	Associates with JARID2 and the regulatory regions of target genes to recruit the complex by epigenetic regulation (PRC2/JARID2/ H3K27 axis).	Knockdown inhibits TGF-β-mediated changes in cell morphology and cell motility characteristic of EMT and counteracts TGF-β-dependent changes in the expression of EMT-related genes; In contrast, overexpression enhances these effects.
	lnc-MMP2-2 [187]	Is highly enriched in TGF-β-mediated exosomes and might function by increasing the expression of MMP2 through its enhancer activity.	Knockdown affects lung cancer invasion and vascular permeability.
	ANCR [188]	Inhibits NSCLC cell migration and invasion by downregulating TGF-β1 expression, however TGF-β1 treatment shows no significant effects on ANCR expression but promotes NSCLC cell migration and invasion.	Low expression level indicates shorter postoperative survival time of NSCLC patients, whereas, ectopic expression inhibits NSCLC cell migration, invasion and downregulated TGF-β1 expression, and this effect can be attenuated by TGF-β1.
	LINC01186 [189]	Functions as a mediator of TGF-β signaling, is down-regulated by TGF-β1 regulating EMT by Smad3	Knockdown promotes cell migration and invasion, whereas, overexpression prevents cell metastasis.
	HOXA11-AS [190]	Regulates the expression of various pathways and genes, especially DOCK8 and TGF-β pathway, however, there is no clear mechanism.	Its expression may determine the overall survival and disease-free survival of lung adenocarcinoma patients in TCGA.
Liver cancer	lncRNA SBF2-AS1 [191]	Functions as a ceRNA of miR-140-5p and regulates the expression of TβR1.	Knockdown inhibits the proliferation, migration and invasion of HCC cells and attenuate the development of HCC tumor in vivo.
	SNHG6-003 [166]	Functions as a ceRNA of miR-26a/b and thereby modulates the expression of transforming growth factor-β-activated kinase 1 (TAK1).	Ectopic expression in HCC cells promotes cell proliferation and induces drug resistance, whereas knockdown promotes apoptosis. High expression of SNHG6-003 closely correlated with tumor progression and shorter survival in HCC patients.
	LINC00974 [192]	Interacts with KRT19, as a sponge of miR-642, activating the Notch and TGF-β pathways.	Knockdown inhibits cell proliferation and invasion with an activation of apoptosis and cell cycle arrest both in vitro and in vitro.
	lncRNA-ATB [132]	Upregulated by TGF-b and can induce EMT and invasion by acting as ceRNA of miR-200 family and increasing ZEB1/2; promotes organ colonization by binding IL-11 mRNA, autocrine induction of IL-11, and triggering STAT3 signaling.	Promotes the invasion-metastasis cascade in hepatocellular carcinoma.

Table 1. *Cont.*

Cancer Type	LncRNA	Function and Mechanism of Action	Example of Key Findings or Experiments
Pancreatic cancer	PVT1 [193]	Regulates EMT process via TGF-β1/Smad signaling.	Acts as an oncogene in pancreatic cancer, knockdown of PVT1 inhibits viability, adhesion, migration and invasion.
	MEG8 [194]	MEG8, which shares the DLK1-DIO3 locus with MEG3, can induce the recruitment of EZH2 protein to miR-34a and miR-203 genes for histone H3 methylation and transcriptional repression, inducing EMT-related cell morphological changes and increases cell motility in the absence of TGF-β by activating the gene expression program required for EMT.	Plays critical role in TGF-β-induced EMT in A549 lung cancer and Panc1 pancreatic cancer cells.
	TUG1 [195]	Regulates EMT process though TGF-β/Smad pathway	Overexpression increases cell proliferation and migration capacities, enhancing the proliferation and migration of pancreatic cancer cells
	MIR100HG [196]	Induced by TGF-β, and this gene contains miR-100, miR-125b and let-7a in its intron, through SMAD2/3. These miRNAs regulate a multitude of genes involved in the inhibition of p53 and DNA damage response pathways.	Plays prominent role in metastasis of pancreatic cancer.
	lncRNA-ATB [197]	No clear mechanism.	Low expression levels are correlated with lymph node metastases neural invasion, and clinical stage and worse overall survival prognoses of patients.
Renal cancer	BX357664 [197,198]	Blocks the TGF-β1/p38/HSP27 pathway.	Upregulation reduces migration, invasion, and proliferation capabilities in RCC cells.
	lncRNA-ATB [199]	No clear mechanism.	Its expression is correlated with metastases and promotes cell migration and invasion in renal cell carcinoma. Knockdown could inhibit cell proliferation, trigger apoptosis, reduce epithelial-to-mesenchymal transition program and suppress cell migration and invasion.
Colorectal cancer	lncRNA-ATB [199–201]	Upregulated by TGF-β, and suppresses E-cadherin expression and promoting EMT process.	High expression is significantly associated with greater tumor size, depth of tumor invasion, lymphatic invasion, vascular invasion, and lymph node metastasis.
	LINC01133 [202]	Downregulated by TGF-β and its expression is positively correlated with E-cadherin, and negatively correlated with Vimentin. Directly interacting with SRSF6, which promotes EMT and metastasis in CRC cells.	Inhibits EMT and metastasis in colorectal cancer (CRC) cells and low LIINC01133 expression in tumors with poor survival in CRC samples.
Bone tumor	MEG3 [203]	Represses Notch and TGF-β signaling pathway by inhibiting Notch1, Hes1, TGF-β and N-cadherin expression, and increasing E-cadherin.	Overexpression represses cell proliferation and migration ability.
	MALAT1 [204]	Induced by TGF-β and its overexpression decreases E-cadherin level, however, this effect was partially reversed by EZH2 knockdown.	Overexpression promotes cell metastasis, a potential diagnostic and prognostic factor in osteosarcoma.
	SNHG7 [205]	Inhibits tumor suppressor miR-34s signals and the targets of miR-34a, including proliferation-related Notch1, apoptosis-related BCL-2, cell cycle-related CDK6, and EMT-related SMAD4.	Knockdown delays the tumor growth in osteosarcoma tissues.
Thyroid cancer	MALAT1 [206]	Induced by TGF-β and supports a role for MALAT1 in EMT in thyroid tumors.	Functions as an oncogene and as a tumor suppressor in different types of thyroid tumors.
	ROR [207]	Functions as a competing endogenous RNA (ceRNA) of sponging miR-145.	Expression is induced by TGF-β in cells undergoing EMT.
	SPRY4-IT1 [208]	Knockdown increases the levels of TGF-β1 and p-Smad2/3 and this effect could be rescued by the interference of TGF-β1.	A novel prognostic factor, which correlates with poor prognosis and exhibits that silenced SPRY4-IT1 inhibited the proliferative and migratory abilities of TC cells.
	ANRIL [209]	Reduces p15INK4B expression through inhibiting TGF-β/Smad signaling pathway.	Promotes invasion and metastasis of TC cells, and the silencing of ANRIL inhibits the invasion and metastasis of TPC-1 cells.

6. TGF-β and Metastasis

6.1. TGF-β-induced Metastasis in Tissues

Tumor metastasis is the primary cause of cancer lethality; it is a progressive, multifactorial and multistep dynamic process, including detachment from a primary tumor, invasion into surrounding tissues, invasion into blood circulation/lymphatic circulation, survival in the circulatory system, extravasation from blood vessels, distal colonization, etc. [210]. The migration of cancer cells is pivotal to early metastasis, and changes in tumor cells (acquired by EMT) and the microenvironment are the two main factors that help cancer stem cells (CSCs) to escape from the primary site [211]. In the tumor microenvironment, TGF-β, Chemokine 4/12 (CXCL4/12), interleukin-6 (IL-6) and tumor necrosis factor-α (TNF-α), etc. can enhance EMT. At the same time, tumor cells secrete more epithelial growth factors, fibroblast growth factor (FGF) and insulin-like growth factor (IGF), leading to a hypoxic, acidic, high interstitial fluid pressure (IFP) state in the microenvironment, which activates cancer associated fibroblasts (CAFs) to produce more matrix metalloproteinases (MMPs) and remodel the tumor extracellular matrices (ECM) [212].

TGF-β induces metastasis to bone, liver, lung and other tissues of specific cancer types, such as breast, lung, gastric and prostate cancers (Figure 6A) [213]. Metastatic cancer cells have been shown to disturb the tight balance of bone transformation by osteoblasts and osteoclasts, conferring a receptive outgrowth microenvironment [214]. Tumor cells secrete cytokines and parathyroid hormone-related protein (PTHrP), which is the main inducer of osteoclast formation (also interleukin (IL)-1/-6/-11, etc.), and its expression is specific to the bone metastasis microenvironment [215]. TGF-β released in the active form upon osteoclastic bone resorption enhances PTHrP signaling in osteoblasts, resulting in osteoblasts expressing receptor activator of NFκB ligand (RANKL) while reducing osteoprotegerin (OPG) expression [216]. The high RANKL/OPG ratio enhances the osteolytic activity, which is associated with the release of high levels of active TGF-β. This TGF-β can further upregulate PTHrP expression by cancer cells, thereby forming a positive feedback loop called a "vicious cycle" [217].

Many specific studies have analyzed the role of TGF-β in lung and liver metastases. In a previous study, increased TGF-β levels were shown to lead to lung metastases in the MMTV/PyVmT transgenic model of metastatic breast cancer [218]. In addition, in breast cancer cells, TGF-β induced ANGPTL4 via the SMAD signaling pathway, and this cytokine could disrupt lung capillary walls and seed pulmonary metastases [219]. Because there are inherent differences in the microvasculature of these two tissues, lung metastasis requires robust extravasation functions provided by ANGPTl4 and other factors and additional lung colonizing functions achieved by ID1/ID3 [220]. Therefore, the vasculature disruptive mechanism provides a selective invasive advantage in the lung but not bone. Another study showed that the WNT signaling inhibitor Dickkopf 1 (DKK1) is a key factor for this metastatic preference; it reduces the recruitment of macrophages and neutrophils by the WNT/PCP-RAC1-JNK pathway and inhibits the level of tumor-derived TGF-β to inhibit lung metastasis. Thus, tumor cells that highly secrete DKK1 tend to metastasize to bone, while those with low DKK1 secretion tend to metastasize to the lung [221].

Statistical analysis of gene expression profiles revealed that TGF-β signaling is the most significant gene pathway in liver metastases of colorectal cancer [222]. Exosomes (small membrane vesicles with a size ranging from 40 to 100 nm) from pancreatic cancer induce Kupffer cells to release more TGF-β, which in turn activates the fibrotic pathway and forms a proinflammatory environment that supports pancreatic cancer metastasis [223].

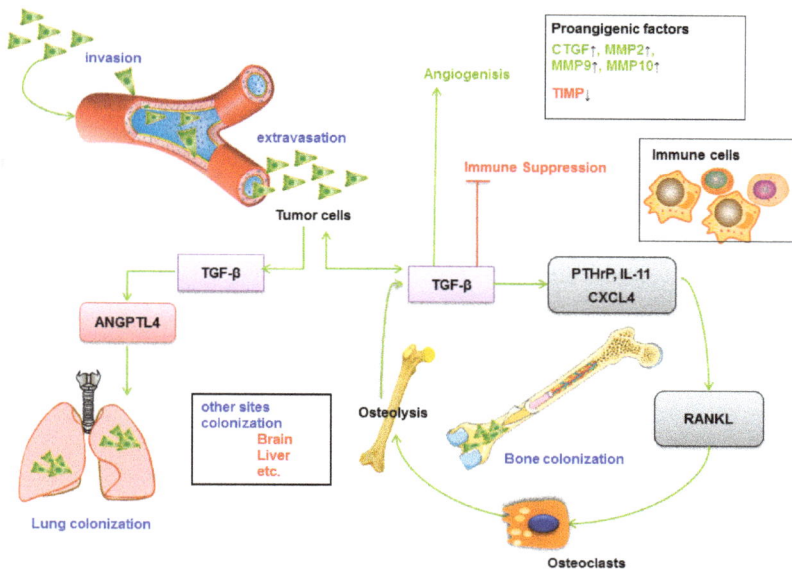

Figure 6. TGF-β mediates metastasis. TGF-β produced by cancer cells can alter the bone microenvironment by inducing the expression of osteolytic factors like PTHrP and IL11. The osteoclast bone resorption via RANKL production by osteoblasts results in more release of TGF-β, which in turn acts on tumor cells thereby creating a positive feedback loop called a "vicious cycle". Chemokine receptor CXCL4 and angiogenesis inducer connective tissue growth factor (CTGF) are also key modulators induced by TGF-β in this process. Another important factor of this metastatic preference is the Wnt antagonist Dickkopf 1 (DKK1); cells that highly secrete DKK1 tend to metastasize to bone, while low DKK1-secreting tumor cells tend to metastasize to the lung. TGF-β leads to transcriptional upregulation of proangiogenic factors, including CTGF, matrix metalloprotease (MMP)2, MMP-9, and MMP10, or inhibition of TIMP to mediate the formation of new blood vessels. TGF-β inhibits the proliferation of T cells and B cells and inhibits the production of immune factors by B lymphocytes. In addition, TGF-β-induced angiopoietin-like 4 (ANGPTL4) via the SMAD signaling pathway is proangiogenic and can disrupt lung capillary walls and seed pulmonary metastases. Inhibitory signals are indicated with inhibitory red arrows; Stimulatory signals are indicated with green arrows.

6.2. TGF-β Promotes Angiogenesis

Regardless of the primary or secondary tumor, angiogenesis occurs once the tumor is more than 1–2 mm in diameter; therefore, a rich blood supply is necessary to provide nutrients and oxygen for tumor growth and metastasis [224]. Tumor cells secrete a variety of growth factors, including TGF-β, to accelerate the development of cancer by inducing angiogenesis (Figure 6B) [225].

On endothelial cells TGF-β can bind to the type I receptor, activin receptor-like kinase 1 (ALK-1), prompting downstream signaling involving intracellular and nuclear proteins (SMADs and Id1) and leading to a proangiogenic response [226]. Furthermore, TGF-β1 and hypoxia are potent inducers of vascular endothelial growth factor (VEGF) expression in tumor cells, and oncogenes, especially RAS, can also combine with the tumor microenvironment, providing the foundation for tumor cell invasion and angiogenesis [227]. Using a mouse mammary carcinoma model, researchers confirmed that VEGF expression in peri-necrotic areas is synergized by both hypoxia and TGF-β1, further showing that this cooperation is achieved through hypoxia-inducible factor (HIF)-1α physically associating with SMAD3 [228]. For example, TGF-β increases the expression of VEGF-C by coordinating with sine oculis homeobox homolog 1 (SIX1) in tumor cells, promoting tumor lymph angiogenesis and lymph node metastasis [229]. Likewise, TGF-β1 could promote macrophages to secrete more VEGF via the TβRII/SMAD3 signaling pathway in oral squamous cell carcinoma [230]. TGF-β and VEGF form a feedback loop through the Semaphorin3A/NEM axis; in general, abrogated VEGF inhibits the endothelial cell paracrine TGF-β1 and endothelial SMAD2/3 activation; in turn, TGF-β1 further stimulates endothelial Semaphorin3A expression [231]. Studies have shown that in RAS-transformed epithelial tumors, TGF-β significantly increases the expression of VEGF/VEGF-R, which has a powerful effect on capillary formation and migration of endothelial cells, thereby promoting angiogenesis in tumor cells [232]. TGF-β mediates the formation of new blood vessels by promoting connective tissue growth factor (CTGF) and angiogenic regulatory enzymes, such as matrix metalloproteinases (MMP-2, MMP-9, MMP-10, etc.) or by inhibiting tissue inhibitor of metalloproteinases (TIMP) [233].

6.3. TGF-β Promotes Immune Evasion

Under physiological conditions, the immune system is the most important element of human defense against cancer, and T lymphocytes and natural killer cells can recognize and specifically clear tumor cells [4,234]. However, tumor cells evade this immune surveillance through immune evasion of TGF-β, but the cellular mechanism by which TGF-β induces T cell dysfunction remains unclear. TGF-β may inhibit the proliferation of T cells and B cells and inhibit the production of immune factors by B lymphocytes (Figure 6C). In transgenic mouse studies, CD4+ and CD8+ T lymphocytes showed that expression of dominant-negative TβR2 was more effective in clearing thymoma and melanoma cells than in wild-type mice, which indicates that T lymphocytes are central targets for the negative regulation of TGF-β [235]. Interestingly, T cell production of TGF-β1 was shown to be a requirement for tumors to evade immune surveillance independent of TGF-β produced by tumors [236].

For example, regulation of tumor metastasis by TGF-β/SMAD signaling was found to be achieved by impairing the activity of tumor-infiltrating T cells [237,238], i.e., the infiltration level of CD3+, CD4+ and CD8+ cells and the proliferation and activity of T cells (such as the secretion of granzyme, FAS ligand (FASL), perforin and interferon (IFN)-γ) [239,240]. Regulatory T-cells (Tregs), whose excessive function inhibit antitumor immune responses, are another vital factor for TGF-β-mediated immune evasion by suppressing the proliferation and activation of CD8+ cytotoxic T-cells [241]. Effector Treg cells express high amounts of integrin αvβ8, which enables them to activate latent TGF-β, and tumor-derived TGF-β, in turn, induces FoxP3 expression and generates induced Treg cells [242]. Tregs can produce cell surface docking receptors for latent TGF-β, called glycoprotein A repetitions predominant (GARP). Further experiments revealed that overexpression of GARP leads to more TGF-β-releasing Treg cells and enhanced TGF-β signaling, tumor growth and metastasis in immunodeficient mice [243].

Additionally, TGF-β blocks immune surveillance by inhibiting migration and inducing apoptosis of antigen-presenting cells, such as dendritic cells (DCs), whose function is to mature and stimulate T cells during the immune response. Studies have shown that tumor-derived TGF-β significantly inhibits the proliferation of human CD4+ T cells activated by dendritic cells [244–246]. TGF-β also has an impact on myeloid cell functions. TGF-β in the tumor microenvironment polarizes tumor-associated macrophages (TAMs) from the pretumor (M2) phenotype to the antitumor (M1) phenotype. TGF-β inhibits neutrophil activity (i.e., degranulation) [247]. Furthermore, tumor-derived TGF-β polarizes the tumor-associated neutrophil (TAN) phenotype from N1 to N2 and induces a population of protumor TANs [248].

Moreover, the development of natural killer (NK) cells and T helper 1 (Th1) differentiation depend on TGF-β signaling. Functional studies have demonstrated that selective deletion of SMAD4 in NK cells impedes NK cell homeostasis and maturation, thereby reducing murine cytomegalovirus clearance [249]. A TGF-β-regulated transcription factor, T-bet, is responsible for Th1 differentiation and survival of activated CD4+ T cells via mediating CD122 expression and IL-15 signaling in Th1 cells [250].

7. Targeting the TGF-β Signaling Pathway in Cancer

Several TGF-β signaling inhibitors have been developed to curtail the aberrant TGF-β signaling characteristics of tumors. There are several types of TGF-β drugs in (pre)clinical development: ligand traps, antisense oligonucleotides (AONs), neutralizing antibodies, receptor domain-immunoglobulin fusions and receptor kinase inhibitors (Figure 7) [251]. While these agents have shown promise in the clinic, the complexity and pleiotropic nature of TGF-β tumor regulation render TGF-β targeted therapy a challenge. Related to this, careful administration/dosing of TGF-β therapies as well as judicious patient selection is needed to overcome on-target and off-target toxic side-effects [252].

Antisense oligonucleotides have been incorporated into immune cells to block translation or to degrade TGF-β mRNA. Belagenpumatucel-L (Lucanix) is a therapeutic vaccine for non-small cell lung cancer (NSCLC), which inhibits the expression of TGF-β2, thereby reducing the immunosuppressive effect of TGF-β2 and thus enhancing its antitumor effect [253]. Trabedersen (AP 12009) is an antisense oligonucleotide that specifically targets human TGF-β2 mRNA and has been used in the treatment of metastatic melanoma and pancreatic cancer [254]. Its clinical development has been put on hold (Clinical Trials: NCT00761280).

Multiple TβRI kinase inhibitors have been developed for the treatment of cancer and are under clinical trials. For example, a phase I study of Galunisertib (LY2157299), a TβR1 kinase inhibitor, showed an acceptable tolerability and safety profile in Japanese patients with advanced solid tumors [255] and is currently under clinical development in combination with immunotherapy, anti-PD-1 antibodies, including nivolumab and durvalumab (Clinical Trials: NCT02423343). SM16 (a TβRI kinase inhibitor) and 1D11 (a TGF-β neutralizing antibody) synergistically inhibited metastasis in combination with the agonistic OX40 antibody [256]. SD208, an inhibitor of TβR1, blocks the TGF-β/SMAD pathway, Matrigel invasion and expression of TGF-β target genes (PTHrP, IL-11, CTGF, and RUNX2, etc.) and was effective at preventing the development of bone metastases and decreasing the progression of established osteolytic lesions in melanoma and glioma models [257,258]. However, another study showed that SD-208 could not significantly reduce tumor growth and angiogenesis in SW-48 cells, a human colorectal cancer model, and thus, its efficiency still needs to be assessed [259].

Blockade of TGF-β ligand and receptor binding is a crucial mechanism for TGF-β inhibitory targeting agents. In a clinical phase I trial, PF-03446962, an anti-ALK1 antibody that displays (dose-dependent) antiangiogenic activity, was tested. This antibody was administered to patients who were resistant to anti-VEGF/VEGFR therapy [260]. However, in phase II, trials were stopped due to ineffectiveness (NCT01620970). Another TGF-β monoclonal antibody, Fresolimumab (GC-1008), was demonstrated to be safe and well tolerated in Phase I and Phase II trials [261]. Fresolimumab is an anti-TGF-β neutralizing antibody capable of neutralizing all human isoforms of TGF-β [262]. John et al. have shown that some melanoma or renal cancer patients with multiple doses of fresolimumab treatment experienced cutaneous lesions during phase 1 clinical trials [195]. Moreover, a phase I study noted that the maximum tolerated dose for the anti-TβRII monoclonal antibody LY3022859 could not be determined, but dose escalation beyond 25 mg was considered unsafe in patients with advanced solid tumors [263].

Finally, an interesting paper showed that a TGF-β inhibitor in combination with a PD-L1 inhibitor (Atezolizumab) may be able to remodel the matrix microenvironment and allow T cells to enter the interior of the tumor [237]. Furthermore, for chimeric antigen receptor (CAR) T cell therapies, CAR-T cells can reverse the immunosuppressive effect of TGF-β [264]. This finding suggests that combination therapies rather than mono-TGF-β therapies could prove to be the most effective approach in the future.

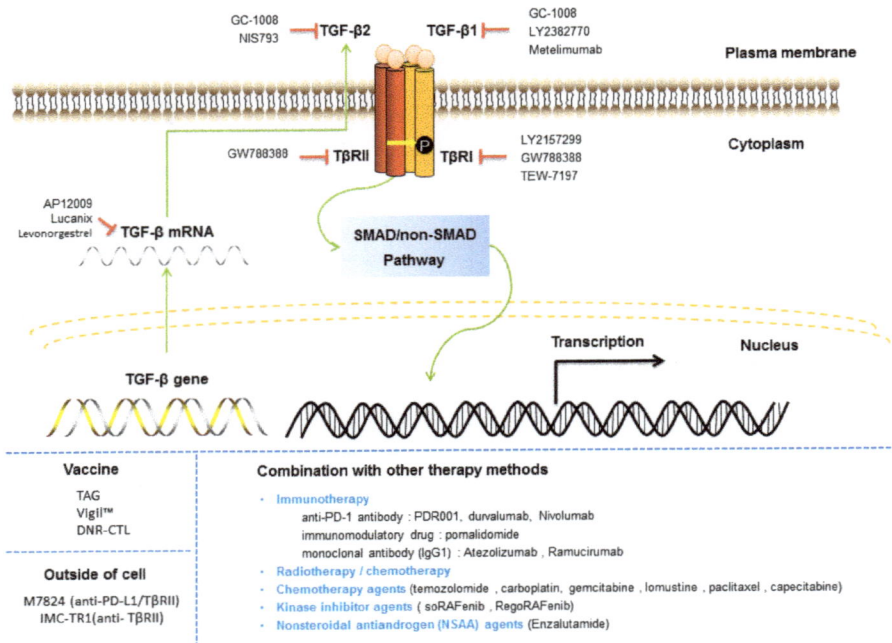

Figure 7. Targeting the TGF-β signaling pathway in cancer. Various molecular mechanisms by which TGF-β signaling is targeted for therapeutic gain are depicted. TGF-β inhibitory targeting agents (Table 1), including TβRI kinase inhibitors, AON targeting ligand or receptor gene expression, and antibodies interfering with ligand-receptor interactions, have been developed to curtail excessive TGF-β pathway activation. Inhibitory signals are indicated with inhibitory red arrows; Stimulatory signals are indicated with green arrows.

8. Concluding Remarks

TGF-β acts as a tumor suppressor during the early phase of cancer progression and as a tumor promotor in advanced stages. The latter role, in particular, has been targeted as a potential therapeutic approach to inhibiting tumor growth and dissemination. However, developing effective treatments is severely hampered by the biphasic role of TGF-β in cancer and its function in many physiological processes, including the cardiovascular system. There has been some progress in the treatment of liver cancer and other types of cancer, but there is a pressing need for a greater understanding of pathological TGF-β mechanisms as well as greater clarity regarding protocols of therapy administration and patient selection.

With the success of immune checkpoint inhibitors for cancer therapy and considering the potent immune suppressive effects of TGF-β, it may be of particular interest to see whether TGF-β targeting agents can increase the efficiency and range of immune therapeutic agents. Such trials are underway, and the results are eagerly awaited. As highlighted in this review, other approaches could include the targeting of E3 ubiquitin ligases and deubiquitinating enzymes and miRNAs and lncRNAs, which control the stability of TGF-β receptors and SMAD proteins.

Author Contributions: Writing—original draft preparation, Y.H.; writing—review and editing, D.B.; supervision, P.t.D.

Funding: Our studies on TGF-β signaling in cancer are supported by Cancer Genomics Centre Netherlands (CGC.nl). YH is supported by CSC scholarship.

Int. J. Mol. Sci. **2019**, *20*, 2767

Acknowledgments: Our studies on TGF-β signaling in cancer are supported by Cancer Genomics Centre Netherlands (CGC.nl). YH is supported by CSC scholarship.

Conflicts of Interest: The authors declare no conflict of interest.

References

1. Faguet, G.B. A brief history of cancer: Age-old milestones underlying our current knowledge database. *Int. J. Cancer* **2015**, *136*, 2022–2036. [CrossRef] [PubMed]

2. Hu, Y.; Fu, L. Targeting cancer stem cells: A new therapy to cure cancer patients. *Am. J. Cancer Res.* **2012**, *2*, 340–356. [PubMed]

3. Farkona, S.; Diamandis, E.P.; Blasutig, I.M. Cancer immunotherapy: The beginning of the end of cancer? *BMC Med.* **2016**, *14*, 73. [CrossRef] [PubMed]

4. David, C.J.; Massague, J. Contextual determinants of TGF-β action in development, immunity and cancer. *Nat. Rev. Mol. Cell Biol.* **2018**, *19*, 419–435. [CrossRef] [PubMed]

5. Exposito-Villen, A.; Aranega, E.A.; Franco, D. Functional role of non-coding RNAs during epithelial-To-mesenchymal transition. *Noncoding RNA* **2018**, *4*, 14. [CrossRef]

6. Derynck, R.; Budi, E.H. Specificity, versatility, and control of TGF-β family signaling. *Sci. Signal* **2019**, *12*, eaav5183. [CrossRef] [PubMed]

7. De Larco, J.E.; Todaro, G.J. Growth factors from murine sarcoma virus-transformed cells. *Proc. Natl. Acad. Sci. USA* **1978**, *75*, 4001–4005. [CrossRef]

8. Derynck, R.; Jarrett, J.A.; Chen, E.Y.; Eaton, D.H.; Bell, J.R.; Assoian, R.K.; Roberts, A.B.; Sporn, M.B.; Goeddel, V.D. Human transforming growth factor-β complementary DNA sequence and expression in normal and transformed cells. *Nature* **1985**, *316*, 701–705. [CrossRef]

9. Sha, X.; Yang, L.; Gentry, E.L. Identification and analysis of discrete functional domains in the pro region of pre-pro-transforming growth factor β 1. *J. Cell Biol.* **1991**, *114*, 827–839. [CrossRef]

10. Shi, M.; Zhu, J.; Wang, R.; Chen, X.; Mi, L.; Walz, T.; Springer, T.A. Latent TGF-β structure and activation. *Nature* **2011**, *474*, 343–349. [CrossRef]

11. Cheifetz, S.; Hernandez, H.; Laiho, M.; ten Dijke, P.; Iwata, K.K.; Massague, J. Distinct transforming factor-β (TGF-β) receptor subsets as determinants of cellular responsiveness to three TGF-β isoforms. *J. Biol. Chem.* **1990**, *265*, 20533–20538. [PubMed]

12. Dong, X.; Zhao, B.; Iacob, R.E.; Zhu, J.; Koksal, A.C.; Lu, C.; Engen, J.R.; Springer, T.A. Force interacts with macromolecular structure in activation of TGF-β. *Nature* **2017**, *542*, 55–59. [CrossRef] [PubMed]

13. Massague, J. Receptors for the TGF-β family. *Cell* **1992**, *69*, 1067–1070. [CrossRef]

14. Heldin, C.H.; Moustakas, A. Signaling receptors for TGF-β family members. *Cold Spring Harb. Perspect. Biol.* **2016**, *8*, a022053. [CrossRef] [PubMed]

15. Upadhyay, A.; Moss-Taylor, L.; Kim, M.J.; Ghosh, A.C.; O'Connor, M.B. TGF-β family signaling in drosophila. *Cold Spring Harb. Perspect. Biol.* **2017**, *9*. [CrossRef] [PubMed]

16. Savage-Dunn, C.; Padgett, R.W. The TGF-β family in caenorhabditis elegans. *Cold Spring Harb. Perspect. Biol.* **2017**, *9*, a022178. [CrossRef] [PubMed]

17. Derynck, R.; Gelbart, W.M.; Harland, R.M.; Heldin, C.H.; Kern, S.E.; Massagué, J.; Melton, D.A.; Mlodzik, M.; Padgett, R.W.; Roberts, A.B.; et al. Nomenclature: Vertebrate mediators of TGF-β family signals. *Cell* **1996**, *87*, 173. [CrossRef]

18. Wrana, J.L. Signaling by the TGF-β superfamily. *Cold Spring Harb. Perspect. Biol.* **2013**, *5*, a011197. [CrossRef] [PubMed]

19. Hill, C.S. Transcriptional control by the SMADs. *Cold Spring Harb. Perspect. Biol.* **2016**, *8*, a022079. [CrossRef]

20. Miyazawa, K.; Miyazono, K. Regulation of TGF-β family signaling by inhibitory SMADs. *Cold Spring Harb. Perspect. Biol.* **2017**, *9*, a022095. [CrossRef]

21. Zhang, Y.E. Non-SMAD signaling pathways of the TGF-β family. *Cold Spring Harb. Perspect. Biol.* **2017**, *9*, a022129. [CrossRef] [PubMed]

22. Zhang, Y.E. Non-SMAD pathways in TGF-β signaling. *Cell Res.* **2009**, *19*, 128–139. [CrossRef] [PubMed]

23. Ravichandran, K.S. Signaling via Shc family adapter proteins. *Oncogene* **2001**, *20*, 6322–6330. [CrossRef] [PubMed]

24. Yamashita, M.; Fatyol, K.; Jin, C.; Wang, X.; Liu, Z.; Zhang, Y.E. TRAF6 mediates SMAD-independent activation of JNK and p38 by TGF-β. *Mol. Cell.* **2008**, *31*, 918–924. [CrossRef] [PubMed]

25. Sorrentino, A.; Thakur, N.; Grimsby, S.; Marcusson, A.; von Bulow, V.; Schuster, N.; Zhang, S.; Heldin, C.H.; Landstrom, M. The type I TGF-β receptor engages TRAF6 to activate TAK1 in a receptor kinase-independent manner. *Nat. Cell Biol.* **2008**, *10*, 1199–1207. [CrossRef] [PubMed]

26. Ozdamar, B.; Bose, R.; Barrios-Rodiles, M.; Wang, H.R.; Zhang, Y.; Wrana, J.L. Regulation of the polarity protein Par6 by TGF-β receptors controls epithelial cell plasticity. *Science* **2005**, *307*, 1603–1609. [CrossRef] [PubMed]

27. Wilkes, M.C.; Murphy, S.J.; Garamszegi, N.; Leof, E.B. Cell-type-specific activation of PAK2 by transforming growth factor β independent of SMAD2 and SMAD3. *Mol. Cell Biol.* **2003**, *23*, 8878–8889. [CrossRef]

28. Bakin, A.V.; Tomlinson, A.K.; Bhowmick, N.A.; Moses, H.L.; Arteaga, C.L. Phosphatidylinositol 3-kinase function is required for transforming growth factor β-mediated epithelial to mesenchymal transition and cell migration. *J. Biol. Chem.* **2000**, *275*, 36803–36810. [CrossRef]

29. Zhang, Y.; Alexander, P.B.; Wang, X.F. TGF-β family signaling in the control of cell proliferation and survival. *Cold Spring Harb. Perspect. Biol.* **2017**, *9*, a022145. [CrossRef]

30. Azar, R.; Alard, A.; Susini, C.; Bousquet, C.; Pyronnet, S. 4E-BP1 is a target of SMAD4 essential for TGF-β-mediated inhibition of cell proliferation. *EMBO J.* **2009**, *28*, 3514–3522. [CrossRef]

31. Baghdassarian, N.; Ffrench, M. Cyclin-dependent kinase inhibitors (CKIs) and hematological malignancies. *Hematol. Cell Ther.* **1996**, *38*, 313–323. [CrossRef] [PubMed]

32. Xu, J.; Acharya, S.; Sahin, O.; Zhang, Q.; Saito, Y.; Yao, J.; Wang, H.; Li, P.; Zhang, L.; Lowery, F.J.; et al. 14-3-3zeta turns TGF-β's function from tumor suppressor to metastasis promoter in breast cancer by contextual changes of SMAD partners from p53 to Gli2. *Cancer Cell.* **2015**, *27*, 177–192. [CrossRef] [PubMed]

33. Iavarone, A.; Massague, J. Repression of the CDK activator Cdc25A and cell-cycle arrest by cytokine TGF-β in cells lacking the CDK inhibitor p15. *Nature* **1997**, *387*, 417–422. [CrossRef] [PubMed]

34. Ling, M.T.; Wang, X.; Tsao, S.W.; Wong, Y.C. Down-regulation of Id-1 expression is associated with TGF-β1-induced growth arrest in prostate epithelial cells. *Biochim. Biophys. Acta* **2002**, *1570*, 145–152. [CrossRef]

35. Chen, C.R.; Kang, Y.; Siegel, P.M.; Massague, J. E2F4/5 and p107 as SMAD cofactors linking the TGF-β receptor to c-myc repression. *Cell* **2002**, *110*, 19–32. [CrossRef]

36. Ozaki, I.; Hamajima, H.; Matsuhashi, S.; Mizuta, T. Regulation of TGF-β1-induced pro-apoptotic signaling by growth factor receptors and extracellular matrix receptor integrins in the liver. *Front. Physiol.* **2011**, *2*, 78. [CrossRef] [PubMed]

37. Wiener, Z.; Band, A.M.; Kallio, P.; Hogstrom, J.; Hyvonen, V.; Kaijalainen, S.; Ritvos, O.; Haglund, C.; Kruuna, O.; Robine, S.; et al. Oncogenic mutations in intestinal adenomas regulate Bim-mediated apoptosis induced by TGF-β. *Proc. Natl. Acad. Sci. USA* **2014**, *111*, E2229–E2236. [CrossRef]

38. Zhao, X.; Liu, Y.; Du, L.; He, L.; Ni, B.; Hu, J.; Zhu, D.; Chen, Q. Threonine 32 (Thr32) of FoxO3 is critical for TGF-β-induced apoptosis via Bim in hepatocarcinoma cells. *Protein Cell* **2015**, *6*, 127–138. [CrossRef]

39. Spender, L.C.; O'Brien, D.I.; Simpson, D.; Dutt, D.; Gregory, C.D.; Allday, M.J.; Clark, L.J.; Inman, G.J. TGF-β induces apoptosis in human B cells by transcriptional regulation of BIK and BCL-XL. *Cell Death Differ.* **2009**, *16*, 593–602. [CrossRef]

40. Ohgushi, M.; Kuroki, S.; Fukamachi, H.; O'Reilly, L.A.; Kuida, K.; Strasser, A.; Yonehara, S. Transforming growth factor β-dependent sequential activation of SMAD, Bim, and caspase-9 mediates physiological apoptosis in gastric epithelial cells. *Mol. Cell. Biol.* **2005**, *25*, 10017–10028. [CrossRef]

41. Shima, Y.; Nakao, K.; Nakashima, T.; Kawakami, A.; Nakata, K.; Hamasaki, K.; Kato, Y.; Eguchi, K.; Ishii, N. Activation of caspase-8 in transforming growth factor-β-induced apoptosis of human hepatoma cells. *Hepatology* **1999**, *30*, 1215–1222. [CrossRef] [PubMed]

42. Takekawa, M.; Tatebayashi, K.; Itoh, F.; Adachi, M.; Imai, K.; Saito, H. SMAD-dependent GADD45β expression mediates delayed activation of p38 MAP kinase by TGF-β. *EMBO J.* **2002**, *21*, 6473–6482. [CrossRef] [PubMed]

43. Yan, F.; Wang, Y.; Wu, X.; Peshavariya, H.M.; Dusting, G.J.; Zhang, M.; Jiang, F. Nox4 and redox signaling mediate TGF-β-induced endothelial cell apoptosis and phenotypic switch. *Cell Death Dis.* **2014**, *5*, e1010. [CrossRef] [PubMed]

44. Cardenas, H.; Vieth, E.; Lee, J.; Segar, M.; Liu, Y.; Nephew, K.P.; Matei, D. TGF-β induces global changes in DNA methylation during the epithelial-to-mesenchymal transition in ovarian cancer cells. *Epigenetics* **2014**, *9*, 1461–1472. [CrossRef] [PubMed]

45. Evanno, E.; Godet, J.; Piccirilli, N.; Guilhot, J.; Milin, S.; Gombert, J.M.; Fouchaq, B.; Roche, J. Tri-methylation of H3K79 is decreased in TGF-β1-induced epithelial-to-mesenchymal transition in lung cancer. *Clin. Epigenetics* **2017**, *9*, 80. [CrossRef] [PubMed]

46. Cassar, L.; Nicholls, C.; Pinto, A.R.; Chen, R.; Wang, L.; Li, H.; Liu, J.P. TGF-β receptor mediated telomerase inhibition, telomere shortening and breast cancer cell senescence. *Protein Cell* **2017**, *8*, 39–54. [CrossRef] [PubMed]

47. Kiyono, K.; Suzuki, H.I.; Matsuyama, H.; Morishita, Y.; Komuro, A.; Kano, M.R.; Sugimoto, K.; Miyazono, K. Autophagy is activated by TGF-β and potentiates TGF-β-mediated growth inhibition in human hepatocellular carcinoma cells. *Cancer Res.* **2009**, *69*, 8844–8852. [CrossRef]

48. Suzuki, H.I.; Kiyono, K.; Miyazono, K. Regulation of autophagy by transforming growth factor-β (TGF-β) signaling. *Autophagy* **2010**, *6*, 645–647. [CrossRef]

49. Korkut, A.; Zaidi, S.; Kanchi, R.S.; Rao, S.; Gough, N.R.; Schultz, A.; Li, X.; Lorenzi, P.L.; Berger, A.C.; Robertson, G.; et al. A pan-cancer analysis reveals high-frequency genetic alterations in mediators of signaling by the TGF-β superfamily. *Cell Syst.* **2018**, *7*, 422–437.e7. [CrossRef]

50. Hahn, S.A.; Schutte, M.; Hoque, A.T.; Moskaluk, C.A.; da Costa, L.T.; Rozenblum, E.; Weinstein, C.L.; Fischer, A.; Yeo, C.J.; Hruban, R.H.; et al. DPC4, a candidate tumor suppressor gene at human chromosome 18q21.1. *Science* **1996**, *271*, 350–353. [CrossRef]

51. Markowitz, S.; Wang, J.; Myeroff, L.; Parsons, R.; Sun, L.; Lutterbaugh, J.; Fan, R.S.; Zborowska, E.; Kinzler, K.W.; Vogelstein, B.; et al. Inactivation of the type II TGF-β receptor in colon cancer cells with microsatellite instability. *Science* **1995**, *268*, 1336–1338. [CrossRef] [PubMed]

52. Macias-Silva, M.; Abdollah, S.; Hoodless, P.A.; Pirone, R.; Attisano, L.; Wrana, J.L. MADR2 is a substrate of the TGF-β receptor and its phosphorylation is required for nuclear accumulation and signaling. *Cell* **1996**, *87*, 1215–1224. [CrossRef]

53. Yang, L.; Pang, Y.; Moses, H.L. TGF-β and immune cells: An important regulatory axis in the tumor microenvironment and progression. *Trends Immunol.* **2010**, *31*, 220–227. [CrossRef] [PubMed]

54. Pinto, M.; Oliveira, C.; Cirnes, L.; Machado, J.C.; Ramires, M.; Nogueira, A.; Carneiro, F.; Seruca, R. Promoter methylation of TGF-β receptor I and mutation of TGF-β receptor II are frequent events in MSI sporadic gastric carcinomas. *J. Pathol.* **2003**, *200*, 32–38. [CrossRef]

55. Bruna, A.; Darken, R.S.; Rojo, F.; Ocana, A.; Penuelas, S.; Arias, A.; Paris, R.; Tortosa, A.; Mora, J.; Baselga, J.; et al. High TGF-β-SMAD activity confers poor prognosis in glioma patients and promotes cell proliferation depending on the methylation of the PDGF-B gene. *Cancer Cell* **2007**, *11*, 147–160. [CrossRef] [PubMed]

56. Caja, L.; Dituri, F.; Mancarella, S.; Caballero-Diaz, D.; Moustakas, A.; Giannelli, G.; Fabregat, I. TGF-β and the tissue microenvironment: Relevance in fibrosis and cancer. *Int. J. Mol. Sci.* **2018**, *19*, 1294. [CrossRef] [PubMed]

57. Peinado, H.; Quintanilla, M.; Cano, A. Transforming growth factor β-1 induces snail transcription factor in epithelial cell lines: Mechanisms for epithelial mesenchymal transitions. *J. Biol. Chem.* **2003**, *278*, 21113–21123. [CrossRef]

58. Lamouille, S.; Xu, J.; Derynck, R. Molecular mechanisms of epithelial-mesenchymal transition. *Nat. Rev. Mol. Cell Biol.* **2014**, *15*, 178–196. [CrossRef]

59. Nieto, M.A.; Huang, R.Y.; Jackson, R.A.; Thiery, J.P. Emt: 2016. *Cell* **2016**, *166*, 21–45. [CrossRef]

60. Tsai, J.H.; Yang, J. Epithelial-mesenchymal plasticity in carcinoma metastasis. *Genes Dev.* **2013**, *27*, 2192–2206. [CrossRef]

61. van Staalduinen, J.; Baker, D.; Dijke, P.t.; van Dam, H. Epithelial-mesenchymal-transition-inducing transcription factors: New targets for tackling chemoresistance in cancer? *Oncogene* **2018**, *37*, 6195–6211. [CrossRef] [PubMed]

62. Jolly, M.K.; Ware, K.E.; Gilja, S.; Somarelli, J.A.; Levine, H. EMT and MET: Necessary or permissive for metastasis? *Mol. Oncol.* **2017**, *11*, 755–769. [CrossRef] [PubMed]

63. Zheng, X.; Carstens, J.L.; Kim, J.; Scheible, M.; Kaye, J.; Sugimoto, H.; Wu, C.C.; LeBleu, V.S.; Kalluri, R. Epithelial-to-mesenchymal transition is dispensable for metastasis but induces chemoresistance in pancreatic cancer. *Nature* **2015**, *527*, 525–530. [CrossRef] [PubMed]

64. Fischer, K.R.; Durrans, A.; Lee, S.; Sheng, J.; Li, F.; Wong, S.T.; Choi, H.; el Rayes, T.; Ryu, S.; Troeger, J.; et al. Epithelial-to-mesenchymal transition is not required for lung metastasis but contributes to chemoresistance. *Nature* **2015**, *527*, 472–476. [CrossRef] [PubMed]

65. Aiello, N.M.; Brabletz, T.; Kang, Y.; Nieto, M.A.; Weinberg, R.A.; Stanger, B.Z. Upholding a role for EMT in pancreatic cancer metastasis. *Nature* **2017**, *547*, E7–E8. [CrossRef] [PubMed]

66. Li, W.; Kang, Y. Probing the fifty shades of EMT in metastasis. *Trends Cancer* **2016**, *2*, 65–67. [CrossRef] [PubMed]

67. Pastushenko, I.; Brisebarre, A.; Sifrim, A.; Fioramonti, M.; Revenco, T.; Boumahdi, S.; van Keymeulen, A.; Brown, D.; Moers, V.; Lemaire, S.; et al. Identification of the tumour transition states occurring during EMT. *Nature* **2018**, *556*, 463–468. [CrossRef]

68. Nikitorowicz-Buniak, J.; Denton, C.P.; Abraham, D.; Stratton, R. Partially evoked epithelial-mesenchymal transition (EMT) is associated with increased TGF-β signaling within lesional scleroderma skin. *PLoS ONE* **2015**, *10*, e0134092.

69. Puram, S.V.; Parikh, A.S.; Tirosh, I. Single cell RNA-seq highlights a role for a partial EMT in head and neck cancer. *Mol. Cell Oncol.* **2018**, *5*, e1448244. [CrossRef]

70. Deckers, M.; van Dinther, M.; Buijs, J.; Que, I.; Lowik, C.; van der Pluijm, G.; Dijke, P.t. The tumor suppressor SMAD4 is required for transforming growth factor β-induced epithelial to mesenchymal transition and bone metastasis of breast cancer cells. *Cancer Res.* **2006**, *66*, 2202–2209. [CrossRef]

71. Vincent, T.; Neve, E.P.; Johnson, J.R.; Kukalev, A.; Rojo, F.; Albanell, J.; Pietras, K.; Virtanen, I.; Philipson, L.; Leopold, P.L.; et al. A SNAIL1-SMAD3/4 transcriptional repressor complex promotes TGF-β mediated epithelial-mesenchymal transition. *Nat. Cell Biol.* **2009**, *11*, 943–950. [CrossRef] [PubMed]

72. Hoot, K.E.; Lighthall, J.; Han, G.; Lu, S.L.; Li, A.; Ju, W.; Kulesz-Martin, M.; Bottinger, E.; Wang, X.J. Keratinocyte-specific SMAD2 ablation results in increased epithelial-mesenchymal transition during skin cancer formation and progression. *J. Clin. Investig.* **2008**, *118*, 2722–2732. [CrossRef] [PubMed]

73. Yu, J.; Lei, R.; Zhuang, X.; Li, X.; Li, G.; Lev, S.; Segura, M.F.; Zhang, X.; Hu, G. MicroRNA-182 targets SMAD7 to potentiate TGF-β-induced epithelial-mesenchymal transition and metastasis of cancer cells. *Nat. Commun.* **2016**, *7*, 13884. [CrossRef] [PubMed]

74. Kuang, J.; Li, L.; Guo, L.; Su, Y.; Wang, Y.; Xu, Y.; Wang, X.; Meng, S.; Lei, L.; Xu, L.; et al. RNF8 promotes epithelial-mesenchymal transition of breast cancer cells. *J. Exp. Clin. Cancer Res.* **2016**, *35*, 88. [CrossRef] [PubMed]

75. Zavadil, J.; Bitzer, M.; Liang, D.; Yang, Y.C.; Massimi, A.; Kneitz, S.; Piek, E.; Bottinger, E.P. Genetic programs of epithelial cell plasticity directed by transforming growth factor-β. *Proc. Natl. Acad. Sci. USA* **2001**, *98*, 6686–6691. [CrossRef] [PubMed]

76. Lee, M.K.; Pardoux, C.; Hall, M.C.; Lee, P.S.; Warburton, D.; Qing, J.; Smith, S.M.; Derynck, R. TGF-β activates Erk MAP kinase signalling through direct phosphorylation of ShcA. *EMBO J.* **2007**, *26*, 3957–3967. [CrossRef] [PubMed]

77. Davies, M.; Robinson, M.; Smith, E.; Huntley, S.; Prime, S.; Paterson, I. Induction of an epithelial to mesenchymal transition in human immortal and malignant keratinocytes by TGF-β1 involves MAPK, SMAD and AP-1 signalling pathways. *J. Cell Biochem.* **2005**, *95*, 918–931. [CrossRef] [PubMed]

78. Lin, T.; Zeng, L.; Liu, Y.; DeFea, K.; Schwartz, M.A.; Chien, S.; Shyy, J.Y. Rho-ROCK-LIMK-cofilin pathway regulates shear stress activation of sterol regulatory element binding proteins. *Circ. Res.* **2003**, *92*, 1296–1304. [CrossRef] [PubMed]

79. Landstrom, M. The TAK1-TRAF6 signalling pathway. *Int. J. Biochem. Cell Biol.* **2010**, *42*, 585–589. [CrossRef]

80. Song, J.; Landstrom, M. TGF-β activates PI3K-AKT signaling via TRAF6. *Oncotarget* **2017**, *8*, 99205–99206. [CrossRef]

81. Xue, G.; Restuccia, D.F.; Lan, Q.; Hynx, D.; Dirnhofer, S.; Hess, D.; Ruegg, C.; Hemmings, B.A. Akt/PKB-mediated phosphorylation of TWIST1 promotes tumor metastasis via mediating cross-talk between PI3K/Akt and TGF-β signaling axes. *Cancer Dis.* **2012**, *2*, 248–259. [CrossRef] [PubMed]

82. Ma, L.; Teruya-Feldstein, J.; Weinberg, R.A. Tumour invasion and metastasis initiated by microRNA-10b in breast cancer. *Nature* **2007**, *449*, 682–688. [CrossRef] [PubMed]

83. Werden, S.J.; Sphyris, N.; Sarkar, T.R.; Paranjape, A.N.; LaBaff, A.M.; Taube, J.H.; Hollier, B.G.; Ramirez-Pena, E.Q.; Soundararajan, R.; den Hollander, P.; et al. Phosphorylation of serine 367 of FOXC2 by p38 regulates ZEB1 and breast cancer metastasis, without impacting primary tumor growth. *Oncogene* **2016**, *35*, 5977–5988. [CrossRef] [PubMed]

84. Chen, L.; Li, Y.C.; Wu, L.; Yu, G.T.; Zhang, W.F.; Huang, C.F.; Sun, Z.J. TRAF6 regulates tumour metastasis through EMT and CSC phenotypes in head and neck squamous cell carcinoma. *J. Cell. Mol. Med.* **2018**, *22*, 1337–1349. [CrossRef] [PubMed]

85. Gonzalez, D.M.; Medici, D. Signaling mechanisms of the epithelial-mesenchymal transition. *Sci. Signal* **2014**, *7*, re8. [CrossRef] [PubMed]

86. Nishita, M.; Hashimoto, M.K.; Ogata, S.; Laurent, M.N.; Ueno, N.; Shibuya, H.; Cho, K.W. Interaction between Wnt and TGF-β signalling pathways during formation of Spemann's organizer. *Nature* **2000**, *403*, 781–785. [CrossRef] [PubMed]

87. Medici, D.; Hay, E.D.; Goodenough, D.A. Cooperation between snail and LEF-1 transcription factors is essential for TGF-β1-induced epithelial-mesenchymal transition. *Mol. Biol. Cell.* **2006**, *17*, 1871–1879. [CrossRef] [PubMed]

88. Jamora, C.; Lee, P.; Kocieniewski, P.; Azhar, M.; Hosokawa, R.; Chai, Y.; Fuchs, E. A signaling pathway involving TGF-β2 and snail in hair follicle morphogenesis. *PLoS Biol.* **2005**, *3*, e11. [CrossRef]

89. Murillo-Garzon, V.; Gorrono-Etxebarria, I.; Akerfelt, M.; Puustinen, M.C.; Sistonen, L.; Nees, M.; Carton, J.; Waxman, J.; Kypta, R.M. Frizzled-8 integrates Wnt-11 and transforming growth factor-β signaling in prostate cancer. *Nat. Commun.* **2018**, *9*, 1747. [CrossRef]

90. Kanei-Ishii, C.; Ninomiya-Tsuji, J.; Tanikawa, J.; Nomura, T.; Ishitani, T.; Kishida, S.; Kokura, K.; Kurahashi, T.; Ichikawa-Iwata, E.; Kim, Y.; et al. Wnt-1 signal induces phosphorylation and degradation of c-Myb protein via TAK1, HIPK2, and NLK. *Genes Dev.* **2004**, *18*, 816–829. [CrossRef]

91. Sjolund, J.; Bostrom, A.K.; Lindgren, D.; Manna, S.; Moustakas, A.; Ljungberg, B.; Johansson, M.; Fredlund, E.; Axelson, H. The notch and TGF-β signaling pathways contribute to the aggressiveness of clear cell renal cell carcinoma. *PLoS ONE* **2011**, *6*, e23057. [CrossRef] [PubMed]

92. Zavadil, J.; Cermak, L.; Soto-Nieves, N.; Bottinger, E.P. Integration of TGF-β/SMAD and Jagged1/Notch signalling in epithelial-to-mesenchymal transition. *EMBO J.* **2004**, *23*, 1155–1165. [CrossRef] [PubMed]

93. Zhang, Y.; Feng, X.H.; Derynck, R. SMAD3 and SMAD4 cooperate with c-Jun/c-Fos to mediate TGF-β-induced transcription. *Nature* **1998**, *394*, 909–913. [CrossRef] [PubMed]

94. Huber, M.A.; Kraut, N.; Beug, H. Molecular requirements for epithelial-mesenchymal transition during tumor progression. *Curr. Opin. Cell Biol.* **2005**, *17*, 548–558. [CrossRef] [PubMed]

95. Olsen, S.N.; Wronski, A.; Castano, Z.; Dake, B.; Malone, C.; de Raedt, T.; Enos, M.; DeRose, Y.S.; Zhou, W.; Guerra, S.; et al. Loss of RasGAP tumor suppressors underlies the aggressive nature of luminal b breast cancers. *Cancer Discov.* **2017**, *7*, 202–217. [CrossRef] [PubMed]

96. Bowman, T.; Broome, M.A.; Sinibaldi, D.; Wharton, W.; Pledger, W.J.; Sedivy, J.M.; Irby, R.; Yeatman, T.; Courtneidge, S.A.; Jove, R. Stat3-mediated Myc expression is required for Src transformation and PDGF-induced mitogenesis. *Proc. Natl. Acad. Sci. USA* **2001**, *98*, 7319–7324. [CrossRef] [PubMed]

97. Lamouille, S.; Subramanyam, D.; Blelloch, R.; Derynck, R. Regulation of epithelial-mesenchymal and mesenchymal-epithelial transitions by microRNAs. *Curr. Opin. Cell Biol.* **2013**, *25*, 200–207. [CrossRef] [PubMed]

98. Gregory, P.A.; Bert, A.G.; Paterson, E.L.; Barry, S.C.; Tsykin, A.; Farshid, G.; Vadas, M.A.; Khew-Goodall, Y.; Goodall, G.J. The miR-200 family and miR-205 regulate epithelial to mesenchymal transition by targeting ZEB1 and SIP1. *Nat. Cell Biol.* **2008**, *10*, 593–601. [CrossRef]

99. Diepenbruck, M.; Tiede, S.; Saxena, M.; Ivanek, R.; Kalathur, R.K.R.; Luond, F.; Meyer-Schaller, N.; Christofori, G. miR-1199-5p and Zeb1 function in a double-negative feedback loop potentially coordinating EMT and tumour metastasis. *Nat. Commun.* **2017**, *8*, 1168. [CrossRef]

100. Kim, N.H.; Kim, H.S.; Li, X.Y.; Lee, I.; Choi, H.S.; Kang, S.E.; Cha, S.Y.; Ryu, J.K.; Yoon, D.; Fearon, E.R.; et al. A p53/miRNA-34 axis regulates Snail1-dependent cancer cell epithelial-mesenchymal transition. *J. Cell. Biol.* **2011**, *195*, 417–433. [CrossRef]

101. Hahn, S.; Jackstadt, R.; Siemens, H.; Hunten, S.; Hermeking, H. SNAIL and miR-34a feed-forward regulation of ZNF281/ZBP99 promotes epithelial-mesenchymal transition. *EMBO J.* **2013**, *32*, 3079–3095. [CrossRef] [PubMed]

102. Majid, S.; Dar, A.A.; Saini, S.; Shahryari, V.; Arora, S.; Zaman, M.S.; Chang, I.; Yamamura, S.; Tanaka, Y.; Chiyomaru, T.; et al. miRNA-34b inhibits prostate cancer through demethylation, active chromatin modifications, and AKT pathways. *Clin. Cancer Res.* **2013**, *19*, 73–84. [CrossRef] [PubMed]

103. Siemens, H.; Neumann, J.; Jackstadt, R.; Mansmann, U.; Horst, D.; Kirchner, T.; Hermeking, H. Detection of miR-34a promoter methylation in combination with elevated expression of c-Met and β-catenin predicts distant metastasis of colon cancer. *Clin. Cancer Res.* **2013**, *19*, 710–720. [CrossRef] [PubMed]

104. Genovese, G.; Ergun, A.; Shukla, S.A.; Campos, B.; Hanna, J.; Ghosh, P.; Quayle, S.N.; Rai, K.; Colla, S.; Ying, H.; et al. microRNA regulatory network inference identifies miR-34a as a novel regulator of TGF-β signaling in glioblastoma. *Cancer Discov.* **2012**, *2*, 736–749. [CrossRef] [PubMed]

105. Tian, X.J.; Zhang, H.; Xing, J. Coupled reversible and irreversible bistable switches underlying TGF-β-induced epithelial to mesenchymal transition. *Biophys. J.* **2013**, *105*, 1079–1089. [CrossRef] [PubMed]

106. Gregory, P.A.; Bracken, C.P.; Smith, E.; Bert, A.G.; Wright, J.A.; Roslan, S.; Morris, M.; Wyatt, L.; Farshid, G.; Lim, Y.Y.; et al. An autocrine TGF-β/ZEB/miR-200 signaling network regulates establishment and maintenance of epithelial-mesenchymal transition. *Mol. Biol. Cell* **2011**, *22*, 1686–1698. [CrossRef] [PubMed]

107. Kumar, M.S.; Armenteros-Monterroso, E.; East, P.; Chakravorty, P.; Matthews, N.; Winslow, M.M.; Downward, J. HMGA2 functions as a competing endogenous RNA to promote lung cancer progression. *Nature* **2014**, *505*, 212–217. [CrossRef]

108. Liao, Y.C.; Wang, Y.S.; Guo, Y.C.; Lin, W.L.; Chang, M.H.; Juo, S.H. Let-7g improves multiple endothelial functions through targeting transforming growth factor-β and SIRT-1 signaling. *J. Am. Coll. Cardiol.* **2014**, *63*, 1685–1694. [CrossRef]

109. Han, X.; Yan, S.; Weijie, Z.; Feng, W.; Liuxing, W.; Mengquan, L.; Qingxia, F. Critical role of miR-10b in transforming growth factor-β1-induced epithelial-mesenchymal transition in breast cancer. *Cancer Gene Ther.* **2014**, *21*, 60–67. [CrossRef]

110. Ma, L.; Reinhardt, F.; Pan, E.; Soutschek, J.; Bhat, B.; Marcusson, E.G.; Teruya-Feldstein, J.; Bell, G.W.; Weinberg, R.A. Therapeutic silencing of miR-10b inhibits metastasis in a mouse mammary tumor model. *Nat. Biotechnol.* **2010**, *28*, 341–347. [CrossRef]

111. Song, L.; Liu, L.; Wu, Z.; Li, Y.; Ying, Z.; Lin, C.; Wu, J.; Hu, B.; Cheng, S.Y.; Li, M.; et al. TGF-β induces miR-182 to sustain NF-kappaB activation in glioma subsets. *J. Clin. Investig.* **2012**, *122*, 3563–3578. [CrossRef] [PubMed]

112. Taylor, M.A.; Sossey-Alaoui, K.; Thompson, C.L.; Danielpour, D.; Schiemann, W.P. TGF-β upregulates miR-181a expression to promote breast cancer metastasis. *J. Clin. Investig.* **2013**, *123*, 150–163. [CrossRef] [PubMed]

113. Parikh, A.; Lee, C.; Joseph, P.; Marchini, S.; Baccarini, A.; Kolev, V.; Romualdi, C.; Fruscio, R.; Shah, H.; Wang, F.; et al. microRNA-181a has a critical role in ovarian cancer progression through the regulation of the epithelial-mesenchymal transition. *Nat. Commun.* **2014**, *5*, 2977. [CrossRef] [PubMed]

114. Bu, P.; Wang, L.; Chen, K.Y.; Rakhilin, N.; Sun, J.; Closa, A.; Tung, K.L.; King, S.; Varanko, A.K.; Xu, Y.; et al. miR-1269 promotes metastasis and forms a positive feedback loop with TGF-β. *Nat. Commun.* **2015**, *6*, 6879. [CrossRef] [PubMed]

115. Xia, H.; Ooi, L.L.; Hui, K.M. MicroRNA-216a/217-induced epithelial-mesenchymal transition targets PTEN and SMAD7 to promote drug resistance and recurrence of liver cancer. *Hepatology* **2013**, *58*, 629–641. [CrossRef] [PubMed]

116. Kong, W.; Yang, H.; He, L.; Zhao, J.J.; Coppola, D.; Dalton, W.S.; Cheng, J.Q. MicroRNA-155 is regulated by the transforming growth factor β/SMAD pathway and contributes to epithelial cell plasticity by targeting RhoA. *Mol. Cell. Biol.* **2008**, *28*, 6773–6784. [CrossRef]

117. Yin, K.; Yin, W.; Wang, Y.; Zhou, L.; Liu, Y.; Yang, G.; Wang, J.; Lu, J. MiR-206 suppresses epithelial mesenchymal transition by targeting TGF-β signaling in estrogen receptor positive breast cancer cells. *Oncotarget* **2016**, *7*, 24537–24548. [CrossRef] [PubMed]

118. Suzuki, H.I. MicroRNA Control of TGF-β Signaling. *Int. J. Mol. Sci.* **2018**, *19*, 1901. [CrossRef]

119. Zaravinos, A. The regulatory role of MicroRNAs in EMT and cancer. *J. Oncol.* **2015**, *2015*, 865816. [CrossRef]

120. Romano, G.; Kwong, L.N. miRNAs, melanoma and microenvironment: An intricate network. *Int. J. Mol. Sci.* **2017**, *18*, 2354. [CrossRef]

121. Musavi Shenas, M.H.; Eghbal-Fard, S.; Mehrisofiani, V.; Yazdani, N.A.; Farzam, O.R.; Marofi, F.; Yousefi, M. MicroRNAs and signaling networks involved in epithelial-mesenchymal transition. *J. Cell Physiol.* **2019**, *234*, 5775–5785. [CrossRef]

122. Lin, C.W.; Kao, S.H.; Yang, P.C. The miRNAs and epithelial-mesenchymal transition in cancers. *Curr. Pharm. Des.* **2014**, *20*, 5309–5318. [CrossRef]

123. Shields, E.J.; Petracovici, A.F.; Bonasio, R. lncRedibly versatile: Biochemical and biological functions of long noncoding RNAs. *Biochem. J.* **2019**, *476*, 1083–1104. [CrossRef]

124. Gutschner, T.; Diederichs, S. The hallmarks of cancer: A long non-coding RNA point of view. *RNA Biol.* **2012**, *9*, 703–719. [CrossRef]

125. Schmitt, A.M.; Chang, H.Y. Long noncoding RNAs in cancer pathways. *Cancer Cell* **2016**, *29*, 452–463. [CrossRef]

126. Lu, H.; Yang, D.; Zhang, L.; Lu, S.; Ye, J.; Li, M.; Hu, W. Linc-pint inhibits early stage pancreatic ductal adenocarcinoma growth through TGF-β pathway activation. *Oncol. Lett.* **2019**, *17*, 4633–4639. [CrossRef]

127. Shin, T.J.; Lee, K.H.; Cho, H.M.; Cho, J.Y. Concise approach for screening long non-coding RNAs functionally linked to human breast cancer associated genes. *Exp. Mol. Pathol.* **2019**, *108*, 89–96. [CrossRef]

128. Shi, T.; Gao, G.; Cao, Y. Long noncoding RNAs as novel biomarkers have a promising future in cancer diagnostics. *Dis. Markers* **2016**, *2016*, 9085195. [CrossRef]

129. Sakai, S.; Ohhata, T.; Kitagawa, K.; Uchida, C.; Aoshima, T.; Niida, H.; Suzuki, T.; Inoue, Y.; Miyazawa, K.; Kitagawa, M. Long noncoding RNA ELIT-1 acts as a SMAD3 cofactor to facilitate TGF-β/SMAD signaling and promote epithelial-mesenchymal transition. *Cancer Res.* **2019**.

130. He, L.; Thomson, J.M.; Hemann, M.T.; Hernando-Monge, E.; Mu, D.; Goodson, S.; Powers, S.; Cordon-Cardo, C.; Lowe, S.W.; Hannon, G.J.; et al. A microRNA polycistron as a potential human oncogene. *Nature* **2005**, *435*, 828–833. [CrossRef]

131. Li, W.; Kang, Y. A new Lnc in metastasis: Long noncoding RNA mediates the prometastatic functions of TGF-β. *Cancer Cell* **2014**, *25*, 557–559. [CrossRef]

132. Yuan, J.H.; Yang, F.; Wang, F.; Ma, J.Z.; Guo, Y.J.; Tao, Q.F.; Liu, F.; Pan, W.; Wang, T.T.; Zhou, C.C.; et al. A long noncoding RNA activated by TGF-β promotes the invasion-metastasis cascade in hepatocellular carcinoma. *Cancer Cell* **2014**, *25*, 666–681. [CrossRef]

133. Grelet, S.; Link, L.A.; Howley, B.; Obellianne, C.; Palanisamy, V.; Gangaraju, V.K.; Diehl, J.A.; Howe, P.H. A regulated PNUTS mRNA to lncRNA splice switch mediates EMT and tumour progression. *Nat. Cell Biol.* **2017**, *19*, 1105–1115. [CrossRef]

134. Mondal, T.; Subhash, S.; Vaid, R.; Enroth, S.; Uday, S.; Reinius, B.; Mitra, S.; Mohammed, A.; James, A.R.; Hoberg, E.; et al. MEG3 long noncoding RNA regulates the TGF-β pathway genes through formation of RNA-DNA triplex structures. *Nat. Commun.* **2015**, *6*, 7743. [CrossRef]

135. Mitra, R.; Chen, X.; Greenawalt, E.J.; Maulik, U.; Jiang, W.; Zhao, Z.; Eischen, C.M. Decoding critical long non-coding RNA in ovarian cancer epithelial-to-mesenchymal transition. *Nat. Commun.* **2017**, *8*, 1604. [CrossRef]

136. Yan, X.; Zhang, D.; Wu, W.; Wu, S.; Qian, J.; Hao, Y.; Yan, F.; Zhu, P.; Wu, J.; Huang, G.; et al. Mesenchymal stem cells promote hepatocarcinogenesis via lncRNA-MUF interaction with ANXA2 and miR-34a. *Cancer Res.* **2017**, *77*, 6704–6716. [CrossRef]

137. Zhang, J.; Han, C.; Ungerleider, N.; Chen, W.; Song, K.; Wang, Y.; Kwon, H.; Ma, W.; Wu, T. A transforming growth factor-β and H19 signaling axis in tumor-initiating hepatocytes that regulates hepatic carcinogenesis. *Hepatology* **2018**, *69*, 1549–1563. [CrossRef]

138. Matouk, I.J.; Raveh, E.; Abu-lail, R.; Mezan, S.; Gilon, M.; Gershtain, E.; Birman, T.; Gallula, J.; Schneider, T.; Barkali, M.; et al. Oncofetal H19 RNA promotes tumor metastasis. *Biochim. Biophys. Acta* **2014**, *1843*, 1414–1426. [CrossRef]

139. Zhou, W.; Ye, X.L.; Xu, J.; Cao, M.G.; Fang, Z.Y.; Li, L.Y.; Guan, G.H.; Liu, Q.; Qian, Y.H.; Xie, D. The lncRNA H19 mediates breast cancer cell plasticity during EMT and MET plasticity by differentially sponging miR-200b/c and let-7b. *Sci. Signal* **2017**, *10*, eaak9557. [CrossRef]

140. Luo, M.; Li, Z.; Wang, W.; Zeng, Y.; Liu, Z.; Qiu, J. Long non-coding RNA H19 increases bladder cancer metastasis by associating with EZH2 and inhibiting E-cadherin expression. *Cancer Lett.* **2013**, *333*, 213–221. [CrossRef]

141. Rodriguez-Mateo, C.; Torres, B.; Gutierrez, G.; Pintor-Toro, J.A. Downregulation of Lnc-Spry1 mediates TGF-β-induced epithelial-mesenchymal transition by transcriptional and posttranscriptional regulatory mechanisms. *Cell Death Differ.* **2017**, *24*, 785–797. [CrossRef]

142. Song, Y.X.; Sun, J.X.; Zhao, J.H.; Yang, Y.C.; Shi, J.X.; Wu, Z.H.; Chen, X.W.; Gao, P.; Miao, Z.F.; Wang, Z.N. Non-coding RNAs participate in the regulatory network of CLDN4 via ceRNA mediated miRNA evasion. *Nat. Commun.* **2017**, *8*, 289. [CrossRef]

143. Battistelli, C.; Cicchini, C.; Santangelo, L.; Tramontano, A.; Grassi, L.; Gonzalez, F.J.; de Nonno, V.; Grassi, G.; Amicone, L.; Tripodi, M. The snail repressor recruits EZH2 to specific genomic sites through the enrollment of the lncRNA HOTAIR in epithelial-to-mesenchymal transition. *Oncogene* **2017**, *36*, 942–955. [CrossRef]

144. Zhang, Y.; Li, J.; Jia, S.; Wang, Y.; Kang, Y.; Zhang, W. Down-regulation of lncRNA-ATB inhibits epithelial-mesenchymal transition of breast cancer cells by increasing miR-141-3p expression. *Biochem. Cell Biol.* **2018**, *97*, 85–90. [CrossRef]

145. Li, G.Y.; Wang, W.; Sun, J.Y.; Xin, B.; Zhang, X.; Wang, T.; Zhang, Q.F.; Yao, L.B.; Han, H.; Fan, D.M.; et al. Long non-coding RNAs AC026904.1 and UCA1: A "one-two punch" for TGF-β-induced SNAI2 activation and epithelial-mesenchymal transition in breast cancer. *Theranostics* **2018**, *8*, 2846–2861. [CrossRef]

146. Wu, W.; Chen, F.; Cui, X.; Yang, L.; Chen, J.; Zhao, J.; Huang, D.; Liu, J.; Yang, L.; Zeng, J.; et al. LncRNA NKILA suppresses TGF-β-induced epithelial-mesenchymal transition by blocking NF-kappaB signaling in breast cancer. *Int. J. Cancer* **2018**, *143*, 2213–2224. [CrossRef]

147. Li, Z.; Dong, M.; Fan, D.; Hou, P.; Li, H.; Liu, L.; Lin, C.; Liu, J.; Su, L.; Wu, L.; et al. LncRNA ANCR down-regulation promotes TGF-β-induced EMT and metastasis in breast cancer. *Oncotarget* **2017**, *8*, 67329–67343. [CrossRef]

148. Richards, E.J.; Zhang, G.; Li, Z.P.; Permuth-Wey, J.; Challa, S.; Li, Y.; Kong, W.; Dan, S.; Bui, M.M.; Coppola, D.; et al. Long non-coding RNAs (LncRNA) regulated by transforming growth factor (TGF) β: LncRNA-hit-mediated TGF-β-induced epithelial to mesenchymal transition in mammary epithelia. *J. Biol. Chem.* **2015**, *290*, 6857–6867. [CrossRef]

149. Pan, Y.; Li, C.; Chen, J.; Zhang, K.; Chu, X.; Wang, R.; Chen, L. The emerging roles of long noncoding RNA ROR (lincRNA-ROR) and its possible mechanisms in human cancers. *Cell Physiol. Biochem.* **2016**, *40*, 219–229. [CrossRef]

150. Eades, G.; Wolfson, B.; Zhang, Y.; Li, Q.; Yao, Y.; Zhou, Q. lincRNA-RoR and miR-145 regulate invasion in triple-negative breast cancer via targeting ARF6. *Mol. Cancer Res.* **2015**, *13*, 330–338. [CrossRef]

151. Fang, Y.; Wang, J.; Wu, F.; Song, Y.; Zhao, S.; Zhang, Q. Long non-coding RNA HOXA-AS2 promotes proliferation and invasion of breast cancer by acting as a miR-520c-3p sponge. *Oncotarget* **2017**, *8*, 46090–46103. [CrossRef]

152. Wu, Z.J.; Li, Y.; Wu, Y.Z.; Wang, Y.; Nian, W.Q.; Wang, L.L.; Li, L.C.; Luo, H.L.; Wang, D.L. Long non-coding RNA CCAT2 promotes the breast cancer growth and metastasis by regulating TGF-β signaling pathway. *Eur. Rev. Med. Pharmacol. Sci.* **2017**, *21*, 706–714.

153. Zhang, C.Y.; Yu, M.S.; Li, X.; Zhang, Z.; Han, C.R.; Yan, B. Overexpression of long non-coding RNA MEG3 suppresses breast cancer cell proliferation, invasion, and angiogenesis through AKT pathway. *Tumour Biol.* **2017**, *39*, 1010428317701311. [CrossRef]

154. Zhang, W.; Shi, S.; Jiang, J.; Li, X.; Lu, H.; Ren, F. LncRNA MEG3 inhibits cell epithelial-mesenchymal transition by sponging miR-421 targeting E-cadherin in breast cancer. *Biomed. Pharmacother.* **2017**, *91*, 312–319. [CrossRef]

155. Fu, M.; Huang, Z.; Zang, X.; Pan, L.; Liang, W.; Chen, J.; Qian, H.; Xu, W.; Jiang, P.; Zhang, X. Long noncoding RNA LINC00978 promotes cancer growth and acts as a diagnostic biomarker in gastric cancer. *Cell Prolif.* **2018**, *51*, e12425. [CrossRef]

156. Zuo, Z.K.; Gong, Y.; Chen, X.H.; Ye, F.; Yin, Z.M.; Gong, Q.N.; Huang, J.S. TGF-β1-Induced LncRNA UCA1 upregulation promotes gastric cancer invasion and migration. *DNA Cell Biol.* **2017**, *36*, 159–167. [CrossRef]

157. Saito, T.; Kurashige, J.; Nambara, S.; Komatsu, H.; Hirata, H.; Ueda, M.; Sakimura, S.; Uchi, R.; Takano, Y.; Shinden, Y.; et al. A long non-coding RNA activated by transforming growth factor-β is an independent prognostic marker of gastric cancer. *Ann. Surg. Oncol.* **2015**, *22* (Suppl. 3), S915–S922. [CrossRef]

158. Chen, Y.; Wei, G.; Xia, H.; Tang, Q.; Bi, F. Long noncoding RNAATB promotes cell proliferation, migration and invasion in gastric cancer. *Mol. Med. Rep.* **2018**, *17*, 1940–1946.

159. Lei, K.; Liang, X.; Gao, Y.; Xu, B.; Xu, Y.; Li, Y.; Tao, Y.; Shi, W.; Liu, J. Lnc-ATB contributes to gastric cancer growth through a MiR-141-3p/TGF-β2 feedback loop. *Biochem. Biophys. Res. Commun.* **2017**, *484*, 514–521. [CrossRef]

160. Zhang, Q.; Chen, B.; Liu, P.; Yang, J. XIST promotes gastric cancer (GC) progression through TGF-β1 via targeting miR-185. *J. Cell Biochem.* **2018**, *119*, 2787–2796. [CrossRef]

161. Han, Q.; Zhang, W.; Meng, J.; Ma, L.; Li, A. LncRNA-LET inhibits cell viability, migration and EMT while induces apoptosis by up-regulation of TIMP2 in human granulosa-like tumor cell line KGN. *Biomed. Pharmacother.* **2018**, *100*, 250–256. [CrossRef]

162. Li, J.; Huang, Y.; Deng, X.; Luo, M.; Wang, X.; Hu, H.; Liu, C.; Zhong, M. Long noncoding RNA H19 promotes transforming growth factor-β-induced epithelial-mesenchymal transition by acting as a competing endogenous RNA of miR-370-3p in ovarian cancer cells. *Onco. Targets Ther.* **2018**, *11*, 427–440. [CrossRef]

163. Li, Q.; Zhang, C.; Chen, R.; Xiong, H.; Qiu, F.; Liu, S.; Zhang, M.; Wang, F.; Wang, Y.; Zhou, X.; et al. Disrupting MALAT1/miR-200c sponge decreases invasion and migration in endometrioid endometrial carcinoma. *Cancer Lett.* **2016**, *383*, 28–40. [CrossRef]

164. Hu, J.; Zhang, L.; Mei, Z.; Jiang, Y.; Yi, Y.; Liu, L.; Meng, Y.; Zhou, L.; Zeng, J.; Wu, H.; et al. Interaction of E3 ubiquitin ligase MARCH7 with long noncoding RNA MALAT1 and autophagy-related protein ATG7 promotes autophagy and invasion in ovarian cancer. *Cell Physiol. Biochem.* **2018**, *47*, 654–666. [CrossRef]

165. Ma, J.; Xue, M. LINK-A lncRNA promotes migration and invasion of ovarian carcinoma cells by activating TGF-β pathway. *Biosci. Rep.* **2018**, *38*, BSR20180936. [CrossRef]

166. Cao, C.; Zhang, T.; Zhang, D.; Xie, L.; Zou, X.; Lei, L.; Wu, D.; Liu, L. The long non-coding RNA, SNHG6-003, functions as a competing endogenous RNA to promote the progression of hepatocellular carcinoma. *Oncogene* **2017**, *36*, 1112–1122. [CrossRef]

167. Liang, H.; Zhao, X.; Wang, C.; Sun, J.; Chen, Y.; Wang, G.; Fang, L.; Yang, R.; Yu, M.; Gu, Y.; et al. Systematic analyses reveal long non-coding RNA (PTAF)-mediated promotion of EMT and invasion-metastasis in serous ovarian cancer. *Mol. Cancer* **2018**, *17*, 96. [CrossRef]

168. Zhuang, J.; Lu, Q.; Shen, B.; Huang, X.; Shen, L.; Zheng, X.; Huang, R.; Yan, J.; Guo, H. TGF-β1 secreted by cancer-associated fibroblasts induces epithelial-mesenchymal transition of bladder cancer cells through lncRNA-ZEB2NAT. *Sci. Rep.* **2015**, *5*, 11924. [CrossRef]

169. Fan, Y.; Shen, B.; Tan, M.; Mu, X.; Qin, Y.; Zhang, F.; Liu, Y. TGF-β-induced upregulation of malat1 promotes bladder cancer metastasis by associating with suz12. *Clin. Cancer Res.* **2014**, *20*, 1531–1541. [CrossRef]

170. Gou, L.; Liu, M.; Xia, J.; Wan, Q.; Jiang, Y.; Sun, S.; Tang, M.; Zhou, L.; He, T.; Zhang, Y. BMP9 promotes the proliferation and migration of bladder cancer cells through up-regulating lncRNA UCA1. *Int. J. Mol. Sci.* **2018**, *19*, 1116. [CrossRef]

171. Zhai, X.; Xu, W. Long noncoding RNA ATB promotes proliferation, migration, and invasion in bladder cancer by suppressing MicroRNA-126. *Oncol. Res.* **2018**, *26*, 1063–1072. [CrossRef]

172. Jin, Y.; Cui, Z.; Li, X.; Jin, X.; Peng, J. Upregulation of long non-coding RNA PlncRNA-1 promotes proliferation and induces epithelial-mesenchymal transition in prostate cancer. *Oncotarget* **2017**, *8*, 26090–26099. [CrossRef]

173. Wang, S.H.; Zhang, M.D.; Wu, X.C.; Weng, M.Z.; Zhou, D.; Quan, Z.W. Overexpression of LncRNA-ROR predicts a poor outcome in gallbladder cancer patients and promotes the tumor cells proliferation, migration, and invasion. *Tumour Biol.* **2016**, *37*, 12867–12875. [CrossRef]

174. Zhao, B.; Lu, Y.L.; Yang, Y.; Hu, L.B.; Bai, Y.; Li, R.Q.; Zhang, G.Y.; Li, J.; Bi, C.W.; Yang, L.B.; et al. Overexpression of lncRNA ANRIL promoted the proliferation and migration of prostate cancer cells via regulating let-7a/TGF-β1/ SMAD signaling pathway. *Cancer Biomark.* **2018**, *21*, 613–620. [CrossRef]

175. Xu, S.; Yi, X.M.; Tang, C.P.; Ge, J.P.; Zhang, Z.Y.; Zhou, W.Q. Long non-coding RNA ATB promotes growth and epithelial-mesenchymal transition and predicts poor prognosis in human prostate carcinoma. *Oncol. Rep.* **2016**, *36*, 10–22. [CrossRef]

176. Wang, H.; Wang, G.; Gao, Y.; Zhao, C.; Li, X.; Zhang, F.; Jiang, C.; Wu, B. Lnc-SNHG1 activates the TGFBR2/SMAD3 and RAB11A/Wnt/β-catenin pathway by sponging MiR-302/372/373/520 in invasive pituitary tumors. *Cell Physiol. Biochem.* **2018**, *48*, 1291–1303. [CrossRef]

177. Li, Z.; Liu, H.; Zhong, Q.; Wu, J.; Tang, Z. LncRNA UCA1 is necessary for TGF-β-induced epithelial-mesenchymal transition and stemness via acting as a ceRNA for SLUG in glioma cells. *FEBS Open Bio.* **2018**, *8*, 1855–1865. [CrossRef]

178. Zhang, C.; Hao, Y.; Wang, Y.; Xu, J.; Teng, Y.; Yang, X. TGF-β/SMAD4-regulated LncRNA-LINP1 inhibits epithelial-mesenchymal transition in lung cancer. *Int. J. Biol. Sci.* **2018**, *14*, 1715–1723. [CrossRef]

179. Lu, Z.; Li, Y.; Che, Y.; Huang, J.; Sun, S.; Mao, S.; Lei, Y.; Li, N.; Sun, N.; He, J. The TGF-β-induced lncRNA TBILA promotes non-small cell lung cancer progression in vitro and in vivo via cis-regulating HGAL and activating S100A7/JAB1 signaling. *Cancer Lett.* **2018**, *432*, 156–168. [CrossRef]

180. Wang, X.; Zhang, G.; Cheng, Z.; Dai, L.; Jia, L.; Jing, X.; Wang, H.; Zhang, R.; Liu, M.; Jiang, T.; et al. Knockdown of LncRNA-XIST suppresses proliferation and TGF-β1-induced EMT in NSCLC through the notch-1 pathway by regulation of miR-137. *Genet. Test Mol. Biomark.* **2018**, *22*, 333–342. [CrossRef]

181. Li, C.; Wan, L.; Liu, Z.; Xu, G.; Wang, S.; Su, Z.; Zhang, Y.; Zhang, C.; Liu, X.; Lei, Z.; et al. Long non-coding RNA XIST promotes TGF-β-induced epithelial-mesenchymal transition by regulating miR-367/141-ZEB2 axis in non-small-cell lung cancer. *Cancer Lett.* **2018**, *418*, 185–195. [CrossRef]

182. Kawasaki, N.; Miwa, T.; Hokari, S.; Sakurai, T.; Ohmori, K.; Miyauchi, K.; Miyazono, K.; Koinuma, D. Long noncoding RNA NORAD regulates transforming growth factor-β signaling and epithelial-to-mesenchymal transition-like phenotype. *Cancer Sci.* **2018**, *109*, 2211–2220. [CrossRef]

183. Cao, Y.; Luo, X.; Ding, X.; Cui, S.; Guo, C. LncRNA ATB promotes proliferation and metastasis in A549 cells by down-regulation of microRNA-494. *J. Cell Biochem.* **2018**, *119*, 6935–6942. [CrossRef]

184. Lu, W.; Zhang, H.; Niu, Y.; Wu, Y.; Sun, W.; Li, H.; Kong, J.; Ding, K.; Shen, H.M.; Wu, H.; et al. Long non-coding RNA linc00673 regulated non-small cell lung cancer proliferation, migration, invasion and epithelial mesenchymal transition by sponging miR-150-5p. *Mol. Cancer* **2017**, *16*, 118. [CrossRef]

185. Lu, Z.; Li, Y.; Wang, J.; Che, Y.; Sun, S.; Huang, J.; Chen, Z.; He, J. Long non-coding RNA NKILA inhibits migration and invasion of non-small cell lung cancer via NF-kappaB/Snail pathway. *J. Exp. Clin. Cancer Res.* **2017**, *36*, 54. [CrossRef]

186. Terashima, M.; Tange, S.; Ishimura, A.; Suzuki, T. MEG3 long noncoding RNA contributes to the epigenetic regulation of epithelial-mesenchymal transition in lung cancer cell lines. *J. Biol. Chem.* **2017**, *292*, 82–99. [CrossRef]

187. Wu, D.M.; Deng, S.H.; Liu, T.; Han, R.; Zhang, T.; Xu, Y. TGF-β-mediated exosomal lnc-MMP2-2 regulates migration and invasion of lung cancer cells to the vasculature by promoting MMP2 expression. *Cancer Med.* **2018**, *7*, 5118–5129. [CrossRef]

188. Wang, S.; Lan, F.; Xia, Y. lncRA ANCR inhibits non-small cell lung cancer cell migration and invasion by inactivating TGF-β pathway. *Med. Sci. Monit.* **2018**, *24*, 6002–6009. [CrossRef]

189. Hao, Y.; Yang, X.; Zhang, D.; Luo, J.; Chen, R. Long noncoding RNA LINC01186, regulated by TGF-β/SMAD3, inhibits migration and invasion through epithelial-mesenchymal-transition in lung cancer. *Gene* **2017**, *608*, 1–12. [CrossRef]

190. Zhang, Y.; He, R.Q.; Dang, Y.W.; Zhang, X.L.; Wang, X.; Huang, S.N.; Huang, W.T.; Jiang, M.T.; Gan, X.N.; Xie, Y.; et al. Comprehensive analysis of the long noncoding RNA HOXA11-AS gene interaction regulatory network in NSCLC cells. *Cancer Cell Int.* **2016**, *16*, 89. [CrossRef]

191. Li, Y.; Liu, G.; Li, X.; Dong, H.; Xiao, W.; Lu, S. Long non-coding RNA SBF2-AS1 promotes hepatocellular carcinoma progression through regulation of miR-140-5p-TGFBR1 pathway. *Biochem. Biophys. Res. Commun.* **2018**, *503*, 2826–2832. [CrossRef]

192. Tang, J.; Zhuo, H.; Zhang, X.; Jiang, R.; Ji, J.; Deng, L.; Qian, X.; Zhang, F.; Sun, B. A novel biomarker Linc00974 interacting with KRT19 promotes proliferation and metastasis in hepatocellular carcinoma. *Cell Death Dis.* **2014**, *5*, e1549. [CrossRef]

193. Zhang, X.; Feng, W.; Zhang, J.; Ge, L.; Zhang, Y.; Jiang, X.; Peng, W.; Wang, D.; Gong, A.; Xu, M. Long noncoding RNA PVT1 promotes epithelialmesenchymal transition via the TGF-β/SMAD pathway in pancreatic cancer cells. *Oncol. Rep.* **2018**, *40*, 1093–1102.

194. Terashima, M.; Ishimura, A.; Wanna-Udom, S.; Suzuki, T. MEG8 long noncoding RNA contributes to epigenetic progression of the epithelial-mesenchymal transition of lung and pancreatic cancer cells. *J. Biol. Chem.* **2018**, *293*, 18016–18030. [CrossRef]

195. Qin, C.F.; Zhao, F.L. Long non-coding RNA TUG1 can promote proliferation and migration of pancreatic cancer via EMT pathway. *Eur. Rev. Med. Pharmacol. Sci.* **2017**, *21*, 2377–2384.

196. Ottaviani, S.; Stebbing, J.; Frampton, A.E.; Zagorac, S.; Krell, J.; de Giorgio, A.; Trabulo, S.M.; Nguyen, V.T.M.; Magnani, L.; Feng, H.; et al. TGF-β induces miR-100 and miR-125b but blocks let-7a through LIN28B controlling PDAC progression. *Nat. Commun.* **2018**, *9*, 1845. [CrossRef]

197. Qu, S.; Yang, X.; Song, W.; Sun, W.; Li, X.; Wang, J.; Zhong, Y.; Shang, R.; Ruan, B.; Zhang, Z.; et al. Downregulation of lncRNA-ATB correlates with clinical progression and unfavorable prognosis in pancreatic cancer. *Tumour Biol.* **2016**, *37*, 3933–3938. [CrossRef]

198. Liu, Y.; Qian, J.; Li, X.; Chen, W.; Xu, A.; Zhao, K.; Hua, Y.; Huang, Z.; Zhang, J.; Liang, C.; et al. Long noncoding RNA BX357664 regulates cell proliferation and epithelial-to-mesenchymal transition via inhibition of TGF-β1/p38/HSP27 signaling in renal cell carcinoma. *Oncotarget* **2016**, *7*, 81410–81422.

199. Xiong, J.; Liu, Y.; Jiang, L.; Zeng, Y.; Tang, W. High expression of long non-coding RNA lncRNA-ATB is correlated with metastases and promotes cell migration and invasion in renal cell carcinoma. *Jpn. J. Clin. Oncol.* **2016**, *46*, 378–384. [CrossRef]

200. Yue, B.; Qiu, S.; Zhao, S.; Liu, C.; Zhang, D.; Yu, F.; Peng, Z.; Yan, D. LncRNA-ATB mediated E-cadherin repression promotes the progression of colon cancer and predicts poor prognosis. *J. Gastroenterol. Hepatol.* **2016**, *31*, 595–603. [CrossRef]

201. Iguchi, T.; Uchi, R.; Nambara, S.; Saito, T.; Komatsu, H.; Hirata, H.; Ueda, M.; Sakimura, S.; Takano, Y.; Kurashige, J.; et al. A long noncoding RNA, lncRNA-ATB, is involved in the progression and prognosis of colorectal cancer. *Anticancer Res.* **2015**, *35*, 1385–1388.

202. Kong, J.; Sun, W.; Li, C.; Wan, L.; Wang, S.; Wu, Y.; Xu, E.; Zhang, H.; Lai, M. Long non-coding RNA LINC01133 inhibits epithelial-mesenchymal transition and metastasis in colorectal cancer by interacting with SRSF6. *Cancer Lett.* **2016**, *380*, 476–484. [CrossRef]

203. Zhang, S.Z.; Cai, L.; Li, B. MEG3 long non-coding RNA prevents cell growth and metastasis of osteosarcoma. *Bratisl Lek Listy* **2017**, *118*, 632–636. [CrossRef]

204. Huo, Y.; Li, Q.; Wang, X.; Jiao, X.; Zheng, J.; Li, Z.; Pan, X. MALAT1 predicts poor survival in osteosarcoma patients and promotes cell metastasis through associating with EZH2. *Oncotarget* **2017**, *8*, 46993–47006. [CrossRef]

205. Deng, Y.; Zhao, F.; Zhang, Z.; Sun, F.; Wang, M. Long noncoding RNA SNHG7 promotes the tumor growth and epithelial-to-mesenchymal transition via regulation of mir-34a signals in osteosarcoma. *Cancer Biother. Radiopharm.* **2018**, *33*, 365–372. [CrossRef]

206. Zhang, R.; Hardin, H.; Huang, W.; Chen, J.; Asioli, S.; Righi, A.; Maletta, F.; Sapino, A.; Lloyd, R.V. MALAT1 long non-coding RNA expression in thyroid tissues: Analysis by in situ hybridization and real-time PCR. *Endocr. Pathol.* **2017**, *28*, 7–12. [CrossRef]

207. Zhang, R.; Hardin, H.; Huang, W.; Buehler, D.; Lloyd, R.V. Long non-coding RNA linc-ROR is upregulated in papillary thyroid carcinoma. *Endocr. Pathol.* **2018**, *29*, 1–8. [CrossRef]

208. Zhou, H.; Sun, Z.; Li, S.; Wang, X.; Zhou, X. LncRNA SPRY4-IT was concerned with the poor prognosis and contributed to the progression of thyroid cancer. *Cancer Gene Ther.* **2018**, *25*, 39–46. [CrossRef]

209. Zhao, J.J.; Hao, S.; Wang, L.L.; Hu, C.Y.; Zhang, S.; Guo, L.J.; Zhang, G.; Gao, B.; Jiang, Y.; Tian, W.G.; et al. Long non-coding RNA ANRIL promotes the invasion and metastasis of thyroid cancer cells through TGF-β/SMAD signaling pathway. *Oncotarget* **2016**, *7*, 57903–57918. [CrossRef]

210. Lambert, A.W.; Pattabiraman, D.R.; Weinberg, R.A. Emerging biological principles of metastasis. *Cell* **2017**, *168*, 670–691. [CrossRef]

211. Chiang, A.C.; Massague, J. Molecular basis of metastasis. *N. Engl. J. Med.* **2008**, *359*, 2814–2823. [CrossRef]

212. Jung, H.Y.; Fattet, L.; Yang, J. Molecular pathways: Linking tumor microenvironment to epithelial-mesenchymal transition in metastasis. *Clin. Cancer Res.* **2015**, *21*, 962–968. [CrossRef]

213. Nguyen, D.X.; Bos, P.D.; Massague, J. Metastasis: From dissemination to organ-specific colonization. *Nat. Rev. Cancer* **2009**, *9*, 274–284. [CrossRef]

214. Roodman, G.D. Biology of osteoclast activation in cancer. *J. Clin. Oncol.* **2001**, *19*, 3562–3571. [CrossRef]

215. Yin, J.J.; Selander, K.; Chirgwin, J.M.; Dallas, M.; Grubbs, B.G.; Wieser, R.; Massague, J.; Mundy, G.R.; Guise, T.A. TGF-β signaling blockade inhibits PTHrP secretion by breast cancer cells and bone metastases development. *J. Clin. Investig.* **1999**, *103*, 197–206. [CrossRef]

216. Kang, Y.; Siegel, P.M.; Shu, W.; Drobnjak, M.; Kakonen, S.M.; Cordon-Cardo, C.; Guise, T.A.; Massague, J. A multigenic program mediating breast cancer metastasis to bone. *Cancer Cell* **2003**, *3*, 537–549. [CrossRef]

217. Buijs, J.T.; Stayrook, K.R.; Guise, T.A. The role of TGF-β in bone metastasis: Novel therapeutic perspectives. *Bonekey Rep.* **2012**, *1*, 96. [CrossRef]

218. Biswas, S.; Guix, M.; Rinehart, C.; Dugger, T.C.; Chytil, A.; Moses, H.L.; Freeman, M.L.; Arteaga, C.L. Inhibition of TGF-β with neutralizing antibodies prevents radiation-induced acceleration of metastatic cancer progression. *J. Clin. Investig.* **2007**, *117*, 1305–1313. [CrossRef]

219. Padua, D.; Zhang, X.H.; Wang, Q.; Nadal, C.; Gerald, W.L.; Gomis, R.R.; Massague, J. TGF-β primes breast tumors for lung metastasis seeding through angiopoietin-like 4. *Cell* **2008**, *133*, 66–77. [CrossRef]

220. Gupta, G.P.; Perk, J.; Acharyya, S.; de Candia, P.; Mittal, V.; Todorova-Manova, K.; Gerald, W.L.; Brogi, E.; Benezra, R.; Massague, J. ID genes mediate tumor reinitiation during breast cancer lung metastasis. *Proc. Natl. Acad. Sci. USA* **2007**, *104*, 19506–19511. [CrossRef]

221. Zhuang, X.; Zhang, H.; Li, X.; Li, X.; Cong, M.; Peng, F.; Yu, J.; Zhang, X.; Yang, Q.; Hu, G. Differential effects on lung and bone metastasis of breast cancer by Wnt signalling inhibitor DKK1. *Nat. Cell. Biol.* **2017**, *19*, 1274–1285. [CrossRef]

222. Jung, B.; Staudacher, J.J.; Beauchamp, D. Transforming growth factor β superfamily signaling in development of colorectal cancer. *Gastroenterology* **2017**, *152*, 36–52. [CrossRef]

223. Steinbichler, T.B.; Dudas, J.; Riechelmann, H.; Skvortsova, I.I. The role of exosomes in cancer metastasis. *Semin. Cancer Biol.* **2017**, *44*, 170–181. [CrossRef]

224. Carmeliet, P.; Jain, R.K. Angiogenesis in cancer and other diseases. *Nature* **2000**, *407*, 249–257. [CrossRef]

225. Roberts, A.B.; Thompson, N.L.; Heine, U.; Flanders, C.; Sporn, M.B. Transforming growth factor-β: Possible roles in carcinogenesis. *Br. J. Cancer* **1988**, *57*, 594–600. [CrossRef]

226. Goumans, M.J.; Valdimarsdottir, G.; Itoh, S.; Rosendahl, A.; Sideras, P.; Dijke, P.t. Balancing the activation state of the endothelium via two distinct TGF-β type I receptors. *EMBO J.* **2002**, *21*, 1743–1753. [CrossRef]

227. Breier, G.; Blum, S.; Peli, J.; Groot, M.; Wild, C.; Risau, W.; Reichmann, E. Transforming growth factor-β and RAS regulate the VEGF/VEGF-receptor system during tumor angiogenesis. *Int. J. Cancer* **2002**, *97*, 142–148. [CrossRef]

228. Sanchez-Elsner, T.; Botella, L.M.; Velasco, B.; Corbi, A.; Attisano, L.; Bernabeu, C. Synergistic cooperation between hypoxia and transforming growth factor-β pathways on human vascular endothelial growth factor gene expression. *J. Biol. Chem.* **2001**, *276*, 38527–38535. [CrossRef]

229. Liu, D.; Li, L.; Zhang, X.X.; Wan, D.Y.; Xi, B.X.; Hu, Z.; Ding, W.C.; Zhu, D.; Wang, X.L.; Wang, W.; et al. SIX1 promotes tumor lymphangiogenesis by coordinating TGF-β signals that increase expression of VEGF-C. *Cancer Res.* **2014**, *74*, 5597–5607. [CrossRef]

230. Sun, H.; Miao, C.; Liu, W.; Qiao, X.; Yang, W.; Li, L.; Li, C. TGF-β1/TβRII/SMAD3 signaling pathway promotes VEGF expression in oral squamous cell carcinoma tumor-associated macrophages. *Biochem. Biophys. Res. Commun.* **2018**, *497*, 583–590. [CrossRef]

231. Groppa, E.; Brkic, S.; Bovo, E.; Reginato, S.; Sacchi, V.; di Maggio, N.; Muraro, M.G.; Calabrese, D.; Heberer, M.; Gianni-Barrera, R.; et al. VEGF dose regulates vascular stabilization through Semaphorin3A and the Neuropilin-1 + monocyte/TGF-β1 paracrine axis. *EMBO Mol. Med.* **2015**, *7*, 1366–1384. [CrossRef]

232. Rak, J.; Mitsuhashi, Y.; Sheehan, C.; Tamir, A.; Viloria-Petit, A.; Filmus, J.; Mansour, S.J.; Ahn, N.G.; Kerbel, R.S. Oncogenes and tumor angiogenesis: Differential modes of vascular endothelial growth factor up-regulation in ras-transformed epithelial cells and fibroblasts. *Cancer Res.* **2000**, *60*, 490–498.

233. Miyazono, K.; Ehata, S.; Koinuma, D. Tumor-promoting functions of transforming growth factor-β in progression of cancer. *Ups. J. Med. Sci.* **2012**, *117*, 143–152. [CrossRef]

234. Chen, W.; Ten Dijke, P. Immunoregulation by members of the TGF-β superfamily. *Nat. Rev. Immunol.* **2016**, *16*, 723–740. [CrossRef]

235. Gorelik, L.; Flavell, R.A. Abrogation of TGF-β signaling in T cells leads to spontaneous T cell differentiation and autoimmune disease. *Immunity* **2000**, *12*, 171–181. [CrossRef]

236. Donkor, M.K.; Sarkar, A.; Savage, P.A.; Franklin, R.A.; Johnson, L.K.; Jungbluth, A.A.; Allison, J.P.; Li, M.O. T cell surveillance of oncogene-induced prostate cancer is impeded by T cell-derived TGF-β1 cytokine. *Immunity* **2011**, *35*, 123–134. [CrossRef]

237. Mariathasan, S.; Turley, S.J.; Nickles, D.; Castiglioni, A.; Yuen, K.; Wang, Y.; Kadel, E.E., III; Koeppen, H.; Astarita, J.L.; Cubas, R.; et al. TGF-β attenuates tumour response to PD-L1 blockade by contributing to exclusion of T cells. *Nature* **2018**, *554*, 544–548. [CrossRef]

238. Tauriello, D.V.F.; Palomo-Ponce, S.; Stork, D.; Berenguer-Llergo, A.; Badia-Ramentol, J.; Iglesias, M.; Sevillano, M.; Ibiza, S.; Canellas, A.; Hernando-Momblona, X.; et al. TGF-β drives immune evasion in genetically reconstituted colon cancer metastasis. *Nature* **2018**, *554*, 538–543. [CrossRef]

239. Gorelik, L.; Flavell, R.A. Immune-mediated eradication of tumors through the blockade of transforming growth factor-β signaling in T cells. *Nat. Med.* **2001**, *7*, 1118–1122. [CrossRef]

240. Thomas, D.A.; Massague, J. TGF-β directly targets cytotoxic T cell functions during tumor evasion of immune surveillance. *Cancer Cell* **2005**, *8*, 369–380. [CrossRef]

241. Fu, S.; Zhang, N.; Yopp, A.C.; Chen, D.; Mao, M.; Chen, D.; Zhang, H.; Ding, Y.; Bromberg, J.S. TGF-β induces Foxp3 + T-regulatory cells from CD4 + CD25 - precursors. *Am. J. Transplant.* **2004**, *4*, 1614–1627. [CrossRef]

242. Worthington, J.J.; Kelly, A.; Smedley, C.; Bauche, D.; Campbell, S.; Marie, J.C.; Travis, M.A. Integrin alphavβ8-mediated TGF-β activation by effector regulatory T cells is essential for suppression of T-Cell-mediated inflammation. *Immunity* **2015**, *42*, 903–915. [CrossRef]

243. Metelli, A.; Wu, B.X.; Fugle, C.W.; Rachidi, S.; Sun, S.; Zhang, Y.; Wu, J.; Tomlinson, S.; Howe, P.H.; Yang, Y.; et al. Surface expression of TGF-β docking receptor GARP promotes oncogenesis and immune tolerance in breast cancer. *Cancer Res.* **2016**, *76*, 7106–7117. [CrossRef]

244. Imai, K.; Minamiya, Y.; Koyota, S.; Ito, M.; Saito, H.; Sato, Y.; Motoyama, S.; Sugiyama, T.; Ogawa, J. Inhibition of dendritic cell migration by transforming growth factor-β1 increases tumor-draining lymph node metastasis. *J. Exp. Clin. Cancer Res.* **2012**, *31*, 3. [CrossRef]

245. Ito, M.; Minamiya, Y.; Kawai, H.; Saito, S.; Saito, H.; Nakagawa, T.; Imai, K.; Hirokawa, M.; Ogawa, J. Tumor-derived TGF-β-1 induces dendritic cell apoptosis in the sentinel lymph node. *J. Immunol.* **2006**, *176*, 5637–5643. [CrossRef]

246. Kobie, J.J.; Wu, R.S.; Kurt, R.A.; Lou, S.; Adelman, M.K.; Whitesell, L.J.; Ramanathapuram, L.V.; Arteaga, C.L.; Akporiaye, E.T. Transforming growth factor β inhibits the antigen-presenting functions and antitumor activity of dendritic cell vaccines. *Cancer Res.* **2003**, *63*, 1860–1864.

247. Allavena, P.; Sica, A.; Garlanda, C.; Mantovani, A. The Yin-Yang of tumor-associated macrophages in neoplastic progression and immune surveillance. *Immunol. Rev.* **2008**, *222*, 155–161. [CrossRef]

248. Fridlender, Z.G.; Sun, J.; Kim, S.; Kapoor, V.; Cheng, G.; Ling, L.; Worthen, G.S.; Albelda, S.M. Polarization of tumor-associated neutrophil phenotype by TGF-β: "N1" versus "N2" TAN. *Cancer Cell* **2009**, *16*, 183–194. [CrossRef]

249. Wang, Y.; Chu, J.; Yi, P.; Dong, W.; Saultz, J.; Wang, Y.; Wang, H.; Scoville, S.; Zhang, J.; Wu, L.C.; et al. SMAD4 promotes TGF-β-independent NK cell homeostasis and maturation and antitumor immunity. *J. Clin. Investig.* **2018**, *128*, 5123–5136. [CrossRef]

250. Li, M.O.; Sanjabi, S.; Flavell, R.A. Transforming growth factor-β controls development, homeostasis, and tolerance of T cells by regulatory T cell-dependent and -independent mechanisms. *Immunity* **2006**, *25*, 455–471. [CrossRef]

251. Akhurst, R.J. Targeting TGF-β signaling for therapeutic gain. *Cold Spring Harb. Perspect. Biol.* **2017**, *9*, a022301. [CrossRef]

252. Colak, S.; Ten Dijke, P. Targeting TGF-β signaling in cancer. *Trends Cancer* **2017**, *3*, 56–71. [CrossRef]

253. Zappa, C.; Mousa, S.A. Non-small cell lung cancer: Current treatment and future advances. *Transl. Lung Cancer Res.* **2016**, *5*, 288–300. [CrossRef]

254. Jaschinski, F.; Rothhammer, T.; Jachimczak, P.; Seitz, C.; Schneider, A.; Schlingensiepen, K.H. The antisense oligonucleotide trabedersen (AP 12009) for the targeted inhibition of TGF-β2. *Curr. Pharm. Biotechnol.* **2011**, *12*, 2203–2213. [CrossRef]

255. Fujiwara, Y.; Nokihara, H.; Yamada, Y.; Yamamoto, N.; Sunami, K.; Utsumi, H.; Asou, H.; Takahash, I.O.; Ogasawara, K.; Gueorguieva, I.; et al. Phase 1 study of galunisertib, a TGF-β receptor I kinase inhibitor, in Japanese patients with advanced solid tumors. *Cancer Chemother. Pharmacol.* **2015**, *76*, 1143–1152. [CrossRef]

256. Garrison, K.; Hahn, T.; Lee, W.C.; Ling, L.E.; Weinberg, A.D.; Akporiaye, E.T. The small molecule TGF-β signaling inhibitor SM16 synergizes with agonistic OX40 antibody to suppress established mammary tumors and reduce spontaneous metastasis. *Cancer Immunol. Immunother.* **2012**, *61*, 511–521. [CrossRef]

257. Mohammad, K.S.; Javelaud, D.; Fournier, P.G.; Niewolna, M.; McKenna, C.R.; Peng, X.H.; Duong, V.; Dunn, L.K.; Mauviel, A.; Guise, T.A. TGF-β RI kinase inhibitor SD-208 reduces the development and progression of melanoma bone metastases. *Cancer Res.* **2011**, *71*, 175–184. [CrossRef]

258. Uhl, M.; Aulwurm, S.; Wischhusen, J.; Weiler, M.; Ma, J.Y.; Almirez, R.; Mangadu, R.; Liu, Y.W.; Platten, M.; Herrlinger, U.; et al. SD-208, a novel transforming growth factor β receptor I kinase inhibitor, inhibits growth and invasiveness and enhances immunogenicity of murine and human glioma cells in vitro and in vivo. *Cancer Res.* **2004**, *64*, 7954–7961. [CrossRef]

259. Akbari, A.; Amanpour, S.; Muhammadnejad, S.; Ghahremani, M.H.; Ghaffari, S.H.; Dehpour, A.R.; Mobini, G.R.; Shidfar, F.; Abastabar, M.; Khoshzaban, A.; et al. Evaluation of antitumor activity of a TGF-β receptor I inhibitor (SD-208) on human colon adenocarcinoma. *Daru* **2014**, *22*, 47. [CrossRef]

260. Doi, T.; Lee, K.H.; Kim, T.M.; Ohtsu, A.; Kim, T.Y.; Ikeda, M.; Yoh, K.; Stampino, C.G.; Hirohashi, T.; Suzuki, A.; et al. A phase I study of the human anti-activin receptor-like kinase 1 antibody PF-03446962 in Asian patients with advanced solid tumors. *Cancer Med.* **2016**, *5*, 1454–1463. [CrossRef]

261. Morris, J.C.; Tan, A.R.; Olencki, T.E.; Shapiro, G.I.; Dezube, B.J.; Reiss, M.; Hsu, F.J.; Berzofsky, J.A.; Lawrence, D.P. Phase I study of GC1008 (fresolimumab): A human anti-transforming growth factor-β (TGF-β) monoclonal antibody in patients with advanced malignant melanoma or renal cell carcinoma. *PLoS ONE* **2014**, *9*, e90353. [CrossRef]

262. Lacouture, M.E.; Morris, J.C.; Lawrence, D.P.; Tan, A.R.; Olencki, T.E.; Shapiro, G.I.; Dezube, B.J.; Berzofsky, J.A.; Hsu, F.J.; Guitart, J. Cutaneous keratoacanthomas/squamous cell carcinomas associated with neutralization of transforming growth factor β by the monoclonal antibody fresolimumab (GC1008). *Cancer Immunol. Immunother.* **2015**, *64*, 437–446. [CrossRef]

263. Tolcher, A.W.; Berlin, J.D.; Cosaert, J.; Kauh, J.; Chan, E.; Piha-Paul, S.A.; Amaya, A.; Tang, S.; Driscoll, K.; Kimbung, R.; et al. A phase 1 study of anti-TGF-β receptor type-II monoclonal antibody LY3022859 in patients with advanced solid tumors. *Cancer Chemother. Pharmacol.* **2017**, *79*, 673–680. [CrossRef]

264. Chang, Z.L.; Lorenzini, M.H.; Chen, X.; Tran, U.; Bangayan, N.J.; Chen, Y.Y. Rewiring T-cell responses to soluble factors with chimeric antigen receptors. *Nat. Chem. Biol.* **2018**, *14*, 317–324. [CrossRef]

International Journal of
Molecular Sciences

MDPI

Review

Nano-Strategies to Target Breast Cancer-Associated Fibroblasts: Rearranging the Tumor Microenvironment to Achieve Antitumor Efficacy

Marta Truffi [1], Serena Mazzucchelli [1], Arianna Bonizzi [1], Luca Sorrentino [1], Raffaele Allevi [1], Renzo Vanna [2], Carlo Morasso [2] and Fabio Corsi [1,2,3,*]

[1] Department of Biomedical and Clinical Sciences "L. Sacco", Università degli studi di Milano, via G. B. Grassi 74, 20157 Milano, Italy; marta.truffi@unimi.it (M.T.); serena.mazzucchelli@unimi.it (S.M.); arianna.bonizzi@unimi.it (A.B.); luca.sorrentino1@unimi.it (L.S.); raffaele.allevi@unimi.it (R.A.)

[2] Nanomedicine and Molecular Imaging Lab, Istituti Clinici Scientifici Maugeri IRCCS, via Maugeri 4, 27100 Pavia, Italy; renzo.vanna@icsmaugeri.it (R.V.); carlo.morasso@icsmaugeri.it (C.M.)

[3] Breast Unit, Surgery Department, Istituti Clinici Scientifici Maugeri IRCCS, via Maugeri 4, 27100 Pavia, Italy

* Correspondence: fabio.corsi@unimi.it; Tel.: +39-0250319850

Received: 4 February 2019; Accepted: 8 March 2019; Published: 13 March 2019

Abstract: Cancer-associated fibroblasts (CAF) are the most abundant cells of the tumor stroma and they critically influence cancer growth through control of the surrounding tumor microenvironment (TME). CAF-orchestrated reactive stroma, composed of pro-tumorigenic cytokines and growth factors, matrix components, neovessels, and deregulated immune cells, is associated with poor prognosis in multiple carcinomas, including breast cancer. Therefore, beyond cancer cells killing, researchers are currently focusing on TME as strategy to fight breast cancer. In recent years, nanomedicine has provided a number of smart delivery systems based on active targeting of breast CAF and immune-mediated overcome of chemoresistance. Many efforts have been made both to eradicate breast CAF and to reshape their identity and function. Nano-strategies for CAF targeting profoundly contribute to enhance chemosensitivity of breast tumors, enabling access of cytotoxic T-cells and reducing immunosuppressive signals. TME rearrangement also includes reorganization of the extracellular matrix to enhance permeability to chemotherapeutics, and nano-systems for smart coupling of chemo- and immune-therapy, by increasing immunogenicity and stimulating antitumor immunity. The present paper reviews the current state-of-the-art on nano-strategies to target breast CAF and TME. Finally, we consider and discuss future translational perspectives of proposed nano-strategies for clinical application in breast cancer.

Keywords: cancer-associated fibroblasts; tumor microenvironment; nanoparticles; breast cancer; antitumor efficacy

1. Introduction

In recent years, the focus of cancer research has shifted from cancer to cancer-related stroma [1,2]. Indeed, tumor microenvironment (TME) plays a key role in several processes related to cancer progression, including the acquisition of an invasive phenotype, cell migration capability, chemoresistance, protection from antitumor immune response and neoangiogenesis [3]. The crosstalk between cancer and TME has such a pivotal relevance, that nowadays the epithelial-to-mesenchymal transition (EMT) is recognized as the central moment of cancer progression from in situ to invasive disease [4,5]. EMT refers to the milieu of biological processes by which cancer cells gradually lose their epithelial hallmarks and acquire mesenchymal properties related to invasion of surrounding tissues and remodeling of the extracellular matrix (ECM) [6]. The final result of EMT is the capability of cancer cells to metastasize in distant sites where, again, proliferation of cancer cells up to clinically detectable metastases depends on the organ microenvironment—the "seed and soil" hypothesis [7]. Thus, cancer stroma not only support, but are rather protagonists of cancer progression (Figure 1). Therefore, it is not surprising that currently an increasing proportion of research is focusing on development of anti-cancer strategies targeted toward TME [8]. Moreover, another more pragmatic reason is exciting the interest on anti-stromal therapies: TME is much more genetically stable than cancer [9,10]. A paradigmatic example is breast cancer, which is genetically and phenotypically heterogeneous [11]. Therefore, targeting breast cancer cells or pathways is rather difficult, considering that they change continuously between patients and in the same patient. A double-acting strategy targeting both breast cancer cells and TME might be a key for success in the treatment of breast cancer, but innovative drug delivery systems are needed. Nanomedicine could respond to this clinical unmet need, providing smart delivery systems based on active targeting and internalization both in cancer and in TME cells [12,13]. Due to their size and versatility, nanoparticles have attracted interest in the field of anticancer medicine, and they have shown ability to deliver consistent amounts of drugs and control their release at the tumor site [14–18]. Interesting feature of nanoparticles and nano-systems is the possibility to chemically or genetically modify their surface with a variety of targeting moieties or active ligands in order to trigger specific direction and recognition of the biological target [19–21]. The aim of the present paper is to review the current state-of-the-art and future translational perspectives on nano-strategies to target breast cancer microenvironment.

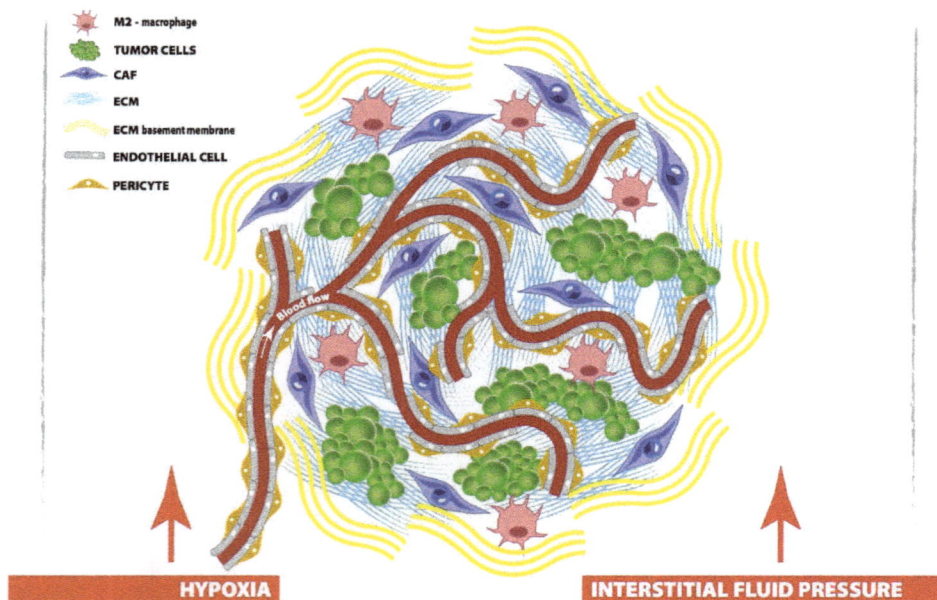

Figure 1. In desmoplastic cancer, tumor cells share their niche with CAF, macrophages, blood vessels and perivascular cells, all of which contribute to arrange the TME. First, CAF actively remodel cellular and matrix components of the TME; M2-polarized macrophages promote metastases and generate an immunosuppressive habit; endothelial cells and pericytes favor tumor angiogenesis and handle oxygen supply. Finally, in the niche, interstitial ECM supports tumor architecture and regulates drug penetration. Surrounding the tumor, ECM basement membrane acts as a barrier toward migration. Both ECM and blood vessels contribute to increase interstitial fluid pressure and tumor hypoxia, which are big obstacles to tumor treatment.

2. Biological Hallmarks to Target Breast CAF

Cancer-associated fibroblasts (CAF) are the most abundant cells of the tumor stroma and they critically influence cancer growth and progression through control of the surrounding TME (Figure 2). CAF produce and secrete a variety of growth factors and cytokines, including transforming growth factor β (TGFβ), vascular endothelial growth factor (VEGF), platelet-derived growth factor (PDGF), interleukins, as well as ECM components, particularly fibrillar collagens and fibronectins, and metalloproteinases (MMP) which support tumor growth, generate a physical barrier against drugs and immune infiltration and facilitate cancer invasion [1,22–25]. This contributes to generate the so-called reactive stroma and to induce a desmoplastic reaction in TME, which has been associated with poor prognosis in multiple carcinomas, including breast cancer [26–28]. Therefore, it is increasingly evident that effective anticancer therapies should tackle not only cancer cells but even such a tumor fortress composed by TME and orchestrated by CAF.

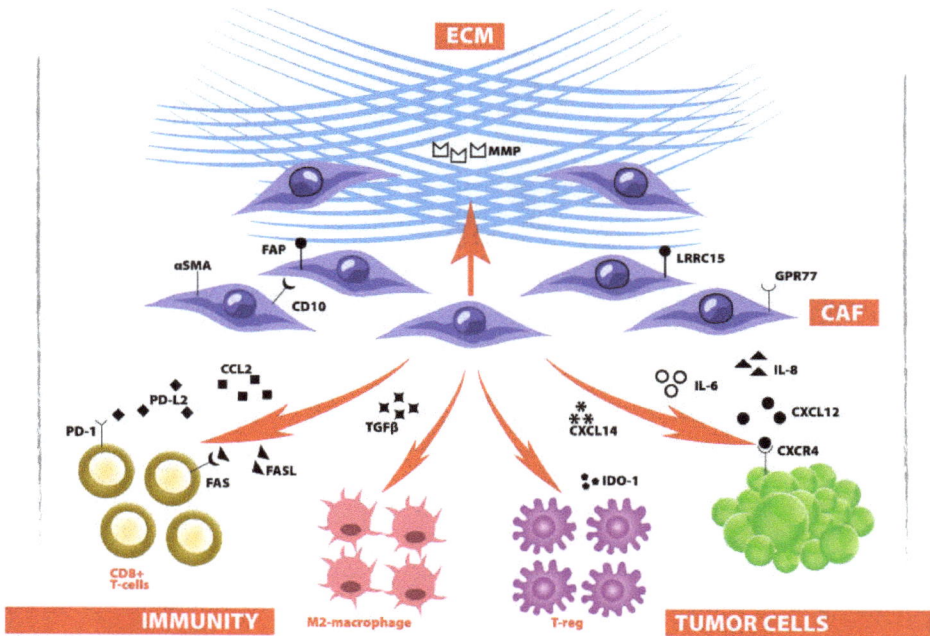

Figure 2. CAF and their essential role in the TME: CAF secrete cytokines to promote cancer growth and shape antitumor immunity; CAF control ECM composition to avoid drug penetration, favor tumor migration and mediate exclusion of antitumor immunity from the tumor.

Previous studies have identified some phenotypical markers to detect and target pro-tumorigenic CAF in breast cancer. Among them, we can find markers of mesenchymal origin, such as αSMA or cell surface proteins associated with crucial biological functions of CAF, like fibroblast activation protein (FAP), which enzymatically remodel ECM and induce cancer cells migration [29], CD10, a zinc-dependent metalloproteinase [30], or G protein-coupled receptor 77 (GPR77), which activates VEGF expression and angiogenesis in hypoxic breast TME [31]. Leucine-rich repeat containing 15 (LRRC15) membrane protein was also found highly expressed on CAF in many solid tumors [32]. Exploitation of these markers to drive localization of novel therapeutics would be useful to eradicate CAF, and in some cases even to reprogram their biological functions.

Additionally, CAF express large amounts of FAS ligand (FASL), which induces apoptosis of FAS-expressing CD8+ T cells, and programmed cell death 1 ligand 2 (PD-L2), which induces T cell anergy by interacting with the immune checkpoint molecule PD-1, thus avoiding anticancer immunity in the host [33]. CAF also secrete a lot of chemokine ligands (CXCL12/SDF1, CXCL14, CCL2, and others) that promote the proliferation of cancer cells and encourage the recruitment of tumor-associated macrophages, thereby contributing to immunosuppression [34–37]. In particular, the SDF1–CXCR4 interaction, together with the heat shock factor 1 (HSF1), generate an autocrine loop in CAF that drives the transcription of pro-tumorigenic cytokines and growth factors and supports rapid tumor growth [38,39]. Crucial signaling behind stromal communication with breast cancer cells is represented by the transcription factor STAT1, which is able to enlarge tumorigenicity and chemoresistance. Ablation of STAT1 in CAF may decrease cancer cell proliferation and reduce α-SMA+ reactive fibroblasts and ductal carcinoma in situ (DCIS)-like lesions in a mouse model of early breast cancer progression [38–40].

In recent years, some biological hallmarks of CAF have been exploited to design and study novel therapeutics and nano-therapeutics to remodel the TME and enhance the therapeutic activity of chemotherapy [8,41]. Some others have revealed insufficient specificity for CAF targeting or still require further preclinical research to achieve accessibility and exploitation for therapeutic purposes. Here, we resume and list some nano-systems that have been successfully explored in preclinical setting to target breast CAF and rearrange the breast TME (Table 1).

Table 1. Nano-strategies to target CAF and remodel the TME in preclinical models of breast cancer.

Nanoparticle	Payload	Effect on	Breast Cancer Model	Antitumor Effect	Ref.
Z@FRT-scFv	ZnF$_{16}$Pc	FAP + CAF	Orthotopic 4T1 tumor	CAF eradication by PDT	[42,43]
CAP-NP	PTX (or other hydrophobic drugs)	FAP + CAF and surrounding cells	MCF-7 xenograft	Enhanced local drug accumulation	[44]
C$_{16}$-N/losartan hydrogel	Losartan	Angiotensin	E0771 and 4T1 mouse models	CAF inhibition + ECM remodeling	[45,46]
DOX-HPEG-PH20-NP	rHuPH20 + DOX	Hyaluronic acid + cancer cells	4T1 syngeneic breast tumor	ECM remodeling + chemotherapy	[47]
PTX-SNPs	Sulfatide + PTX	Tenascin-C + cancer cells	Murine breast cancer EMT6	Enhanced chemotherapy	[48]
1-NP	Laminin-mimic peptide 1	TME	MDA-MB-231 tumor model	Reduced metastases by artificial ECM formation	[49]
HNP liposomes	Marimastat + HA-PTX prodrug	MMP + CD44+ cancer cells	orthotopic 4T1 tumor	ECM remodeling + chemotherapy	[50]
BCPN	OXA prodrug + NLG919	IDO-1 + cancer cells	orthotopic 4T1 tumor	Immunotherapy + chemotherapy	[51]
DOX/IND-liposome	Indoximod + DOX	IDO-1 + cancer cells	orthotopic 4T1 tumor model	Immunotherapy + chemotherapy	[52]
HA-DOX/PHIS/R848	Resiquimod + DOX	DC + CD44+ cancer cells	4T1 tumor-bearing mice	Immunotherapy + chemotherapy	[53]
PLGA NP	Tumor antigens	DC	Tumor and blood samples from breast cancer patients	Immune-stimulation	[54]
LPD	Plasmid encoding IL-10 trap	TME	Orthotopic 4T1 triple-negative model	Immunotherapy	[55]

3. Active Nano-Systems for Breast CAF Disruption and Regulation

CAF are key actors in the restricted penetration of drug and nanodrug in the tumor tissue [56]. Indeed, CAF contribute to the biosynthesis and remodeling of the ECM and to the high tumor interstitial fluid pressure [57]. Therefore, the development of treatment's strategy able to eradicate CAF could result in reduced collagen content in the ECM, leading to improved drug and nanodrug accumulation and diffusion. Many efforts have been made to the exploitation of CAF as a potential target for cancer therapy [8]. In particular, they have been involved the fibroblast activation protein (FAP), which is an integral membrane serine protease of the dipeptidyl peptidase subfamily selectively expressed by CAF. FAP is undetectable in the stroma of normal tissue [29], suggesting that the exploitation of FAP as a selective target of CAF could lead to specific and active delivery of cytotoxic drugs into these cells. Currently, three different nano-approaches have been investigated with the aim to specifically target CAF and improve therapeutic efficacy of anticancer treatments. They could be attributed to three different subcategories that will be described in detail in the following paragraphs.

3.1. Nanoparticles for Photodynamic Therapy

In recent years, a novel antitumor approach called photodynamic therapy (PDT) has been exploited for cancer treatment with the aim to enhance nanoparticle's tumor uptake and improve their therapeutic efficacy. This strategy consists in exposing the cells to non-toxic dose of light in presence of light-sensitive molecules, known as photosensitizer (PS). Conventional PDT uses non-targeting photosensitizer molecules and inflicts direct damage on tumor cells mainly generating reactive oxygen species (ROS) or acts indirectly by disrupting the vasculature via endothelial cells damage [58].

In this scenario, a very exciting idea is the possibility to perform a CAF-targeted PDT to modulate the microenvironment and promote the anticancer therapy avoiding systemic toxicities. Indeed, it is increasingly evident the crucial role of CAF in cancer progression due to their capability to isolate cancer cells from drugs and T-cytotoxic cell [59].

In this research area the most important studies have been performed by Zhen group, which exploited ferritin nanocages to deliver a photosensitizer to CAF in context of breast cancer [42,43]. They achieved CAF-targeted delivery of the photosensitizer thanks to the surface conjugation of an anti-FAP single chain variable fragment (scFv) antibody. Moreover, the combination of FAP-targeted delivery of the photosensitizer with a localized photoirradiation of the tumor, allowed the selective eradication of CAF. This treatment destroyed ECM and suppressed $C-X-C$ motif chemokine ligand 12 (CXCL12) secretion by CAF, resulting in the significant improvement of CD8+ T-cell infiltration. This study suggests a novel approach to modulate TME through selective killing of CAF [42]. In another study, Zhen and colleagues combined the CAF-targeted PDT treatment with the administration of quantum dots (QDs), demonstrating that PDT treatment could be useful to improve QDs penetration. Indeed, this approach allowed CAF eradication in irradiated tumors and, in the meantime, promoted increase in nanoparticle's tumor uptake, due to reduced amount of secreted collagen in the surrounding ECM [43]. Results from this study points out a potential role for CAF eradication in improving tumor drug penetration and documents a promising effective combination with chemotherapy. Finally, CAF-targeted PDT represents a great advance to modulate TME for optimal breast cancer management.

3.2. Nanoparticles for Cytotoxic Delivery

Another approach to disrupt CAF is represented by the exploitation of CAF destroying nanoparticles. They have been designed to deliver chemotherapeutics or drugs with specific cytotoxicity against CAF. Ji's group proposed a double acting nanomaterial, which exploits the CAF targeting capability and an efficient cell penetration to improve chemotherapeutic drug delivery for the treatment of human prostate cancer model [60]. The researchers designed and synthesized a novel cell-penetrating peptide (CPP) based on an amphiphilic peptide (C2KKG2R9) linked by the hydrophobic tail to a cholesterol molecule monomer. Monomers of this CPP linked to a cholesterol molecule self-assembled into a core-shell structured peptide nanoparticles (PNP). PNP was then loaded with the antitumor drug doxorubicin (DOX) and the surface of the resulting DOX-loaded PNP (PNP-D) was modified with an anti-FAP monoclonal antibody to specifically recognize CAF. The anti-FAP antibody displayed CAF specificity, while the presence of CPP/cholesterol and DOX enhanced drug penetration and cytotoxicity, respectively [60].

In order to target CAF, another approach exploited a novel cleavable amphiphilic peptide (CAP), which is specifically responsive to FAPα, expressed on CAF surface [44]. In aqueous solution, CAP monomers readily self-assembled into nanofibers, which could be loaded with hydrophobic chemotherapeutic drugs, to obtain drug-loaded spherical nanoparticles (CAP-NP). Once they reached the tumor stroma, CAP-NP rapidly disassembled upon cleavage by FAPα and released the drug. This strategy was effective in the treatment of prostate, breast and pancreatic tumor models containing or recruiting FAPα+ CAF; it disrupted the stromal barrier and enhanced local drug accumulation. In case of breast cancer model, it has to be noted that MCF-7 breast tumor cell line was FAPα−negative but ultimately formed tumors that exhibited positive FAPα expression. Despite effective, the proposed approach may be possible with hydrophobic drugs only, as demonstrated in case of treatment of breast cancer-bearing mice with CAP-NP loaded with paclitaxel [44].

Recently, a peptide derivative nanofiber (C16-GNNQQNYKD-OH) has been designed and synthesized. It is able to self-assemble into long filaments entrapping losartan molecules inside a hydrogel exploitable for localized drug delivery [45]. Following a topical intratumor injection in mice model of aggressive triple negative breast cancer, this hydrogel induced inhibition of collagen I synthesis by CAF. Moreover, its combination with chemotherapeutic drug resulted in the amplification of the therapeutic efficacy [45].

Otherwise, in a work published by Chen and colleagues, nanoliposomes loaded with Navitoclax have been exploited for the target delivery of a small molecule inhibitor in hepatocellular carcinoma. CAF targeting was achieved by functionalization of the liposome with FH peptide (FH-SSLNav), which displayed an extremely high affinity to tenascin-C [61]. Tenascin-C is a tumor-specific extracellular matrix highly expressed in most solid tumors and mainly secreted by CAF. Tumor exhibited a high-level expression of tenascin-C, which was found to colocalize with CAF. Once in CAF, Navitoclax was released by nanoliposomes in slow rate and inhibited the Bcl-2 proteins, thus inducing apoptosis in CAF at very low dosage. FH-SSLNav revealed efficacy in CAF-induced death, while being much less effective in tumor cells or healthy tissues. Moreover, the treatment led to improved nanoparticles penetration in solid tumor due to modulation of the TME [62].

3.3. Regulation of CAF Function

The third approach to modulate TME through direct CAF targeting deals with regulation of CAF function. Indeed, beyond complete ablation of CAFs, some preclinical research has attempted the metabolic targeting of tumor stroma, with the aim to reprogram CAF identity and enhance therapeutic performance of currently available chemotherapeutics.

Major studies for regulation of CAF function have been pursued by Huang group in desmoplastic tumors (i.e., bladder and pancreatic carcinomas) and they assumed to be effective also in other cancer subtypes (i.e., breast cancer) as long as they are desmoplastic [63–65]. The first strategy exploits the suppression of Wnt16 in CAF. Indeed, Wnt16 is one of the major mitogenic growth factors and its downregulation improves the antitumor effect of cisplatin in resistant cancers. By developing a Lipid-calcium phosphate (LCP) nanoparticle loaded with cisplatin, the authors observed that the off-target exposure of these nanoparticles induced Wnt16 secretion by CAF. This resulted in stroma reconstitution and onset of cisplatin-resistance, which was reverted by the administration of LCP NP loaded with the anti-Wnt16 siRNA [63]. Wnt16 downregulation was also obtained with LCP NP loaded with the prodrug quercetin, demonstrating its capability to downregulate Wnt16 levels, reduce the number of CAF, normalize the collagen content in tumor tissue and improve nanoparticles delivery into the tumor [64]. Another cunning approach exploits off-targeting capability of LCP nanoparticle coated with Protamine (LPD) to mediate the transfection of CAF with a cytotoxic protein. A plasmid DNA encoding the secretable TNF-related apoptosis-induced ligand (sTRAIL) was formulated in LCP NP and intravenously administered in mice bearing a desmoplastic tumor. As a result of off-target distribution, LCP NP accumulated in CAFs and induced production of sTRAIL. The produced protein was secreted and induced apoptosis in surrounding cells. Moreover, the production of sTRAIL induced also a change in CAF activation, reverting them to a quiescent state, which resulted in inhibition of tumor growth [65]. Despite the low specificity of this delivery strategy, no significant side-effects have been reported, since noticeable morphological changes were not detected in organs where LCP-sTRAIL were distributed. Indeed, a reason can be attributed to the low level of sTRAIL plasmid expression in these organs [65].

Interesting crucial pathway to be targeted when one aims at regulating CAF function is represented by Sonic Hedgehog. A strategy to reduce proliferation and number of CAF was studied in a xenograft mouse model of breast cancer, by using a Sonic Hedgehog inhibitor, vismodegib. Vismogedib targeted CAF with the aim to improve drug delivery. Results from this study suggested that CAF depletion was effective in remodelling the TME, thus promoting fluid stress alleviation. Moreover, the combination of vismodegib with two clinically approved nanoparticles, such as Abraxane and Doxil, improved the treatment efficacy and overall survival in selected patients [66].

Superparamagnetic iron oxide nanoparticles (SPION) were tested to modulate differentiation of human pancreatic stellate cells (hPSCs) into CAF-like myofibroblasts. Indeed, Relaxin-2 is a hormone belonging to the insulin superfamily, able to resolve fibrosis by stimulating MMPs production and inhibiting Smad2/3 phosphorilation, through its binding with the Relaxin Family Peptide Receptor type 1. However, this protein displays small size, short circulation half-life and undesired systemic vasodilatation, which limit its direct application. To solve this issue, the endogenous hormone relaxin-2 was conjugated to SPION, displaying inhibition of Smad2 signaling pathway and blocking hPSCs differentiation into CAF. The inhibition of hPSCs differentiation reduced cytoskeleton marker expression, ECM production and TGF-β-induced contractility of hPCSs. In addition, when combined with gemcitabine, the treatment enhanced the therapeutic efficacy of chemotherapy, resulting in a significant blockade of tumor growth [67].

4. Nano-Strategies to Modulate CAF-Instructed Tumor Microenvironment

4.1. Remodeling the Extracellular Environment

The ability to control local remodelling of ECM is a critical function of CAF and a feature of paramount importance during the desmoplastic reaction occurring in many breast carcinomas. CAF synthesize and secrete ECM components, cytokines and growth factors that create a favorable support for tumor progression and a physical barrier to several drugs [28]. Dynamic stromal alterations may induce tissue stiffening and increased tension, which have been associated with poor outcome in patients with solid tumors, including breast cancer [26]. Moreover, collagen-rich stroma may induce EMT and promote migration and invasion of breast tumor cells, thus supporting a crucial relevance for tumor environment when approaching novel therapeutic strategies [68]. To face this issue, some preclinical research has counteracted the excessive production of stromal collagen and hyaluronan either through regulation of their secretion by CAF or enzymatically disrupting ECM.

Losartan, an angiotensin inhibitor, was assembled with C_{16}-N peptide hydrogels and injected in a murine model of triple negative breast cancer, where it inhibited collagen I synthesis by CAF [46]. While acting as a local and sustainable depot of drug, the formulation significantly improved the intratumoral accumulation and penetration of PEGylated doxorubicin-loaded liposomes, which were administered as combine therapy. Through physical action on CAF-controlled ECM, C_{16}-N/losartan improved drug delivery and vascular perfusion in breast tumor, thereby reducing growth of primary tumor and lung metastasis, as compared to single chemotherapy. In another study, recombinant human hyaluronidase, an enzymatic agent degrading hyaluronic acid, was conjugated to the surface of PLGA NP and pegylated to avoid rapid clearance [47]. Due to ECM degradation, rHuPH20-conjugation improved the intratumoral accumulation of the nanoparticles in 4T1 syngeneic mouse breast tumors by fourfold, as compared to bare nanoparticles. When loaded with doxorubicin, the hyaluronidase-nanoconjugate was able to reduce breast tumor growth in vitro and in vivo, by showing improved efficacy at low dosage as compared to unconjugated nanodrug [47]. ECM-reducing treatments may also increase blood vessels density and revert tumor hypoxia, thus normalizing tumor vasculature and enhancing nanodrug perfusion through EPR [46]. With this perspective, Gong et al. showed enhanced tumor permeation of nanomicelles for PDT when coupled with hyaluronidase for treatment of 4T1-bearing Balb/c mice. The anti-ECM/PDT combined therapy induced increased tumor uptake of the nanomedicine and amplified antitumor efficacy in both breast primary tumor and metastatic lymph nodes [69]. Promising targets for exploitation in nano-mediated tackling of the TME also include tenascin-C, an ECM glycoprotein mainly produced by CAF and highly expressed in breast tumors. The tenascin-C ligand sulfatide was used as targeting moiety for lipid perfluorooctylbromide nanoparticles in order to produce a breast cancer delivery nanovehicle for paclitaxel. Nanoparticles administration in EMT6 mouse model achieved increased accumulation of paclitaxel in breast cancer tissue and remarkable tumor inhibition as compared to free drug or untargeted nanoparticles [48].

If ECM disruption improves anticancer drug penetration, reduces hypoxia and overcomes chemoresistance, on the other hand reduced tumor ECM may weaken the physical barrier that confine the tumor at its site of onset, thus allowing formation of neovessels and favoring cancer cells migration and metastasis [70]. For this reason, some research has been devoted to production of artificial ECM. Inspired by the self-assembled formation of natural ECM, Hu et al. developed nanofibers-transforming nanoparticles that mimicked laminin, a component of ECM [49]. Building block for the artificial ECM was laminin-mimic peptide 1 made from (i) hydrophobic bis-pyrene unit for nanoparticle formation, (ii) peptide scaffold for fibers structure, (iii) specific targeting peptide sequence (i.e., RGD/YIGSR) for binding to cancer cells, as well as natural laminin does. Intravenously injected nanoparticles (1-NP) in MDA-MB-231-bearing mice accumulated at the tumor site through combined EPR effect and active targeting. Upon binding to the receptor, 1-NP transformed into nanofibers of artificial ECM that surrounded the tumor mass for over 72 hours post-injection. 1-NP efficiently restrained migration of MDA-MB-231 cells in vitro and induced remarkable inhibition of lung metastatic rate in murine model of highly metastatic breast cancer [49]. In order to block the metastatic spread of cancer cells, mimics of metalloproteinase (MMP) substrates have also been used to counteract the activity of MMPs highly expressed in a variety of tumors, including breast cancer [71]. An example is represented by marimastat (MATT), a broad-spectrum synthetic enzyme inhibitor that achieves great inhibition of collagenases, gelatinases and MMPs even at a nanomolar concentration [72]. (MATT)-loaded thermosensitive liposomes (LTSLs) were assembled with hyaluronic acid-paclitaxel (HA-PTX) prodrug to achieve dual targeting of extracellular TME and cancer cells [50]. Hybrid nanoparticles (HNPs) released their payload upon mild hyperthermia in the TME, where HA-PTX could enter and kill cancer cells, while MATT inhibited MMP activity, reduced activation of CAF and slowed down cancer cells migration. As a result, treatment of 4T1-tumor-bearing mice with HNPs reduced tumor volume, inhibited angiogenesis and decreased lung metastases.

4.2. Orchestrate an Anti-Breast Cancer Immune Response

CAF contribute to shape the immune cells in tumors by secreting proinflammatory cytokines and chemokines, notably TGFβ, IL-6 and CCL2, to recruit immunosuppressive cells into the tumor stroma and reject effector T cells [73,74]. The immunosuppressive TME drastically limits the promises of effective immunotherapeutics and checkpoint inhibitors, which have risen new hope for the treatment of several malignant tumors [75]. Overcoming immune suppression in the tumor is of fundamental importance for effective cancer treatment. As already mentioned, therapeutic nano-strategies aimed at targeting and ablating stromal CAF provide great benefit for tumor eradication, by both enabling access of cytotoxic T-cells and by reducing immunosuppressive signals in the TME (see Section 3). However, CAF eradication may not be enough, as pro-tumorigenic cytokines and chemokines are produced by other cell types beyond CAF, here including tumor cells, pericytes, endothelial cells, and adipocytes, which could somehow compensate for CAF absence [76]. The common features of some components of the TME may suggest that targeting mediators of the intercellular communication among CAF, tumor cells, vascular endothelial cells, neutrophils, dendritic cells, T-cells, and macrophages would lead to successful anticancer applications and complement other treatment options. To this purpose, some preclinical studies have proposed to directly modulate the immunosuppressive factors in the TME by nano-mediated delivery of immunotherapy.

As an example, Feng et al. have generated a dual-activatable binary cooperative prodrug nanoparticle, termed BCPN, which contains self-assembled PEG-grafted oxaliplatin prodrug and a disulphide bond-cross-linked homodimer of NLG919 [51]. Such a nanoparticle was proposed as a platform for codelivery of anticancer drug (OXA) and potent inhibitor of IDO-1 (NLG919), an enzyme implicated in tumor immunosuppression through its ability to limit T-cell function and engage mechanisms of immune tolerance [77]. Due to high sensitivity to the acidic pH of the TME, BCPN showed a negative to positive surface charge reversion for improved tumor penetration. The OXA prodrug and NLG919 dimer were both released following the intracellular reductive microenvironment. OXA induced adaptive antitumor immunogenicity by triggering immunogenic cell death of the tumor cell, while NLG919 inactivated IDO-1, thereby inhibiting intratumoral infiltration of T-regs. Once administered to 4T1-breast-tumor-bearing mice, BCPN promoted sustained and enhanced antitumor immune response as compared to combine treatment with free OXA and NLG919, thus resulting in long-term tumor regression [51]. The same strategy was adopted by Lu et al., who designed a dual-delivery liposome for anticancer drug and IDO-1 inhibitor [52]. They constructed a phospholipid-conjugated indoximod (IDO-1 inhibitor, IND) prodrug that self-assembled into a lipid bilayer liposome, and performed doxorubicin loading as a second step. NPs injected in an orthotopic 4T1 tumor model dramatically improved the pharmacokinetics and tumor concentration of the two drugs induced immunogenic cell death as a result of DOX activity on cancer cells and recruitment of CD8+ cytotoxic T lymphocytes, with disappearance of T-regs, thus showing effective immune response. The authors further combined the DOX/IND-liposome treatment with an anti-PD-1 antibody and showed synergistic efficacy with immune checkpoint blocking agents. Additional example of combined nano-immunotherapy and -chemotherapy is represented by dual pH-responsive multifunctional nanoparticle for co-loading of the Toll-like receptor agonist resiquimod (R848) and the chemotherapeutic drug doxorubicin (HA-DOX/PHIS/R848) [53]. Tumor active targeting was achieved by doxorubicin conjugation to hyaluronic acid, a ligand for breast cancer cells expressing CD44. R848 is a Toll-like receptor agonist that promote maturation of DCs. The pH-dependent hydrophobic/hydrophilic transformation of NP triggered the disintegration of the core, with release of R848 to exert stimulation and maturation of dendritic cells, while DOX was specifically internalized by cancer cells by receptor-mediated endocytosis. Immunoregulation of DCs activity of NP was assessed in vitro as production of type I interferon and proinflammatory cytokines. Then, intravenous injection of HA-DOX/PHIS/R848 NPs in a murine model revealed significant inhibition of tumor growth compared to both free DOX and free R848, thus demonstrating the synergistic effects of the combined nano-drugs on breast cancer [53].

An interesting study on lung carcinoma documents the exploitation and modulation of neutrophils infiltration to mediate the transport of NPs across the tumor vessel barrier [78]. Gold nanorods were linked to anti-CD11b antibodies (GNRs-CD11b), which target activated neutrophils in the peripheral blood. After photosensitization, acute inflammation can be induced at the tumor site, thus promoting tumor infiltration by neutrophils. The neutrophils, loaded with NPs, infiltrated the tumor, significantly increased NP accumulation and enhanced efficacy of photothermal therapy.

Another strategy could be the adoption of nanoparticles as vehicle for tumor-associated antigens to stimulate antitumor T-cell-mediated immune response. Indeed, PLGA NP were loaded with tumor lysate obtained from fresh breast tumor resections and used to stimulate dendritic cells isolated from human peripheral blood mononuclear cells [54]. Tumor lysate-loaded NPs triggered a more efficient maturation of monocyte-derived DC compared to either tumor lysate or NPs alone, as assessed by immunophenotyping and cytokine release. They also showed a capacity to stimulate naive autologous T helper cells through matured DC.

Quite recently, Shen et al. produced a nano-delivery system for a gene encoding an antibody-like protein to trap IL-10 in the TME [55]. Indeed, high expression of IL-10 was associated with poor survival in pancreatic and triple-negative breast cancer patients [79]. Therefore, transient expression of IL-10 trap through NP intratumoral delivery could change cytokines and tumor-infiltrating lymphocytes within the TME, resulting in significant antitumor efficacy. The plasmid encoding IL-10 trap was encapsulated into liposome-protamine-DNA (LPD) and administered to 4T1-bearing mice. IL-10 trap alone significantly reduced tumor growth and enhanced median survival over one month, indicating efficacy in immunosuppressive triple negative breast cancer [55].

5. Prospective Advancement for Clinical Translation

TME plays a key role in most of tumour progression processes, including proliferation, invasion, and neoangiogenesis [3]. Stroma is not significant only from a biological point of view, but it has also strong clinical implications. Indeed, CAF can affect chemosensitivity, protecting cancer cells from cytotoxic drugs with obvious limitations in anticancer efficacy, particularly in breast cancer subtypes in which an intense cross-talk between tumor cells and microenvironment is present, such as basal-like malignancies [80]. An unsolved issue in clinical management of breast cancer is the possibility to enhance antitumor immunity against cancer cells, and immune-modulating treatments has recently gained a strong relevance both in adjuvant and in neoadjuvant settings [81,82]. However, CAF are also capable to protect malignant cells from T-cells antitumor response, thus probably limiting the great potential of such immune-therapies, as well as antitumor vaccination strategies [33,83,84]. If we consider such relevance of CAF and the great promise of CAF-targeted strategies, it is surprising that stroma-targeted therapy has been poorly explored in clinical trials. Two main reasons explain this lacking. First, it is not clear how to specifically target CAF. FAP has been considered a promising target for CAF, being expressed in over 90% of these stromal cells [85]. However, on the other hand, FAP is also over-expressed in multipotent bone marrow stem cells, thus explaining the overwhelming myelotoxicity observed when this strategy was assessed in pilot clinical studies [86,87]. Preliminary clinical trials have substantially demonstrated no efficacy of anti-FAP strategy, both by direct targeting using the monoclonal antibody sibrotuzumab in lung and colorectal cancer [88,89], and by inhibiting the enzymatic function of FAP using talabostat [90,91]. A fortunate pre-clinical exception is the use of a DNA-based FAP vaccine, which showed an excellent anti-stromal CD8-mediated immunity, with subsequent restoring of chemosensitivity in a mouse model of breast cancer [92]. However, significant myelotoxicity due to over-expression of FAP in the bone marrow has currently limited the clinical exploitation of FAP for CAF targeting. Two surface markers specifically linked to pro-tumorigenic CAF have been identified: CD10 and GPR77 [93,94]. These targets promise a higher specificity for CAF, and a monoclonal antibody toward GPR77 has demonstrated enhanced chemosensitivity and reduction of tumor stem cells in a patient-derived model of breast cancer [30]. Secondly, it is not clear which CAF have to be depleted. A tout-court depletion of CAF, indeed, might paradoxically promote cancer progression, since CAF within certain subsets or within certain stages of tumorigenesis express an anti-cancer profile [95]. Recently, it has been suggested that quiescent fibroblasts, upon development of cancer, might differentiate into cancer-restraining (F1 subtype) and cancer-promoting (F2 subtype) CAF [22]. Furthermore, "secretory CAF" have been identified as a third possible differentiation, leading itself to cancer survival vs. cancer apoptosis pathways, based on the secretome context. For example, some F3 CAF produce ECM-degrading proteases which facilitate motility and invasion of cancer cells [96]. However, on the other hand, it has been observed that CAF, upon secretion of TGFβ, inhibit cancer development in early stages but the same pathway may stimulate cancer in advanced stages [97]. Furthermore, tumor-restraining CAF may increase antitumor immunity, by secretion of immunomodulatory cytokines such as IL-10, TNF, and IL-6, aiding in recruitment of macrophages and T lymphocytes, thus converting an immune-suppressive into immune-stimulating cancer microenvironment [98]. Recently, novel approaches including single-cell RNA sequencing have revealed that distinct subtypes of CAF co-exist within TME, with different phenotypes and functions [99]. Thus, the concept that cancer-related stroma is not heterogeneous like cancer cells has been profoundly revised, and precision medicine is required also in CAF-targeted anticancer strategies.

Turning CAF "from foes to friends" may, therefore, establish as the most appropriate treatment to target the stroma, since possible pitfalls of complete depletion of CAF [8], which may limit clinical translation of this strategy, might be avoided. Under this innovative perspective, nanomedicine represents an added value for CAF-targeted anticancer treatments. Indeed, nanoparticles are optimal drug delivery systems since they present several advantages in cancer therapy. First, they allow their cargo to follow the so-called EPR effect, by which molecules of certain sizes tend to accumulate in tumor tissue much more than they do in normal tissues. This property not only leads to increased drug accumulation in the tumor mass, but also avoids off targeting and associated side effects, which could rise hope on reduced myelotoxicity when targeting FAP by appropriate nano-vehicles. Second, nanoparticles can be loaded with specific drugs focused on particular cell pathways and possibly acting on different subcellular districts. Precision medicine should rely not only on accurate targets, such as in the case of targeted monoclonal antibodies which can exert their anti-stromal activity by immune-mediated CAF depletion. Additionally, accurate effects are needed, for example to convert a pro-tumorigenic into an anti-cancer CAF profile by acting on CAF-mediated signaling, such as the JAK1-STAT3 pathway [100], or by genetically modifying CAF themselves to make them quiescent or even to allow in situ production of pro-apoptotic mediators, as recently proposed [65]. An accurate targeting might be ensured by novel nano-drugs, but the question is which target should be used for properly addressing such strategies in activated CAF and avoiding cytotoxic effects on normal/quiescent or F1 fibroblasts. As previously stated, FAP showed encouraging results, but its over-expression in bone marrow may also represent a relevant drawback. CD10 and GPR77 could be more specific and accurate as hallmarks of activated cancer-promoting CAF, and further research should address their usefulness as targets for nano-delivery of cytotoxic drugs or, possibly, of CAF-reprogramming strategies. Another aspect to be considered is the availability of concurrent drugs delivery and multiple contemporary anticancer effects thanks to nano-therapy, making possible to act both on cancer cells and on TME at the same time, thus maximizing the anticancer effects and reducing chemoresistance. Notably, a double-threaded link is present between nanomedicine and TME. On one hand, nano-delivery may overcome current clinical limitations and acts on stromal cells to enhance chemosensitivity and immune response. However, on the other hand, TME strongly affects the capability of nanoparticles to reach the tumor sites, thus impacting on anticancer efficacy of delivered drugs, particularly in the complex stroma of breast cancer [101]. A combinatorial therapy may therefore act on CAF and other stromal compounds, allowing cytotoxic drugs specifically delivered by nanoparticles to enhance their effect.

Several potentialities are promised by nanomedicine for specific and multi-acting cancer treatments, but the pivotal issue is whether nanomedicine is clinically transferable in the next future. A first great limitation is the lack of suitable preclinical models which accurately resemble spatially and biologically human breast cancer and its complex interaction with surrounding microenvironment. Clinical translation cannot be independent from this point, as widely demonstrated in current research, since the great majority of promising drugs and therapeutic strategies developed in preclinical studies do not reach clinical relevance [102]. Furthermore, preclinical reliable models of TME is even more difficult, since spatial issues might be fundamental to properly resemble the real biological situation [103]. Other major questions arise about safety, production standardization, and costs as well as potential toxicities before the use of targeted nanoparticles in clinical practice, and research in this field needs to be further accelerated.

Author Contributions: Conceptualization, M.T. and F.C.; writing—review and editing, M.T., S.M., A.B., L.S., F.C., R.V., and C.M.; figures, R.A. and M.T.; supervision, F.C. and M.T.

Funding: This paper was funded by Associazione Italiana per la Ricerca sul Cancro (AIRC IG 20172 to F.C.).

Int. J. Mol. Sci. **2019**, *20*, 1263

Acknowledgments: M.T. thanks Università degli studi di Milano for research fellowship. A.B. is grateful to Associazione Italiana per la Ricerca sul Cancro (AIRC IG 20172 to F.C.) for fellowship.

Conflicts of Interest: The authors declare no conflict of interest.

Abbreviations

α-SMA	Alpha-smooth muscle actin
Bcl-2	B-cell lymphoma 2
BCPN	Binary cooperative prodrug nanoparticle
CAF	Cancer-associated fibroblasts
CAP	Cleavable amphiphilic peptide
CCL2	Chemokine C-C motif ligand 2
CD	Cluster of differentiation
CPP	Cell-penetrating peptide
CXCL	C-X-C motif ligand
DC	Dendritic cells
DOX	Doxorubicin
ECM	Extracellular matrix
EMT	Epithelial to mesenchymal transition
EPR	Enhanced permeability and retention
FAP	Fibroblast activation protein
GNRs	Gold nanorods
HA	Hyaluronic acid
HNP	Hybrid nanoparticle
hPSCs	Human pancreatic stellate cells
IDO-1	Indoleamine 2,3-Dioxygenase 1
IL	Interleukin
LCP	Lipid-calcium phosphate
LPD	Liposome-protamine-DNA
LTSLs	Loaded thermosensitive liposomes
MATT	Marimastat
MMP	Metalloproteinases
NP	Nanoparticles
OXA	Oxaliplatin
PEG	Polyethylene glycol
PDGF	Platelet-derived growth factor
PDT	Photodynamic therapy
PLGA	Poly(lactic-co-glycolic) acid
PNP	Peptide nanoparticles
PS	Photosensitizer
PTX	Paclitaxel
QDs	Quantum dots
ROS	Reactive oxygen species
SPION	Superparamagnetic iron oxide nanoparticles
siRNA	Short interfering RNA
TGFβ	Transforming growth factor beta
TME	Tumor microenvironment
TNF	Tumor necrosis factor
TRAIL	TNF-related apoptosis-induced ligand
VEGF	Vascular endothelial growth factor

References

1. Luo, H.; Tu, G.; Liu, Z.; Liu, M. Cancer-associated fibroblasts: A multifaceted driver of breast cancer progression. *Cancer Lett.* **2015**, *361*, 155–163. [CrossRef] [PubMed]

2. Buchsbaum, R.J.; Oh, S.Y. Breast Cancer-Associated Fibroblasts: Where We Are and Where We Need to Go. *Cancers* **2016**, *8*, 19. [CrossRef] [PubMed]

3. Mao, Y.; Keller, E.T.; Garfield, D.H.; Shen, K.; Wang, J. Stromal cells in tumor microenvironment and breast cancer. *Cancer Metastasis Rev.* **2013**, *32*, 303–315. [CrossRef]

4. Hay, E.D. An overview of epithelio-mesenchymal transformation. *Acta Anat.* **1995**, *154*, 8–20. [CrossRef] [PubMed]

5. Hu, M.; Yao, J.; Carroll, D.K.; Weremowicz, S.; Chen, H.; Carrasco, D.; Richardson, A.; Violette, S.; Nikolskaya, T.; Nikolsky, Y.; et al. Regulation of in situ to invasive breast carcinoma transition. *Cancer Cell* **2008**, *13*, 394–406. [CrossRef] [PubMed]

6. Lamouille, S.; Xu, J.; Derynck, R. Molecular mechanisms of epithelial-mesenchymal transition. *Nat. Rev. Mol. Cell Biol.* **2014**, *15*, 178–196. [CrossRef]

7. Paget, S. The distribution of secondary growths in cancer of the breast. 1889. *Cancer Metastasis Rev.* **1989**, *8*, 98–101.

8. Chen, X.; Song, E. Turning foes to friends: Targeting cancer-associated fibroblasts. *Nat. Rev. Drug Discov.* **2018**, *18*, 99–115. [CrossRef]

9. Trimboli, A.J.; Cantemir-Stone, C.Z.; Li, F.; Wallace, J.A.; Merchant, A.; Creasap, N.; Thompson, J.C.; Caserta, E.; Wang, H.; Chong, J.L.; et al. Pten in stromal fibroblasts suppresses mammary epithelial tumours. *Nature* **2009**, *461*, 1084–1091. [CrossRef]

10. Tchou, J.; Conejo-Garcia, J. Targeting the tumor stroma as a novel treatment strategy for breast cancer: Shifting from the neoplastic cell-centric to a stroma-centric paradigm. *Adv. Pharm.* **2012**, *65*, 45–61.

11. Stingl, J.; Caldas, C. Molecular heterogeneity of breast carcinomas and the cancer stem cell hypothesis. *Nat. Rev. Cancer* **2007**, *7*, 791–799. [CrossRef] [PubMed]

12. Wu, D.; Si, M.; Xue, H.-Y.; Wong, H.-L. Nanomedicine applications in the treatment of breast cancer: Current state of the art. *Int. J. Nanomed.* **2017**, *12*, 5879–5892. [CrossRef] [PubMed]

13. Tong, R.; Langer, R. Nanomedicines Targeting the Tumor Microenvironment. *Cancer J.* **2015**, *21*, 314–321. [CrossRef] [PubMed]

14. Gurunathan, S.; Kang, M.-H.; Qasim, M.; Kim, J.-H. Nanoparticle-Mediated Combination Therapy: Two-in-One Approach for Cancer. *Int. J. Mol. Sci.* **2018**, *19*, 3264. [CrossRef]

15. Batra, H.; Pawar, S.; Bahl, D. Curcumin in combination with anti-cancer drugs: A nanomedicine review. *Pharm. Res.* **2018**, *139*, 91–105. [CrossRef]

16. Tang, H.; Zhao, W.; Yu, J.; Li, Y.; Zhao, C. Recent Development of pH-Responsive Polymers for Cancer Nanomedicine. *Molecules* **2018**, *24*, 4. [CrossRef] [PubMed]

17. Truffi, M.; Fiandra, L.; Sorrentino, L.; Monieri, M.; Corsi, F.; Mazzucchelli, S. Ferritin nanocages: A biological platform for drug delivery, imaging and theranostics in cancer. *Pharm. Res.* **2016**, *107*, 57–65. [CrossRef] [PubMed]

18. Bottai, G.; Truffi, M.; Corsi, F.; Santarpia, L. Progress in nonviral gene therapy for breast cancer and what comes next? *Expert Opin. Biol. Ther.* **2017**, *17*, 595–611. [CrossRef]

19. Colombo, M.; Rizzuto, M.A.; Pacini, C.; Pandolfi, L.; Bonizzi, A.; Truffi, M.; Monieri, M.; Catrambone, F.; Giustra, M.; Garbujo, S.; et al. Half-Chain Cetuximab Nanoconjugates Allow Multitarget Therapy of Triple Negative Breast Cancer. *Bioconjug. Chem.* **2018**, *29*, 3817–3832. [CrossRef]

20. Truffi, M.; Colombo, M.; Sorrentino, L.; Pandolfi, L.; Mazzucchelli, S.; Pappalardo, F.; Pacini, C.; Allevi, R.; Bonizzi, A.; Corsi, F.; et al. Multivalent exposure of trastuzumab on iron oxide nanoparticles improves antitumor potential and reduces resistance in HER2-positive breast cancer cells. *Sci. Rep.* **2018**, *8*, 6563. [CrossRef]

21. Truffi, M.; Colombo, M.; Peñaranda-Avila, J.; Sorrentino, L.; Colombo, F.; Monieri, M.; Collico, V.; Zerbi, P.; Longhi, E.; Allevi, R.; et al. Nano-targeting of mucosal addressin cell adhesion molecule-1 identifies bowel inflammation foci in murine model. *Nanomedicine* **2017**, *12*, 1547–1560. [CrossRef] [PubMed]

22. Kalluri, R. The biology and function of fibroblasts in cancer. *Nat. Rev. Cancer* **2016**, *16*, 582–598. [CrossRef] [PubMed]

23. LeBleu, V.S.; Kalluri, R. A peek into cancer-associated fibroblasts: Origins, functions and translational impact. *Dis. Model. Mech.* **2018**, *11*. [CrossRef] [PubMed]

24. Qiao, A.; Gu, F.; Guo, X.; Zhang, X.; Fu, L. Breast cancer-associated fibroblasts: Their roles in tumor initiation, progression and clinical applications. *Front. Med.* **2016**, *10*, 33–40. [CrossRef] [PubMed]

25. Alkasalias, T.; Moyano-Galceran, L.; Arsenian-Henriksson, M.; Lehti, K. Fibroblasts in the Tumor Microenvironment: Shield or Spear? *Int. J. Mol. Sci.* **2018**, *19*, 1532. [CrossRef] [PubMed]

26. Calvo, F.; Ege, N.; Grande-Garcia, A.; Hooper, S.; Jenkins, R.P.; Chaudhry, S.I.; Harrington, K.; Williamson, P.; Moeendarbary, E.; Charras, G.; et al. Mechanotransduction and YAP-dependent matrix remodelling is required for the generation and maintenance of cancer-associated fibroblasts. *Nat. Cell Biol.* **2013**, *15*, 637–646. [CrossRef] [PubMed]

27. Yamashita, M.; Ogawa, T.; Zhang, X.; Hanamura, N.; Kashikura, Y.; Takamura, M.; Yoneda, M.; Shiraishi, T. Role of stromal myofibroblasts in invasive breast cancer: Stromal expression of alpha-smooth muscle actin correlates with worse clinical outcome. *Breast Cancer* **2012**, *19*, 170–176. [CrossRef] [PubMed]

28. Bochet, L.; Lehuédé, C.; Dauvillier, S.; Wang, Y.Y.; Dirat, B.; Laurent, V.; Dray, C.; Guiet, R.; Maridonneau-Parini, I.; Le Gonidec, S.; et al. Adipocyte-derived fibroblasts promote tumor progression and contribute to the desmoplastic reaction in breast cancer. *Cancer Res.* **2013**, *73*, 5657–5668. [CrossRef]

29. Kelly, T. Fibroblast activation protein-alpha and dipeptidyl peptidase IV (CD26): Cell-surface proteases that activate cell signaling and are potential targets for cancer therapy. *Drug Resist. Updat.* **2005**, *8*, 51–58. [CrossRef]

30. Su, S.; Chen, J.; Yao, H.; Liu, J.; Yu, S.; Lao, L.; Wang, M.; Luo, M.; Xing, Y.; Chen, F.; et al. CD10+GPR77+ Cancer-Associated Fibroblasts Promote Cancer Formation and Chemoresistance by Sustaining Cancer Stemness. *Cell* **2008**, *172*, 841–856.e16.

31. De Francesco, E.M.; Sims, A.H.; Maggiolini, M.; Sotgia, F.; Lisanti, M.P.; Clarke, R.B. GPER mediates the angiocrine actions induced by IGF1 through the HIF-1α/VEGF pathway in the breast tumor microenvironment. *Breast Cancer Res.* **2017**, *19*, 129. [CrossRef]

32. Purcell, J.W.; Tanlimco, S.G.; Hickson, J.; Fox, M.; Sho, M.; Durkin, L.; Uziel, T.; Powers, R.; Foster, K.; McGonigal, T.; et al. LRRC15 Is a Novel Mesenchymal Protein and Stromal Target for Antibody-Drug Conjugates. *Cancer Res.* **2018**, *78*, 4059–4072. [CrossRef] [PubMed]

33. Lakins, M.A.; Ghorani, E.; Munir, H.; Martins, C.P.; Shields, J.D. Cancer-associated fibroblasts induce antigen-specific deletion of CD8 + T Cells to protect tumour cells. *Nat. Commun.* **2018**, *9*, 948. [CrossRef] [PubMed]

34. Tsuyada, A.; Chow, A.; Wu, J.; Somlo, G.; Chu, P.; Loera, S.; Luu, T.; Li, A.X.; Wu, X.; Ye, W.; et al. CCL2 mediates cross-talk between cancer cells and stromal fibroblasts that regulates breast cancer stem cells. *Cancer Res.* **2012**, *72*, 2768–2779. [CrossRef] [PubMed]

35. Eck, S.M.; Côté, A.L.; Winkelman, W.D.; Brinckerhoff, C.E. CXCR4 and matrix metalloproteinase-1 are elevated in breast carcinoma-associated fibroblasts and in normal mammary fibroblasts exposed to factors secreted by breast cancer cells. *Mol. Cancer Res.* **2009**, *7*, 1033–1044. [CrossRef] [PubMed]

36. Comito, G.; Giannoni, E.; Segura, C.P.; Barcellos-de-Souza, P.; Raspollini, M.R.; Baroni, G.; Lanciotti, M.; Serni, S.; Chiarugi, P. Cancer-associated fibroblasts and M2-polarized macrophages synergize during prostate carcinoma progression. *Oncogene* **2014**, *33*, 2423–2431. [CrossRef]

37. Raschioni, C.; Bottai, G.; Sagona, A.; Errico, V.; Testori, A.; Gatzemeier, W.; Corsi, F.; Tinterri, C.; Roncalli, M.; Santarpia, L.; et al. CXCR4/CXCL12 Signaling and Protumor Macrophages in Primary Tumors and Sentinel Lymph Nodes Are Involved in Luminal B Breast Cancer Progression. *Dis. Mark.* **2018**, *2018*, 5018671. [CrossRef]

38. Scherz-Shouval, R.; Santagata, S.; Mendillo, M.L.; Sholl, L.M.; Ben-Aharon, I.; Beck, A.H.; Dias-Santagata, D.; Koeva, M.; Stemmer, S.M.; Whitesell, L.; et al. The reprogramming of tumor stroma by HSF1 is a potent enabler of malignancy. *Cell* **2014**, *158*, 564–578. [CrossRef]

39. Orimo, A.; Gupta, P.B.; Sgroi, D.C.; Arenzana-Seisdedos, F.; Delaunay, T.; Naeem, R.; Carey, V.J.; Richardson, A.L.; Weinberg, R.A. Stromal fibroblasts present in invasive human breast carcinomas promote tumor growth and angiogenesis through elevated SDF-1/CXCL12 secretion. *Cell* **2005**, *121*, 335–348. [CrossRef] [PubMed]

40. Zellmer, V.R.; Schnepp, P.M.; Fracci, S.L.; Tan, X.; Howe, E.N.; Zhang, S. Tumor-induced Stromal STAT1 Accelerates Breast Cancer via Deregulating Tissue Homeostasis. *Mol. Cancer Res.* **2017**, *15*, 585–597. [CrossRef]

41. Chen, Q.; Liu, G.; Liu, S.; Su, H.; Wang, Y.; Li, J.; Luo, C. Remodeling the Tumor Microenvironment with Emerging Nanotherapeutics. *Trends Pharm. Sci.* **2018**, *39*, 59–74. [CrossRef]

42. Zhen, Z.; Tang, W.; Wang, M.; Zhou, S.; Wang, H.; Wu, Z.; Hao, Z.; Li, Z.; Liu, L.; Xie, J. Protein Nanocage Mediated Fibroblast-Activation Protein Targeted Photoimmunotherapy To Enhance Cytotoxic T Cell Infiltration and Tumor Control. *Nano Lett.* **2017**, *17*, 862–869. [CrossRef] [PubMed]

43. Li, L.; Zhou, S.; Lv, N.; Zhen, Z.; Liu, T.; Gao, S.; Xie, J.; Ma, Q. Photosensitizer-Encapsulated Ferritins Mediate Photodynamic Therapy against Cancer-Associated Fibroblasts and Improve Tumor Accumulation of Nanoparticles. *Mol. Pharm.* **2018**, *15*, 3595–3599. [CrossRef] [PubMed]

44. Ji, T.; Zhao, Y.; Ding, Y.; Wang, J.; Zhao, R.; Lang, J.; Qin, H.; Liu, X.; Shi, J.; Tao, N.; et al. Transformable Peptide Nanocarriers for Expeditious Drug Release and Effective Cancer Therapy via Cancer-Associated Fibroblast Activation. *Angew. Chem. Int. Ed. Engl.* **2016**, *55*, 1050–1055. [CrossRef] [PubMed]

45. Hu, C.; Liu, X.; Ran, W.; Meng, J.; Zhai, Y.; Zhang, P.; Yin, Q.; Yu, H.; Zhang, Z.; Li, Y. Regulating cancer associated fibroblasts with losartan-loaded injectable peptide hydrogel to potentiate chemotherapy in inhibiting growth and lung metastasis of triple negative breast cancer. *Biomaterials* **2017**, *144*, 60–72. [CrossRef] [PubMed]

46. Chauhan, V.P.; Martin, J.D.; Liu, H.; Lacorre, D.A.; Jain, S.R.; Kozin, S.V.; Stylianopoulos, T.; Mousa, A.S.; Han, X.; Adstamongkonkul, P.; et al. Angiotensin inhibition enhances drug delivery and potentiates chemotherapy by decompressing tumour blood vessels. *Nat. Commun.* **2013**, *4*, 2516. [CrossRef] [PubMed]

47. Zhou, H.; Fan, Z.; Deng, J.; Lemons, P.K.; Arhontoulis, D.C.; Bowne, W.B.; Cheng, H. Hyaluronidase Embedded in Nanocarrier PEG Shell for Enhanced Tumor Penetration and Highly Efficient Antitumor Efficacy. *Nano Lett.* **2016**, *16*, 3268–3277. [CrossRef] [PubMed]

48. Li, X.; Qin, F.; Yang, L.; Mo, L.; Li, L.; Hou, L. Sulfatide-containing lipid perfluorooctylbromide nanoparticles as paclitaxel vehicles targeting breast carcinoma. *Int. J. Nanomed.* **2014**, *9*, 3971–3985. [CrossRef]

49. Hu, X.X.; He, P.P.; Qi, G.B.; Gao, Y.J.; Lin, Y.X.; Yang, C.; Yang, P.P.; Hao, H.; Wang, L.; Wang, H. Transformable Nanomaterials as an Artificial Extracellular Matrix for Inhibiting Tumor Invasion and Metastasis. *ACS Nano* **2017**, *11*, 4086–4096. [CrossRef]

50. Lv, Y.; Xu, C.; Zhao, X.; Lin, C.; Yang, X.; Xin, X.; Zhang, L.; Qin, C.; Han, X.; Yang, L.; et al. Nanoplatform Assembled from a CD44-Targeted Prodrug and Smart Liposomes for Dual Targeting of Tumor Microenvironment and Cancer Cells. *ACS Nano* **2018**, *12*, 1519–1536. [CrossRef]

51. Feng, B.; Zhou, F.; Hou, B.; Wang, D.; Wang, T.; Fu, Y.; Ma, Y.; Yu, H.; Li, Y. Binary Cooperative Prodrug Nanoparticles Improve Immunotherapy by Synergistically Modulating Immune Tumor Microenvironment. *Adv. Mater.* **2018**, *30*, e180300. [CrossRef]

52. Lu, J.; Liu, X.; Liao, Y.P.; Wang, X.; Ahmed, A.; Jiang, W.; Ji, Y.; Meng, H.; Nel, A.E. Breast Cancer Chemo-immunotherapy through Liposomal Delivery of an Immunogenic Cell Death Stimulus Plus Interference in the IDO-1 Pathway. *ACS Nano* **2018**, *12*, 11041–11061. [CrossRef] [PubMed]

53. Liu, Y.; Qiao, L.; Zhang, S.; Wan, G.; Chen, B.; Zhou, P.; Zhang, N.; Wang, Y. Dual pH-responsive multifunctional nanoparticles for targeted treatment of breast cancer by combining immunotherapy and chemotherapy. *Acta Biomater.* **2018**, *66*, 310–324. [CrossRef] [PubMed]

54. Iranpour, S.; Nejati, V.; Delirezh, N.; Biparva, P.; Shirian, S. Enhanced stimulation of anti-breast cancer T cells responses by dendritic cells loaded with poly lactic-co-glycolic acid (PLGA) nanoparticle encapsulated tumor antigens. *J. Exp. Clin. Cancer Res.* **2016**, *35*, 168. [CrossRef] [PubMed]

55. Shen, L.; Li, J.; Liu, Q.; Song, W.; Zhang, X.; Tiruthani, K.; Hu, H.; Das, M.; Goodwin, T.J.; Liu, R.; et al. Local Blockade of Interleukin 10 and C-X-C Motif Chemokine Ligand 12 with Nano-Delivery Promotes Antitumor Response in Murine Cancers. *ACS Nano* **2018**, *12*, 9830–9841. [CrossRef] [PubMed]

56. Li, X.-Y.; Hu, S.-Q.; Xiao, L. The cancer-associated fibroblasts and drug resistance. *Eur. Rev. Med. Pharmacol. Sci.* **2015**, *19*, 2112–2119. [PubMed]

57. Kalluri, R.; Zeisberg, M. Fibroblasts in cancer. *Nat. Rev. Cancer* **2006**, *6*, 392–401. [CrossRef]

58. Agostinis, P.; Berg, K.; Cengel, K.A.; Foster, T.H.; Girotti, A.W.; Gollnick, S.O.; Hahn, S.M.; Hamblin, M.R.; Juzeniene, A.; Kessel, D.; et al. Photodynamic therapy of cancer: An update. *CA Cancer J. Clin.* **2011**, *61*, 250–281. [CrossRef] [PubMed]

59. Joyce, J.A.; Fearon, D.T. T cell exclusion, immune privilege, and the tumor microenvironment. *Science* **2015**, *348*, 74–80. [CrossRef]

60. Ji, T.; Ding, Y.; Zhao, Y.; Wang, J.; Qin, H.; Liu, X.; Lang, J.; Zhao, R.; Zhang, Y.; Shi, J.; et al. Peptide assembly integration of fibroblast-targeting and cell-penetration features for enhanced antitumor drug delivery. *Adv. Mater.* **2015**, *27*, 1865–1873. [CrossRef]

61. Mertens, J.C.; Fingas, C.D.; Christensen, J.D.; Smoot, R.L.; Bronk, S.F.; Werneburg, N.W.; Gustafson, M.P.; Dietz, A.B.; Roberts, L.R.; Sirica, A.E.; et al. Therapeutic effects of deleting cancer-associated fibroblasts in cholangiocarcinoma. *Cancer Res.* **2013**, *73*, 897–907. [CrossRef]

62. Chen, B.; Wang, Z.; Sun, J.; Song, Q.; He, B.; Zhang, H.; Wang, X.; Dai, W.; Zhang, Q. A tenascin C targeted nanoliposome with navitoclax for specifically eradicating of cancer-associated fibroblasts. *Nanomedicine* **2016**, *12*, 131–141. [CrossRef] [PubMed]

63. Miao, L.; Wang, Y.; Lin, C.M.; Xiong, Y.; Chen, N.; Zhang, L.; Kim, W.Y.; Huang, L. Nanoparticle modulation of the tumor microenvironment enhances therapeutic efficacy of cisplatin. *J. Control. Release* **2015**, *217*, 27–41. [CrossRef] [PubMed]

64. Hu, K.; Miao, L.; Goodwin, T.J.; Li, J.; Liu, Q.; Huang, L. Quercetin Remodels the Tumor Microenvironment to Improve the Permeation, Retention, and Antitumor Effects of Nanoparticles. *ACS Nano* **2017**, *11*, 4916–4925. [CrossRef]

65. Miao, L.; Liu, Q.; Lin, C.M.; Luo, C.; Wang, Y.; Liu, L.; Yin, W.; Hu, S.; Kim, W.Y.; Huang, L. Targeting Tumor-Associated Fibroblasts for Therapeutic Delivery in Desmoplastic Tumors. *Cancer Res.* **2017**, *77*, 719–731. [CrossRef] [PubMed]

66. Mpekris, F.; Papageorgis, P.; Polydorou, C.; Voutouri, C.; Kalli, M.; Pirentis, A.P.; Stylianopoulos, T. Sonic-hedgehog pathway inhibition normalizes desmoplastic tumor microenvironment to improve chemo- and nanotherapy. *J. Control. Release* **2017**, *261*, 105–112. [CrossRef] [PubMed]

67. Mardhian, D.F.; Storm, G.; Bansal, R.; Prakash, J. Nano-targeted relaxin impairs fibrosis and tumor growth in pancreatic cancer and improves the efficacy of gemcitabine in vivo. *J. Control. Release* **2018**, *290*, 1–10. [CrossRef] [PubMed]

68. Provenzano, P.P.; Inman, D.R.; Eliceiri, K.W.; Knittel, J.G.; Yan, L.; Rueden, C.T.; White, J.G.; Keely, P.J. Collagen density promotes mammary tumor initiation and progression. *BMC Med.* **2008**, *6*, 11. [CrossRef] [PubMed]

69. Gong, H.; Chao, Y.; Xiang, J.; Han, X.; Song, G.; Feng, L.; Liu, J.; Yang, G.; Chen, Q.; Liu, Z. Hyaluronidase to Enhance Nanoparticle-Based Photodynamic Tumor Therapy. *Nano Lett.* **2016**, *16*, 2512–2521. [CrossRef]

70. Lu, P.; Weaver, V.M.; Werb, Z. The extracellular matrix: A dynamic niche in cancer progression. *J. Cell Biol.* **2012**, *196*, 395–406. [CrossRef]

71. Turpeenniemi-Hujanen, T. Gelatinases (MMP-2 and -9) and their natural inhibitors as prognostic indicators in solid cancers. *Biochimie* **2005**, *87*, 287–297. [CrossRef]

72. Sparano, J.A.; Bernardo, P.; Stephenson, P.; Gradishar, W.J.; Ingle, J.N.; Zucker, S.; Davidson, N.E. Randomized phase III trial of marimastat versus placebo in patients with metastatic breast cancer who have responding or stable disease after first-line chemotherapy: Eastern Cooperative Oncology Group trial E2196. *J. Clin. Oncol.* **2004**, *22*, 4683–4690. [CrossRef] [PubMed]

73. Flavell, R.A.; Sanjabi, S.; Wrzesinski, S.H.; Licona-Limón, P. The polarization of immune cells in the tumour environment by TGFbeta. *Nat. Rev. Immunol.* **2010**, *10*, 554–567. [CrossRef] [PubMed]

74. Yang, X.; Lin, Y.; Shi, Y.; Li, B.; Liu, W.; Yin, W.; Dang, Y.; Chu, Y.; Fan, J.; He, R. FAP Promotes Immunosuppression by Cancer-Associated Fibroblasts in the Tumor Microenvironment via STAT3-CCL2 Signaling. *Cancer Res.* **2016**, *76*, 4124–4135. [CrossRef] [PubMed]

75. Turley, S.J.; Cremasco, V.; Astarita, J.L. Immunological hallmarks of stromal cells in the tumour microenvironment. *Nat. Rev. Immunol.* **2015**, *15*, 669–682. [CrossRef] [PubMed]

76. Balkwill, F.R.; Capasso, M.; Hagemann, T. The tumor microenvironment at a glance. *J. Cell Sci.* **2012**, *125*, 5591–5596. [CrossRef] [PubMed]

77. Mellor, A.L.; Lemos, H.; Huang, L. Indoleamine 2,3-Dioxygenase and Tolerance: Where Are We Now? *Front. Immunol.* **2017**, *8*, 1360. [CrossRef] [PubMed]

78. Chu, D.; Dong, X.; Zhao, Q.; Gu, J.; Wang, Z. Photosensitization Priming of Tumor Microenvironments Improves Delivery of Nanotherapeutics via Neutrophil Infiltration. *Adv. Mater.* **2017**, *29*, 1701021. [CrossRef]

79. Sato, T.; Terai, M.; Tamura, Y.; Alexeev, V.; Mastrangelo, M.J.; Selvan, S.R. Interleukin 10 in the tumor microenvironment: A target for anticancer immunotherapy. *Immunol. Res.* **2011**, *51*, 170–182. [CrossRef]

80. Velaei, K.; Samadi, N.; Barazvan, B.; Rad, J.S. Tumor microenvironment-mediated chemoresistance in breast cancer. *Breast* **2016**, *30*, 92–100. [CrossRef]

81. Solinas, C.; Gombos, A.; Latifyan, S.; Piccart-Gebhart, M.; Kok, M.; Buisseret, L. Targeting immune checkpoints in breast cancer: An update of early results. *ESMO Open* **2017**, *2*, e000255. [CrossRef]

82. Pelekanou, V.; Carvajal-Hausdorf, D.E.; Altan, M.; Wasserman, B.; Carvajal-Hausdorf, C.; Wimberly, H.; Brown, J.; Lannin, D.; Pusztai, L.; Rimm, D.L. Effect of neoadjuvant chemotherapy on tumor-infiltrating lymphocytes and PD-L1 expression in breast cancer and its clinical significance. *Breast Cancer Res.* **2017**, *19*, 91. [CrossRef] [PubMed]

83. Mariathasan, S.; Turley, S.J.; Nickles, D.; Castiglioni, A.; Yuen, K.; Wang, Y.; Kadel, E.E., III; Koeppen, H.; Astarita, J.L.; Cubas, R. TGFβ attenuates tumour response to PD-L1 blockade by contributing to exclusion of T cells. *Nature* **2018**, *554*, 544–548. [CrossRef] [PubMed]

84. Kraman, M.; Bambrough, P.J.; Arnold, J.N.; Roberts, E.W.; Magiera, L.; Jones, J.O.; Gopinathan, A.; Tuveson, D.A.; Fearon, D.T. Suppression of antitumor immunity by stromal cells expressing fibroblast activation protein-alpha. *Science* **2010**, *330*, 827–830. [CrossRef] [PubMed]

85. Huber, M.A.; Schubert, R.D.; Peter, R.U.; Kraut, N.; Park, J.E.; Rettig, W.J.; Garin-Chesa, P. Fibroblast activation protein: Differential expression and serine protease activity in reactive stromal fibroblasts of melanocytic skin tumors. *J. Investig. Dermatol.* **2003**, *120*, 182–188. [CrossRef]

86. Chung, K.M.; Hsu, S.C.; Chu, Y.R.; Lin, M.Y.; Jiaang, W.T.; Chen, R.H.; Chen, X. Fibroblast activation protein (FAP) is essential for the migration of bone marrow mesenchymal stem cells through RhoA activation. *PLoS ONE* **2014**, *9*, e88772. [CrossRef] [PubMed]

87. Roberts, E.W.; Deonarine, A.; Jones, J.O.; Denton, A.E.; Feig, C.; Lyons, S.K.; Espeli, M.; Kraman, M.; McKenna, B.; Wells, R.J.; et al. Depletion of stromal cells expressing fibroblast activation protein-α from skeletal muscle and bone marrow results in cachexia and anemia. *J. Exp. Med.* **2013**, *210*, 1137–1151. [CrossRef]

88. Scott, A.M.; Wiseman, G.; Welt, S.; Adjei, A.; Lee, F.T.; Hopkins, W.; Divgi, C.R.; Hanson, L.H.; Mitchell, P.; Gansen, D.N.; et al. A Phase I dose-escalation study of sibrotuzumab in patients with advanced or metastatic fibroblast activation protein-positive cancer. *Clin. Cancer Res.* **2003**, *9*, 1639–1647.

89. Hofheinz, R.D.; Al-Batran, S.E.; Hartmann, F.; Hartung, G.; Jäger, D.; Renner, C.; Tanswell, P.; Kunz, U.; Amelsberg, A.; Kuthan, H.; et al. Stromal antigen targeting by a humanised monoclonal antibody: An early phase II trial of sibrotuzumab in patients with metastatic colorectal cancer. *Onkologie* **2003**, *26*, 44–48. [CrossRef]

90. Eager, R.M.; Cunningham, C.C.; Senzer, N.; Richards, D.A.; Raju, R.N.; Jones, B.; Uprichard, M.; Nemunaitis, J. Phase II trial of talabostat and docetaxel in advanced non-small cell lung cancer. *Clin. Oncol.* **2009**, *21*, 464–472. [CrossRef]

91. Narra, K.; Mullins, S.R.; Lee, H.O.; Strzemkowski-Brun, B.; Magalong, K.; Christiansen, V.J.; McKee, P.A.; Egleston, B.; Cohen, S.J.; Weiner, L.M.; et al. Phase II trial of single agent Val-boroPro (Talabostat) inhibiting Fibroblast Activation Protein in patients with metastatic colorectal cancer. *Cancer Biol. Ther.* **2007**, *6*, 1691–1699. [CrossRef]

92. Loeffler, M.; Krüger, J.A.; Niethammer, A.G.; Reisfeld, R.A. Targeting tumor-associated fibroblasts improves cancer chemotherapy by increasing intratumoral drug uptake. *J. Clin. Investig.* **2006**, *116*, 1955–1962. [CrossRef] [PubMed]

93. LeBien, T.W.; McCormack, R.T. The common acute lymphoblastic leukemia antigen (CD10)—Emancipation from a functional enigma. *Blood* **1989**, *73*, 625–635. [PubMed]

94. Gerard, N.P.; Lu, B.; Liu, P.; Craig, S.; Fujiwara, Y.; Okinaga, S.; Gerard, C. An anti-inflammatory function for the complement anaphylatoxin C5a-binding protein, C5L2. *J. Biol. Chem.* **2005**, *280*, 39677–39680. [CrossRef] [PubMed]

95. Ishii, G.; Ochiai, A.; Neri, S. Phenotypic and functional heterogeneity of cancer-associated fibroblast within the tumor microenvironment. *Adv. Drug Deliv. Rev.* **2016**, *99*, 186–196. [CrossRef]

96. Boire, A.; Covic, L.; Agarwal, A.; Jacques, S.; Sherifi, S.; Kuliopulos, A. PAR1 is a matrix metalloprotease-1 receptor that promotes invasion and tumorigenesis of breast cancer cells. *Cell* **2005**, *120*, 303–313. [CrossRef] [PubMed]

97. Engle, S.J.; Hoying, J.B.; Boivin, G.P.; Ormsby, I.; Gartside, P.S.; Doetschman, T. Transforming growth factor beta1 suppresses nonmetastatic colon cancer at an early stage of tumorigenesis. *Cancer Res.* **1999**, *59*, 3379–3386.

98. Ene–Obong, A.; Clear, A.J.; Watt, J.; Wang, J.; Fatah, R.; Riches, J.C.; Marshall, J.F.; Chin–Aleong, J.; Chelala, C.; Gribben, J.G.; et al. Activated pancreatic stellate cells sequester CD8+ T cells to reduce their infiltration of the juxtatumoral compartment of pancreatic ductal adenocarcinoma. *Gastroenterology* **2013**, *145*, 1121–1132. [CrossRef] [PubMed]

99. Bartoschek, M.; Oskolkov, N.; Bocci, M.; Lövrot, J.; Larsson, C.; Sommarin, M.; Madsen, C.D.; Lindgren, D.; Pekar, G.; Karlsson, G.; et al. Spatially and functionally distinct subclasses of breast cancer-associated fibroblasts revealed by single cell RNA sequencing. *Nat. Commun.* **2018**, *9*, 5150. [CrossRef] [PubMed]

100. Hurwitz, H.I.; Uppal, N.; Wagner, S.A.; Bendell, J.C.; Beck, J.T.; Wade, S.M., III; Nemunaitis, J.J.; Stella, P.J.; Pipas, J.M.; Wainberg, Z.A.; et al. Randomized, Double-Blind, Phase II Study of Ruxolitinib or Placebo in Combination with Capecitabine in Patients with Metastatic Pancreatic Cancer for Whom Therapy with Gemcitabine Has Failed. *J. Clin. Oncol.* **2015**, *33*, 4039–4047. [CrossRef] [PubMed]

101. Song, G.; Darr, D.B.; Santos, C.M.; Ross, M.; Valdivia, A.; Jordan, J.L.; Midkiff, B.R.; Cohen, S.; Nikolaishvili-Feinberg, N.; Miller, C.R.; et al. Effects of tumor microenvironment heterogeneity on nanoparticle disposition and efficacy in breast cancer tumor models. *Clin. Cancer Res.* **2014**, *20*, 6083–6095. [CrossRef]

102. Day, C.-P.; Merlino, G.; van Dyke, T. Preclinical mouse cancer models: A maze of opportunities and challenges. *Cell* **2015**, *163*, 39–53. [CrossRef] [PubMed]

103. Sethi, P.; Jyoti, A.; Swindell, E.P.; Chan, R.; Langner, U.W.; Feddock, J.M.; Nagarajan, R.; O'Halloran, T.V.; Upreti, M. 3D tumor tissue analogs and their orthotopic implants for understanding tumor-targeting of microenvironment-responsive nanosized chemotherapy and radiation. *Nanomedicine* **2011**, *11*, 2013–2023. [CrossRef] [PubMed]

International Journal of
Molecular Sciences

MDPI

Review

Activated Fibroblast Program Orchestrates Tumor Initiation and Progression; Molecular Mechanisms and the Associated Therapeutic Strategies

Go J. Yoshida [1],*, Arata Azuma [2],*, Yukiko Miura [2] and Akira Orimo [1],*

[1] Department of Molecular Pathogenesis, Juntendo University Faculty of Medicine, 2-1-1 Hongo, Bunkyo-ku, Tokyo, 113-8421, Japan

[2] Department of Pulmonary Medicine and Oncology, Graduate School of Medicine, Nippon Medical School, 1-1-5, Sendagi, Bunkyo-ku, Tokyo 1138603, Japan; s7081@nms.ac.jp

* Correspondence: go-yoshida@juntendo.ac.jp (G.J.Y.); a-azuma@nms.ac.jp (A.A.); aorimo@juntendo.ac.jp (A.O.); Tel.: +81-3-5802-1039 (G.J.Y.); Fax: +81-3-5684-1646 (G.J.Y.)

Received: 25 April 2019; Accepted: 3 May 2019; Published: 7 May 2019

Abstract: Neoplastic epithelial cells coexist in carcinomas with various non-neoplastic stromal cells, together creating the tumor microenvironment. There is a growing interest in the cross-talk between tumor cells and stromal fibroblasts referred to as carcinoma-associated fibroblasts (CAFs), which are frequently present in human carcinomas. CAF populations extracted from different human carcinomas have been shown to possess the ability to influence the hallmarks of cancer. Indeed, several mechanisms underlying CAF-promoted tumorigenesis are elucidated. Activated fibroblasts in CAFs are characterized as alpha-smooth muscle actin-positive myofibroblasts and actin-negative fibroblasts, both of which are competent to support tumor growth and progression. There are, however, heterogeneous CAF populations presumably due to the diverse sources of their progenitors in the tumor-associated stroma. Thus, molecular markers allowing identification of bona fide CAF populations with tumor-promoting traits remain under investigation. CAFs and myofibroblasts in wound healing and fibrosis share biological properties and support epithelial cell growth, not only by remodeling the extracellular matrix, but also by producing numerous growth factors and inflammatory cytokines. Notably, accumulating evidence strongly suggests that anti-fibrosis agents suppress tumor development and progression. In this review, we highlight important tumor-promoting roles of CAFs based on their analogies with wound-derived myofibroblasts and discuss the potential therapeutic strategy targeting CAFs.

Keywords: angiogenesis; cancer-associated fibroblasts; extracellular matrix; fibrosis; heterogeneity; interstitial fluid pressure; metabolic reprogramming; transforming growth factor-β; tumor stiffness

1. Significant Roles of Fibrosis in Cancer Development

1.1. Contributions of Fibrosis to Cancer Development

Injured epithelial tissues are repaired by the formation of granulation tissues rich in α-smooth muscle actin (α-SMA)-positive myofibroblasts (a hallmark of activated fibroblasts), platelets, newly formed blood vessels, macrophages, and other inflammatory cells and extracellular matrix (ECM). The transforming growth factor-β (TGF-β) signal pathway is involved in the emergence of myofibroblasts, which contribute to the production of matrix metalloproteinase (MMP) and ECM proteins, such as collagen I, fibronectin and hyaluronic acid [1–4]. The damaged tissues are then degraded and ECM proteins are simultaneously generated *de novo* [2–5]. Sustained activation of myofibroblasts promotes dysfunctional repair mechanisms, leading to accumulation of fibrotic ECM which is rich in collagen fibers and resistant to MMP-mediated degradation [1,6,7]. The fibrotic ECM inhibits epithelial cell polarity and stimulates epithelial cell proliferation, which in turn results in conditions allowing tumor formation and development [8,9].

In fact, a growing body of evidence suggests that the presence of fibrotic lesions significantly increases the risk of cancer in numerous tissues, including the lungs, liver and breast [8–11]. Idiopathic pulmonary fibrosis (IPF), which is a progressive and fatal lung disease of unknown etiology, is associated with a higher incidence of lung cancers as compared with the general population [12]. IPF is characterized by scar tissue accumulation in the lung interstitium. The injury to type II alveolar epithelial cells triggers production of TGF-β that leads to mitogenesis of macrophages, platelets and myofibroblasts in the injured areas, leading to the formation of fibroblastic foci. Fibroblastic foci containing myofibroblasts at the leading edge of lung fibrosis are an indicator of poor prognosis and decreased survival [13].

The secreted protein acidic and rich in cysteine (SPARC) family of proteins regulate ECM assembly and growth factor signaling to modulate interactions between cells and the extracellular environment [14,15]. SPARC (also known as osteonectin, an acidic extracellular matrix glycoprotein) binds to soluble procollagen and prevents procollagen from interacting with cellular receptors, such as discoidin domain receptor 2 and integrins [15,16]. In the absence of SPARC, procollagen accumulates at the cell surface and is inefficiently incorporated into the ECM, resulting in the production of thin collagen fibers. SPARC is thus required for procollagen to be dissociated from the cell surface and incorporated into the ECM.

SPARC is exclusively expressed in IPF patients, never in healthy individuals [9,17]. SPARC expression is also tightly correlated with increased collagen deposition. Inhibition of SPARC expression significantly attenuates fibrosis in various animal models of disease [15]. SPARC is also localized in the cytoplasm of the actively-migrating myofibroblasts within the fibroblastic foci [17]. SPARC expression and TGF-β signaling are reciprocally regulated; TGF-β induces SPARC expression via canonical Smad2/3 signaling in lung fibroblasts and SPARC which, in turn, activates TGF-β signaling [18]. TGF-β also induces plasminogen activator inhibitor-1 (PAI-1) expression via Smad2/3 signaling in lung fibroblasts. Moreover, SPARC-activated integrin promotes Akt activation that inhibits glycogen synthase kinase-3β (GSK-3β) by serine-9/21 phosphorylation, leading to β-catenin activation and PAI-1 expression [17]. As PAI-1 prevents lung fibroblasts from undergoing apoptosis induced by plasminogen, ectopic SPARC expression in IPF apparently mediates the progression of interstitial fibrosis by inhibiting apoptosis in lung myofibroblasts via β-catenin activation and PAI-1 expression in collaboration with the TGF-β signal pathway. Taken together, the observations of these cellular mechanisms by which SPARC promotes the activation of fibroblasts in culture and its fibrosis-promoting ability in vivo encourage investigators to seek therapeutic strategies for blocking SPARC activity. Such research may lead to the eradication of fibrotic diseases.

In contrast to the fibrosis-promoting SPARC function, the roles of stromal SPARC in human carcinomas appear to be far more complex and even contradictory according to previous reports. Enhanced SPARC expression in the tumor-associated stroma correlates with a poor prognosis for patients with non-small cell lung cancers (NSCLC) [19] and pancreatic adenocarcinomas [20], but not for those with bladder cancers [21]. Chemical agent-induced bladder carcinomas have been shown to grow and progress more significantly in SPARC$^{-/-}$ mice than in control SPARC$^{+/+}$ mice [21]. Murine carcinoma-associated fibroblasts (CAFs) extracted from SPARC$^{-/-}$ bladder carcinomas also exhibit enhanced inflammatory phenotypes via NF-κB and AP-1 signaling, thereby promoting tumor growth and metastasis, indicating a tumor-suppressive role of SPARC in bladder CAFs. Collectively, these observations indicate cell-context dependent roles of stromal SPARC in different tumors.

Furthermore, non-alcoholic steatohepatitis (NASH), characterized by fat accumulation, inflammation and liver cell damage, leads to advanced fibrosis and cirrhosis, thereby increasing the risk of developing hepatocellular carcinoma (HCC) [22,23]. Diabetes mellitus (DM) with insulin resistance has also been demonstrated to be an independent risk factor for HCC development in NASH patients [23,24]. Activation of insulin-like growth factor 1 (IGF1) signaling stimulates cellular proliferation by activating the mitogen-activated protein kinase (MAPK) pathway and increases the transcription of c-Fos and c-Jun proto-oncogenes [25–27]. Moreover, phosphatase and tensin homologue deleted on chromosome 10 (PTEN) is a crucial negative regulator of the insulin signal pathway mediated by suppression of the phosphatidylinositol-3 kinase (PI3K)-Akt signal pathway. It has been shown that concomitant down-regulation of PTEN and up-regulation of c-Met occurs in HCC, leading to poor clinical outcomes [28]. Loss of PTEN function leads to the accumulation of phosphatidylinositol-3,4,5-triphosphate (PIP3), which mimics the effects of PI3K activation and triggers the activation of its downstream effectors, PDK1, Akt and Rac1/CDC42. Taken together, these observations demonstrate that NASH induces activation of an oncogenic signal transduction series of events in the non-cancerous liver to initiate tumor development.

1.2. Epithelial-Mesenchymal Transition (EMT) and Endothelial-Mesenchymal Transition (EndoMT) in Fibrosis and Tumor Stroma

As the saying "tumors: wounds that do not heal" goes, myofibroblasts in wounds and fibrosis mimic CAFs within a tumor [4,29,30] (Figure 1). Epithelial cells frequently transdifferentiate into mesenchymal cells through EMT during wound healing and fibrosis. EndoMT, another form of cellular transition, has also emerged as a mechanism underlying pathological fibrosis development [31–33]. Lineage-tagging experiments using a murine fibrosis model of renal injury indicate that about 30% of the cells involved are derived from tubular epithelial cells via EMT, while about 35% arise from EndoMT [34].

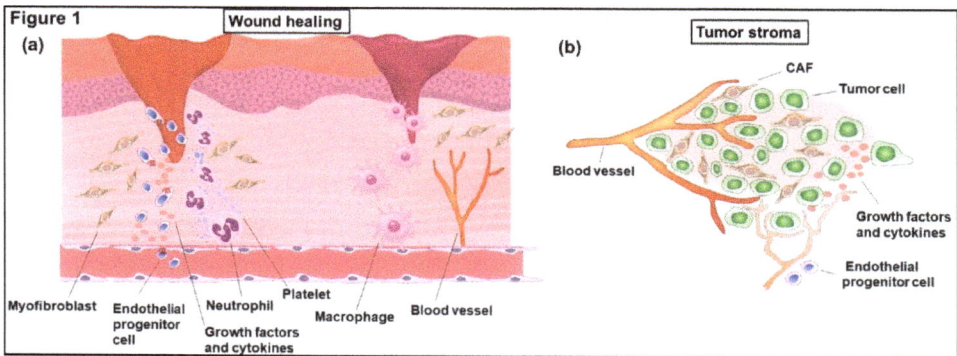

Figure 1. Schematic representation of both wound healing and tumor stroma. Platelets, inflammatory immune cells including neutrophils and macrophages, vascular endothelial cells and activated fibroblasts (myofibroblasts and carcinoma-associated fibroblasts (CAFs)) are recruited into granulation tissues during wound healing (**a**) and tumor stroma (**b**).

EndoMT is a complex biological process in which endothelial cells lose their molecular markers, such as vascular endothelial cadherin (VE cadherin), and acquire the myofibroblastic phenotype expressing mesenchymal markers including α-SMA, type I collagen, and vimentin. These cells also gain motility and are thus capable of migrating into the surrounding tissues [31]. TGF-β treatment induces the downstream signaling pathway to significantly upregulate Snail1 expression in endothelial cells via EndoMT [35–37]. This observation strongly suggests that EMT and EndoMT share the similar molecular mechanisms. Remarkably, TGF-β1-induced EndoMT occurs independently of Smad2/3 phosphorylation via non-canonical TGF-β signaling [31]. Furthermore, several important kinases including the c-Abl protein kinase (c-Abl), protein kinase C δ (PKC-δ) and GSK-3β, have been shown to play pivotal roles in this Smad-independent TGF-β pathway. In the absence of GSK-3β phosphorylation, the kinase activity of GSK-3β is promoted and induces proteasomal degradation of Snail1, thereby abrogating EndoMT. Both c-Abl and PKC-δ are required for GSK-3β phosphorylation to induce EndoMT [38]. From the perspective of preventing tissue fibrosis, the inhibition of GSK-3β serine-9 phosphorylation by a specific inhibitor of PKC-δ, i.e., rottlerin, or by c-Abl, widely known as imatinib, degrades Snail1 and thereby inhibits EndoMT [31]. Thus, rottlerin and imatinib both effectively suppress acquisition of the myofibroblastic phenotype and pathological fibrotic changes.

Myofibroblasts reportedly promote the induction and maintenance of EMT of epithelial cells at wound edges [4,39]. When this physiological EMT process is disrupted, wounds cannot heal. For instance, wound re-epithelialization of dermal tissue is compromised in mice lacking functional Slug, which is one of the transcription factors involved in TGF-β-induced EMT [40,41]. To achieve EMT at the wound edge, myofibroblasts secrete extracellular proteolytic enzymes, such as MMPs, which cleave ECM components and release potent TGF-β (latent form) and other EMT-inducing cytokines [6,42]. This is an intriguing parallel with the role EMT in cancer development and progression; while cancer cells are regulated by cytokines, such as TGF-β, these cytokines become tethered within the ECM, such that it remains ready for mobilization in response to certain triggers.

Stromal myofibroblasts are also observed in proximity to carcinoma cells associated with the EMT phenotype [43]. CAFs in this context participate in the release and bioavailability by secreting extracellular proteases and ECM-remodeling enzymes. It was previously shown that normal colonic fibroblasts differentiate into α-SMA-positive CAFs and secrete larger amounts of MMP2 and urokinase-type plasminogen activator (uPA) associated with various cancer cells [44,45]. These proteolytic enzymes have been suggested to cleave various ECM components such as decorin, which covalently and potently binds to TGF-β and prevents the potential ligand from binding to the TGF-β receptor in adjacent cancer cells [46]. ECM components act as a reservoir for various cytokines; since decorin is able to bind TGF-β1, proteolytic degradation of decorin results in the release of this sequestered TGF-β ligand [47]. These lines of evidence all strongly suggest that paracrine signaling from CAFs and myofibroblasts in a wound regulates epithelial-mesenchymal plasticity in nearby epithelial cells to further promote tumor progression and fibrosis, respectively.

2. Fibrosis-Induced Tumor Progression

2.1. Origin and Differentiation into CAFs

Although CAFs represent a major cellular component of the tumor stroma, a precise molecular definition of CAFs is as yet lacking. Attempts to define CAFs are usually aimed at identifying morphological features and expression patterns of the following proteins: α-SMA, asporin, collagen 11-α1 (COL11A1), fibroblast-activating protein (FAP), platelet-derived growth factor receptor (PDGFR) α/β, fibroblast-specific protein 1 (FSP1, also called S100A4), podoplanin, SPARC, S100A4, tenascin-C, microfibrillar-associated protein 5 (MFAP5), and vimentin [33,48–50]. However, none of these markers are specific to CAFs. Lack of the appropriate molecular markers for identifying tumor-promoting CAFs thus makes it difficult to elucidate the biology of these fibroblasts. Such a precise understanding would be the first, and most fundamental, step toward developing a cell type-specific targeting approach.

CAFs produce growth factors and inflammatory cytokines that are capable not only of regulating fibroblast activation in an autocrine fashion, but also of controlling the behaviors of cancer cells as well as other stromal cells, along with remodeling the ECM in a paracrine manner [33,51,52]. CAFs transdifferentiate from their progenitors, such as resident fibroblasts, endothelial cells, preadipocytes and bone marrow-derived mesenchymal stem cells (MSCs) during tumor progression [50,53–55]. MSCs are known to differentiate into CAFs in culture [56]. Injection of MSCs with carcinoma cells into immunodeficient mice also results in enhanced tumor growth and metastasis presumably through differentiation into tumor-promoting CAFs [57]. However, how differences among cells of origin for CAFs impact their biological functions has yet to be elucidated. A recent elegant study demonstrated the unique roles of CAFs originating from bone marrow in breast carcinomas [58]. Using MMTV-PyMT transgenic mice and adaptive bone marrow transplantation techniques, Raz et al. showed bone marrow-derived CAFs extracted from breast tumors to have proangiogenic traits and that, when implanted into recipient mice, these traits were significantly more marked than in those locally arising from the mammary gland. They also found decreased PDGFRα expression to allow bone marrow-derived CAFs to be distinguished from other CAF populations in breast tumors, highlighting distinct fibroblast populations present in the tumor.

Myofibroblastic CAFs are induced *de novo* from their progenitors including normal fibroblasts, when treated with TGF-β, platelet-derived growth factor (PDGF), Wnt7a, exosomes and microRNAs in culture [49,50,59–61]. Activation of fibroblasts with a pro-inflammatory state also occurs in otherwise non-activated fibroblasts treated with interleukin (IL)-1β, IL-6, leukemia inhibitory factor (LIF) and osteopontin [62,63]. However, it remains unknown whether these *de novo* generated CAFs continue to maintain their activated, tumor-promoting traits after a series of passages in culture or incubation with carcinoma cells within a tumor mass.

From the perspective of epigenetic modifications in myofibroblasts, LIF induces constitutive activation of the Janus kinase 1 (JAK1)/STAT3 signaling pathway mediated by post-translational regulation of STAT3 acetylation by p300 [64–66]. The acetylated STAT3 causes an epigenetic-dependent loss of expression of the Src homology region 2 domain-containing phosphatase-1 (SHP-1) tyrosine phosphatase, which is a negative regulator of the JAK/STAT pathway [67,68]. Silencing of SHP-1 gene expression by promoter methylation leads to sustained phosphorylation of JAK1 kinase and the STAT3 transcription factor that maintain the contractile and invasive abilities of CAFs [65]. Blockage of both JAK signal and DNA methyltransferase activities results, both in vitro and in vivo, in the reversal of the invasive phenotype of CAFs.

Table 1 shows the common activated signal pathways in both wound-induced activated fibroblasts and CAFs. This table comparatively details the biological roles of growth factors and cytokines in wound-healing and tumor stroma settings.

Table 1. The critical signal pathways activated in both wound-induced fibroblasts and CAFs.

Signal Pathway	Wound-Induced Fibroblasts	CAFs
Epithelial growth factor (EGF)	EGF stimulation increases the phosphorylation of myosin light chain (MLC) subunit of myosin that promotes cell contractility in various different cell types. Activation of PKC with the PKC-δ isoform mediates the cell contraction by EGF-stimulated MLC phosphorylation in murine fibroblast cells [69].	Resistance to the epidermal growth factor receptor (EGFR) tyrosine kinase inhibitor (TKI) is partially medicated by CAFs in tumors through paracrine factors secreted from these fibroblasts [70].
Fibroblast growth factor (FGF)	FGFs have the biological activity of stimulating the proliferation of fibroblasts and angiogenesis [71]. FGFs exert multiple functions through binding to and activation of fibroblast growth factor receptors (FGFRs), and the main signaling through the stimulation of FGFRs is the RAS/MAPK signal pathway.	CAFs secrete increased levels of FGF-1/-3 and promote cancer cell growth and angiogenesis through the activation of FGFR4, which is followed by the activation of extracellular signal-regulated kinase (ERK) and the modulation of MMP-7 expression [72]. In addition, FGF-1 and FGF-3 act as primary autocrine mediators of epithelial-stromal interactions in the tumor progression.

Table 1. *Cont.*

Signal Pathway	Wound-Induced Fibroblasts	CAFs
JAK/STAT	Synovial fibroblasts mediate chronic inflammation and joint destruction in patients suffered from rheumatoid arthritis (RA). Increased levels of IL-6, TNF-α and IL-1β production activate STAT3 signaling that in turn boosts expression levels of these cytokines in an autocrine fashion in synovial fibroblasts, promoting chronic inflammation [73]. STAT3 activation also induces receptor activator of nuclear factor kappa B ligand (RANKL) expression that stimulates osteoclastogenesis and thus promotes the joint destruction [73].	CAFs release high levels of IL-6 and CCL2 upon STAT3 activation in co-culture system with cancer cells, promoting the self-renewal and spheroid forming potentials of cancer stem cells [74]. Furthermore, the leukemia inhibitory factor (LIF)-induced JAK1/STAT3 signaling pathway mediates expression of the invasive CAF phenotype [75].
PDGF	PDGFs induce fibroblast activation and fibrosis. PDGF-BB stimulates polarization and provides enhancement and directionality for collagen-driven human dermal fibroblast migration. Akt processes both migratory and proliferative signals from PDGF receptors [76].	Breast tumor cells produce PDGF-CC to activate stromal fibroblasts that in turn confer the basal and estrogen receptor α-negative phenotypes into cancer cells, rendering them unresponsive to endocrine treatment [77].
PGE$_2$-Wnt	Dermal fibroblasts expressing a low level of Dickkopf 1, a Wnt signaling antagonist, exhibit enhancement of the canonical Wnt/β-catenin signal pathway with accumulation of prostaglandin E$_2$ (PGE$_2$) [78]. The PGE$_2$ signaling also increases nuclear β-catenin signaling in fibroblasts.	Autocrine activity of PGE$_2$ regulates the production of angiogenic factors by fibroblasts, which are key to the vascularization of both primary and metastatic tumor growth [79]. Simultaneous activation of PGE$_2$ and Wnt signals in transgenic mice causes gastric cancer with an abundance of vascular endothelial growth factor-A (VEGF-A) expressing CAFs, derived from bone marrow [80].
TGF-β	Upon TGF-β stimulation, fibroblasts are activated and undergo phenotypic transition into myofibroblasts, the key effector cells under fibrotic conditions. The myofibroblast phenotype is characterized by the formation of gap junctions and by the acquisition of a contractile apparatus with associated contractile proteins. In healing wounds, myofibroblasts are required for tissue repair prior to their elimination due to the induction of apoptosis, but constitutively activated myofibroblasts promote fibrosis [81].	Increased TGF-β production by tumor cells gives rise to the desmoplastic stroma in murine tumor models [82,83]. TGF-β potently suppresses immunity, induces angiogenesis and promotes cancer cell migration and invasion by stimulating EMT. Moreover, cancer cell-derived TGF-β activates TGF-β signaling in CAFs, inducing the up-regulation of monocarboxylate transporter 4 (MCT4) (a marker of glycolysis) and BNIP3 (a marker of autophagy) and the loss of caveolin-1 (CAV1) [84].

2.2. Emerging Roles of CAFs for Therapeutic Resistance

Recent emerging evidence supports crucial roles of CAFs for therapeutic resistance, as exemplified by innate and adaptive resistance in various human carcinomas (Figure 2).

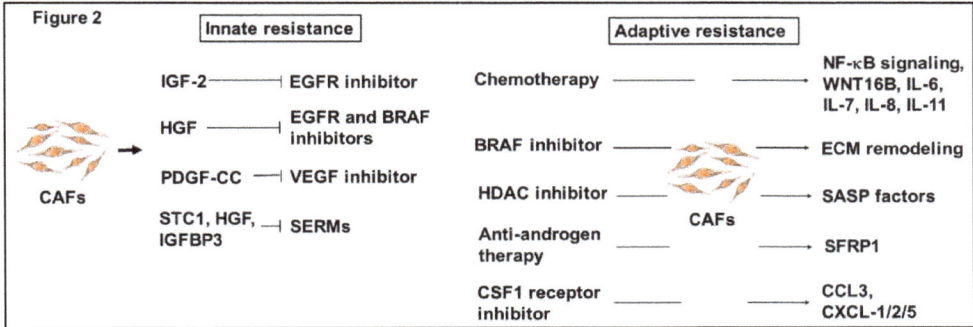

Figure 2. Emerging roles of CAFs for therapeutic resistance. CAFs play crucial roles in innate resistance to anti-cancer drugs (left). CAF-released insulin-like growth factor 2 (IGF2) provides cancer cells with tumor-initiating ability and EGFR-TKI-resistance. HGF produced by CAFs blunts the efficacy of BRAF and EGFR inhibitors in BRAF-mutant melanoma cells and lung cancer cells via MAPK and PI3K/AKT signal pathways. CAF-produced PDGF-CC attenuates the efficacy of anti-VEGF therapy via increasing neo-angiogenesis. Breast tumor cell-derived PDGF-CC also enables CAFs to produce stanniocalcin 1 (STC1), HGF and IGFBP3 that contribute to promoting conversion of luminal cancer cells into basal cancer cells, resulting in resistance to treatment with selective estrogen receptor modulators (SERMs). Upon therapeutic insult, CAFs acquire adaptive resistance (right). Chemotherapy induces pro-inflammatory phenotypes in CAFs via activation of NF-κB signaling, resulting in enhanced production of Wnt family member wingless-type MMTV integration site family member 16B (WNT16B), IL-6 and IL-8 from these fibroblasts that provides breast cancer cells with chemoresistant ability. Increased levels of IL-7 and IL-11 production are also induced in CAFs by chemotherapy, rendering cancer cells tumor-initiating and apoptosis-resistant. Treatment of CAFs with the BRAF inhibitor induces ECM remodeling, resulting in activation of integrin β1/focal adhesion kinase (FAK)/Src and ERK signaling in melanoma cells. The histone deacetylase (HDAC) inhibitor treatment enables CAFs to produce the senescence-associated secretory phenotype (SASP) factors. Exposure to anti-androgen therapy encourages CAFs to produce SFRP1 that promotes prostate cancer neuroendocrine differentiation. Treatment of different human carcinomas with the CSF1 receptor inhibitor targeting TAMs allows CAFs to boost CCL3 and CXCL-1/2/5 productions, resulting in the recruitment of MDSCs into tumors and thus promoting tumor growth and progression.

2.2.1. Innate Resistance of CAFs to Anti-Cancer Drugs

CAFs produce inherently increased levels of growth factors and inflammatory cytokines that attenuate the efficacy of anti-cancer treatment. For example, CAFs produce an abundance of insulin-like growth factor 2 (IGF2) that renders cholangiocarcinoma and pancreatic cancer cells resistant to EGFR tyrosine kinase inhibitors (TKI) by activating the insulin receptor (IR)/insulin-like growth factor 1 receptor (IGF1R) signaling axis [85,86]. CAF-produced IGF2 also reportedly promotes invasion and metastasis of colon cancer cells [87]. Moreover, stromal IGF2 induces NANOG expression and thus boosts the cancer-initiating properties of lung cancer cells through IR/IGF1R signaling followed by activation of the AKT-PI3K pathway [88].

Mechanisms underlying the stroma-mediated innate resistance to the BRAF inhibitor have also been addressed using BRAF-mutant melanoma cells in other studies. Hepatocyte growth factor (HGF) released from fibroblasts contributes to resistance to the BRAF inhibitor, presumably via the downstream signaling of the MAPK and PI3K/AKT pathways [89]. While treatment with BRAF-

and MEK-inhibitors is not sufficient to overcome HGF-induced resistance, BRAF- and MET (the HGF receptor)-inhibitors suppress the majority of HGF-induced drug resistance in BRAF-mutant melanoma [89,90]. Similarly, HGF-producing human fibroblastic cells confer resistance to gefitinib, a TKI selective to EGFR in lung cancer cells through the PI3K/Akt signal pathway [91]. Importantly, anti-HGF neutralizing antibody and the natural HGF inhibitor NK4 significantly overcome the gefitinib resistance in culture and in tumor xenografts raised by lung cancer cells admixed with HGF-producing human fibroblasts in mice treated with gefitinib.

Inhibition of VEGF-A is effective in treating several human carcinomas [92]. However, tumors often show resistance to anti-VEGF treatment. Importantly, CAFs have been shown to mediate the resistance to anti-VEGF therapy and the molecular mechanism was elucidated in murine lymphoma models [93]. Murine CAFs were isolated from subcutaneous tumors developed by lymphoma cell lines resistant to the antiangiogenic therapy with VEGF inhibitors. The increased level of PDGF-CC produced by CAFs resulted in rendering tumor cells resistant to anti-VEGF therapy by stimulation of neoangiogenesis, when lymphoma cells otherwise sensitive to anti-VEGF therapy were co-implanted with these fibroblasts into recipient mice [93].

Human breast cancers of the luminal subtypes expressing female hormone receptors are effectively treated with selective estrogen receptor modulators (SERMs), such as tamoxifen, while tumors of the basal-like subtype that do not express hormone receptors fail to have effective targeted therapies. Roswall et al. have recently demonstrated that CAFs play crucial roles in regulating the phenotypic conversion of luminal breast cancers into basal-like cancers, which show the worst overall survival among various human breast cancer subtypes [77]. Human breast cancer cells produce PDGF-CC that acts onto the cognate PDGF receptors expressed on closely apposed CAFs to activate these fibroblasts. The resulting activated CAFs produce stanniocalcin 1 (STC1), HGF and insulin growth factor binding protein 3 (IGFBP3), all of which downregulate the expression levels of the luminal markers including FOXA1, estrogen receptor and GATA3, resulting in the conversion of the luminal tumors into basal-like tumors [77]. Notably, the luminal phenotype and sensitivity to endocrine therapy were shown to be restored in otherwise resistant tumors, not only by genetic targeting of the PDGF-C gene in the MMTV-PyMT murine basal-like breast cancer model, but also by treatment with the neutralizing PDGF-CC antibody of patient-derived triple-negative breast tumor xenografts transplanted orthotopically in immunodeficient mice.

2.2.2. Adaptive Resistance of CAFs to Anti-Cancer Drugs

CAFs have been shown to prime chronic inflammation, as exemplified by recruitment of protumorigenic macrophages in an NFκB signal-dependent fashion, resulting in the promotion of tumor growth and angiogenesis [94]. With stress exposures, such as chemotherapy and radiation, these fibroblasts also acquire the pro-inflammatory phenotype via further activating NFκB signaling, resulting in increased survival signals of cancer cells. The activated NFκB signaling in the therapy-treated CAFs enhances production of different cytokines including Wnt family member wingless-type MMTV integration site family member 16B (WNT16B), IL-6 and IL-8, leading to the induction of chemoresistance in breast cancer cells [95,96]. CAFs treated with cisplatin also boost IL-11 production to activate the STAT3 signal pathway and upregulate the expressions of the anti-apoptotic proteins such as Bcl-2 and survivin for prostate cancer cells to acquire resistance to apoptosis [97]. Moreover, CAFs extracted from freshly resected human colorectal cancer specimens after chemotherapy reportedly show higher IL-7 production than those without chemotherapy [98]. This stromal IL-7 provides CD44-positive colon cancer cells with further increased tumor-initiating ability, thereby promoting tumor cell growth both in vitro and in vivo [98].

Furthermore, chemotherapy-induced stromal chronic inflammation is responsible for angiogenesis and ECM remodeling, and subsequently provides tumor cells with a physical barrier against the cytotoxic agents administered [33,99,100]. Recent investigations have shown that conventional chemotherapy and radiotherapy can lead to increased tumor stiffness involving the stroma response [49,101,102].

Remarkably, treatment with PLX-4270 (BRAF inhibitor) activates tumor-associated fibroblasts to induce ECM remodeling that activates integrin β1/FAK/Src signaling in melanoma cells [101]. This signal activation then renders melanoma cells resistant to PLX-4270 via ERK activation.

Treatment of stromal fibroblasts with high concentrations of HDAC inhibitors, such as SAHA, TSA and vorinostat, causes the senescence-associated secretory phenotype (SASP) mediated by the direct activation of the NFκB signal [103,104]. Treating fibroblasts with HDAC inhibitors results in significant paracrine stimulation of tumor growth, which suggests that high-dose HDAC inhibitors would likely impact the stromal compartment adversely in a therapeutic setting.

Intriguingly, androgen deprivation therapy increases the population of CD105 (endoglin)-positive CAFs, which contribute to neuroendocrine differentiation of epithelial prostate carcinoma cells [105]. CAF-derived secreted frizzled-related protein 1 (SFRP1), which is driven by the CD105-mediated signal pathway, is both necessary and sufficient to induce the neuroendocrine differentiation of prostate carcinoma in a paracrine manner. These series of observations raise concern regarding the undesirable side effects of therapeutic agents; paracrine signaling from treatment-primed CAFs might influence the regrowth and malignancy of nearby tumor cells.

CAFs play a key role in driving the drug resistance not only by raising particular gene expression patterns and signaling pathways as mentioned above, but also by stimulating recruitment of immunosuppressive cells into the tumor. Tumor-associated macrophages (TAMs) are non-neoplastic cells abundant in stroma of different human tumors and exert either pro-tumoral or tumoricidal functions in response to cytokine exposure [106,107]. A growing body of evidence indicates that pro-tumoral TAMs support tumor growth and progression by influencing tumor hallmarks [107]. Colony-stimulating factor 1 (CSF1) receptor signaling is a key regulator of TAM recruitment, differentiation and survival. Treatment with CSF1 receptor inhibitors targeting TAMs clearly reduces tumor growth in murine tumor models, though the anti-tumor effect was shown to be very limited in patients [106]. The molecular mechanisms underlying the tumor progression elicited by substantial depletion of TAMs remain, however, unknown. Importantly, a recent study revealed this to be due to increased CCL3 and CXCL-1/2/5 productions from CAFs treated with the CSF1 receptor inhibitor [108]. These CAF-produced chemokines then stimulate the recruitment of polymorpho-nuclear myeloid-derived suppressor cells (PMN-MDSCs) into tumors, resulting in the promotion of tumor growth and progression. These findings therefore demonstrate that CAFs mediate neutralization of the anti-tumor effect exerted by CSF1 receptor inhibitors via recruitment of PMN-MDSCs into tumors.

2.3. Cross-Talk between CAFs and Tumor Microenvironment

The wound-healing program is strongly dependent on the cross-talk between various stromal cells and myofibroblasts at the wound site [109,110] (Figure 1). For instance, myofibroblasts induce angiogenesis from preexisting parental vessels or from the circulating endothelial precursor cells (EPCs) recruited at the wound site via the secretion of a potent preangiogenic chemokine, CXCL12, also known as stromal cell-derived factor-1 (SDF-1) [111,112]. Chemotactically-attracted EPCs then transdifferentiate into endothelial cells in the presence of VEGF, which is also secreted by myofibroblasts.

In certain contexts, myofibroblastic CAFs induce neo-angiogenesis via secretion of preangiogenic factors including CXCL12, VEGF, PDGF, TGF-β and HGF in a wide range of cancers [49,50,113–117]. Most desmoplastic tumors are highly vascularized, wherein shifting the switch toward an angiogenesis-promoting phenotype occurs [118,119]. Interestingly, as in the case of wound healing, the CAF niches depend on the CXCL12/CXCR4 axis and VEGF production to stimulate the formation of neovasculature at the invasive front of breast cancer [115,120–122]. As the CXCR4 receptor for CXCL12 is expressed on both the tumor cells and EPCs [123,124], CAF-produced CXCL12 stimulates tumor growth and neoangiogenesis via acting CXCR4 expressed on these cells. The production of VEGF from tumor cells and CAFs also boosts neoangiogenesis in breast cancer tissues. Collectively, niches of myofibroblasts in wounds and CAFs are both likely to support angiogenesis, apparently through similar signaling pathways.

It was recently shown that hypermethylated in cancer 1 (HIC1), which is a tumor suppressor gene located at 17p13.3, resides exclusively within CpG islands, frequently showing hypermethylation in several tumors such as breast, lung and prostate carcinomas [125–128]. A recent study found that HIC1-depleted breast cancer cells markedly produce CXCL14 that activates Akt and ERK1/2 signal pathways, by acting through its cognate receptor GPR85 on resident fibroblasts in a paracrine manner, resulting in the induction of a phenotypic conversion into CAFs [127]. The activated CAFs, in turn, boost the production of chemokine CCL17 that acts on its cognate receptor CCR4, located on breast cancer cells, to drive metastasis. Collectively, the HIC1-CXCL14-CCL17 positive-feedback loop reciprocally mediating interactions between breast tumor cells and myofibroblastic CAFs contributes to the malignant potentials of breast tumors.

CAFs remodel the ECM components and thereby regulate tumor stiffness [129]. It was recently shown that squamous cancer cells activate EGFR in response to tumor stiffness, which leads to actomyosin contractility and collective invasion [130]. From a mechanistic standpoint, enhanced tyrosine kinase activity of EGFR results in Ca^{2+}-dependent regulation of Cdc42 small GTPase activity in tumor cells, which in turn leads to phosphorylation of MLC2. The MLC kinase regulates actomyosin-dependent ECM remodeling due to CAFs. Surprisingly, two Ca^{2+} channel blockers, the phenylalkylamine verapamil and the nondihydropiridine diltiazem, which have been used for treating hypertension and arrhythmia for decades, show the therapeutic efficacy for preventing collective cancer invasion, an effect achieved by significantly down-regulating the phosphorylation of MLC2.

It is noteworthy that mechanical force-mediated ECM remodeling by CAFs depends on actomyosin contractility generated through the Rho-associated protein kinase (ROCK) signal pathway [129]. The IL-6/JAK1/signal transducer and activator of transcription 3 (STAT3) axis also controls actomyosin contractility by regulating the levels of phosphorylated-MLC2 in both melanoma cells and CAFs. In striking contrast to melanoma cells, in which the IL-6-gp130/JAK1-ROCK axis is required for the amoeboid-like individual tumor migration, this signaling pathway is not required for the migration of squamous carcinoma cells themselves, but is required for CAFs to remodel the matrix, which is necessary for promoting the collective invasion of these carcinoma cells [64].

EMT of epithelial tumor cells, the process by which the number of tumor-initiating cells (TICs) is increased [131,132], is apparently regulated in cooperation with CAFs. Recent studies also highlight the importance of epithelial-mesenchymal plasticity to be determined by several transcription factors, including ZEB1, Snail and Twist. The resulting tumor cells with the hybrid epithelial/mesenchymal trait mediated by partial EMT are considered to enhance the tumor-initiating, invasive and metastatic properties as well, along with promoting chemoresistance [132–135]. These phenotypic changes are also presumably induced by CAF-regulated ECM components, exosomes and soluble factors.

3. Metabolic Reprogramming of CAFs During Cancer Progression

3.1. Metabolic Symbiosis Between Cancer Cells and CAFs

Features of desmoplastic tumor stroma resemble those of wound healing and involution during gestation. Mammography measures and compares the different types of breast tissue visible on a mammogram, which is an X-ray image of the breasts routinely used to screen for breast cancer in clinics. High breast density represents a greater amount of glandular and connective tissue than fat and has an association with higher risk for breast cancer development [136]. The expression level of CD36, a cell surface receptor for fatty acids, is downregulated in fibroblasts extracted from noncancerous breast tissues with high mammographic density as well as in breast CAFs [137]. CD36 expression is also known to be required for human mammary fibroblasts to transdifferentiate into preadipocytes in culture. Consistently, adipocytes were shown to regenerate from myofibroblasts in a murine skin wound healing model [138]. These findings demonstrate that down-regulation of CD36 expression in the stroma results in increased fibrosis in the breast, resulting in high mammographic density, presumably via attenuated transdifferentiation into adipocytes from mammary fibroblasts.

AMP-activated protein kinase (AMPK) reportedly regulates the translocation of the fatty acid transporter CD36 from intracellular stores to the plasma membrane [139], thereby promoting fatty acid uptake into skeletal muscle. CD36 also contributes to the activation of mitochondrial fatty acid β-oxidation (FAO) which in turn influences their metabolic plasticity in ovarian and oral carcinoma cells, leading to greater lymph node metastasis [140,141].

CD36 is involved in caveolae, a subset of lipid rafts forming part of the cell membrane microdomain enriched in cholesterol and signaling proteins. Decreased expression of CAV1, another component of caveolae within the tumor microenvironment, is also consistently associated with poor clinical outcomes in patients with a wide variety of malignancies [142]. CAV1-deficient fibroblasts also show concomitantly decreased CD36 expression, stabilization of hypoxia-induced factor-1α (HIF-1α), activation of TGF-β signal transduction and induction of myofibroblast differentiation [142,143]. These CD36-deficient fibroblasts likewise undergo a metabolic shift from mitochondrial oxidative phosphorylation to aerobic glycolysis, promoting the metabolic plasticity of these fibroblasts [142,144,145] (Figure 3). These findings suggest that altered caveolae function in the tumor microenvironment induces tumor metabolic heterogeneity, leading to the manifestation of malignant features. These mechanistic insights into how the alteration of caveolae is induced and maintained in CAFs are currently under investigation.

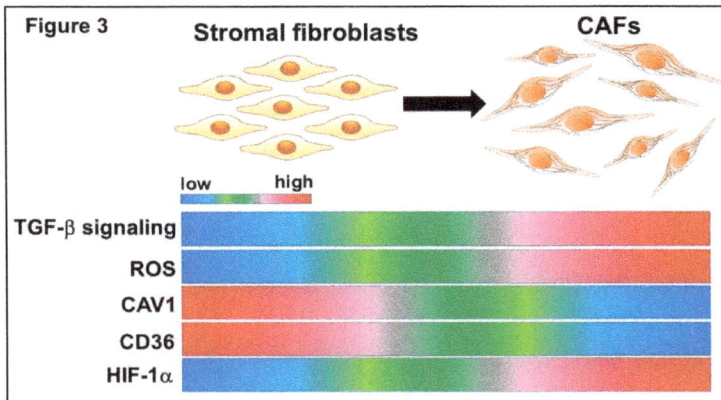

Figure 3. Metabolic reprogramming in CAFs. CD36 and caveolin 1 (CAV1) are components of caveolae, a subset of lipid rafts found in the cell membrane microdomain enriched for cholesterol and signaling proteins. These CD36 and CAV1 expressions are downregulated in CAFs. The attenuated CAV1 expression concomitantly decreases CD36 expression, stabilizes hypoxia-induced factor-1α (HIF-1α), activates TGF-β signal transduction and induces myofibroblast differentiation in fibroblasts. This attenuated CD36 expression also shows a metabolic shift from mitochondrial oxidative phosphorylation to aerobic glycolysis, promoting metabolic plasticity in these fibroblasts. Tumor-derived reactive oxygen species (ROS) are responsible for down-regulation of CAV1 in CAFs. Loss of CAV1 in CAFs also results in ROS elevations, which in turn stabilize HIF-1α.

Initially, the Warburg effect was believed to be confined to specific tumor cell types [146–148]. However, the emerging concept of a "reverse Warburg effect" has recently attracted considerable research attention [145,149,150]. Tumor-derived reactive oxygen species (ROS) are responsible for down-regulation of CAV1 in CAFs [151–153] (Figure 3). Loss of CAV1 in CAFs also results in ROS elevations, which in turn stabilize HIF-1α. In other words, malignant cells induce a "pseudo-hypoxic" microenvironment for CAFs [145,154]. Because the transcription factor HIF-1α promotes glycolysis and provides cancer cells with lactate and glutamate, elevated ROS production in tumor cells indirectly induces the uptake of intermediate metabolites of the tricarboxylic acid (TCA) cycle in mitochondria (Figure 4). Of note, CAFs consume more glucose and secrete more lactate than normal fibroblasts [154–156]. Furthermore, CAFs depend significantly on autophagy that may lead to resistance to chemotherapy [145,155,157]. Collectively, fibroblasts surrounding epithelial tumor cells undergo metabolic reprogramming, which results in a metabolic phenotype resembling that induced by the Warburg effect. Importantly, metabolic symbiosis between epithelial cancer cells and CAFs requires a cell population to express a different MCT subtype [145,158–161]. Epithelial tumor cells express MCT1, which contributes to uptake of the lactate provided by CAV1-deficient CAFs, which in turn express MCT4, a marker of both aerobic glycolysis and lactate efflux (Figure 4).

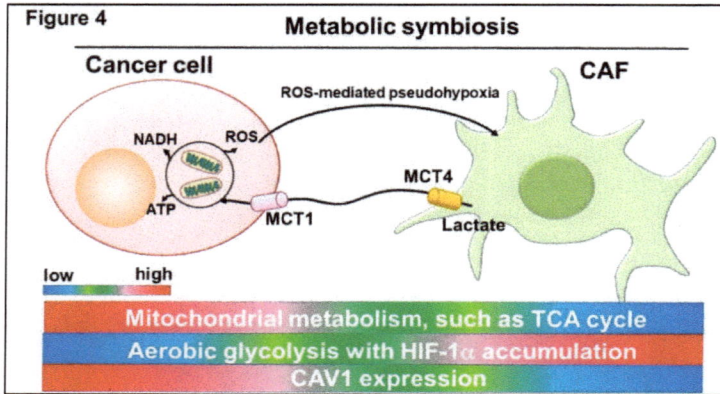

Figure 4. Metabolic symbiosis between cancer cells and CAFs requires the expression of a different MCT subtype. Monocarboxylate transporter 1 (MCT1)-expressing cancer cells induce ROS-mediated pseudohypoxia for MCT4-expressing CAFs, causing HIF-1α accumulation in the nucleus. CAFs depend on aerobic glycolysis and secrete lactate via MCT4. Cancer cells exhibit robust lactate uptake via MCT1, allowing them to generate large amounts of ATP via the mitochondrial TCA cycle. Tumor cells then efficiently produce metabolic intermediates, such as NADH by utilizing lactate derived from CAFs. ROS are a major hallmark of cancer tissues that drives robust metabolism in adjacent proliferating MCT1-positive cancer cells, which are abundant in mitochondria, mediated by the paracrine transfer of mitochondrial fuels, such as lactate, pyruvate and ketone bodies.

3.2. Signal Pathways Involved in Metabolic Reprogramming of CAFs

Accumulating evidence strongly suggests that p62 (also known as sequestosome 1) is involved in metabolic reprogramming of activated fibroblasts in fibrosis and tumor stroma [162–166]. p62 is a multifunctional adaptor protein and a specific substrate for autophagy. p62 is thus selectively incorporated into autophagosomes through the direct binding with LC3 (microtubule-associated protein light chain 3) to be degraded by autophagy [167,168].

Hepatic stellate cells (HSCs), which can transdifferentiate into myofibroblasts in response to certain stimuli, play critical roles in liver fibrosis and HCC development [163]. A study showed that vitamin D receptor (VDR) signaling exerts the anti-fibrotic and anti-inflammatory effects in HSCs [163]. p62 also mediates the anti-fibrotic function by a direct interaction with VDR and the retinoid X receptor that promotes their heterodimerization, a process critical for target gene recruitment [163]. Moreover, Duran et al. have shown that loss of p62 expression in HSCs enhances their myofibroblastic differentiation, thereby impairing suppression of fibrosis and inflammation by VDR agonists in chemical agent-induced murine fibrosis and tumor models. Consistent with the aforementioned observations, these findings demonstrate decreased p62 expression to be crucial for myofibroblast differentiation to exert their actions of supporting fibrosis and tumor growth via attenuated VDR signaling. However, the molecular mechanisms underlying the observed down-regulation of p62 expression in activated fibroblasts remain as yet unknown.

Significant down-regulation of p62 expression also underlies the metabolic reprogramming in CAFs mediated by the mammalian target of rapamycin (mTOR) complex 1/Myc cascade controlling IL-6 secretion [166]. Reduced activity of the mTOR complex 1 in p62-deficient fibroblasts accounts for c-Myc down-regulation and the subsequent up-regulation of IL-6, resulting in the promotion of inflammation and tumorigenesis. The lack of p62 in CAFs promotes resistance to glutamine deprivation by directly regulating ATF4 stability via its p62-mediated polyubiquitination [162]. Interestingly, selective autophagy mainly regulated by p62 does not account for the capacity of p62-deficient CAFs to withstand glutamine starvation [162,166]. Up-regulation of ATF4 due to p62 deficiency in the tumor stroma enhances glucose carbon flux through a pyruvate carboxylase-asparagine synthase cascade, which in turn results in asparagine generation as a compensatory source of the nitrogen required for proliferation of both cancer cells and CAFs. It has been shown both in vitro and in vivo that p62-deficient stromal fibroblasts produce non-essential amino acids which are crucial for proliferation in the absence of glutamine by maintaining the TCA cycle in mitochondria, explaining how the p62-deficient tumor stroma stably provides asparagine in an ATF4-dependent manner [162,164,169]. In addition, CSL/RBPJκ, a transcriptional suppressor which is converted into an activator by Notch, plays the role of a negative regulator for CAFs [165]. CSL interacts with p62 and their expression levels are downregulated in murine dermal CAFs through autophagy, indicating that autophagy downmodulates CSL protein expression via p62 in CAFs [165].

4. Targeting Tumor Stroma Fibroblasts to Attenuate Tumor Progression

4.1. Tumor Stiffness and Enhanced Interstitial Fluid Pressure

As noted above, both tumor cells and CAFs secrete a number of factors which promote angiogenesis, the most widely-accepted of which are members of the VEGF family. Angiogenesis and lymphatic co-options correlate with tumor progression and poor patient outcomes, and are the primary contributors to the altered fluid flow and interstitial fluid pressure (IFP) in the tumor microenvironment [170–172]. Vessels developing in the tumor microenvironment are generally irregular and have major gaps in the endothelial cell layer, reducing the degree of coverage by myofibroblasts and pericytes [173–176]. Furthermore, myofibroblastic CAFs induce not only increases in the numbers of fibrotic foci, but also the contraction of the interstitial space [33,177]. The increased vessel number in conjunction with increased hydraulic conductivity or the relative ease with which fluid moves across the vessel wall, is responsible for the irregular and increased influx of fluid into the tumor stroma. Indeed, rising IFP is frequently reported in solid tumors, such as breast carcinoma, glioblastoma and malignant melanoma [178–180]. This increased IFP is due not only to fluid failing to properly drain out of the interstitial space, but also to a number of other physiological changes in the tumor microenvironment, including both an increased number of tumor cells and more ECM deposition in the tumor stroma.

Recent studies support tumor microenvironment stiffness as a therapeutic target aimed at preventing cancer development and progression [181–184]. The tumor stromal region typically consists of excessive amounts of fibrous collagen, which can be cross-linked by soluble mediators, such as lysyl oxidase (LOX), thereby increasing the stiffness of the tumor microenvironment [185–187]. In turn, this increased tumor stiffness is considered to profoundly influence tumor progression inducing activated oncogenic signal pathways driven by activated FAK, Akt, β-catenin, and PI3K, as well as the inhibition of tumor suppressor molecules, such as PTEN. Targeting tumor stiffness via the inhibition of LOX enzymatic activity has been demonstrated to decrease metastatic dissemination of breast and colorectal tumor cells in vivo [188,189]. Treatment with LOX-blocking antibody in combination with gemcitabine also shows attenuated metastases of early-stage pancreatic tumors in Pdx1-Cre

KrasG12D/+ Trp53R172H/+ (KPC) mice, however, the effects are not observed in late-stage tumors, presumably due to the presence of considerable levels of already established cross-linked collagen [190].

In pancreatic cancers, tumor cells secrete Hedgehog (Hh) ligands to act on a patched (PTCH1) receptor expressed in CAFs. Hh signaling is thus activated by the ligand binding to PTCH1 that relieves an inhibitory effect on Smoothened (SMO) in these fibroblasts [191]. Hh signaling in CAFs coordinates the acquisition of a poorly-vascularized, desmoplastic microenvironment which impairs drug delivery in pancreatic adenocarcinoma [192–194]. Suppression of the stromal Hh signal by the inhibitor of SMO improves the delivery of gemcitabine via transiently increasing vascular density of pancreatic tumors in KPC mice [194].

Increased tumor stiffness impacts not only tumor cells, but also similarly exerts its effects on the surrounding stromal cells, wherein tumor stiffness activates normal fibroblasts to acquire CAF phenotypes and maintains them by the nuclear localization of yes-associated protein (YAP) in the Hippo signal pathway [195–198]. Actomyosin contractility and Src function are required for YAP activation by stiff matrices. Conversely, YAP depletion reduces the ability of CAFs to form fibrous collagen networks and to promote angiogenesis in vivo. YAP regulates expression levels of several cytoskeleton-related molecules including ANLN and DIAPH3 and then stabilizes MLC2/MYL9. Matrix stiffness further enhances YAP activation, thereby establishing a feed-forward self-reinforcing loop which helps to maintain the CAF phenotype [198]. Increased YAP1 activity in CAFs thus also induces a stiff ECM associated with the Rho-ROCK axis, thereby activating both Src and YAP signaling in a self-stimulating manner.

4.2. Therapeutic Strategy Against Activated Tumor Stroma

As noted above, CAFs compromise the effects of cancer therapies not only by producing large amounts of tumor-promoting growth factors and inflammatory cytokines, but also by recruiting other stromal cell types including immunosuppressive inflammatory cells into tumors. Nonetheless, as clonal somatic genetic alterations are rarely harbored in CAFs of different human carcinomas and these fibroblasts are anticipated to be less likely than carcinoma cells to acquire resistance to therapy [199], CAFs are speculated to be a promising therapeutic target [50,200,201].

Treatment with chemotherapy significantly eradicates chemosensitive tumors. However, a considerable number of CAFs often survive in the remnant tumors after treatment. The surviving CAFs acquire innate and adaptive therapeutic resistance that are accompanied by stromal inflammation and increased collagen accumulation, leading to iatrogenic tumor stiffness and the development of chemoresistant tumors. Treatment with several drugs in combination with chemotherapy shows promising results compromising the CAF-induced drug resistance in murine tumor models (Figure 5).

Aberrant IFP elevation disrupts the distribution of systemically administered anti-cancer drugs and thereby compromises the treatment of solid tumors [171,172,202,203]. Hyaluronidases are enzymes that catalyze the degradation of hyaluronic acid (HA), a glycosaminoglycan distributed widely throughout various different tissues. Pegylated recombinant hyaluronidase, known as PEGPH20, contributes to the significant decrease in IFP, thus improving gemcitabine sensitivity in pancreatic ductal adenocarcinoma (PDAC) [204–206]. PEGPH20 reportedly reduces IFP in the PDAC microenvironment and expands the tumor vasculature to improve perfusion, which increases access for anti-tumor immune cells and therapeutic agents. A randomized phase II study was also performed using a total of 279 patients with previously untreated metastatic pancreatic ductal adenocarcinoma treated with chemotherapy alone or chemotherapy plus PEGPH20 [207]. The results demonstrated that PEGPH20 treatment is more beneficial in patients with HA-high tumors than in those with HA-low tumors. The level of HA in the tumor-associated stroma was also shown to be a promising biomarker for identifying patients who may benefit from PEGPH20 treatment [207].

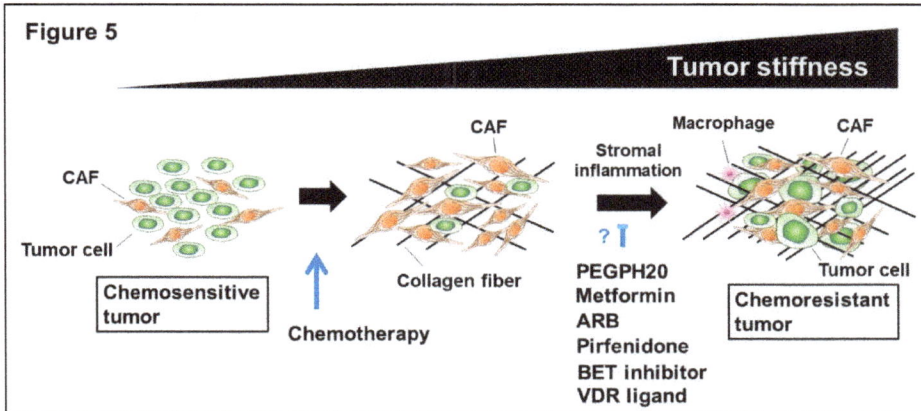

Figure 5. Schematic representation of chemoresistant tumor formation by CAFs and the potential treatment.

Metformin is an oral drug used in the management of patients with type II DM. Metformin administration was first reported to likely be associated with a reduced risk of cancer in the DM patients more than a decade ago [208]. Recently, mounting evidence has pointed to its anti-cancer effects in various malignancies [209–211]. This is a typical example of drug re-positioning [212]. Metformin has been reported to play a suppressor role in inflammatory and fibrosis-related diseases, such as atherosclerosis, cardiac fibrosis, renal fibrosis, interstitial pulmonary fibrosis and endometriosis [213–216]. Inhibition of TGF-β signal pathway, monocyte-to-macrophage differentiation, and NFκB-mediated inflammatory factors is involved in the molecular mechanisms underlying these non-malignant diseases. Importantly, similar mechanisms of metformin action have been suggested to contribute to suppression of the stromal reaction in tumors. In lung cancer, metformin was demonstrated to suppress pulmonary interstitial fibrosis during gefitinib therapy [217]. In ovarian cancer patients, cisplatin administration increases IL-6-producing myofibroblastic CAFs populations through activation of NFκB signal pathway in the tumor-associated stroma [218]. Pretreatment with metformin actually inhibits the desmoplastic stromal reaction via attenuation of the NFκB signal and IL-6 secretion from CAFs. This explains how the IL-6 receptor antagonist, which has conventionally been used for treating rheumatoid arthritis, might serve as an anti-cancer agent in clinical settings.

Renin-angiotensin system inhibitors, which have been prescribed for the treatment of cardiovascular diseases, receive considerable attention in oncology [219]. Angiotensin II (AngII) /AngII type I receptor (AT1R) axis plays pivotal roles in promoting tumor growth and progression. The treatment of CAFs with losartan, which is a selective AT1R blocker (ARB), reportedly attenuates activated fibroblastic state, as exemplified by TGF-β signaling and α-SMA expression, as well as ECM production in culture [220]. Importantly, in mice orthotopically bearing breast cancer cells chemotherapy in combination with losartan inhibits tumor growth and increases survival of the mice more significantly than monotherapy does [219–221].

Pirfenidone is an orally active synthetic anti-fibrotic agent structurally similar to pyridine 2,4-dicarboxylate [222]. This drug was recently approved for the treatment of patients with IPF. Pirfenidone exerts anti-fibrotic effects through inhibition of TGF-β and Hh signaling in lung fibroblasts of IPF patients [223]. Miura et al. recently reported that pirfenidone is likely to reduce the risk of lung cancer development in patients with IPF [224]. The retrospective analysis also demonstrated lung cancer incidence to be significantly lower in a pirfenidone-treated group than in a non-pirfenidone-treated group. Indeed, this anti-fibrotic agent induces apoptotic cell death of CAFs residing among NSCLC cells [225] and decreases the expression of collagen triple helix repeat containing 1 (CTHRC1), which is associated with tumor aggressiveness and poor clinical outcomes for NSCLC patients [226]. Importantly, simultaneous co-administration of pirfenidone with chemotherapy inhibits tumor growth and metastasis of breast and pancreatic carcinoma cells, presumably due to attenuation of the TGF-β signal pathway, activated fibroblastic state and ECM protein production in CAFs [227–229].

Bromodomain-containing protein 4 (BRD4), a bromodomain and extra-terminal (BET) family member is an important epigenetic reader. BRD4 is critical for the activated fibroblastic state and enhancer-mediated profibrotic gene expression in both HSCs and CAFs [230,231]. Several investigations also show that treatment with a small molecule inhibitor of BRD4 significantly attenuates fibrosis and tumorigenesis via inhibition of stromal TGF-β signaling in murine liver fibrosis models [230,232] as well as patient-derived pancreas and skin squamous cell carcinoma models [231,233].

Experimental evidence supports that the VDR signaling suppresses TGF-β-Smad2/3 signaling to attenuate the fibrotic reactions in fibrosis and tumor stroma [234]. Treatment with VDR ligands thus inhibits the activated state of HSCs and CAFs in murine liver fibrosis [235] and pancreatic carcinoma [236] models. This anti-fibrotic effect is due to the inhibition of the Smad3 recruitment into the binding sites of cis-regulatory regions in the profibrotic genes.

Taken together, these findings indicate that several anti-fibrosis drugs have major potential for impairing and even blocking tumor-promoting CAFs, resulting in attenuation of tumor growth and progression in experimental animal models.

5. Closing Remarks

In this review, we have described the close relationship between tumorigenesis and fibrosis, both of which are accompanied by the expansion of activated fibroblast populations. Several aspects of the cellular mechanisms underlying CAF-promoted tumorigenesis and therapy-resistance have been elucidated. However, there are CAF populations that have surprisingly been shown to suppress tumor growth and progression in different murine tumor models including those of the pancreas, bladder and colon [237–240]. These studies indicate that activation of Hh signaling in CAFs by tumor cell-produced Hh ligand suppresses the growth of tumors via bone morphogenetic protein (BMP) signaling in tumor cells, suggesting the presence of CAF populations with tumor-suppressive functions. In marked contrast, a very recent study using murine models of TNBC found that Hh signaling in CAFs promotes cancer stem cell plasticity and chemoresistance in cancer cells via elevation of stromal FGF5 production [182]. These contradictory observations raise the possibility of cancer cell context-dependent differences in stromal Hh signaling. Although tumor-suppressive or -promoting functions may be inherent in fibroblasts within tumors due to their multiple cells-of-origin, their complex interactions with other stromal cells and carcinoma cells with genetically and epigenetically diverse alterations would also presumably be crucial for generating CAF heterogeneity during tumor progression. Thus, CAFs can reasonably be described as a "cell state" rather than a "cell type" [241].

Several growth factors and cytokines have been identified as inducing CAF differentiation in progenitors, some of which consist of the feedback loop between cancer cells and CAFs in the tumor microenvironment. CAFs also stably maintain their transcriptome and metabolic profiling in an autocrine fashion. It is noteworthy that activated and tumor-promoting traits in these fibroblasts are retained during in vitro propagations, despite a lack of ongoing interactions with carcinoma cells, suggesting the key roles of epigenetic alterations in CAFs, as exemplified by DNA methylation [65]. Given that CAFs are responsible for high IFP in the tumor microenvironment, therapies aimed at preventing iatrogenic tumor stiffness hold great promise. Furthermore, it is remarkable that pirfenidone, one of the anti-fibrotic drugs, not only prevents IPF-associated lung cancer development but also inhibits the distant metastasis of difficult-to-cure breast carcinoma. However, the molecular mechanisms which would allow anti-cancer therapies to precisely target CAFs have yet to be elucidated. The importance of targeting the tumor stroma as well as tumor cells themselves has attracted increasing academic attention as researchers strive to achieve the precision medicine.

Funding: We gratefully acknowledge Grants-in-Aid for Scientific Research from the Ministry of Education, Culture, Sports, Science and Technology, Japan (Research Project Number: 18K07207 and 15K14385 to A.O.) and a Grant-in-Aid (S1311011) from the Foundation of Strategic Research Projects in Private Universities from the MEXT, Japan (A.O.)

Acknowledgments: The authors apologize to researchers whose work is not cited due to space limitations. We appreciate Ms. Chie Kataoka for proofreading and editing this manuscript.

Conflicts of Interest: The authors have no conflicts of interest to declare.

Abbreviations

AngII	angiotensin II
ARB	angiotensin II type I receptor blocker
AT1R	angII type I receptor
α-SMA	α-smooth muscle actin
BRD4	Bromodomain-containing protein 4
CAFs	carcinoma-associated fibroblasts
CAV1	caveolin 1
c-Abl	c-Abl protein kinase
CSF1	colony-stimulating factor 1
DM	diabetes mellitus
ECM	extracellular matrix
EGF	epithelial growth factor
EGFR	epidermal growth factor receptor
EMT	epithelial-mesenchymal transition
EndoMT	endothelial-mesenchymal transition
EPCs	endothelial precursor cells
ERK	extracellular signal-regulated kinase
FAK	focal adhesion kinase
FGF	fibroblast growth factor
FGFRs	fibroblast growth factor receptors
GSK-3β	glycogen synthase kinase-3β
HA	hyaluronic acid
HCC	hepatocellular carcinoma
HGF	hepatocyte growth factor
Hh	Hedgehog
HIC1	hypermethylated in cancer 1
HIF-1α	hypoxia-induced factor-1α
HSCs	hepatic stellate cells
IFP	interstitial fluid pressures
IGF1R	insulin-like growth factor 1 receptor
IGF2	insulin-like growth factor 2
IL	interleukin
IPF	idiopathic pulmonary fibrosis
IR	insulin receptor
JAK1	Janus kinase 1
LIF	leukemia inhibitory factor
LOX	lysyl oxidase
MAPK	mitogen-activate protein kinase
MCT	monocarboxylate transporter
MLC	myosin light chain
MMP	matrix metalloproteinase
MSCs	mesenchymal stem cells
mTOR	mammalian target of rapamycin
NASH	non-alcoholic steatohepatitis

NSCLC	non-small cell lung cancer
PAI-1	plasminogen activator inhibitor-1
PDAC	pancreatic ductal adenocarcinoma
PDGF	platelet-derived growth factor
PDGFR	platelet-derived growth factor receptor
PGE$_2$	prostaglandin E$_2$
PI3K	phosphatidylinositol-3 kinase
PIP3	phosphatidylinositol-3,4,5-triphosphate
PKC-δ	protein kinase C δ
KPC	Pdx1-Cre KrasG12D/+ Trp53R172H/+
PMN-MDSCs	polymorpho-nuclear myeloid-derived suppressor cells
PTEN	phosphatase and tensin homologue deleted on chromosome 10
ROCK	Rho-associated protein kinase
ROS	reactive oxygen species
SASP	senescence-associated secretory phenotype
SHP-1	src homology region 2 domain-containing phosphatase-1
SDF-1	stromal cell-derived factor-1
SMO	Smoothened
SPARC	secreted protein acidic and rich in cysteine
STAT3	signal transducer and activator of transcription 3
STC1	stanniocalcin 1
TAMs	tumor-associated macrophages
TCA	tricarboxylic acid
TGF-β	transforming growth factor-β
TKI	tyrosine kinase inhibitor
TNBC	triple-negative breast carcinoma
TIC	tumor-initiating cells
uPA	urokinase-type plasminogen activator
VDR	vitamin D receptor
VEGF	vascular endothelial growth factor
WNT16B	Wnt family member wingless-type MMTV integration site family member 16B
YAP	yes-associated protein

References

1. Hinz, B. Formation and function of the myofibroblast during tissue repair. *J. Invest. Dermatol.* **2007**, *127*, 526–537. [CrossRef]
2. Thannickal, V.J.; Lee, D.Y.; White, E.S.; Cui, Z.; Larios, J.M.; Chacon, R.; Horowitz, J.C.; Day, R.M.; Thomas, P.E. Myofibroblast differentiation by transforming growth factor-beta1 is dependent on cell adhesion and integrin signaling via focal adhesion kinase. *J. Biol. Chem.* **2003**, *278*, 12384–12389. [CrossRef] [PubMed]
3. Scharenberg, M.A.; Pippenger, B.E.; Sack, R.; Zingg, D.; Ferralli, J.; Schenk, S.; Martin, I.; Chiquet-Ehrismann, R. TGF-beta-induced differentiation into myofibroblasts involves specific regulation of two MKL1 isoforms. *J. Cell Sci.* **2014**, *127*, 1079–1091. [CrossRef]
4. Foster, D.S.; Jones, R.E.; Ransom, R.C.; Longaker, M.T.; Norton, J.A. The evolving relationship of wound healing and tumor stroma. *JCI Insight* **2018**, *3*, e99911. [CrossRef]
5. Carthy, J.M. TGFbeta signaling and the control of myofibroblast differentiation: Implications for chronic inflammatory disorders. *J. Cell Physiol.* **2018**, *233*, 98–106. [CrossRef] [PubMed]
6. Wipff, P.J.; Rifkin, D.B.; Meister, J.J.; Hinz, B. Myofibroblast contraction activates latent TGF-beta1 from the extracellular matrix. *J. Cell Biol.* **2007**, *179*, 1311–1323. [CrossRef] [PubMed]
7. Wynn, T.A.; Ramalingam, T.R. Mechanisms of fibrosis: Therapeutic translation for fibrotic disease. *Nat. Med.* **2012**, *18*, 1028–1040. [CrossRef] [PubMed]
8. Uehara, T.; Ainslie, G.R.; Kutanzi, K.; Pogribny, I.P.; Muskhelishvili, L.; Izawa, T.; Yamate, J.; Kosyk, O.; Shymonyak, S.; Bradford, B.U.; et al. Molecular mechanisms of fibrosis-associated promotion of liver carcinogenesis. *Toxicol. Sci.* **2013**, *132*, 53–63. [CrossRef] [PubMed]

9. Lederer, D.J.; Martinez, F.J. Idiopathic Pulmonary Fibrosis. *N. Engl. J. Med.* **2018**, *378*, 1811–1823. [CrossRef] [PubMed]

10. Zhao, M.; Dumur, C.I.; Holt, S.E.; Beckman, M.J.; Elmore, L.W. Multipotent adipose stromal cells and breast cancer development: Think globally, act locally. *Mol. Carcinog* **2010**, *49*, 923–927. [CrossRef] [PubMed]

11. Cox, T.R.; Erler, J.T. Molecular pathways: Connecting fibrosis and solid tumor metastasis. *Clin. Cancer Res.* **2014**, *20*, 3637–3643. [CrossRef] [PubMed]

12. Yoon, J.H.; Nouraie, M.; Chen, X.; Zou, R.H.; Sellares, J.; Veraldi, K.L.; Chiarchiaro, J.; Lindell, K.; Wilson, D.O.; Kaminski, N.; et al. Characteristics of lung cancer among patients with idiopathic pulmonary fibrosis and interstitial lung disease—Analysis of institutional and population data. *Respir. Res.* **2018**, *19*, 195. [CrossRef] [PubMed]

13. Tomassetti, S.; Gurioli, C.; Ryu, J.H.; Decker, P.A.; Ravaglia, C.; Tantalocco, P.; Buccioli, M.; Piciucchi, S.; Sverzellati, N.; Dubini, A.; et al. The impact of lung cancer on survival of idiopathic pulmonary fibrosis. *Chest* **2015**, *147*, 157–164. [CrossRef] [PubMed]

14. Wong, S.L.; Sukkar, M.B. The SPARC protein: An overview of its role in lung cancer and pulmonary fibrosis and its potential role in chronic airways disease. *Br. J. Pharmacol.* **2017**, *174*, 3–14. [CrossRef] [PubMed]

15. Trombetta-Esilva, J.; Bradshaw, A.D. The Function of SPARC as a Mediator of Fibrosis. *Open Rheumatol. J.* **2012**, *6*, 146–155. [CrossRef] [PubMed]

16. Hohenester, E.; Sasaki, T.; Giudici, C.; Farndale, R.W.; Bachinger, H.P. Structural basis of sequence-specific collagen recognition by SPARC. *Proc. Natl. Acad. Sci. USA* **2008**, *105*, 18273–18277. [CrossRef]

17. Chang, W.; Wei, K.; Jacobs, S.S.; Upadhyay, D.; Weill, D.; Rosen, G.D. SPARC suppresses apoptosis of idiopathic pulmonary fibrosis fibroblasts through constitutive activation of beta-catenin. *J. Biol. Chem.* **2010**, *285*, 8196–8206. [CrossRef]

18. Francki, A.; McClure, T.D.; Brekken, R.A.; Motamed, K.; Murri, C.; Wang, T.; Sage, E.H. SPARC regulates TGF-beta1-dependent signaling in primary glomerular mesangial cells. *J. Cell Biochem.* **2004**, *91*, 915–925. [CrossRef]

19. Koukourakis, M.I.; Giatromanolaki, A.; Brekken, R.A.; Sivridis, E.; Gatter, K.C.; Harris, A.L.; Sage, E.H. Enhanced expression of SPARC/osteonectin in the tumor-associated stroma of non-small cell lung cancer is correlated with markers of hypoxia/acidity and with poor prognosis of patients. *Cancer Res.* **2003**, *63*, 5376–5380.

20. Infante, J.R.; Matsubayashi, H.; Sato, N.; Tonascia, J.; Klein, A.P.; Riall, T.A.; Yeo, C.; Iacobuzio-Donahue, C.; Goggins, M. Peritumoral fibroblast SPARC expression and patient outcome with resectable pancreatic adenocarcinoma. *J. Clin. Oncol.* **2007**, *25*, 319–325. [CrossRef]

21. Said, N.; Frierson, H.F.; Sanchez-Carbayo, M.; Brekken, R.A.; Theodorescu, D. Loss of SPARC in bladder cancer enhances carcinogenesis and progression. *J. Clin. Investig.* **2013**, *123*, 751–766. [CrossRef]

22. Baffy, G.; Brunt, E.M.; Caldwell, S.H. Hepatocellular carcinoma in non-alcoholic fatty liver disease: An emerging menace. *J. Hepatol.* **2012**, *56*, 1384–1391. [CrossRef]

23. Lade, A.; Noon, L.A.; Friedman, S.L. Contributions of metabolic dysregulation and inflammation to nonalcoholic steatohepatitis, hepatic fibrosis, and cancer. *Curr. Opin. Oncol.* **2014**, *26*, 100–107. [CrossRef]

24. Monga, S.P. beta-Catenin Signaling and Roles in Liver Homeostasis, Injury, and Tumorigenesis. *Gastroenterology* **2015**, *148*, 1294–1310. [CrossRef]

25. Bakiri, L.; Hamacher, R.; Grana, O.; Guio-Carrion, A.; Campos-Olivas, R.; Martinez, L.; Dienes, H.P.; Thomsen, M.K.; Hasenfuss, S.C.; Wagner, E.F. Liver carcinogenesis by FOS-dependent inflammation and cholesterol dysregulation. *J. Exp. Med.* **2017**, *214*, 1387–1409. [CrossRef]

26. Chun, Y.S.; Huang, M.; Rink, L.; Von Mehren, M. Expression levels of insulin-like growth factors and receptors in hepatocellular carcinoma: A retrospective study. *World J. Surg. Oncol.* **2014**, *12*, 231. [CrossRef]

27. Enguita-German, M.; Fortes, P. Targeting the insulin-like growth factor pathway in hepatocellular carcinoma. *World J. Hepatol.* **2014**, *6*, 716–737. [CrossRef]

28. Xu, Z.; Hu, J.; Cao, H.; Pilo, M.G.; Cigliano, A.; Shao, Z.; Xu, M.; Ribback, S.; Dombrowski, F.; Calvisi, D.F.; et al. Loss of Pten synergizes with c-Met to promote hepatocellular carcinoma development via mTORC2 pathway. *Exp. Mol. Med.* **2018**, *50*, e417. [CrossRef]

29. Dvorak, H.F. Tumors: Wounds that do not heal. Similarities between tumor stroma generation and wound healing. *N. Engl. J. Med.* **1986**, *315*, 1650–1659.

30. Schafer, M.; Werner, S. Cancer as an overhealing wound: An old hypothesis revisited. *Nat. Rev. Mol. Cell Biol.* **2008**, *9*, 628–638. [CrossRef]
31. Piera-Velazquez, S.; Li, Z.; Jimenez, S.A. Role of endothelial-mesenchymal transition (EndoMT) in the pathogenesis of fibrotic disorders. *Am. J. Pathol.* **2011**, *179*, 1074–1080. [CrossRef] [PubMed]
32. Piera-Velazquez, S.; Mendoza, F.A.; Jimenez, S.A. Endothelial to Mesenchymal Transition (EndoMT) in the Pathogenesis of Human Fibrotic Diseases. *J. Clin. Med.* **2016**, *5*, 45. [CrossRef] [PubMed]
33. LeBleu, V.S.; Kalluri, R. A peek into cancer-associated fibroblasts: Origins, functions and translational impact. *Dis Model. Mech* **2018**, *11*, dmm029447. [CrossRef] [PubMed]
34. Kalluri, R.; Weinberg, R.A. The basics of epithelial-mesenchymal transition. *J. Clin. Investig.* **2009**, *119*, 1420–1428. [CrossRef] [PubMed]
35. Kokudo, T.; Suzuki, Y.; Yoshimatsu, Y.; Yamazaki, T.; Watabe, T.; Miyazono, K. Snail is required for TGFbeta-induced endothelial-mesenchymal transition of embryonic stem cell-derived endothelial cells. *J. Cell Sci.* **2008**, *121*, 3317–3324. [CrossRef]
36. Medici, D.; Potenta, S.; Kalluri, R. Transforming growth factor-beta2 promotes Snail-mediated endothelial-mesenchymal transition through convergence of Smad-dependent and Smad-independent signalling. *Biochem. J.* **2011**, *437*, 515–520. [CrossRef]
37. Batlle, E.; Sancho, E.; Franci, C.; Dominguez, D.; Monfar, M.; Baulida, J.; Garcia De Herreros, A. The transcription factor snail is a repressor of E-cadherin gene expression in epithelial tumour cells. *Nat. Cell Biol.* **2000**, *2*, 84–89. [CrossRef]
38. Li, Z.; Jimenez, S.A. Protein kinase Cdelta and c-Abl kinase are required for transforming growth factor beta induction of endothelial-mesenchymal transition in vitro. *Arthritis Rheumatol.* **2011**, *63*, 2473–2483. [CrossRef]
39. Hinz, B.; Phan, S.H.; Thannickal, V.J.; Prunotto, M.; Desmouliere, A.; Varga, J.; De Wever, O.; Mareel, M.; Gabbiani, G. Recent developments in myofibroblast biology: Paradigms for connective tissue remodeling. *Am. J. Pathol.* **2012**, *180*, 1340–1355. [CrossRef]
40. Hudson, L.G.; Newkirk, K.M.; Chandler, H.L.; Choi, C.; Fossey, S.L.; Parent, A.E.; Kusewitt, D.F. Cutaneous wound reepithelialization is compromised in mice lacking functional Slug (Snai2). *J. Dermatol. Sci.* **2009**, *56*, 19–26. [CrossRef]
41. Savagner, P.; Kusewitt, D.F.; Carver, E.A.; Magnino, F.; Choi, C.; Gridley, T.; Hudson, L.G. Developmental transcription factor slug is required for effective re-epithelialization by adult keratinocytes. *J. Cell Physiol.* **2005**, *202*, 858–866. [CrossRef]
42. Arpino, V.; Brock, M.; Gill, S.E. The role of TIMPs in regulation of extracellular matrix proteolysis. *Matrix Biol.* **2015**, *44–46*, 247–254. [CrossRef]
43. Vered, M.; Dayan, D.; Yahalom, R.; Dobriyan, A.; Barshack, I.; Bello, I.O.; Kantola, S.; Salo, T. Cancer-associated fibroblasts and epithelial-mesenchymal transition in metastatic oral tongue squamous cell carcinoma. *Int. J. Cancer* **2010**, *127*, 1356–1362. [CrossRef]
44. Hassona, Y.; Cirillo, N.; Heesom, K.; Parkinson, E.K.; Prime, S.S. Senescent cancer-associated fibroblasts secrete active MMP-2 that promotes keratinocyte dis-cohesion and invasion. *Br. J. Cancer* **2014**, *111*, 1230–1237. [CrossRef]
45. Tian, B.; Chen, X.; Zhang, H.; Li, X.; Wang, J.; Han, W.; Zhang, L.Y.; Fu, L.; Li, Y.; Nie, C.; et al. Urokinase plasminogen activator secreted by cancer-associated fibroblasts induces tumor progression via PI3K/AKT and ERK signaling in esophageal squamous cell carcinoma. *Oncotarget* **2017**, *8*, 42300–42313. [CrossRef]
46. Yamaguchi, Y.; Mann, D.M.; Ruoslahti, E. Negative regulation of transforming growth factor-beta by the proteoglycan decorin. *Nature* **1990**, *346*, 281–284. [CrossRef]
47. Imai, K.; Hiramatsu, A.; Fukushima, D.; Pierschbacher, M.D.; Okada, Y. Degradation of decorin by matrix metalloproteinases: Identification of the cleavage sites, kinetic analyses and transforming growth factor-beta1 release. *Biochem. J.* **1997**, *322 Pt 3*, 809–814. [CrossRef]
48. Gascard, P.; Tlsty, T.D. Carcinoma-associated fibroblasts: Orchestrating the composition of malignancy. *Genes Dev.* **2016**, *30*, 1002–1019. [CrossRef]
49. Mezawa, Y.; Orimo, A. The roles of tumor- and metastasis-promoting carcinoma-associated fibroblasts in human carcinomas. *Cell Tissue Res.* **2016**, *365*, 675–689. [CrossRef]
50. Pietras, K.; Ostman, A. Hallmarks of cancer: Interactions with the tumor stroma. *Exp. Cell Res.* **2010**, *316*, 1324–1331. [CrossRef]

51. Heneberg, P. Paracrine tumor signaling induces transdifferentiation of surrounding fibroblasts. *Crit. Rev. Oncol. Hematol.* **2016**, *97*, 303–311. [CrossRef]

52. Tao, L.; Huang, G.; Song, H.; Chen, Y.; Chen, L. Cancer associated fibroblasts: An essential role in the tumor microenvironment. *Oncol. Lett.* **2017**, *14*, 2611–2620. [CrossRef]

53. Polanska, U.M.; Orimo, A. Carcinoma-associated fibroblasts: Non-neoplastic tumour-promoting mesenchymal cells. *J. Cell Physiol.* **2013**, *228*, 1651–1657. [CrossRef]

54. Kalluri, R. The biology and function of fibroblasts in cancer. *Nat. Rev. Cancer* **2016**, *16*, 582–598. [CrossRef]

55. Ohlund, D.; Elyada, E.; Tuveson, D. Fibroblast heterogeneity in the cancer wound. *J. Exp. Med.* **2014**, *211*, 1503–1523. [CrossRef]

56. Mishra, P.J.; Mishra, P.J.; Humeniuk, R.; Medina, D.J.; Alexe, G.; Mesirov, J.P.; Ganesan, S.; Glod, J.W.; Banerjee, D. Carcinoma-associated fibroblast-like differentiation of human mesenchymal stem cells. *Cancer Res.* **2008**, *68*, 4331–4339. [CrossRef]

57. Karnoub, A.E.; Dash, A.B.; Vo, A.P.; Sullivan, A.; Brooks, M.W.; Bell, G.W.; Richardson, A.L.; Polyak, K.; Tubo, R.; Weinberg, R.A. Mesenchymal stem cells within tumour stroma promote breast cancer metastasis. *Nature* **2007**, *449*, 557–563. [CrossRef]

58. Raz, Y.; Cohen, N.; Shani, O.; Bell, R.E.; Novitskiy, S.V.; Abramovitz, L.; Levy, C.; Milyavsky, M.; Leider-Trejo, L.; Moses, H.L.; et al. Bone marrow-derived fibroblasts are a functionally distinct stromal cell population in breast cancer. *J. Exp. Med.* **2018**, *215*, 3075–3093. [CrossRef]

59. Midgley, A.C.; Rogers, M.; Hallett, M.B.; Clayton, A.; Bowen, T.; Phillips, A.O.; Steadman, R. Transforming growth factor-beta1 (TGF-beta1)-stimulated fibroblast to myofibroblast differentiation is mediated by hyaluronan (HA)-facilitated epidermal growth factor receptor (EGFR) and CD44 co-localization in lipid rafts. *J. Biol. Chem.* **2013**, *288*, 14824–14838. [CrossRef]

60. Bonner, J.C. Regulation of PDGF and its receptors in fibrotic diseases. *Cytokine Growth Factor Rev.* **2004**, *15*, 255–273. [CrossRef]

61. Avgustinova, A.; Iravani, M.; Robertson, D.; Fearns, A.; Gao, Q.; Klingbeil, P.; Hanby, A.M.; Speirs, V.; Sahai, E.; Calvo, F.; et al. Tumour cell-derived Wnt7a recruits and activates fibroblasts to promote tumour aggressiveness. *Nat. Commun.* **2016**, *7*, 10305. [CrossRef]

62. Biffi, G.; Oni, T.E.; Spielman, B.; Hao, Y.; Elyada, E.; Park, Y.; Preall, J.; Tuveson, D.A. IL1-Induced JAK/STAT Signaling Is Antagonized by TGFbeta to Shape CAF Heterogeneity in Pancreatic Ductal Adenocarcinoma. *Cancer Discov.* **2019**, *9*, 282–301. [CrossRef]

63. Sharon, Y.; Raz, Y.; Cohen, N.; Ben-Shmuel, A.; Schwartz, H.; Geiger, T.; Erez, N. Tumor-derived osteopontin reprograms normal mammary fibroblasts to promote inflammation and tumor growth in breast cancer. *Cancer Res.* **2015**, *75*, 963–973. [CrossRef]

64. Sanz-Moreno, V.; Gaggioli, C.; Yeo, M.; Albrengues, J.; Wallberg, F.; Viros, A.; Hooper, S.; Mitter, R.; Feral, C.C.; Cook, M.; et al. ROCK and JAK1 signaling cooperate to control actomyosin contractility in tumor cells and stroma. *Cancer Cell* **2011**, *20*, 229–245. [CrossRef]

65. Albrengues, J.; Bertero, T.; Grasset, E.; Bonan, S.; Maiel, M.; Bourget, I.; Philippe, C.; Herraiz Serrano, C.; Benamar, S.; Croce, O.; et al. Epigenetic switch drives the conversion of fibroblasts into proinvasive cancer-associated fibroblasts. *Nat. Commun.* **2015**, *6*, 10204. [CrossRef]

66. Shien, K.; Papadimitrakopoulou, V.A.; Ruder, D.; Behrens, C.; Shen, L.; Kalhor, N.; Song, J.; Lee, J.J.; Wang, J.; Tang, X.; et al. JAK1/STAT3 Activation through a Proinflammatory Cytokine Pathway Leads to Resistance to Molecularly Targeted Therapy in Non-Small Cell Lung Cancer. *Mol. Cancer Ther.* **2017**, *16*, 2234–2245. [CrossRef]

67. Zhang, Q.; Wang, H.Y.; Marzec, M.; Raghunath, P.N.; Nagasawa, T.; Wasik, M.A. STAT3- and DNA methyltransferase 1-mediated epigenetic silencing of SHP-1 tyrosine phosphatase tumor suppressor gene in malignant T lymphocytes. *Proc. Natl. Acad. Sci. USA* **2005**, *102*, 6948–6953. [CrossRef]

68. Lee, H.; Zhang, P.; Herrmann, A.; Yang, C.; Xin, H.; Wang, Z.; Hoon, D.S.; Forman, S.J.; Jove, R.; Riggs, A.D.; et al. Acetylated STAT3 is crucial for methylation of tumor-suppressor gene promoters and inhibition by resveratrol results in demethylation. *Proc. Natl. Acad. Sci. USA* **2012**, *109*, 7765–7769. [CrossRef]

69. Iwabu, A.; Smith, K.; Allen, F.D.; Lauffenburger, D.A.; Wells, A. Epidermal growth factor induces fibroblast contractility and motility via a protein kinase C delta-dependent pathway. *J. Biol. Chem.* **2004**, *279*, 14551–14560. [CrossRef]

70. Mink, S.R.; Vashistha, S.; Zhang, W.; Hodge, A.; Agus, D.B.; Jain, A. Cancer-associated fibroblasts derived from EGFR-TKI-resistant tumors reverse EGFR pathway inhibition by EGFR-TKIs. *Mol. Cancer Res.* **2010**, *8*, 809–820. [CrossRef]

71. Yun, Y.R.; Won, J.E.; Jeon, E.; Lee, S.; Kang, W.; Jo, H.; Jang, J.H.; Shin, U.S.; Kim, H.W. Fibroblast growth factors: Biology, function, and application for tissue regeneration. *J. Tissue Eng.* **2010**, *2010*, 218142. [CrossRef]

72. Bai, Y.P.; Shang, K.; Chen, H.; Ding, F.; Wang, Z.; Liang, C.; Xu, Y.; Sun, M.H.; Li, Y.Y. FGF-1/-3/FGFR4 signaling in cancer-associated fibroblasts promotes tumor progression in colon cancer through Erk and MMP-7. *Cancer Sci.* **2015**, *106*, 1278–1287. [CrossRef]

73. Mori, T.; Miyamoto, T.; Yoshida, H.; Asakawa, M.; Kawasumi, M.; Kobayashi, T.; Morioka, H.; Chiba, K.; Toyama, Y.; Yoshimura, A. IL-1beta and TNFalpha-initiated IL-6-STAT3 pathway is critical in mediating inflammatory cytokines and RANKL expression in inflammatory arthritis. *Int. Immunol.* **2011**, *23*, 701–712. [CrossRef]

74. Tsuyada, A.; Chow, A.; Wu, J.; Somlo, G.; Chu, P.; Loera, S.; Luu, T.; Li, A.X.; Wu, X.; Ye, W.; et al. CCL2 mediates cross-talk between cancer cells and stromal fibroblasts that regulates breast cancer stem cells. *Cancer Res.* **2012**, *72*, 2768–2779. [CrossRef]

75. Albrengues, J.; Bourget, I.; Pons, C.; Butet, V.; Hofman, P.; Tartare-Deckert, S.; Feral, C.C.; Meneguzzi, G.; Gaggioli, C. LIF mediates proinvasive activation of stromal fibroblasts in cancer. *Cell Rep.* **2014**, *7*, 1664–1678. [CrossRef]

76. Li, W.; Fan, J.; Chen, M.; Guan, S.; Sawcer, D.; Bokoch, G.M.; Woodley, D.T. Mechanism of human dermal fibroblast migration driven by type I collagen and platelet-derived growth factor-BB. *Mol. Biol. Cell* **2004**, *15*, 294–309. [CrossRef]

77. Roswall, P.; Bocci, M.; Bartoschek, M.; Li, H.; Kristiansen, G.; Jansson, S.; Lehn, S.; Sjolund, J.; Reid, S.; Larsson, C.; et al. Microenvironmental control of breast cancer subtype elicited through paracrine platelet-derived growth factor-CC signaling. *Nat. Med.* **2018**, *24*, 463–473. [CrossRef]

78. Kabashima, K.; Sakabe, J.; Yoshiki, R.; Tabata, Y.; Kohno, K.; Tokura, Y. Involvement of Wnt signaling in dermal fibroblasts. *Am. J. Pathol.* **2010**, *176*, 721–732. [CrossRef]

79. Inada, M.; Takita, M.; Yokoyama, S.; Watanabe, K.; Tominari, T.; Matsumoto, C.; Hirata, M.; Maru, Y.; Maruyama, T.; Sugimoto, Y.; et al. Direct Melanoma Cell Contact Induces Stromal Cell Autocrine Prostaglandin E2-EP4 Receptor Signaling That Drives Tumor Growth, Angiogenesis, and Metastasis. *J. Biol. Chem.* **2015**, *290*, 29781–29793. [CrossRef]

80. Guo, X.; Oshima, H.; Kitmura, T.; Taketo, M.M.; Oshima, M. Stromal fibroblasts activated by tumor cells promote angiogenesis in mouse gastric cancer. *J. Biol. Chem.* **2008**, *283*, 19864–19871. [CrossRef]

81. Biernacka, A.; Dobaczewski, M.; Frangogiannis, N.G. TGF-beta signaling in fibrosis. *Growth Factors* **2011**, *29*, 196–202. [CrossRef]

82. Lohr, M.; Schmidt, C.; Ringel, J.; Kluth, M.; Muller, P.; Nizze, H.; Jesnowski, R. Transforming growth factor-beta1 induces desmoplasia in an experimental model of human pancreatic carcinoma. *Cancer Res.* **2001**, *61*, 550–555.

83. Berking, C.; Takemoto, R.; Schaider, H.; Showe, L.; Satyamoorthy, K.; Robbins, P.; Herlyn, M. Transforming growth factor-beta1 increases survival of human melanoma through stroma remodeling. *Cancer Res.* **2001**, *61*, 8306–8316.

84. Guido, C.; Whitaker-Menezes, D.; Capparelli, C.; Balliet, R.; Lin, Z.; Pestell, R.G.; Howell, A.; Aquila, S.; Ando, S.; Martinez-Outschoorn, U.; et al. Metabolic reprogramming of cancer-associated fibroblasts by TGF-beta drives tumor growth: Connecting TGF-beta signaling with "Warburg-like" cancer metabolism and L-lactate production. *Cell Cycle* **2012**, *11*, 3019–3035. [CrossRef]

85. Vaquero, J.; Lobe, C.; Tahraoui, S.; Claperon, A.; Mergey, M.; Merabtene, F.; Wendum, D.; Coulouarn, C.; Housset, C.; Desbois-Mouthon, C.; et al. The IGF2/IR/IGF1R Pathway in Tumor Cells and Myofibroblasts Mediates Resistance to EGFR Inhibition in Cholangiocarcinoma. *Clin. Cancer Res.* **2018**, *24*, 4282–4296. [CrossRef]

86. Ireland, L.; Santos, A.; Ahmed, M.S.; Rainer, C.; Nielsen, S.R.; Quaranta, V.; Weyer-Czernilofsky, U.; Engle, D.D.; Perez-Mancera, P.A.; Coupland, S.E.; et al. Chemoresistance in Pancreatic Cancer Is Driven by Stroma-Derived Insulin-Like Growth Factors. *Cancer Res.* **2016**, *76*, 6851–6863. [CrossRef]

87. Unger, C.; Kramer, N.; Unterleuthner, D.; Scherzer, M.; Burian, A.; Rudisch, A.; Stadler, M.; Schlederer, M.; Lenhardt, D.; Riedl, A.; et al. Stromal-derived IGF2 promotes colon cancer progression via paracrine and autocrine mechanisms. *Oncogene* **2017**, *36*, 5341–5355. [CrossRef]

88. Chen, W.J.; Ho, C.C.; Chang, Y.L.; Chen, H.Y.; Lin, C.A.; Ling, T.Y.; Yu, S.L.; Yuan, S.S.; Chen, Y.J.; Lin, C.Y.; et al. Cancer-associated fibroblasts regulate the plasticity of lung cancer stemness via paracrine signalling. *Nat. Commun.* **2014**, *5*, 3472. [CrossRef]

89. Straussman, R.; Morikawa, T.; Shee, K.; Barzily-Rokni, M.; Qian, Z.R.; Du, J.; Davis, A.; Mongare, M.M.; Gould, J.; Frederick, D.T.; et al. Tumour micro-environment elicits innate resistance to RAF inhibitors through HGF secretion. *Nature* **2012**, *487*, 500–504. [CrossRef]

90. Wilson, T.R.; Fridlyand, J.; Yan, Y.; Penuel, E.; Burton, L.; Chan, E.; Peng, J.; Lin, E.; Wang, Y.; Sosman, J.; et al. Widespread potential for growth-factor-driven resistance to anticancer kinase inhibitors. *Nature* **2012**, *487*, 505–509. [CrossRef]

91. Wang, W.; Li, Q.; Yamada, T.; Matsumoto, K.; Matsumoto, I.; Oda, M.; Watanabe, G.; Kayano, Y.; Nishioka, Y.; Sone, S.; et al. Crosstalk to stromal fibroblasts induces resistance of lung cancer to epidermal growth factor receptor tyrosine kinase inhibitors. *Clin. Cancer Res.* **2009**, *15*, 6630–6638. [CrossRef]

92. Ferrara, N.; Kerbel, R.S. Angiogenesis as a therapeutic target. *Nature* **2005**, *438*, 967–974. [CrossRef]

93. Crawford, Y.; Kasman, I.; Yu, L.; Zhong, C.; Wu, X.; Modrusan, Z.; Kaminker, J.; Ferrara, N. PDGF-C mediates the angiogenic and tumorigenic properties of fibroblasts associated with tumors refractory to anti-VEGF treatment. *Cancer Cell* **2009**, *15*, 21–34. [CrossRef]

94. Erez, N.; Truitt, M.; Olson, P.; Arron, S.T.; Hanahan, D. Cancer-Associated Fibroblasts Are Activated in Incipient Neoplasia to Orchestrate Tumor-Promoting Inflammation in an NF-kappaB-Dependent Manner. *Cancer Cell* **2010**, *17*, 135–147. [CrossRef]

95. Su, S.; Chen, J.; Yao, H.; Liu, J.; Yu, S.; Lao, L.; Wang, M.; Luo, M.; Xing, Y.; Chen, F.; et al. CD10(+)GPR77(+) Cancer-Associated Fibroblasts Promote Cancer Formation and Chemoresistance by Sustaining Cancer Stemness. *Cell* **2018**, *172*, 841–856. [CrossRef]

96. Sun, Y.; Campisi, J.; Higano, C.; Beer, T.M.; Porter, P.; Coleman, I.; True, L.; Nelson, P.S. Treatment-induced damage to the tumor microenvironment promotes prostate cancer therapy resistance through WNT16B. *Nat. Med.* **2012**, *18*, 1359–1368. [CrossRef]

97. Tao, L.; Huang, G.; Wang, R.; Pan, Y.; He, Z.; Chu, X.; Song, H.; Chen, L. Cancer-associated fibroblasts treated with cisplatin facilitates chemoresistance of lung adenocarcinoma through IL-11/IL-11R/STAT3 signaling pathway. *Sci. Rep.* **2016**, *6*, 38408. [CrossRef]

98. Lotti, F.; Jarrar, A.M.; Pai, R.K.; Hitomi, M.; Lathia, J.; Mace, A.; Gantt, G.A., Jr.; Sukhdeo, K.; DeVecchio, J.; Vasanji, A.; et al. Chemotherapy activates cancer-associated fibroblasts to maintain colorectal cancer-initiating cells by IL-17A. *J. Exp. Med.* **2013**, *210*, 2851–2872. [CrossRef]

99. Erdogan, B.; Webb, D.J. Cancer-associated fibroblasts modulate growth factor signaling and extracellular matrix remodeling to regulate tumor metastasis. *Biochem. Soc. Trans.* **2017**, *45*, 229–236. [CrossRef]

100. Karagiannis, G.S.; Poutahidis, T.; Erdman, S.E.; Kirsch, R.; Riddell, R.H.; Diamandis, E.P. Cancer-associated fibroblasts drive the progression of metastasis through both paracrine and mechanical pressure on cancer tissue. *Mol. Cancer Res.* **2012**, *10*, 1403–1418. [CrossRef]

101. Hirata, E.; Girotti, M.R.; Viros, A.; Hooper, S.; Spencer-Dene, B.; Matsuda, M.; Larkin, J.; Marais, R.; Sahai, E. Intravital imaging reveals how BRAF inhibition generates drug-tolerant microenvironments with high integrin beta1/FAK signaling. *Cancer Cell* **2015**, *27*, 574–588. [CrossRef]

102. Qayyum, M.A.; Insana, M.F. Stromal responses to fractionated radiotherapy. *Int. J. Radiat. Biol.* **2012**, *88*, 383–392. [CrossRef]

103. Pazolli, E.; Alspach, E.; Milczarek, A.; Prior, J.; Piwnica-Worms, D.; Stewart, S.A. Chromatin remodeling underlies the senescence-associated secretory phenotype of tumor stromal fibroblasts that supports cancer progression. *Cancer Res.* **2012**, *72*, 2251–2261. [CrossRef]

104. Nguyen, A.H.; Elliott, I.A.; Wu, N.; Matsumura, C.; Vogelauer, M.; Attar, N.; Dann, A.; Ghukasyan, R.; Toste, P.A.; Patel, S.G.; et al. Histone deacetylase inhibitors provoke a tumor supportive phenotype in pancreatic cancer associated fibroblasts. *Oncotarget* **2017**, *8*, 19074–19088. [CrossRef]

105. Kato, M.; Placencio-Hickok, V.R.; Madhav, A.; Haldar, S.; Tripathi, M.; Billet, S.; Mishra, R.; Smith, B.; Rohena-Rivera, K.; Agarwal, P.; et al. Heterogeneous cancer-associated fibroblast population potentiates neuroendocrine differentiation and castrate resistance in a CD105-dependent manner. *Oncogene* **2018**. [CrossRef]

106. Quail, D.F.; Joyce, J.A. Molecular Pathways: Deciphering Mechanisms of Resistance to Macrophage-Targeted Therapies. *Clin. Cancer Res.* **2017**, *23*, 876–884. [CrossRef]

107. Cassetta, L.; Pollard, J.W. Targeting macrophages: Therapeutic approaches in cancer. *Nat. Rev. Drug Discov.* **2018**. [CrossRef]

108. Kumar, V.; Donthireddy, L.; Marvel, D.; Condamine, T.; Wang, F.; Lavilla-Alonso, S.; Hashimoto, A.; Vonteddu, P.; Behera, R.; Goins, M.A.; et al. Cancer-Associated Fibroblasts Neutralize the Anti-tumor Effect of CSF1 Receptor Blockade by Inducing PMN-MDSC Infiltration of Tumors. *Cancer Cell* **2017**, *32*, 654–668. [CrossRef]

109. Darby, I.A.; Hewitson, T.D. Fibroblast differentiation in wound healing and fibrosis. *Int. Rev. Cytol.* **2007**, *257*, 143–179.

110. Darby, I.A.; Laverdet, B.; Bonte, F.; Desmouliere, A. Fibroblasts and myofibroblasts in wound healing. *Clin. Cosmet. Investig. Dermatol.* **2014**, *7*, 301–311.

111. Penn, M.S. SDF-1:CXCR4 axis is fundamental for tissue preservation and repair. *Am. J. Pathol.* **2010**, *177*, 2166–2168. [CrossRef]

112. Schmidt, A.; Brixius, K.; Bloch, W. Endothelial precursor cell migration during vasculogenesis. *Circ. Res.* **2007**, *101*, 125–136. [CrossRef]

113. Sewell-Loftin, M.K.; Bayer, S.V.H.; Crist, E.; Hughes, T.; Joison, S.M.; Longmore, G.D.; George, S.C. Cancer-associated fibroblasts support vascular growth through mechanical force. *Sci. Rep.* **2017**, *7*, 12574. [CrossRef]

114. Fukumura, D.; Xavier, R.; Sugiura, T.; Chen, Y.; Park, E.C.; Lu, N.; Selig, M.; Nielsen, G.; Taksir, T.; Jain, R.K.; et al. Tumor induction of VEGF promoter activity in stromal cells. *Cell* **1998**, *94*, 715–725. [CrossRef]

115. Orimo, A.; Gupta, P.B.; Sgroi, D.C.; Arenzana-Seisdedos, F.; Delaunay, T.; Naeem, R.; Carey, V.J.; Richardson, A.L.; Weinberg, R.A. Stromal fibroblasts present in invasive human breast carcinomas promote tumor growth and angiogenesis through elevated SDF-1/CXCL12 secretion. *Cell* **2005**, *121*, 335–348. [CrossRef]

116. Calon, A.; Espinet, E.; Palomo-Ponce, S.; Tauriello, D.V.; Iglesias, M.; Cespedes, M.V.; Sevillano, M.; Nadal, C.; Jung, P.; Zhang, X.H.; et al. Dependency of colorectal cancer on a TGF-beta-driven program in stromal cells for metastasis initiation. *Cancer Cell* **2012**, *22*, 571–584. [CrossRef]

117. De Wever, O.; Westbroek, W.; Verloes, A.; Bloemen, N.; Bracke, M.; Gespach, C.; Bruyneel, E.; Mareel, M. Critical role of N-cadherin in myofibroblast invasion and migration in vitro stimulated by colon-cancer-cell-derived TGF-beta or wounding. *J. Cell Sci.* **2004**, *117*, 4691–4703. [CrossRef]

118. Ng, C.F.; Frieboes, H.B. Model of vascular desmoplastic multispecies tumor growth. *J. Theor. Biol.* **2017**, *430*, 245–282. [CrossRef]

119. Nishida, N.; Yano, H.; Nishida, T.; Kamura, T.; Kojiro, M. Angiogenesis in cancer. *Vasc. Health Risk Manag.* **2006**, *2*, 213–219. [CrossRef]

120. Izumi, D.; Ishimoto, T.; Miyake, K.; Sugihara, H.; Eto, K.; Sawayama, H.; Yasuda, T.; Kiyozumi, Y.; Kaida, T.; Kurashige, J.; et al. CXCL12/CXCR4 activation by cancer-associated fibroblasts promotes integrin beta1 clustering and invasiveness in gastric cancer. *Int. J. Cancer* **2016**, *138*, 1207–1219. [CrossRef]

121. Sun, X.; Cheng, G.; Hao, M.; Zheng, J.; Zhou, X.; Zhang, J.; Taichman, R.S.; Pienta, K.J.; Wang, J. CXCL12 / CXCR4 / CXCR7 chemokine axis and cancer progression. *Cancer Metastasis Rev.* **2010**, *29*, 709–722. [CrossRef]

122. Teng, F.; Tian, W.Y.; Wang, Y.M.; Zhang, Y.F.; Guo, F.; Zhao, J.; Gao, C.; Xue, F.X. Cancer-associated fibroblasts promote the progression of endometrial cancer via the SDF-1/CXCR4 axis. *J. Hematol. Oncol.* **2016**, *9*, 8. [CrossRef]

123. Jin, F.; Brockmeier, U.; Otterbach, F.; Metzen, E. New insight into the SDF-1/CXCR4 axis in a breast carcinoma model: Hypoxia-induced endothelial SDF-1 and tumor cell CXCR4 are required for tumor cell intravasation. *Mol. Cancer Res.* **2012**, *10*, 1021–1031. [CrossRef]

124. Mego, M.; Cholujova, D.; Minarik, G.; Sedlackova, T.; Gronesova, P.; Karaba, M.; Benca, J.; Cingelova, S.; Cierna, Z.; Manasova, D.; et al. CXCR4-SDF-1 interaction potentially mediates trafficking of circulating tumor cells in primary breast cancer. *BMC Cancer* **2016**, *16*, 127. [CrossRef]

125. Fujii, H.; Biel, M.A.; Zhou, W.; Weitzman, S.A.; Baylin, S.B.; Gabrielson, E. Methylation of the HIC-1 candidate tumor suppressor gene in human breast cancer. *Oncogene* **1998**, *16*, 2159–2164. [CrossRef]

126. Wang, X.; Wang, Y.; Xiao, G.; Wang, J.; Zu, L.; Hao, M.; Sun, X.; Fu, Y.; Hu, G.; Wang, J. Hypermethylated in cancer 1(HIC1) suppresses non-small cell lung cancer progression by targeting interleukin-6/Stat3 pathway. *Oncotarget* **2016**, *7*, 30350–30364. [CrossRef]

127. Wang, Y.; Weng, X.; Wang, L.; Hao, M.; Li, Y.; Hou, L.; Liang, Y.; Wu, T.; Yao, M.; Lin, G.; et al. HIC1 deletion promotes breast cancer progression by activating tumor cell/fibroblast crosstalk. *J. Clin. Investig.* **2018**. [CrossRef]

128. Zheng, J.; Wang, J.; Sun, X.; Hao, M.; Ding, T.; Xiong, D.; Wang, X.; Zhu, Y.; Xiao, G.; Cheng, G.; et al. HIC1 modulates prostate cancer progression by epigenetic modification. *Clin. Cancer Res.* **2013**, *19*, 1400–1410. [CrossRef]

129. Gaggioli, C.; Hooper, S.; Hidalgo-Carcedo, C.; Grosse, R.; Marshall, J.F.; Harrington, K.; Sahai, E. Fibroblast-led collective invasion of carcinoma cells with differing roles for RhoGTPases in leading and following cells. *Nat. Cell Biol.* **2007**, *9*, 1392–1400. [CrossRef]

130. Grasset, E.M.; Bertero, T.; Bozec, A.; Friard, J.; Bourget, I.; Pisano, S.; Lecacheur, M.; Maiel, M.; Bailleux, C.; Emelyanov, A.; et al. Matrix Stiffening and EGFR Cooperate to Promote the Collective Invasion of Cancer Cells. *Cancer Res.* **2018**, *78*, 5229–5242. [CrossRef]

131. Mani, S.A.; Guo, W.; Liao, M.J.; Eaton, E.N.; Ayyanan, A.; Zhou, A.Y.; Brooks, M.; Reinhard, F.; Zhang, C.C.; Shipitsin, M.; et al. The epithelial-mesenchymal transition generates cells with properties of stem cells. *Cell* **2008**, *133*, 704–715. [CrossRef]

132. Shibue, T.; Weinberg, R.A. EMT, CSCs, and drug resistance: The mechanistic link and clinical implications. *Nat. Rev. Clin. Oncol.* **2017**, *14*, 611–629. [CrossRef]

133. Saitoh, M. Involvement of partial EMT in cancer progression. *J. Biochem.* **2018**, *164*, 257–264. [CrossRef]

134. Nieto, M.A.; Huang, R.Y.; Jackson, R.A.; Thiery, J.P. Emt: 2016. *Cell* **2016**, *166*, 21–45. [CrossRef]

135. Krebs, A.M.; Mitschke, J.; Lasierra Losada, M.; Schmalhofer, O.; Boerries, M.; Busch, H.; Boettcher, M.; Mougiakakos, D.; Reichardt, W.; Bronsert, P.; et al. The EMT-activator Zeb1 is a key factor for cell plasticity and promotes metastasis in pancreatic cancer. *Nat. Cell Biol.* **2017**, *19*, 518–529. [CrossRef]

136. Byrne, C.; Schairer, C.; Wolfe, J.; Parekh, N.; Salane, M.; Brinton, L.A.; Hoover, R.; Haile, R. Mammographic features and breast cancer risk: Effects with time, age, and menopause status. *J. Natl. Cancer Inst.* **1995**, *87*, 1622–1629. [CrossRef]

137. DeFilippis, R.A.; Chang, H.; Dumont, N.; Rabban, J.T.; Chen, Y.Y.; Fontenay, G.V.; Berman, H.K.; Gauthier, M.L.; Zhao, J.; Hu, D.; et al. CD36 repression activates a multicellular stromal program shared by high mammographic density and tumor tissues. *Cancer Discov.* **2012**, *2*, 826–839. [CrossRef]

138. Plikus, M.V.; Guerrero-Juarez, C.F.; Ito, M.; Li, Y.R.; Dedhia, P.H.; Zheng, Y.; Shao, M.; Gay, D.L.; Ramos, R.; Hsi, T.C.; et al. Regeneration of fat cells from myofibroblasts during wound healing. *Science* **2017**, *355*, 748–752. [CrossRef]

139. Thomson, D.M.; Winder, W.W. AMP-activated protein kinase control of fat metabolism in skeletal muscle. *Acta Physiol.* **2009**, *196*, 147–154. [CrossRef]

140. Ladanyi, A.; Mukherjee, A.; Kenny, H.A.; Johnson, A.; Mitra, A.K.; Sundaresan, S.; Nieman, K.M.; Pascual, G.; Benitah, S.A.; Montag, A.; et al. Adipocyte-induced CD36 expression drives ovarian cancer progression and metastasis. *Oncogene* **2018**, *37*, 2285–2301. [CrossRef]

141. Pascual, G.; Avgustinova, A.; Mejetta, S.; Martin, M.; Castellanos, A.; Attolini, C.S.; Berenguer, A.; Prats, N.; Toll, A.; Hueto, J.A.; et al. Targeting metastasis-initiating cells through the fatty acid receptor CD36. *Nature* **2017**, *541*, 41–45. [CrossRef]

142. Martinez-Outschoorn, U.E.; Sotgia, F.; Lisanti, M.P. Caveolae and signalling in cancer. *Nat. Rev. Cancer* **2015**, *15*, 225–237. [CrossRef]

143. Santi, A.; Kugeratski, F.G.; Zanivan, S. Cancer Associated Fibroblasts: The Architects of Stroma Remodeling. *Proteomics* **2018**, *18*, e1700167. [CrossRef]

144. Whitaker-Menezes, D.; Martinez-Outschoorn, U.E.; Lin, Z.; Ertel, A.; Flomenberg, N.; Witkiewicz, A.K.; Birbe, R.C.; Howell, A.; Pavlides, S.; Gandara, R.; et al. Evidence for a stromal-epithelial "lactate shuttle" in human tumors: MCT4 is a marker of oxidative stress in cancer-associated fibroblasts. *Cell Cycle* **2011**, *10*, 1772–1783. [CrossRef]

145. Yoshida, G.J. Metabolic reprogramming: The emerging concept and associated therapeutic strategies. *J. Exp. Clin. Cancer Res.* **2015**, *34*, 111. [CrossRef]

146. Liberti, M.V.; Locasale, J.W. The Warburg Effect: How Does it Benefit Cancer Cells? *Trends Biochem. Sci.* **2016**, *41*, 211–218. [CrossRef]

147. Schwartz, L.; Supuran, C.T.; Alfarouk, K.O. The Warburg Effect and the Hallmarks of Cancer. *Anticancer Agents Med. Chem.* **2017**, *17*, 164–170. [CrossRef]

148. Warburg, O. On the origin of cancer cells. *Science* **1956**, *123*, 309–314. [CrossRef]

149. Fu, Y.; Liu, S.; Yin, S.; Niu, W.; Xiong, W.; Tan, M.; Li, G.; Zhou, M. The reverse Warburg effect is likely to be an Achilles' heel of cancer that can be exploited for cancer therapy. *Oncotarget* **2017**, *8*, 57813–57825. [CrossRef]

150. Pavlides, S.; Whitaker-Menezes, D.; Castello-Cros, R.; Flomenberg, N.; Witkiewicz, A.K.; Frank, P.G.; Casimiro, M.C.; Wang, C.; Fortina, P.; Addya, S.; et al. The reverse Warburg effect: Aerobic glycolysis in cancer associated fibroblasts and the tumor stroma. *Cell Cycle* **2009**, *8*, 3984–4001. [CrossRef]

151. Jezierska-Drutel, A.; Rosenzweig, S.A.; Neumann, C.A. Role of oxidative stress and the microenvironment in breast cancer development and progression. *Adv. Cancer Res.* **2013**, *119*, 107–125.

152. Mougeolle, A.; Poussard, S.; Decossas, M.; Lamaze, C.; Lambert, O.; Dargelos, E. Oxidative stress induces caveolin 1 degradation and impairs caveolae functions in skeletal muscle cells. *PLoS ONE* **2015**, *10*, e0122654. [CrossRef]

153. Wang, S.; Wang, N.; Zheng, Y.; Zhang, J.; Zhang, F.; Wang, Z. Caveolin-1: An Oxidative Stress-Related Target for Cancer Prevention. *Oxid. Med. Cell Longev.* **2017**, *2017*, 7454031. [CrossRef]

154. Avagliano, A.; Granato, G.; Ruocco, M.R.; Romano, V.; Belviso, I.; Carfora, A.; Montagnani, S.; Arcucci, A. Metabolic Reprogramming of Cancer Associated Fibroblasts: The Slavery of Stromal Fibroblasts. *Biomed. Res. Int.* **2018**, *2018*, 6075403. [CrossRef]

155. Pavlides, S.; Vera, I.; Gandara, R.; Sneddon, S.; Pestell, R.G.; Mercier, I.; Martinez-Outschoorn, U.E.; Whitaker-Menezes, D.; Howell, A.; Sotgia, F.; et al. Warburg meets autophagy: Cancer-associated fibroblasts accelerate tumor growth and metastasis via oxidative stress, mitophagy, and aerobic glycolysis. *Antioxid. Redox Signal.* **2012**, *16*, 1264–1284. [CrossRef]

156. Xing, Y.; Zhao, S.; Zhou, B.P.; Mi, J. Metabolic reprogramming of the tumour microenvironment. *FEBS J.* **2015**, *282*, 3892–3898. [CrossRef]

157. Capparelli, C.; Guido, C.; Whitaker-Menezes, D.; Bonuccelli, G.; Balliet, R.; Pestell, T.G.; Goldberg, A.F.; Pestell, R.G.; Howell, A.; Sneddon, S.; et al. Autophagy and senescence in cancer-associated fibroblasts metabolically supports tumor growth and metastasis via glycolysis and ketone production. *Cell Cycle* **2012**, *11*, 2285–2302. [CrossRef]

158. Allen, E.; Mieville, P.; Warren, C.M.; Saghafinia, S.; Li, L.; Peng, M.W.; Hanahan, D. Metabolic Symbiosis Enables Adaptive Resistance to Anti-angiogenic Therapy that Is Dependent on mTOR Signaling. *Cell Rep.* **2016**, *15*, 1144–1160. [CrossRef]

159. Porporato, P.E.; Dhup, S.; Dadhich, R.K.; Copetti, T.; Sonveaux, P. Anticancer targets in the glycolytic metabolism of tumors: A comprehensive review. *Front. Pharmacol.* **2011**, *2*, 49. [CrossRef]

160. Semenza, G.L. Tumor metabolism: Cancer cells give and take lactate. *J. Clin. Investig.* **2008**, *118*, 3835–3837. [CrossRef]

161. Sonveaux, P.; Vegran, F.; Schroeder, T.; Wergin, M.C.; Verrax, J.; Rabbani, Z.N.; De Saedeleer, C.J.; Kennedy, K.M.; Diepart, C.; Jordan, B.F.; et al. Targeting lactate-fueled respiration selectively kills hypoxic tumor cells in mice. *J. Clin. Investig.* **2008**, *118*, 3930–3942. [CrossRef]

162. Linares, J.F.; Cordes, T.; Duran, A.; Reina-Campos, M.; Valencia, T.; Ahn, C.S.; Castilla, E.A.; Moscat, J.; Metallo, C.M.; Diaz-Meco, M.T. ATF4-Induced Metabolic Reprograming Is a Synthetic Vulnerability of the p62-Deficient Tumor Stroma. *Cell Metab.* **2017**, *26*, 817–829. [CrossRef]

163. Duran, A.; Hernandez, E.D.; Reina-Campos, M.; Castilla, E.A.; Subramaniam, S.; Raghunandan, S.; Roberts, L.R.; Kisseleva, T.; Karin, M.; Diaz-Meco, M.T.; et al. p62/SQSTM1 by Binding to Vitamin D Receptor Inhibits Hepatic Stellate Cell Activity, Fibrosis, and Liver Cancer. *Cancer Cell* **2016**, *30*, 595–609. [CrossRef]

164. Reina-Campos, M.; Shelton, P.M.; Diaz-Meco, M.T.; Moscat, J. Metabolic reprogramming of the tumor microenvironment by p62 and its partners. *Biochim. Biophys. Acta Rev. Cancer* **2018**, *1870*, 88–95. [CrossRef]

165. Goruppi, S.; Jo, S.H.; Laszlo, C.; Clocchiatti, A.; Neel, V.; Dotto, G.P. Autophagy Controls CSL/RBPJkappa Stability through a p62/SQSTM1-Dependent Mechanism. *Cell Rep.* **2018**, *24*, 3108–3114. [CrossRef]

166. Valencia, T.; Kim, J.Y.; Abu-Baker, S.; Moscat-Pardos, J.; Ahn, C.S.; Reina-Campos, M.; Duran, A.; Castilla, E.A.; Metallo, C.M.; Diaz-Meco, M.T.; et al. Metabolic reprogramming of stromal fibroblasts through p62-mTORC1 signaling promotes inflammation and tumorigenesis. *Cancer Cell* **2014**, *26*, 121–135. [CrossRef]

167. Mizushima, N.; Komatsu, M. Autophagy: Renovation of cells and tissues. *Cell* **2011**, *147*, 728–741. [CrossRef]

168. Moscat, J.; Diaz-Meco, M.T. p62 at the crossroads of autophagy, apoptosis, and cancer. *Cell* **2009**, *137*, 1001–1004. [CrossRef]

169. Huang, J.; Diaz-Meco, M.T.; Moscat, J. The macroenviromental control of cancer metabolism by p62. *Cell Cycle* **2018**, *17*, 2110–2121. [CrossRef]

170. Jain, R.K.; Tong, R.T.; Munn, L.L. Effect of vascular normalization by antiangiogenic therapy on interstitial hypertension, peritumor edema, and lymphatic metastasis: Insights from a mathematical model. *Cancer Res.* **2007**, *67*, 2729–2735. [CrossRef]

171. Munson, J.M.; Shieh, A.C. Interstitial fluid flow in cancer: Implications for disease progression and treatment. *Cancer Manag. Res.* **2014**, *6*, 317–328. [CrossRef]

172. Omidi, Y.; Barar, J. Targeting tumor microenvironment: Crossing tumor interstitial fluid by multifunctional nanomedicines. *Bioimpacts* **2014**, *4*, 55–67.

173. Kerbel, R.S. Tumor angiogenesis. *N. Engl. J. Med.* **2008**, *358*, 2039–2049. [CrossRef]

174. Christian, S.; Winkler, R.; Helfrich, I.; Boos, A.M.; Besemfelder, E.; Schadendorf, D.; Augustin, H.G. Endosialin (Tem1) is a marker of tumor-associated myofibroblasts and tumor vessel-associated mural cells. *Am. J. Pathol.* **2008**, *172*, 486–494. [CrossRef]

175. Otranto, M.; Sarrazy, V.; Bonte, F.; Hinz, B.; Gabbiani, G.; Desmouliere, A. The role of the myofibroblast in tumor stroma remodeling. *Cell Adhes. Migr.* **2012**, *6*, 203–219. [CrossRef]

176. Ribeiro, A.L.; Okamoto, O.K. Combined effects of pericytes in the tumor microenvironment. *Stem Cells Int.* **2015**, *2015*, 868475. [CrossRef]

177. Liao, Z.; Tan, Z.W.; Zhu, P.; Tan, N.S. Cancer-associated fibroblasts in tumor microenvironment—Accomplices in tumor malignancy. *Cell Immunol.* **2018**. [CrossRef]

178. Boucher, Y.; Salehi, H.; Witwer, B.; Harsh, G.R.T.; Jain, R.K. Interstitial fluid pressure in intracranial tumours in patients and in rodents. *Br. J. Cancer* **1997**, *75*, 829–836. [CrossRef]

179. Nathanson, S.D.; Nelson, L. Interstitial fluid pressure in breast cancer, benign breast conditions, and breast parenchyma. *Ann. Surg. Oncol.* **1994**, *1*, 333–338. [CrossRef]

180. Simonsen, T.G.; Gaustad, J.V.; Leinaas, M.N.; Rofstad, E.K. High interstitial fluid pressure is associated with tumor-line specific vascular abnormalities in human melanoma xenografts. *PLoS ONE* **2012**, *7*, e40006. [CrossRef]

181. Bhome, R.; Al Saihati, H.A.; Goh, R.W.; Bullock, M.D.; Primrose, J.N.; Thomas, G.J.; Sayan, A.E.; Mirnezami, A.H. Translational aspects in targeting the stromal tumour microenvironment: From bench to bedside. *New Horiz Transl. Med.* **2016**, *3*, 9–21. [CrossRef]

182. Cazet, A.S.; Hui, M.N.; Elsworth, B.L.; Wu, S.Z.; Roden, D.; Chan, C.L.; Skhinas, J.N.; Collot, R.; Yang, J.; Harvey, K.; et al. Targeting stromal remodeling and cancer stem cell plasticity overcomes chemoresistance in triple negative breast cancer. *Nat. Commun.* **2018**, *9*, 2897. [CrossRef]

183. Giussani, M.; Merlino, G.; Cappelletti, V.; Tagliabue, E.; Daidone, M.G. Tumor-extracellular matrix interactions: Identification of tools associated with breast cancer progression. *Semin. Cancer Biol.* **2015**, *35*, 3–10. [CrossRef]

184. Son, B.; Lee, S.; Youn, H.; Kim, E.; Kim, W.; Youn, B. The role of tumor microenvironment in therapeutic resistance. *Oncotarget* **2017**, *8*, 3933–3945. [CrossRef]

185. Levental, K.R.; Yu, H.; Kass, L.; Lakins, J.N.; Egeblad, M.; Erler, J.T.; Fong, S.F.; Csiszar, K.; Giaccia, A.; Weninger, W.; et al. Matrix crosslinking forces tumor progression by enhancing integrin signaling. *Cell* **2009**, *139*, 891–906. [CrossRef]

186. Peng, C.; Liu, J.; Yang, G.; Li, Y. Lysyl oxidase activates cancer stromal cells and promotes gastric cancer progression: Quantum dot-based identification of biomarkers in cancer stromal cells. *Int. J. Nanomed.* **2018**, *13*, 161–174. [CrossRef]

187. Wang, T.H.; Hsia, S.M.; Shieh, T.M. Lysyl Oxidase and the Tumor Microenvironment. *Int. J. Mol. Sci.* **2016**, *18*, 62. [CrossRef]

188. Reynaud, C.; Ferreras, L.; Di Mauro, P.; Kan, C.; Croset, M.; Bonnelye, E.; Pez, F.; Thomas, C.; Aimond, G.; Karnoub, A.E.; et al. Lysyl Oxidase Is a Strong Determinant of Tumor Cell Colonization in Bone. *Cancer Res.* **2017**, *77*, 268–278. [CrossRef]

189. Cox, T.R.; Gartland, A.; Erler, J.T. Lysyl Oxidase, a Targetable Secreted Molecule Involved in Cancer Metastasis. *Cancer Res.* **2016**, *76*, 188–192. [CrossRef]

190. Miller, B.W.; Morton, J.P.; Pinese, M.; Saturno, G.; Jamieson, N.B.; McGhee, E.; Timpson, P.; Leach, J.; McGarry, L.; Shanks, E.; et al. Targeting the LOX/hypoxia axis reverses many of the features that make pancreatic cancer deadly: Inhibition of LOX abrogates metastasis and enhances drug efficacy. *EMBO Mol. Med.* **2015**, *7*, 1063–1076. [CrossRef]

191. Kelleher, F.C. Hedgehog signaling and therapeutics in pancreatic cancer. *Carcinogenesis* **2011**, *32*, 445–451. [CrossRef]

192. Erkan, M.; Hausmann, S.; Michalski, C.W.; Fingerle, A.A.; Dobritz, M.; Kleeff, J.; Friess, H. The role of stroma in pancreatic cancer: Diagnostic and therapeutic implications. *Nat. Rev. Gastroenterol. Hepatol.* **2012**, *9*, 454–467. [CrossRef]

193. Mathew, E.; Zhang, Y.; Holtz, A.M.; Kane, K.T.; Song, J.Y.; Allen, B.L.; Pasca di Magliano, M. Dosage-dependent regulation of pancreatic cancer growth and angiogenesis by hedgehog signaling. *Cell Rep.* **2014**, *9*, 484–494. [CrossRef]

194. Olive, K.P.; Jacobetz, M.A.; Davidson, C.J.; Gopinathan, A.; McIntyre, D.; Honess, D.; Madhu, B.; Goldgraben, M.A.; Caldwell, M.E.; Allard, D.; et al. Inhibition of Hedgehog signaling enhances delivery of chemotherapy in a mouse model of pancreatic cancer. *Science* **2009**, *324*, 1457–1461. [CrossRef]

195. Jiang, Z.; Zhou, C.; Cheng, L.; Yan, B.; Chen, K.; Chen, X.; Zong, L.; Lei, J.; Duan, W.; Xu, Q.; et al. Inhibiting YAP expression suppresses pancreatic cancer progression by disrupting tumor-stromal interactions. *J. Exp. Clin. Cancer Res.* **2018**, *37*, 69. [CrossRef]

196. Kharaishvili, G.; Simkova, D.; Bouchalova, K.; Gachechiladze, M.; Narsia, N.; Bouchal, J. The role of cancer-associated fibroblasts, solid stress and other microenvironmental factors in tumor progression and therapy resistance. *Cancer Cell Int.* **2014**, *14*, 41. [CrossRef]

197. Warren, J.S.A.; Xiao, Y.; Lamar, J.M. YAP/TAZ Activation as a Target for Treating Metastatic Cancer. *Cancers* **2018**, *10*, 115. [CrossRef]

198. Calvo, F.; Ege, N.; Grande-Garcia, A.; Hooper, S.; Jenkins, R.P.; Chaudhry, S.I.; Harrington, K.; Williamson, P.; Moeendarbary, E.; Charras, G.; et al. Mechanotransduction and YAP-dependent matrix remodelling is required for the generation and maintenance of cancer-associated fibroblasts. *Nat. Cell Biol.* **2013**, *15*, 637–646. [CrossRef]

199. Campbell, I.; Polyak, K.; Haviv, I. Clonal mutations in the cancer-associated fibroblasts: The case against genetic coevolution. *Cancer Res.* **2009**, *69*, 6765–6768, discussion 6769. [CrossRef]

200. Togo, S.; Polanska, U.M.; Horimoto, Y.; Orimo, A. Carcinoma-associated fibroblasts are a promising therapeutic target. *Cancers* **2013**, *5*, 149–169. [CrossRef]

201. Hanahan, D.; Coussens, L.M. Accessories to the crime: Functions of cells recruited to the tumor microenvironment. *Cancer Cell* **2012**, *21*, 309–322. [CrossRef]

202. Tredan, O.; Galmarini, C.M.; Patel, K.; Tannock, I.F. Drug resistance and the solid tumor microenvironment. *J. Natl. Cancer Inst.* **2007**, *99*, 1441–1454. [CrossRef]

203. Wu, M.; Frieboes, H.B.; Chaplain, M.A.; McDougall, S.R.; Cristini, V.; Lowengrub, J.S. The effect of interstitial pressure on therapeutic agent transport: Coupling with the tumor blood and lymphatic vascular systems. *J. Theor. Biol.* **2014**, *355*, 194–207. [CrossRef]

204. Gourd, E. PEGPH20 for metastatic pancreatic ductal adenocarcinoma. *Lancet Oncol.* **2018**, *19*, e81. [CrossRef]

205. Michl, P.; Gress, T.M. Improving drug delivery to pancreatic cancer: Breaching the stromal fortress by targeting hyaluronic acid. *Gut* **2012**, *61*, 1377–1379. [CrossRef]

206. Provenzano, P.P.; Cuevas, C.; Chang, A.E.; Goel, V.K.; Von Hoff, D.D.; Hingorani, S.R. Enzymatic targeting of the stroma ablates physical barriers to treatment of pancreatic ductal adenocarcinoma. *Cancer Cell* **2012**, *21*, 418–429. [CrossRef]

207. Hingorani, S.R.; Zheng, L.; Bullock, A.J.; Seery, T.E.; Harris, W.P.; Sigal, D.S.; Braiteh, F.; Ritch, P.S.; Zalupski, M.M.; Bahary, N.; et al. HALO 202: Randomized Phase II Study of PEGPH20 Plus Nab-Paclitaxel/Gemcitabine Versus Nab-Paclitaxel/Gemcitabine in Patients With Untreated, Metastatic Pancreatic Ductal Adenocarcinoma. *J. Clin. Oncol.* **2018**, *36*, 359–366. [CrossRef]

208. Evans, J.M.; Donnelly, L.A.; Emslie-Smith, A.M.; Alessi, D.R.; Morris, A.D. Metformin and reduced risk of cancer in diabetic patients. *BMJ* **2005**, *330*, 1304–1305. [CrossRef]

209. Kourelis, T.V.; Siegel, R.D. Metformin and cancer: New applications for an old drug. *Med. Oncol.* **2012**, *29*, 1314–1327. [CrossRef]

210. Leone, A.; Di Gennaro, E.; Bruzzese, F.; Avallone, A.; Budillon, A. New perspective for an old antidiabetic drug: Metformin as anticancer agent. *Cancer Treat. Res.* **2014**, *159*, 355–376.

211. Saini, N.; Yang, X. Metformin as an anti-cancer agent: Actions and mechanisms targeting cancer stem cells. *Acta Biochim. Biophys. Sin.* **2018**, *50*, 133–143. [CrossRef]

212. Yoshida, G.J. Therapeutic strategies of drug repositioning targeting autophagy to induce cancer cell death: From pathophysiology to treatment. *J. Hematol. Oncol.* **2017**, *10*, 67. [CrossRef]

213. Nesti, L.; Natali, A. Metformin effects on the heart and the cardiovascular system: A review of experimental and clinical data. *Nutr. Metab. Cardiovasc. Dis.* **2017**, *27*, 657–669. [CrossRef]

214. Rangarajan, S.; Bone, N.B.; Zmijewska, A.A.; Jiang, S.; Park, D.W.; Bernard, K.; Locy, M.L.; Ravi, S.; Deshane, J.; Mannon, R.B.; et al. Metformin reverses established lung fibrosis in a bleomycin model. *Nat. Med.* **2018**, *24*, 1121–1127. [CrossRef]

215. Shao, R.; Li, X.; Feng, Y.; Lin, J.F.; Billig, H. Direct effects of metformin in the endometrium: A hypothetical mechanism for the treatment of women with PCOS and endometrial carcinoma. *J. Exp. Clin. Cancer Res.* **2014**, *33*, 41. [CrossRef]

216. Yi, H.; Huang, C.; Shi, Y.; Cao, Q.; Zhao, Y.; Zhang, L.; Chen, J.; Pollock, C.A.; Chen, X.M. Metformin attenuates folic-acid induced renal fibrosis in mice. *J. Cell Physiol.* **2018**, *233*, 7045–7054. [CrossRef]

217. Li, L.; Huang, W.; Li, K.; Zhang, K.; Lin, C.; Han, R.; Lu, C.; Wang, Y.; Chen, H.; Sun, F.; et al. Metformin attenuates gefitinib-induced exacerbation of pulmonary fibrosis by inhibition of TGF-beta signaling pathway. *Oncotarget* **2015**, *6*, 43605–43619. [CrossRef]

218. Xu, S.; Yang, Z.; Jin, P.; Yang, X.; Li, X.; Wei, X.; Wang, Y.; Long, S.; Zhang, T.; Chen, G.; et al. Metformin Suppresses Tumor Progression by Inactivating Stromal Fibroblasts in Ovarian Cancer. *Mol. Cancer Ther.* **2018**, *17*, 1291–1302. [CrossRef]

219. Pinter, M.; Jain, R.K. Targeting the renin-angiotensin system to improve cancer treatment: Implications for immunotherapy. *Sci. Transl. Med.* **2017**, *9*, eaan5616. [CrossRef]

220. Chauhan, V.P.; Martin, J.D.; Liu, H.; Lacorre, D.A.; Jain, S.R.; Kozin, S.V.; Stylianopoulos, T.; Mousa, A.S.; Han, X.; Adstamongkonkul, P.; et al. Angiotensin inhibition enhances drug delivery and potentiates chemotherapy by decompressing tumour blood vessels. *Nat. Commun.* **2013**, *4*, 2516. [CrossRef]

221. Vennin, C.; Murphy, K.J.; Morton, J.P.; Cox, T.R.; Pajic, M.; Timpson, P. Reshaping the Tumor Stroma for Treatment of Pancreatic Cancer. *Gastroenterology* **2018**, *154*, 820–838. [CrossRef]

222. Azuma, A. Pirfenidone treatment of idiopathic pulmonary fibrosis. *Ther. Adv. Respir. Dis.* **2012**, *6*, 107–114. [CrossRef]

223. Didiasova, M.; Singh, R.; Wilhelm, J.; Kwapiszewska, G.; Wujak, L.; Zakrzewicz, D.; Schaefer, L.; Markart, P.; Seeger, W.; Lauth, M.; et al. Pirfenidone exerts antifibrotic effects through inhibition of GLI transcription factors. *FASEB J.* **2017**, *31*, 1916–1928. [CrossRef]

224. Miura, Y.; Saito, T.; Tanaka, T.; Takoi, H.; Yatagai, Y.; Inomata, M.; Nei, T.; Saito, Y.; Gemma, A.; Azuma, A. Reduced incidence of lung cancer in patients with idiopathic pulmonary fibrosis treated with pirfenidone. *Respir. Investig.* **2018**, *56*, 72–79. [CrossRef]

225. Mediavilla-Varela, M.; Boateng, K.; Noyes, D.; Antonia, S.J. The anti-fibrotic agent pirfenidone synergizes with cisplatin in killing tumor cells and cancer-associated fibroblasts. *BMC Cancer* **2016**, *16*, 176. [CrossRef]

226. Ke, Z.; He, W.; Lai, Y.; Guo, X.; Chen, S.; Li, S.; Wang, Y.; Wang, L. Overexpression of collagen triple helix repeat containing 1 (CTHRC1) is associated with tumour aggressiveness and poor prognosis in human non-small cell lung cancer. *Oncotarget* **2014**, *5*, 9410–9424. [CrossRef]

227. Takai, K.; Le, A.; Weaver, V.M.; Werb, Z. Targeting the cancer-associated fibroblasts as a treatment in triple-negative breast cancer. *Oncotarget* **2016**, *7*, 82889–82901. [CrossRef]

228. Kozono, S.; Ohuchida, K.; Eguchi, D.; Ikenaga, N.; Fujiwara, K.; Cui, L.; Mizumoto, K.; Tanaka, M. Pirfenidone inhibits pancreatic cancer desmoplasia by regulating stellate cells. *Cancer Res.* **2013**, *73*, 2345–2356. [CrossRef]

229. Polydorou, C.; Mpekris, F.; Papageorgis, P.; Voutouri, C.; Stylianopoulos, T. Pirfenidone normalizes the tumor microenvironment to improve chemotherapy. *Oncotarget* **2017**, *8*, 24506–24517. [CrossRef]

230. Ding, N.; Hah, N.; Yu, R.T.; Sherman, M.H.; Benner, C.; Leblanc, M.; He, M.; Liddle, C.; Downes, M.; Evans, R.M. BRD4 is a novel therapeutic target for liver fibrosis. *Proc. Natl. Acad. Sci. USA* **2015**, *112*, 15713–15718. [CrossRef]

231. Yamamoto, K.; Tateishi, K.; Kudo, Y.; Hoshikawa, M.; Tanaka, M.; Nakatsuka, T.; Fujiwara, H.; Miyabayashi, K.; Takahashi, R.; Tanaka, Y.; et al. Stromal remodeling by the BET bromodomain inhibitor JQ1 suppresses the progression of human pancreatic cancer. *Oncotarget* **2016**, *7*, 61469–61484. [CrossRef]

232. Middleton, S.A.; Rajpal, N.; Cutler, L.; Mander, P.; Rioja, I.; Prinjha, R.K.; Rajpal, D.; Agarwal, P.; Kumar, V. BET Inhibition Improves NASH and Liver Fibrosis. *Sci. Rep.* **2018**, *8*, 17257. [CrossRef]

233. Kim, D.E.; Procopio, M.G.; Ghosh, S.; Jo, S.H.; Goruppi, S.; Magliozzi, F.; Bordignon, P.; Neel, V.; Angelino, P.; Dotto, G.P. Convergent roles of ATF3 and CSL in chromatin control of cancer-associated fibroblast activation. *J. Exp. Med.* **2017**, *214*, 2349–2368. [CrossRef]

234. Hah, N.; Sherman, M.H.; Yu, R.T.; Downes, M.; Evans, R.M. Targeting Transcriptional and Epigenetic Reprogramming in Stromal Cells in Fibrosis and Cancer. *Cold Spring Harb. Symp. Quant. Biol.* **2015**, *80*, 249–255. [CrossRef]

235. Ding, N.; Yu, R.T.; Subramaniam, N.; Sherman, M.H.; Wilson, C.; Rao, R.; Leblanc, M.; Coulter, S.; He, M.; Scott, C.; et al. A vitamin D receptor/SMAD genomic circuit gates hepatic fibrotic response. *Cell* **2013**, *153*, 601–613. [CrossRef]

236. Sherman, M.H.; Yu, R.T.; Engle, D.D.; Ding, N.; Atkins, A.R.; Tiriac, H.; Collisson, E.A.; Connor, F.; Van Dyke, T.; Kozlov, S.; et al. Vitamin D receptor-mediated stromal reprogramming suppresses pancreatitis and enhances pancreatic cancer therapy. *Cell* **2014**, *159*, 80–93. [CrossRef]

237. Shin, K.; Lim, A.; Zhao, C.; Sahoo, D.; Pan, Y.; Spiekerkoetter, E.; Liao, J.C.; Beachy, P.A. Hedgehog signaling restrains bladder cancer progression by eliciting stromal production of urothelial differentiation factors. *Cancer Cell* **2014**, *26*, 521–533. [CrossRef]

238. Ozdemir, B.C.; Pentcheva-Hoang, T.; Carstens, J.L.; Zheng, X.; Wu, C.C.; Simpson, T.R.; Laklai, H.; Sugimoto, H.; Kahlert, C.; Novitskiy, S.V.; et al. Depletion of carcinoma-associated fibroblasts and fibrosis induces immunosuppression and accelerates pancreas cancer with reduced survival. *Cancer Cell* **2014**, *25*, 719–734. [CrossRef]

239. Rhim, A.D.; Oberstein, P.E.; Thomas, D.H.; Mirek, E.T.; Palermo, C.F.; Sastra, S.A.; Dekleva, E.N.; Saunders, T.; Becerra, C.P.; Tattersall, I.W.; et al. Stromal elements act to restrain, rather than support, pancreatic ductal adenocarcinoma. *Cancer Cell* **2014**, *25*, 735–747. [CrossRef]

240. Gerling, M.; Buller, N.V.; Kirn, L.M.; Joost, S.; Frings, O.; Englert, B.; Bergstrom, A.; Kuiper, R.V.; Blaas, L.; Wielenga, M.C.; et al. Stromal Hedgehog signalling is downregulated in colon cancer and its restoration restrains tumour growth. *Nat. Commun.* **2016**, *7*, 12321. [CrossRef]

241. Madar, S.; Goldstein, I.; Rotter, V. 'Cancer associated fibroblasts'—More than meets the eye. *Trends Mol. Med.* **2013**, *19*, 447–453. [CrossRef]

International Journal of
Molecular Sciences

MDPI

Review

Molecular Mechanisms of Pulmonary Fibrogenesis and Its Progression to Lung Cancer: A Review

Tomonari Kinoshita [1] and Taichiro Goto [2,*]

[1]   Division of General Thoracic Surgery, Department of Surgery, Keio University School of Medicine,
      35 Shinanomachi, Shinjuku, Tokyo 1608582, Japan; kinotomo0415@gmail.com
[2]   Lung Cancer and Respiratory Disease Center, Yamanashi Central Hospital, Kofu, Yamanashi 4008506, Japan
*   Correspondence: taichiro@1997.jukuin.keio.ac.jp; Tel.: +81-55-253-7111

Received: 31 January 2019; Accepted: 20 March 2019; Published: 22 March 2019

Abstract: Idiopathic pulmonary fibrosis (IPF) is defined as a specific form of chronic, progressive fibrosing interstitial pneumonia of unknown cause, occurring primarily in older adults, and limited to the lungs. Despite the increasing research interest in the pathogenesis of IPF, unfavorable survival rates remain associated with this condition. Recently, novel therapeutic agents have been shown to control the progression of IPF. However, these drugs do not improve lung function and have not been tested prospectively in patients with IPF and coexisting lung cancer, which is a common comorbidity of IPF. Optimal management of patients with IPF and lung cancer requires understanding of pathogenic mechanisms and molecular pathways that are common to both diseases. This review article reflects the current state of knowledge regarding the pathogenesis of pulmonary fibrosis and summarizes the pathways that are common to IPF and lung cancer by focusing on the molecular mechanisms.

Keywords: idiopathic pulmonary fibrosis; lung cancer; pathogenesis; common pathways

1. Introduction

Idiopathic pulmonary fibrosis is a progressive and usually fatal lung disease characterized by fibroblast proliferation and extracellular matrix remodeling, which results in irreversible distortion of the lung's architecture. Although its cause remains to be elucidated fully, advances in cellular and molecular biology have greatly expanded our understanding of the biological processes involved in its initiation and progression [1]. It is widely accepted that environmental and occupational factors, smoking, viral infections, and traction injury to the peripheral lung can cause chronic damage to the alveolar epithelium [2]. Based on recent in vitro and in vivo studies of IPF, the novel therapeutic reagents pirfenidone and nintedanib were developed to slow the progression of this complex disease [3–5]. However, these drugs do not improve lung function and patients often remain with poor pulmonary function [6,7]. Furthermore, neither drug has been tested prospectively in patients with coexisting IPF and lung cancer [8]. In previous studies, 22% of patients with IPF developed primary lung cancers, corresponding with a five-fold greater risk than that in the general population [8–12]. Similarly, primary lung cancer risk is more than 20 times higher in patients who undergo lung transplantation for IPF than in the general population [13,14]. These observations warrant efforts to identify pathways that are common to both disorders. Questions regarding the proper and ideal management of patients who suffer from both IPF and lung cancer are also raised. It is assumed that pathogenetic similarities between IPF and lung cancer are a starting point for investigations of disease pathogenesis and the resulting insights will improve therapeutic approaches. This review article summarizes the current knowledge of the pathogenesis of pulmonary fibrosis and outlines the common molecular pathways between IPF and lung cancer.

2. The Pathogenesis of Pulmonary Fibrosis

Although knowledge of the pathogenesis of IPF remains incomprehensive, numerous research papers have contributed to the understanding of this disease. In particular, some environmental and microbial exposures have been associated with the initiation of IPF. Various individual genetic and epigenetic factors have also been related to the development of fibrosis, and potential contributions of variants and interactions with putative external factors have been presumed but not clarified. Repeated microinjury to alveolar epithelial tissues has been revealed as the first trigger of an aberrant repair process in which several lung cells develop abnormal behaviors that promote the fibrotic process. IPF is currently considered an epithelium-driven disease wherein dysfunctional aging lung epithelia are exposed to recurrent microinjuries that sabotage regeneration and lead to aberrant epithelial–mesenchymal crosstalk, creating an imbalance between profibrotic and antifibrotic mediators. Concurrently, environments that are supportive of elevated fibroblast and myofibroblast activities are maintained, and the normal repair mechanisms are replaced with chronic fibrosis. This review article details the current evidence of molecular contributions to the pathogenesis of IPF. The currently accepted mechanisms of pulmonary fibrosis are shown in Table 1 and Figure 1.

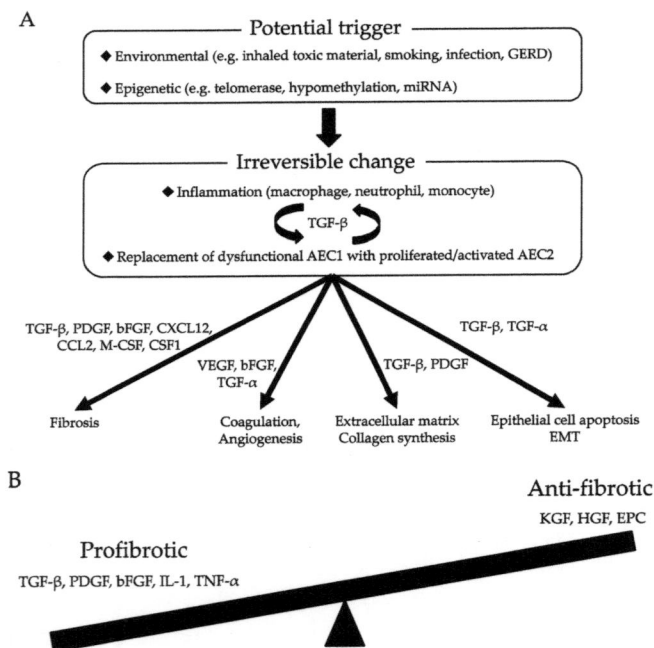

Figure 1. (**A**) Molecular mechanisms of pulmonary fibrosis; (**B**) Imbalance of profibrotic and antifibrotic mediators lead to defective regeneration and aberrant remodeling, resulting in the pathological transformation of pulmonary fibrosis. GERD, gastroesophageal reflux disease; AEC, alveolar epithelial cell; EMT, epithelial mesenchymal transition; EPC, Endothelial progenitor cell.

Table 1. Functions of representative molecules contributing to idiopathic pulmonary fibrosis.

Molecules	Profibrotic/Antifibrotic	Function in IPF
TGF-β	Profibrotic	Extracellular matrix production Epithelial mesenchymal transition Epithelial cell apoptosis and migration Recruitment of fibrocytes and immune cells Fibroblast activation, myofibroblast proliferation Induction of growth factor production Induction of pro-angiogenic mediator production
PDGF	Profibrotic	Extracellular matrix production Fibroblast proliferation
FGF	Profibrotic	Fibroblast activation Endothelial cell proliferation Epithelial cell proliferation
TGF-α	Profibrotic	Endothelial cell proliferation Fibroblast proliferation
VEGF	Profibrotic	Angiogenesis in injured lung
KGF	Antifibrotic	Maintenance and repair of injured lung
HGF	Antifibrotic	Maintenance and repair of injured lung

2.1. Dysfunctional Epithelia Trigger Aberrant Wound Healing Processes

It is assumed that fibrosis advances over long periods of time in patients with IPF. Thus, at the time of diagnosis, modifications of lung structure have already been established by the disease and pathological features, such as various stages of epithelial damage, alveolar epithelial cell (AEC) 2s hyperplasia, dense fibrosis, and abnormally proliferating mesenchymal cells, are found. At this time, it is not possible to determine the course of events that have led to lung damage; however, it is accepted that dysfunctional epithelia are key to the pathogenesis of IPF [15].

Under normal conditions of lung injury, AEC1s are replaced with proliferating and differentiating AEC2 cells and stem cells, which restore alveolar integrity by stimulating coagulation, the formation of new vessels, activation and migration of fibroblasts, and synthesis and proper alignment of collagen. Chemokines, such as transforming growth factor (TGF)-β1, platelet-derived growth factor (PDGF), vascular endothelial growth factor (VEGF), and fibroblast growth factor (FGF), are central to these processes. Conversely, continued lung injury or loss of normal restorative capacity invokes an inflammatory phase of the wound healing process. The associated increases in the expression levels of interleukin-1 (IL-1) and tumor necrosis factor-alpha (TNF-α) create a biochemical environment that favors chronic flaws of regeneration and tissue remodeling [16].

2.2. Growth Factors Associated with the Initial Stages of Pulmonary Fibrogenesis

2.2.1. TGF-β

TGF-βs are multifunctional cytokines that are present as three isoforms: TGF-β1, TGF-β2, and TGF-β3. Although the biological activities of these isoforms are indiscrete, TGF-β1 plays a predominant role in pulmonary fibrosis [17]. The three TGF-β receptors, type I (TGFRI), type II (TGFRII), and type III (TGFRIII), have the potential to bind to all three TGF-βs with high affinity. However, TGF-β is the best characterized promoter of extracellular matrix (ECM) production and is considered the strongest chemotactic factor for immune cells, such as monocytes and macrophages. In these cell types, TGF-β activates the release of cytokines, such as PDGF, IL-1β, basic FGF (bFGF), and TNF-α, and autoregulates its own expression. Increases in TGF-β production are consistently observed in epithelial cells and macrophages from lung tissues of patients with IPF [18] and in rodents with bleomycin-induced pulmonary fibrosis [19]. Smad proteins are known as mediators of TGF-β signaling from the membrane to the nucleus [20]. Activated TGF-β receptors induce phosphorylation of Smad2 and Smad3, and complexes of these with other Smad proteins are translocated into the nucleus to regulate transcriptional responses. Studies show that the deficiency of Smad3 attenuates bleomycin-induced pulmonary fibrosis in mice [21] and that the inhibitory Smad7 prevents the phosphorylation of Smad2 and Smad3 via activated TGF-β receptors [22,23].

TGF-β1 is considered the most important mediator of IPF. AEC2s produce TGF-β1 following actin–myosin-mediated cytoskeletal contractions that are induced by the unfolded protein response (UPR) following αvβ6 integrin activation. The αvβ6 integrin/TGF-β1 pathway is a constitutively expressed molecular sensing mechanism that is primed to recognize injurious stimuli. TGF-β1 is a strong profibrotic mediator that promotes the epithelial–mesenchymal transition (EMT); epithelial cell apoptosis; epithelial cell migration; other profibrotic mediator production; circulating fibrocyte recruitment; fibroblast activation and proliferation and transformation into myofibroblasts; and VEGF, connective-tissue growth factor, and other pro-angiogenic mediator production [24].

2.2.2. PDGF

PDGF is a potent chemoattractant for mesenchymal cells and induces the proliferation of fibroblasts and the synthesis of ECM. Activated homologous A and B subunits of PDGF can form three dimeric PDGF isoforms. Alveolar macrophages with IPF produce higher volumes of PDGF-B mRNA and protein [25,26]. AEC2s and mesenchymal cells also express abnormal levels of PDGF in animal models [27]. Moreover, PDGF-B transgenic mice develop lung disease with diffusely emphysematous lung lesions and inflammation/fibrosis in focal areas [28]. In agreement, intratracheal instillation of recombinant human PDGF-B into rats produces fibrotic lesions that are concentrated around large airways and blood vessels [29]. In another study, gene transfer of an extracellular domain of the PDGF receptor ameliorated bleomycin-induced pulmonary fibrosis in a mouse model [30]. Insulin-like growth factor (IGF)-1 also promoted fibroblast proliferation synergistically with PGDF [31]. Accordingly, alveolar macrophages from patients with IPF expressed IGF-1 mRNA and protein at greater levels than those in normal alveolar macrophages [31,32].

2.2.3. FGF

bFGF is a stimulator of fibroblast and endothelial cell proliferation that has been correlated with the proliferative aspects of fibrosis. In particular, bFGF expression is up-regulated at various periods of wound healing, and recombinant bFGF has been shown to accelerate wound healing. Accordingly, anti-bFGF antibody inhibited the formation of granulated tissue and normal wound repair. Alveolar macrophages are a predominant source of bFGF in intra-alveolar fibrotic areas following acute lung injury [33]. In a study of IPF, mast cells were found to be the predominant bFGF-producing cells, and bFGF levels were associated with bronchoalveolar lavage cellularity and with the severity of gas exchange abnormalities [34].

2.2.4. TGF-α

TGF-α induces proliferation in endothelial cells, epithelial cells, and fibroblasts, and is present in fibrotic areas [35]. In proliferative fibrotic lesions in rats with asbestos- or bleomycin-induced pulmonary fibrosis, AECs and macrophages had elevated expression levels of TGF-α [36]. Similarly, in transgenic mice expressing human TGF-α, proliferative fibrotic responses in interstitial and pleural surfaces were epithelial cell specific [37]. These results indicate that TGF-α is involved in cell proliferation under fibrotic conditions following lung injury.

2.2.5. Keratinocyte Growth Factor (KGF)

KGF is produced by mesenchymal cells, and the KGF receptor is expressed in the epithelial tissues of developing lungs. In rats, KGF accelerated the functional differentiation of AEC2s, and the intratracheal instillation of KGF significantly improved bleomycin-induced pulmonary fibrosis [38]. These data suggest that KGF participates in the maintenance and repair of alveolar epithelium and has potential in the treatment of lung injury and pulmonary fibrosis.

2.2.6. Hepatocyte Growth Factor (HGF)

HGF is produced by mesenchymal cells and has been identified as a potent mitogen for mature hepatocytes. The HGF receptor is a c-Met proto-oncogene product that is predominantly expressed in various types of epithelial cells. HGF levels are higher in bronchoalveolar lavage fluid and serum from patients with IPF than in serum from healthy people [39,40]. HGF is also highly expressed by hyperplastic AECs and macrophages in lung tissues of patients with IPF. In in vitro studies of epithelial cells, HGF promoted DNA synthesis in AEC2s [41]. The administration of HGF also inhibited fibrotic changes in mice with bleomycin-induced lung injury [42]. Promisingly, the combination of HGF and interferon-γ (IFN-γ) enhanced the migratory activity of A549 cells by up-regulating the c-Met/HGF receptor [43]. Based on these observations, HGF treatments may offer a novel strategy for promoting the repair of inflammatory lung damage for patients with pulmonary fibrosis.

2.3. Changes in AEC2s that Lead to Aberrant Tissue Repair

Repetitive exposures of alveolar epithelium to microinjuries, such as infection, smoking, toxic environmental inhalants, and gastroesophageal reflux, contribute to AEC1 damage. AEC2s normally regenerate damaged cells, but when dysfunctional, their ability to reestablish homeostasis is impaired. This condition is considered indicative of the pathogenesis of IPF [44,45].

2.3.1. UPR

High cellular activity leads to protein over-expression, and if unchecked, it can cause endoplasmic reticulum (ER) stress. The correcting protective pathway is stimulated by the imbalance between cellular demand for protein synthesis and the capacity of the ER to dispose of unfolded or damaged proteins. This protective pathway is known as UPR, and it re-establishes ER homeostasis. To this end, this pathway inhibits protein translation, targets proteins for degradation, and induces apoptosis when overwhelmed. The activation of UPR stimulates the expression of profibrotic mediators, such as TGF-β1, PDGF, C-X-C motif chemokine 12 (CXCL12), and chemokine C-C motif ligand 2 (CCL2), and thus, can lead to apoptosis [46].

2.3.2. Epithelial–Mesenchymal Transition (EMT)

EMT is a molecular reprograming process, and in AEC2s, it is induced by UPR and enhanced by profibrotic mediators and signaling pathways. Under these conditions, epithelial cells express mesenchymal cell-associated genes, detach from basement membranes, and migrate and down-regulate their typical markers. The most used marker of these transitioning cells is alpha smooth-muscle actin (αSMA). However, EMT occurs during development and in cancerous and fibrotic tissues, but it is not involved in the restoration of tissues through wound healing processes [46].

2.3.3. Wnt-β-Catenin Signaling

Other key pathways of IPF are related to the deregulation of embryological programs, such as Wnt-β-catenin signaling, which has been associated with EMT and fibrogenesis following activation by TGF-β1, sonic hedgehog, gremlin-1, and phosphatase and tensin homolog. Deregulation of these pathways confers resistance to apoptosis and offers proliferative advantages to cells [47].

2.4. Endothelium and Coagulation

Damage to alveolar structures and the loss of AECs with basement membranes involves alveolar vessels and leads to increased vascular permeability. Wound clots form during this early phase of wound healing responses, and sequentially, new vessels are formed through the proliferation of endothelial cells and endothelial progenitor cells (EPCs). Patients with IPF with failure of re-endothelization have significantly decreased numbers of EPCs, likely resulting in dysfunctional alveolar–capillary barriers, profibrotic responses, and compensatively augmented VEGF expression.

This series of endothelial changes could stimulate fibrotic processes and abnormalities of vessel functions, contributing to cardio–respiratory declines and advanced disease. Furthermore, endothelial cells may undergo a mesenchymal transition with similar consequences as those of EMT [48].

Endothelial and epithelial damage also activates coagulation cascades during the early phases of wound healing. Coagulation proteinases have several cellular effects on wound healing. In particular, the tissue factor-dependent pathway is central to the pathogenesis of IPF and promotes a pro-coagulation state with increased levels of inhibitors of plasminogen activation, active fibrinolysis, and protein C. Under these pro-coagulation conditions, degradation of ECM is decreased, resulting in profibrotic effects and the induction of fibroblast differentiation into myofibroblasts via proteinase-activated receptors [16].

2.5. Immunogenic Changes that Lead to Pulmonary Fibrosis

The pathobiology of IPF is led by aberrant epithelial–mesenchymal signaling, but inflammation may also play an important role because inflammatory cells are involved in normal wound healing from early phases. Initially, macrophages produce cytokines that induce inflammatory responses and participate in the transition to healing environments by recruiting fibroblasts, epithelial cells, and endothelial cells. If injury persists, neutrophils and monocytes are recruited, and the production of reactive oxygen species exacerbates epithelial damage. The resulting imbalances between antioxidants and pro-oxidants may also promote apoptosis of epithelial cells and activation of pathways that impair function. Finally, monocytes and macrophages produce PDGF, CCL2, macrophage colony stimulating factor, and colony stimulating factor 1. These proteins may also have direct profibrotic effects [44,49].

The roles of lymphocytes in IPF are still unclear. However, some lymphocytic cytokines are considered profibrotic due to their direct effects on the activities of fibroblast and myofibroblast. Th-1, Th-2, and Th-17 T-cells have been clearly associated with the pathogenesis of IPF. The Th1 T-cell subset produces IL-1α, TNF-α, PDGF, and TGF-β1 and has net profibrotic effects. Th2 and Th17 responses appear more important in the pathogenesis of IPF. In particular, the typical Th2 interleukin IL-4 induces IL-5, IL-13, and TGF-β1 expression, leading to the recruitment of macrophages, mast cells, eosinophils, and mesenchymal cells and the direct activation of fibroblasts. Additionally, fibroblasts from patients with IPF are hyperresponsive to IL-13, which has a positive effect on fibroblast activity and enhances the production of ECM. The Th17 T-cell subset indirectly promotes fibrosis by increasing TGF-β1 levels. Th17 cells are also positively regulated by TGF-β1, suggesting the presence of a positive feedback loop [16]. Numbers of regulatory T-cells are reportedly lower in bronchoalveolar lavage fluid and peripheral blood samples from patients with IPF than in those of healthy subjects. Regulatory T-cells (Tregs) play a crucial role in immune tolerance and the prevention of autoimmunity; deficiencies in numbers and functions of these T-cells play an important role in the initial phases of pathogenesis of IPF. The function of Treg in IPF is severely impaired due to reduced number of infiltrating Tregs in addition to dysfunction of Tregs. Interestingly, the compromised Treg function in bronchoalveolar lavage is associated with parameters of the disease severity of IPF, indicating a causal relationship between the development of IPF and impaired immune regulation mediated by Tregs [50]. Previous studies have demonstrated low IFN-γ levels in the lungs of patients with IPF. IFN-γ inhibits fibroblastic activity and abolishes Th2 responses. However, further studies are required to characterize the roles of inflammation in the pathobiology of IPF. Currently, the early stages of IPF are poorly understood, as are the mechanisms of disease progression [49,51]. Nonetheless, pirfenidone (5-methyl-1-phenyl-2-[1H]-pyridone) was designed to have anti-inflammatory and antifibrotic effects and was efficacious in the clinical setting [6].

2.6. Interactions Between ECM and Mesenchymal Cells, Fibrocytes, Fibroblasts, and Myofibroblasts

Contributions of mesenchymal cells, and particularly fibroblasts and myofibroblasts, are crucial for the pathogenesis of IPF. These cells are recruited, activated, and induced to differentiate and proliferate in the abnormal biochemical environments that are created by activated epithelial and endothelial cells. Although the initial trigger and source of mesenchymal cell recruitment remain unclear, the current published consensus defines fibroblasts and myofibroblasts as the key cell types for IPF. Circulating fibrocytes, pulmonary fibroblasts, and myofibroblasts have also been identified among mesenchymal cells that are involved in IPF [52]. The most recent studies of these processes are summarized in a well-integrated review [53].

3. Common Characteristics of IPF and Lung Cancer

Multiple studies compare IPF with cancer to provide insights into the pathogenesis of both diseases, for which survival rates are low. Arguments against the similarities of cancer and IPF include the presence of homogeneity, metastases, and laterality in cancers. However, cytogenetic heterogeneity has been shown in myofibroblasts, which do not metastasize to other organs. In addition, simultaneous involvement of both lungs is a definitive indication of IPF. However, this is primarily based on the generally accepted assumption that tumors are almost always monoclonal and grow in only one lung before metastasizing and invading other organs. From an anatomical viewpoint, patients with IPF mainly exhibit fibrosis in the lung periphery and in the lower lobes, which are sites of lung tumors in a high percentage of cases [54]. Additionally, patients with lung transplants due to IPF have much higher rates of lung cancer, as stated above [13,14]. These observations warrant further studies regarding the molecular connections between these two lung diseases. Furthermore, epigenetic and genetic abnormalities, changed relationships between cells, uncontrolled proliferation, and abnormal activation of specific signal transduction pathways are pathogenic features of both diseases [55,56]. Principal fibrogenic molecules, signal transduction pathways and immune cells that potentially participate both in two diseases are shown in Table 2.

Table 2. Principal factors participating both in lung cancer and idiopathic pulmonary fibrosis.

Mediators		IPF	Lung Cancer
Abnormal mRNA	let-7	down-regulated	down-regulated
	miR-21	up-regulated	up-regulated
	miR-29	down-regulated	down-regulated
	miR-30	down-regulated	down-regulated
	miR-155	up-regulated	up-regulated
	miR-200	down-regulated	down-regulated
Cell-free DNA	-	up-regulated	up-regulated
Glycoprotein	Thy-1	down-regulated	down-regulated
Connexin	Cx43	down-regulated	down-regulated
Growth Factors	TGF-β	up-regulated	up-regulated
	PDGF	up-regulated	up-regulated
	VEGF	up-regulated	up-regulated
Migration	FGF	up-regulated	up-regulated
	laminin	up-regulated	up-regulated
	fascin	up-regulated	up-regulated
Pathways	heat shock protein 27	up-regulated	up-regulated
	Wnt pathway	up-regulated	up-regulated
	PI3K/Akt pathway	up-regulated	up-regulated
Immune Cells	FAM	up-regulated	up-regulated
	MDSC	up-regulated	up-regulated
	Treg	down-regulated	up-regulated

FAM, fibrosis-associated macrophage. MDSC, myeloid-derived suppressor cell. Treg, regulatory T-cell.

3.1. Epigenetic and Genetic Abnormalities

Hypomethylation of oncogenes and methylation of tumor suppressor genes are established pathogenic mechanisms for most tumors. Epigenetic responses to environmental exposures, including smoking and dietary factors, and aging have recently been identified in patients with IPF. Recent studies also demonstrated changes to global methylation patterns in patients with IPF that are reciprocal to those in patients with lung cancers [57]. Under the conditions of IPF, hypermethylation of the CD90/Thy-1 promoter region decreases the expression of the glycoprotein Thy-1, which is normally expressed by fibroblasts [58,59]. The loss of this molecule in patients with IPF also correlates with invasive behaviors of cancers and the transition from fibroblasts into myofibroblasts. Hence, pharmaceutical inhibition of the methylation of *Thy-1* gene may restore Thy-1 expression, suggesting a new therapeutic approach for this disease. Specific gene mutations have also been considered important to the origin and progression of cancer [60]. Similarly, expression of the oncogene p53, fragile histidine triads, microsatellite instability, and loss of heterozygosity were observed in approximately half of the cases of IPF, frequently in the peripheral honeycombed lung regions that are specifically characteristic of IPF [60–63]. Additionally, mutations that are generally related to cancer occurrence and development, including those affecting telomere shortening and telomerase expression, have been observed in familial IPF [64–66]. Recently, circulating and cell-free DNA has been considered as a diagnostic and prognostic biomarker of cancer [67]. In these studies, free circulating concentrations of DNA increased in patients with cancer and IPF compared with that in patients with other fibrotic lung diseases [68]. In addition to circulating DNA, abnormal expression levels of mRNA were correlated with the pathogenesis of both diseases. These studies suggest that short non-protein-coding RNAs regulate carcinogenesis related genes that are involved in growth, invasion, and metastasis; these features are characteristic of cancer cells [69–71]. Recent papers show that 10% of mRNAs are aberrantly expressed in patients with IPF [72–74]. Among them, let-7, miR-29, miR-30, and miR-200 were down-regulated, whereas miR-21 and miR-155 were up-regulated. These changes corresponded with groups of genes that are associated with fibrosis, regulation of ECM, induction of EMT and apoptosis. Some of these mRNAs may also affect and be affected by TGF-β expression, potentially speeding functional deterioration in patients with IPF.

3.2. Abnormal Cell–Cell Communication

Intercellular channels provide metabolic and electrical coupling of cells and are formed by proteins of the connexins (Cxs) family. Cxs are necessary for the synchronization of cell proliferation and tissue repair [75]. Among them, Cx43 is the most abundant on fibroblast membranes and is involved in tissue repair and wound healing. At wound sites, the repression of Cx43 promotes repair of injured skin tissues with increased cell proliferation and migration of keratinocytes and fibroblasts. Accordingly, down-regulation of Cx43 is related to increased expression levels of TGF-β and production of collagen and acceleration of the differentiation of myofibroblast, which likely promotes healing. These changes contribute to the loss of control over the proliferation of fibroblasts that characterizes abnormal repair and fibrosis. This contention is supported by observations of low expression of Cx43 in fibroblasts derived from keloids and hypertrophic scars than in those derived from normal skin tissues [76]. Although low expression levels of Cxs are often correlated with the progression of cancer and the loss of intercellular communication [77], human lung carcinoma cell lines with high expression of Cx43 showed reduced proliferation [78]. Reduced expression of Cx43 was reported in primary lung fibroblasts from patients with IPF, and reduced intercellular communication was also identified in these cells [79]. Limited cell–cell communications are often reported in fibroblasts from patients with IPF and in cancer cells, reflecting common defects of contact inhibition and uncontrolled proliferation.

3.3. Abnormal Activation of Signaling Pathways

The Wnt/β-catenin signaling pathway regulates molecules that are related to tissue invasion, such as matrilysin, laminin, and cyclin-D1. However, arguably, the most important function of Wnt/β-catenin pathway is to mediate crosstalk with TGF-β. This pathway is abnormally activated in some tumors, as shown in lung cancer and mesothelioma [80]. Wnt/β-catenin pathway activation was also shown recently in fibroproliferative disorders of liver and kidney tissues [81]. The Wnt/β-catenin pathway is strongly activated in the lung tissues of patients with IPF [82], potentially reflecting the activities of TGF-β [83]. Specifically, TGF-β potentially activates extracellular signal-regulated protein kinases 1 and 2 (ERK1/2), and the target genes of this pathway activate other signaling pathways, including the phosphatidylinositol 3-kinase (PI3K)/Akt pathway, which regulates proliferation and apoptosis. The roles of PI3K in proliferation and differentiation into myofibroblasts have been demonstrated following stimulation with TGF-β [84]. In cancer cells, the activation of PI3K pathway participates in the demise of regulatory controls over cell proliferation. Therapeutic inhibitors have been developed using the PI3K pathway as a target, and their effects on tumor growth and survival is being assessed in many cancers [85]. Oral administration of a PI3K pathway inhibitors significantly prevented bleomycin-induced pulmonary fibrosis in rats [86]. Hence, clinical trials of such inhibitors are eagerly awaited for patients with IPF.

Tyrosine kinases are key mediators of multiple signaling pathways in healthy cells with demonstrated roles in cell growth, differentiation, adhesion, and motility and in the regulation of cell death. Tyrosine kinase activity is controlled by specific transmembrane receptors that mediate the activity of various ligands. Conversely, abnormal activities of these kinases have been associated with development, progression, and spread of several types of cancer [87]. Recently, activities of tyrosine kinase receptors were investigated in wound healing process and fibrogenesis.

TGF-β, PDGF, VEGF, and FGF are common mediators of carcinogenesis and fibrogenesis. Among them, VEGF may directly or indirectly promote cell survival and proliferation by activating ERK1/2 and PI3K. Accordingly, elevated expression levels of VEGF mRNA were shown in EPCs from patients with IPF. Furthermore, antifibrotic strategies using multiple inhibitors of tyrosine kinase receptors have been evaluated in a rat model of bleomycin-induced fibrosis; PDGF, VEGF, and FGF inhibitors produced significant improvement in fibrosis [48,88–90]. In support of these in vitro and in vivo observations, the multiple tyrosine kinase inhibitor nintedanib showed highly favorable results for the treatment of IPF [7].

3.4. Abnormal Migration and Invasion Activities

TGF-β is the most important mediator of the pathogenesis and carcinogenesis of IPF. In tumor microenvironments, TGF-β, predominantly from cancer-derived epithelial cells, induces myofibroblast recruitment at the invasive front of the cancer tissue and protects myofibroblasts from apoptosis. These cells encircle tumor tissues and produce TGF-β. With inflammatory mediators and metalloproteinases, myofibroblasts break basement membranes of surrounding tissues to facilitate tumor invasion [91,92]. Likewise, in IPF, myofibroblasts maintain proliferation through autocrine production of TGF-β, leading to their uncontrolled proliferation [93]. Moreover, related, antifibrotic prostaglandin E2 is down-regulated in myofibroblasts from IPF tissues [94]. TGF-β1 promotes the nuclear localization of myocardin-related transcription factor-A (MRTF-A), which regulates the differentiation and survival of fibroblasts, resulting in enhanced lung fibrosis [95–98]. MRTF-A has been targeted as a mediator of tumor progression and metastasis [99–101].

In cancer cells, the capacity to invade surrounding tissue strongly correlates with the expression of various molecules, including laminin, heat shock protein 27, and fascin [102–104]. In IPF, epithelial cells around fibroblast foci also express these molecules [105]. However, these molecules are exclusively expressed by bronchiolar basal cells, which are located as a layer between luminal epithelial cell and myofibroblast layers. Hence, these molecules are likely contributors to the migration of cells and the invasion of bronchiolar basal cells into myofibroblasts and luminal epithelium and are expressed at the invasive front of tumors.

Matrix metalloproteases and integrins are strongly associated with invasion and migration of cells [106]. Integrins activate cancer cells through the KRAS/RelB/NF-κB pathway and lead to the development of stem cell-like properties, such as independent growth and drug resistance. These properties provide cell–cell communications between inflammatory cells, fibroblasts, and parenchymal cells through ECM. Under conditions of IPF, integrin promotes initiation, maintenance, and resolution of tissue fibrosis. Accordingly, integrin expression was reportedly high in myofibroblasts and AECs after lung injury. Integrin is also considered a strong regulator of TGF-β during the progression of lung fibrosis. A clinical study of the humanized antibody STX-100 has been conducted for IPF [107]. Other inhibitors, such as specific antibodies against αvβ6, have also been investigated in clinical trials, and these antibodies were tested in preclinical models of fibrosis and in the murine model of bleomycin-induced pulmonary fibrosis.

3.5. Inflammatory Environment

Inflammatory reaction is described by some reports as a promoting factor in the development and progression step of tumorigenesis [108]. As described above, some kinds of macrophages produce cytokines which contribute to the inflammatory responses such as fibrosis-associated macrophages. This macrophage behaves as an M2 phenotype macrophage expressing arginase and CD206 [109]. M2 macrophages have been broadly identified as trigger cells towards tumor progression [110–112]. Myeloid-derived suppressor cells are associated with poor prognosis in malignancies and their accumulation in IPF is also correlated with disease progression [113]. On the other hand, infiltrating T lymphocytes play a crucial role in tumor progression and suppression, although their roles in IPF are still unclear [114]. Infiltrating Tregs are significantly correlated with the tumor progression whereas deficiency in numbers and functions of Tregs is observed in the initial step of IPF (Table 2) [50,115]. Further studies regarding the role of Treg in the IPF-related cancer are awaited.

4. Conclusions

In conclusion, cancer and fibrosis are both severe lung diseases, and they share biological pathways. Although the specific genetic and cellular mechanisms are not yet fully understood, several signaling pathways and microenvironments have been shown to disrupt tissue architecture and lead to dysfunction. Conversely, it is clear that lung tumorigenesis and fibrosis display highly heterogeneous behaviors, warranting personalized therapeutic approaches. Lung fibrosis may eventually be attenuated by therapies that are developed after considering mechanisms that are common to cancer and IPF.

Author Contributions: Writing—Original Draft Preparation, T.K.; Writing—Review & Editing, T.G.

Funding: This research received no external funding.

Acknowledgments: The authors greatly appreciate Yoshihiro Miyashita, Yumiko Kakizaki, and Toshiharu Tsutsui for their helpful scientific discussions.

Conflicts of Interest: The authors declare no conflict of interest.

References

1. Raghu, G.; Collard, H.R.; Egan, J.J.; Martinez, F.J.; Behr, J.; Brown, K.K.; Colby, T.V.; Cordier, J.F.; Flaherty, K.R.; Lasky, J.A.; et al. An official ats/ers/jrs/alat statement: Idiopathic pulmonary fibrosis: Evidence-based guidelines for diagnosis and management. *Am. J. Respir. Crit. Care Med.* **2011**, *183*, 788–824. [CrossRef] [PubMed]
2. Selman, M.; King, T.E.; Pardo, A.; American Thoracic Society; European Respiratory Society; American College of Chest Physicians. Idiopathic pulmonary fibrosis: Prevailing and evolving hypotheses about its pathogenesis and implications for therapy. *Ann. Intern. Med.* **2001**, *134*, 136–151. [CrossRef] [PubMed]
3. Iwata, T.; Yoshida, S.; Fujiwara, T.; Wada, H.; Nakajima, T.; Suzuki, H.; Yoshino, I. Effect of Perioperative Pirfenidone Treatment in Lung Cancer Patients With Idiopathic Pulmonary Fibrosis. *Ann. Thorac. Surg.* **2016**, *102*, 1905–1910. [CrossRef]
4. Iwata, T.; Yoshida, S.; Nagato, K.; Nakajima, T.; Suzuki, H.; Tagawa, T.; Mizobuchi, T.; Ota, S.; Nakatani, Y.; Yoshino, I. Experience with perioperative pirfenidone for lung cancer surgery in patients with idiopathic pulmonary fibrosis. *Surg. Today* **2015**, *45*, 1263–1270. [CrossRef]

5. Iwata, T.; Yoshino, I.; Yoshida, S.; Ikeda, N.; Tsuboi, M.; Asato, Y.; Katakami, N.; Sakamoto, K.; Yamashita, Y.; Okami, J.; et al. A phase II trial evaluating the efficacy and safety of perioperative pirfenidone for prevention of acute exacerbation of idiopathic pulmonary fibrosis in lung cancer patients undergoing pulmonary resection: West Japan Oncology Group 6711 L (PEOPLE Study). *Respir. Res.* **2016**, *17*, 90. [CrossRef]

6. Costabel, U.; Albera, C.; Lancaster, L.H.; Lin, C.Y.; Hormel, P.; Hulter, H.N.; Noble, P.W. An Open-Label Study of the Long-Term Safety of Pirfenidone in Patients with Idiopathic Pulmonary Fibrosis (RECAP). *Respiration* **2017**, *94*, 408–415. [CrossRef]

7. Richeldi, L.; du Bois, R.M.; Raghu, G.; Azuma, A.; Brown, K.K.; Costabel, U.; Cottin, V.; Flaherty, K.R.; Hansell, D.M.; Inoue, Y.; et al. Efficacy and safety of nintedanib in idiopathic pulmonary fibrosis. *N. Engl. J. Med.* **2014**, *370*, 2071–2082. [CrossRef] [PubMed]

8. Karampitsakos, T.; Tzilas, V.; Tringidou, R.; Steiropoulos, P.; Aidinis, V.; Papiris, S.A.; Bouros, D.; Tzouvelekis, A. Lung cancer in patients with idiopathic pulmonary fibrosis. *Pulm. Pharmacol. Ther.* **2017**, *45*, 1–10. [CrossRef] [PubMed]

9. Ozawa, Y.; Suda, T.; Naito, T.; Enomoto, N.; Hashimoto, D.; Fujisawa, T.; Nakamura, Y.; Inui, N.; Nakamura, H.; Chida, K. Cumulative incidence of and predictive factors for lung cancer in IPF. *Respirology* **2009**, *14*, 723–728. [CrossRef]

10. Goto, T.; Maeshima, A.; Akanabe, K.; Oyamada, Y.; Kato, R. Acute exacerbation of idiopathic pulmonary fibrosis of microscopic usual interstitial pneumonia pattern after lung cancer surgery. *Ann. Thorac. Cardiovasc. Surg.* **2011**, *17*, 573–576. [CrossRef]

11. Goto, T.; Maeshima, A.; Oyamada, Y.; Kato, R. Idiopathic pulmonary fibrosis as a prognostic factor in non-small cell lung cancer. *Int. J. Clin. Oncol.* **2014**, *19*, 266–273. [CrossRef]

12. Goto, T. Measuring Surgery Outcomes of Lung Cancer Patients with Concomitant Pulmonary Fibrosis: A Review of the Literature. *Cancers* **2018**, *10*, 223. [CrossRef]

13. Hendriks, L.E.; Drent, M.; van Haren, E.H.; Verschakelen, J.A.; Verleden, G.M. Lung cancer in idiopathic pulmonary fibrosis patients diagnosed during or after lung transplantation. *Respir. Med. Case Rep.* **2012**, *5*, 37–39. [CrossRef] [PubMed]

14. Daniels, C.E.; Jett, J.R. Does interstitial lung disease predispose to lung cancer? *Curr. Opin. Pulm. Med.* **2005**, *11*, 431–437. [CrossRef] [PubMed]

15. Liu, Y.M.; Nepali, K.; Liou, J.P. Idiopathic Pulmonary Fibrosis: Current Status, Recent Progress, and Emerging Targets. *J. Med. Chem.* **2017**, *60*, 527–553. [CrossRef] [PubMed]

16. Betensley, A.; Sharif, R.; Karamichos, D. A Systematic Review of the Role of Dysfunctional Wound Healing in the Pathogenesis and Treatment of Idiopathic Pulmonary Fibrosis. *J. Clin. Med.* **2016**, *6*, 2. [CrossRef] [PubMed]

17. Coker, R.K.; Laurent, G.J.; Shahzeidi, S.; Lympany, P.A.; du Bois, R.M.; Jeffery, P.K.; McAnulty, R.J. Transforming growth factors-beta(1), -beta(2), and -beta(3) stimulate fibroblast procollagen production in vitro but are differentially expressed during bleomycin-induced lung fibrosis. *Am. J. Pathol.* **1997**, *150*, 981–991.

18. Khalil, N.; O'Connor, R.N.; Unruh, H.W.; Warren, P.W.; Flanders, K.C.; Kemp, A.; Bereznay, O.H.; Greenberg, A.H. Increased production and immunohistochemical localization of transforming growth factor-beta in idiopathic pulmonary fibrosis. *Am. J. Respir. Cell Mol. Biol.* **1991**, *5*, 155–162. [CrossRef] [PubMed]

19. Raghow, B.; Irish, P.; Kang, A.H. Coordinate regulation of transforming growth factor beta gene expression and cell proliferation in hamster lungs undergoing bleomycin-induced pulmonary fibrosis. *J. Clin. Investig.* **1989**, *84*, 1836–1842. [CrossRef]

20. Heldin, C.H.; Miyazono, K.; ten Dijke, P. TGF-beta signalling from cell membrane to nucleus through SMAD proteins. *Nature* **1997**, *390*, 465–471. [CrossRef]

21. Zhao, J.; Shi, W.; Wang, Y.L.; Chen, H.; Bringas, P., Jr.; Datto, M.B.; Frederick, J.P.; Wang, X.F.; Warburton, D. Smad3 deficiency attenuates bleomycin-induced pulmonary fibrosis in mice. *Am. J. Physiol. Lung Cell Mol. Physiol.* **2002**, *282*, L585–L593. [CrossRef] [PubMed]

22. Nakao, A.; Afrakhte, M.; Moren, A.; Nakayama, T.; Christian, J.L.; Heuchel, R.; Itoh, S.; Kawabata, M.; Heldin, N.E.; Heldin, C.H.; et al. Identification of Smad7, a TGFbeta-inducible antagonist of TGF-beta signalling. *Nature* **1997**, *389*, 631–635. [CrossRef]

23. Hayashi, H.; Abdollah, S.; Qiu, Y.; Cai, J.; Xu, Y.Y.; Grinnell, B.W.; Richardson, M.A.; Topper, J.N.; Gimbrone, M.A., Jr.; Wrana, J.L.; et al. The MAD-related protein Smad7 associates with the TGFbeta receptor and functions as an antagonist of TGFbeta signaling. *Cell* **1997**, *89*, 1165–1173. [CrossRef]

24. Grimminger, F.; Gunther, A.; Vancheri, C. The role of tyrosine kinases in the pathogenesis of idiopathic pulmonary fibrosis. *Eur. Respir. J.* **2015**, *45*, 1426–1433. [CrossRef] [PubMed]

25. Antoniades, H.N.; Bravo, M.A.; Avila, R.E.; Galanopoulos, T.; Neville-Golden, J.; Maxwell, M.; Selman, M. Platelet-derived growth factor in idiopathic pulmonary fibrosis. *J. Clin. Investig.* **1990**, *86*, 1055–1064. [CrossRef] [PubMed]

26. Martinet, Y.; Rom, W.N.; Grotendorst, G.R.; Martin, G.R.; Crystal, R.G. Exaggerated spontaneous release of platelet-derived growth factor by alveolar macrophages from patients with idiopathic pulmonary fibrosis. *N. Engl. J. Med.* **1987**, *317*, 202–209. [CrossRef] [PubMed]

27. Liu, J.Y.; Morris, G.F.; Lei, W.H.; Hart, C.E.; Lasky, J.A.; Brody, A.R. Rapid activation of PDGF-A and -B expression at sites of lung injury in asbestos-exposed rats. *Am. J. Respir. Cell Mol. Biol.* **1997**, *17*, 129–140. [CrossRef] [PubMed]

28. Hoyle, G.W.; Li, J.; Finkelstein, J.B.; Eisenberg, T.; Liu, J.Y.; Lasky, J.A.; Athas, G.; Morris, G.F.; Brody, A.R. Emphysematous lesions, inflammation, and fibrosis in the lungs of transgenic mice overexpressing platelet-derived growth factor. *Am. J. Pathol.* **1999**, *154*, 1763–1775. [CrossRef]

29. Yi, E.S.; Lee, H.; Yin, S.; Piguet, P.; Sarosi, I.; Kaufmann, S.; Tarpley, J.; Wang, N.S.; Ulich, T.R. Platelet-derived growth factor causes pulmonary cell proliferation and collagen deposition in vivo. *Am. J. Pathol.* **1996**, *149*, 539–548.

30. Yoshida, M.; Sakuma-Mochizuki, J.; Abe, K.; Arai, T.; Mori, M.; Goya, S.; Matsuoka, H.; Hayashi, S.; Kaneda, Y.; Kishimoto, T. In vivo gene transfer of an extracellular domain of platelet-derived growth factor beta receptor by the HVJ-liposome method ameliorates bleomycin-induced pulmonary fibrosis. *Biochem. Biophys. Res. Commun.* **1999**, *265*, 503–508. [CrossRef]

31. Rom, W.N.; Basset, P.; Fells, G.A.; Nukiwa, T.; Trapnell, B.C.; Crysal, R.G. Alveolar macrophages release an insulin-like growth factor I-type molecule. *J. Clin. Investig.* **1988**, *82*, 1685–1693. [CrossRef] [PubMed]

32. Bitterman, P.B.; Adelberg, S.; Crystal, R.G. Mechanisms of pulmonary fibrosis. Spontaneous release of the alveolar macrophage-derived growth factor in the interstitial lung disorders. *J. Clin. Investig.* **1983**, *72*, 1801–1813. [CrossRef] [PubMed]

33. Henke, C.; Marineili, W.; Jessurun, J.; Fox, J.; Harms, D.; Peterson, M.; Chiang, L.; Doran, P. Macrophage production of basic fibroblast growth factor in the fibroproliferative disorder of alveolar fibrosis after lung injury. *Am. J. Pathol.* **1993**, *143*, 1189–1199.

34. Inoue, Y.; King, T.E., Jr.; Tinkle, S.S.; Dockstader, K.; Newman, L.S. Human mast cell basic fibroblast growth factor in pulmonary fibrotic disorders. *Am. J. Pathol.* **1996**, *149*, 2037–2054. [PubMed]

35. Liu, J.Y.; Morris, G.F.; Lei, W.H.; Corti, M.; Brody, A.R. Up-regulated expression of transforming growth factor-alpha in the bronchiolar-alveolar duct regions of asbestos-exposed rats. *Am. J. Pathol.* **1996**, *149*, 205–217. [PubMed]

36. Madtes, D.K.; Busby, H.K.; Strandjord, T.P.; Clark, J.G. Expression of transforming growth factor-alpha and epidermal growth factor receptor is increased following bleomycin-induced lung injury in rats. *Am. J. Respir. Cell Mol. Biol.* **1994**, *11*, 540–551. [CrossRef]

37. Korfhagen, T.R.; Swantz, R.J.; Wert, S.E.; McCarty, J.M.; Kerlakian, C.B.; Glasser, S.W.; Whitsett, J.A. Respiratory epithelial cell expression of human transforming growth factor-alpha induces lung fibrosis in transgenic mice. *J. Clin. Investig.* **1994**, *93*, 1691–1699. [CrossRef] [PubMed]

38. Sugahara, K.; Iyama, K.; Kuroda, M.J.; Sano, K. Double intratracheal instillation of keratinocyte growth factor prevents bleomycin-induced lung fibrosis in rats. *J. Pathol.* **1998**, *186*, 90–98. [CrossRef]

39. Sakai, T.; Satoh, K.; Matsushima, K.; Shindo, S.; Abe, S.; Abe, T.; Motomiya, M.; Kawamoto, T.; Kawabata, Y.; Nakamura, T.; et al. Hepatocyte growth factor in bronchoalveolar lavage fluids and cells in patients with inflammatory chest diseases of the lower respiratory tract: Detection by RIA and in situ hybridization. *Am. J. Respir. Cell Mol. Biol.* **1997**, *16*, 388–397. [CrossRef]

40. Maeda, J.; Ueki, N.; Hada, T.; Higashino, K. Elevated serum hepatocyte growth factor/scatter factor levels in inflammatory lung disease. *Am. J. Respir. Crit. Care Med.* **1995**, *152*, 1587–1591. [CrossRef]

41. Shiratori, M.; Michalopoulos, G.; Shinozuka, H.; Singh, G.; Ogasawara, H.; Katyal, S.L. Hepatocyte growth factor stimulates DNA synthesis in alveolar epithelial type II cells in vitro. *Am. J. Respir. Cell Mol. Biol.* **1995**, *12*, 171–180. [CrossRef]

42. Yaekashiwa, M.; Nakayama, S.; Ohnuma, K.; Sakai, T.; Abe, T.; Satoh, K.; Matsumoto, K.; Nakamura, T.; Takahashi, T.; Nukiwa, T. Simultaneous or delayed administration of hepatocyte growth factor equally represses the fibrotic changes in murine lung injury induced by bleomycin. A morphologic study. *Am. J. Respir. Crit. Care Med.* **1997**, *156*, 1937–1944. [CrossRef]

43. Nagahori, T.; Dohi, M.; Matsumoto, K.; Saitoh, K.; Honda, Z.I.; Nakamura, T.; Yamamoto, K. Interferon-gamma upregulates the c-Met/hepatocyte growth factor receptor expression in alveolar epithelial cells. *Am. J. Respir. Cell Mol. Biol.* **1999**, *21*, 490–497. [CrossRef] [PubMed]

44. Coward, W.R.; Saini, G.; Jenkins, G. The pathogenesis of idiopathic pulmonary fibrosis. *Ther. Adv. Respir. Dis.* **2010**, *4*, 367–388. [CrossRef] [PubMed]

45. Evans, C.M.; Fingerlin, T.E.; Schwarz, M.I.; Lynch, D.; Kurche, J.; Warg, L.; Yang, I.V.; Schwartz, D.A. Idiopathic Pulmonary Fibrosis: A Genetic Disease That Involves Mucociliary Dysfunction of the Peripheral Airways. *Physiol. Rev.* **2016**, *96*, 1567–1591. [CrossRef] [PubMed]

46. Wolters, P.J.; Collard, H.R.; Jones, K.D. Pathogenesis of idiopathic pulmonary fibrosis. *Annu. Rev. Pathol.* **2014**, *9*, 157–179. [CrossRef]

47. King, T.E., Jr.; Pardo, A.; Selman, M. Idiopathic pulmonary fibrosis. *Lancet* **2011**, *378*, 1949–1961. [CrossRef]

48. Malli, F.; Koutsokera, A.; Paraskeva, E.; Zakynthinos, E.; Papagianni, M.; Makris, D.; Tsilioni, I.; Molyvdas, P.A.; Gourgoulianis, K.I.; Daniil, Z. Endothelial progenitor cells in the pathogenesis of idiopathic pulmonary fibrosis: An evolving concept. *PLoS ONE* **2013**, *8*, e53658. [CrossRef] [PubMed]

49. Zolak, J.S.; de Andrade, J.A. Idiopathic pulmonary fibrosis. *Immunol. Allergy Clin. N. Am.* **2012**, *32*, 473–485. [CrossRef]

50. Kotsianidis, I.; Nakou, E.; Bouchliou, I.; Tzouvelekis, A.; Spanoudakis, E.; Steiropoulos, P.; Sotiriou, I.; Aidinis, V.; Margaritis, D.; Tsatalas, C.; et al. Global impairment of CD4+CD25+FOXP3+ regulatory T cells in idiopathic pulmonary fibrosis. *Am. J. Respir. Crit. Care Med.* **2009**, *179*, 1121–1130. [CrossRef]

51. King, T.E., Jr.; Bradford, W.Z.; Castro-Bernardini, S.; Fagan, E.A.; Glaspole, I.; Glassberg, M.K.; Gorina, E.; Hopkins, P.M.; Kardatzke, D.; Lancaster, L.; et al. A phase 3 trial of pirfenidone in patients with idiopathic pulmonary fibrosis. *N. Engl. J. Med.* **2014**, *370*, 2083–2092. [CrossRef]

52. Hinz, B. Mechanical aspects of lung fibrosis: A spotlight on the myofibroblast. *Proc. Am. Thorac. Soc.* **2012**, *9*, 137–147. [CrossRef] [PubMed]

53. Sgalla, G.; Iovene, B.; Calvello, M.; Ori, M.; Varone, F.; Richeldi, L. Idiopathic pulmonary fibrosis: Pathogenesis and management. *Respir. Res.* **2018**, *19*, 32. [CrossRef] [PubMed]

54. Antoniou, K.M.; Tomassetti, S.; Tsitoura, E.; Vancheri, C. Idiopathic pulmonary fibrosis and lung cancer: A clinical and pathogenesis update. *Curr. Opin. Pulm. Med.* **2015**, *21*, 626–633. [CrossRef] [PubMed]

55. Vancheri, C.; Failla, M.; Crimi, N.; Raghu, G. Idiopathic pulmonary fibrosis: A disease with similarities and links to cancer biology. *Eur. Respir. J.* **2010**, *35*, 496–504. [CrossRef] [PubMed]

56. Vancheri, C. Common pathways in idiopathic pulmonary fibrosis and cancer. *Eur. Respir. Rev.* **2013**, *22*, 265–272. [CrossRef]

57. Rabinovich, E.I.; Kapetanaki, M.G.; Steinfeld, I.; Gibson, K.F.; Pandit, K.V.; Yu, G.; Yakhini, Z.; Kaminski, N. Global methylation patterns in idiopathic pulmonary fibrosis. *PLoS ONE* **2012**, *7*, e33770. [CrossRef] [PubMed]

58. Sanders, Y.Y.; Kumbla, P.; Hagood, J.S. Enhanced myofibroblastic differentiation and survival in Thy-1(-) lung fibroblasts. *Am. J. Respir. Cell Mol. Biol.* **2007**, *36*, 226–235. [CrossRef] [PubMed]

59. Sanders, Y.Y.; Pardo, A.; Selman, M.; Nuovo, G.J.; Tollefsbol, T.O.; Siegal, G.P.; Hagood, J.S. Thy-1 promoter hypermethylation: A novel epigenetic pathogenic mechanism in pulmonary fibrosis. *Am. J. Respir. Cell Mol. Biol.* **2008**, *39*, 610–618. [CrossRef]

60. Kuwano, K.; Kunitake, R.; Kawasaki, M.; Nomoto, Y.; Hagimoto, N.; Nakanishi, Y.; Hara, N. P21Waf1/Cip1/Sdi1 and p53 expression in association with DNA strand breaks in idiopathic pulmonary fibrosis. *Am. J. Respir. Crit. Care Med.* **1996**, *154*, 477–483. [CrossRef] [PubMed]

61. Hojo, S.; Fujita, J.; Yamadori, I.; Kamei, T.; Yoshinouchi, T.; Ohtsuki, Y.; Okada, H.; Bandoh, S.; Yamaji, Y.; Takahara, J.; et al. Heterogeneous point mutations of the p53 gene in pulmonary fibrosis. *Eur. Respir. J.* **1998**, *12*, 1404–1408. [CrossRef] [PubMed]

62. Uematsu, K.; Yoshimura, A.; Gemma, A.; Mochimaru, H.; Hosoya, Y.; Kunugi, S.; Matsuda, K.; Seike, M.; Kurimoto, F.; Takenaka, K.; et al. Aberrations in the fragile histidine triad (FHIT) gene in idiopathic pulmonary fibrosis. *Cancer Res.* **2001**, *61*, 8527–8533. [PubMed]

63. Demopoulos, K.; Arvanitis, D.A.; Vassilakis, D.A.; Siafakas, N.M.; Spandidos, D.A. MYCL1, FHIT, SPARC, p16(INK4) and TP53 genes associated to lung cancer in idiopathic pulmonary fibrosis. *J. Cell. Mol. Med.* **2002**, *6*, 215–222. [CrossRef] [PubMed]

64. Cronkhite, J.T.; Xing, C.; Raghu, G.; Chin, K.M.; Torres, F.; Rosenblatt, R.L.; Garcia, C.K. Telomere shortening in familial and sporadic pulmonary fibrosis. *Am. J. Respir. Crit. Care Med.* **2008**, *178*, 729–737. [CrossRef] [PubMed]

65. Diaz de Leon, A.; Cronkhite, J.T.; Katzenstein, A.L.; Godwin, J.D.; Raghu, G.; Glazer, C.S.; Rosenblatt, R.L.; Girod, C.E.; Garrity, E.R.; Xing, C.; et al. Telomere lengths, pulmonary fibrosis and telomerase (TERT) mutations. *PLoS ONE* **2010**, *5*, e10680. [CrossRef] [PubMed]

66. Liu, T.; Chung, M.J.; Ullenbruch, M.; Yu, H.; Jin, H.; Hu, B.; Choi, Y.Y.; Ishikawa, F.; Phan, S.H. Telomerase activity is required for bleomycin-induced pulmonary fibrosis in mice. *J. Clin. Investig.* **2007**, *117*, 3800–3809. [CrossRef]

67. Schwarzenbach, H.; Hoon, D.S.; Pantel, K. Cell-free nucleic acids as biomarkers in cancer patients. *Nat. Rev. Cancer* **2011**, *11*, 426–437. [CrossRef]

68. Casoni, G.L.; Ulivi, P.; Mercatali, L.; Chilosi, M.; Tomassetti, S.; Romagnoli, M.; Ravaglia, C.; Gurioli, C.; Gurioli, C.; Zoli, W.; et al. Increased levels of free circulating DNA in patients with idiopathic pulmonary fibrosis. *Int. J. Biol. Markers* **2010**, *25*, 229–235. [CrossRef]

69. Lovat, F.; Valeri, N.; Croce, C.M. MicroRNAs in the pathogenesis of cancer. *Semin Oncol.* **2011**, *38*, 724–733. [CrossRef]

70. Oak, S.R.; Murray, L.; Herath, A.; Sleeman, M.; Anderson, I.; Joshi, A.D.; Coelho, A.L.; Flaherty, K.R.; Toews, G.B.; Knight, D.; et al. A micro RNA processing defect in rapidly progressing idiopathic pulmonary fibrosis. *PLoS ONE* **2011**, *6*, e21253. [CrossRef]

71. Pandit, K.V.; Milosevic, J.; Kaminski, N. MicroRNAs in idiopathic pulmonary fibrosis. *Transl. Res.* **2011**, *157*, 191–199. [CrossRef] [PubMed]

72. Correll, K.A.; Edeen, K.E.; Redente, E.F.; Zemans, R.L.; Edelman, B.L.; Danhorn, T.; Curran-Everett, D.; Mikels-Vigdal, A.; Mason, R.J. TGF beta inhibits HGF, FGF7, and FGF10 expression in normal and IPF lung fibroblasts. *Physiol. Rep.* **2018**, *6*, e13794. [CrossRef] [PubMed]

73. Roach, K.M.; Feghali-Bostwick, C.A.; Amrani, Y.; Bradding, P. Lipoxin A4 Attenuates Constitutive and TGF-beta1-Dependent Profibrotic Activity in Human Lung Myofibroblasts. *J. Immunol.* **2015**, *195*, 2852–2860. [CrossRef] [PubMed]

74. Samara, K.D.; Trachalaki, A.; Tsitoura, E.; Koutsopoulos, A.V.; Lagoudaki, E.D.; Lasithiotaki, I.; Margaritopoulos, G.; Pantelidis, P.; Bibaki, E.; Siafakas, N.M.; et al. Upregulation of citrullination pathway: From Autoimmune to Idiopathic Lung Fibrosis. *Respir. Res.* **2017**, *18*, 218. [CrossRef] [PubMed]

75. Losa, D.; Chanson, M.; Crespin, S. Connexins as therapeutic targets in lung disease. *Expert Opin. Ther. Targets* **2011**, *15*, 989–1002. [CrossRef]

76. Mori, R.; Power, K.T.; Wang, C.M.; Martin, P.; Becker, D.L. Acute downregulation of connexin43 at wound sites leads to a reduced inflammatory response, enhanced keratinocyte proliferation and wound fibroblast migration. *J. Cell Sci.* **2006**, *119*, 5193–5203. [CrossRef]

77. Cesen-Cummings, K.; Fernstrom, M.J.; Malkinson, A.M.; Ruch, R.J. Frequent reduction of gap junctional intercellular communication and connexin43 expression in human and mouse lung carcinoma cells. *Carcinogenesis* **1998**, *19*, 61–67. [CrossRef]

78. Zhang, Z.Q.; Zhang, W.; Wang, N.Q.; Bani-Yaghoub, M.; Lin, Z.X.; Naus, C.C. Suppression of tumorigenicity of human lung carcinoma cells after transfection with connexin43. *Carcinogenesis* **1998**, *19*, 1889–1894. [CrossRef]

79. Trovato-Salinaro, A.; Trovato-Salinaro, E.; Failla, M.; Mastruzzo, C.; Tomaselli, V.; Gili, E.; Crimi, N.; Condorelli, D.F.; Vancheri, C. Altered intercellular communication in lung fibroblast cultures from patients with idiopathic pulmonary fibrosis. *Respir. Res.* **2006**, *7*, 122. [CrossRef]

80. Mazieres, J.; He, B.; You, L.; Xu, Z.; Jablons, D.M. Wnt signaling in lung cancer. *Cancer Lett.* **2005**, *222*, 1–10. [CrossRef]

81. Bowley, E.; O'Gorman, D.B.; Gan, B.S. Beta-catenin signaling in fibroproliferative disease. *J. Surg. Res.* **2007**, *138*, 141–150. [CrossRef]

82. Chilosi, M.; Poletti, V.; Zamo, A.; Lestani, M.; Montagna, L.; Piccoli, P.; Pedron, S.; Bertaso, M.; Scarpa, A.; Murer, B.; et al. Aberrant Wnt/beta-catenin pathway activation in idiopathic pulmonary fibrosis. *Am. J. Pathol.* **2003**, *162*, 1495–1502. [CrossRef]

83. Caraci, F.; Gili, E.; Calafiore, M.; Failla, M.; La Rosa, C.; Crimi, N.; Sortino, M.A.; Nicoletti, F.; Copani, A.; Vancheri, C. TGF-beta1 targets the GSK-3beta/beta-catenin pathway via ERK activation in the transition of human lung fibroblasts into myofibroblasts. *Pharmacol. Res.* **2008**, *57*, 274–282. [CrossRef] [PubMed]

84. Conte, E.; Fruciano, M.; Fagone, E.; Gili, E.; Caraci, F.; Iemmolo, M.; Crimi, N.; Vancheri, C. Inhibition of PI3K prevents the proliferation and differentiation of human lung fibroblasts into myofibroblasts: The role of class I P110 isoforms. *PLoS ONE* **2011**, *6*, e24663. [CrossRef]

85. Guerreiro, A.S.; Fattet, S.; Kulesza, D.W.; Atamer, A.; Elsing, A.N.; Shalaby, T.; Jackson, S.P.; Schoenwaelder, S.M.; Grotzer, M.A.; Delattre, O.; et al. A sensitized RNA interference screen identifies a novel role for the PI3K p110gamma isoform in medulloblastoma cell proliferation and chemoresistance. *Mol. Cancer Res.* **2011**, *9*, 925–935. [CrossRef]

86. Wei, X.; Han, J.; Chen, Z.Z.; Qi, B.W.; Wang, G.C.; Ma, Y.H.; Zheng, H.; Luo, Y.F.; Wei, Y.Q.; Chen, L.J. A phosphoinositide 3-kinase-gamma inhibitor, AS605240 prevents bleomycin-induced pulmonary fibrosis in rats. *Biochem. Biophys. Res. Commun.* **2010**, *397*, 311–317. [CrossRef] [PubMed]

87. Grimminger, F.; Schermuly, R.T.; Ghofrani, H.A. Targeting non-malignant disorders with tyrosine kinase inhibitors. *Nat. Rev. Drug Discov.* **2010**, *9*, 956–970. [CrossRef]

88. Chaudhary, N.I.; Roth, G.J.; Hilberg, F.; Muller-Quernheim, J.; Prasse, A.; Zissel, G.; Schnapp, A.; Park, J.E. Inhibition of PDGF, VEGF and FGF signalling attenuates fibrosis. *Eur. Respir. J.* **2007**, *29*, 976–985. [CrossRef] [PubMed]

89. Ando, M.; Miyazaki, E.; Ito, T.; Hiroshige, S.; Nureki, S.I.; Ueno, T.; Takenaka, R.; Fukami, T.; Kumamoto, T. Significance of serum vascular endothelial growth factor level in patients with idiopathic pulmonary fibrosis. *Lung* **2010**, *188*, 247–252. [CrossRef]

90. Rhee, C.K.; Lee, S.H.; Yoon, H.K.; Kim, S.C.; Lee, S.Y.; Kwon, S.S.; Kim, Y.K.; Kim, K.H.; Kim, T.J.; Kim, J.W. Effect of nilotinib on bleomycin-induced acute lung injury and pulmonary fibrosis in mice. *Respiration* **2011**, *82*, 273–287. [CrossRef] [PubMed]

91. Desmouliere, A.; Guyot, C.; Gabbiani, G. The stroma reaction myofibroblast: A key player in the control of tumor cell behavior. *Int. J. Dev. Biol.* **2004**, *48*, 509–517. [CrossRef]

92. Micke, P.; Ostman, A. Tumour-stroma interaction: Cancer-associated fibroblasts as novel targets in anti-cancer therapy? *Lung Cancer* **2004**, *45* (Suppl. 2), S163–S175. [CrossRef]

93. Fletcher, C.D. Myofibroblastic tumours: An update. *Verh. Dtsch. Ges. Pathol.* **1998**, *82*, 75–82. [PubMed]

94. Vancheri, C.; Sortino, M.A.; Tomaselli, V.; Mastruzzo, C.; Condorelli, F.; Bellistri, G.; Pistorio, M.P.; Canonico, P.L.; Crimi, N. Different expression of TNF-alpha receptors and prostaglandin E(2)Production in normal and fibrotic lung fibroblasts: Potential implications for the evolution of the inflammatory process. *Am. J. Respir. Cell Mol. Biol.* **2000**, *22*, 628–634. [CrossRef] [PubMed]

95. Zhou, Y.; Huang, X.; Hecker, L.; Kurundkar, D.; Kurundkar, A.; Liu, H.; Jin, T.H.; Desai, L.; Bernard, K.; Thannickal, V.J. Inhibition of mechanosensitive signaling in myofibroblasts ameliorates experimental pulmonary fibrosis. *J. Clin. Investig.* **2013**, *123*, 1096–1108. [CrossRef]

96. Sisson, T.H.; Ajayi, I.O.; Subbotina, N.; Dodi, A.E.; Rodansky, E.S.; Chibucos, L.N.; Kim, K.K.; Keshamouni, V.G.; White, E.S.; Zhou, Y.; et al. Inhibition of myocardin-related transcription factor/serum response factor signaling decreases lung fibrosis and promotes mesenchymal cell apoptosis. *Am. J. Pathol.* **2015**, *185*, 969–986. [CrossRef]

97. Huang, X.; Yang, N.; Fiore, V.F.; Barker, T.H.; Sun, Y.; Morris, S.W.; Ding, Q.; Thannickal, V.J.; Zhou, Y. Matrix stiffness-induced myofibroblast differentiation is mediated by intrinsic mechanotransduction. *Am. J. Respir. Cell Mol. Biol.* **2012**, *47*, 340–348. [CrossRef]

98. Bernau, K.; Ngam, C.; Torr, E.E.; Acton, B.; Kach, J.; Dulin, N.O.; Sandbo, N. Megakaryoblastic leukemia-1 is required for the development of bleomycin-induced pulmonary fibrosis. *Respir. Res.* **2015**, *16*, 45. [CrossRef]

99. Medjkane, S.; Perez-Sanchez, C.; Gaggioli, C.; Sahai, E.; Treisman, R. Myocardin-related transcription factors and SRF are required for cytoskeletal dynamics and experimental metastasis. *Nat. Cell Biol.* **2009**, *11*, 257–268. [CrossRef]

100. Kim, T.; Hwang, D.; Lee, D.; Kim, J.H.; Kim, S.Y.; Lim, D.S. MRTF potentiates TEAD-YAP transcriptional activity causing metastasis. *EMBO J.* **2017**, *36*, 520–535. [CrossRef]

101. Kishi, T.; Mayanagi, T.; Iwabuchi, S.; Akasaka, T.; Sobue, K. Myocardin-related transcription factor A (MRTF-A) activity-dependent cell adhesion is correlated to focal adhesion kinase (FAK) activity. *Oncotarget* **2016**, *7*, 72113–72130. [CrossRef] [PubMed]

102. Moriya, Y.; Niki, T.; Yamada, T.; Matsuno, Y.; Kondo, H.; Hirohashi, S. Increased expression of laminin-5 and its prognostic significance in lung adenocarcinomas of small size. An immunohistochemical analysis of 102 cases. *Cancer* **2001**, *91*, 1129–1141. [CrossRef]

103. Garrido, C.; Schmitt, E.; Cande, C.; Vahsen, N.; Parcellier, A.; Kroemer, G. HSP27 and HSP70: Potentially oncogenic apoptosis inhibitors. *Cell Cycle* **2003**, *2*, 579–584. [CrossRef] [PubMed]

104. Pelosi, G.; Pastorino, U.; Pasini, F.; Maissoneuve, P.; Fraggetta, F.; Iannucci, A.; Sonzogni, A.; De Manzoni, G.; Terzi, A.; Durante, E.; et al. Independent prognostic value of fascin immunoreactivity in stage I nonsmall cell lung cancer. *Br. J. Cancer* **2003**, *88*, 537–547. [CrossRef]

105. Chilosi, M.; Zamo, A.; Doglioni, C.; Reghellin, D.; Lestani, M.; Montagna, L.; Pedron, S.; Ennas, M.G.; Cancellieri, A.; Murer, B.; et al. Migratory marker expression in fibroblast foci of idiopathic pulmonary fibrosis. *Respir. Res.* **2006**, *7*, 95. [CrossRef]

106. Kidera, Y.; Tsubaki, M.; Yamazoe, Y.; Shoji, K.; Nakamura, H.; Ogaki, M.; Satou, T.; Itoh, T.; Isozaki, M.; Kaneko, J.; et al. Reduction of lung metastasis, cell invasion, and adhesion in mouse melanoma by statin-induced blockade of the Rho/Rho-associated coiled-coil-containing protein kinase pathway. *J. Exp. Clin. Cancer Res.* **2010**, *29*, 127. [CrossRef]

107. Cancer Genome Atlas Research Network. Comprehensive molecular profiling of lung adenocarcinoma. *Nature* **2014**, *511*, 543–550. [CrossRef]

108. Behr, J.; Kolb, M.; Cox, G. Treating IPF–all or nothing? A PRO-CON debate. *Respirology* **2009**, *14*, 1072–1081. [CrossRef]

109. Mantovani, A.; Biswas, S.K.; Galdiero, M.R.; Sica, A.; Locati, M. Macrophage plasticity and polarization in tissue repair and remodelling. *J. Pathol.* **2013**, *229*, 176–185. [CrossRef]

110. Mills, C.D.; Lenz, L.L.; Harris, R.A. A Breakthrough: Macrophage-Directed Cancer Immunotherapy. *Cancer Res.* **2016**, *76*, 513–516. [CrossRef] [PubMed]

111. Gordon, S.; Martinez, F.O. Alternative activation of macrophages: Mechanism and functions. *Immunity* **2010**, *32*, 593–604. [CrossRef] [PubMed]

112. Kinoshita, T.; Kudo-Saito, C.; Muramatsu, R.; Fujita, T.; Saito, M.; Nagumo, H.; Sakurai, T.; Noji, S.; Takahata, E.; Yaguchi, T.; et al. Determination of poor prognostic immune features of tumour microenvironment in non-smoking patients with lung adenocarcinoma. *Eur. J. Cancer* **2017**, *86*, 15–27. [CrossRef]

113. Fernandez, I.E.; Greiffo, F.R.; Frankenberger, M.; Bandres, J.; Heinzelmann, K.; Neurohr, C.; Hatz, R.; Hartl, D.; Behr, J.; Eickelberg, O. Peripheral blood myeloid-derived suppressor cells reflect disease status in idiopathic pulmonary fibrosis. *Eur. Respir. J.* **2016**, *48*, 1171–1183. [CrossRef] [PubMed]

114. Kinoshita, T.; Muramatsu, R.; Fujita, T.; Nagumo, H.; Sakurai, T.; Noji, S.; Takahata, E.; Yaguchi, T.; Tsukamoto, N.; Kudo-Saito, C.; et al. Prognostic value of tumor-infiltrating lymphocytes differs depending on histological type and smoking habit in completely resected non-small-cell lung cancer. *Ann. Oncol.* **2016**, *27*, 2117–2123. [CrossRef]

115. Kinoshita, T.; Ishii, G.; Hiraoka, N.; Hirayama, S.; Yamauchi, C.; Aokage, K.; Hishida, T.; Yoshida, J.; Nagai, K.; Ochiai, A. Forkhead box P3 regulatory T cells coexisting with cancer associated fibroblasts are correlated with a poor outcome in lung adenocarcinoma. *Cancer Sci.* **2013**, *104*, 409–415. [CrossRef] [PubMed]

International Journal of
Molecular Sciences

MDPI

Review

The Role of Tumor Necrosis Factor α in the Biology of Uterine Fibroids and the Related Symptoms

Michał Ciebiera [1,*], Marta Włodarczyk [2], Magdalena Zgliczyńska [3], Krzysztof Łukaszuk [4,5], Błażej Męczekalski [6], Christopher Kobierzycki [7], Tomasz Łoziński [8] and Grzegorz Jakiel [9]

1 Second Department of Obstetrics and Gynecology, The Center of Postgraduate Medical Education, 01-809 Warsaw, Poland
2 Department of Biochemistry and Clinical Chemistry, Department of Pharmacogenomics, Medical University of Warsaw, 02-097 Warsaw, Poland; mdwlodarczyk@gmail.com
3 Students' Scientific Association at the I Department of Obstetrics and Gynecology, Medical University of Warsaw, 02-015 Warsaw, Poland; zgliczynska.magda@gmail.com
4 Department of Obstetrics and Gynecological Nursing, Faculty of Health Sciences, Medical University of Gdansk, 80-210 Gdansk, Poland; krzysztof.lukaszuk@invicta.pl
5 INVICTA Fertility and Reproductive Center, 80-172 Gdansk, Poland
6 Department of Gynecological Endocrinology, Poznan University of Medical Sciences, 60-513 Poznan, Poland; blazejmeczekalski@yahoo.com
7 Division of Histology and Embryology, Department of Human Morphology and Embryology, Wroclaw Medical University, 50-368 Wroclaw, Poland; ch.kobierzycki@gmail.com
8 Department of Obstetrics and Gynecology Pro-Familia Hospital, 35-001 Rzeszów, Poland; tomasz.lozinski@pro-familia.pl
9 First Department of Obstetrics and Gynecology, The Center of Postgraduate Medical Education, 01-004 Warsaw, Poland; grzegorz.jakiel1@o2.pl
* Correspondence: michal.ciebiera@gmail.com; Tel.: +48-607-155-177

Received: 18 October 2018; Accepted: 28 November 2018; Published: 4 December 2018

Abstract: Uterine fibroids (UFs) are the most common benign tumors of the female genital tract. The incidence of UFs has been estimated at 25–80% depending on selected population. The pathophysiology of UFs remains poorly understood. The transformation of smooth muscle cells of the uterus into abnormal, immortal cells, capable of clonal division, is the main component of all pathways leading to UF tumor formation and tumor necrosis factor α (TNF-α) is believed to be one of the key factors in this field. TNF-α is a cell signaling protein involved in systemic inflammation and is one of the cytokines responsible for the acute phase reaction. This publication presents current data about the role of tumor necrosis factor α in the biology of UFs and the related symptoms. TNF-α is an extremely important cytokine associated with the biology of UFs, UF-related symptoms and complaints. Its concentration has been proven to be elevated in women with clinically symptomatic UFs. The presented data suggest the presence of an "inflammation-like" state in women with UFs where TNF-α is a potent inflammation inducer. The origin of numerous symptoms reported by women with UFs can be traced back to the TNF-α influence. Nevertheless, our knowledge on this subject remains limited and TNF-α dependent pathways in UF pathophysiology should be investigated further.

Keywords: uterine fibroid; leiomyoma; tumor; tumor necrosis factor α; cytokine; growth factor; inflammation; clinical symptoms; pathophysiology; therapy

1. Introduction

1.1. Uterine Fibroids—An Overview

Uterine fibroids (UFs) are the most common benign tumors of the female genital tract. The incidence of UFs has been estimated at 25–80%, depending on the populations and multiple risk factors [1–4]. A significant percentage of UF-positive women are symptom-free but UFs cause clinical symptoms of sufficient intensity to impair normal daily functioning in about one-third of the affected subjects [2,4–6]. The most common symptoms include excessive bleeding and secondary anemia, pelvic discomfort or pain, bowel and bladder dysfunctions, infertility, and obstetric pathologies [1,2,7,8]. Symptomatic UFs are the leading cause of a decreased quality of patient life [6,9], and the main reason behind various surgeries, chief among them hysterectomy [10,11].

1.2. Uterine Fibroids—Growth Factors and Steroid Control

Despite intensive research, the pathophysiology of UFs remains poorly understood. The transformation of smooth muscle cells of the uterus into abnormal, immortal cells, capable of clonal division, is the main component of all pathways leading to fibroid tumor formation. The second component is tumor growth through uncontrollable cell division, as well as production and accumulation of the extracellular matrix (ECM) [2,7,12,13] (Figure 1).

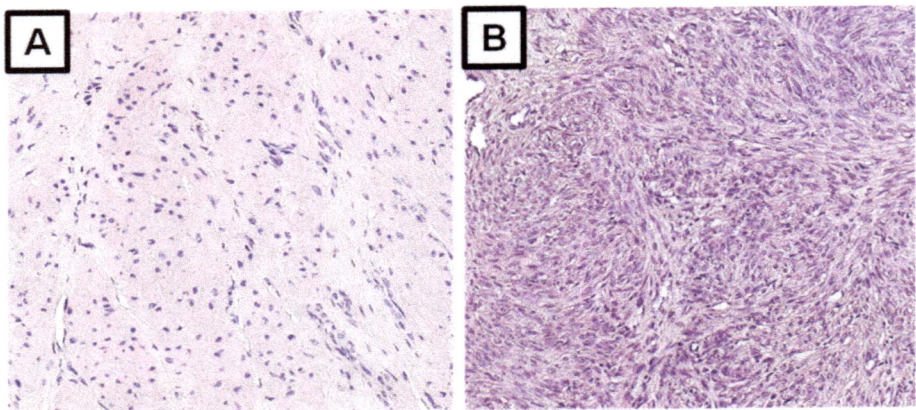

Figure 1. Microscopic slides, histologic specimens of the myometrium (**A**) and a uterine fibroid (**B**). Hematoxylin and eosin stain, 200× magnification.

According to various authors, UF metabolism is affected through steroid hormones, growth factors, cytokines, and chemokines [7,14–18]. Paracrine signaling plays an important role in cellular transformation of the myometrium. The initiators of UF formation are not completely understood, but estrogen and progesterone are believed to be the major promoters of their growth [12,16]. The hormonal effect of these hormones on UFs is related to the aforementioned molecules [7,16]. The influence of steroids on the growth factor expression suggests that these factors represent the ultimate effectors of steroid action [7]. Also, the literature offers some reports suggesting non-genomic interactions between growth factors and hormonal pathways [7,19].

Steroid action is mediated through the interactions of estrogen (ER) and progesterone (PR) receptors with different DNA response elements, which regulate the transcription of selected genes [12,20]. Estrogen and its receptors play an important role in myometrial tissue metabolism and UF growth [12,18,21]. The effect of estrogen on these tumors is evident as these tumors do not appear before the menarche and their size decreases after menopause. It is currently believed that reduced apoptotic potential with increased proliferative potential is associated with the progesterone component rather than with estrogens [12,16,22–24]. The main mechanism of action of progesterone in tumorigenesis is its effect on the increase in the concentration of selected growth factors [12,23–25]. Disturbances in growth factors (e.g., transforming growth factor β (TGF-β)) [26–29] and cytokine secretion may be the cause of UF-derived symptoms [7,20,30]. Some of the UF properties might also be regulated by different miRNAs in order to alter their effect on structural homeostasis of female genital tract [31]. Alas, only a small aspect of the complex UF pathogenesis network is known and more evidence is necessary. Extensive worldwide research is ongoing [8].

1.3. Uterine Fibroids and the Extracellular Matrix

UF growth is determined by the rate of cell proliferation, differentiation, apoptosis, angiogenesis and ECM deposition [13,20,32]. As mentioned above, UFs are considered to be a type of a fibrotic disorder with excessive ECM production [13,20,33,34]. Fibrosis arises through two pathways: recruitment of the inflammatory cells and activation of the fibroblasts [13,35,36]. Fibroid tumor tissue contains approximately 50% more ECM than the adjacent myometrial tissue. In addition, the architecture of collagen fibrils in UFs has been found to be abnormal [36]. The main fibroid ECM components include collagen (type I and type III), fibronectin, and proteoglycans [13,14,32,34,37]. Normal ECM undergoes a continuous balanced rebuild process, which contributes to the maintenance of its proper amount and density. ECM matrix enzymes are regulated by special tissue inhibitors of metalloproteinases [13,38]. Peptide growth factors can have a regulatory effect only if they bind to their specific receptors and induce signal transmission inside the cell [7,39,40]. This condition is possible only when the factor is released from the complexes with matrix components. ECM accumulation is affected by several factors, e.g., TGF-β, activin A and platelet-derived growth factor (PDGF), TNF-α, followed by estrogen and progesterone [20], and by selected microRNAs [13]. According to Islam et al. ECM can be treated as a reservoir of growth factors and cytokines which protects them from being degraded when staying in the ECM microenvironment [13]. When degraded by matrix metalloproteinases (MMPs), ECM releases soluble forms of various growth factors and cytokines, allowing them to play their molecular roles [13].

1.4. Uterine Fibroids and Cytokines

Cytokines are low-molecular-weight proteins which are produced and released by immune system cells [41]. They have a wide range of biological effects and act over short distances, either in an autocrine or paracrine manner [7,42]. Cytokines affect almost all known biological process, including embryonic development, disease pathogenesis, as well as specific and non-specific responses to various antigens and stimuli [41]. They are responsible for intracellular signal transmission by binding to specific surface receptors [41–43]. Numerous cytokines have been identified to play a significant role in myometrial and UF biology [35,44,45]. According to Ciarmela et al. interleukins (IL) such as IL-1, IL-6, IL-11, IL-13, IL-15, interferon (IFN)-γ, TNF-α are involved in crucial pathways in the pathophysiology of UFs. The abovementioned cytokines have an effect on the inflammation, neoangiogenesis and the regulation of tissue remodeling [7]. These cytokines may also be responsible for UF-related symptoms, i.e., pain, infertility, and obstetric pathologies [35,46].

1.5. Tumor Necrosis Factor α—An Overview and Pathways

Although various molecules are involved in UF biology, it appears that TNF-α may be one of the most important myometrium-associated cytokines [39]. TNF-α is a cell signaling protein involved in systemic inflammation and is one of the cytokines responsible for the acute phase reaction. TNF-α has a dual biological nature as it might cause several undesired effects. TNF-α is a pleiotropic cytokine which has been identified as the key regulator of the inflammatory response [43,47]. It also plays a major role in the cell cycle, being the controller of growth, differentiation, and apoptosis [13,43,47]. TNF-α has multiple biological functions throughout the human body, including fever and acute phase stimulation, promotion of the adhesion molecule expression, phagocytosis stimulation, appetite suppression, and modulation of insulin resistance [43]. TNF-α can be an antineoplastic and antiangiogenic agent which stimulates the immune system to fight cancer cells [39,48]. It has been found that TNF-α is an important gene expression regulator. The interaction of chemokines and their receptors may cause the amplification of cellular signaling pathways and induce the expression of proteins responsible for proliferative cells or changes in their normal metabolism [35,39,49]. Despite the anti-cancer properties of TNF-α, elevated levels of TNF-α are not always capable of destroying all abnormal cells and, paradoxically, they can cause severe symptoms related to tumor occurrence [47,50]. Dysregulation of TNF-α production and distribution has been demonstrated in various human diseases, including cancers, dermatoses, and inflammatory bowel diseases (Table 1) [35,43,51,52].

Table 1. Examples of human diseases with dysregulated TNF-α production [30,50,53–59].

Field	Examples
Rheumatology	Rheumatoid arthritis Psoriatic Arthritis Ankylosing Spondylitis
Dermatology	Plaque psoriasis
Ophtalmology	Uveitis
Psychiatry	Depression
Gastroenterology	Crohn's Disease Ulcerative Colitis
Urology	Renal cell carcinoma
Gynecology	Ovarian cancer Uterine fibroids
Neurology	Alzheimer's Disease

TNF-α is produced mainly by activated macrophages but it can also be produced by other cell types, e.g., lymphocytes and neutrophils [43,60]. TNF-α is produced as a 233-amino acid-long type II transmembrane protein arranged in homotrimers [61]. The soluble homotrimeric form is then released via proteolysis by the ADAM metallopeptidase domain 17 (ADAM17) and has a triangular shape [62,63]. Importantly, both the secreted and the membrane forms are biologically active, although their specific functions remain the subject of some controversy among the researchers [60,64].

TNF-α is also found in smooth muscle cells as a response to tissue injury or upon immune responses to various stimuli [65–67]. TNF-α uses two types of receptors: TNF-α receptor type 1 (TNFR1) and TNFR2 (Figure 2).

TNFR1 is expressed in most tissues, whereas TNFR2 is found primarily in the immune system cells [43]. After binding to the receptor, the TNF-α molecule may activate one of the three potential effects: (1) activation of the nuclear factor kappa-light-chain-enhancer of activated B cells (NF-κB), which is a transcription factor involved in cell survival, proliferation and the inflammatory response (this include NF-κB-inducing kinase (NIK) [68] and I kappa B kinase (IKK) [69]); (2) activation of the mitogen-activated protein kinases (MAPK) pathways (through c-jun N-terminal kinase (JNK) [70] and on the other hand through receptor interacting protein (RIP) kinases family [71] and MAP kinase kinase (MEKK) [72]), involved in cell differentiation and proliferation; and (3) induction of death signaling (Figure 2) [43,60,73,74]. Most of the mentioned pathways are tumor necrosis factor receptor-associated factor (TRAF)2 dependent as presented in Figure 2 [75].

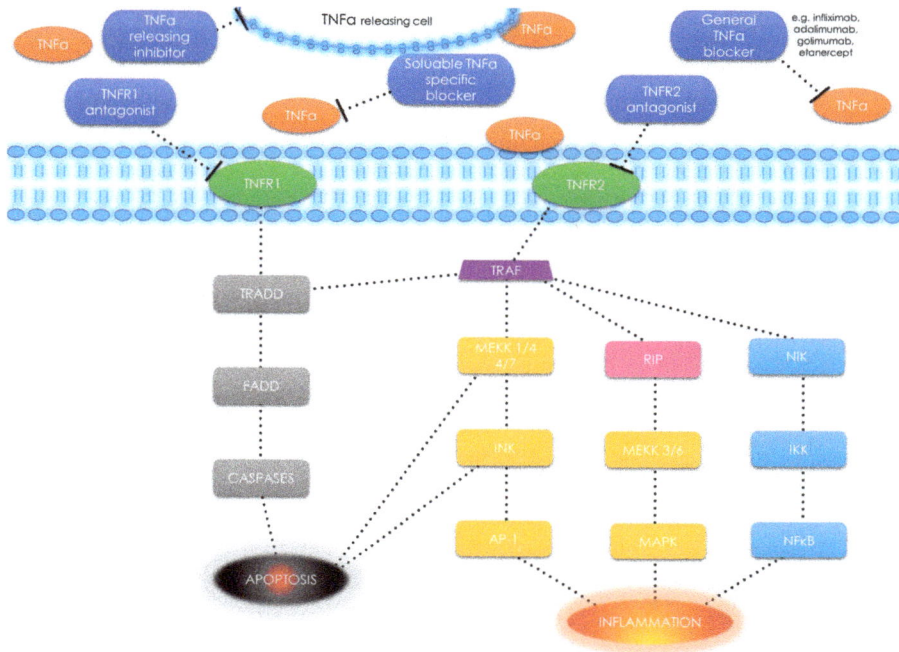

Figure 2. TNF-α receptors, pathways and different types of signals—schematic diagram. TNF-α has an ability to induce apoptosis, cell survival or inflammation depending on selected pathway. Tumor necrosis factor α (TNF-α); tumor necrosis factor α receptor (TNFR); tumor necrosis factor receptor type 1-associated death domain (TRADD); Fas-associated protein with death domain (FADD); tumor necrosis factor receptor-associated factor (TRAF); mitogen-activated protein kinase kinase kinase (MEKK); c-jun N-terminal kinase (JNK); activator protein 1 (AP-1); receptor interacting protein (RIP); mitogen-activated protein kinases (MAPK); nuclear factor κ-light-chain-enhancer of activated B cells (NF-κB); NF-κ-B-inducing kinase (NIK); I κB kinase (IKK).; T-bar as a drug - binding point interaction.

TNFRs form trimers when reached by the ligand. This binding leads to the dissociation of the silencer of death domains (SODD) inhibitory protein [76]. When SODD is finally dissociated, tumor necrosis factor receptor type 1-associated death domain (TRADD) protein binds into free death domain [66,73,77,78]. The TRADD protein binding may activate three different important pathways [66,73,78]. The first pathway is the activation of the NF-κB protein complex, which leads to changes in the DNA transcription, cytokine production and cell survival (Figure 2). The NF-κB pathway is also responsible for cell proliferation, inflammatory response, and anti-apoptotic factors [79]. The NF-κB pathway has some serious connections with the UF-related pathways: the focal adhesion kinase (FAK) signaling activated by TGF-β [29], and with activin A [80]. IL-1α and TNF-α are the factors that block the differentiation of human myoblasts with the activation of the TGF-β-activated kinase (TAK)-1 pathway. As described by Trendelenburg et al. this pathway can be modulated (tracing to p38 and NF-κB) with the use of different factors like drugs or genetic modifications [80]. Additionally, TGF-β increases the expression of apoptosis-related p53 and Bax proteins mediated by their major regulator—TNF-α [81]. Apoptosis is a mechanism by which cells undergo programmed death. This process can be mediated through the receptor or the mitochondrial pathways. The extrinsic pathway involves binding of a ligand e.g., soluble protein like TNF-α to its cognate cell surface receptor [20,82]. The second pathway depends on MAPK activation (Figure 2). TNF-α activates JNKs, which are responsive to stress stimuli. JNKs are involved in various processes, e.g., apoptosis, cell degeneration, differentiation and proliferation, inflammatory conditions, and cytokine production [83,84]. JNKs and p38 MAPKs can exert antagonistic effects on cell proliferation and survival. This crosstalk is an important regulatory mechanism in numerous cellular responses [85]. The third and the weakest pathway is responsible for induction of death signaling (Figure 2). In this pathway, TRADD binds FADD, which is a stimulus to caspase-8 concentration raise and results in subsequent autoproteolytic activation and effector caspases cleaving [86]. These three TNF-α pathways have many conflicting effects. Some of them enhance the transcription of the anti-apoptotic proteins, while others have a positive effect on the inhibitory proteins which interfere with death signaling. This delicate balance can shift both ways, depending on cell type, concurrent stimulation of other cytokines, and the influence of reactive oxygen species [43,73].

2. Material and Methods

A review of the publications on the role of TNF-α in UF biology and UF-derived symptoms is presented. Authors conducted their search in PubMed of the National Library of Medicine and Google Scholar. Databases were extensively searched for all original and review articles/book chapters using keywords (one or in combinations): uterine fibroid; uterine leiomyoma; tumor necrosis factor α published in English until October 2018. Moreover additional articles in bibliographies of reviewed articles were searched. Overall, most relevant articles were reviewed and included as appropriate.

3. Discussion

3.1. Uterine Fibroids and Inflammation

Inflammation plays a major role in tumorigenesis and inflammation-related factors are key players in the development of many benign and malignant neoplasms, especially due to high increase in the occurrence of a mutation and the proliferation rate of the mutated cells [35,87]. As stated by Wegienka, local chronic inflammation constitutes a microenvironment which enables UF development and UFs are the effect of the presence of various inflammatory-derived molecules in the myometrium [88,89]. UF is thought to be an inflammatory-related fibrotic disorder. According to the available data, UF pathophysiological pathways depend on the activated macrophages, whose number in the UF tissue is increased as compared to the adjacent normal myometrium [90]. The abovementioned cells play a key role in the reparative processes and myofibroblast recruitment [35,91]. Menstruation, infections, mechanical injuries, and oxidative stress may be the causes of the inflammation in the uterus [13]. Myofibroblasts produce ECM as a response to the influence of various cytokines and growth factors. In normal conditions, this process can repair damaged tissues. However, it may also lead to fibrosis if there is no control or the deregulation of the controlling pathways is too high [13,32,92–94].

MED12 somatic mutation is the most often detected DNA mutation in human UFs (70–80%) [95–98]. This mutation is a driver for stimulating the development of UFs and influencing genomic instability [99,100]. MED12 is also related to endocrine and growth factor pathways [12]. A possible impact of altered signaling in UF pathophysiology on inflammatory responses and genomic repair mechanisms is being extensively researched [101]. UFs growth is mostly hormone-dependent and progesterone is believed to be the most important hormone in this complicated process [12,23,24]. The main mechanism of progesterone action in UF tumorigenesis is its effect on the increase in the concentration of selected growth factors [7,29]. Factors which have been proven to have a serious impact on UF tumorigenesis include TGFs, activin A, vascular endothelial growth factors (VEGFs), and other pro-inflammatory agents [14,15,29,102] (Figure 3).

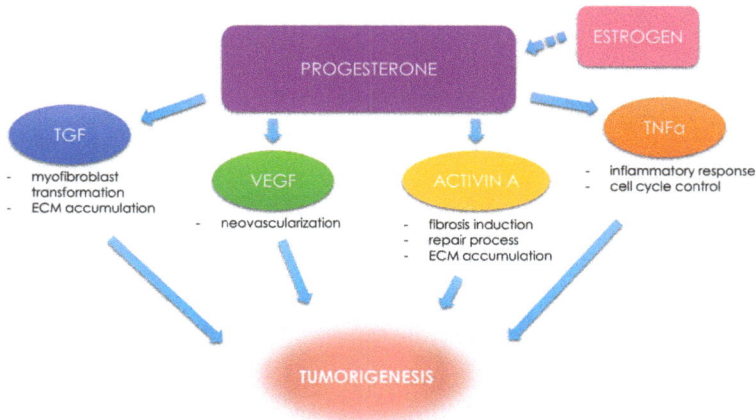

Figure 3. Progesterone and progesterone-related factors. The role of selected factors in UF tumorigenesis. Transforming growth factor (TGF); vascular endothelial growth factor (VEGF); tumor necrosis factor α (TNF-α); extracellular matrix (ECM). Estrogen as a factor preparing the tumor to be stimulated by progesterone (dotted arrow).

According to the available sources, elevated levels of TGF-β play an important role in UF growth and clinical symptom progression [7,26,27,29,103], which appears to be one of the key factors in myofibroblast transformation and fibrosis progression [17,29,35,37]. Activin A is yet another factor, by no means less important, which was found to be involved in cell proliferation, differentiation, death signaling, and metabolism [81,104]. Activin A and the related proteins, e.g., anti-Müllerian hormone (AMH) or bone morphogenetic proteins (BMPs), belong to the TGF-β superfamily [105]. Recent studies have proven the role of activin A in inflammation processes [35,81,105], wound repair [106], and fibrosis [81,107]. Activin A produced by macrophages is responsible for cell transformation, which leads to tumor occurrence [35,108]. It also plays various physiological roles in a wide range of other tissues [109]. As suggested by different authors, elevated activin A concentrations might also be responsible for excessive ECM production [81,104]. In a study by Ciarmela et al. mRNA levels of activin A were found to be more expressed in UF tissue as compared to the adjacent myometrium [109]. Islam et al. found that mRNA levels of collagen, fibronectin and versican variants in ECM were increased by activin A [104]. The same authors also found that activin A induced phosphorylation of the Smad proteins in UF cells [104], which is one of the pathways of profibrotic signaling [13,29]. The potential pathological connections between TGF-β, activin A and TNF-α will be described later in this article.

3.2. ECM and Inflammation in Uterine Fibroids

ECM found in UF differs from the normal ECM in a well-formed myometrial tissue [13,14,110]. The presence of inflammatory cells in UFs contributes to excessive ECM production, tissue remodeling, and tumor growth [35]. ECM accumulation is regulated by growth factors, cytokines and steroid hormones. Increased production and accumulation of abnormal ECM results in further tumor volume gain [13,14,110]. MMPs are the enzymes which are responsible for matrix lysis and remodeling [29,111–113]. They are regulated by tissue inhibitors of metalloproteinases (TIMPs) [38]. Some growth factors increase the concentration of TIMPs, which slows down the conversion of the entire ECM and results in its excessive accumulation [29,103,114]. Modified MMP activity is then insufficient to degrade the appropriate amount of ECM to maintain tissue [29,111,115]. Proteolytic enzymes which are produced as a response to inflammation are responsible for several major UF developing processes, e.g., angiogenesis and proliferation of fibroblasts [20,35].

TGF-β, activin-A, and TNF-α are able to increase the synthesis of ECM components through the activation of multiple signaling pathways, e.g., the Smad proteins and various kinases. Cytokines (e.g., interleukins and TNF-α) and growth factors (e.g., VEGF), fibroblast growth factors (FGFs), endothelial growth factor (EGF), and different TGFs are among those angiogenic factors which have been extensively described due to their ability to regulate the expression of proteases and their inhibitors, and enhance cell proliferation and migration [20,112,115,116]. As stated by Chegini (2010), angiogenesis depends on the specific balance between promoters and inhibitors [20]. In this place, TNF-α plays the role of the angiogenic suppressor [20], but its role is much more complex.

3.3. Tumor Necrosis Factor α in Uterine Fibroids

TNF-α is produced by macrophages whose significant numbers can be found in UFs [35,43,60]. Increased TNF-α expression has been found in UF tumors as compared to the adjacent normal myometrium [49]. According to Nair et al. TNF-α secreted by adipocytes enhances the proliferation of UFs [117]. What is currently known is that polymorphisms in the genes encoding IL-1β, IL-6, and TNF-α have been associated with an increased risk of these tumors [20,118–120].

According to available data, fibroid tissue and the adjacent unchanged myometrial tissue demonstrated that TNF-α is abundantly present in the cytoplasm of tumor cells [49,121], which is consistent with the findings of Plewka et al. from 2013 [122]. In that study, Plewka et al. investigated the expression of different inflammatory mediators, e.g., IL-1β, TNF-α or cyclooxygenase 2 (COX-2), in the normal myometrium and UFs in women of reproductive age. These authors found significantly higher TNF-α immunoreactivity in UFs as compared to normal uterine smooth muscle tissue [122].

In the work of Kurachi et al. staining for TNF-α in UF tissue obtained in the proliferative phase was more abundant than in the secretory phase of the menstrual cycle [49]. This relation was not found in the normal myometrium [49]. The addition of progesterone resulted in a decrease in immunoreactive TNF-α expression as compared to control cultures, while a similar addition of estradiol did not affect the TNF-α expression [49]. In cultured UF cells, the treatment with progesterone inhibited the expression of insulin-like growth factor 1 (IGF-1) and the TNF-α, and augmented the expression of apoptosis-inhibiting Bcl-2 protein [123], which is consistent with other reports in the literature. Progesterone is a steroid hormone with anti-inflammatory and mitogenic activity [101,124]. Natural killer (NK) cell activity is suppressed under the influence of progesterone [125]. Progesterone regulates uterine NK cells through a glucocorticoid receptor mediated process and steroidal antiprogestogens (e.g., mifepristone) could abolish the inhibitory effect of this hormone [126]. Elevated levels of progesterone have an effect on increased TGF-β and decreased TNF-α production [7,29,101]. Progesterone, being the main steroid hormone responsible for the formation and growth of UFs, should lead to a decrease in TNF-α concentration. However, paradoxically, the remaining pathways, many of which are still undefined, make TNF-α concentrations higher in the presence of these tumors, as shown by other authors [39] and our recent work [30].

The picture becomes more complex when we bear in mind that TNF-α is a potent stimulator of aromatase activity, what results in enhanced conversion of androstenedione to estrone [127]. TNF-α influences key genes and enzymes involved in estrogen metabolism, making it more hormonally active and carcinogenic [128]. Estrogens have been known to play a major role in UF pathophysiology as described above, including also pathways like Ras-Raf-MEK, which are more in common with TNF-α [18].

The reciprocal feedback of progesterone and TNF-α seems more complex when considered along with activin A. As mentioned before, activin A is one of the major factors involved in UF pathogenesis, with a direct pro-fibrotic effect on UF cells by the expression of ECM protein (via Smad pathway) induction [104]. The regulation of activin A biological function in fibrosis-related processes is complex due to the influence of several molecules [81]. The effect of TNF-α in the paths in which activin A is involved is vital, but the pathophysiology of this factor involves also TGF-β1, IL-1, IL-1β, IL-13, angiotensin, and others [81,129,130]. In a study by Protic et al. activin A mRNA expression was upregulated by TNF-α in myometrial and UF cells [35]. The same effect was also found in other cell types [129].

Activins are produced in the gonads, pituitary gland, placenta, and other organs. In the ovary, activin increases FSH binding and FSH-induced aromatization. Activin A enhances the activity of aromatase enzyme and simultaneously suppresses progesterone production [131]. In a study by Shukovski and Findlay (1990), activin A was found to delay the process of luteinization [132]. Hillier et al., concluded that activin A is responsible for promoting estrogen synthesis and simultaneously suppresses the synthesis of progesterone [131]. According to Ciarmela et al., activin A should be considered as a steroid-regulated factor involved in myometrial functionality and that the disruption of its signaling may contribute to tumor growth [109]. The abovementioned observations both, the complexity of these processes and the limitations of the available data. The presented information suggests that UFs have their own complicated path of connections and conjugations, in which TNF-α plays a role as an activin A upregulator, where activin A affects progesterone, which in turn inhibits TNF-α. It seems reasonable to assume that includes one or more additional factors which allow for TNF-α to bypass the inhibitory effect of progesterone and induce its related inflammatory reactions. These observations are also supported by some of the effects of ulipristal acetate (UPA) on UF tumors [133]. Despite the fact that UPA has a potent effect on one of the major UF pathophysiological pathways—progesterone and is TGF-β dependent [27], it also increases the activity of alkaline phosphatase, upregulates caspases, and downregulates TNF-α expression [134]. According to Ciarmela et al. UPA was also found to inhibit the expression and functions of activin A and activin receptor in UF tumor cells [135]. In our opinion, more research about the described pathway and its molecular connections is necessary to better understand it in the context of UFs.

Wang et al. found that TNF-α upregulates the mRNA levels of MMP-2 in UF cell cultures. The same observations were made with protein levels, whereas this effect was insignificant in the normal myometrium [136]. Their finding might play a major role in releasing soluble forms of various growth factors and cytokines from ECM, as mentioned before [13]. Such dependence may also have an effect in the form of a self-winding dependency circle—TNF-α, which releases various cytokines also releases further TNF-α molecules and, at some point, this can become a cascade difficult to control.

Islam et al. found that activin A expression was increased under the influence of TNF-α [16], which might solve the mystery how this cytokine can stimulate ECM production, as activin A is an important profibrotic factor in UF pathology [13,81]. The role of TNF-α in UF formation and growth seems even more justified as it was proven that TNF-α has a potent influence on extracellular signal–regulated kinases (ERKs) [13,136]. The abovementioned ERK pathway plays an important role in integrating external signals from the presence of mitogens and is a part of the Ras-Raf-MEK-ERK signal transduction cascade. This cascade participates in various processes such as cell cycle progression, cell migration, survival, differentiation, proliferation, and transcription [137]. The activity of the Ras-Raf-MEK-ERK cascade is increased in more than one-third of human neoplasms, and inhibition of its components might be an effective anti-tumor strategy [137].

3.4. Obesity, Inflammation and Tumor Necrosis Factor α in Uterine Fibroids

Obesity is a condition in which excessive body fat may interfere with the maintenance of an optimal state of health. Obesity is predominantly caused by excessive food intake, lack of physical activity, and genetic susceptibility [138,139]. According to the available data, obesity is considered to be a major risk factor for UFs [3,8,26,140,141]. The reasons for this dependence are in this case sought in the metabolic function of adipose tissue [141,142]. Most adrenal androgens are metabolized to estrogen with aromatase in the adipose tissue [142,143]. Pathophysiological factors attributed to the occurrence of UF include reduced production of sex hormone-binding globulin (SHBG) in obese women [141,144]. This protein binds a large proportion of the circulating sex hormones, not allowing their hormonal activity on sensitive tissues, which affects the delicate hormonal balance of the body [144,145].

Obesity is characterized by a chronic state of inflammation and this might be a cause of abnormal tissue regeneration, as is the case in UFs [35]. Excessive adipose tissue may release various inflammatory mediators. These relations may predispose to a pro-inflammatory state of the tissues and enhanced oxidative stress [146]. Excessive fat accumulation is associated also with increased levels of reactive oxygen species (ROS), which inhibit cell apoptosis and increase ECM deposition [147,148].

Macrophages are important components of the adipose tissue, which actively participate in its activities [35,60,149]. Other immune system cells, like lymphocytes taking part in the metabolism of the adipose tissue, can also lead to immune deregulation [149]. As described in the available studies, adipose tissue produces and releases a variety of pro- and anti-inflammatory factors, including adipokines, as well as various cytokines and chemokines, such as different interleukins and chemoattractant proteins [35,146,150]. Human adipose tissue secretes TNF-α, which may be a good explanation of the relation between obesity and inflammation [151]. In a study by Hotamisligil et al. TNF-α was found to be highly expressed in the adipose tissue of obese people [152], whereas Zaragosi et al. described the same dependence for activin A [35,153]. These inflammation enhancing factors which are accumulated and released by adipose tissue and have a direct effect on the myometrium and myofibroblasts may result in excessive production of the ECM components, tissue remodeling, and UF occurrence [35,81].

Further studies are necessary to show the effect of body weight change (gain or loss) on TNF-α and other pro-inflammatory factors levels, and whether this can affect the symptoms associated with UFs.

3.5. Tumor Necrosis Factor α, Uterine Fibroids and the Related Symptoms—Overview

TNF-α may be considered as a chemokine which could influence numerous clinical symptoms associated with UFs [20,35]. Abdominal and pelvic pain, infertility or gastrointestinal complaints are just a few ailments which may be caused not only by tumor pressure but also by the paracrine and endocrine influence of the tumor. In this section some links between UFs-derived complaints and TNF-α are described.

3.5.1. Pain

Symptomatic UFs can cause chronic pelvic pain. Interestingly, this pain cannot always be attributed to the presence of the mass or position of the tumor. While the subject of chronic pain in patients with UFs has been examined only to a small extent, in the case of endometriosis this topic is relatively well-researched [154]. TNF-α levels in the peritoneal fluid are higher in women with endometriosis (higher stages of endometriosis correspond to higher TNF-α levels) [155,156]. The role of TNF-a as a pain inducer has been well-documented [157,158]. TNF-α stimulates the production of prostaglandins (PG) E2 and F2α [154,159]. PGE2, for example, may play a role in the resolution of the inflammation. Some of its effects on the human body include vomiting, fever, diarrhea, and excessive uterine contraction [160]. Pharmacological therapies, including painkillers or hormones for pain management are used to relieve these symptoms [20]. The literature offers some data about a successful use of anti-TNF-α drug—etanercept—in the treatment of endometriotic implants but, to the best of our knowledge, there is no evidence about the use of this drug in UFs treatment [161]. In our opinion, more research about the potential role of TNF-α on UF-derived pain is needed.

3.5.2. Infertility

The connection between UFs and infertility, together with the potential impact of TNF-α in this regard, should be highlighted.

Infertility is defined as the inability to become pregnant or carry a pregnancy to full term. There are many causes of infertility and many different methods of treatment [162]. The immune mechanisms are among the possible causes of some forms of infertility. Cytokines, like TNF-α, selected ILs and others which trigger a Th1 type immune response, are suspected to play a major role in infertility [163]. According to Wang et al. women with a history of infertility have significantly increased serum TNF-α concentrations in comparison to fertile controls, and TNF-α can potentially serve as an infertility marker [164]. According to Falconer et al. inflammation affects the ovaries and, consequently, the oocyte quality, with less favorable outcomes of in vitro fertilization [165]. TNF-α stimulates the apoptosis of human trophoblast cells and inhibits the proliferation of human trophoblast cells in vitro [166,167]. According to some authors, TNF-α interferes with development of the placenta and trophoblast invasion of the spiral arteries [166,168]. Abnormal production of TNF-α and cytokines which are similar in function has been suggested to cause fetal growth restriction [169]. In a very interesting study by Azizieh and Raghubaty, elevated TNF-α serum levels were correlated with severe pregnancy complications, e.g., recurrent miscarriages, premature rupture of membranes, preeclampsia, and intrauterine growth restriction. These authors concluded that TNF-α must be an important factor in the pathogenesis of these complications [168]. According to Austrian experts, TNF-α concentration, distribution and stimulation period determine whether TNF-α would have a beneficial or adverse effect on the female reproductive system [166]. TNF-α was also found to differ between women with and without the polycystic ovary syndrome (PCOS). According to a meta-analysis by Gao et al. TNF-α serum levels in women with PCOS are elevated as compared to PCOS-free controls. Those authors concluded that TNF-α serum levels might be related to insulin resistance and androgen excess [170].

It is officially accepted that UFs decrease the fertility potential [171,172]. According to Pritts et al. UF-positive women have decreased rates of implantations and live births, and increased rates of spontaneous miscarriages [171]. Some cytokines are thought to be responsible for implantation success and optimal embryonic development [173]. TNF-α may just be one of these cytokines and there is a chance that too high concentrations of TNF-α can have a significant impact on live birth occurrence. In our opinion, reducing the TNF-α induced inflammation in the uterus, e.g., by removing the UFs, might result in successes of the reproductive medicine. Unfortunately, data research on this topic remain limited and further research is necessary. Our study, which demonstrated elevated serum TNF-α levels in UF-positive women, was a step in that direction [30]. The next stage would be to investigate the impact of surgical or drug therapies (e.g., UPA or alternative agents) on the concentrations of selected cytokines and growth factors and find how to achieve the best effect, i.e., successful conception [172–174]. Studies about UPA and infertility are ongoing and the first answers are expected soon [175].

3.5.3. Gastrointestinal Issues

TNF-α plays a major role in gastroenterology. Not all gastrointestinal symptoms associated with UFs are only related to pressure on the neighboring organs. Some of them may also have a paracrine or endocrine background. There is increasing evidence for the involvement of the immune system and the related cytokines in various gastrointestinal disorders, altered cytokine expression, and abnormal presence of immune cells, resulting in the occurrence of the clinical symptoms. Several diseases and symptoms depend on the TNF-α-related pathways [176]. TNF-α is overexpressed in patients with colitis ulcerosa and Crohn's disease, and the degree of mucosal inflammation is positively correlated with chemokine secretion patterns [177]. TNF-α is also a known factor in the initiation and amplification of the inflammatory responses to Helicobacter pylori infection [178–180]. TNF-α is a potent inhibitor of gastric acid secretion [179], and its improper secretion might be a causable factor of many cases of dyspeptic disorders in women with UFs.

Therefore, TNF-α targets were proposed as potential treatment options and today anti-TNF-α drugs are the gold standard in inflammatory bowel disease therapy [181,182]. These diseases are treated with the use of TNF-α inhibitors, e.g., monoclonal antibodies like infliximab, adalimumab, and certolizumab [183]. There is also the abovementioned decoy circulating receptor drug, etanercept, which binds to TNF-α [184]. There are some data indicating prolonged disease stabilization in patients treated with the use of TNF-α inhibitors, but most studies found no difference in the efficacy of that treatment when compared to placebo [185]. However, there is a difference between benign UFs and malignant tumors, and further laboratory tests should be performed to evaluate the effects of these drugs on UFs.

3.6. Tumor Necrosis Factor α, Uterine Fibroids and the Related Symptoms—Management

UFs are just one serious health problem worldwide. Pharmacotherapy is an important chapter in UF management. Prophylaxis of UFs is practically non-existent, while treatment is often costly and expensive. According to Soave et al. current anti-UF strategies that preserve uterus and fertility are not capable yet of controlling clinical symptoms and tumor progression and they are mostly ineffective in the long term outcome [6]. Anti-UF agents should be chosen according to the size and location of UFs, age, dominant symptoms, childbearing plans and treatment availability [186]. Potent drugs like gonadotropin-releasing hormone (GnRH) analogs or selective progesterone receptor modulators (SPRMs) are an option for patients who need symptom relief preoperatively, who are approaching menopause or do not accept surgical intervention. GnRH analogs have been used in women with UFs to reduce bleeding and tumor volume, but their use is limited to short term due to their hypoestrogenic side effects [186]. The other drugs recently used in UF therapy are SPRMs, drugs which interact with progesterone receptors. SPRMs [187] both treat symptoms and eliminate or delay the need of surgery [6,188].

In the case of other anti-UF agents, their poor effectiveness remains the greatest issue. Innovative forms of UF pharmacotherapy are still under intensive investigation. As stated in recent review from 2017 vitamin D, paricalcitol, epigallocatechin gallate, elagolix, aromatase inhibitors (AIs) and cabergoline might find their place in UF therapy as safe and effective alternatives or co-drugs [174]. After reviewing the current literature authors found that anti-TNF-α may be also a promising option in the nearest future. Further research on this topic should focus on individually tailored strategies, where for example anti-TNF-α drugs would be used in women with UF-dependent clinical symptoms that occur due to TNF-α pathways dysregulation. In light of the above, it would seem prudent to assume that drugs directed against TNF-α would be applicable for the relief of UF-derived symptoms (Figure 4).

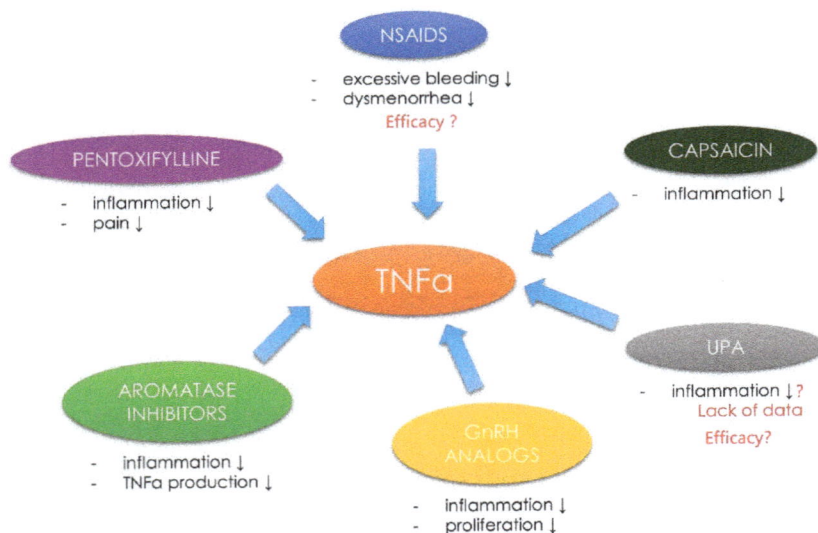

Figure 4. Drugs with proven and potential effect against TNF-α in UF therapy. Nonsteroidal anti-inflammatory drugs (NSAIDS); tumor necrosis factor α (TNF-α); ulipristal acetate (UPA); gonadotropin-releasing hormone (GnRH); ↓ as decrease.

However, this is not so obvious due to the failed attempts to use anti-TNF-α drugs (like infliximab) in case of endometriosis-related pain due to the lack of clinical effects [189,190]. Better results were obtained with pentoxyfilline, a methylated xanthine derivative, a competitive nonselective phosphodiesterase inhibitor [191]. Pentoxyfilline inhibits TNF-α production in vitro and reduces the inflammatory action of this molecule [192,193]. According to a study by Kamencic and Thiel, postoperative treatment with pentoxyfilline reduced the pain in patients with endometriosis [193]. To the best of our knowledge, there are no studies about the use of pentoxyfilline in UF therapy. However, this might be an interesting idea for future studies on UFs (Figure 4).

Nonsteroidal anti-inflammatory drugs (NSAIDs) are widely used to treat endometriosis-TNF-α-related symptoms. Nevertheless, data proving that they significantly reduce endometriosis-related pain remain insufficient [194,195]. According to some sources, NSAIDs can be used to reduce excessive bleeding and dysmenorrhea symptoms, but there are not enough data about their efficacy [196] (Figure 4).

According to less known studies, capsaicin and its analogs are able to block TNF-α-induced NF-κB activation in a dose-dependent manner. This substance can therefore play a chemopreventive role in tumor growth and the derived symptoms. However, no studies on that topic are available [197] (Figure 4).

The use of AIs in therapy has given rise to some expectations [174]. This therapy also includes some TNF-α-dependent pathways. AIs decrease the production of estrogens by blocking or inactivating aromatase [174]. UF tissue expresses aromatase in higher amounts than normal myometrial tissue [143,198]. Some connections between aromatase and TNF-α have been confirmed. Cytokines, like ILs, prostaglandins and TNF-α can stimulate aromatase activity [127,199]. For example, TNF-α induces aromatase expression via the adipose specific promoters [200]. To et al. found that TNF-α regulates estrogen biosynthesis within the breast, affecting the activity of the key estrogen-derived enzymes, i.e., aromatase or estrone sulfatase. These authors stated that TNF-α targeting might be useful as a novel approach to anti-tumor therapies [201]. As described in our recent review on alternative agents in UF therapy, AIs are potent drugs in UF management which reduce UF volume and improve the associated symptoms, probably also by the reduction of TNF-α production (Figure 4). Unfortunately, data about AIs use in UF treatment are incomplete and more studies are still necessary [174]. Last but not least, GnRH analogs are drugs used in UFs management which affect the TNF-α pathways. GnRH analogs act by binding to GnRH receptors, which leads to hypogonadotropic hypogonadal state [202]. GnRH agonists have a direct and indirect effect on the tissue in reducing proliferation and improving clinical symptoms [202]. Some of their effects might be explained by TNF-α-related pathways. In a study by Taniguchi et al. GnRH analog treatment attenuated TNF-α-induced cell proliferation in the endometrial stromal cells [203]. More evidence has come also from other Japanese authors who showed that GnRH analogs treatment attenuated the expression of IL-8 by reducing TNF-α-induced NF-κB activation [204]. In an interesting study about spinal cord inflammatory response, Guzman-Soto et al. found that GnRH analog—leurolide has an effect on the activation/expression levels transcription nuclear factor NF-κB and the proinflammatory cytokines like TNF-α [205]. Further research about the connections between GnRH analogs and TNF-α pathways is needed (Figure 4).

UPA, one of the main drugs related to the treatment of UFs [6,133,188] cannot be omitted. Data on UPA and its effect on inflammation are incomplete. We know many paths of action of UPA, both biochemical and genetic [133,206,207]. In the studies published by our team, we obtained preliminary results that UPA may influence some growth factors, e.g., TGF-β3 [27]. Research on other factors is ongoing and the results will be presented in the nearest future. In their study on the effect of UPA on the human endometrium, Whitaker et al. found that there were significant changes in insulin growth factor binding protein 1, IL-15, HOXA10 mRNAs expression as compared to controls. What is more, cell proliferation in UFs from the UPA group was lower than in women in the proliferative phase [208]. It might suggest some connection between UPA and inflammation.

UPA has a potent effect on progesterone-TGF-β pathway [27], it also increases the activity of alkaline phosphatase, upregulates caspases and downregulates TNF-α expression [134]. There are no studies on the connections between TNF-α and UPA (Figure 4).

3.7. Tumor Necrosis Factor α—Novel Concepts in Diagnosis and Therapy

In our opinion, there is a possibility that TNF-α serum levels may become a marker used for clinical verification in the case of problematic differentiation (UF or different tumor) [30], or to determine the risk of clinical symptoms, as can be partially done with endometriosis [209].

According to a study by our group, women can be diagnosed with UFs solely using the serum TNF-α level cut-off point [30]. Studies on TNF-α levels in patients with diseases similar to UFs, such as uterine sarcoma or adenomyosis, may open up a new chapter in gynecological diagnostics (Figure 5).

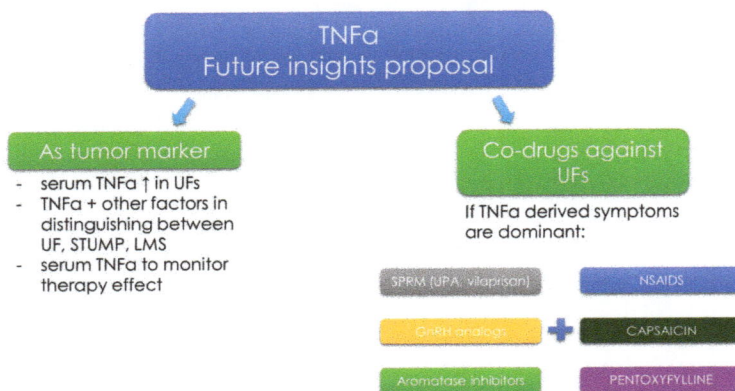

Figure 5. Potential future directions in UF diagnosis and therapy with the use of TNF-α. Tumor necrosis factor α (TNF-α); uterine fibroid (UF); smooth muscle tumor of uncertain malignant potential (STUMP); leiomyosarcoma (LMS); selective progesterone receptor modulator (SPRM); ulipristal acetate (UPA); nonsteroidal anti-inflammatory drugs (NSAIDS); ↑ as increase.

If higher levels of TNF-α are confirmed in the serum of patients with uterine sarcoma, as compared to UF-positive patients, TNF-α could be considered as a non-specific marker which could indicate factors such as what type of surgery should be chosen, if morcellation can be performed, and where to refer the patient. This is important, especially in the current state of surgical management of UFs [210]. TNF-α might be useful marker to estimate the risk of UF occurrence, or for the evaluation of treatment effectiveness (e.g., in UPA therapy). Due to the complexity of the pathophysiological pathways in which it takes part, we believe that TNF-α will not be a specific marker for UFs, but our results may be a starting point for further studies. It is possible that the consideration of other biochemical parameters, such as 25-hydroxyvitamin D or TGF-β3 serum levels, in addition to serum TNF-α levels, could increase specificity [26,28,211].

Even if the concept was wrong, it should also be considered whether, for example, TNF alpha could be used as a potential therapy efficacy marker, as with our idea about TGF-β3 [26,28]. Since patients with UFs have elevated TNF-α serum levels, it seems logical to use them as indicators of patient response to various treatment methods. In our opinion, some of the research on UFs should focus on checking which growth factors and cytokines are most frequently associated with specific clinical conditions caused by these tumors. In the case of confirming the dependence of specific symptoms with specific growth factors (e.g., like TNF-α with pain [157,158], therapies for individual patients could be selected more effectively. However, the necessary prerequisite is to investigate various effects of selected drugs on the growth factors, which will allow to choose the best therapy (Figure 5). At this point, again, the validity of the concept of co-drugs in the treatment of UFs should be emphasized. If safe drug connections were found, a broader spectrum of symptoms could be better eliminated, e.g., one drug decreases TGF-β3 levels and slows down the ECM formation, whereas another drug decreases TNF-α level and has a beneficial effect on pain and infertility (Figure 5). Further extensive research in this field is necessary.

4. Conclusions

TNF-α is an extremely important cytokine associated with the biology of UFs, UF-related symptoms and complaints. Its concentration has been proven to be elevated in women with clinically symptomatic UFs. The presented data suggest the presence of an "inflammation-like" state in women with UFs where TNF-α is a potent inflammation inducer. The origin of numerous symptoms reported by women with UFs can be traced back to the TNF-α dependent pathways.

Nevertheless, our knowledge on this subject remains limited. It seems vital to study the pathophysiological pathways dependent on TNF-α, in particular its associations with progesterone and activin A. Hopefully, the results of that research will be the decisive factor in selecting the appropriate, individually tailored UF treatment methods. It is possible that TNF-α will prove useful as an additional clinical marker for the diagnosis and therapy of UFs. The importance of anti-TNF-α drugs in the treatment of UFs and how drugs with proven anti-UF action affect the TNF-α dependent symptoms should be investigated further.

Author Contributions: M.C., M.W., M.Z., K.Ł., B.M., C.K., T.Ł. and G.J. analyzed the data and wrote the paper. M.C. and M.Z. draw the figures. M.C, K.Ł., B.M. and G.J. supervised the work. M.C., M.W., M.Z., K.Ł., B.M., C.K., T.Ł. and G.J. accepted the final version of the paper.

Funding: This study was funded by The Center of Postgraduate Medical Education. Grant number 501-1-21-27-18.

Conflicts of Interest: The authors declare no conflicts of interest.

Abbreviations

ADAM17	ADAM metallopeptidase domain 17
AMH	anti-Müllerian hormone
AP-1	activator protein 1
BMP	bone morphogenetic protein
COX-2	cyclooxygenase 2
ECM	extracellular matrix
EGF	endothelial growth factor
ER	estrogen receptor
ERK	extracellular signal–regulated kinase
FADD	Fas-associated protein with death domain
FAK	focal adhesion kinase
FGF	fibroblast growth factor
GnRH	gonadotropin-releasing hormone
IFN	Interferon
IGF-1	insulin-like growth factor 1
IKK	I kappa B kinase
IL	Interleukin
JNK	c-jun N-terminal kinase
LMS	leiomyosarcoma
MAPK	mitogen-activated protein kinases
MEKK	mitogen-activated protein kinase kinase kinase
MMP	matrix metalloproteinase
NF-κB	nuclear factor kappa-light-chain-enhancer of activated B cells
NIK	NF-kappa-B-inducing kinase
NK	natural killer

NSAID	nonsteroidal anti-inflammatory drug
PCOS	polycystic ovary syndrome
PDGF	platelet-derived growth factor
PG	Prostaglandin
PR	progesterone receptor
RIP	receptor interacting protein
ROS	reactive oxygen species
SHBG	sex hormone-binding globulin
SODD	silencer of death domains
SPRM	selective progesterone receptor modulator
STUMP	smooth muscle tumor of uncertain malignant potential
TAK	TGF-β-activated kinase
TGF-β	transforming growth factor β
TIMP	tissue inhibitor of metalloproteinase
TNFR	TNF-α receptor
TNF-α	tumor necrosis factor α
TRADD	tumor necrosis factor receptor type 1-associated death domain
TRAF	tumor necrosis factor receptor-associated factor
UF	uterine fibroid
UPA	ulipristal acetate
VEGF	vascular endothelial growth factor

References

1. Stewart, E.A. Uterine fibroids. *Lancet* **2001**, *357*, 293–298. [CrossRef]
2. Stewart, E.A.; Laughlin-Tommaso, S.K.; Catherino, W.H.; Lalitkumar, S.; Gupta, D.; Vollenhoven, B. Uterine fibroids. *Nat. Rev. Dis. Primers* **2016**, *2*, 16043. [CrossRef] [PubMed]
3. Stewart, E.A.; Cookson, C.L.; Gandolfo, R.A.; Schulze-Rath, R. Epidemiology of uterine fibroids: A systematic review. *BJOG* **2017**, *124*, 1501–1512. [CrossRef] [PubMed]
4. Parker, W.H. Etiology, symptomatology, and diagnosis of uterine myomas. *Fertil Steril* **2007**, *87*, 725–736. [CrossRef] [PubMed]
5. Metwally, M.; Farquhar, C.M.; Li, T.C. Is another meta-analysis on the effects of intramural fibroids on reproductive outcomes needed? *Reprod. Biomed. Online* **2011**, *23*, 2–14. [CrossRef] [PubMed]
6. Soave, I.; Marci, R. Uterine leiomyomata: The snowball effect. *Curr. Med. Res. Opin* **2017**, *33*, 1909–1911. [CrossRef] [PubMed]
7. Ciarmela, P.; Islam, M.S.; Reis, F.M.; Gray, P.C.; Bloise, E.; Petraglia, F.; Vale, W.; Castellucci, M. Growth factors and myometrium: Biological effects in uterine fibroid and possible clinical implications. *Hum. Reprod. Update* **2011**, *17*, 772–790. [CrossRef] [PubMed]
8. Al-Hendy, A.; Myers, E.R.; Stewart, E. Uterine fibroids: Burden and unmet medical need. *Semin. Reprod. Med.* **2017**, *35*, 473–480. [CrossRef] [PubMed]
9. Soliman, A.M.; Margolis, M.K.; Castelli-Haley, J.; Fuldeore, M.J.; Owens, C.D.; Coyne, K.S. Impact of uterine fibroid symptoms on health-related quality of life of us women: Evidence from a cross-sectional survey. *Curr. Med. Res. Opin* **2017**, *33*, 1971–1978. [CrossRef] [PubMed]
10. Cardozo, E.R.; Clark, A.D.; Banks, N.K.; Henne, M.B.; Stegmann, B.J.; Segars, J.H. The estimated annual cost of uterine leiomyomata in the united states. *Am. J. Obstet. Gynecol.* **2012**, *206*, e211–e219. [CrossRef]
11. Soliman, A.M.; Yang, H.; Du, E.X.; Kelkar, S.S.; Winkel, C. The direct and indirect costs of uterine fibroid tumors: A systematic review of the literature between 2000 and 2013. *Am. J. Obstet. Gynecol.* **2015**, *213*, 141–160. [CrossRef] [PubMed]
12. Bulun, S.E. Uterine fibroids. *N. Engl. J. Med.* **2013**, *369*, 1344–1355. [CrossRef] [PubMed]
13. Islam, M.S.; Ciavattini, A.; Petraglia, F.; Castellucci, M.; Ciarmela, P. Extracellular matrix in uterine leiomyoma pathogenesis: A potential target for future therapeutics. *Hum. Reprod. Update* **2018**, *24*, 59–85. [CrossRef] [PubMed]

14. Sozen, I.; Arici, A. Interactions of cytokines, growth factors, and the extracellular matrix in the cellular biology of uterine leiomyomata. *Fertil Steril* **2002**, *78*, 1–12. [CrossRef]
15. Dixon, D.; He, H.; Haseman, J.K. Immunohistochemical localization of growth factors and their receptors in uterine leiomyomas and matched myometrium. *Environ. Health Perspect.* **2000**, *108* (Suppl. 5), 795–802. [CrossRef] [PubMed]
16. Islam, M.S.; Protic, O.; Stortoni, P.; Grechi, G.; Lamanna, P.; Petraglia, F.; Castellucci, M.; Ciarmela, P. Complex networks of multiple factors in the pathogenesis of uterine leiomyoma. *Fertil Steril* **2013**, *100*, 178–193. [CrossRef] [PubMed]
17. Joseph, D.S.; Malik, M.; Nurudeen, S.; Catherino, W.H. Myometrial cells undergo fibrotic transformation under the influence of transforming growth factor beta-3. *Fertil Steril* **2010**, *93*, 1500–1508. [CrossRef]
18. Borahay, M.A.; Asoglu, M.R.; Mas, A.; Adam, S.; Kilic, G.S.; Al-Hendy, A. Estrogen receptors and signaling in fibroids: Role in pathobiology and therapeutic implications. *Reprod. Sci.* **2017**, *24*, 1235–1244. [CrossRef]
19. Nierth-Simpson, E.N.; Martin, M.M.; Chiang, T.C.; Melnik, L.I.; Rhodes, L.V.; Muir, S.E.; Burow, M.E.; McLachlan, J.A. Human uterine smooth muscle and leiomyoma cells differ in their rapid 17beta-estradiol signaling: Implications for proliferation. *Endocrinology* **2009**, *150*, 2436–2445. [CrossRef]
20. Chegini, N. Proinflammatory and profibrotic mediators: Principal effectors of leiomyoma development as a fibrotic disorder. *Semin. Reprod. Med.* **2010**, *28*, 180–203. [CrossRef]
21. Maruo, T.; Ohara, N.; Wang, J.; Matsuo, H. Sex steroidal regulation of uterine leiomyoma growth and apoptosis. *Hum. Reprod. Update* **2004**, *10*, 207–220. [CrossRef] [PubMed]
22. Barbarisi, A.; Petillo, O.; Di Lieto, A.; Melone, M.A.; Margarucci, S.; Cannas, M.; Peluso, G. 17-beta estradiol elicits an autocrine leiomyoma cell proliferation: Evidence for a stimulation of protein kinase-dependent pathway. *J. Cell. Physiol.* **2001**, *186*, 414–424. [CrossRef]
23. Ishikawa, H.; Ishi, K.; Serna, V.A.; Kakazu, R.; Bulun, S.E.; Kurita, T. Progesterone is essential for maintenance and growth of uterine leiomyoma. *Endocrinology* **2010**, *151*, 2433–2442. [CrossRef] [PubMed]
24. Chill, H.H.; Safrai, M.; Reuveni Salzman, A.; Shushan, A. The rising phoenix-progesterone as the main target of the medical therapy for leiomyoma. *Biomed. Res. Int.* **2017**, *2017*, 4705164. [CrossRef] [PubMed]
25. Walker, C.L.; Stewart, E.A. Uterine fibroids: The elephant in the room. *Science* **2005**, *308*, 1589–1592. [CrossRef] [PubMed]
26. Ciebiera, M.; Wlodarczyk, M.; Slabuszewska-Jozwiak, A.; Nowicka, G.; Jakiel, G. Influence of vitamin D and transforming growth factor beta3 serum concentrations, obesity, and family history on the risk for uterine fibroids. *Fertil Steril* **2016**, *106*, 1787–1792. [CrossRef] [PubMed]
27. Ciebiera, M.; Wlodarczyk, M.; Wrzosek, M.; Slabuszewska-Jozwiak, A.; Nowicka, G.; Jakiel, G. Ulipristal acetate decreases transforming growth factor beta3 serum and tumor tissue concentrations in patients with uterine fibroids. *Fertil Steril* **2018**, *109*, 501–507. [CrossRef]
28. Halder, S.; Al-Hendy, A. Hypovitaminosis D and high serum transforming growth factor beta-3: Important biomarkers for uterine fibroids risk. *Fertil Steril* **2016**, *106*, 1648–1649. [CrossRef]
29. Ciebiera, M.; Wlodarczyk, M.; Wrzosek, M.; Meczekalski, B.; Nowicka, G.; Lukaszuk, K.; Ciebiera, M.; Slabuszewska-Jozwiak, A.; Jakiel, G. Role of transforming growth factor beta in uterine fibroid biology. *Int J. Mol. Sci.* **2017**, *18*, 2435. [CrossRef]
30. Ciebiera, M.; Wlodarczyk, M.; Wrzosek, M.; Wojtyla, C.; Blazej, M.; Nowicka, G.; Lukaszuk, K.; Jakiel, G. TNF-alpha serum levels are elevated in women with clinically symptomatic uterine fibroids. *Int. J. Immunopathol. Pharmacol.* **2018**, *32*, 2058738418779461. [CrossRef]
31. Kim, Y.J.; Kim, Y.Y.; Shin, J.H.; Kim, H.; Ku, S.Y.; Suh, C.S. Variation in microRNA expression profile of uterine leiomyoma with endometrial cavity distortion and endometrial cavity non-distortion. *Int. J. Mol. Sci.* **2018**, *19*, 2524. [CrossRef] [PubMed]
32. Leppert, P.C.; Jayes, F.L.; Segars, J.H. The extracellular matrix contributes to mechanotransduction in uterine fibroids. *Obstet. Gynecol. Int.* **2014**, *2014*, 783289. [CrossRef] [PubMed]
33. Malik, M.; Norian, J.; McCarthy-Keith, D.; Britten, J.; Catherino, W.H. Why leiomyomas are called fibroids: The central role of extracellular matrix in symptomatic women. *Semin. Reprod. Med.* **2010**, *28*, 169–179. [CrossRef] [PubMed]
34. Rafique, S.; Segars, J.H.; Leppert, P.C. Mechanical signaling and extracellular matrix in uterine fibroids. *Semin. Reprod. Med.* **2017**, *35*, 487–493. [CrossRef] [PubMed]

35. Protic, O.; Toti, P.; Islam, M.S.; Occhini, R.; Giannubilo, S.R.; Catherino, W.H.; Cinti, S.; Petraglia, F.; Ciavattini, A.; Castellucci, M.; et al. Possible involvement of inflammatory/reparative processes in the development of uterine fibroids. *Cell Tissue Res.* **2016**, *364*, 415–427. [CrossRef] [PubMed]

36. Kisseleva, T.; Brenner, D.A. Mechanisms of fibrogenesis. *Exp. Biol. Med.* **2008**, *233*, 109–122. [CrossRef] [PubMed]

37. Arici, A.; Sozen, I. Transforming growth factor-beta3 is expressed at high levels in leiomyoma where it stimulates fibronectin expression and cell proliferation. *Fertil Steril* **2000**, *73*, 1006–1011. [CrossRef]

38. Brew, K.; Dinakarpandian, D.; Nagase, H. Tissue inhibitors of metalloproteinases: Evolution, structure and function. *Biochim. Biophys. Acta* **2000**, *1477*, 267–283. [CrossRef]

39. Wolanska, M.; Taudul, E.; Bankowska-Guszczyn, E.; Kinalski, M. Tumor necrosis factor in uterine leiomyomas at various stages of tumor growth. *Ginekol. Pol.* **2010**, *81*, 431–434.

40. Lee, J.W.; Juliano, R. Mitogenic signal transduction by integrin- and growth factor receptor-mediated pathways. *Mol. Cells* **2004**, *17*, 188–202.

41. Dinarello, C.A. Historical insights into cytokines. *Eur. J. Immunol.* **2007**, *37* (Suppl. 1), S34–S45. [CrossRef] [PubMed]

42. Turner, M.D.; Nedjai, B.; Hurst, T.; Pennington, D.J. Cytokines and chemokines: At the crossroads of cell signalling and inflammatory disease. *Biochim. Biophys. Acta* **2014**, *1843*, 2563–2582. [CrossRef] [PubMed]

43. Locksley, R.M.; Killeen, N.; Lenardo, M.J. The TNF and TNF receptor superfamilies: Integrating mammalian biology. *Cell* **2001**, *104*, 487–501. [CrossRef]

44. Markowska, A.; Mardas, M.; Gajdzik, E.; Zagrodzki, P.; Markowska, J. Oxidative stress markers in uterine fibroids tissue in pre- and postmenopausal women. *Clin. Exp. Obstet. Gynecol.* **2015**, *42*, 725–729. [PubMed]

45. Sivarajasingam, S.P.; Imami, N.; Johnson, M.R. Myometrial cytokines and their role in the onset of labour. *J. Endocrinol.* **2016**, *231*, R101–R119. [CrossRef] [PubMed]

46. Nisenblat, V.; Bossuyt, P.M.; Shaikh, R.; Farquhar, C.; Jordan, V.; Scheffers, C.S.; Mol, B.W.; Johnson, N.; Hull, M.L. Blood biomarkers for the non-invasive diagnosis of endometriosis. *Cochrane Database Syst. Rev.* **2016**, CD012179. [CrossRef]

47. Bradley, J.R. TNF-mediated inflammatory disease. *J. Pathol.* **2008**, *214*, 149–160. [CrossRef]

48. Balkwill, F. Tumor necrosis factor or tumor promoting factor? *Cytokine Growth Factor Rev.* **2002**, *13*, 135–141. [CrossRef]

49. Kurachi, O.; Matsuo, H.; Samoto, T.; Maruo, T. Tumor necrosis factor-alpha expression in human uterine leiomyoma and its down-regulation by progesterone. *J. Clin. Endocrinol. Metab.* **2001**, *86*, 2275–2280. [CrossRef]

50. Postal, M.; Lapa, A.T.; Sinicato, N.A.; de Oliveira Pelicari, K.; Peres, F.A.; Costallat, L.T.; Fernandes, P.T.; Marini, R.; Appenzeller, S. Depressive symptoms are associated with tumor necrosis factor alpha in systemic lupus erythematosus. *J. Neuroinflamm.* **2016**, *13*, 5. [CrossRef]

51. Victor, F.C.; Gottlieb, A.B. TNF-alpha and apoptosis: Implications for the pathogenesis and treatment of psoriasis. *J. Drugs Dermatol.* **2002**, *1*, 264–275. [PubMed]

52. Brynskov, J.; Foegh, P.; Pedersen, G.; Ellervik, C.; Kirkegaard, T.; Bingham, A.; Saermark, T. Tumour necrosis factor alpha converting enzyme (TACE) activity in the colonic mucosa of patients with inflammatory bowel disease. *Gut* **2002**, *51*, 37–43. [CrossRef] [PubMed]

53. Strober, W.; Fuss, I.J. Proinflammatory cytokines in the pathogenesis of inflammatory bowel diseases. *Gastroenterology* **2011**, *140*, 1756–1767. [CrossRef] [PubMed]

54. Decourt, B.; Lahiri, D.K.; Sabbagh, M.N. Targeting tumor necrosis factor alpha for Alzheimer's disease. *Curr. Alzheimer Res.* **2017**, *14*, 412–425. [CrossRef]

55. Chuang, M.J.; Sun, K.H.; Tang, S.J.; Deng, M.W.; Wu, Y.H.; Sung, J.S.; Cha, T.L.; Sun, G.H. Tumor-derived tumor necrosis factor-alpha promotes progression and epithelial-mesenchymal transition in renal cell carcinoma cells. *Cancer Sci.* **2008**, *99*, 905–913. [CrossRef] [PubMed]

56. Victor, F.C.; Gottlieb, A.B.; Menter, A. Changing paradigms in dermatology: Tumor necrosis factor alpha (TNF-alpha) blockade in psoriasis and psoriatic arthritis. *Clin. Dermatol.* **2003**, *21*, 392–397. [CrossRef] [PubMed]

57. Rifkin, L.M.; Birnbaum, A.D.; Goldstein, D.A. TNF inhibition for ophthalmic indications: Current status and outlook. *BioDrugs* **2013**, *27*, 347–357. [CrossRef]

58. Matsuno, H.; Yudoh, K.; Katayama, R.; Nakazawa, F.; Uzuki, M.; Sawai, T.; Yonezawa, T.; Saeki, Y.; Panayi, G.S.; Pitzalis, C.; et al. The role of TNF-alpha in the pathogenesis of inflammation and joint destruction in rheumatoid arthritis (RA): A study using a human ra/scid mouse chimera. *Rheumatology* **2002**, *41*, 329–337. [CrossRef]

59. Gupta, M.; Babic, A.; Beck, A.H.; Terry, K. TNF-alpha expression, risk factors, and inflammatory exposures in ovarian cancer: Evidence for an inflammatory pathway of ovarian carcinogenesis? *Hum. Pathol.* **2016**, *54*, 82–91. [CrossRef]

60. Olszewski, M.B.; Groot, A.J.; Dastych, J.; Knol, E.F. TNF trafficking to human mast cell granules: Mature chain-dependent endocytosis. *J. Immunol.* **2007**, *178*, 5701–5709. [CrossRef]

61. Tang, P.; Hung, M.C.; Klostergaard, J. Human pro-tumor necrosis factor is a homotrimer. *Biochemistry* **1996**, *35*, 8216–8225. [CrossRef] [PubMed]

62. Black, R.A.; Rauch, C.T.; Kozlosky, C.J.; Peschon, J.J.; Slack, J.L.; Wolfson, M.F.; Castner, B.J.; Stocking, K.L.; Reddy, P.; Srinivasan, S.; et al. A metalloproteinase disintegrin that releases tumour-necrosis factor-alpha from cells. *Nature* **1997**, *385*, 729–733. [CrossRef] [PubMed]

63. Moss, M.L.; Jin, S.L.; Milla, M.E.; Bickett, D.M.; Burkhart, W.; Carter, H.L.; Chen, W.J.; Clay, W.C.; Didsbury, J.R.; Hassler, D.; et al. Cloning of a disintegrin metalloproteinase that processes precursor tumour-necrosis factor-alpha. *Nature* **1997**, *385*, 733–736. [CrossRef] [PubMed]

64. Palladino, M.A.; Bahjat, F.R.; Theodorakis, E.A.; Moldawer, L.L. Anti-TNF-alpha therapies: The next generation. *Nat. Rev. Drug Discov.* **2003**, *2*, 736–746. [CrossRef] [PubMed]

65. Dubravec, D.B.; Spriggs, D.R.; Mannick, J.A.; Rodrick, M.L. Circulating human peripheral blood granulocytes synthesize and secrete tumor necrosis factor alpha. *Proc. Natl. Acad. Sci. USA* **1990**, *87*, 6758–6761. [CrossRef] [PubMed]

66. Chen, G.; Goeddel, D.V. TNF-R1 signaling: A beautiful pathway. *Science* **2002**, *296*, 1634–1635. [CrossRef]

67. Reactome. TNF Signaling. Available online: http://www.reactome.org/content/detail/R-HSA-75893 (accessed on 7 October 2018).

68. Sun, S.C. Non-canonical NF-kappaB signaling pathway. *Cell Res.* **2011**, *21*, 71–85. [CrossRef] [PubMed]

69. Karin, M.; Delhase, M. The I kappa B kinase (IKK) and NF-kappa B: Key elements of proinflammatory signalling. *Semin. Immunol.* **2000**, *12*, 85–98. [CrossRef]

70. Oltmanns, U.; Issa, R.; Sukkar, M.B.; John, M.; Chung, K.F. Role of c-Jun N-terminal kinase in the induced release of GM-CSF, RANTES and Il-8 from human airway smooth muscle cells. *Br. J. Pharmacol.* **2003**, *139*, 1228–1234. [CrossRef]

71. Festjens, N.; Vanden Berghe, T.; Cornelis, S.; Vandenabeele, P. RIP1, a kinase on the crossroads of a cell's decision to live or die. *Cell Death Differ.* **2007**, *14*, 400–410. [CrossRef]

72. Riches, D.W.; Chan, E.D.; Winston, B.W. TNF-alpha-induced regulation and signalling in macrophages. *Immunobiology* **1996**, *195*, 477–490. [CrossRef]

73. Wajant, H.; Pfizenmaier, K.; Scheurich, P. Tumor necrosis factor signaling. *Cell Death Differ.* **2003**, *10*, 45–65. [CrossRef] [PubMed]

74. Kant, S.; Swat, W.; Zhang, S.; Zhang, Z.Y.; Neel, B.G.; Flavell, R.A.; Davis, R.J. TNF-stimulated MAP kinase activation mediated by a Rho family GTPase signaling pathway. *Genes Dev.* **2011**, *25*, 2069–2078. [CrossRef] [PubMed]

75. Wajant, H.; Scheurich, P. Tumor necrosis factor receptor-associated factor (TRAF) 2 and its role in tnf signaling. *Int. J. Biochem. Cell Biol.* **2001**, *33*, 19–32. [CrossRef]

76. Takada, H.; Chen, N.J.; Mirtsos, C.; Suzuki, S.; Suzuki, N.; Wakeham, A.; Mak, T.W.; Yeh, W.C. Role of sodd in regulation of tumor necrosis factor responses. *Mol. Cell. Biol.* **2003**, *23*, 4026–4033. [CrossRef] [PubMed]

77. Pobezinskaya, Y.L.; Liu, Z. The role of TRADD in death receptor signaling. *Cell Cycle* **2012**, *11*, 871–876. [CrossRef]

78. Liu, Y.; Lu, D.; Sheng, J.; Luo, L.; Zhang, W. Identification of TRADD as a potential biomarker in human uterine leiomyoma through Itraq based proteomic profiling. *Mol. Cell. Probes* **2017**, *36*, 15–20. [CrossRef] [PubMed]

79. Gilmore, T.D. Introduction to NF-kappaB: Players, pathways, perspectives. *Oncogene* **2006**, *25*, 6680–6684. [CrossRef] [PubMed]

80. Trendelenburg, A.U.; Meyer, A.; Jacobi, C.; Feige, J.N.; Glass, D.J. Tak-1/p38/NFkappaB signaling inhibits myoblast differentiation by increasing levels of Activin A. *Skelet Muscle* **2012**, *2*, 3. [CrossRef]

81. Protic, O.; Islam, M.S.; Greco, S.; Giannubilo, S.R.; Lamanna, P.; Petraglia, F.; Ciavattini, A.; Castellucci, M.; Hinz, B.; Ciarmela, P. Activin A in inflammation, tissue repair, and fibrosis: Possible role as inflammatory and fibrotic mediator of uterine fibroid development and growth. *Semin. Reprod. Med.* **2017**, 35, 499–509. [CrossRef]

82. Hengartner, M.O. The biochemistry of apoptosis. *Nature* **2000**, 407, 770–776. [CrossRef]

83. Vlahopoulos, S.; Zoumpourlis, V.C. JNK: A key modulator of intracellular signaling. *Biochemistry* **2004**, 69, 844–854. [CrossRef] [PubMed]

84. Ip, Y.T.; Davis, R.J. Signal transduction by the c-Jun N-terminal kinase (JNK)—From inflammation to development. *Curr. Opin Cell Biol.* **1998**, 10, 205–219. [CrossRef]

85. Wagner, E.F.; Nebreda, A.R. Signal integration by jnk and p38 MAPK pathways in cancer development. *Nat. Rev. Cancer* **2009**, 9, 537–549. [CrossRef] [PubMed]

86. Gaur, U.; Aggarwal, B.B. Regulation of proliferation, survival and apoptosis by members of the tnf superfamily. *Biochem. Pharmacol.* **2003**, 66, 1403–1408. [CrossRef]

87. Chow, M.T.; Moller, A.; Smyth, M.J. Inflammation and immune surveillance in cancer. *Semin. Cancer Biol.* **2012**, 22, 23–32. [CrossRef] [PubMed]

88. Wegienka, G. Are uterine leiomyoma a consequence of a chronically inflammatory immune system? *Med. Hypotheses* **2012**, 79, 226–231. [CrossRef]

89. Wegienka, G.; Baird, D.D.; Cooper, T.; Woodcroft, K.J.; Havstad, S. Cytokine patterns differ seasonally between women with and without uterine leiomyomata. *Am. J. Reprod. Immunol.* **2013**, 70, 327–335. [CrossRef]

90. Miura, S.; Khan, K.N.; Kitajima, M.; Hiraki, K.; Moriyama, S.; Masuzaki, H.; Samejima, T.; Fujishita, A.; Ishimaru, T. Differential infiltration of macrophages and prostaglandin production by different uterine leiomyomas. *Hum. Reprod.* **2006**, 21, 2545–2554. [CrossRef]

91. Wynn, T.A.; Barron, L. Macrophages: Master regulators of inflammation and fibrosis. *Semin. Liver Dis.* **2010**, 30, 245–257. [CrossRef]

92. Leppert, P.C.; Baginski, T.; Prupas, C.; Catherino, W.H.; Pletcher, S.; Segars, J.H. Comparative ultrastructure of collagen fibrils in uterine leiomyomas and normal myometrium. *Fertil Steril* **2004**, 82 (Suppl. 3), 1182–1187. [CrossRef] [PubMed]

93. Fujisawa, C.; Castellot, J.J., Jr. Matrix production and remodeling as therapeutic targets for uterine leiomyoma. *J. Cell Commun. Signal.* **2014**, 8, 179–194. [CrossRef]

94. Feng, L.; Jayes, F.L.; Johnson, L.N.C.; Schomberg, D.W.; Leppert, P.C. Biochemical pathways and myometrial cell differentiation leading to nodule formation containing collagen and fibronectin. *Curr. Protein Pept. Sci.* **2017**, 18, 155–166. [CrossRef] [PubMed]

95. Makinen, N.; Mehine, M.; Tolvanen, J.; Kaasinen, E.; Li, Y.; Lehtonen, H.J.; Gentile, M.; Yan, J.; Enge, M.; Taipale, M.; et al. Med12, the mediator complex subunit 12 gene, is mutated at high frequency in uterine leiomyomas. *Science* **2011**, 334, 252–255. [CrossRef] [PubMed]

96. Heinonen, H.R.; Sarvilinna, N.S.; Sjoberg, J.; Kampjarvi, K.; Pitkanen, E.; Vahteristo, P.; Makinen, N.; Aaltonen, L.A. Med12 mutation frequency in unselected sporadic uterine leiomyomas. *Fertil Steril* **2014**, 102, 1137–1142. [CrossRef] [PubMed]

97. Makinen, N.; Heinonen, H.R.; Moore, S.; Tomlinson, I.P.; van der Spuy, Z.M.; Aaltonen, L.A. MED12 exon 2 mutations are common in uterine leiomyomas from South African patients. *Oncotarget* **2011**, 2, 966–969. [CrossRef] [PubMed]

98. Halder, S.K.; Laknaur, A.; Miller, J.; Layman, L.C.; Diamond, M.; Al-Hendy, A. Novel MED12 gene somatic mutations in women from the southern United States with symptomatic uterine fibroids. *Mol. Genet. Genom.* **2015**, 290, 505–511. [CrossRef] [PubMed]

99. Mittal, P.; Shin, Y.H.; Yatsenko, S.A.; Castro, C.A.; Surti, U.; Rajkovic, A. MED12 gain-of-function mutation causes leiomyomas and genomic instability. *J. Clin. Investig.* **2015**, 125, 3280–3284. [CrossRef] [PubMed]

100. Elkafas, H.; Qiwei, Y.; Al-Hendy, A. Origin of uterine fibroids: Conversion of myometrial stem cells to tumor-initiating cells. *Semin. Reprod. Med.* **2017**, 35, 481–486. [CrossRef] [PubMed]

101. El Andaloussi, A.; Chaudhry, Z.; Al-Hendy, A.; Ismail, N. Uterine fibroids: Bridging genomic defects and chronic inflammation. *Semin. Reprod. Med.* **2017**, 35, 494–498. [CrossRef] [PubMed]

102. Tal, R.; Segars, J.H. The role of angiogenic factors in fibroid pathogenesis: Potential implications for future therapy. *Hum. Reprod. Update* **2014**, 20, 194–216. [CrossRef]

103. Halder, S.K.; Goodwin, J.S.; Al-Hendy, A. 1,25-dihydroxyvitamin D3 reduces TGF-beta3-induced fibrosis-related gene expression in human uterine leiomyoma cells. *J. Clin. Endocrinol. Metab.* **2011**, *96*, E754–E762. [CrossRef]

104. Islam, M.S.; Catherino, W.H.; Protic, O.; Janjusevic, M.; Gray, P.C.; Giannubilo, S.R.; Ciavattini, A.; Lamanna, P.; Tranquilli, A.L.; Petraglia, F.; et al. Role of Activin-A and myostatin and their signaling pathway in human myometrial and leiomyoma cell function. *J. Clin. Endocrinol. Metab.* **2014**, *99*, E775–E785. [CrossRef] [PubMed]

105. Werner, S.; Alzheimer, C. Roles of activin in tissue repair, fibrosis, and inflammatory disease. *Cytokine Growth Factor Rev.* **2006**, *17*, 157–171. [CrossRef] [PubMed]

106. Mukhopadhyay, A.; Chan, S.Y.; Lim, I.J.; Phillips, D.J.; Phan, T.T. The role of the activin system in keloid pathogenesis. *Am. J. Physiol. Cell Physiol.* **2007**, *292*, C1331–C1338. [CrossRef] [PubMed]

107. Wada, W.; Kuwano, H.; Hasegawa, Y.; Kojima, I. The dependence of transforming growth factor-beta-induced collagen production on autocrine factor Activin A in hepatic stellate cells. *Endocrinology* **2004**, *145*, 2753–2759. [CrossRef]

108. Sierra-Filardi, E.; Puig-Kroger, A.; Blanco, F.J.; Nieto, C.; Bragado, R.; Palomero, M.I.; Bernabeu, C.; Vega, M.A.; Corbi, A.L. Activin a skews macrophage polarization by promoting a proinflammatory phenotype and inhibiting the acquisition of anti-inflammatory macrophage markers. *Blood* **2011**, *117*, 5092–5101. [CrossRef]

109. Ciarmela, P.; Bloise, E.; Gray, P.C.; Carrarelli, P.; Islam, M.S.; De Pascalis, F.; Severi, F.M.; Vale, W.; Castellucci, M.; Petraglia, F. Activin-A and myostatin response and steroid regulation in human myometrium: Disruption of their signalling in uterine fibroid. *J. Clin. Endocrinol. Metab.* **2011**, *96*, 755–765. [CrossRef]

110. Norian, J.M.; Malik, M.; Parker, C.Y.; Joseph, D.; Leppert, P.C.; Segars, J.H.; Catherino, W.H. Transforming growth factor beta3 regulates the versican variants in the extracellular matrix-rich uterine leiomyomas. *Reprod. Sci.* **2009**, *16*, 1153–1164. [CrossRef]

111. Verma, R.P.; Hansch, C. Matrix metalloproteinases (MMPs): Chemical-biological functions and QSARs. *Bioorg. Med. Chem.* **2007**, *15*, 2223–2268. [CrossRef]

112. Dou, Q.; Tarnuzzer, R.W.; Williams, R.S.; Schultz, G.S.; Chegini, N. Differential expression of matrix metalloproteinases and their tissue inhibitors in leiomyomata: A mechanism for gonadotrophin releasing hormone agonist-induced tumour regression. *Mol. Hum. Reprod.* **1997**, *3*, 1005–1014. [CrossRef] [PubMed]

113. Kamel, M.; Wagih, M.; Kilic, G.S.; Diaz-Arrastia, C.R.; Baraka, M.A.; Salama, S.A. Overhydroxylation of lysine of collagen increases uterine fibroids proliferation: Roles of lysyl hydroxylases, lysyl oxidases, and matrix metalloproteinases. *Biomed. Res. Int.* **2017**, *2017*, 5316845. [CrossRef] [PubMed]

114. Halder, S.K.; Osteen, K.G.; Al-Hendy, A. 1,25-dihydroxyvitamin D3 reduces extracellular matrix-associated protein expression in human uterine fibroid cells. *Biol. Reprod.* **2013**, *89*, 150. [CrossRef] [PubMed]

115. Ma, C.; Chegini, N. Regulation of matrix metalloproteinases (MMPs) and their tissue inhibitors in human myometrial smooth muscle cells by TGF-beta1. *Mol. Hum. Reprod.* **1999**, *5*, 950–954. [CrossRef] [PubMed]

116. Halder, S.K.; Osteen, K.G.; Al-Hendy, A. Vitamin D3 inhibits expression and activities of matrix metalloproteinase-2 and -9 in human uterine fibroid cells. *Hum. Reprod.* **2013**, *28*, 2407–2416. [CrossRef] [PubMed]

117. Nair, S.; Al-Hendy, A. Adipocytes enhance the proliferation of human leiomyoma cells via TNF-alpha proinflammatory cytokine. *Reprod. Sci.* **2011**, *18*, 1186–1192. [CrossRef]

118. Hsieh, Y.Y.; Chang, C.C.; Tsai, F.J.; Lin, C.C.; Yeh, L.S.; Tsai, C.H. Tumor necrosis factor-alpha-308 promoter and p53 codon 72 gene polymorphisms in women with leiomyomas. *Fertil Steril* **2004**, *82* (Suppl. 3), 1177–1181. [CrossRef]

119. Litovkin, K.V.; Domenyuk, V.P.; Bubnov, V.V.; Zaporozhan, V.N. Interleukin-6 -174g/c polymorphism in breast cancer and uterine leiomyoma patients: A population-based case control study. *Exp. Oncol.* **2007**, *29*, 295–298.

120. Pietrowski, D.; Thewes, R.; Sator, M.; Denschlag, D.; Keck, C.; Tempfer, C. Uterine leiomyoma is associated with a polymorphism in the interleukin 1-beta gene. *Am. J. Reprod. Immunol.* **2009**, *62*, 112–117. [CrossRef]

121. Martel, K.M.; Ko, A.C.; Christman, G.M.; Stribley, J.M. Apoptosis in human uterine leiomyomas. *Semin. Reprod. Med.* **2004**, *22*, 91–103. [CrossRef]

122. Plewka, A.; Madej, P.; Plewka, D.; Kowalczyk, A.; Miskiewicz, A.; Wittek, P.; Leks, T.; Bilski, R. Immunohistochemical localization of selected pro-inflammatory factors in uterine myomas and myometrium in women of various ages. *Folia Histochem. Cytobiol.* **2013**, *51*, 73–83. [CrossRef] [PubMed]

123. Maruo, T.; Matsuo, H.; Shimomura, Y.; Kurachi, O.; Gao, Z.; Nakago, S.; Yamada, T.; Chen, W.; Wang, J. Effects of progesterone on growth factor expression in human uterine leiomyoma. *Steroids* **2003**, *68*, 817–824. [CrossRef]

124. Manta, L.; Suciu, N.; Toader, O.; Purcarea, R.M.; Constantin, A.; Popa, F. The etiopathogenesis of uterine fibromatosis. *J. Med. Life* **2016**, *9*, 39–45. [PubMed]

125. Wilkens, J.; Male, V.; Ghazal, P.; Forster, T.; Gibson, D.A.; Williams, A.R.; Brito-Mutunayagam, S.L.; Craigon, M.; Lourenco, P.; Cameron, I.T.; et al. Uterine nk cells regulate endometrial bleeding in women and are suppressed by the progesterone receptor modulator asoprisnil. *J. Immunol.* **2013**, *191*, 2226–2235. [CrossRef]

126. Guo, W.; Li, P.; Zhao, G.; Fan, H.; Hu, Y.; Hou, Y. Glucocorticoid receptor mediates the effect of progesterone on uterine natural killer cells. *Am. J. Reprod. Immunol.* **2012**, *67*, 463–473. [CrossRef] [PubMed]

127. Macdiarmid, F.; Wang, D.; Duncan, L.J.; Purohit, A.; Ghilchick, M.W.; Reed, M.J. Stimulation of aromatase activity in breast fibroblasts by tumor necrosis factor alpha. *Mol. Cell. Endocrinol.* **1994**, *106*, 17–21. [CrossRef]

128. Kamel, M.; Shouman, S.; El-Merzebany, M.; Kilic, G.; Veenstra, T.; Saeed, M.; Wagih, M.; Diaz-Arrastia, C.; Patel, D.; Salama, S. Effect of tumour necrosis factor-alpha on estrogen metabolic pathways in breast cancer cells. *J. Cancer* **2012**, *3*, 310–321. [CrossRef]

129. Hubner, G.; Werner, S. Serum growth factors and proinflammatory cytokines are potent inducers of activin expression in cultured fibroblasts and keratinocytes. *Exp. Cell Res.* **1996**, *228*, 106–113. [CrossRef]

130. Shao, L.E.; Frigon, N.L., Jr.; Yu, A.; Palyash, J.; Yu, J. Contrasting effects of inflammatory cytokines and glucocorticoids on the production of activin a in human marrow stromal cells and their implications. *Cytokine* **1998**, *10*, 227–235. [CrossRef]

131. Hillier, S.G.; Miro, F. Inhibin, activin, and follistatin. Potential roles in ovarian physiology. *Ann. N. Y. Acad. Sci.* **1993**, *687*, 29–38. [CrossRef]

132. Shukovski, L.; Findlay, J.K. Activin-a inhibits oxytocin and progesterone production by preovulatory bovine granulosa cells in vitro. *Endocrinology* **1990**, *126*, 2222–2224. [CrossRef]

133. Ali, M.; Al-Hendy, A. Selective progesterone receptor modulators for fertility preservation in women with symptomatic uterine fibroids. *Biol. Reprod.* **2017**, *97*, 337–352. [CrossRef] [PubMed]

134. Yoshida, S.; Ohara, N.; Xu, Q.; Chen, W.; Wang, J.; Nakabayashi, K.; Sasaki, H.; Morikawa, A.; Maruo, T. Cell-type specific actions of progesterone receptor modulators in the regulation of uterine leiomyoma growth. *Semin. Reprod. Med.* **2010**, *28*, 260–273. [CrossRef] [PubMed]

135. Ciarmela, P.; Carrarelli, P.; Islam, M.S.; Janjusevic, M.; Zupi, E.; Tosti, C.; Castellucci, M.; Petraglia, F. Ulipristal acetate modulates the expression and functions of activin a in leiomyoma cells. *Reprod. Sci.* **2014**, *21*, 1120–1125. [CrossRef] [PubMed]

136. Wang, Y.; Feng, G.; Wang, J.; Zhou, Y.; Liu, Y.; Shi, Y.; Zhu, Y.; Lin, W.; Xu, Y.; Li, Z. Differential effects of tumor necrosis factor-alpha on matrix metalloproteinase-2 expression in human myometrial and uterine leiomyoma smooth muscle cells. *Hum. Reprod.* **2015**, *30*, 61–70. [CrossRef] [PubMed]

137. Roskoski, R., Jr. Erk1/2 map kinases: Structure, function, and regulation. *Pharmacol. Res.* **2012**, *66*, 105–143. [CrossRef] [PubMed]

138. Haslam, D.W.; James, W.P. Obesity. *Lancet* **2005**, *366*, 1197–1209. [CrossRef]

139. Albuquerque, D.; Nobrega, C.; Manco, L.; Padez, C. The contribution of genetics and environment to obesity. *Br. Med. Bull.* **2017**, *123*, 159–173. [CrossRef]

140. Baird, D.D.; Dunson, D.B.; Hill, M.C.; Cousins, D.; Schectman, J.M. Association of physical activity with development of uterine leiomyoma. *Am. J. Epidemiol.* **2007**, *165*, 157–163. [CrossRef]

141. Shikora, S.A.; Niloff, J.M.; Bistrian, B.R.; Forse, R.A.; Blackburn, G.L. Relationship between obesity and uterine leiomyomata. *Nutrition* **1991**, *7*, 251–255.

142. Shozu, M.; Murakami, K.; Inoue, M. Aromatase and leiomyoma of the uterus. *Semin. Reprod. Med.* **2004**, *22*, 51–60. [CrossRef] [PubMed]

143. Sumitani, H.; Shozu, M.; Segawa, T.; Murakami, K.; Yang, H.J.; Shimada, K.; Inoue, M. In situ estrogen synthesized by aromatase p450 in uterine leiomyoma cells promotes cell growth probably via an autocrine/intracrine mechanism. *Endocrinology* **2000**, *141*, 3852–3861. [CrossRef] [PubMed]

144. Caldwell, J.D.; Jirikowski, G.F. Sex hormone binding globulin and aging. *Horm. Metab. Res.* **2009**, *41*, 173–182. [CrossRef] [PubMed]

145. Simo, R.; Saez-Lopez, C.; Barbosa-Desongles, A.; Hernandez, C.; Selva, D.M. Novel insights in SHBG regulation and clinical implications. *Trends Endocrinol. Metab.* **2015**, *26*, 376–383. [CrossRef] [PubMed]

146. Ellulu, M.S.; Patimah, I.; Khaza'ai, H.; Rahmat, A.; Abed, Y. Obesity and inflammation: The linking mechanism and the complications. *Arch. Med. Sci.* **2017**, *13*, 851–863. [CrossRef] [PubMed]

147. Alpay, Z.; Saed, G.M.; Diamond, M.P. Female infertility and free radicals: Potential role in adhesions and endometriosis. *J. Soc. Gynecol. Investig.* **2006**, *13*, 390–398. [CrossRef]

148. Ilaria, S.; Marci, R. From obesity to uterine fibroids: An intricate network. *Curr. Med. Res. Opin* **2018**, 1–3. [CrossRef]

149. Fantuzzi, G. Adipose tissue, adipokines, and inflammation. *J. Allergy Clin. Immunol.* **2005**, *115*, 911–919. [CrossRef]

150. Lafontan, M. Fat cells: Afferent and efferent messages define new approaches to treat obesity. *Annu. Rev. Pharmacol. Toxicol.* **2005**, *45*, 119–146. [CrossRef]

151. Rodriguez-Hernandez, H.; Simental-Mendia, L.E.; Rodriguez-Ramirez, G.; Reyes-Romero, M.A. Obesity and inflammation: Epidemiology, risk factors, and markers of inflammation. *Int. J. Endocrinol.* **2013**, *2013*, 678159. [CrossRef]

152. Hotamisligil, G.S.; Shargill, N.S.; Spiegelman, B.M. Adipose expression of tumor necrosis factor-alpha: Direct role in obesity-linked insulin resistance. *Science* **1993**, *259*, 87–91. [CrossRef] [PubMed]

153. Zaragosi, L.E.; Wdziekonski, B.; Villageois, P.; Keophiphath, M.; Maumus, M.; Tchkonia, T.; Bourlier, V.; Mohsen-Kanson, T.; Ladoux, A.; Elabd, C.; et al. Activin A plays a critical role in proliferation and differentiation of human adipose progenitors. *Diabetes* **2010**, *59*, 2513–2521. [CrossRef] [PubMed]

154. Howard, F.M. Endometriosis and mechanisms of pelvic pain. *J. Minim. Invasive Gynecol.* **2009**, *16*, 540–550. [CrossRef] [PubMed]

155. Eisermann, J.; Gast, M.J.; Pineda, J.; Odem, R.R.; Collins, J.L. Tumor necrosis factor in peritoneal fluid of women undergoing laparoscopic surgery. *Fertil Steril* **1988**, *50*, 573–579. [CrossRef]

156. Calhaz-Jorge, C.; Costa, A.P.; Barata, M.; Santos, M.C.; Melo, A.; Palma-Carlos, M.L. Tumour necrosis factor alpha concentrations in the peritoneal fluid of infertile women with minimal or mild endometriosis are lower in patients with red lesions only than in patients without red lesions. *Hum. Reprod.* **2000**, *15*, 1256–1260. [CrossRef]

157. Sommer, C.; Kress, M. Recent findings on how proinflammatory cytokines cause pain: Peripheral mechanisms in inflammatory and neuropathic hyperalgesia. *Neurosci. Lett.* **2004**, *361*, 184–187. [CrossRef]

158. Dogru, H.Y.; Ozsoy, A.Z.; Karakus, N.; Delibas, I.B.; Isguder, C.K.; Yigit, S. Association of genetic polymorphisms in tnf and mif gene with the risk of primary dysmenorrhea. *Biochem. Genet.* **2016**, *54*, 457–466. [CrossRef]

159. Chen, D.B.; Yang, Z.M.; Hilsenrath, R.; Le, S.P.; Harper, M.J. Stimulation of prostaglandin (PG) F2 alpha and PGE2 release by tumour necrosis factor-alpha and interleukin-1 alpha in cultured human luteal phase endometrial cells. *Hum. Reprod.* **1995**, *10*, 2773–2780. [CrossRef]

160. Wiemer, A.J.; Hegde, S.; Gumperz, J.E.; Huttenlocher, A. A live imaging cell motility screen identifies prostaglandin E2 as a T cell stop signal antagonist. *J. Immunol.* **2011**, *187*, 3663–3670. [CrossRef]

161. Ceyhan, S.T.; Onguru, O.; Fidan, U.; Ide, T.; Yaman, H.; Kilic, S.; Baser, I. Comparison of aromatase inhibitor (letrozole) and immunomodulators (infliximab and etanercept) on the regression of endometriotic implants in a rat model. *Eur. J. Obstet. Gynecol. Reprod. Biol.* **2011**, *154*, 100–104. [CrossRef]

162. Gurunath, S.; Pandian, Z.; Anderson, R.A.; Bhattacharya, S. Defining infertility—A systematic review of prevalence studies. *Hum. Reprod. Update* **2011**, *17*, 575–588. [CrossRef] [PubMed]

163. Iwabe, T.; Harada, T.; Terakawa, N. Role of cytokines in endometriosis-associated infertility. *Gynecol. Obstet. Investig.* **2002**, *53* (Suppl. 1), 19–25. [CrossRef] [PubMed]

164. Wang, C.; Ng, S.C.; Kwak-Kim, J.; Gilman-Sachs, A.; Beer, A.; Beaman, K. Increased tumor necrosis factor-alpha level in infertility patient. *Clin. Appl. Immunol. Rev.* **2002**, *3*, 6. [CrossRef]

165. Falconer, H.; Sundqvist, J.; Gemzell-Danielsson, K.; von Schoultz, B.; D'Hooghe, T.M.; Fried, G. Ivf outcome in women with endometriosis in relation to tumour necrosis factor and anti-mullerian hormone. *Reprod. Biomed. Online* **2009**, *18*, 582–588. [CrossRef]

166. Haider, S.; Knofler, M. Human tumour necrosis factor: Physiological and pathological roles in placenta and endometrium. *Placenta* **2009**, *30*, 111–123. [CrossRef] [PubMed]

167. Giannubilo, S.R.; Landi, B.; Pozzi, V.; Sartini, D.; Cecati, M.; Stortoni, P.; Corradetti, A.; Saccucci, F.; Tranquilli, A.L.; Emanuelli, M. The involvement of inflammatory cytokines in the pathogenesis of recurrent miscarriage. *Cytokine* **2012**, *58*, 50–56. [CrossRef] [PubMed]

168. Azizieh, F.Y.; Raghupathy, R.G. Tumor necrosis factor-alpha and pregnancy complications: A prospective study. *Med. Princ. Pract.* **2015**, *24*, 165–170. [CrossRef] [PubMed]

169. Briana, D.D.; Malamitsi-Puchner, A. Reviews: Adipocytokines in normal and complicated pregnancies. *Reprod. Sci.* **2009**, *16*, 921–937. [CrossRef] [PubMed]

170. Gao, L.; Gu, Y.; Yin, X. High serum tumor necrosis factor-alpha levels in women with polycystic ovary syndrome: A meta-analysis. *PLoS ONE* **2016**, *11*, e0164021. [CrossRef] [PubMed]

171. Pritts, E.A.; Parker, W.H.; Olive, D.L. Fibroids and infertility: An updated systematic review of the evidence. *Fertil Steril* **2009**, *91*, 1215–1223. [CrossRef] [PubMed]

172. Van Heertum, K.; Barmat, L. Uterine fibroids associated with infertility. *Womens Health* **2014**, *10*, 645–653. [CrossRef] [PubMed]

173. Purohit, P.; Vigneswaran, K. Fibroids and infertility. *Curr. Obstet. Gynecol. Rep.* **2016**, *5*, 81–88. [CrossRef] [PubMed]

174. Ciebiera, M.; Lukaszuk, K.; Meczekalski, B.; Ciebiera, M.; Wojtyla, C.; Slabuszewska-Jozwiak, A.; Jakiel, G. Alternative oral agents in prophylaxis and therapy of uterine fibroids-an up-to-date review. *Int. J. Mol. Sci.* **2017**, *18*, 2586. [CrossRef] [PubMed]

175. Association Pour Le Developpement En Fecondation In Vitro. Impact of Esmya on Fertility to Infertile Women with Fibroids Managed with Assisted Reproduction Techniques (NACRE). Available online: https://clinicaltrials.gov/ct2/show/NCT03349190 (accessed on 8 October 2018).

176. Tansey, M.G.; Szymkowski, D.E. The TNF superfamily in 2009: New pathways, new indications, and new drugs. *Drug Discov. Today* **2009**, *14*, 1082–1088. [CrossRef] [PubMed]

177. Reinecker, H.C.; Steffen, M.; Witthoeft, T.; Pflueger, I.; Schreiber, S.; MacDermott, R.P.; Raedler, A. Enhanced secretion of tumour necrosis factor-alpha, IL-6, and IL-1 beta by isolated lamina propria mononuclear cells from patients with ulcerative colitis and Crohn's disease. *Clin. Exp. Immunol.* **1993**, *94*, 174–181. [CrossRef]

178. Yamaoka, Y.; Kita, M.; Kodama, T.; Sawai, N.; Kashima, K.; Imanishi, J. Induction of various cytokines and development of severe mucosal inflammation by caga gene positive helicobacter pylori strains. *Gut* **1997**, *41*, 442–451. [CrossRef]

179. Wolfe, M.M.; Nompleggi, D.J. Cytokine inhibition of gastric acid secretion—A little goes a long way. *Gastroenterology* **1992**, *102*, 2177–2178. [CrossRef]

180. Tahara, T.; Shibata, T.; Okubo, M.; Ishizuka, T.; Kawamura, T.; Yamashita, H.; Nakamura, M.; Nakagawa, Y.; Nagasaka, M.; Arisawa, T.; et al. Association between interleukin-1beta and tumor necrosis factor-alpha polymorphisms and symptoms of dyspepsia. *Mol. Med. Rep.* **2015**, *11*, 3888–3893. [CrossRef]

181. Berns, M.; Hommes, D.W. Anti-tnf-alpha therapies for the treatment of crohn's disease: The past, present and future. *Expert Opin. Investig. Drugs* **2016**, *25*, 129–143. [CrossRef]

182. Ungar, B.; Levy, I.; Yavne, Y.; Yavzori, M.; Picard, O.; Fudim, E.; Loebstein, R.; Chowers, Y.; Eliakim, R.; Kopylov, U.; et al. Optimizing anti-TNF-alpha therapy: Serum levels of infliximab and adalimumab are associated with mucosal healing in patients with inflammatory bowel diseases. *Clin. Gastroenterol. Hepatol.* **2016**, *14*, 550–557. [CrossRef]

183. Kawalec, P.; Mikrut, A.; Wisniewska, N.; Pilc, A. Tumor necrosis factor-alpha antibodies (infliximab, adalimumab and certolizumab) in Crohn's disease: Systematic review and meta-analysis. *Arch. Med. Sci.* **2013**, *9*, 765–779. [CrossRef] [PubMed]

184. Scott, L.J. Etanercept: A review of its use in autoimmune inflammatory diseases. *Drugs* **2014**, *74*, 1379–1410. [CrossRef] [PubMed]

185. Korneev, K.V.; Atretkhany, K.N.; Drutskaya, M.S.; Grivennikov, S.I.; Kuprash, D.V.; Nedospasov, S.A. Tlr-signaling and proinflammatory cytokines as drivers of tumorigenesis. *Cytokine* **2017**, *89*, 127–135. [CrossRef] [PubMed]

186. Donnez, J.; Dolmans, M.M. Uterine fibroid management: From the present to the future. *Hum. Reprod. Update* **2016**, *22*, 665–686. [CrossRef] [PubMed]

187. Singh, S.S.; Belland, L.; Leyland, N.; von Riedemann, S.; Murji, A. The past, present, and future of selective progesterone receptor modulators in the management of uterine fibroids. *Am. J. Obstet. Gynecol.* **2018**, *218*, 563–572. [CrossRef]

188. Murji, A.; Whitaker, L.; Chow, T.L.; Sobel, M.L. Selective progesterone receptor modulators (SPRMs) for uterine fibroids. *Cochrane Database Syst. Rev.* **2017**, *4*, CD010770. [CrossRef]

189. Koninckx, P.R.; Craessaerts, M.; Timmerman, D.; Cornillie, F.; Kennedy, S. Anti-TNF-alpha treatment for deep endometriosis-associated pain: A randomized placebo-controlled trial. *Hum. Reprod.* **2008**, *23*, 2017–2023. [CrossRef]

190. Lu, D.; Song, H.; Shi, G. Anti-TNF-alpha treatment for pelvic pain associated with endometriosis. *Cochrane Database Syst. Rev.* **2013**, CD008088. [CrossRef]

191. Essayan, D.M. Cyclic nucleotide phosphodiesterases. *J. Allergy Clin. Immunol.* **2001**, *108*, 671–680. [CrossRef]

192. Deree, J.; Martins, J.O.; Melbostad, H.; Loomis, W.H.; Coimbra, R. Insights into the regulation of tnf-alpha production in human mononuclear cells: The effects of non-specific phosphodiesterase inhibition. *Clinics* **2008**, *63*, 321–328. [CrossRef]

193. Kamencic, H.; Thiel, J.A. Pentoxifylline after conservative surgery for endometriosis: A randomized, controlled trial. *J. Minim. Invasive Gynecol.* **2008**, *15*, 62–66. [CrossRef] [PubMed]

194. Brown, J.; Crawford, T.J.; Allen, C.; Hopewell, S.; Prentice, A. Nonsteroidal anti-inflammatory drugs for pain in women with endometriosis. *Cochrane Database Syst. Rev.* **2017**, *1*, CD004753. [CrossRef] [PubMed]

195. Ferrero, S.; Evangelisti, G.; Barra, F. Current and emerging treatment options for endometriosis. *Expert Opin. Pharmacother.* **2018**, *19*, 1109–1125. [CrossRef] [PubMed]

196. Chwalisz, K.; Taylor, H. Current and emerging medical treatments for uterine fibroids. *Semin. Reprod. Med.* **2017**, *35*, 510–522. [CrossRef] [PubMed]

197. Luqman, S.; Meena, A.; Marler, L.E.; Kondratyuk, T.P.; Pezzuto, J.M. Suppression of tumor necrosis factor-alpha-induced nuclear factor kappaB activation and aromatase activity by capsaicin and its analog capsazepine. *J. Med. Food* **2011**, *14*, 1344–1351. [CrossRef] [PubMed]

198. Ishikawa, H.; Reierstad, S.; Demura, M.; Rademaker, A.W.; Kasai, T.; Inoue, M.; Usui, H.; Shozu, M.; Bulun, S.E. High aromatase expression in uterine leiomyoma tissues of african-american women. *J. Clin. Endocrinol. Metab* **2009**, *94*, 1752–1756. [CrossRef] [PubMed]

199. Purohit, A.; Singh, A.; Ghilchik, M.W.; Reed, M.J. Inhibition of tumor necrosis factor alpha-stimulated aromatase activity by microtubule-stabilizing agents, paclitaxel and 2-methoxyestradiol. *Biochem. Biophys. Res. Commun.* **1999**, *261*, 214–217. [CrossRef] [PubMed]

200. To, S.Q.; Knower, K.C.; Clyne, C.D. Origins and actions of tumor necrosis factor alpha in postmenopausal breast cancer. *J. Interferon Cytokine Res.* **2013**, *33*, 335–345. [CrossRef] [PubMed]

201. To, S.Q.; Knower, K.C.; Clyne, C.D. NFkappaB and MAPK signalling pathways mediate tnfalpha-induced early growth response gene transcription leading to aromatase expression. *Biochem. Biophys. Res. Commun.* **2013**, *433*, 96–101. [CrossRef] [PubMed]

202. Khan, K.N.; Kitajima, M.; Hiraki, K.; Fujishita, A.; Nakashima, M.; Ishimaru, T.; Masuzaki, H. Cell proliferation effect of gnrh agonist on pathological lesions of women with endometriosis, adenomyosis and uterine myoma. *Hum. Reprod.* **2010**, *25*, 2878–2890. [CrossRef]

203. Taniguchi, F.; Higaki, H.; Azuma, Y.; Deura, I.; Iwabe, T.; Harada, T.; Terakawa, N. Gonadotropin-releasing hormone analogues reduce the proliferation of endometrial stromal cells but not endometriotic cells. *Gynecol. Obstet. Investig.* **2013**, *75*, 9–15. [CrossRef] [PubMed]

204. Sakamoto, Y.; Harada, T.; Horie, S.; Iba, Y.; Taniguchi, F.; Yoshida, S.; Iwabe, T.; Terakawa, N. Tumor necrosis factor-alpha-induced interleukin-8 (IL-8) expression in endometriotic stromal cells, probably through nuclear factor-kappa B activation: Gonadotropin-releasing hormone agonist treatment reduced il-8 expression. *J. Clin. Endocrinol. Metab.* **2003**, *88*, 730–735. [CrossRef] [PubMed]

205. Guzman-Soto, I.; Salinas, E.; Quintanar, J.L. Leuprolide acetate inhibits spinal cord inflammatory response in experimental autoimmune encephalomyelitis by suppressing NF-kappaB activation. *Neuroimmunomodulation* **2016**, *23*, 33–40. [CrossRef]

206. Courtoy, G.E.; Donnez, J.; Marbaix, E.; Dolmans, M.M. In vivo mechanisms of uterine myoma volume reduction with ulipristal acetate treatment. *Fertil Steril* **2015**, *104*, 426–434. [CrossRef] [PubMed]

207. Courtoy, G.E.; Donnez, J.; Ambroise, J.; Arriagada, P.; Luyckx, M.; Marbaix, E.; Dolmans, M.M. Gene expression changes in uterine myomas in response to ulipristal acetate treatment. *Reprod. Biomed. Online* **2018**, *37*, 224–233. [CrossRef] [PubMed]

208. Whitaker, L.H.; Murray, A.A.; Matthews, R.; Shaw, G.; Williams, A.R.; Saunders, P.T.; Critchley, H.O. Selective progesterone receptor modulator (sprm) ulipristal acetate (upa) and its effects on the human endometrium. *Hum. Reprod.* **2017**, *32*, 531–543. [CrossRef]

209. Bedaiwy, M.A.; Falcone, T.; Sharma, R.K.; Goldberg, J.M.; Attaran, M.; Nelson, D.R.; Agarwal, A. Prediction of endometriosis with serum and peritoneal fluid markers: A prospective controlled trial. *Hum. Reprod.* **2002**, *17*, 426–431. [CrossRef]

210. Parker, W.; Berek, J.S.; Pritts, E.; Olive, D.; Kaunitz, A.M.; Chalas, E.; Clarke-Pearson, D.; Goff, B.; Bristow, R.; Taylor, H.S.; et al. An open letter to the food and drug administration regarding the use of morcellation procedures in women having surgery for presumed uterine myomas. *J. Minim. Invasive Gynecol.* **2016**, *23*, 303–308. [CrossRef]

211. Brakta, S.; Diamond, J.S.; Al-Hendy, A.; Diamond, M.P.; Halder, S.K. Role of vitamin d in uterine fibroid biology. *Fertil Steril* **2015**, *104*, 698–706. [CrossRef]

International Journal of
Molecular Sciences

MDPI

Article

Protein S is Protective in Acute Lung Injury by Inhibiting Cell Apoptosis

Prince Baffour Tonto [1,†], Taro Yasuma [1,2,†], Tetsu Kobayashi [3,†],
Corina N. D'Alessandro-Gabazza [1,*,†], Masaaki Toda [1,†], Haruko Saiki [3], Hajime Fujimoto [3],
Kentaro Asayama [3], Kentaro Fujiwara [3], Kota Nishihama [2], Tomohito Okano [3],
Atsuro Takeshita [1,2] and Esteban C. Gabazza [1]

[1] Department of Immunology, Mie University, Graduate School of Medicine Mie, Edobashi 2-174, Tsu,
  Mie 514-8507, Japan; 316MS02@m.mie-u.ac.jp (P.B.T.); t-yasuma0630@clin.medic.mie-u.ac.jp (T.Y.);
  t-masa@doc.medic.mie-u.ac.jp (M.T.); johnpaul0114@yahoo.co.jp (A.T.);
  gabazza@doc.medic.mie-u.ac.jp (E.C.G.)
[2] Department of Diabetes and Endocrinology, Mie University, Graduate School of Medicine Mie,
  Edobashi 2-174, Tsu, Mie 514-8507, Japan; kn2480@gmail.com
[3] Department of Pulmonary and Critical Care Medicine, Mie University, Graduate School of Medicine Mie,
  Edobashi 2-174, Tsu, Mie 514-8507, Japan; ktetsu@clin.medic.mie-u.ac.jp (T.K.);
  mchharuko@city-hosp.matsusaka.mie.jp (H.S.); genfujimoto1974@yahoo.co.jp (H.F.);
  longwan@yahoo.co.jp (K.A.); kentaro-fu@clin.medic.mie-u.ac.jp (K.F.); okat-omo-525@live.jp (T.O.)
* Correspondence: immunol@doc.medic.mie-u.ac.jp; Tel.: +81-59-231-5037; Fax: +81-59-231-5225
† These authors equally contributed for the completion of this work.

Received: 10 February 2019; Accepted: 27 February 2019; Published: 2 March 2019

Abstract: Acute lung injury is a fatal disease characterized by inflammatory cell infiltration, alveolar-capillary barrier disruption, protein-rich edema, and impairment of gas exchange. Protein S is a vitamin K-dependent glycoprotein that exerts anticoagulant, immunomodulatory, anti-inflammatory, anti-apoptotic, and neuroprotective effects. The aim of this study was to evaluate whether human protein S inhibits cell apoptosis in acute lung injury. Acute lung injury in human protein S transgenic and wild-type mice was induced by intratracheal instillation of lipopolysaccharide. The effect of human protein S on apoptosis of lung tissue cells was evaluated by Western blotting. Inflammatory cell infiltration, alveolar wall thickening, myeloperoxidase activity, and the expression of inflammatory cytokines were reduced in human protein S transgenic mice compared to the wild-type mice after lipopolysaccharide instillation. Apoptotic cells and caspase-3 activity were reduced while phosphorylation of extracellular signal-regulated kinase was enhanced in the lung tissue from human protein S transgenic mice compared to wild-type mice after lipopolysaccharide instillation. The results of this study suggest that human protein S is protective in lipopolysaccharide-induced acute lung injury by inhibiting apoptosis of lung cells.

Keywords: acute lung injury; protein S; apoptosis; signal pathway; Erk1/2; lipopolysaccharide

1. Introduction

Acute lung injury (ALI) and its severe form, acute respiratory distress syndrome (ARDS), are non-cardiogenic pulmonary edema, clinically defined as a severe dysfunction of gas exchange and chest radiographic abnormalities in the absence of heart failure [1,2]. Several conditions including severe sepsis, trauma, and ischemia/reperfusion injury may cause ALI/ARDS [1]. Multiple therapeutic modalities are being used including lung-protective ventilation strategies, prone position and fluid-conservative therapy, but most of them are ineffective in controlling the disease [1,3,4]. The mortality rate of patients with ARDS remains very high ranging between 35% and 46% [3–5].

Therefore, development of new therapeutic strategies is urgently required. The pathogenesis of the disease is not completely clear but apoptosis of alveolar epithelial cells is considered to play a critical role [6,7]. Enhanced apoptosis of lung epithelial cells leads to damage and rupture of the pulmonary epithelial/endothelial barrier [8]. Lipopolysaccharide (LPS) that accumulates during sepsis by Gram-negative bacteria is an important cause of lung epithelial cell apoptosis [9,10]. The relevance of cell apoptosis in the mechanism of ALI has been demonstrated by studies showing upregulation of the Fas/FasL system with activation of pro-apoptotic signal pathways in the lung from patients with ALI [7,11]. Inhibition of caspases, which are essential enzymes for the process of cell apoptosis, has been reported to prolong the survival rate in experimental LPS-induced acute lung injury, and inhibition of pro-apoptotic signaling pathways attenuates LPS-induced apoptosis and release of inflammatory cytokines in rat type II alveolar epithelial cells [12,13]. Apart from supporting the role of apoptosis in ALI, these studies point to apoptosis of lung cells as a potential target for therapeutic intervention in ALI/ARDS.

Protein S (PS) is a vitamin K-dependent glycoprotein that has anticoagulant action by enhancing several-fold the inhibitory activity of APC on blood coagulation, by directly blocking the activity of prothrombinase complex, tenase, or by stimulating the inhibition of the tissue factor inhibitory pathway [14,15]. Apart from its anticoagulant activity, PS can also independently exert anti-inflammatory effect [16]. Previously we demonstrated that administration of PS in mice can attenuate cell infiltration and the expression of inflammatory cytokines in LPS-induced ALI [16]. PS also has anti-apoptotic activity [17,18]. PS reduces bleomycin-induced pulmonary fibrosis by inhibiting apoptosis of lung epithelial cells and ameliorates streptozotocin-induced diabetes by suppressing apoptosis of pancreatic β-cells [17,18]. However, the strong anti-apoptotic activity of PS can be also detrimental; for example, previous studies have shown that administration of PS worsens acute liver injury induced by alcohol and chronic liver injury and liver fibrosis caused by carbon tetrachloride [19,20].

In the present study, we hypothesized that overexpression of PS will attenuate ALI not only by its anti-inflammatory activity but also by its inhibitory activity on cell apoptosis. To demonstrate this hypothesis, in this study we evaluated and compared the inflammatory and apoptosis-related markers between wild-type mice and mice overexpressing human PS in the lungs.

2. Results

2.1. Systemic and Lung Upregulation of Protein S (PS) During Lipopolysaccharide (LPS)-Induced ALI

As expected, the antigen concentration of PS was significantly increased in plasma and lung tissue samples from both hPS-TG/SAL and hPS-TG/LPS mice compared to samples from their WT counterparts. The plasma concentration of PS was significantly higher in the hPS-TG/LPS group than in the hPS-TG/SAL group. The mRNA expression of hPS was positive in both hPS-TG groups but it was not detected in both WT groups (Figure 1). No difference in the mRNA expression of mouse PS was observed between groups.

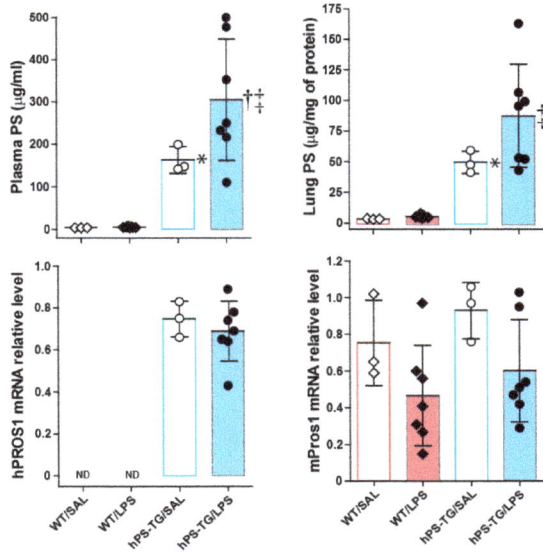

Figure 1. Systemic and local upregulation of Protein S (PS) during lipopolysaccharide (LPS)-induced acute lung injury (ALI). Mice were allocated in four groups including wild type mice treated with saline (WT/SAL, $n = 3$) or LPS (WT/LPS, $n = 7$), and human PS transgenic (TG) mice treated with saline (hPS TG/SAL, $n = 3$) or LPS (hPS TG/LPS, $n = 7$). The level of PS was measured by enzyme immunoassay and the gene expression by reverse transcription polymerase chain reaction (RT-PCR). Data are expressed as the mean ± SD. * $p < 0.05$: hPS-TG/SAL vs WT/SAL; † $p < 0.05$: hPS-TG/LPS vs hPS-TG/SAL; ‡ $p < 0.05$: hPS-TG/LPS vs WT/LPS. ND, not detected.

2.2. Less Lung Cell Infiltration in Mice Overexpressing Human Protein S (hPS)

The total number of infiltrating cells and the total number of neutrophils in BALF were significantly increased in WT/LPS and hPS-TG/LPS group compared to WT/SAL and hPS-TG/SAL groups, respectively. The total number of cells and the total number of neutrophils were decreased in the hPS-TG/LPS group compared to the WT/LPS group but the differences were not statistically significant (Figure 2A). There were not significant differences in the count of macrophages and lymphocytes among groups (Figure 2A).

The number of infiltrating inflammatory cells in lung tissue was significantly increased in lung tissue from hPS-TG/LPS and WT/LPS mice compared to mice receiving intratracheal saline, but it was significantly decreased in hPS-TG/LPS mice compared to WT/LPS mice (Figure 2B,C).

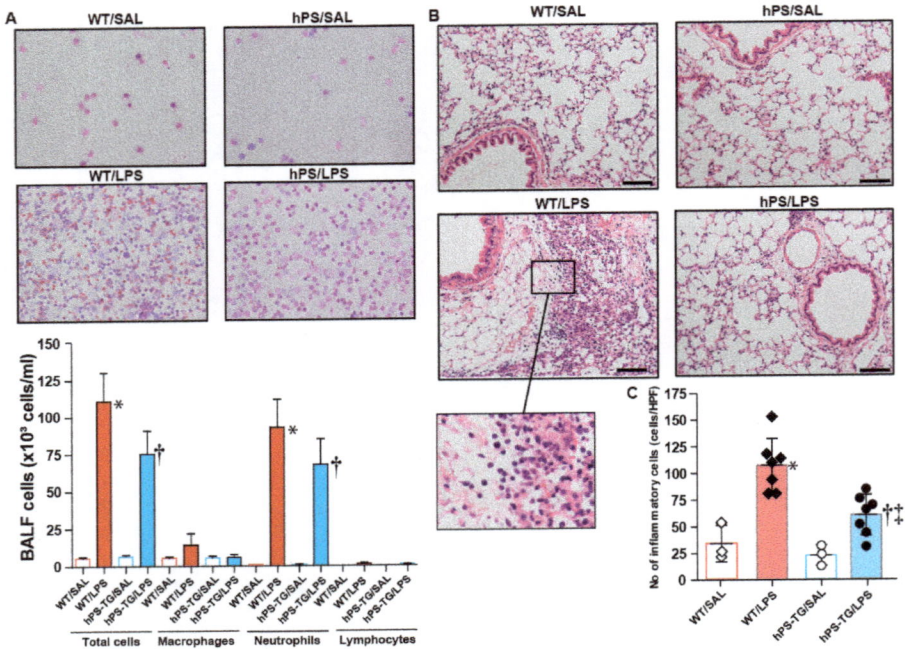

Figure 2. Less lung cell infiltration in mice overexpressing hPS. Mice were allocated in four groups including wild type mice treated with saline (WT/SAL, *n* = 3) or LPS (WT/LPS, *n* = 7), and human PS transgenic (TG) mice treated with saline (hPS TG/SAL, *n* = 3) or LPS (hPS TG/LPS, *n* = 7). Cells in bronchoalveolar lavage fluid were counted using automatic cell counter and stained for differential counting (**A**). Lung tissue samples were stained with hematoxylin & eosin (**B**). The number of cells was counted using the WindROOF image processing software (**C**). Scale bars indicate 100 μm. Data are expressed as the mean ± SD. * *p* < 0.05: WT/LPS vs WT/SAL; † *p* < 0.05: hPS-TG/LPS vs hPS-TG/SAL; ‡ *p* < 0.05: hPS-TG/LPS vs WT/LPS.

2.3. The Coagulation System Was Not Affected by hPS Overexpression

No significant difference in the plasma concentration of TAT was found between WT/LPS and hPS-TG/LPS groups or between hPS-TG/SAL and hPS-TG/LPS groups (Figure 3). The BALF concentration of TAT was significantly higher in hPS-TG/LPS group compared to hPS-TG/SAL group but not between WT/SAL and WT/LPS groups. The lung tissue concentration of TAT was significantly increased in WT/LPS and hPS-TG/LPS groups compared to WT/SAL and hPS-TG/SAL groups, respectively (Figure 3). No significant difference in the plasma, BALF and lung tissue concentrations of TAT was observed between WT/LPS and hPS-TG/LPS groups (Figure 3).

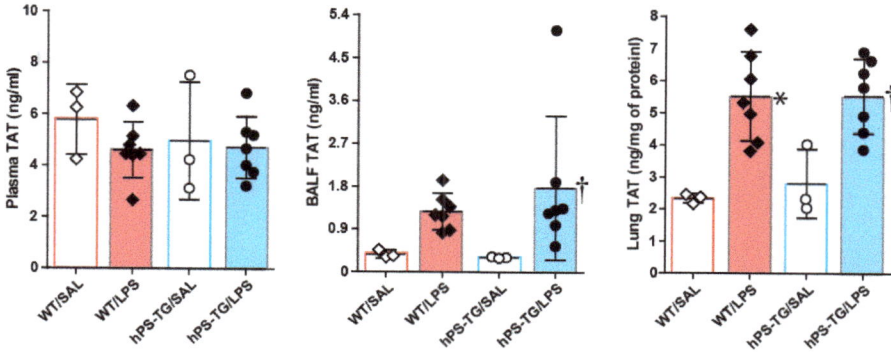

Figure 3. Activation of the coagulation system was inhibited by hPS overexpression. Mice were allocated in four groups including wild type mice treated with saline (WT/SAL, $n = 3$) or LPS (WT/LPS, $n = 7$), and human PS transgenic (TG) mice treated with saline (hPS TG/SAL, $n = 3$) or LPS (hPS TG/LPS, $n = 7$). The concentration of thrombin–antithrombin (TAT) was measured using an enzyme immunoassay. Data are expressed as the mean $\pm$ SD. * $p < 0.05$: WT/LPS vs WT/SAL; † $p < 0.05$: hPS-TG/LPS vs hPS-TG/SAL.

2.4. Suppression of Pro-Inflammatory Markers by hPS Overexpression

The plasma levels of MCP-1, MPO, TNF-α and IL-6 were significantly higher in WT/LPS mice than in WT/SAL mice but there was no difference between hPS-TG/LPS and hPS-TG/SAL groups. The plasma levels of MCP-1 and MPO were significantly decreased in hPS-TG/LPS group compared to their WT counterpart (Figure 4A).

The BALF levels of MPO, TNF-α and IL-6 were significantly higher in the WT/LPS group than in the WT/SAL group but no difference was observed between hPS-TG/LPS and hPS-TG/SAL groups. The BALF levels of MPO, TNF-α and IL-6 were significantly reduced in the hPS-TG/LPS group compared to its WT counterpart (Figure 4B). The BALF level of MCP-1 was not significantly different between groups.

The lung tissue levels of MCP-1, MPO, TNF-α and IL-6 were significantly increased in the WT/LPS group compared to the WT/SAL group but they were significantly decreased in the hPS-TG/LPS group compared to the WT/LPS group (Figure 4C). No difference was found between hPS-TG/LPS and hPS-TG//SAL groups.

The protein concentration of total protein in BALF was significantly ($p < 0.05$) enhanced in both WT/LPS (339.1 ± 52.0 µg/mL), and hPS-TG/LPS (321.1 ± 84.3 µg/mL) groups compared to the WT/SAL (193.1 ± 37.2 µg/mL) and hPS-TG/SAL (129.0 ± 27.2 µg/mL) groups but there was no significant difference between WT/LPS and hPS-TG/LPS groups.

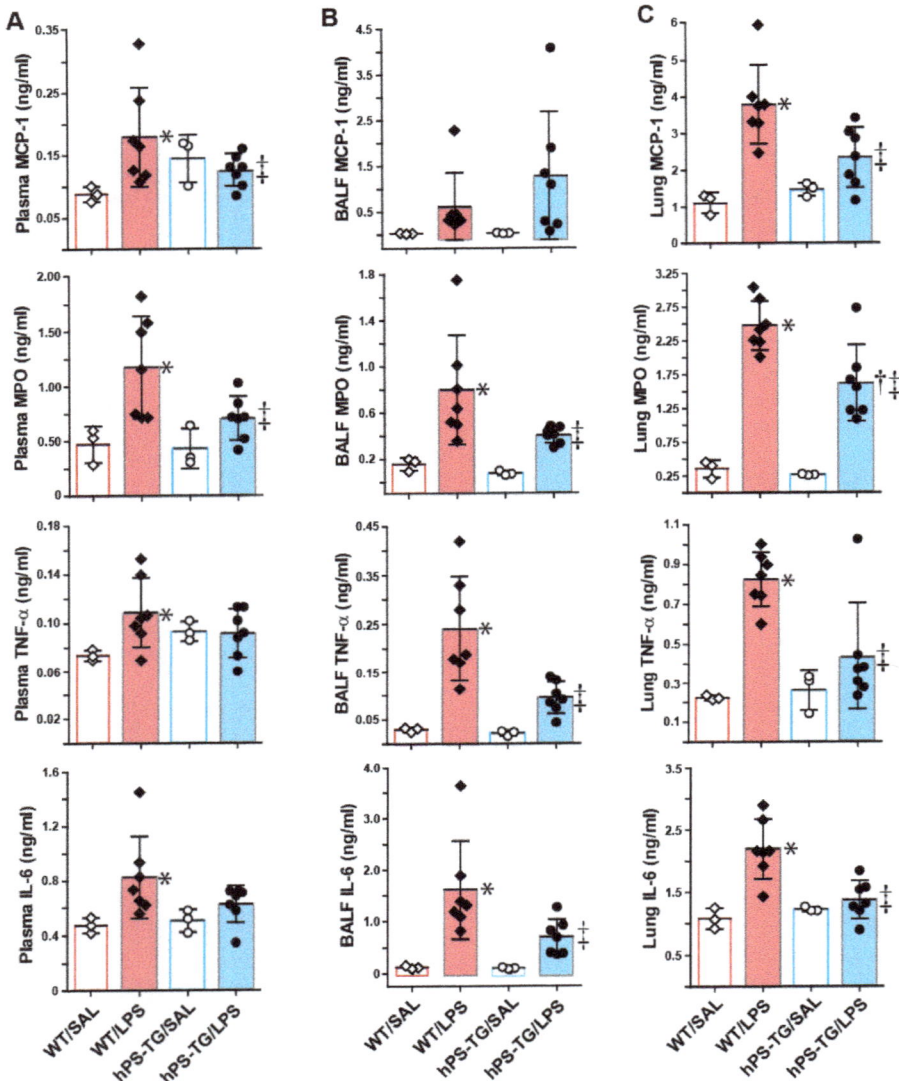

Figure 4. Suppression of pro-inflammatory markers by hPS overexpression. Mice were allocated in four groups including wild type mice treated with saline (WT/SAL, *n* = 3) or LPS (WT/LPS, *n* = 7), and human PS transgenic (TG) mice treated with saline (hPS TG/SAL, *n* = 3) or LPS (hPS TG/LPS, *n* = 7). The plasma (**A**), bronchoalveolar lavage fluid (**B**) and lung tissue (**C**) concentrations of cytokines and chemokines were measured using enzyme immunoassays and myeloperoxidase was measured using a colorimetric assay as described under materials and methods. Data are expressed as the mean ± SD. * *p* < 0.05: WT/LPS vs WT/SAL; † *p* < 0.05: hPS-TG/LPS vs hPS-TG/SAL; ‡ *p* < 0.05: hPS-TG/LPS vs WT/LPS.

2.5. Decreased Lung Apoptotic Cells in Mice Overexpressing hPS

The number of TUNEL (+) cells in lung tissue was significantly increased in the WT/LPS group compared to the WT/SAL group. There was not significant statistical difference between hPS-TG/LPS and hPS-TG/SAL groups in the number of TUNEL (+) cells. The number of TUNEL (+) cells in lung tissue was significantly increased in WT/LPS group compared to hPS-TG/LPS group (Figure 5A,B).

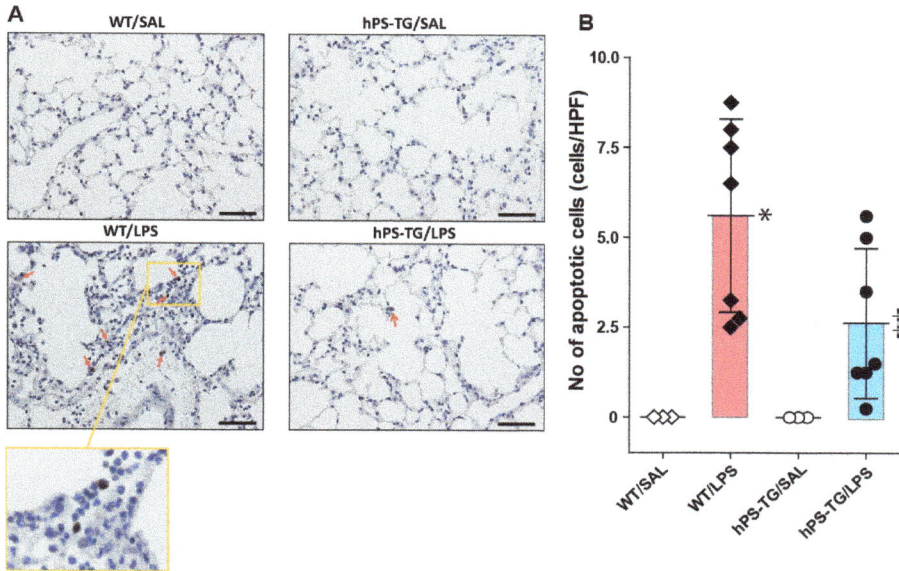

Figure 5. Decreased lung apoptotic cells in mice overexpressing hPS. Mice were allocated in four groups including wild type mice treated with saline (WT/SAL, *n* = 3) or LPS (WT/LPS, *n* = 7), and human PS transgenic (TG) mice treated with saline (hPS TG/SAL, *n* = 3) or LPS (hPS TG/LPS, *n* = 7). Staining of terminal deoxynucleotidyl transferase dUTP nick-end labeling (TUNEL) was performed as described under materials and methods (**A**) and the number of apoptotic cells was counted using WindROOF image processing software (**B**). Arrows indicate apoptotic cells. Scale bars indicate 100 μm. Data are expressed as the mean ± SD. *$p < 0.05$: WT/LPS vs WT/SAL; ‡ $p < 0.05$: hPS-TG/LPS vs WT/LPS.

2.6. The Expression of Apoptotic Factors Is Regulated by hPS

The mRNA expression of the markers of apoptosis inhibition Bcl2 and Bcl-xl and both the Bcl2/Bax and Bcl-xl/Bax ratios significantly increased in the hPS-TG/LPS group compared to the WT/LPS group (Figure 6). The mRNA expression of Bcl-xl and both the Bcl2/Bax and Bcl-xl/Bax ratios significantly decreased in the WT/LPS group compared to the WT/SAL group (Figure 6). The mRNA expression of Bcl2 decreased, but not at a statistically significant level, in the WT/LPS group compared to the WT/SAL group (Figure 6).

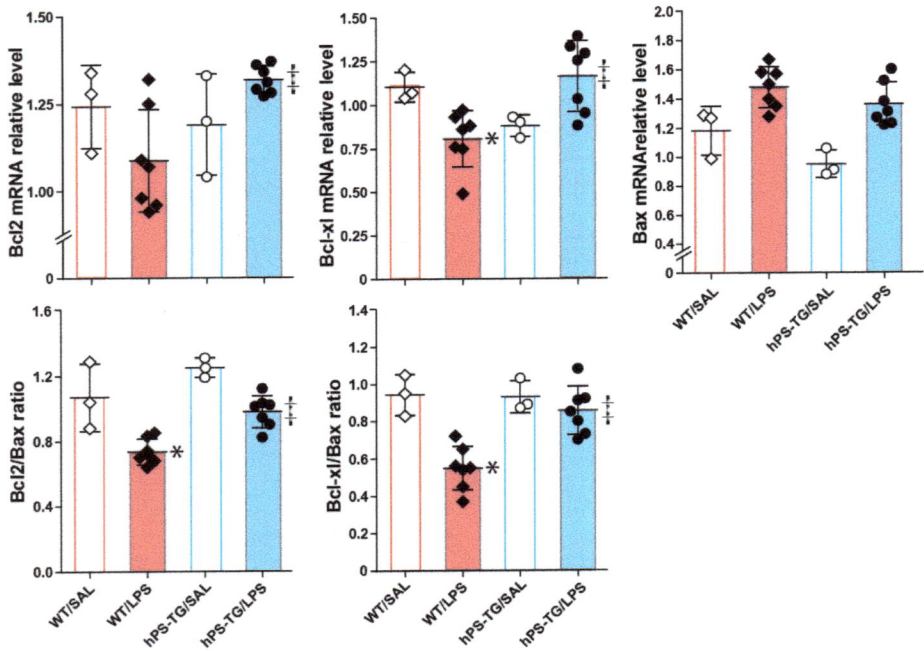

Figure 6. Regulation of the expression of apoptotic factors by hPS. Mice were allocated in four groups including wild-type mice treated with saline (WT/SAL, *n* = 3) or LPS (WT/LPS, *n* = 7), and human PS transgenic (TG) mice treated with saline (hPS TG/SAL, *n* = 3) or LPS (hPS TG/LPS, *n* = 7). Total RNA was extracted from the whole lung tissue, reverse-transcribed and then PCR was performed using gene-specific primers. The mRNA expression of the anti-apoptotic factors Bcl2 and Bcl-xl and of the apro-apoptotic factor was evaluated by PCR. Data are expressed as the mean ± SD. *$p < 0.05$: WT/LPS vs WT/SAL; ‡ $p < 0.05$: hPS-TG/LPS vs WT/LPS.

2.7. Decreased Activation of Caspase-3 in Mice Overexpressing hPS

Western blotting of lung tissue showed that the cleaved form of caspase-3, a marker of cell apoptosis, is increased in the WT/LPS group compared to the WT/SAL and hPS-TG/LPS groups. There was no difference between hPS-TG/LPS and hPS-TG/SAL groups (Figure 7A).

The activation of Erk1/2 was significantly increased in the hPS-TG/LPS group compared to the WT/LPS group but no statistical difference was found between other groups (Figure 7B).

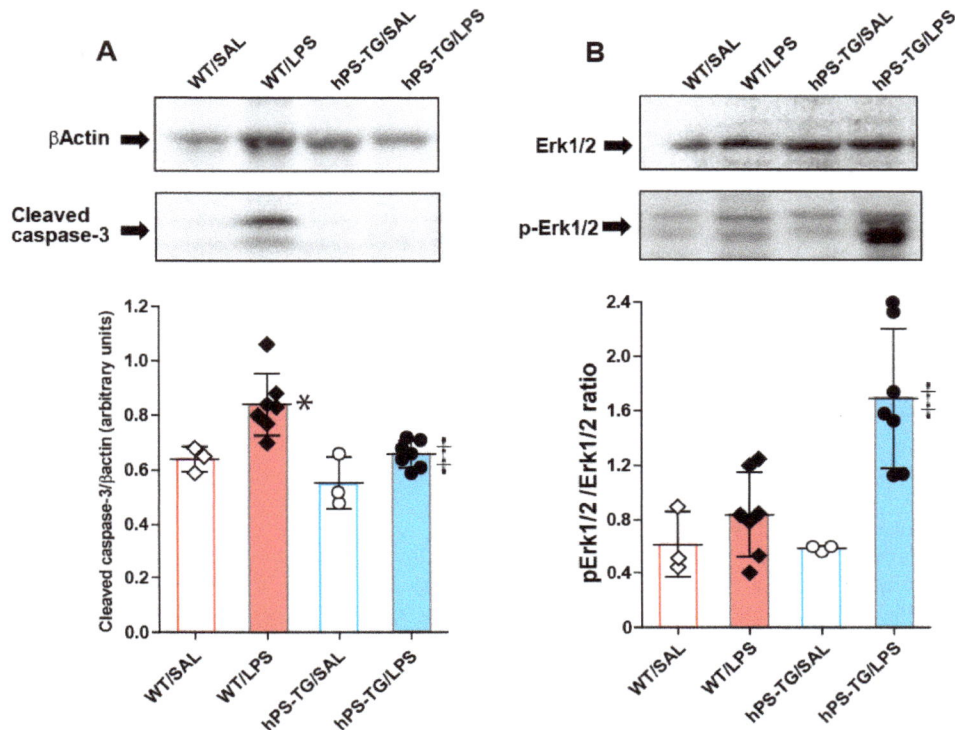

Figure 7. Decreased cleavage of caspase-3 and increased activation of Erk1/2 in mice overexpressing hPS. Mice were allocated in four groups including wild-type mice treated with saline (WT/SAL, *n* = 3) or LPS (WT/LPS, *n* = 7), and human PS transgenic (TG) mice treated with saline (hPS TG/SAL, *n* = 3) or LPS (hPS TG/LPS, *n* = 7). The presence of cleaved caspase-3 (**A**) and p-Erk1/2 (**B**) in lung tissue was evaluated by Western blotting. Quantification was undertaken using ImageJ (https://imagej.nih.gov/ij/index.html). Data are expressed as the mean ± SD. * *p* < 0.05: WT/LPS vs WT/SAL; ‡ *p* < 0.05: hPS-TG/LPS vs WT/LPS. p-Erk1/2, phosphorylated-Erk1/2.

3. Discussion

The results of the present study show that overexpression of PS in the lungs attenuates ALI by inhibiting apoptosis of lung cells. PS is a 69 KDa glycoprotein expressed by multiple cells including hepatocytes, vascular endothelial cells, lymphocytes and lung cells [14]. Besides its anticoagulant effect, PS can also inhibit the inflammatory response, the complement system, and prolong cell survival. A previous study has shown that administration of hPS ameliorates LPS-induced ALI in mice by suppressing the mRNA and protein expressions of pro-inflammatory cytokines and chemokines from alveolar epithelial cells and macrophages, suggesting the therapeutic benefits of the anti-inflammatory activity of hPS in ALI [16]. Consistent with these findings, here we found significantly decreased inflammatory cells in lung tissue and significantly decreased lung concentrations of MCP-1, TNF-α and IL-6 with reduced release of MPO in mice overexpressing hPS compared to their WT counterparts after inhalation of LPS, further supporting the favorable effect of hPS-mediated inhibition of inflammation. It is worth noting that there was no significant difference in the BALF concentration of protein between WT and hPS-TG treated with LPS, suggesting that hPS overexpression exerts no effect on the epithelial/endothelial barrier. In addition, the beneficial effect of hPS was not due to its anticoagulant effect because the level of TAT, a marker of coagulation activation, was not decreased in the hPS-TG/LPS group compared to the WT/LPS group.

Previous studies have shown that apoptosis of endothelial and alveolar epithelial cells plays a critical role in the pathogenesis of ALI [21–23]. Enhanced cell apoptosis may be detrimental because abnormal accumulation of apoptotic cells can perpetuate the inflammatory response by the increased release of inflammatory mediators, proteases, reactive oxygen species or lysozymes [24]. Therefore, inhibition of cell apoptosis may prevent tissue damage, dampen the immune response and accelerate tissue resolution and healing during acute inflammatory conditions including ALI [24]. Previous studies have shown that PS may directly inhibit the cell process of apoptosis by binding to the TAM (Axl, Mer, Tyro3) family of receptor tyrosine kinases. Interaction of PS with TAM receptors activates the signaling pathways of phosphoinosotide 3-kinase(PI3k)/Akt and the mitogen-activated protein kinase/extracellular signal-regulated kinase(MAPK/Erk) that promote cell growth and inhibit cell apoptosis [25–28]. In addition, we previously demonstrated that hPS inhibits apoptosis by activating the PI3K/Akt pathway in a variety of cells including lung epithelial cells [17–20]. However, to date whether hPS also suppresses apoptosis of lung cells in ALI remains unknown. In the present study, we demonstrated that the number of lung apoptotic cells and the cleavage of caspase-3 are decreased and the activation of Erk1/2 is increased in mice overexpressing hPS compared to WT after induction of ALI by intratracheal instillation of LPS. In addition, the expression of the anti-apoptotic factors Bcl2, Bcl-xl and the Bcl2/Bax and Bcl2-xl/Bax ratios were significantly enhanced in mice overexpressing hPS compared to their WT counterpart after treatment with LPS. These observations suggest that hPS inhibits apoptosis of lung cells in ALI. It is worth noting here that inhibition of apoptosis by hPS may be detrimental in some acute or chronic diseases in which its target cells play a pathological role. For example, there is evidence showing that hPS may worsen acute alcoholic liver injury by inhibiting apoptosis of activated natural killer T cells, enhance liver fibrosis by prolonging the survival of fibrogenic stellate cells or promote cancer progression and metastasis by providing a survival advantage to cancer cells [19,20,29]. However, beneficial effects of hPS have been reported in animal models of acute and chronic disorders of the lungs including ALI and pulmonary fibrosis [16,17]. The differential expression level of hPS receptors on the cell surface may explain these organ-dependent differential effects of hPS [30].

The low power of the experiment due to the small number of animals allocated in each experimental group is a limitation of the present study and, therefore, the reported statistical differences and related clinical implications should be interpreted with caution. However, this is the first report showing that hPS inhibits apoptosis of lung cells in ALI, thereby suggesting the potential beneficial effect of hPS in this devastating disease. Further studies should be undertaken to corroborate these findings.

4. Materials and Methods

4.1. Experimental Animals

Characterization of the homozygous human (h) PS transgenic (TG) mouse on a C57BL/6 background has been previously reported [19]. Wild-type (WT) C57BL/6 mice were used as controls. Male mice (8- to 10-week old) were used in the experiments. All animals were maintained in a specific pathogen-free environment and subjected to a 12 h light:dark cycle in the animal house of Mie University. Genotyping of hPS mice was performed by polymerase chain reaction (PCR) using DNA extracted from the tails and by measuring the plasma concentration of PS antigen as previously described [19]. The Committee for Animal Investigation of Mie University approved the protocols(Approval No: 24-50, Date: 2016/02/01) of the study and all animal procedures were performed in accordance with the institutional guidelines of Mie University and following the internationally approved principles of laboratory animal care published by the National Institute of Health (https://olaw.nih.gov/).

4.2. Acute Lung Injury (ALI) Induction

Lung damage was induced by intratracheal instillation of LPS (5 mg/kg; Sigma-Aldrich, St. Louis, MO, USA). Mice were categorized into the following groups: WT/LPS ($n = 7$) and hPS-TG/LPS ($n = 7$) groups that received intratracheal instillation of LPS dissolved in physiological saline (SAL) (75 µL), and WT/SAL ($n = 3$) and hPS-TG/SAL ($n = 3$) groups that received intratracheal instillation of saline (75 µL). The instillation of LPS and saline was performed under anesthesia by intraperitoneal injection of 62.5 mg/kg sodium pentobarbital. Mice were sacrificed 24h after LPS instillation by an overdose of intraperitoneal pentobarbital.

4.3. Collection of Samples

After mouse euthanasia, bronchoalveolar lavage fluid (BALF) was sampled as previously described and blood samples were collected by heart puncture and placed in tubes containing 10 U/mL heparin [17]. The total number of cells in BALF was measured using a nucleocounter from ChemoMetec (Allerød, Denmark). The BALF supernatant was separated by centrifugation and stored at −80 °C until use for biochemical analysis. For differential cell counting BALF was centrifuged using a cytospin and the cells were stained with May–Grunwald–Giemsa (Merck, Darmstadt, Germany).

4.4. Histological Examination

Mice were thoracotomized under profound anesthesia, the pulmonary circulation flushed with saline and then the lungs were excised. One of the lungs was perfused with 10% neutral buffered formalin, fixed in formalin for 24 h and then embedded in paraffin. Tissue specimens of 5 µm were prepared, stained with hematoxylin-eosin and then examined under a light microscopy (Olympus BX53 microscope; Tokyo, Japan). An investigator blinded to the treatment group counted the number of inflammatory cells in a blinded fashion; at least five microscopic fields were counted per mouse. Cell counting in the histological sections was performed using the Olympus BX53 microscope with a plan objective, combined with an Olympus DP70 digital camera (Tokyo, Japan) and the WinROOF image processing software (Mitani Corp., Fukui, Japan) for Windows.

4.5. Western Blotting

Activation of ERK in lung tissue was evaluated by Western blotting following standard methods using specific antibodies purchased from Cell Signaling (Danvers, MA, USA). Apoptosis was evaluated by measuring the cleaved form of caspase-3 (Cell Signaling) by Western blotting using specific antibodies.

4.6. Immunohistochemistry

DNA fragmentation in lung tissue was evaluated by the terminal deoxynucleotidyl transferase-mediated dUTP nick end labeling (TUNEL) method. TUNEL staining was performed at the Biopathology institute Corporation using terminal deoxynucleotidyl transferase enzyme (Millipore Sigma, St. Louis, MO, USA), anti-digoxigenin peroxidase (Millipore Sigma, St. Louis, MO, USA) and 3,3′-diaminobenzidine. The total number of TUNEL (+) cells was counted using the WinROOF image processing software (Mitani Corporation, Osaka, Japan).

4.7. Biochemical Analysis

Lung tissue was homogenized in the presence of a protease inhibitor cocktail (Nacalai Tesque Inc.; Kyoto, Japan) using Tomy Micro Smash™ MS-100R (Tomy Digital Biology Co., Tokyo, Japan), the preparation was then centrifuged at 15,000 rpm for 10 min and supernatants collected and stored at −80 °C until analysis. The concentration of PS was measured using polyclonal antihuman PS antibody (Dako, Santa Clara, CA, USA) as capture antibody and biotin-labeled monoclonal antihuman PS antibody (Haematologic Technologies Inc., Esset, VT) as second antibody; absolute values

Int. J. Mol. Sci. **2019**, *20*, 1082

were extrapolated from a standard curve drawn using PS antigen (Enzyme Research Laboratories, South Bend, IN, USA). The detection limit of this PS immunoassay is 0.8 ng/mL. The polyclonal anti-hPS antibody used for this immunoassay cross-reacts with mouse PS [19]. The concentration of total protein was measured using a commercial kit (BCATM protein assay kit; Pierce, Rockford, IL, USA) following the manufacturer's instructions. The concentrations of TNF-α, IL-6 and MCP-1 were measured using commercial enzyme immunoassay (EIA) kits from BD Biosciences Pharmingen (San Diego, CA, USA) following the manufacturer's instructions. Thrombin–antithrombin complex (TAT), a marker of coagulation activation, was measured using enzyme immunoassay kits from Cedarlane Laboratories (Hornby, ON, Canada) following the manufacturer's instructions. The concentration of myeloperoxidase (MPO) was measured by a colorimetric assay using the synthetic substrate 2,2′-azinobis (ethylbenzyl-thiazoline-6-sulfonic acid) diammonium salt.

4.8. Gene Expression Analysis

Total RNA was harvested from the whole lung tissue with TRIzol reagent (Invitrogen, Carlsbad, CA), according to the manufacturer's instructions. The first-strand cDNA was synthesized from 2 µg of total RNA with oligo-dT primer and SuperScript II RNase H Reverse Transcriptase (Invitrogen), and PCR was performed using gene-specific primers. The RNA concentration was measured by ultraviolet (UV) absorption at 260:280 nm using an Ultrospec 1100 pro ultraviolet/visible (UV/Vis) spectrophotometer (Amersham Biosciences, NJ, USA). Reverse transcription of RNA into cDNA was done using a ReverTra Ace quantitative reverse transcription PCR (qPCR RT) kit (TOYOBO, Osaka, Japan) and then the DNA was amplified by PCR using Quick Taq HS DyeMix (TOYOBO). The primers used in the experiments were as follows: mouse glyceraldehyde 3-phosphate dehydrogenase (GAPDH), forward: 5′-CCCTTATTGACCTCAACTACATGGT-3′, reverse: 5′-GAGGGGCCATCCACAGTCTTCTG-3′; Mouse PS, forward: 5′-TGCTCAGTTCAGCATAGC TACA-3′, reverse: 5′-CTGATCCGAGCACAGAGATACC-3′; human PS, forward: 5′-AGGGCTCCTA CTATCCTGGTTCTG-3′, reverse: 5′-GCCATTATAAAAGGCATTCACTGG-3′; B-cell lymphoma2 (Bcl2), forward: 5′-AGCTGCACCTGACGCCCTT-3′, reverse: 5′-GTTCAGGTACTCAGTCATCCAC-3′; Bcl-extra large (Bcl-xl), forward: 5′-AGGTTCCTAAGCTTCGCAATTC-3′, reverse: 5′-TGTTTAGCGATT CTCTTCCAGG-3′; Bcl2-associated x (Bax), forward: 5′-CGGCGAATTGGAGATGAACTG-3′, reverse: 5′-GCAAAGTAGAAGAGGGCAACC-3′; as described [19]. The gene expression was normalized by the transcription level of glyceraldehyde-3-phosphate dehydrogenase.

4.9. Statistical Analysis

Data are expressed as the mean ± standard deviation of the means (S.D.). The statistical difference between variables was calculated by analysis of variance with post hoc analysis using the Fisher's least significant difference test. Normality of the data was calculated using the W/S test [30]. Statistical analyses were performed using the StatView 4.1 package software for the Macintosh (Abacus Concepts, Berkeley, CA, USA). Statistical significance was considered as $p < 0.05$.

Author Contributions: Conceptualization, T.K., E.C.G.; Formal analysis, K.N., K.A., H.S., K.F.; Investigation, T.Y., C.N.D'.A.-G., P.B.T., M.T.; Methodology, A.T., H.F., T.O.

Funding: This work was financially supported in part by a grant from the Ministry of Education, Culture, Sports, Science, and Technology of Japan (Kakenhi No 15K09170).

Acknowledgments: We would like to acknowledge the important contribution of Liqiang Qin from Taizhou Hospital, Wenzhou University Hospital, in the present work.

References

1. Fan, E.; Brodie, D.; Slutsky, A.S. Acute Respiratory Distress Syndrome: Advances in Diagnosis and Treatment. *JAMA* **2018**, *319*, 698–710. [CrossRef] [PubMed]

2. Umbrello, M.; Formenti, P.; Bolgiaghi, L.; Chiumello, D. Current Concepts of ARDS: A Narrative Review. *Int. J. Mol. Sci.* **2016**, *18*, 64. [CrossRef] [PubMed]

3. Laffey, J.G.; Madotto, F.; Bellani, G.; Pham, T.; Fan, E.; Brochard, L.; Amin, P.; Arabi, Y.; Bajwa, E.K.; Bruhn, A.; et al. Geo-economic variations in epidemiology, patterns of care, and outcomes in patients with acute respiratory distress syndrome: Insights from the LUNG SAFE prospective cohort study. *Lancet Respir. Med.* **2017**, *5*, 627–638. [CrossRef]

4. Mutlu, G.M.; Budinger, G.R. Incidence and outcomes of acute lung injury. *N. Engl. J. Med.* **2006**, *354*, 416–417; author reply 416–417. [PubMed]

5. Bellani, G.; Laffey, J.G.; Pham, T.; Fan, E.; Brochard, L.; Esteban, A.; Gattinoni, L.; van Haren, F.; Larsson, A.; McAuley, D.F.; et al. Epidemiology, Patterns of Care, and Mortality for Patients With Acute Respiratory Distress Syndrome in Intensive Care Units in 50 Countries. *JAMA* **2016**, *315*, 788–800. [CrossRef] [PubMed]

6. Dias-Freitas, F.; Metelo-Coimbra, C.; Roncon-Albuquerque, R., Jr. Molecular mechanisms underlying hyperoxia acute lung injury. *Respir. Med.* **2016**, *119*, 23–28. [CrossRef] [PubMed]

7. Lee, K.S.; Choi, Y.H.; Kim, Y.S.; Baik, S.H.; Oh, Y.J.; Sheen, S.S.; Park, J.H.; Hwang, S.C.; Park, K.J. Evaluation of bronchoalveolar lavage fluid from ARDS patients with regard to apoptosis. *Respir. Med.* **2008**, *102*, 464–469. [CrossRef] [PubMed]

8. Ji, Q.; Sun, Z.; Yang, Z.; Zhang, W.; Ren, Y.; Chen, W.; Yao, M.; Nie, S. Protective effect of ginsenoside Rg1 on LPS-induced apoptosis of lung epithelial cells. *Mol. Immunol.* **2018**. [CrossRef] [PubMed]

9. Fang, Y.; Gao, F.; Hao, J.; Liu, Z. microRNA-1246 mediates lipopolysaccharide-induced pulmonary endothelial cell apoptosis and acute lung injury by targeting angiotensin-converting enzyme 2. *Am. J. Transl. Res.* **2017**, *9*, 1287–1296. [PubMed]

10. Lin, W.C.; Chen, C.W.; Huang, Y.W.; Chao, L.; Chao, J.; Lin, Y.S.; Lin, C.F. Kallistatin protects against sepsis-related acute lung injury via inhibiting inflammation and apoptosis. *Sci. Rep.* **2015**, *5*, 12463. [CrossRef] [PubMed]

11. Albertine, K.H.; Soulier, M.F.; Wang, Z.; Ishizaka, A.; Hashimoto, S.; Zimmerman, G.A.; Matthay, M.A.; Ware, L.B. Fas and fas ligand are up-regulated in pulmonary edema fluid and lung tissue of patients with acute lung injury and the acute respiratory distress syndrome. *Am. J. Pathol.* **2002**, *161*, 1783–1796. [CrossRef]

12. Kawasaki, M.; Kuwano, K.; Hagimoto, N.; Matsuba, T.; Kunitake, R.; Tanaka, T.; Maeyama, T.; Hara, N. Protection from lethal apoptosis in lipopolysaccharide-induced acute lung injury in mice by a caspase inhibitor. *Am. J. Pathol.* **2000**, *157*, 597–603. [CrossRef]

13. Ma, X.; Xu, D.; Ai, Y.; Ming, G.; Zhao, S. Fas inhibition attenuates lipopolysaccharide-induced apoptosis and cytokine release of rat type II alveolar epithelial cells. *Mol. Biol. Rep.* **2010**, *37*, 3051–3056. [CrossRef] [PubMed]

14. Dahlback, B. Vitamin K-Dependent Protein S: Beyond the Protein C Pathway. *Semin. Thromb. Hemost.* **2018**, *44*, 176–184. [CrossRef] [PubMed]

15. Van der Meer, J.H.; van der Poll, T.; van't Veer, C. TAM receptors, Gas6, and protein S: Roles in inflammation and hemostasis. *Blood* **2014**, *123*, 2460–2469. [CrossRef] [PubMed]

16. Takagi, T.; Taguchi, O.; Aoki, S.; Toda, M.; Yamaguchi, A.; Fujimoto, H.; Boveda-Ruiz, D.; Gil-Bernabe, P.; Ramirez, A.Y.; Naito, M.; et al. Direct effects of protein S in ameliorating acute lung injury. *J. Thromb. Haemost.* **2009**, *7*, 2053–2063. [CrossRef] [PubMed]

17. Urawa, M.; Kobayashi, T.; D'Alessandro-Gabazza, C.N.; Fujimoto, H.; Toda, M.; Roeen, Z.; Hinneh, J.A.; Yasuma, T.; Takei, Y.; Taguchi, O.; et al. Protein S is protective in pulmonary fibrosis. *J. Thromb. Haemost.* **2016**, *14*, 1588–1599. [CrossRef] [PubMed]

18. Yasuma, T.; Yano, Y.; D'Alessandro-Gabazza, C.N.; Toda, M.; Gil-Bernabe, P.; Kobayashi, T.; Nishihama, K.; Hinneh, J.A.; Mifuji-Moroka, R.; Roeen, Z.; et al. Amelioration of Diabetes by Protein S. *Diabetes* **2016**, *65*, 1940–1951. [CrossRef] [PubMed]

19. Chelakkot-Govindalayathil, A.L.; Mifuji-Moroka, R.; D'Alessandro-Gabazza, C.N.; Toda, M.; Matsuda, Y.; Gil-Bernabe, P.; Roeen, Z.; Yasuma, T.; Yano, Y.; Gabazza, E.C.; et al. Protein S exacerbates alcoholic hepatitis by stimulating liver natural killer T cells. *J. Thromb. Haemost.* **2015**, *13*, 142–154. [CrossRef] [PubMed]

20. Totoki, T.; D'Alessandro-Gabazza, C.N.; Toda, M.; Tonto, P.B.; Takeshita, A.; Yasuma, T.; Nishihama, K.; Iwasa, M.; Horiki, N.; Takei, Y.; et al. Protein S Exacerbates Chronic Liver Injury and Fibrosis. *Am. J. Pathol.* **2018**, *188*, 1195–1203. [CrossRef] [PubMed]

21. Barlos, D.; Deitch, E.A.; Watkins, A.C.; Caputo, F.J.; Lu, Q.; Abungu, B.; Colorado, I.; Xu, D.Z.; Feinman, R. Trauma-hemorrhagic shock-induced pulmonary epithelial and endothelial cell injury utilizes different programmed cell death signaling pathways. *Am. J. Physiol. Lung Cell. Mol. Physiol.* **2009**, *296*, L404–L417. [CrossRef] [PubMed]

22. Perl, M.; Lomas-Neira, J.; Chung, C.S.; Ayala, A. Epithelial cell apoptosis and neutrophil recruitment in acute lung injury—A unifying hypothesis? What we have learned from small interfering RNAs. *Mol. Med.* **2008**, *14*, 465–475. [CrossRef] [PubMed]

23. Zhou, X.; Dai, Q.; Huang, X. Neutrophils in acute lung injury. *Front. Biosci.* **2012**, *17*, 2278–2283. [CrossRef]

24. Robb, C.T.; Regan, K.H.; Dorward, D.A.; Rossi, A.G. Key mechanisms governing resolution of lung inflammation. *Semin. Immunopathol.* **2016**, *38*, 425–448. [CrossRef] [PubMed]

25. Brown, J.E.; Krodel, M.; Pazos, M.; Lai, C.; Prieto, A.L. Cross-phosphorylation, signaling and proliferative functions of the Tyro3 and Axl receptors in Rat2 cells. *PLoS ONE* **2012**, *7*, e36800. [CrossRef] [PubMed]

26. Cook, S.J.; Stuart, K.; Gilley, R.; Sale, M.J. Control of cell death and mitochondrial fission by ERK1/2 MAP kinase signalling. *FEBS J.* **2017**, *284*, 4177–4195. [CrossRef] [PubMed]

27. Fraineau, S.; Monvoisin, A.; Clarhaut, J.; Talbot, J.; Simonneau, C.; Kanthou, C.; Kanse, S.M.; Philippe, M.; Benzakour, O. The vitamin K-dependent anticoagulant factor, protein S, inhibits multiple VEGF-A-induced angiogenesis events in a Mer- and SHP2-dependent manner. *Blood* **2012**, *120*, 5073–5083. [CrossRef] [PubMed]

28. Tian, Y.; Zhang, Z.; Miao, L.; Yang, Z.; Yang, J.; Wang, Y.; Qian, D.; Cai, H.; Wang, Y. Anexelekto (AXL) Increases Resistance to EGFR-TKI and Activation of AKT and ERK1/2 in Non-Small Cell Lung Cancer Cells. *Oncol. Res.* **2016**, *24*, 295–303. [CrossRef] [PubMed]

29. Suleiman, L.; Négrier, C.; Boukerche, H. Protein S: A multifunctional anticoagulant vitamin K-dependent protein at the crossroads of coagulation, inflammation, angiogenesis, and cancer. *Crit. Rev. Oncol. Hematol.* **2013**, *88*, 637–654. [CrossRef] [PubMed]

30. Kanji, G.K.; SAGE Research Methods Core. *100 Statistical Tests*, 3rd ed.; SAGE Publications, Incorporated Distributor: London, UK; Thousand Oaks, CA, USA, 2006; 256p.

International Journal of
Molecular Sciences

MDPI

Article

Non-Canonical Regulation of Type I Collagen through Promoter Binding of SOX2 and Its Contribution to Ameliorating Pulmonary Fibrosis by Butylidenephthalide

Hong-Meng Chuang [1,2], Li-Ing Ho [3], Mao-Hsuan Huang [1,2], Kun-Lun Huang [4], Tzyy-Wen Chiou [5], Shinn-Zong Lin [1,6], Hong-Lin Su [2,*,†] and Horng-Jyh Harn [1,7,*,†]

[1] Buddhist Tzu Chi Bioinnovation Center, Tzu Chi Foundation, Hualien 970, Taiwan; kavin273@gmail.com (H.-M.C.); spleo0825@gmail.com (M.-H.H.); shinnzong@yahoo.com.tw (S.-Z.L.)
[2] Department of Life Sciences, Agricultural Biotechnology Center, National Chung Hsing University, Taichung 402, Taiwan
[3] Division of Respiratory Therapy, Department of Chest Medicine, Taipei Veterans General Hospital, Taipei 112, Taiwan; breathho@gmail.com
[4] Hyperbaric Oxygen Therapy Center, Division of Pulmonary and Critical Care Medicine, Graduate Institute of Aerospace and Undersea Medicine, Department of Internal Medicine, Tri-Service General Hospital, National Defense Medical Center, Taipei 114, Taiwan; kun@mail.ndmctsgh.edu.tw
[5] Department of Life Science and Graduate Institute of Biotechnology, National Dong Hwa University, Hualien 974, Taiwan; twchiou@mail.ndhu.edu.tw
[6] Department of Neurosurgery, Buddhist Tzu Chi General Hospital, Tzu Chi University, Hualien 970, Taiwan
[7] Department of Pathology, Buddhist Tzu Chi General Hospital, Tzu Chi University, Hualien 970, Taiwan
[*] Correspondence: duke1945@tzuchi.com.tw (H.-J.H.); suhonglin@dragon.nchu.edu.tw (H.-L.S.); Tel.: +886-3-8561825 (ext. 15615) (H.-J.H.); +886-4-22840416 (ext. 417) (H.-L.S.); Fax: +886-3-8573710 (H.-J.H.); +886-3-8630262 (H.-L.S.)
[†] These authors contributed equally to this study.

Received: 12 September 2018; Accepted: 29 September 2018; Published: 4 October 2018

Abstract: Pulmonary fibrosis is a fatal respiratory disease that gradually leads to dyspnea, mainly accompanied by excessive collagen production in the fibroblast and myofibroblast through mechanisms such as abnormal alveolar epithelial cells remodeling and stimulation of the extracellular matrix (ECM). Our results show that a small molecule, butylidenephthalide (BP), reduces type I collagen (COL1) expression in Transforming Growth Factor beta (TGF-β)-induced lung fibroblast without altering downstream pathways of TGF-β, such as Smad phosphorylation. Treatment of BP also reduces the expression of transcription factor Sex Determining Region Y-box 2 (SOX2), and the ectopic expression of SOX2 overcomes the inhibitory actions of BP on COL1 expression. We also found that serial deletion of the SOX2 binding site on 3′COL1 promoter results in a marked reduction in luciferase activity. Moreover, chromatin immunoprecipitation, which was found on the SOX2 binding site of the COL1 promoter, decreases in BP-treated cells. In an in vivo study using a bleomycin-induced pulmonary fibrosis C57BL/6 mice model, mice treated with BP displayed reduced lung fibrosis and collagen deposition, recovering in their pulmonary ventilation function. The reduction of SOX2 expression in BP-treated lung tissues is consistent with our findings in the fibroblast. This is the first report that reveals a non-canonical regulation of COL1 promoter via SOX2 binding, and contributes to the amelioration of pulmonary fibrosis by BP treatment.

Keywords: pulmonary fibrosis; butylidenephthalide; SOX2; type I collagen; bleomycin

1. Introduction

Pulmonary fibrosis is a disease that occurs when scar-like tissue accumulates extracellular matrix (ECM) components (such as collagen, elastin, and fibronectin) in the pulmonary interstitial tissue. Such prolonged fibrogenesis makes the pulmonary tissue lose elasticity, resulting in a gradual loss of ability to contract, relax, and exchange gas. Statistics show that patients with idiopathic pulmonary fibrosis (IPF) have a two- to three- year median duration of survival from the time of diagnosis [1]. When definitively diagnosed, most patients' pulmonary functions have already deteriorated, showing symptoms of dyspnea, leading to chronic hypoxia [2]. The efficacy of current medications is not commensurate with other treatments such as pulmonary transplantation. For example, despite clinical data showing little efficacy with anti-inflammatory agents, inflammatory effects after injury play an important role in fibrogenesis. However, the biggest disadvantage of pulmonary transplantation is the time it takes to find a suitable transplant. Thus, this unmet medical need requires the development of new targets [3].

In a classical model of fibrosis, TGF-β pathways play a pivotal role [4] in inducing (1) secretion of growth factors that facilitate the proliferation of connective tissues [5]; (2) epithelial-mesenchymal transition (EMT) in alveolar type I or alveolar type II cells such as a myofibroblast source [6]; and (3) regulation of metalloproteinase activity which results in tissue remodeling [7]. Inhibition of TGF-β using approved clinical treatments for IPF (Pirfenidone) causes common side effects such as nausea and photosensitivity [8]. In TGF-β signaling, receptor phosphorylates JNK, Akt, and p38 regulate cell proliferation and differentiation, affecting downstream pathways like Smad and non-Smad pathways. Phosphorylation of Smad2/3 regulates ECM deposition and EMT induction, whereas, the non-Smad pathway improves the proliferation of fibroblast [9]. A typical signaling pathway in TGF-β enhances the expression of collagen and other fibrogenic genes, and collagen deposition is the most common pathological finding in patients with pulmonary fibrosis. Fibrillar-type collagens, including type I, II, III, and V, also express during tissue fibrosis [10], with collagen I predominating in areas of mature fibrosis [11].

There has been evidence indicating the requirement of SOX2 expression in induced skin fibrosis, recruited into fibrotic lesions in response to bleomycin treatments [12]. There is also evidence of elevated SOX2 expression in IPF patients [13]. As such, we postulate a possibility of the requirement of SOX2 in pulmonary fibrosis. SOX2 exhibits pluripotent qualities, allowing cells to reprogram, but also induces tumorgenecity. In general, SOX2 functions and pathways are too complicated, making it difficult to explain its relationship with fibrogenesis. One possibility is that the specific promoter consensus sequences of SOX2, reported previously, drive expressions such as fibroblast growth factor 4 (*Fgf4*) [14]. Another author suggests that in bronchiolar Clara cells, SOX2 forms a complex with Smad3 and influences cell proliferation [15]. Nevertheless, these studies focus on the importance of SOX2 but not on the regulatory activity of its transcription. To address this, we have found in our previous studies that Butylidenephthalide (BP), also known as (3E)-3-butylidene-2-benzofuran-1-one in IUPAC, showed anti-fibrotic effects in the liver, with its mechanism enhancing Bone morphogenetic protein 7 (BMP7) expression and reducing TGF-β in hepatic stellate cells (HSC) [16]. The HSC functions as a fibroblast after EMT, and can be reduced through BP treatment. In addition, BP decreases the expression of SOX2 by reducing its metastasis and invasiveness in glioblastoma multiform (GBM) cell [17].

Since our previous study demonstrated the effect of BP in treating liver fibrosis, it is possible that BP is effective for pulmonary fibrosis. To this end, we sought to examine the anti-fibrotic effect of BP on pulmonary fibrosis. Our data indicates that while BP reduced the expression of Collagen I, it altered neither the TGF-β-induced downstream pathways, nor the EMT status, and the expression of Collagen I was recovered in the exogenous overexpression of SOX2 in lung fibroblasts. Moreover, a consensus the SOX2 binding region was found in the promoter region of *COL1A1*, where BP reduction of SOX2 and its binding activity on the collagen I promoter leads to the inhibition of collagen deposition. The relationship between SOX2 and *COL1A1* was further examined by using chromatin immunoprecipitation. In the bleomycin-induced mouse model, treatment groups were found to not only ameliorate the fibrosis score, but also the pulmonary function. All in all, these findings reveal a BP-induced non-canonical regulation of SOX2, and a potential candidate for the amelioration of pulmonary fibrosis.

2. Results

2.1. BP Treatment in Human Lung Fibroblasts Attenuated Collagen Expression Driven by Exogenous TGF-β

The most extensive evidence on the origin of the myofibroblast is from fibroblast transdifferentiation. Intensive studies have shown that TGF-β1 would induce a similar progression of pulmonary fibrosis in mice [18], where treatment of TGF-β1 in isolated lung fibroblasts induces fibrotic reaction such as collagen and cytokines expression [19]. To evaluate the effects of BP in lung fibrosis, we established an in vitro model using normal human lung fibroblast cell lines (NHLF). Previous studies have shown that TGF-β1 is an effective cytokine to induce various types of collagen [20]. After stimulating with exogenous TGF-β treatment, cells were treated with or without BP (0 to 30 µg/mL) for 24 h. As shown in Figure 1A, the mRNA levels of type I collagen α (*COL1A1*) were dramatically increased by TGF-β (1.0 vs. 4.9), but decreased in BP treatment groups. In addition, BP reduces mRNA and protein levels of type I collagen in a time-dependent manner (Figure 1C,D). The protein and mRNA levels in SOX2 were found to have similar trends (response to dosage) as those of type I collagen (Figure 1A,B). These results suggest that BP could reduce collagen production, and that it has a potential for treating lung fibrosis.

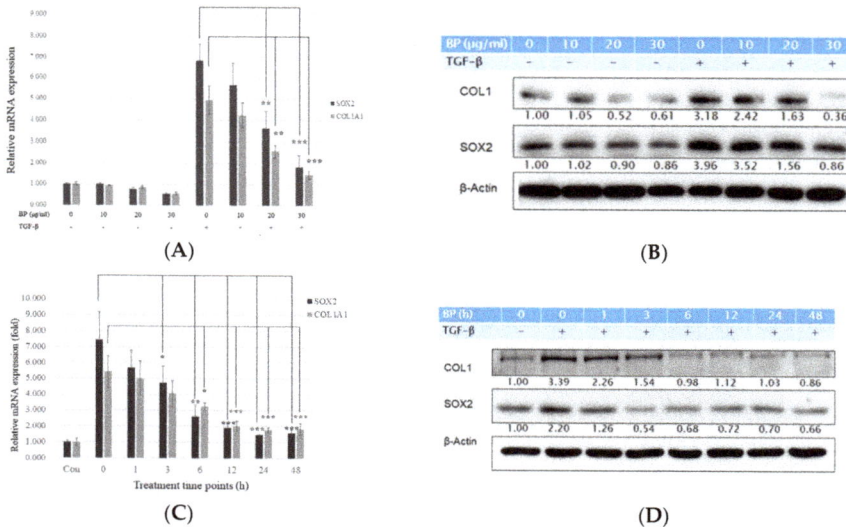

Figure 1. BP reduced collagen I production in lung fibroblast pre-stimulated with recombinant TGF-β1. TGF-β (5 ng/mL) stimulate for 12 h and (**A**) mRNA and (**B**) protein expression levels of BP treatment in several dosages (0, 10, 20, and 30 µg/mL) for 24 h on type I collagen, and SOX2 expressions. (**C**) mRNA and (**D**) protein expression levels of BP treatment for several time points (0, 1, 3, 6, 12, 24, and 48 h) in the dosage of 30 µg/mL on type I collagen and SOX2 expressions. The same amount of DMSO was added in 0 µg/mL groups as a vehicle control. Data showed 3 independent qPCR experiments and presented are mean ± SD. * denotes a significant decrease with the 0 µg/ml group of $p < 0.05$; ** $p < 0.01$; *** $p < 0.001$ by student's *t*-test.

2.2. BP Did Not Regulate the Smad and Non-Smad Pathways to Block the Effects of TGF-β

Previous reports have concluded that the role of TGF-β through the phosphorylation of Smad2/3 leads to the stimulation of ECM production in pulmonary fibroblasts [21]. In this study, TGF-β1 stimulation in normal lung fibroblast cell lines (NHLF) successfully upregulated in the expression of collagen and phosphorylation of Smad 2/3. Remarkably, BP did not reduce the phosphorylation of the downstream messenger of TGF-β, Smad2/3 (Figure 2A). Previous studies suggest that TGF-β-induced EMT results in the loss of E-cadherin expression, breaking the tight junctions in the epithelial type cells. However, in this study, the expression of E-cadherin was not changed by BP, as shown in Figure 2A, suggesting that BP would not alter the EMT status in lung fibroblasts. In this study, the NHLF treated TGF-β1 was sufficient to induce non-Smad pathways such as JNK, Akt, and p38 phosphorylation (Figure 2B). This treatment did not reduce the phosphorylation in these pathways, suggesting that BP would not alter the canonic TGF-β signal pathways.

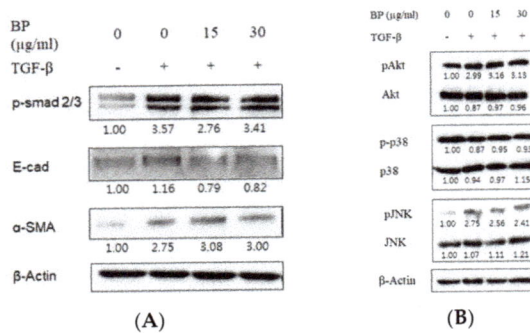

Figure 2. Effect of BP in canonical TGF-β downstream Smad and non-Smad pathways. Representative images of western blot analyses for TGF-β-induced (**A**) phosphorylated Smad2/3 for Smad pathway and its downstream EMT markers, and (**B**) phosphorylated Akt, p38 and JNK for non-Smad pathways and the effect of BP treatment.

2.3. SOX2-Overexpression Prevented Collagen Reduction in BP Treatment

Although SOX2 plays an important role in fibroblast differentiation, the relationship between SOX2 and fibrosis is still unclear. Weina et al. proposed that SOX2 might be involved in the TGF-β signaling pathway and correlates it with melanoma aggressiveness and metastasis in human melanoma cells [22]. Our previous study has shown that BP reduced SOX2 expression and stemness in glioblastoma cancer stem cells [17]. In this study, SOX2 expression was reduced in a time-and dose-dependent manner in BP-treated NHLFs. In order to reveal the relationship between SOX2 and BP-reduced collagen production, we constructed a SOX2 coding sequence on the pcDNA3.1 expression vector. NHLF transfected pcDNA3.1/SOX2 increased mRNA (Figure 3A) and protein (Figure 3B) levels of SOX2. However, in SOX2-overexpressed cells, BP treatment did not decrease type I collagen in both mRNA and protein levels, while the vector control group still decreased SOX2 and collagen. These finding suggested that the expression of collagen might be correlated by SOX2.

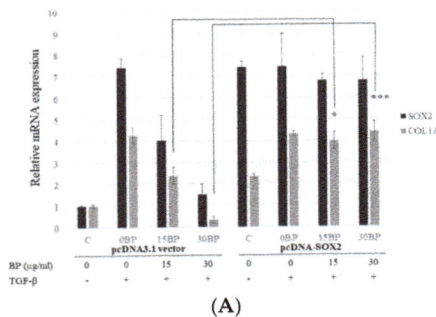

(A)

Figure 3. *Cont.*

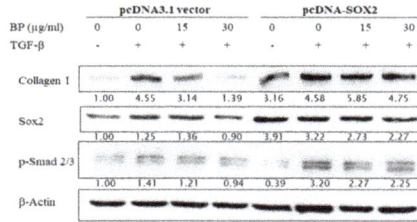

(**B**)

Figure 3. Effects of exogenous overexpression of SOX2 on BP treatments in fibroblasts. Stable expression lines of pcDNA3.1/SOX2 and its vector control were isolated by G418 and treated with or without BP for 24 h. (**A**) Analysis of the Col I and SOX2 expressions were done on the mRNA levels by qPCR; and (**B**) analysis of the Col I, phospho-Smad2/3, and SOX2 expressions were done on the protein level by Western blot analysis (representative images). Data showed 3 independent qPCR experiments and presented are mean ± SD. * denotes a significant decrease with the - group of $p < 0.05$; *** $p < 0.001$ by student's *t*-test.

2.4. BP Decreased SOX2 Specific Promoter Binding on Collagen

The regulation of collagen expression has been extensively studied, and components such as SP1 and AP2 contributing to its activity are already known. However, the consensus binding of SOX2 has still not been elucidated. To further elucidate whether the expression of collagen is reduced by BP, we constructed the type I collagen promoter sequence from −1091 to +3 of transcription starting site (TSS), as described in a previous report [23]. According to previous reports [24], the *COL1A1* promoter has a consensus SOX2 binding sequence, (T/A)TTGTT. Thus, we divided the promoter sequence into (1) a full length of *COL1A1* promoter; (2) a SOX2 binding site mutant construct; (3) a SOX2-AP2-SP1 binding site containing promoter; (4) a AP2-SP1 binding site containing promoter; (5) a SP1 binding site containing promoter; and (6) a TATA box only promoter (Figure 4A). The luciferase activity showed a significant decrease in both SOX2 binding site mutant and removal promoters, suggesting that the consensus SOX2 binding sequence, (T/A)TTGTT, could regulate the expression of *COL1A1*. To address whether BP might reduce the expression through the promoter regulation, we evaluated the luciferase activity in the condition of BP treatment (Figure 4B). The elevated luciferase activity in TGF-β-induced cells is significantly reduced by BP treatment (15, 30 μg/mL), and there was no difference between the control and BP treatment groups. To further define the binding characteristics of SOX2, we used chromatin immunoprecipitation to isolate the specific binding of SOX2 of the promoter sequence in TGF-β-induced NHLF cells. We found the consensus SOX2 binding site, (T/A)TTGTT, containing *COL1A1* promoter existed in the anti-SOX2 bound amplicons. In the result of qPCR analysis, SOX2 binding amplicon was markedly decreased while NHLF was being treated by BP (16.12 vs. 6.56) (Figure 4C). These finding suggest that the expression of type I collagen could regulated by the promoter-binding ability of SOX2 in response to BP treatment.

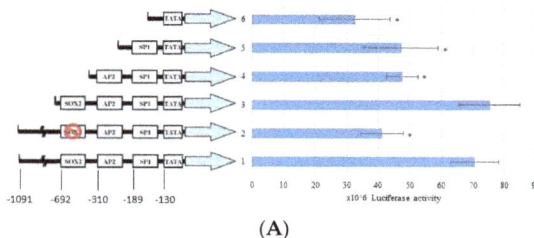

(**A**)

Figure 4. *Cont.*

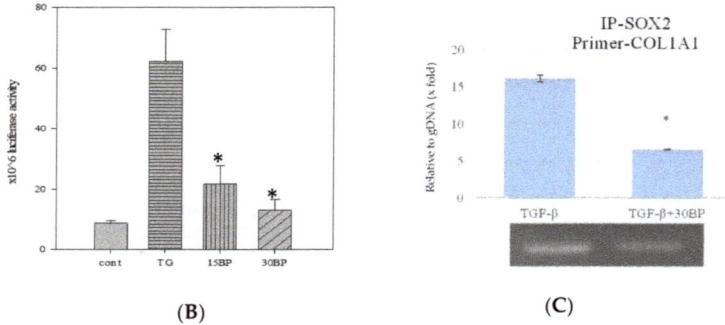

(B)

(C)

Figure 4. Effects of BP and SOX2 on the activity of type I collagen promoter deletion constructs in NHLF. Luciferase reporter gene constructs including truncated sequences of the *COL1A1* promoter and (**A**) Responsive elements exhibiting the previous described localization of consensus motifs of transcription factors SOX2, AP2, SP1, and TATA box are indicated. (**B**) The luciferase activity was determined 45 h after transfection of pGL3/COL1A1 in NHLF cells and BP treatment. Numbers represent the mean of 4 independent experiments. (**C**) ChIP analysis of SOX2 binding on the full length *COL1A1* promoter in BP-treated groups, and CT values (upper) were obtained from qPCR analysis and the representative image (bottom) from three independent experiments of gel electrophoresis, respectively. * denotes a significant decrease with the full-length group of $p < 0.05$ by ANOVA followed by *post hoc* analysis of Dunnett's test.

2.5. BP Reduced bleomycin-Induced Pulmonary Fibrosis in Mice

To determine whether BP regulates the fibrotic phenotype in vivo, we delivered BP with oral administration to bleomycin-treated mice of lung injury (Figure 5A). BP treatment of C57B/L6 mice did not result in weight loss and death at both high (50 mg/kg) and low (10 mg/kg) dosages (data not shown). In terms of histology analysis, the increased cell infiltration in bleomycin-induced groups was reduced by dose-dependent BP treatment (Figure 5B). The enlarged figure (shown in the vehicle group) demonstrates a noted increase of inflammatory cells according to their morphology. The collagen deposition was indicated in the blue color of the tissues by Masson-trichrome staining, and more severe fibrosis in the vehicle group mice was observed as compared to that in BP treated groups (Figure 5C). Based on the lung thickening and the distortion structure of the histological feature, the Ashcroft score was found to significantly reduced in 10 mg/kg ($p < 0.05$) and 50 mg/kg ($p < 0.05$) of BP treatment groups (Figure 5D).

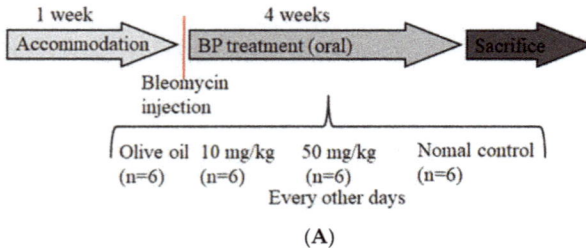

(A)

Figure 5. *Cont.*

(B)

(C)

(D)

(E)

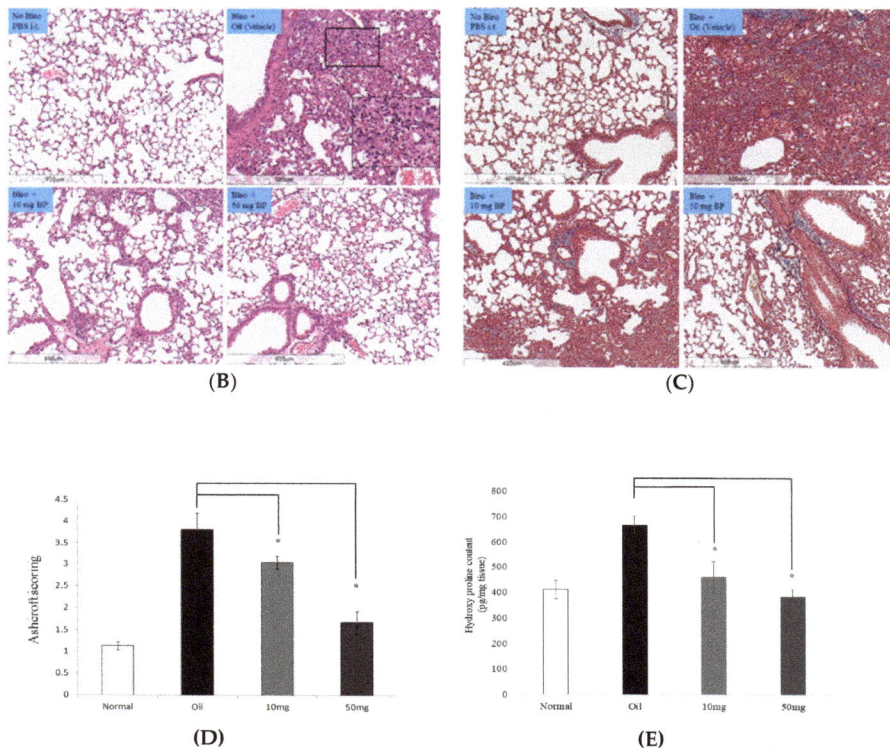

Figure 5. BP treatment in bleomycin-induced mice. (**A**) Schematic illustration experimental design. Mice were divided into four groups, and bleomycin was instilled on day 0 in all three experimental groups. The control group instilled the same volume of saline. Treatment groups were administered 0 (vehicle), 50 and 10 mg/kg doses of BP every other day for 30 days. Histological findings revealed lung inflammation and fibrosis by both (**B**) H&E (200 × magnification) and (**C**) Masson's staining (100× magnification). (**D**) The Ashcroft score was measured with all field of slides, each view counted and represented with mean ± SEM from six mice per group. (**E**) The hydroxyl proline content was obtained from tissues of each lobes of lung. * denotes a significant decrease of $p < 0.05$ by ANOVA followed by post hoc analysis of Dunnett's test.

Data from BAL fluid investigation showed IL-1β, IL-6, Macrophage inflammatory protein-1β (MIP-1β), and TNFα (which was reported as pulmonary fibrosis-related cytokines in BAL [25]) significantly decreased in BP treatment groups (Figure 6A–D. $p < 0.05$). These results suggest that BP is effective in ameliorating pulmonary fibrosis in bleomycin-induced mice mediated both by reducing collagen fiber and decreasing fibrotic related cytokines.

2.6. BP Restored Pulmonary Function in Bleomycin-Treated Mice

Bleomycin-induced pulmonary fibrosis is one of the most convincing animal models for investigating potential therapies, as its progression is consistent with patients who have developed lung fibrosis from bleomycin treatment. In our study, we administered bleomycin intratracheally to observe lung histopathology and pulmonary functions. In the treatment groups, we performed two doses of BP (50 and 10 mg/kg), and used olive oil as a vehicle control. On day 30, pulmonary functions were measured by whole body plethysmography. The respiratory rate and tidal volumes did not change in each of the groups. However, the Penh, EEP, relaxation times, and minute volume all significantly decreased ($p < 0.05$) (Table 1.) in BP treatment groups compared to the vehicle group and even the control group. Penh is used as an empirical index of airway resistance [26], and the result suggests that lung stiffness decreased and airway functions improved after BP treatment.

Table 1. Parameters of respiratory function in BP-treated groups were compared with vehicle control in bleomycin-induced mice. Data represent the mean ± standard error of the mean of $n = 6$ mice per group. * denotes a significant decrease of $p < 0.05$ by ANOVA followed by *post hoc* analysis of Dunnett's test. Data represent the mean standard error of the mean of $n = 6$ mice per group. * denotes a significant decrease of $p < 0.05$ by ANOVA and subsequent post hoc test with Dunnett's test if the *p*-value below 0.01.

Variable	Normal	10 mg/kg	50 mg/kg	Vehicle
Frequency (breaths/m)	452.5 ± 2.8	447.4 ± 75.4	460.1 ± 27.1	393.1 ± 41
Tidal volume (mL)	0.046 ± 0.001	0.046 ± 0.002	0.046 ± 0.001	0.048 ± 0.002
Accumulated volume (mL)	247.23 ± 201.77	141.03 ± 99.02 *	288.96 ± 7.5 *	124.42 ± 48.14
Minute volume (mL/m)	20.51 ± 0.06	19.94 ± 2.62 *	20.11 ± 2.59 *	18.08 ± 1.36
Inspiratory time (s)	0.0569 ± 0.0023	0.0623 ± 0.0086	0.0587 ± 0.008	0.0587 ± 0.0037
Expiratory time (s)	0.092 ± 0.003	0.101 ± 0.029	0.096 ± 0.027	0.078 ± 0.041
Peak inspiratory (mL/s)	1.313 ± 0.008	1.271 ± 0.076	1.269 ± 0.139	1.361 ± 0.048
Peak expiratory (mL/s)	1.103 ± 0.008	1.046 ± 0.07	1.037 ± 0.066	0.848 ± 0.411
Relaxation time (s)	0.0598 ± 0.0043	0.0586 ± 0.0168 *	0.0501 ± 0.0106 *	0.0815 ± 0.0075
End inspiratory pause (ms)	5.514 ± 1.089	5.296 ± 0.916	4.955 ± 0.377	6.047 ± 1.293
End expiratory pause (ms)	7.1129 ± 2.339	21.35 ± 24.67	14.08 ± 9.7 *	37.9 ± 16.21
Enhanced pause (Penh)	0.62 ± 0.002	0.608 ± 0.068 *	0.582 ± 0.105 *	0.742 ± 0.091

2.7. SOX2 and Collagen Expression Reduced in BP-Treated Lung Tissues

To further examine the inhibitory effects of SOX2 in BP-treated mice, we extracted protein lysate from lung tissues and performed western blot analysis. Protein expressions were quantified using immunoblot and its histogram plot show that both SOX2 and Collagen I expressions were markedly reduced in 50 mg/kg (Figure 6E).

3. Discussion

Our findings show evidence of BP as a potential therapeutic treatment in pulmonary fibrosis. Because fibroblasts and/or myofibroblasts in the fibrotic foci respond to the collagen synthesis [27], lung fibroblast cells are used to identify the inhibitory effects of fibrosis. From dose- and time-dependent studies, BP reduces the expression of collagen I in TGF-β1-treated NHLF. However, BP treatment did not alter Smad phosphorylation, nor non-Smad pathways such as the JNK, p38, and Akt. Interestingly, our previous data showed that SOX2 correlates with stemness but not with EMT status [17]. TGF-β signaling also induces SOX2 expression, reiterating the relationship between SOX2 and fibrogenesis. Thus, we conducted an ectopic expression of SOX2 to reactivate the collagen expression in BP-treated groups. In summary, these findings revealed that without the inhibition of the canonical TGF-β signaling pathways, BP decreases collagen deposition via SOX2 regulation.

In previous studies, ChIP-sequence analysis showed that binding activity of SOX2 on COL1A1 promoter decreases when embryonic stem cells (ESC) differentiates into neural progenitor cells (NPC) [28], leading to a reduction in cell development with fibroblast characteristics. These findings suggest that direct regulation of the SOX family proteins is essential for collagen expression, as well as fibrosis. Unfortunately, these studies didn't indicate the specific binding motif. Hence, we constructed a series of promoter deletions in the current study and found reduced reporter activity in the −722~+3 region in the promoter, removing a SOX2 consensus binding site. Furthermore, the promoter activity was significantly reduced in BP-treated TGF-β-induced NHLF (Figure 4B). Collectively, BP reduces the expression of SOX2 and Collagen I (Figure 1A), and regulates COL1A1 promoter activity through SOX2 binding site (Figure 4A), suggesting a relationship in the regulation of SOX2 and COL1A1. In ChIP assay, direct binding of SOX2 was found in the COL1A1 promoter region (Figure 4C). Inhibition of collagen synthesis is able to prevent bleomycin-induced fibrosis in vivo by not only restriction of serine/glycine uptake, but also through the key serine and glycine synthesis enzyme, phosphoglycerate dehydrogenase [29]. This recent study suggested that TGF-β-induced collagen synthesis and bleomycin-induced pulmonary fibrosis could inhibit fibrogenesis through collagen protein synthesis. Similarly, our result showed that a collagen transcription regulation is a potential therapeutic target; however, blocking serine and glycine uptake may influence other translation levels. Based on these findings, our study showed that BP can reduce collagen production and fibrogenesis potentially through SOX2 binding cis-element.

To further confirm our hypothesis of fibrosis reduction through BP, we established a bleomycin-induced pulmonary fibrosis animal model. The bleomycin-induced mice produced excessive collagen in our vehicle control groups, measured using Masson-trichrome staining (Figure 5C) and hydroxyl proline content assay (Figure 5E). The excessive collagen can be repressed with BP treatment, which is consistent with our results in human lung fibroblasts (Figure 1). The change in histopathology also shows more alveoli for better air exchange (Figure 5B), such that the accumulated and minute volumes are significantly restored in BP treatment groups (Table 1). Increased pulmonary fibrosis positive correlates with enhanced pause (Penh) [30,31], and our data showed a significant reduction in Penh compared to vehicle control groups, suggesting restoration of the respiratory function with a non-invasive measurement. Although our current study conducts a SOX2 knockout mice in animal models, the expression of SOX2 still reduces in BP-treated lung tissues (Figure 6E), which supports our findings in the regulation of SOX2 and collagen, thus clarifying the therapeutic effects of BP in lung fibrosis.

Our data showed immune modulation not only in the lung tissues through infiltration of multinucleated giant cells (Figure 5B, bleomycin with vehicle treatment), but also in the findings of BAL fluid (Figure 6), such as the IL-6 and IL-1β, that were both decreased in BP-treated groups. Other studies show that evoked T-helper cell 2 (Th2) is related to the excessive migration of macrophages and fibroblasts and resulted in Macrophage Inflammatory Proteins 1(MIP1) augmentation [32], as well as pulmonary fibrosis, further supporting our data [33]. Furthermore, our study suggests a possible relationship between inflammation and fibrosis, but the cause/result could not be determined in the current study. This might be a possible reason for the inefficient effects of immune-repressive drugs in the treatment of idiopathic pulmonary fibrosis patients [34].

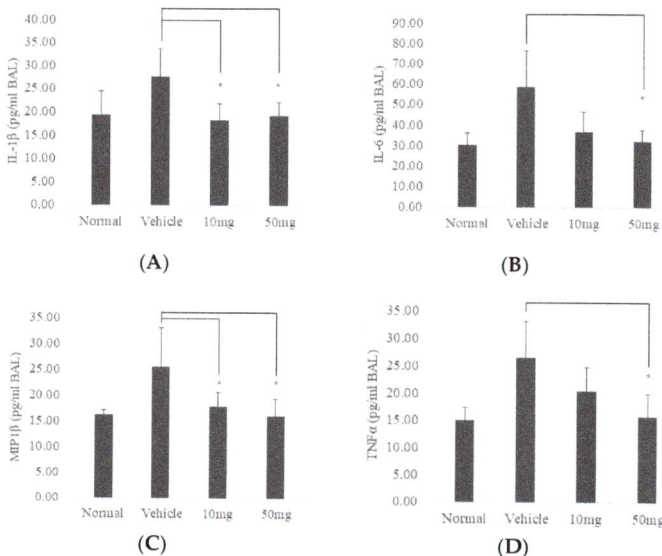

(A)

(B)

(C)

(D)

Figure 6. *Cont.*

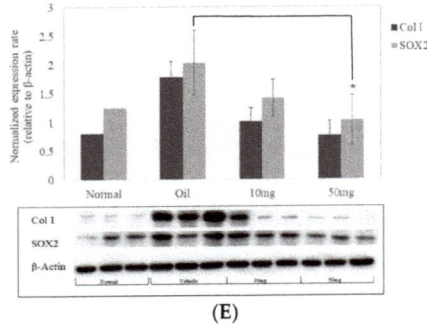

(E)

Figure 6. BP reduces inflammatory response and expressions of collagen I and SOX2 in bleomycin-induced mice. BALF was isolated ($n = 5$) and analyzed by ELISA for changes in (**A**) IL-1β, (**B**) IL-6, (**C**) MIP1β, and (**D**) TNF-α. (**E**) Expressions of collagen I and SOX2 protein from lung tissues in mice ($n = 3$) were quantified by Western blot and standardized by β-Actin. * denotes a significant decrease of $p < 0.05$ by ANOVA followed by *post hoc* analysis of Dunnett's test. Data presented are mean ± SD. (BALF, broncho alveolar lavage fluid; MIP-1β, Macrophage Inflammatory Proteins 1 beta.).

Compared to our previous studies, a consistent down regulation in SOX2 was also observed in glioblastoma stem cell [17]. Chiou et al. reported a DNA methyltransferase 1 (DNMT-1) dependent manner which inhibits SOX2 expression through high-mobility group AT-hook 2 (HMGA2) and miR142-3p [35], where interleukin 6 (IL-6) induces DNMT-1 hypermethylation on the promotor of miR142-3p. Our previous microarray data in GBM observed a downregulation of *HMGA2* and *COL1A1* as well, suggesting that BP reduces these targets in different cell types. One possible explanation might be epigenetic modification, such as methylation or miRNAs. Moreover, the present findings in BAL fluid showed a reduction of IL-6 compared to the vehicle treatment group (Figure 6). Detailed mechanisms about BP-influenced regulation of SOX2 and its relationship between IL-6 and other cytokines are still unknown, and will be studied in the future. In summary, our present data supports SOX2 regulation of *COL1A1* promoter as a potential target in BP treatment of pulmonary fibrosis (Figure 7).

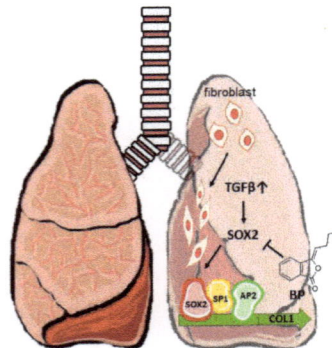

Figure 7. A short schematic mechanism of BP to treat pulmonary fibrosis. In a mouse model of bleomycin-induced pulmonary fibrosis, promoter binding enhances the collagen expression in fibroblast. BP inhibits SOX2 expression and its binding on collagen promoter blocks fibrogenesis in vitro and in vivo. ⊥: Inhibition of the gene expression; Gray upwards arrow ↑: Increased level of TGF-β.

4. Material and Methods

4.1. Chemical and Treatment

Butylidenephthalide (MW: 188.23), purchased from Lancaster Synthesis Ltd. (Newgate, Morecambe, UK), was dissolved in dimethylsulfoxide (DMSO), as previously described [36] for treatments in cell cultures. The same amount of DMSO was added as a vehicle control. For the animal study, BP was dissolved in food grade olive oil (Forlì, Italy). In vitro studies of lung fibrosis were performed in normal Human lung fibroblast (NHLF), which was stimulated into fibrogenesis by recombinant human TGF-β1 (PeproTech, Catalog Number: 100-21, 5 ng/mL) for 12 h.

4.2. Cell Culture and Transfection

NHLF cell was purchased from Lonza (Basel, Switzerland) and maintained in DMEM supplemented with bFGF (Peprotech, London, UK) (5 ng/mL), insulin (5 µg/mL), and 2% FBS (HyClone, Logan, UT, USA). Serum supplement was removed in TGF-β treatment because the presence of serum is likely to diminish the fibrogenic actions of TGF-β. The construct of pcDNA3.1/SOX2 was previously described [17], and NHLF was transfected with pcDNA3.1 vector or pcDNA3.1/SOX2 by FuGENE HD® Transfection Reagent (Promega, Mannheim, Germany). Cells were then exposed to 600 µg/mL G418 (Invitrogen, Carlsbad, CA, USA) in a complete medium containing 2% FBS for stable clone selection over 3 weeks. The transfection efficiency was measured by Western blot for His-tag expression before the treatment of TGF-β and BP.

4.3. RT-PCR and Western Blot Analysis

After the treatment of BP, total RNA was isolated by RNeasy RNA isolation kit (Qiagen, Germantown, MD, USA). The subsequent RNA was quantified into 2 µg samples, and reverse transcription was performed by QuantiTect Reverse Transcription Kit (Qiagen, Germantown, MD, USA). The qPCR reactions were performed by the LightCycler® SYBR Green I Master (Roche, Basel, Switzerland) reagent, and primers are listed in Tab S1. The semi-quantitative RT-PCR was performed according to the following program: 10 min 95 °C followed by 40 cycles of 15 s for 95 °C, 20 s for 60 °C, and 20 s for 72 °C, and followed by 5 min for 72 °C. The qPCR detection program was as follows: 10 min 95 °C followed by 40 cycles of 15 s for 95 °C and 1 m for 60 °C. The ΔCT values were performed from *ACTB* normalization, and further obtained the $\Delta\Delta$CT values from control group. The relative expression rates were obtained from the $2^{-\Delta\Delta CT}$ algorithm. In terms of Western blot, cells were lysed by PRO-PREP™, which was purchased from iNtRON Biotechnology (Gyeonggi-do, Korea), and incubated on ice for 30 min. Cells were centrifuged at 13,000 rpm for 15 min at 4 °C, and the supernatant was then quantified for SDS-polyacrylamide gel electrophoresis. Blots were blocked in 5% skimmed milk for 1 h and hybridized with primary antibodies of His-tag (Abcam, Cambridge, MA, USA), Collagen I (GeneTex, San Antonio, TX, USA), each at a ratio of 1:1000, and SOX2 (GeneTex, San Antonio, TX, USA), ACTB (Sigma, St. Louis, MO, USA), each at a ratio of 1:5000, for overnight hybridization. These data were confirmed with three independent experiments.

4.4. Promoter Construct and Assay

The promoter sequence of *COL1A1* was divided into 4 parts which included a consensus SOX2 binding site (T/A)TTGTT ($-724\sim+3$), an AP2 ($-342\sim+3$), an SP1 ($-221\sim+3$) binding sequence, and a TATA box only ($-177\sim+3$), as described in [23], and these fragments were linked to the luciferase reporter vector pGL3-basic (Promega, Madison, WI, USA). A SOX2del of *COL1A1* was conducted by removing the ATTGTT($-693\sim-687$) in the promoter sequence, and cloned into the same vector. All plasmids were prepared and transfected by FuGENE HD® Transfection Reagent (Promega). The subsequent reaction was catalyzed and measured by Steady-Glo® Luciferase Assay System (Promega) and a luminance ELISA reader (Fluoroskan Ascent FL, Thermo Fisher Scientific, Cleveland,

OH, USA), following the manufacturer's instructions. The luciferase activity was obtained by three independent experiments, with six replicates for each group.

4.5. Chromatin Immunoprecipitation Assay

Total cell lysates (1.5 mg) were isolated from TGF-β-induced NHLF with an additional BP-treated group. The lysates were stored on ice, and sonication was performed with UP100H (Hielscher Ultrasonics GmbH, Teltow, Germany) to break into a 100–200 base pair length. The protein-DNA complexes (150 μg with protease inhibitor) were allowed to hybridize with SOX2 antibodies (10 μg Genetex) at 4 °C overnight. Protein A/G beads (Millipore, Hayward, CA, USA) were added and rotated at 4 °C for 2 h, and washed 4 times, then heated 65 °C to reverse cross links. Real-time PCR was performed to detect *COL1A1* promoter sequence. The internal control used an equal DNA concentration from the control cell lysate.

4.6. Bleomycin-Induced Pulmonary Fibrosis and Tissue Collection

Male C57BL/6 mice were purchased from the National Laboratory Animal Center and animal studies were approved by the China Medical University Institutional Animal Care and Use Committee. Four-week aged mice were housed in a clean enclosure and allowed to accommodate for a week; they had free access to a standard diet. Bleomycin (0.045 U dissolved in 50 μL PBS) or PBS was then injected with intratracheal instillation into the mice, allowing a recovery of 3 days. Bleomycin-treated mice were then randomly grouped into BP treatment (10, 50 mg/kg) and olive oil (in both vehicle and no bleomycin controls) orally every other day (*n* = 6 in all four groups), as illustrated in Figure 5A. On Day 30, animals were sacrificed to collect bronchoalveolar lavage fluid (BALF) and lung tissue. BALF collection is performed using cold PBS (0.5 mL), which was gently injected and pulled out 5 times through the trachea with a syringe, then stored on ice. The BALF samples were performed by magnetic bead-based multiplex immunoassays (Bio-Plex) (BIO-RAD Laboratories, Milano, Italy) following manufactures' instructions. Lung tissues were removed for histopathology and protein extraction, and divided into formalin fixation and −80 °C frozen storage, respectively. Fixed tissues were embedded and followed by H&E or Masson's trichrome stain.

4.7. Pulmonary Function Test in Mice

Bleomycin–treated mice were administered BP orally with two dosages and olive oil as a vehicle control. Pulmonary function was measured by unrestrained Whole Body Plethysmography (WBP) (Buxco Electronics, Inc. Wilmington, NC, USA). The WBP detected atmospheric change to obtain data, such as respiratory rate and Tidal volume, in a specific time range. We assessed non-invasive airway responsive data using the WBP with a single-chamber prior to animal sacrifice, and as the chamber pressure changes, the box flow signal derived the following data: Inspiratory time (Ti); expiratory time (Te); relaxation time (TR), defined by the time for declining 36% of the total expiratory area; peak inspiratory flow (PIF) and peak expiratory flow (PEF); tidal volume (VT); accumulative volume (AV); end inspiratory pause (EIP); end expiratory pause (EEP); and respiratory rate (RR). The Penh (enhanced Pause) is referred to as an empirical parameter to limit airflow. Mice were incubated in the chamber to prevent light and noise, and the data collection period was set at 5~15 min to exclude the accommodation time.

4.8. Pathologic Morphology Staining and Evaluation

After lung tissues were obtained from sacrificed animals, samples were fixed in 3.7% formaldehyde for 2 d. The dehydration, clearance, and infiltration of all tissues were performed by a Histoprocessor (Tissue-Tek; Sakura, Tokyo, Japan). Paraffin embedded tissues were cut at a 4 μm serial section and stained with H&E. The procedure of Masson's trichrome staining was previously described [16]. Briefly, sections were immersed in Bouin's solution and then stained with Mayer's hematoxylin solution, Biebrich scarlet–acid, phosphomolybdic acid–phosphotungstic acid, and aniline

Int. J. Mol. Sci. **2018**, *19*, 3024

blue reagents (Sigma-Aldrich, Steinheim, Germany), respectively, with a ddH$_2$O wash between each reagent. The samples were dried and mounted on glass slides, and sections were examined using a microscope (IX70; Olympus Tokyo, Japan). The Ashcroft scores were evaluated with blinded label and confirmed by a pathologist, who defined the assignment of grades from 0 to 8 [37]. The hydroxyl proline was examined with each group of lung tissues (*n* = 6), and equal weight of lung tissues was minced, following the manufacturer's instructions of the assay kit (Cell Biolabs, San Diego, CA).

4.9. Statistical Analysis

All of the experiments were performed in three or more independent experiments. Statistical analysis of the results (*n* > 5) between groups was analyzed by ANOVA to obtain the *p*-value < 0.05, and a subsequent *post hoc* test with Dunnett's test by SigmaPlot V12.5.

Author Contributions: Writing—Original Draft Preparation, H.-M.C.; Writing—Review & Editing, L.-I.H.; M.-H.H. preparing materials of experiments. Formal Analysis, K.-L.H.; Supervision, T.-W.C.; Project Administration, S.-Z.L.; Conceptualization, H.-L.S. and H.-J.H.

Funding: This study was funded by Buddhist Tzu Chi Bioinnovation Center, Tzu Chi Foundation, Hualien, Taiwan; Ministry of Science and Technology, Taiwan (MOST 106-2320-B-303-001-MY3 and MOST 106-2320-B-303-002-MY3).

Acknowledgments: We are grateful to Tina Emily Shih for English proofreading and Wei-Ju Lin for picture enhancement.

Conflicts of Interest: The authors declare that they have no competing financial interests.

Abbreviations

Term	Definition
ECM	extracellular matrix
SOX2	sex determining region Y (SRY)-box 2
COL1	type I collagen
COL1A1	type I collagen α1
TGF-β	transforming growth factor beta
BP	butylidenephthalide
NHLF	normal lung fibroblast cell lines
E-cad	E-cadherin
i.t.	intratracheal
WBP	whole body plethysmography
BAL	bronchoalveolar lavage

References

1. Raghu, G.; Collard, H.R.; Egan, J.J.; Martinez, F.J.; Behr, J.; Brown, K.K.; Colby, T.V.; Cordier, J.F.; Flaherty, K.R.; Lasky, J.A.; et al. An official ATS/ERS/JRS/ALAT statement: Idiopathic pulmonary fibrosis: Evidence-based guidelines for diagnosis and management. *Am. J. Respir. Crit. Care Med.* **2011**, *183*, 788–824. [CrossRef] [PubMed]

2. Du Bois, R.M. An earlier and more confident diagnosis of idiopathic pulmonary fibrosis. *Eur. Respir. Rev.* **2012**, *21*, 141–146. [CrossRef] [PubMed]

3. Datta, A.; Scotton, C.J.; Chambers, R.C. Novel therapeutic approaches for pulmonary fibrosis. *Br. J. Pharmacol* **2011**, *163*, 141–172. [CrossRef] [PubMed]

4. Meng, X.M.; Nikolic-Paterson, D.J.; Lan, H.Y. TGF-beta: The master regulator of fibrosis. *Nat. Rev. Nephrol.* **2016**, *12*, 325–338. [CrossRef] [PubMed]

5. Cutroneo, K.R.; White, S.L.; Phan, S.H.; Ehrlich, H.P. Therapies for bleomycin induced lung fibrosis through regulation of TGF-beta1 induced collagen gene expression. *J. Cell Physiol.* **2007**, *211*, 585–589. [CrossRef] [PubMed]

6. Willis, B.C.; Liebler, J.M.; Luby-Phelps, K.; Nicholson, A.G.; Crandall, E.D.; du Bois, R.M.; Borok, Z. Induction of epithelial-mesenchymal transition in alveolar epithelial cells by transforming growth factor-beta1: Potential role in idiopathic pulmonary fibrosis. *Am. J. Pathol.* **2005**, *166*, 1321–1332. [CrossRef]

7. Palcy, S.; Bolivar, I.; Goltzman, D. Role of activator protein 1 transcriptional activity in the regulation of gene expression by transforming growth factor beta1 and bone morphogenetic protein 2 in ROS 17/2.8 osteoblast-like cells. *J. Bone Miner. Res.* **2000**, *15*, 2352–2361. [CrossRef] [PubMed]

8. King, T.E., Jr.; Bradford, W.Z.; Castro-Bernardini, S.; Fagan, E.A.; Glaspole, I.; Glassberg, M.K.; Gorina, E.; Hopkins, P.M.; Kardatzke, D.; Lancaster, L.; et al. A phase 3 trial of pirfenidone in patients with idiopathic pulmonary fibrosis. *N. Engl. J. Med.* **2014**, *370*, 2083–2092. [CrossRef] [PubMed]

9. Zhang, Y.E. Non-Smad pathways in TGF-beta signaling. *Cell Res.* **2009**, *19*, 128–139. [CrossRef] [PubMed]

10. Exposito, J.Y.; Valcourt, U.; Cluzel, C.; Lethias, C. The fibrillar collagen family. *Int. J. Mol. Sci.* **2010**, *11*, 407–426. [CrossRef] [PubMed]

11. Kaarteenaho-Wiik, R.; Lammi, L.; Lakari, E.; Kinnula, V.L.; Risteli, J.; Ryhanen, L.; Paakko, P. Localization of precursor proteins and mRNA of type I and III collagens in usual interstitial pneumonia and sarcoidosis. *J. Mol. Histol.* **2005**, *36*, 437–446. [CrossRef] [PubMed]

12. Liu, S.; Herault, Y.; Pavlovic, G.; Leask, A. Skin progenitor cells contribute to bleomycin-induced skin fibrosis. *Arthritis. Rheumatol.* **2014**, *66*, 707–713. [CrossRef] [PubMed]

13. Plantier, L.; Crestani, B.; Wert, S.E.; Dehoux, M.; Zweytick, B.; Guenther, A.; Whitsett, J.A. Ectopic respiratory epithelial cell differentiation in bronchiolised distal airspaces in idiopathic pulmonary fibrosis. *Thorax* **2011**, *66*, 651–657. [CrossRef] [PubMed]

14. Basu-Roy, U.; Ambrosetti, D.; Favaro, R.; Nicolis, S.K.; Mansukhani, A.; Basilico, C. The transcription factor Sox2 is required for osteoblast self-renewal. *Cell Death Differ.* **2010**, *17*, 1345–1353. [CrossRef] [PubMed]

15. Tompkins, D.H.; Besnard, V.; Lange, A.W.; Wert, S.E.; Keiser, A.R.; Smith, A.N.; Lang, R.; Whitsett, J.A. Sox2 is required for maintenance and differentiation of bronchiolar Clara, ciliated, and goblet cells. *PLoS ONE* **2009**, *4*, e8248. [CrossRef] [PubMed]

16. Chuang, H.M.; Su, H.L.; Li, C.; Lin, S.Z.; Yen, S.Y.; Huang, M.H.; Ho, L.I.; Chiou, T.W.; Harn, H.J. The Role of Butylidenephthalide in Targeting the Microenvironment Which Contributes to Liver Fibrosis Amelioration. *Front Pharmacol.* **2016**, *7*, 112–125. [CrossRef] [PubMed]

17. Yen, S.Y.; Chuang, H.M.; Huang, M.H.; Lin, S.Z.; Chiou, T.W.; Harn, H.J. n-Butylidenephthalide Regulated Tumor Stem Cell Genes EZH2/AXL and Reduced Its Migration and Invasion in Glioblastoma. *Int. J. Mol. Sci.* **2017**, *18*, 372. [CrossRef] [PubMed]

18. Bonniaud, P.; Margetts, P.J.; Kolb, M.; Schroeder, J.A.; Kapoun, A.M.; Damm, D.; Murphy, A.; Chakravarty, S.; Dugar, S.; Higgins, L.; et al. Progressive transforming growth factor beta1-induced lung fibrosis is blocked by an orally active ALK5 kinase inhibitor. *Am. J. Respir. Crit. Care Med.* **2005**, *171*, 889–898. [CrossRef] [PubMed]

19. Raghu, G.; Masta, S.; Meyers, D.; Narayanan, A.S. Collagen synthesis by normal and fibrotic human lung fibroblasts and the effect of transforming growth factor-beta. *Am. Rev. Respir. Dis.* **1989**, *140*, 95–100. [CrossRef] [PubMed]

20. Leask, A.; Abraham, D.J. TGF-beta signaling and the fibrotic response. *FASEB J.* **2004**, *18*, 816–827. [CrossRef] [PubMed]

21. Baarsma, H.A.; Engelbertink, L.H.; van Hees, L.J.; Menzen, M.H.; Meurs, H.; Timens, W.; Postma, D.S.; Kerstjens, H.A.; Gosens, R. Glycogen synthase kinase-3 (GSK-3) regulates TGF-beta(1)-induced differentiation of pulmonary fibroblasts. *Br. J. Pharmacol.* **2013**, *169*, 590–603. [CrossRef] [PubMed]

22. Weina, K.; Wu, H.; Knappe, N.; Orouji, E.; Novak, D.; Bernhardt, M.; Huser, L.; Larribere, L.; Umansky, V.; Gebhardt, C.; et al. TGF-beta induces SOX2 expression in a time-dependent manner in human melanoma cells. *Pigment Cell Melanoma Res.* **2016**, *29*, 453–458. [CrossRef] [PubMed]

23. Buttner, C.; Skupin, A.; Rieber, E.P. Transcriptional activation of the type I collagen genes COL1A1 and COL1A2 in fibroblasts by interleukin-4: Analysis of the functional collagen promoter sequences. *J. Cell Physiol.* **2004**, *198*, 248–258. [CrossRef] [PubMed]

24. Kamachi, Y.; Kondoh, H. Sox proteins: Regulators of cell fate specification and differentiation. *Development* **2013**, *140*, 4129–4144. [CrossRef] [PubMed]

25. Emad, A.; Emad, V. Elevated levels of MCP-1, MIP-alpha and MIP-1 beta in the bronchoalveolar lavage (BAL) fluid of patients with mustard gas-induced pulmonary fibrosis. *Toxicology* **2007**, *240*, 60–69. [CrossRef] [PubMed]

26. Schafer, M.J.; White, T.A.; Iijima, K.; Haak, A.J.; Ligresti, G.; Atkinson, E.J.; Oberg, A.L.; Birch, J.; Salmonowicz, H.; Zhu, Y.; et al. Cellular senescence mediates fibrotic pulmonary disease. *Nat. Commun.* **2017**, *8*, 14532. [CrossRef] [PubMed]

27. Wynn, T.A. Common and unique mechanisms regulate fibrosis in various fibroproliferative diseases. *J. Clin. Investig.* **2007**, *117*, 524–529. [CrossRef] [PubMed]

28. Lodato, M.A.; Ng, C.W.; Wamstad, J.A.; Cheng, A.W.; Thai, K.K.; Fraenkel, E.; Jaenisch, R.; Boyer, L.A. SOX2 co-occupies distal enhancer elements with distinct POU factors in ESCs and NPCs to specify cell state. *PLoS Genet.* **2013**, *9*, e1003288. [CrossRef] [PubMed]

29. Hamanaka, R.B.; Nigdelioglu, R.; Meliton, A.Y.; Tian, Y.; Witt, L.J.; O'Leary, E.; Sun, K.A.; Woods, P.S.; Wu, D.; Ansbro, B.; et al. Inhibition of Phosphoglycerate Dehydrogenase Attenuates Bleomycin-induced Pulmonary Fibrosis. *Am. J. Respir. Cell Mol. Biol.* **2018**, *58*, 585–593. [CrossRef] [PubMed]

30. Ganzert, S.; Moller, K.; Steinmann, D.; Schumann, S.; Guttmann, J. Pressure-dependent stress relaxation in acute respiratory distress syndrome and healthy lungs: An investigation based on a viscoelastic model. *Crit. Care* **2009**, *13*, R199. [CrossRef] [PubMed]

31. Hamelmann, E.; Schwarze, J.; Takeda, K.; Oshiba, A.; Larsen, G.L.; Irvin, C.G.; Gelfand, E.W. Noninvasive measurement of airway responsiveness in allergic mice using barometric plethysmography. *Am. J. Respir. Crit. Care Med.* **1997**, *156*, 766–775. [CrossRef] [PubMed]

32. Swaney, J.S.; Chapman, C.; Correa, L.D.; Stebbins, K.J.; Bundey, R.A.; Prodanovich, P.C.; Fagan, P.; Baccei, C.S.; Santini, A.M.; Hutchinson, J.H.; et al. A novel, orally active LPA(1) receptor antagonist inhibits lung fibrosis in the mouse bleomycin model. *Br. J. Pharmacol* **2010**, *160*, 1699–1713. [CrossRef] [PubMed]

33. Singh, B.; Kasam, R.K.; Sontake, V.; Wynn, T.A.; Madala, S.K. Repetitive intradermal bleomycin injections evoke T-helper cell 2 cytokine-driven pulmonary fibrosis. *Am. J. Physiol. Lung Cell. Mol. Physiol.* **2017**, *313*, L796–L806. [CrossRef] [PubMed]

34. Raghu, G.; Anstrom, K.J.; King, T.E., Jr.; Lasky, J.A.; Martinez, F.J. Prednisone, azathioprine, and N-acetylcysteine for pulmonary fibrosis. *N. Engl. J. Med.* **2012**, *366*, 1968–1977. [PubMed]

35. Chiou, G.Y.; Chien, C.S.; Wang, M.L.; Chen, M.T.; Yang, Y.P.; Yu, Y.L.; Chien, Y.; Chang, Y.C.; Shen, C.C.; Chio, C.C.; et al. Epigenetic regulation of the miR142-3p/interleukin-6 circuit in glioblastoma. *Mol. Cell* **2013**, *52*, 693–706. [CrossRef] [PubMed]

36. Tsai, N.M.; Chen, Y.L.; Lee, C.C.; Lin, P.C.; Cheng, Y.L.; Chang, W.L.; Lin, S.Z.; Harn, H.J. The natural compound n-butylidenephthalide derived from Angelica sinensis inhibits malignant brain tumor growth in vitro and in vivo. *J. Neurochem.* **2006**, *99*, 1251–1262. [CrossRef] [PubMed]

37. Ashcroft, T.; Simpson, J.M.; Timbrell, V. Simple method of estimating severity of pulmonary fibrosis on a numerical scale. *J. Clin. Pathol.* **1988**, *41*, 467–470. [CrossRef] [PubMed]

International Journal of
Molecular Sciences

MDPI

Article

Involvement of G-Protein-Coupled Receptor 40 in the Inhibitory Effects of Docosahexaenoic Acid on SREBP1-Mediated Lipogenic Enzyme Expression in Primary Hepatocytes

Seungtae On [†], Hyun Young Kim [†], Hyo Seon Kim, Jeongwoo Park and Keon Wook Kang *

College of Pharmacy and Research Institute of Pharmaceutical Sciences, Seoul National University, Seoul 08826, Korea; seungtae.on@gmail.com (S.O.); hens93@snu.ac.kr (H.Y.K.); hyoseonkim@snu.ac.kr (H.S.K.); jwpark0110@snu.ac.kr (J.P.)
* Correspondence: kwkang@snu.ac.kr
† These authors contributed equally to this work.

Received: 16 April 2019; Accepted: 26 May 2019; Published: 28 May 2019

Abstract: Nonalcoholic fatty liver disease is a frequent liver malady, which can progress to cirrhosis, the end-stage liver disease if proper treatment is not applied. Omega-3 fatty acids, such as docosahexaenoic acid (DHA) and eicosapentaenoic acid, have been clinically proven to lower serum triglyceride levels. Various physiological activities of omega-3 fatty acids are due to their agonistic actions on G-protein-coupled receptor 40 (GPR40) and GPR120. Lipid droplets (LD) accumulation in hepatocytes confirmed that DHA treatment reduced the number of larger (>10 μm^2) LDs, as well as the total area of LDs. Moreover, DHA lowered protein and mRNA expression levels of lipogenic enzymes such as fatty acid synthase (FAS), acetyl-CoA carboxylase and stearoyl-CoA desaturase-1 (SCD-1) in primary hepatocytes incubated with liver X receptor (LXR) agonist T0901317 or high glucose and insulin. DHA also decreased protein expression of nuclear and precursor sterol response-element binding protein (SREBP)-1, a key lipogenesis transcription factor. We further found that exposure of murine primary hepatocytes to DHA for 12 h increased GPR40 and GPR120 mRNA levels. Specific agonists (Compound A for GPR120 and AMG-1638 for GPR40), hepatocytes from GPR120 knock-out mice and GPR40 selective antagonist (GW1100) were used to assess whether DHA's antilipogenic effects are mediated through GPR120 or GPR40. Compound A did not decrease SREBP-1 and FAS protein expression in hepatocytes exposed to T0901317 or high glucose with insulin. Moreover, DHA downregulated lipogenesis enzyme expression in GPR120-null hepatocytes. In contrast, AMG-1638 lowered SREBP-1 and SCD-1 protein levels. Additionally, GW1100, a GPR40 antagonist, reversed the antilipogenic effects of DHA. Collectively, our data demonstrate that DHA downregulates the expression SREBP-1-mediated lipogenic enzymes via GPR40 in primary hepatocytes.

Keywords: GPR40; GPR120; DHA; omega-3 fatty acid; SREBP-1; hepatocytes

1. Introduction

Nonalcoholic fatty liver disease (NAFLD) affects 80–100 million people in the U.S. alone [1]. Most patients in the U.S., who visit primary care providers for routine checkup have elevated levels of alanine aminotransferase (ALT) and aspartate aminotransferase (AST), which are markers of liver damage [2]. Among those with elevated AST and ALT levels, 33 percent are eventually diagnosed with NAFLD [2]. This pattern is not limited to the Western world. In Asia, 25 percent of individuals aged >18 years are diagnosed with NAFLD [3]. Even though this widespread disease has no specific signs or symptoms, it can progress to nonalcoholic steatohepatitis (NASH) and eventually lead to cirrhosis [4]. Although treatment options for NAFLD are limited, thiazolidinediones and vitamin E have been proven to effectively manage symptoms [5].

Omega-3 fatty acid provides an energy source as nutrient, but also act as a remedy for serum hyperlipidemia. Commercially available fish oil supplements are one of the most popular over-the-counter medications for the management of elevated serum triglyceride (TG) levels. In the U.S. alone, about 18.8 million adults are taking fish oil supplements [6]. Omega-3-acid ethyl esters, such as Omacor® and Lovaza®, are widely used as prescription medications for dyslipidemia [7]. Omega-3 fatty acid supplementation decreases liver fat and reduces hepatic steatosis, as well as having beneficial effects on most cardiometabolic risk factors. According to clinical studies, dietary supplementation with omega-3 is efficacious for NAFLD management [8]. However, the mechanism of its TG-lowering effect is not fully understood.

Omega-3 fatty acids, such as docosahexaenoic acid (DHA) and eicosapentaenoic acid (EPA), have been recognized as endogenous ligands for G-protein-coupled receptor 40 (GPR40) and G-protein-coupled receptor 120 (GPR120) [9,10]. In animal models, the administration of selective agonists for GPR40 or GPR120 improves biochemical and pathological indices of hepatic steatosis [11,12]. Nevertheless, the proposed direct effects of omega-3 fatty acids on the G-protein-coupled receptor (GPCR) in hepatocytes are largely uncharacterized and controversial. In the present study, we assessed the effect of DHA on hepatic lipogenesis and attempted to clarify the roles of GPR40 and GPR120 in antilipogenesis effect of DHA in mouse primary hepatocytes.

2. Results

2.1. Downregulation of Sterol Regulatory Element Binding Protein (SREBP)-1-Dependent Lipogenic Enzyme Expression by DHA in Primary Hepatocytes

Public microarray data comparing control mice with mice receiving omega-3 fish oil for 2 weeks were analyzed to assess whether omega-3 fatty acids functionally benefit liver (GSE32706). Gene ontology analyses revealed that the genes, including those involved with lipid metabolic processes and cholesterol biosynthetic processes, were mainly altered for similar functional annotations in the liver tissues of fish-oil-fed mice (Figure S1A). In addition, expression of genes involved in lipid metabolism and fatty acid synthesis was repressed in the fish-oil-fed group (Figure S1A). Moreover, fish oil diet led to a significant reduction in the expression of several genes, including *srebf1*, *acacb*, *fasn*, and *scd1* which are required to control hepatic TG synthesis (Figure S1B).

The SREBP1 transcription factor is a critical regulator of fatty acid homeostasis in hepatocytes. Activation of SREBP1 controls the expression of a range of lipogenic enzymes, such as fatty acid synthase (FAS), acetyl-coenzyme A carboxylase (ACC), and stearoyl-CoA desaturase-1 (SCD-1) [13]. Hence, we analyzed the effects of DHA and EPA (300 μM, each) on the expression of SREBP1 and its downstream lipogenesis enzymes. The levels of FAS, ACC, and SCD-1 were diminished in mouse primary hepatocytes treated with DHA or EPA for 12 h, and the degree of inhibition was more prominent in DHA-treated hepatocytes (Figure 1A). The protein level of the precursor form of SREBP-1 (preSREBP-1) was decreased in primary hepatocytes in response to DHA (Figure 1B). To gain further insight into the antilipogenic properties of DHA, primary hepatocytes were treated with liver X receptor (LXR) agonist T0901317 (T090) to stimulate LXR-dependent SREBP-1 activation. T090-induced protein expression of nuclear SREBP-1 and SCD-1 were decreased significantly following pretreatment of hepatocytes with 300 μM DHA for 12 h (Figure 1C). Although the levels of preSREBP-1, FAS, and ACC were marginally elevated by T090, all the protein expression decreased in response to the DHA treatment (Figure 1D and Figure S2A). Primary hepatocytes were exposed to high glucose with insulin condition to confirm the antilipogenic properties of DHA in metabolic dysfunction. As expected, the amounts of lipogenic proteins involved in TG synthesis, such as preSREBP-1, nSREBP-1, FAS, ACC, and SCD-1, increased under the high glucose with insulin condition, and the enhanced levels were reversed by DHA treatment (Figure 1E and Figure S2B). These data support the notion that DHA ameliorates the SREBP-1-mediated lipogenesis enzyme expression caused by an LXR agonist or the high glucose with insulin condition.

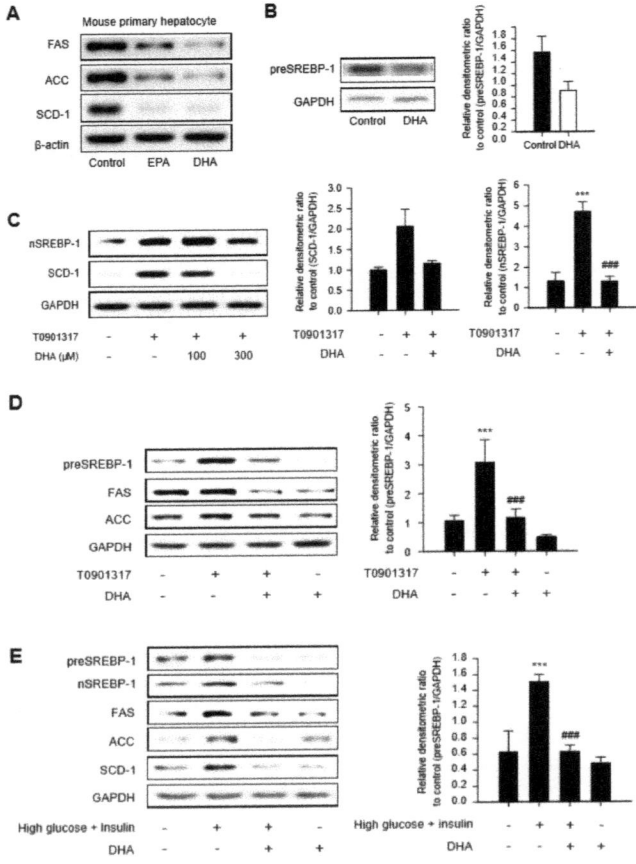

Figure 1. Decrease in lipogenesis enzymes after docosahexaenoic acid (DHA) treatment. (**A**,**B**) The protein expression level of fatty acid synthase (FAS), acetyl-coenzyme A carboxylase (ACC), stearoyl-CoA desaturase-1 (SCD-1) and precursor form of SREBP-1 (preSREBP-1). Mouse primary hepatocytes were treated with DHA 300 μM or eicosapentaenoic acid (EPA) 300 μM for 12 h. Data represent means ± SD ($n = 5$). GAPDH: Glyceraldehyde 3-phosphate dehydrogenase. (**C**) In order to screen an optimal DHA concentration that exerts potent antilipogenic effects, primary hepatocytes were treated with DHA 100 μM and 300 μM for 12 h followed by T0901317 (T090), liver X receptor (LXR) agonist for additional 12 h. Data represent means ± SD ($n = 5$), *** $p < 0.005$ compared with the control group; ### $p < 0.005$ compared with the T090-treated group (**D**) Effects of DHA on the expression of lipogenic proteins stimulated by LXR agonist. Primary hepatocytes were treated with DHA 300 μM for 12 h followed by T090 for additional 12 h. Data represent means ± SD ($n = 5$), *** $p < 0.005$ compared with the control group; ### $p < 0.005$ compared with the T090-treated group (**E**) Effects of DHA on the expression of lipogenic proteins in high glucose with insulin condition. Primary hepatocytes were treated with DHA 300 μM for 12 h followed by 30 mM high-glucose medium for 30 min and further incubation with 200 nM insulin for 24 h. Data represent means ± SD ($n = 3$), *** $p < 0.005$ compared with the control group; ### $p < 0.005$ compared with the high glucose and insulin group.

2.2. DHA-Induced Reduction of Total Area of Lipid Droplets in Hepatocytes

Hepatic steatosis is defined as abnormal retention of lipid droplets (LD) in hepatocytes, which reflects the dysregulation of TG fat [14]. We explored the effect of DHA on LD formation in hepatocytes. BODIPY® staining showed a greater increase in LDs in hepatocytes incubated under the high glucose and insulin condition, but not with T090 (Figure 2A,B). Interestingly, no difference was detected in the number of LDs per cell between hepatocytes treated with high glucose and insulin and those cotreated with DHA (Figure 2C). Nevertheless, the number of LDs > 10 μm^2, representing pathological lipid accumulation, as well as the total area of LDs per cell decreased in response to DHA (Figure 2D,E). One study showed that the decrease in the size of LDs is due mainly to the lower activity of SCD-1 [15]. The immunoblot results showed a dramatic decrease in SCD-1 with DHA treatment (Figure 1C,D), suggesting that DHA mainly targets enlargement of LDs in hepatocytes.

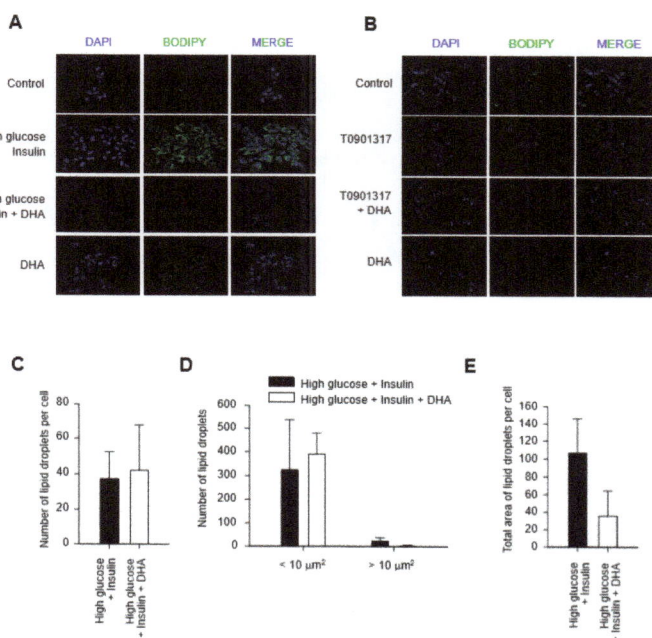

Figure 2. Morphological changes in lipid droplets (LD) after DHA treatment. (**A,B**) LD formation in hepatocytes by T090 treatment or high glucose with insulin condition. Mouse primary hepatocytes were treated with DHA for 12 h followed by treatment of T090 for 12 h or incubation with 30 mM high glucose for 30 min and further incubation with 200 nM insulin for 24 h. LDs were stained with BODIPY® 493/503 and visualized by confocal microscopy. (**C**) Images of 16 random cells from each slide were captured and analyzed by MetaMorph to quantify number of LDs per cell. (**D**) Numbers of LDs with areas ≤10 μm^2 or >10 μm^2 were analyzed. (**E**) Total area of LDs per cell were measured and presented as means ± SEM.

2.3. Limited Role of GPR 120 in Antilipogenic Effects of DHA in Hepatocytes

Long-chain fatty acids, such as DHA, act as endogenous ligands on GPR120 and GPR40 and regulates metabolic and inflammatory homeostasis in adipocytes and macrophages [10,16]. To investigate whether GPR40 and GPR120 are present in hepatocytes, mRNA levels were determined in HepG2 human hepatoma cell line and mouse primary hepatocytes. mRNA levels of GPR40 and GPR120 were detectable in HepG2 and mouse primary hepatocytes (Figure S3A and Figure 3A). Moreover, exposure to 300 μM DHA for 12 h increased the GPR40 and GPR120 mRNA levels

(Figure S3A and Figure 3A). Plasma membrane localization of GPR40 and GPR120 in primary hepatocyte was confirmed by immunocytochemistry (Figure 3B). These data demonstrate that GPR40 and GPR120 are expressed in hepatocytes and HepG2 cells are upregulated after DHA exposure. Compound A (CpdA), a specific GPR120 agonist [11], was used to clarify the function of GPR120 on the antilipogenic effects of DHA. Although the basal SCD-1 level was higher in HepG2 cells [17], CpdA concentration-dependently inhibited SCD-1 expression in T090-treated HepG2 cells and primary hepatocytes (Figure 3C and Figure S3B). However, unlike DHA, CpdA up to 10 μM did not reduce the expression of preSREBP-1 in primary hepatocytes treated with T090 (Figure 3C). Because series of studies proposed direct interaction between LXR and SCD-1 [18,19], we presumed that downregulation of SCD-1 by CpdA may not be dependent on SREBP-1 pathway. Under the high glucose with insulin condition, CpdA did not reduce the protein level of preSREBP-1, nSREBP-1, and FAS in hepatocytes (Figure 3D and Figure S3C). To confirm these findings, we used age-matched GPR120 knock-out (KO) mice. A pathological examination was performed in hematoxylin and eosin (H&E) stained liver tissues from 8-weeks-old GPR120 wild-type and KO mice to rule out preexisting histological conditions (Figure S3D). The decreased lipogenic enzyme expressions by DHA in primary hepatocytes incubated with T090 or high glucose and insulin were not reversed by GPR120 deficiency (Figure 3E,F and Figure S3E,F). These data imply that the inhibitory effects of DHA on the expression of SREBP-1-mediated lipogenic enzymes are minimally associated with GPR120 signaling.

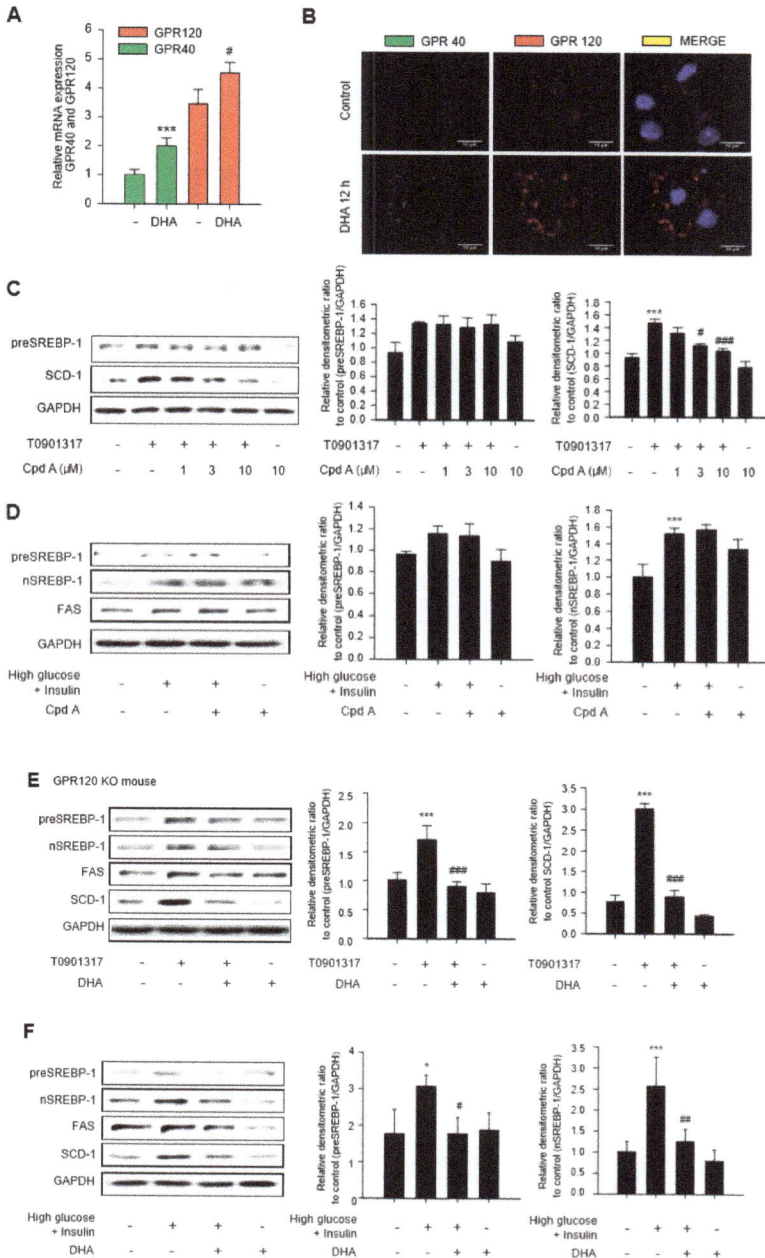

Figure 3. Role of GPR120 in expression of lipogenesis enzymes in hepatocytes. (**A**) Relative mRNA expressions of GPR40 and GPR120 levels were determined by real-time qPCR analyses in mouse primary hepatocytes. Data represent means ± SD ($n = 3$). *** $p < 0.005$ compared with the GPR40 level of the control group; # $p < 0.05$ compared with the GPR120 level of the control group.

(**B**) Immunocytochemistry of GPR40 and GPR120 in primary hepatocytes. Immunofluorescent images of GPR40 and GPR120 at their basal levels and after DHA treatment are shown. (**C**) Primary hepatocytes were treated with compound A (CpdA) (1–10 µM) for 12 h followed by T090, LXR agonist for additional 12 h. Data represent means ± SD (n = 5), *** $p < 0.005$ compared with the control group; # $p < 0.05$, ### $p < 0.005$ compared with the T090-treated group. (**D**) Effects of CpdA on the expression of preSREBP1, nSREBP1, and FAS in primary hepatocytes. Primary hepatocytes were treated with CpdA 10 µM for 12 h followed by 30 mM high-glucose medium for 30 min and further incubation with 200 nM insulin for 24 h. Data represent means ± SD (n = 5), *** $p < 0.005$ compared with the control group. (**E**) Effects of DHA on the expression of preSREBP1, nSREBP1, FAS, and SCD-1 in GPR120 knock-out (KO) primary hepatocytes stimulated with T090. GPR120 KO hepatocytes were treated with DHA for 12 h followed by T090 for additional 12 h. Data represent means ± SD (n = 3), *** $p < 0.005$ compared with the control group; # $p < 0.05$, ### $p < 0.005$ compared with the T090-treated group. (**F**) Effects of DHA on the expression of preSREBP-1, nSREBP-1, FAS, and SCD-1 in GPR120 KO primary hepatocytes stimulated with high glucose and insulin. GPR120 KO primary hepatocytes were treated with DHA 300 µM for 12 h followed by 30 mM high-glucose medium for 30 min and further incubation with 200 nM insulin for 24 h. Data represent means ± SD (n = 3), * $p < 0.05$, *** $p < 0.005$ compared with the control group; # $p < 0.05$, ## $p < 0.01$ compared with the high glucose and insulin group.

2.4. Involvement of GPR40 in Antilipogenesis Effect of DHA

We tested the potential role of GPR40 in antilipogenic effects of DHA in hepatocytes. AMG-1638 is a well-characterized GPR40 agonist in murine species and has shown a specific full agonistic activity compared with other candidates [20]. The enhanced expression of preSREBP-1 and SCD-1 caused by T090 was significantly diminished in hepatocytes incubated with 3 µM AMG-1638 (Figure 4A). Moreover, 3 µM AMG-1638 reduced the protein expression of FAS and SCD-1 as well as preSREBP-1 and nSREBP-1, under the high glucose and insulin condition (Figure 4B). We further assessed the effect of the GPR40 antagonist GW1100 to confirm whether the antilipogenic properties of DHA were related with GPR40 activation. When hepatocytes were co-incubated with both GW1100 and DHA, the inhibitory effects of DHA on the expression of preSREBP-1 and SCD-1 were almost completely reversed by GW1100 in T090-exposed hepatocytes (Figure 4C). Similar results were obtained for GPR120 KO hepatocytes (Figure 4D). Hence, downregulation of SREBP-1-mediated expression of lipogenic enzymes by DHA was under the control of GPR40 in hepatocytes.

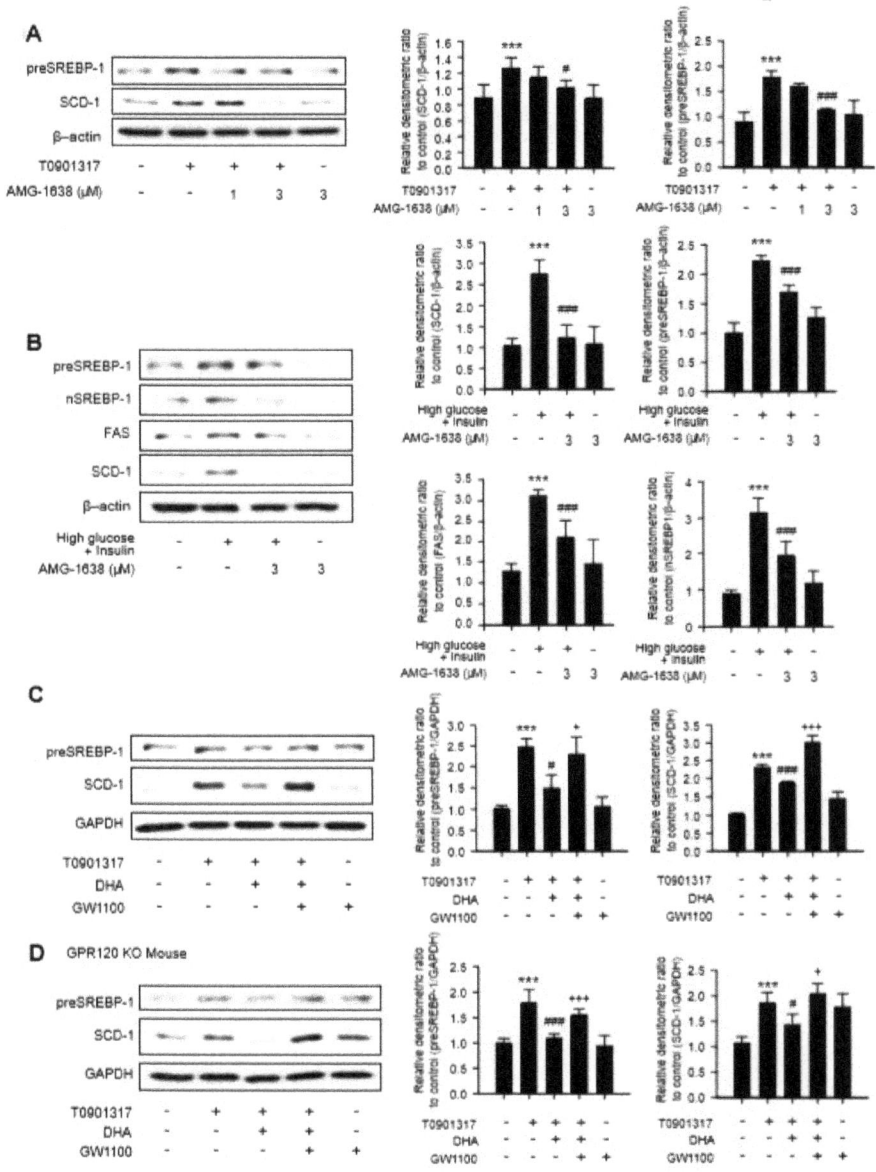

Figure 4. Role of GPR40 in expression of lipogenesis enzymes in hepatocytes. (**A**) Primary hepatocytes were treated with AMG-1638 (1 and 3 μM) for 12 h followed by T090 for additional 12 h, and total cell lysates were subjected to immunoblotting for preSREBP-1 and SCD-1.

Data represent means ± SD (n = 4), ***P < 0.005 compared with the control group; [#] $p < 0.05$, [###] $p < 0.005$ compared with the T090-treated group. (**B**) Inhibitory effects of AMG-1638 on the protein expression of lipogenic enzymes in high glucose and insulin condition. Primary hepatocytes were treated with 3 μM AMG-1638 for 12 h followed by 30 mM high-glucose medium for 30 min and further incubation with 200 nM insulin for 24 h. Data represent means ± SD ($n = 4$), *** $p < 0.005$ compared with the control group, [###] $p < 0.005$ compared with the T090-treated group. (**C**) Primary hepatocytes were treated with DHA and 10 μM GW1100, GPR40 antagonist for 12 h followed by 30 mM high-glucose medium for 30 min and further incubation with 200 nM insulin for 24 h. Data represent means ± SD ($n = 4$), *** $p < 0.005$ compared with the control group; # $p < 0.05$, [###] $p < 0.005$ compared with the T090-treated group; [+] $p < 0.05$, [+++] $p < 0.005$ compared with the T090 and DHA treated group. (**D**) In GPR120 KO hepatocytes, 10 μM GW1100 reversed the antilipogenic effects of DHA. Data represent means ± SD ($n = 4$). *** $p < 0.005$ compared with the control group; [#] $p < 0.05$, [###] $p < 0.005$ compared with the T090-treated group; +$p < 0.05$, [+++] $p < 0.005$ compared with the T090 and DHA treated group.

3. Discussion

DHA and EPA are omega-3 fatty acids and known as ligands for GPR40 and GPR120. Although they both have antilipogenic properties, DHA is associated with a greater reduction in serum TG levels compared with EPA [21]. In addition, DHA has a lipid control benefit over EPA, as it increases the serum level of high-density lipoprotein (HDL) [22]. Many studies have explored the potentially beneficial effects of GPR40 and GPR120 in various cells and organs. However, possibly due to limited expression of GPR40 and GPR120 in hepatocytes compared with other cell types [23,24], few studies have clearly identified the receptor involved in the antilipogenic effects of omega-3 fatty acids in the liver. In this study, we investigated the effects of DHA on lipogenic enzyme expression in primary hepatocytes and sought to clarify the related lipid-sensing GPCR exerting its antilipogenesis effect.

GPCRs are usually downregulated to eschew repercussive effects by persistent stimulation with agonists. Nobili et al. reported that patients administered DHA develop increased levels of GPR120 in hepatocytes [25]. This result is consistent with our finding that GPR120 and GPR40 mRNA levels are increased in mouse primary hepatocytes treated with DHA. Although regulatory mechanisms underlying the expression of long-chain fatty-acid-sensing GPCRs are not fully understood, Abaraviciene et al. showed that the GPR40 protein level is increased in response to 100- and 1000-μM palmitate exposures in rat islet [26]. However, GPR40 mRNA expression was enhanced only by 100 μM palmitate treatment. Therefore, we expect that both transcriptional and post-translational regulation play a role in the enhanced expression of GPR40 and GPR120 after DHA exposure. Moreover, immunohistochemical analyses using GPR40 and GPR120 antibodies confirmed that both GPR40 and GPR120 were present in mouse primary hepatocytes. Although GPR40 and GPR120 have different molecular structures, they share long chain fatty acids as their endogenous ligands, and the identification of highly specific ligands to discriminate the receptors is difficult [11]. Before CpdA was developed by Merck, TUG-891 was the most selective compound for GPR120. Unfortunately, TUG-891 loses its GPR120 selectivity in murine species [27]. Thus, in this study, CpdA was used to target GPR120 in mouse primary hepatocytes. We found that CpdA did not lower the protein expression of SREBP-1-dependent lipogenic enzymes stimulated by the high glucose with insulin milieu. Moreover, DHA reduced protein levels of preSREBP-1 and FAS in primary GPR120 KO hepatocytes exposed to T090 or high glucose and insulin. From this finding, we suggest that GPR120 is not mainly involved in the antilipogenic effects of DHA in hepatocytes. Contrary to our findings, Kang et al. reported that DHA and TUG-891 decreased LXR-mediated lipogenic protein expression in HepG2 and Hep3B hepatoma cell lines, as well as mouse primary hepatocytes, and further revealed that GPR120 is involved in the antilipogenic effect of DHA [28] .This discrepancy seems to be mainly to differences in the concentration range of DHA in cell-based analyses. In our experimental conditions, protein expression of preSREBP-1 and a series of lipogenic enzymes were suppressed by 300 μM DHA treatment. In contrast, others have reported that 10–30 μM DHA efficiently inhibited lipid accumulation by downregulating the SREBP-1 pathway in hepatoma cell lines and primary hepatocytes. DHA seems to have a potent binding affinity to GPR120. A DHA concentration <1 µM evoked internalization of GPR120 in HEK293 cells stably expressing GPR120-enhanced green fluorescent protein (EGFP) [24]. Hence, relatively low concentrations of DHA may preferentially stimulate the GPR120 signal, whereas higher concentration act on GPR40. In fact, 10–20 µM DHA induces calcium influx in Chinese Hamster Ovary (CHO) cells expressing GPR40 [29].

Because the $G\alpha_{q/11}$–coupled receptor GPR40 is expressed mainly in pancreatic β-cells, most of previous studies focused on its potential ability to stimulate insulin secretion. Because GPR40 activation stimulates insulin secretion only in the presence of elevated glucose levels [9,30,31], it has become an attractive potential therapeutic target for glucose homeostasis with little to no hypoglycemic risk. GPR40 is negligibly expressed in the liver [16]. Thus, the functional roles of GPR40 in hepatocytes have received less attention. In this study, we used GPR40 full agonist, AMG-1638, to assess whether GPR40 plays a functional role in hepatocytes. The enhanced lipogenic enzyme levels caused by LXR agonist or high glucose with insulin milieu were diminished by AMG-1638 in primary hepatocytes. Furthermore, GPR40 antagonist GW1100 efficiently abrogated the antilipogenic effects of DHA. These cell-based analyses using specific pharmacological tools suggest that receptor signaling plays an antilipogenic role in primary hepatocytes. Li et al. reported that the GPR40/120 agonist GW9508 improves hepatic steatosis in mice fed a high-cholesterol diet and inhibits LXR ligand-induced expression of lipogenic enzymes in HepG2 cells [12]. They further found that the antilipogenic activity of GW9508 is significantly reversed by GPR40 siRNA, suggesting a pivotal role of GPR40 in the regulation of hepatic lipid accumulation. These data are consistent with our findings.

Taken together, our findings may help to unravel how DHA alleviates fatty acid accumulation in hepatocytes. Furthermore, these findings support the notion of using specific GPCR agonists as an add-on therapy to manage metabolic syndrome and suggest that GPR40 merits further investigation as an adjuvant therapy for NAFLD.

4. Materials and Methods

4.1. Reagent and Antibodies

Antibody recognizing precursor sterol regulatory element binding protein-1 (SREBP-1) was obtained from Santa Cruz Biotechnology (Dallas, TX, USA). Anti-fatty acid synthase (FAS) and antinuclear form of SREBP-1 antibodies were supplied by BD Biosciences (Franklin Lakes, NJ). Stearoyl-CoA desaturase-1 (SCD1), phospho-AMP-activated protein kinase (p-AMPK), and acetyl-CoA carboxylase (ACC) antibodies were obtained from Cell Signaling Technology (Beverly, MA). Anti-glyceraldehyde 3-phosphate dehydrogenase (GAPDH) antibody and T0901317 (T090) were supplied by Calbiochem (San Diego, CA, USA). Anti-β actin antibody, insulin, and glucose were purchased from Sigma-Aldrich (St. Louis, MO). Horseradish peroxidase-conjugated donkey anti-rabbit IgG, and alkaline phosphatase-conjugated donkey anti-mouse IgG were obtained from Jackson Immunoresearch Laboratories (West Grove, PA, USA). Eicosapentaenoic acid (EPA), docosahexaenoic acids (DHA), and compound A (CpdA) were supplied by Cayman Chemical (Ann Arbor, MI, USA). GW1100 and AMG-1638 were kindly donated from LG Chem Ltd. (Seoul, South Korea).

4.2. Cell Culture

Human hepatoma cell line HepG2 was obtained from American Type Culture Collection (ATCC, Manassas, VA, USA). HepG2 cells were grown in low-glucose Dulbecco's modified Eagle's medium (DMEM) with 10% fetal bovine serum (FBS, Gibco, Thermo Fisher Scientific, Waltham, MA, USA) with 50 U/mL penicillin and 50 µg/mL streptomycin. All cell lines were maintained at 37 °C in a humidified incubator with 5% CO_2.

4.3. Animals

GPR120 knock-out (KO) C57BL/6 mice were kindly donated from LG Chem Ltd. The mice were housed in a pathogen-free animal facility under a standard 12 h light/dark cycle. Animal experiments were conducted under the guidelines of the Institutional Animal Use and Care Committee at Seoul National University (SNU-160512-11-1, 12 May, 2016).

4.4. Isolation of Primary Hepatocytes

After anesthetizing 8-weeks-old C57BL/6 mice (DBL, Eumseong, Korea) or GPR120 KO mice with zoletil® and rompun®, 24G catheter was cannulated to liver portal vein. After perfusion with Hank's Balanced Salt solution medium (Life Technologies, Grand Island, NY, USA) supplemented with 0.5 mM ethylene glycol tetraacetic acid and 25 mM 4-(2-hydroxyethyl)-1-piperazineethanesulfonic acid (HEPES), liver tissue was digested with low-glucose DMEM containing 1% penicillin/streptomycin, 15 mM HEPES and 1 mg/mL collagenase from *Clostridium histolyticum* (Sigma-Aldrich, St.Louis, MO, USA). After tearing digested liver tissue, the isolated cells were washed three times with high-glucose DMEM supplemented with 10% FBS, 1% penicillin/streptomycin, 15 mM HEPES, 10 nM dexamethasone. Other cells except primary hepatocytes were all removed by centrifugation.

4.5. mRNA Isolation and Real-Time Quantitative Polymerase Chain Reaction (qPCR)

After washing with sterile phosphate-buffered saline (PBS), total RNA was isolated using TRIzol reagent (Life Technologies, Grand Island, NY). mRNA was reverse transcribed to cDNA using Maxime RT Premix (iNtRON, Seongnam, South Korea). Amplified cDNA was analyzed by Bio-Rad CFX Manager™ Software (Bio-Rad, Hercules, CA, USA) using iTaq Universal SYBR Green Supermix (Bio-Rad) and SYBR Select Master Mix (Life Technologies). Primer sequences used in experiments were:

5'-CTGTGCAGGAATGAGTGGAAG-3' (mouse GPR120-forward)
5'-CTGATGGAGGGTACTGGAAATG-3' (mouse GPR120-reverse)
5'-CTGTGCAGGAATGAGTGGAAG-3' (human GPR120-forward)
5'-CTGATGGAGGGTACTGGAAATG-3' (human GPR120-reverse)
5'-TGGCGCGCCAGCCTGG-3' (mouse GPR120 KO-forward)
5'-CCATATGAAAGCCAGCAGTGCC-3' (mouse GPR120 KO-reverse)
5'-CGCCGTCGGCGCCGTG-3' (mouse GPR120 WT-forward)
5'-CCATATGAAAGCCAGCAGTGCC-3' (mouse GPR120 WT-reverse)
5'-TTTCATAAACCCGGACCTAGGA-3' (mouse GPR40-forward)
5'-CCAGTGACCAGTGGGTTGAGT-3' (mouse GPR40-reverse)
5'- GCCCACTTCTTCCCACTCTA-3' (human GPR40-forward)
5'- AGACCCAGGTGACACAGGAC -3' (human GPR40-reverse)
5'-GTAACCCGTTGAACCCCATT-3' (mouse and human 18s rRNA protein-forward)
5'-CCATCCAATCGGTAGTAGCG-3' (mouse and human 18s rRNA protein-reverse)

4.5. Immunoblot Analysis

After washing the cultured cells with PBS, cells were lysed in lysis buffer containing 20 mM TrisHCl (pH 7.5), 1% Triton X-100, 137 mM sodium chloride, 10% glycerol, 2 mM EDTA, 1 mM sodium orthovanadate, 25 mM β-glycerophosphate, 2 mM sodium pyrophosphate, 1 mM phenyl methyl sulfonyl fluoride, and 1 μg/mL leupeptin. Cells were then incubated in ice for 1 h. The cell lysates were centrifuged at 10,000 *g* for 20 min to remove debris, and the protein samples were loaded on 8–15% SDS-PAGE gel and transferred to nitrocellulose membrane (GE healthcare Life Sciences, Chalfont, Buckinghamshire, UK). Membranes were incubated for 1 h with 5% skim milk (BD Bioscience, San Jose, CA, USA) and reacted with primary antibodies overnight at 4 °C. The membranes were washed and incubated with a second antibody for 1 h at room temperature. Protein expression was visualized with LAS3000-mini (Fujifilm, Tokyo, Japan) using enhanced chemiluminescence (ECL) system reagent

Int. J. Mol. Sci. **2019**, *20*, 2625

(EMD Millipore, Billerica, MA, USA). Densitometric analysis was performed by using Multiguage software (Fujifilm, Tokyo, Japan) and Image J 1.46r.

4.6. Hematoxylin and Eosin Staining

The left lateral lobe of the liver was sliced, and tissue slices were fixed in 10% buffered-neutral formalin. The liver slices were stained with hematoxylin and eosin (H&E).

4.7. Detection of Lipid Droplets (LDs) by Confocal Fluorescence Scanning Microscopy

For BODIPY® 493/503 staining (Thermo Fisher Scientific, Waltham, MA, USA), the 488 nm laser was used and signals were collected with a long pass 505 nm filter. Quantification of LD number and size was performed with MetaMorph, version 7.5 (Molecular Devices, San Jose, CA, USA).

4.8. Statistical Analysis

The analyses were performed with IBM® SPSS® Statistics 23 (IBM SPSS Statistics, IBM Corporation). The significance level was set at p value < 0.05 for all comparisons. Although the size of sample is small, Shapiro–Wilk test showed normality of the data and one-way analysis of variance (ANOVA) was used for multiple comparisons. Tukey's or Dunnett's tests were used as post hoc analysis methods. Statistical analysis for GSE data was performed by using Student's t-test and Benjamini–Hochberg test.

Supplementary Materials: **Supplementary Materials** can be found at http://www.mdpi.com/1422-0067/20/11/2625/s1. **Figure S1.** (A) Functional annotations associated with genes exhibiting lipid metabolism and cholesterol metabolism were altered in the fish-oil-fed mouse. (B) According to GSE32706, genes controlling synthesis of triglyceride including *srebf1, acacb, fasn,* and *scd1* were reduced in the liver from the fish-oil-fed mouse. Data represent means ± SD ($n = 3$), * $p < 0.05$, *** $p < 0.05$ compared with the control diet group. **Figure S2.** (A) Effects of DHA on the expression of lipogenic proteins stimulated by T090. Relative densitometric ratio of FAS and ACC. Data represent means ± SD ($n = 5$), # $p < 0.05$, ### $p < 0.005$ compared with the T090-treated group. (B) Effects of DHA on the expression of lipogenic proteins stimulated by high glucose and insulin. Relative densitometric ratio of nSREBP-1, FAS, ACC, and SCD-1. Data represent means ± SD ($n = 5$), *** $p < 0.05$ compared with the control group; ### $p < 0.005$ compared with the T090-treated group. **Figure S3.** (A) Relative mRNA expressions of GPR40 and GPR120 levels were determined by real-time qPCR analyses in HepG2 cell line. Data represent means ± SD ($n = 4$), *** $p < 0.005$ and ### $p < 0.005$, compared with the GPR40 and GPR120 levels of control group, respectively. (B) Effects of CpdA on the expression of preSREBP-1, and SCD-1 in HepG2 were determined by immunoblottings. (C) Effects of CpdA on the expression of lipogenic proteins stimulated by high glucose and insulin. Relative densitometric ratio of FAS. Data represent means ± SD ($n = 5$), *** $p < 0.05$ compared with the control group. (D) Hematoxylin and eosin (H&E) staining of liver sections from 8-weeks-old wild-type (WT) and GPR120 mice. Original magnification setting of 200× for H&E was used. (E) Effects of DHA on the expression of lipogenic proteins in GPR120 KO primary hepatocytes stimulated by T090. Relative densitometric ratio of nSREBP-1 and FAS. Data represent means ± SD ($n = 3$), *** $p < 0.005$ compared with the control group; ### $p < 0.005$ compared with the T090-treated group. (F) Effects of DHA on the expression of lipogenic proteins in GRP120 KO primary hepatocytes stimulated by high glucose and insulin. Relative densitometric ratio of FAS and SCD-1. Data represent means ± SD ($n = 3$), *** $p < 0.05$ compared with the control group; ## $p < 0.01$, ### $p < 0.005$ compared with the high glucose and insulin group.

Author Contributions: Conceptualization, K.W.K. and S.O.; Investigation, S.O., H.Y.K., H.S.K. and J.P.; Writing-original draft, S.O., H.Y.K. and K.W.K.; Writing-review & editing, H.Y.K. and K.W.K.

Acknowledgments: This work was supported by the National Research Foundation of Korea (NRF) grants funded by the Korean Government (2017M3A9C8028794).

Conflicts of Interest: The authors declare no conflict of interest. The funders had no role in the design of the study; in the collection, analyses, or interpretation of data; in the writing of the manuscript, and in the decision to publish the results.

Abbreviations

ACC	Acetyl-Coenzyme A carboxylase
ALT	Alanine transaminase
AMPK	Adenosine Monophosphate-activated protein kinase
AST	Aspartate transaminase
CpdA	Compound A
DHA	Docosahexaenoic acid
FAS	Fatty acid synthase
FFA	Free fatty acid
GPR120	G-protein-coupled receptor 120
GPR40	G-protein-coupled receptor 40
GADPH	Glyceraldehyde-3-phosphate dehydrogenase
GO	Gene Ontology
LDs	Lipid droplets
LXR	Liver X receptor
NAFLD	Nonalcoholic fatty liver disease
NASH	Nonalcoholic steatohepatitis
ND	Not detected
SCD-1	Stearoyl-Coenzyme A desaturase-1
SREBP-1	Sterol regulatory element binding protein-1
TG	Triglyceride
T090	T0901317

References

1. Perumpail, B.J.; Khan, M.A.; Yoo, E.R.; Cholankeril, G.; Kim, D.; Ahmed, A. Clinical epidemiology and disease burden of nonalcoholic fatty liver disease. *World J. Gastroenterol* **2017**, *23*, 8263–8276. [CrossRef] [PubMed]
2. Bedogni, G.; Nobili, V.; Tiribelli, C. Epidemiology of fatty liver: An update. *World J. Gastroenterol* **2014**, *20*, 9050–9054. [PubMed]
3. Fan, J.G.; Kim, S.U.; Wong, V.W. New trends on obesity and nafld in asia. *J. Hepatol.* **2017**, *67*, 862–873. [CrossRef]
4. Calzadilla Bertot, L.; Adams, L.A. The natural course of non-alcoholic fatty liver disease. *Int. J. Mol. Sci.* **2016**, *17*. [CrossRef]
5. Chalasani, N.; Younossi, Z.; Lavine, J.E.; Charlton, M.; Cusi, K.; Rinella, M.; Harrison, S.A.; Brunt, E.M.; Sanyal, A.J. The diagnosis and management of nonalcoholic fatty liver disease: Practice guidance from the american association for the study of liver diseases. *Hepatology* **2018**, *67*, 328–357. [CrossRef]
6. Clarke, T.C.; Black, L.I.; Stussman, B.J.; Barnes, P.M.; Nahin, R.L. Trends in the use of complementary health approaches among adults: United states, 2002–2012. *Natl. Health Stat. Rep.* **2015**, *10*, 1–16.
7. Backes, J.; Anzalone, D.; Hilleman, D.; Catini, J. The clinical relevance of omega-3 fatty acids in the management of hypertriglyceridemia. *Lipids Health Dis.* **2016**, *15*, 118. [CrossRef]
8. Parker, H.M.; Johnson, N.A.; Burdon, C.A.; Cohn, J.S.; O'Connor, H.T.; George, J. Omega-3 supplementation and non-alcoholic fatty liver disease: A systematic review and meta-analysis. *J. Hepatol.* **2012**, *56*, 944–951. [CrossRef] [PubMed]
9. Briscoe, C.P.; Tadayyon, M.; Andrews, J.L.; Benson, W.G.; Chambers, J.K.; Eilert, M.M.; Ellis, C.; Elshourbagy, N.A.; Goetz, A.S.; Minnick, D.T. The orphan g protein-coupled receptor gpr40 is activated by medium and long chain fatty acids. *J. Biol. Chem.* **2003**, *278*, 11303–11311. [CrossRef]
10. Oh, D.Y.; Talukdar, S.; Bae, E.J.; Imamura, T.; Morinaga, H.; Fan, W.; Li, P.; Lu, W.J.; Watkins, S.M.; Olefsky, J.M. Gpr120 is an omega-3 fatty acid receptor mediating potent anti-inflammatory and insulin-sensitizing effects. *Cell* **2010**, *142*, 687–698. [CrossRef]
11. Oh, D.Y.; Walenta, E.; Akiyama, T.E.; Lagakos, W.S.; Lackey, D.; Pessentheiner, A.R.; Sasik, R.; Hah, N.; Chi, T.J.; Cox, J.M.; et al. A gpr120-selective agonist improves insulin resistance and chronic inflammation in obese mice. *Nat. Med.* **2014**, *20*, 942–947. [CrossRef] [PubMed]

12. Li, M.; Meng, X.; Xu, J.; Huang, X.; Li, H.; Li, G.; Wang, S.; Man, Y.; Tang, W.; Li, J. Gpr40 agonist ameliorates liver x receptor-induced lipid accumulation in liver by activating ampk pathway. *Sci. Rep.* **2016**, *6*, 25237. [CrossRef] [PubMed]

13. Eberle, D.; Hegarty, B.; Bossard, P.; Ferre, P.; Foufelle, F. Srebp transcription factors: Master regulators of lipid homeostasis. *Biochimie* **2004**, *86*, 839–848. [CrossRef] [PubMed]

14. Gluchowski, N.L.; Becuwe, M.; Walther, T.C.; Farese, R.V., Jr. Lipid droplets and liver disease: From basic biology to clinical implications. *Nat. Rev. Gastroenterol. Hepatol.* **2017**, *14*, 343–355. [CrossRef]

15. Shi, X.; Li, J.; Zou, X.; Greggain, J.; Rødkær, S.V.; Færgeman, N.J.; Liang, B.; Watts, J.L. Regulation of lipid droplet size and phospholipid composition by stearoyl-coa desaturase. *J. Lipid Res.* **2013**, *54*, 2504–2514. [CrossRef]

16. Steneberg, P.; Rubins, N.; Bartoov-Shifman, R.; Walker, M.D.; Edlund, H. The ffa receptor gpr40 links hyperinsulinemia, hepatic steatosis, and impaired glucose homeostasis in mouse. *Cell Metab.* **2005**, *1*, 245–258. [CrossRef]

17. Huang, G.-M.; Jiang, Q.-H.; Cai, C.; Qu, M.; Shen, W. Scd1 negatively regulates autophagy-induced cell death in human hepatocellular carcinoma through inactivation of the ampk signaling pathway. *Cancer Lett.* **2015**, *358*, 180–190. [CrossRef]

18. Paton, C.M.; Ntambi, J.M. Biochemical and physiological function of stearoyl-coa desaturase. *Am. J. Physiol. Endocrinol. Metab.* **2009**, *297*, E28–E37. [CrossRef]

19. Wang, Y.; Kurdi-Haidar, B.; Oram, J.F. Lxr-mediated activation of macrophage stearoyl-coa desaturase generates unsaturated fatty acids that destabilize abca1. *J. Lipid Res.* **2004**, *45*, 972–980. [CrossRef]

20. Brown, S.P.; Dransfield, P.J.; Vimolratana, M.; Jiao, X.; Zhu, L.; Pattaropong, V.; Sun, Y.; Liu, J.; Luo, J.; Zhang, J.; et al. Discovery of am-1638: A potent and orally bioavailable gpr40/ffa1 full agonist. *Acs. Med. Chem. Lett.* **2012**, *3*, 726–730. [CrossRef]

21. Jacobson, T.A.; Glickstein, S.B.; Rowe, J.D.; Soni, P.N. Effects of eicosapentaenoic acid and docosahexaenoic acid on low-density lipoprotein cholesterol and other lipids: A review. *J. Clin. Lipidol.* **2012**, *6*, 5–18. [CrossRef]

22. Wei, M.Y.; Jacobson, T.A. Effects of eicosapentaenoic acid versus docosahexaenoic acid on serum lipids: A systematic review and meta-analysis. *Curr. Atheroscler. Rep.* **2011**, *13*, 474–483. [CrossRef]

23. Gotoh, C.; Hong, Y.H.; Iga, T.; Hishikawa, D.; Suzuki, Y.; Song, S.H.; Choi, K.C.; Adachi, T.; Hirasawa, A.; Tsujimoto, G.; et al. The regulation of adipogenesis through gpr120. *Biochem. Biophys. Res. Commun.* **2007**, *354*, 591–597. [CrossRef]

24. Hirasawa, A.; Tsumaya, K.; Awaji, T.; Katsuma, S.; Adachi, T.; Yamada, M.; Sugimoto, Y.; Miyazaki, S.; Tsujimoto, G. Free fatty acids regulate gut incretin glucagon-like peptide-1 secretion through gpr120. *Nat. Med.* **2005**, *11*, 90–94. [CrossRef]

25. Nobili, V.; Carpino, G.; Alisi, A.; De Vito, R.; Franchitto, A.; Alpini, G.; Onori, P.; Gaudio, E. Role of docosahexaenoic acid treatment in improving liver histology in pediatric nonalcoholic fatty liver disease. *Plos ONE* **2014**, *9*, e88005. [CrossRef]

26. Abaraviciene, S.M.; Muhammed, S.J.; Amisten, S.; Lundquist, I.; Salehi, A. Gpr40 protein levels are crucial to the regulation of stimulated hormone secretion in pancreatic islets. Lessons from spontaneous obesity-prone and non-obese type 2 diabetes in rats. *Mol. Cell. Endocrinol.* **2013**, *381*, 150–159. [CrossRef]

27. Hudson, B.D.; Shimpukade, B.; Mackenzie, A.E.; Butcher, A.J.; Pediani, J.D.; Christiansen, E.; Heathcote, H.; Tobin, A.B.; Ulven, T.; Milligan, G. The pharmacology of tug-891, a potent and selective agonist of the free fatty acid receptor 4 (ffa4/gpr120), demonstrates both potential opportunity and possible challenges to therapeutic agonism. *Mol. Pharm.* **2013**, *84*, 710–725. [CrossRef]

28. Kang, S.; Huang, J.; Lee, B.-K.; Jung, Y.-S.; Im, E.; Koh, J.-M.; Im, D.-S. Omega-3 polyunsaturated fatty acids protect human hepatoma cells from developing steatosis through FFA4 (GPR120). *Biochim. Biophys. Acta Mol. Cell Biol. Lipids.* **2018**, *1863*, 105–116. [CrossRef]

29. Itoh, Y.; Kawamata, Y.; Harada, M.; Kobayashi, M.; Fujii, R.; Fukusumi, S.; Ogi, K.; Hosoya, M.; Tanaka, Y.; Uejima, H. Free fatty acids regulate insulin secretion from pancreatic β cells through gpr40. *Nature* **2003**, *422*, 173. [CrossRef]

30. Feng, X.T.; Leng, J.; Xie, Z.; Li, S.L.; Zhao, W.; Tang, Q.L. Gpr40: A therapeutic target for mediating insulin secretion (review). *Int. J. Mol. Med.* **2012**, *30*, 1261–1266. [CrossRef]

31. Syed, I.; Lee, J.; Moraes-Vieira, P.M.; Donaldson, C.J.; Sontheimer, A.; Aryal, P.; Wellenstein, K.; Kolar, M.J.; Nelson, A.T.; Siegel, D.; et al. Palmitic acid hydroxystearic acids activate gpr40, which is involved in their beneficial effects on glucose homeostasis. *Cell Metab.* **2018**, *27*, 419–427.e414. [CrossRef] [PubMed]

International Journal of
Molecular Sciences

MDPI

Review

The Usefulness of Immunohistochemistry in the Differential Diagnosis of Lesions Originating from the Myometrium

Piotr Rubisz [1], Michał Ciebiera [2,*], Lidia Hirnle [1], Magdalena Zgliczyńska [3], Tomasz Łoziński [4], Piotr Dzięgiel [5,6] and Christopher Kobierzycki [5]

[1] First Department of Gynecology and Obstetrics, Wroclaw Medical University, 50-368 Wrocław, Poland; piotr.rubisz@gmail.com (P.R.); lidia.hirnle@umed.wroc.pl (L.H.)
[2] Second Department of Obstetrics and Gynecology, The Center of Postgraduate Medical Education, 01-809 Warsaw, Poland
[3] Students' Scientific Association at the I Department of Obstetrics and Gynecology, Medical University of Warsaw, 02-015 Warsaw, Poland; zgliczynska.magda@gmail.com
[4] Department of Obstetrics and Gynecology Pro-Familia Hospital, 35-001 Rzeszów, Poland; tomasz.lozinski@pro-familia.pl
[5] Division of Histology and Embryology, Department of Human Morphology and Embryology, Wroclaw Medical University, 50-368 Wroclaw, Poland; piotr.dziegiel@umed.wroc.pl (P.D.); ch.kobierzycki@gmail.com (C.K.)
[6] Department of Physiotherapy, University School of Physical Education, 51-612 Wroclaw, Poland
* Correspondence: michal.ciebiera@gmail.com; Tel.: +48-607-155-177

Received: 22 January 2019; Accepted: 1 March 2019; Published: 6 March 2019

Abstract: Uterine leiomyomas (LMs), currently the most common gynecological complaint around the world, are a serious medical, social and economic problem. Accurate diagnosis is the necessary prerequisite of the diagnostic-therapeutic process. Statistically, mistakes may occur more often in case of disease entities with high prevalence rates. Histopathology, based on increasingly advanced immunohistochemistry methods, is routinely used in the diagnosis of neoplastic diseases. Markers of the highest sensitivity and specificity profiles are used in the process. As far as LMs are concerned, the crux of the matter is to identify patients with seemingly benign lesions which turn out to be suspicious (e.g., atypical LM) or malignant (e.g., leiomyosarcoma (LMS)), which is not uncommon. In this study, we present the current state of knowledge about the use of immunohistochemical markers in the differential diagnosis of LM, atypical LM, smooth muscle tumors of uncertain malignant potential (STUMP), and LMS, as well as their clinical predictive value.

Keywords: uterine fibroid; leiomyoma; smooth muscle tumor of uncertain malignant potential; leiomyosarcoma; myometrium; immunohistochemistry; marker; pathology; tumor; diagnosis

1. Introduction

1.1. Epidemiology and Etiopathology of Myometrial Neoplasms

Tumors arising from the smooth muscle cells of the uterus are the most common neoplasms of the female genital tract around the world [1,2], chief among them leiomyomas (LMs), which are benign lesions of the uterus. Their prevalence is age-dependent, reaching 70–80% among women >50 years of age, and has been estimated at approximately 40–60% among women <35 years of age, depending on the population [3,4]. The results of the population observations have been confirmed by hysterectomy preparations—LMs are diagnosed in 77–96% of the patients [5,6], followed by lipoleiomyomas, which are relatively common LM variants (approx. 2.3%) among postmenopausal women who require surgical intervention [6].

Despite the fact that LMs are very common, the etiopathogenesis and pathophysiology of these benign tumors remain unclear [7–10]. The risk factors for the development of LMs include modifiable factors [1], e.g., dietary components [7,11,12], and non-modifiable factors, e.g., genetic [13,14], both in case of spontaneous changes [15] as well as autosomal dominant heredity [16].

According to the available literature, LM growth depends mostly on steroid hormones [13,17–19]. Nowadays, progesterone is believed to be the dominant factor in LM pathophysiology and growth stimulation [13,19,20]. The effect of progesterone on the development of uterine fibroids consists in the overexpression of cytokine-related genes and the increase of selected growth factor concentrations, e.g., transforming growth factors (TGFs), vascular endothelial growth factor (VEGF), tumor necrosis factor-α (TNF-α), and many others [7–9,21]. LMs are benign lesions with favorable prognosis but, due to their prevalence and considerable similarities to malignant tumors which are also derived from the body of the uterus, detailed histopathological evaluation is necessary to exclude dormant leiomyosarcomas and to determine their potential for malignant transformation [22]. According to the available literature, leiomyosarcomas (LMSs) account for 1–2% of all malignant tumors of the uterus [22], and remain the most common sarcomas of the female reproductive tract [22]. Approximately 1 in 100,000 people are diagnosed with LMSs [23]. According to Kho et. al., occult uterine sarcoma occurs in 0.089% of the cases or 1 in 1124 hysterectomies in women undergoing surgery for benign gynecologic indications [24], while Graebe et al., reported an even higher risk –0.22% [25]. Zhao et al. (2015), calculated the risk for unexpected uterine sarcomas at 0.47% [26].

1.2. Differentiation and Advanced Pathological Diagnostics in Myometrial Neoplasm

Determination of the hormone receptor status of the tumor cells allows the prognosis and assessment of the dynamics of the neoplastic process [22,27]. In the case of LMs, identification of patients with a high risk for malignant transformation or dynamic tumor growth may help to decide on the treatment plan and/or the extent of the operative treatment, or pharmacotherapy for non-surgical management [27–30]. It is a well-known fact that numerous regulatory mechanisms of proliferation are disturbed in cancer cells, as is the case with LM, whose growth becomes a self-stimulatory, often tumor volume-dependent process [7,8,13].

Several well-described diagnostic markers are used in the diagnosis of neoplastic lesions, chief among them Ki-67 antigen, p53 protein and steroid receptors: estrogen (ER) and progesterone (PR) [22,31]. These markers are used to assess the biology of the myometrial lesions and for differential diagnosis of tumors with atypical presentation. Nevertheless, histopathological evaluation sometimes remains inconclusive, which is reason enough to continue the search for more reliable markers. Apart from the aforementioned markers, the following, whose diagnostic applicability is yet to be determined, are currently being tested:

(a) proteins that control the cell cycle and have a direct influence on the dynamics of proliferation, e.g., p16 [32–34]; cyclin D1 [35,36]; Bcl-2 [37]; proliferating cell nuclear antigen (PCNA) [38];

(b) TGFs [7]; epidermal growth factor (EGF) [39]; VEGF [40]; insulin-like growth factors (IGFs) [41];

(c) proteins which are responsible for intercellular interactions: mucins [42]; galectins [39,42];

(d) proteins which are responsible for muscle cell contraction: caldesmon [43,44]; calponin [36,44]; α-smooth muscle actin [9,45];

(e) cytoskeletal proteins: vimentin, desmin, nestin, keratin [46,47].

1.3. Classification of Uterine Smooth Muscle Tumors

The differentiation between benign and malignant myometrial lesions continues to present a considerable challenge for pathologists, and sometimes the final diagnosis remains inconclusive [22,48]. A number of criteria have been suggested in an attempt to unify the classification of the uterine tumors [22,48–50]. Based on tumor morphology and biology, four groups of myometrial tumors have been distinguished [51]. The current classification of smooth muscle neoplasms is presented in Table 1 [51].

Table 1. The 2014 classification of smooth muscle tumors according to the World Health Organization (WHO) [51].

Myometrial Neoplasms	
Benign Lesions	
leiomyoma	*leiomyoma with Unusual Presentation*
cellular leiomyoma	diffuse leiomyomatosis
epithelioid leiomyoma	intravenous leiomyomatosis
myxoid leiomyoma	benign metastasizing leiomyoma
atypical leiomyoma	
lipoleiomyoma	
Uncertain Potential	
Smooth Muscle Tumor of Uncertain Malignant Potential (STUMP)	
Malignant	
leiomyosarcoma	
epithelioid variant	
myxoid variant	

At present, a number of criteria are used for the differentiation and the final diagnosis [22,49]. Topographic hematoxylin-eosin (HE) staining is commonly used in anatomic pathology diagnostics to evaluate tissue architecture [22,52]. The criteria presented in Table 2 [53,54] or the Stanford University Criteria presented in Table 3 [55], are applicable for the differentiation between benign and malignant lesions.

Table 2. Differentiation criteria for uterine smooth muscle tumors [53,54].

Benign	Malignant
• Low mitotic index (<5 mitotic figures for 10 High Power Field) • No cell atypia • No tumor cell necrosis (with the exception of ischemic necrosis) • Typical presentation of the smooth muscle cells, with uniform shape and size • No intravascular component • Well-demarcated	• Numerous mitotic figures (≥5 for 10 High-Power Field) • Significant cell atypia • Areas of tumor cell necrosis with island-like presentation

Table 3. Leiomyosarcoma of Deep Soft Tissue Stanford Medicine Surgical Pathology Criteria [55].

Leiomyoma *(requires all below)*	Smooth Muscle Tumor of Uncertain Malignant Potential *(used for any of below)*	Leiomyosarcoma *(requires any one of below)*
Cytologically bland	Bland but 1-4 mitotic figures/50 High-Power Field	Cytologic pleomorphism or atypia
<1 mitotic figure/50 High-Power Field	Multiple recurrences but lacking other atypical features	>4 mitotic figures/50 High-Power Field
No tumor cell necrosis		Coagulative tumor cell necrosis

Among malignant myometrial tumors, LMSs comprise approximately one-third of all uterine sarcomas. The annual incidence rate for LMSs has been calculated at 0.7–1 in 100,000 women [23,56]. LMSs are a relatively rare occurrence but have high malignancy potential and poor prognosis.

The 5-year survival rate for LMSs is 50% in cases without metastases [57]. The presence of distant metastases significantly worsens the already unfavorable prognosis. According to most clinical studies, progression-free survival for patients with distant metastases is 12–15 months [58].

In light of the above, accurate diagnosis and pre-operative histopathological evaluation are essential [29,59]. Nevertheless, we still lack tools to distinguish categorically between benign and malignant lesions. It seems that possible disproportions in the concentrations of selected growth factors among patients with LMSs as compared to LMs might be a turning point as far as improved pre-operative detection of LMS is concerned. However, this subject has been discussed elsewhere [60].

The existence of lesions of uncertain potential, which further complicates the diagnostic process, presents a challenge for both pathologists and clinicians and is cause for concern to the affected patient. It is described as smooth muscle tumor of uncertain malignant potential (STUMP) according to the current classification of myometrial lesions. Tumors from that group have diverse structure and unclear clinical course [48,61,62]. To the best of our knowledge, there have been no reports or observational studies about patients with STUMP after surgical interventions [63]. Due to considerable clinical and morphological similarities between atypical, STUMP, and LMs lesions, advanced histopathological differentiation is necessary [62,64], and only a diagnostic process that includes immunohistochemical (IHC) evaluation may offer that.

Tumor differentiation is additionally complicated by the fact that lesions not derived from the myometrial cells may still arise in the myometrium. Based on the histological structure of a uterus, which consists of endo-, myo- and perimetrium, we know that the endometrium is composed of the epithelium and lamina propria. The latter is a connective tissue of mesenchymal origin as well as the elements of the myometrium. It is vital to bear that in mind during histopathological diagnosis and differentiation between myometrial lesions and tumors such as endometrial stromal sarcoma. The available literature reports suggest using diagnostic panels for the differentiation of the aforementioned lesions [36].

Due to the complexity of differentiating between myometrial changes, physicians and diagnosticians often find differentiation and the process of decision-making about further management challenging. The aim of our study was to systematize the knowledge on the currently used IHC markers and other markers, whose diagnostic potential remains to be fully validated. Also, the up-to-date opinions about the clinical practice for differential diagnosis are presented.

2. Methodology of Obtaining Data and Data Analysis

We conducted a search in PubMed of the National Library of Medicine and Google Scholar. Databases were extensively searched for all original and review articles/book chapters published in English until September 2018 and related to myometrial neoplasms using the following keywords (one or in combinations): uterine fibroid; uterine leiomyoma; uterine leiomyosarcoma; uterine sarcoma; smooth muscle tumor of uncertain malignant potential (STUMP); immunohistochemistry (IHC). Moreover, additional articles from the reference sections of the reviewed articles were searched. Overall, most relevant articles were reviewed and included as appropriate.

3. Available Immunohistochemical Markers

As mentioned above, tumors arising from the smooth muscle cells of the uterus may present a considerable challenge for pathomorphologists. IHC, owing to its steady development and advances, has become one of the main diagnostic tools in anatomic pathology in general. As far as histopathological evaluation of myometrial tumors is concerned, experts agree that no therapeutic intervention should be initiated without additional IHC testing, regardless of the primary diagnosis. Lack of compliance with this recommendation might result in false positive or false negative results and expose the affected patient to the risks associated with a failure to detect a malignant lesion [22,27,65].

IHC markers which may be useful in differentiating between benign and malignant myometrial tumors are presented below. Additionally, marker-dependent implications for clinical management and prognosis for the patients have been discussed.

3.1. Markers with Strong Evidence

3.1.1. Ki-67

The Ki-67 proliferation antigen is the most common IHC marker in laboratory practice [22,31]. Despite being first described by Gerdes et al. as early as 1983, it continues to be the gold standard for the evaluation of the intensity of proliferative processes, and its expression in various neoplastic diseases has prognostic and predictive value [66]. The Ki-67 protein is expressed only in the proliferative phase of a cell (late G1, S, G2, and mitotic phases). Ki-67 antibody (MIB-1), used in the diagnostic process, makes IHC analysis more repeatable and accurate [67]. IHC testing confirmed that Ki-67 expression as a cellular marker for proliferation increases with tumor aggressiveness. Mayerhofer et al. (2004), demonstrated a significantly higher Ki-67 expression in the case of LMS vs. STUMP ($p = 0.0001$) and LMS vs. LM ($p = 0.0002$). At the same time, these authors found no statistically significant differences for STUMP vs. LM ($p = 0.491$) [68]. Mittal and Demopoulos (2001) reported similar results for LMS vs. STUMP and LMS vs. LM, and claimed that Ki-67 may be applicable in differentiating between STUMP vs. LM. Ki-67 expression level exceeded 15% in 11 out of 12 cases of LMS. Expression at the level of 5–10% was observed in 6 out of 7 STUMP cases. Importantly for result interpretation, Ki-67 expression was present in only 1 out of 15 cases of cellular LM, which is consistent with the biology of slow-growing benign lesions [69]. Petrović et al. (2010), found no Ki-67 expression in LM (LMS vs. LM ($p = 0.0001$) and STUMP vs. LM ($p = 0.0001$)), which indicated high diagnostic value of the marker in question. In their study, LMS vs. STUMP did not reach the level of statistical significance [70]. In 2009, Lee et al., reported that Ki-67, both as an isolated marker and in combination with p16 and p53, demonstrated a 92% sensitivity and a 98% specificity in differentiating between LMS and LM (65% LMS; 0% LM > 10% Ki-67 proliferation index $p < 0.001$) [71].

The literature offers reports about a positive correlation between Ki-67 expression and tumor aggressiveness, as well as clinical advancement of the disease in the case of LMS. Akhan et al. (2005), observed prolonged survival among patients with low Ki-67 expression ($p = 0.034$) [72], which is consistent with the findings of Mayerhofer et al. (2004), who demonstrated that rapid tumor growth and shortened disease-free survival are associated with high Ki-67 expression, which is correlated with involvement of the vascular space [68,73]. Lusby et al. (2013), reported Ki-67 overexpression, with an accompanying loss of ER and PR expression in case of LMS, whereas in metastatic tumors Ki-67 expression with VEGF and survival was higher as compared to the primary foci [74]. D'Angelo et al. (2011), confirmed the fact that high Ki-67 expression correlated with worsened long-term prognosis for the patient ($p = 0.01$). These authors also suggested that simultaneous evaluation of the clinical-pathological markers such as tumor size, mitotic index, and IHC Ki-67, Bcl-2 greatly increases statistical significance ($p = 0.001$) [75]. Recently, Demura et al. (2017), reported lower Ki-67 expression in those patients treated with selective progesterone receptor modulators (SPRMs) whose tumor volume significantly decreased, and claimed it was an antiproliferative and a proapoptotic effect of the treatment [32].

In light of the aforementioned data, it seems unquestionable that Ki-67 is a highly useful marker for differentiating between malignant and benign tumors, determining the prognosis for patients with this rare malignancy (LMS), and also planning further oncologic treatment [22]. The diagnostic value of Ki-67 is not to be underestimated; however, certain discrepancies between the reported results explain why Ki-67 has not become a part of the diagnostic panel. A potential panel is presented in Figure 1.

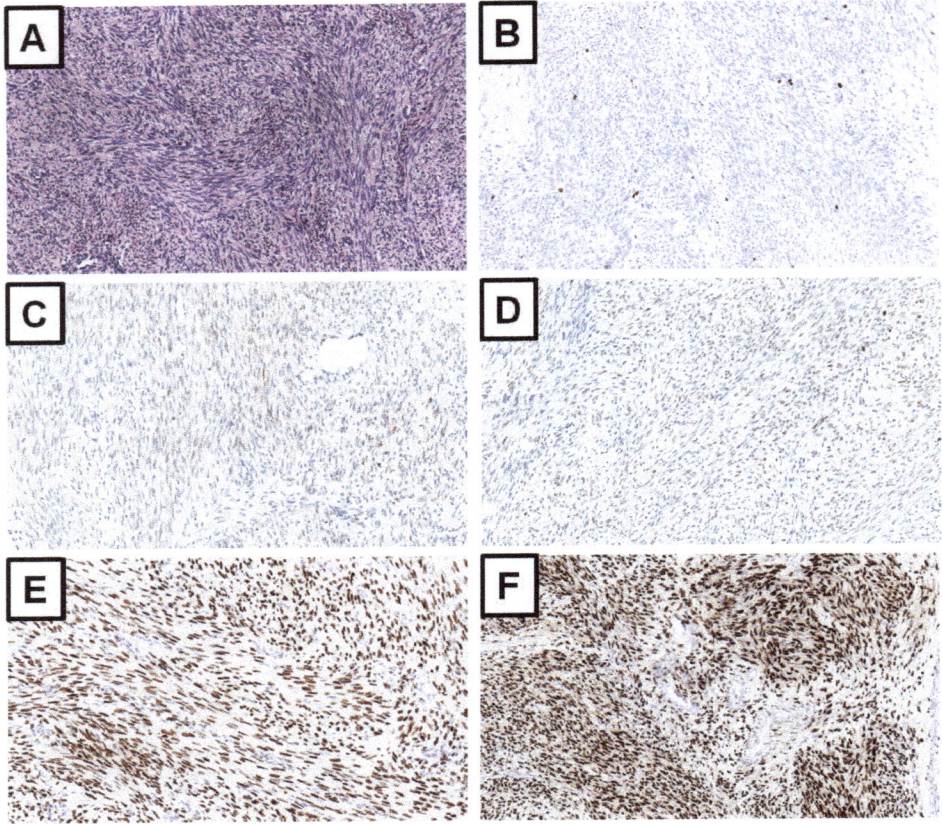

Figure 1. Leiomyoma. (**A**) Topographical staining hematoxylin and eosin. Immunochistochemical expression of (**B**) Ki-67 antigen, (**C**) p53 protein, (**D**) p16 protein, (**E**) estrogen and (**F**) progesteron receptors. Magnification ×20.

3.1.2. Tumor Protein p53 (p53, Cellular Tumor Antigen p53)

According to the available data, the *TP53* gene is the most frequently mutated gene in human cancer. The *TP53* gene encodes more than 15 protein isoforms of various sizes. These p53 proteins are known as the p53 isoforms [76]. The p53 protein plays a crucial role in multicellular organisms, where it prevents cancer formation, thus functioning as a tumor suppressor [76]. The p53 protein is engaged in the regulation of numerous cellular processes. It is responsible for the activation of the mechanisms of DNA repair and induction of apoptosis in response to DNA damage, thus being an example of a suppressor protein. Also, p53 displays features of a transcription factor, so mutations within the DNA-binding domain inhibit the transcription of protein-encoding genes, which are responsible for cell protection against tumor invasion [77,78]. Mutations in the *TP53* gene which encodes the p53 protein are correlated with unfavorable prognosis for the affected patients [70].

The analysis of p53 expression has been widely applied in differential diagnosis of various types of tumors, including lesions derived from uterine smooth muscle cells [78]. It seems that the background for using p53 expression in differential diagnosis is strong. The evidence may be found in studies by O'Neill et al. (2007), and Chen et al. (2008), who revealed that expression of p16, p53 and Ki-67 is stronger in uterine LMS as compared to normal LMs, LM variants, and STUMP [31,79]. Dastranj Tabrizi et al. (2015), and Azimpouran et al. (2016), confirmed the previous findings, even for separate p53 evaluation in differential diagnosis of uterine LMS vs. LM and STUMP [80,81]. Zhou et al. (2015), confirmed the overexpression of the p53 protein in uterine LMS. No significant correlations with age, tumor size, clinical stage were found [82]. Layfield et al. (2000), demonstrated an association between p53 expression and the prognosis for patients with LMS [83]. In an earlier study, Blom et al. (1998), found no such link, but they did find a connection between p53 and the frequency of recurrence in LMS [84]. Similar findings were reported by Maltese et al. (2018), who very recently described that ER status ($p = 0.027$) and p53 expression ($p = 0.015$) predicted the risk for relapse in LMS [85]. Notably, p53 positivity in the case of malignant lesions is usually spread, as opposed to focal reaction in the case of benign changes. However, that fact should be treated merely as additional information and not as conclusive evidence [86]. Dall'Asta et al. (2014), recommended using IHC verification of p16 and p53 overexpression in order to identify patients at an increased risk for disease recurrence, who may benefit from aggressive oncological treatment [62].

The aforementioned authors suggest that the p53 protein is one of the key elements of differentiating between smooth muscle cell-derived tumors and should be routinely used as a part of the diagnostic panel (Figure 1). The protein in question might not only provide better differentiation but also help to determine the malignancy potential and further prognosis for LMS patients.

3.1.3. p16 Protein (Cyclin-Dependent Kinase Inhibitor 2A, Multiple Tumor Suppressor 1)

The p16 tumor suppressor protein, encoded by the *CDKN2A* gene [87], plays a crucial role in cell cycle regulation, especially by decelerating cell progression from G1 to S phase [87]. The p16 protein inhibits cell cycle control using the pRB protein [88]. At present, p53 is the second (after pRB) most often measured suppressor protein. The altered expression profile of p53 is found in various tumors, e.g., laryngeal, esophageal, cervical cancers or malignant melanoma [89]. Diagnostic sensitivity is particularly increased when differentiating between cervical intraepithelial neoplasia (CIN)2 vs. CIN3 of cervical cancers associated with HPV (human papilloma virus) infection [90,91].

As far as myometrial changes are concerned, most authors report that IHC p16 expression increases with tumor aggressiveness. In 2007, O'Neill et al., analyzed the expression of p16, p53 and Ki-67 in myometrial lesions and found their levels to be elevated in uterine LMS as compared to normal LMs, LM variants, and STUMP [79]. In 2008, Gannon et al., and Chen et al., revealed a statistically stronger expression of p16 in LMS than in LM and its various subtypes ($p < 0.001$) [31,33]. As far as the findings of O'Neill are concerned, Chen et al., also stated that the use of a panel of antibodies to p16, p53, and Ki-67 is very helpful in distinguishing LMS from cellular LM and usual LM [31]. These reports are consistent with the results of Bodner-Adler et al. (2005), who found p16 expression in 12% of LMs, 21% of STUMP, and 57% of LMS cases. The aforementioned authors found statistically significant differences regarding p16 expression in LMS in comparison to STUMP lesions ($p < 0.05$) as well as LMS vs. LMs ($p < 0.05$), whereas the STUMP vs. LM difference was statistically insignificant ($p > 0.05$). Staining intensity differed significantly between LMS and LM and between LMS and STUMP ($p < 0.05$), but no statistically significant difference was observed between STUMP and LM ($p > 0.05$). No statistically significant correlations could be found between p16 expression and clinical stage, age, vascular space involvement, and disease recurrence in patients with LMS ($p > 0.05$). Interestingly, p16 positivity had no differential value ($p > 0.05$) as far as the overall survival is concerned [92]. In 2011, Hakverdi et al., reported p16 overexpression in LMS, suggesting that p16 might be a useful IHC marker in distinguishing uterine LMS from LM and its benign variants [93].

More recent studies argue in favor of using IHC panels as markers for the evaluation of uterine LMS, with the loss of ER and PR expression, overexpression of Ki-67, and altered p53, RB, p16 expressions [74]. This is due to fact that in some studies statistical significance was not reached. In a study by Liang et al. (2015), the expression of PR, p16, and phosphorylated histone H3 (pHH3) was found to be significantly different between atypical LMs and LMSs, but there were also no statistically significant differences between atypical LMs and LMs [94]. Finally, Schaefer et al. (2017), analyzed the expression of p16 and p53 on the mRNA and protein level in differential diagnosis between LMS and inflammatory myofibroblastic tumors (IMTs). p16 loss was detected in 5 out of 10 myxoid tumors and 2 out of 11 LMSs, but it was not found in IMTs (p = 0.0005), correlating with CDKN2A deletion (p = 0.014). Strong p16 staining in 6 out of 21 LMSs and 3 out of 26 IMTs did not correlate with changes in CDKN2A. Schaefer et al., concluded that abnormal staining for p53 and p16 loss was observed more frequently in uterine LMSs, with 100% specificity and 70% sensitivity against IMTs [34].

The aforementioned studies emphasize the importance and the possible role of p16 in differentiating between myometrial changes. In our opinion, the p16 protein should be included in the basic diagnostic panels (Figure 1).

3.1.4. Proliferating Cell Nuclear Antigen (PCNA)

PCNA is a proliferating cell nuclear antigen—a sliding DNA clamp and an auxiliary protein—which increases polymerase activity approximately 1000-fold, preventing its separation from DNA [95]. PCNA was originally identified as an antigen expressed in the nuclei of cells during the DNA synthesis phase of the cell cycle [96]. It is involved in processes like DNA replication and repair, chromatin remodeling and epigenetics. High PCNA expression in proliferative cells is the reasons why it is considered to be a cell proliferation marker [95]. In a normal healthy tissue of the uterine muscle, both in the secretory and proliferative phases of the cycle, PCNA expression is low, as opposed to a very high expression, especially in the secretory phase, in the pathologically changed tissues of the uterine muscle, e.g., in LMs [18]. According to Vu et al. (1998), a positive PCNA expression is observed in as many as 874/1000 LMS cells, in contrast to 381/1000 in patients after treatment with gonadotropin-releasing hormone (GnRH) agonists, which is evidence for decreased proliferation in response to pharmacological treatment (p < 0.001) [97].

Also, PCNA allows us to differentiate between malignant and benign lesions derived from smooth muscle cells of the uterus [38]. Higher PCNA expression is observed in LMS as compared to cellular and atypical LM which, together with lowered ER and PR expressions, makes it possible to differentiate between the lesions (p < 0.01) [45]. According to Mittal and Demopoulos (2001), and Zhu et al. (2003), IHC using PCNA can be useful in distinguishing between cellular LM and malignant tumors, and can be treated as an IHC marker [45,69]. Guan et al. (2012), reported similar findings when they found the levels of Ki-67 and PCNA expression to be lower in cellular LMs than in LMSs (p < 0.05) [38].

An analysis of LM tissues from patients undergoing pharmacological treatment also generated interesting observations about PCNA. Ulipristal acetate (UPA) belongs to a group of SPRMs. UPA has partial agonist as well as antagonistic effect on the progesterone receptor. It also binds to the glucocorticoid receptor, but has no relevant affinity to the estrogen, androgen and mineralocorticoid receptors [98]. In a study by Yun et al. (2015), who evaluated changes in proliferating and apoptotic markers of UFs after SPRMs or GnRH agonists, PCNA and caspase-3 protein expression was found to be higher in the LM tissue after SPRMs in comparison to the control group (no difference between the control and GnRH agonists groups) [99]. In a study by Luo et al. (2010), who evaluated the effects of different SPRM—telaprisone on proliferation and apoptosis in cultured LM cells, treatment with this drug also significantly decreased the levels of the proliferation marker PCNA and the anti-apoptotic protein Bcl-2 [100]. Epigallocatechin gallate (EGCG) is the ester of epigallocatechin and gallic acid that can be found in black or green tea [101,102]. EGCG has various biological effects confirmed in laboratory studies, and some of them were also performed on LMs. In 2010, Zhang et al., found that cultured LM cells treated with EGCG showed an inhibition of cell proliferation. An extensive further analysis confirmed a significant decrease in the expression of PCNA and Bcl-2 [102].

All these studies on drugs and proliferation markers confirm the effect of the aforementioned substances on the proliferative processes of LM cells. The effect of the pharmacological treatment on the later histopathological differentiation needs further consideration. The literature offers little, if any, data on the matter. Since medications for LM may affect the IHC markers, mistaking malignant lesions for benign pathologies is a risk which should not be ignored. Therefore, it should be obligatory to inform the histopathologist about each case of pharmacotherapy before surgical intervention for LM. Further research about the effects of pharmacotherapy on IHC proliferation markers is necessary.

PCNA also seems to be a promising IHC marker. In our opinion, and based on the reports in the literature, PCNA improves the detection rates of malignant lesions and may be used as a proliferation marker in cases of smooth muscle cell-derived uterine changes after pharmacological treatment.

3.1.5. Bcl-2 Protein (B-Cell Lymphoma 2)

Bcl-2, a protein encoded in humans by the *BCL2* gene, plays a major role in the regulation of apoptosis. Damage to the *BCL2* gene has been identified as a causative agent in various cancers where overexpression of the anti-apoptotic genes and under-expression of the pro-apoptotic genes might occur [103–105]. Additionally, which is important in the process of carcinogenesis, the Bcl-2 protein may initiate cell replication, thus decreasing the need for growth factors [106].

As far as LM diagnostics and Bcl-2 are concerned, higher Bcl-2 expression was demonstrated in LMs cell as compared to the normal, healthy myometrium [107,108]. In a study by Bodner-Adler et al. (2004), Bcl-2 was expressed more often and more strongly in LMs as compared to LMS and STUMP (Bcl-2 was present in 12 out of 21 LMS, 8 out of 22 STUMP, and 20 out of 25 LMs cases). Statistical significance of Bcl-2 expression was observed between LMS and LMs, and between STUMP and LMs ($p \leq 0.05$), although no statistical significance has not been found between LMS and STUMP [109]. Zhai et al. (1999), also demonstrated Bcl-2 overexpression in benign uterine smooth muscle tumors (LM, cellular LM, and STUMP) as compared to LMS ($p < 0.05$) [110].

According to the available literature, Bcl-2 and its encoding gene seem to be effective tools to determine the malignancy potential of various tumors and patient prognosis. Banas et al. (2017), recently found that Bcl-2 and selected DNA fragmentation factors are significantly under-expressed in uterine LMS, but only lack of DNA fragmentation factor 40 and Bcl-2 negatively influences disease-free survival and the overall survival [111].

In a study by Conconi et al. (2017), the presence of multiple *BCL2* gene copies and their expression in suspicious STUMPs and relapsed tumors was confirmed. These authors concluded that amplification of the *BCL2* gene present in the STUMPs and its multiple copies suggest its potential role as a marker of STUMP malignancy potential [112]. Smaller involvement of the vascular space and longer survival rates were observed in patients with LMS and positive Bcl-2 expression, which allows us to classify that protein as an effective prognostic marker for patients with LMS [113]. According to Lusby et al., high Bcl-2 expression also predicted longer disease-specific survival in women with uterine LMS [74]. The same observations were made by D'Angelo et al. (2011), who concluded that a combination of clinicopathologic parameters including Ki-67 and Bcl-2 protein expression allows us to distinguish groups of LMS with different survival, and that tumors that were Ki-67 positive and Bcl-2 negative had worse prognosis [75]. On the other hand, in 2016 de Graaff et al. reported that high expression of Bcl-2 proteins might contribute to increased chemoresistance of soft tissue LMS [114].

The abovementioned studies presented evidence that Bcl-2 may be used as yet another marker for differentiating between malignant and benign uterine smooth muscle cell tumors. Regardless of the findings of de Graaff et al. [114], Bcl-2 seems to be a valuable prognostic factor for LMS lesions. More data on the matter might be the key component when planning treatment regime for malignant uterine smooth muscle cell tumors. Further research is necessary as the number of reports in the literature is still rather limited.

3.1.6. Other Markers

A number of other markers of myometrial differential diagnosis have been described in the literature. Alas, they do not have strong evidence to support their diagnostic value. However, in our opinion, they should be considered in future studies.

Bodner-Adler et al. (2004), studied matrix metalloproteinase 2 (MMP-2) expression in myometrial lesions and suggested that MMPs play an important role in tumor invasion and metastasis because of their influence on the degradation of the extracellular matrix components. These authors reported stronger MMP-2 expression in patients with LMS as compared to STUMP ($p = 0.025$) and LMs ($p = 0.006$), and concluded that MMP-2 might be a useful IHC parameter to differentiate borderline cases [109]. MMP-2 negative uterine LMSs were also found to have decreased vascular space involvement ($p = 0.04$), and prolonged disease-free survival was observed in MMP-2-negative LMS patients ($p = 0.09$) [115].

In 2006, Bodner-Adler et al. published their results about the expression of thrombospondin 1 (TSP1) in myometrial lesions. According to these authors, TSP1 suppresses angiogenesis by inhibiting endothelial cell proliferation and inducing endothelial cell apoptosis. They observed a stronger expression of TSP1 in LMs as compared to STUMP and LMS ($p < 0.05$), but this significance was not detected between LMS and STUMP [116]. They concluded that a negative correlation between vascular space involvement and TSP1 expression might be a new predictive factor in women with uterine LMS [116].

In 2011, Weissenbacher et al., evaluated the expression of mucin-1, galectin-1 and galectin-3 in tumors derived from the myometrium [42]. Mucins, highly glycosylated proteins, are a component in various secretions [117]. Galectins play an important role in cell adhesion, angiogenesis, metastasis and apoptosis [118]. The expression of Muc-1 was increased in LM and LMS as compared to the normal myometrium. Increased expression of Gal-1 was found in LM in comparison to the healthy myometrium and LMS. The results of Gal-3 expression were statistically insignificant [42].

Very recently, Soltan et al. (2018) published their observations about the expression of galectin-3 and epidermal growth factor receptor (EGFR) in myometrial lesions [39]. EGFR is believed to be one of the most important transmembrane receptors which transduce signals into the cell [119]. In a study by Soltan et al., EGFR overexpression was detected in LMS, while lack of or reduced expression of EGFR was observed in LMs, atypical LMs, and STUMPs. Meanwhile, galectin-3 expression was downregulated in LMS as compared to other myometrial tumors [39].

The minichromosome maintenance protein complex (MCM) proteins, which initiate and limit DNA replication, have also been an area of interest for some researchers. According to a study by Chuang et al. (2012), LMs expressed significantly elevated levels of MCM7 as compared to the myometrium [120]. These proteins remain highly stable but active during all phases (G1, S, G2 and M), which is why it might be possible to differentiate between tumor cells during the phases of inactivity and intensive growth [121]. Unfortunately, data on that topic are very limited.

Major vault protein (MVP)/Catechol-O-methyltransferase (COMT)—Lintel et al. (2018) attempted to assess IHC expression of proteins allowing to distinguish between LM and LMS performed initial proteomic studies to select markers for further evaluation. MVP and COMT had 3.05 and 13.94 times higher expression in LMS relative to LM in ion spectra mass spectrometry, respectively. Subsequently in the IHC, MVP was found to be 50% sensitive and 100% specific when comparing LMS to LM. COMT had a sensitivity of 38% and a specificity of 88%. Immunohistochemical expression of MVP might be suggested as a useful marker in distinguishing LMS from LM in difficult cases [122]. In case of COMT, additional factors may also exert their influence, e.g., unstable concentrations of vit. D (whose influence on COMT has been reported [123]), vit. D supplementation, consumption of large quantities of green tea, or supplementation of its extracts [124].

Cellular retinol-binding protein-1 (CRBP-1) is the carrier protein involved in the transport of retinol from the liver storage site to peripheral tissue [125]. CRBP-1 contributes to the maintenance of the differentiated state of the endometrium through retinol bioavailability regulation. As mentioned by Orlandi et al. (2004), CRBP-1 may help to point the changes which occur in endometrial stroma and therefore be applied as an additional endometrial stromal marker [126]. In 2002, Orlandi et al. found that the expression of CRBP-1 is higher in uterine LMS than in LM and healthy myometrium [127]. In 2007, Zaitseva et al., published their study which demonstrated that CRBP-1 is differentially expressed between myometrium and LM. In this study authors concluded that the expression of CRBP-1 is altered in LMs when compared with healthy myometrium which might be a point for further studies to investigate the importance of these alteration in development of LM [128]. The aforementioned studies suggest that various CRBP-1 expression might represents a new useful marker for the differential diagnosis of smooth muscle tumors of the uterus [127].

Lactate dehydrogenase (LDH) catalyzes the conversion of pyruvate and lactate. It converts pyruvate, the final product of glycolysis, to lactate when oxygen is absent or in short supply, and it performs the reverse reaction during the Cori cycle in the liver. LDH is involved in tumor initiation and metabolism, while cancer cells rely on increased glycolysis resulting in increased lactate production in addition to aerobic respiration even under oxygen-sufficient conditions (in the process called Warburg effect) [129]. Song et al. analyzed the expression and prognostic value of two different subunits of LDH—A and D in LM, cellular LM and LMS. They disclosed significantly stronger IHC expression of LDH-A and LDH-D in patients with LMS vs. LM as well as LMS vs. cellular LM. Moreover, they found that LDH-A-positive LMS patients had a poorer prognosis than LDH-A-negative patients ($p = 0.03$). The authors stated, that overexpressions of LDH-A and LDH-D in LMS patients point to a more aggressive character of the tumor and a positive expression of LDH-A in patients with LMS may have prognostic value in these patients [130].

DNA topoisomerase 2-alpha (TOP2A) is an enzyme that controls and alters the topologic states of DNA during transcription. Baiocchi G et al. (2016) examined protein expression by IHC and gene TOP2A copy number by fluorescence in situ hybridization to verify its prognostic value in malignant (LMSs) and non-malignant (LMs, STUMPs) myometrial lesions; 56.8% of patients with LMS showed high expression of TOP2A. Greater TOP2A levels were found in patients with stage $\geq$II disease compared with stage I and also in high mitotic index tumors (>20/10 HPF). They stated that protein TOP2A expression does not correlate with TOP2A gene expression and does not predict outcome [131].

Studies about new, alternative IHC markers are necessary. The available diagnostic methods are of high quality but, in many cases, it is still impossible to determine the nature of the tumor and these new IHC markers might be used to improve the detection rates for malignant lesions, which in consequence allows for optimal diagnosis and treatment plan.

3.1.7. Non-Myometrial Tumors in Differential Diagnosis

- PEComa—Perivascular Epitheliod Cell Tumor

PEComa is very rare tumor which present co-expression of melanocytic and smooth muscle markers [132]. In a study by Bennett et al., it was clearly shown that all PEComas had expression of HMB-45 antibody, cathepsin K, and at least one muscle marker, with most expressing melan-A (77%) and/or microphthalmia associated transcription factor (MITF) (79%) [133]. What stays in line with the previous observations by Vang et al., with regard to the expression of HMB-45, which is positive for epithelioid mesenchymal tumors of the uterus with an uncertain relationship to pure smooth muscle tumors [134]. Most PEComas can be morphologically distinguished from classical smooth muscle tumors by their distinctive capillary architecture [135]. Future studies about their morphology and IHC patterns for better differential diagnosis are still required.

- Endometrial Stromal Sarcoma (ESS)

In the differential diagnostics of myometrial lesions, in doubtful cases a patomorphologist may also think about neoplasms originating from endometrium, i.e., endometrial stromal tumors. Conklin et al. in 2014 presented new World Health Organization (WHO) classification of endometrial stromal tumors, which recognizes 4 main categories i.e.,: endometrial stromal nodule (ESN), low-grade endometrial stromal sarcoma (LG-ESS), high-grade endometrial stromal sarcoma (HG-ESS), and undifferentiated uterine sarcoma (UUS). These categories are defined by the presence of genetic (distinct translocations) and clinicopathologic (tumor morphology) features. Available genetic diagnostics highlighted the presence of the JAZF1-SUZ12 (formerly JAZF1-JJAZ1) fusion characteristic for ESN and LG-ESSs, the YWHAE-FAM22 translocation identifies HG-ESSs, whereas UUS exhibits no specific molecular pattern [136].

For sure, genetic diagnostics are really precise tests (applicable in particular cases) whereas primarily we search for easily available methods of differentiating in the microscopic examination i.e., IHC. There were separate markers tested for their potential usage. Nucci et al. (2001) pointed out that h-Caldesmon appears to be a more sensitive and specific marker of smooth muscle differentiation in the uterus than desmin and may be a useful tool for distinguishing and classifying uterine mesenchymal tumors [137]. This stays in line with study by Rush et al. (2001) [138]. In the same year Chu et al. (2001) analyzed the expression of CD10, desmin, smooth muscle actin (SMA), ER and inhibin in ESS, LM and LMS. They concluded that in combination with SMA, and desmin, CD10 seems to be a useful IHC marker in the differential diagnosis of ESS from LM and LMS [139].

Zhu et al. (2004) evaluated the potential utility of a panel of antibodies in the differential diagnosis of ESS and cellular LM. They tested expression of desmin, alpha SMA, calponin h1, h-caldesmon, ER, PR, CD10, CD44v3, PCNA, and mast cells in LG-ESS, HG-ESS, cellular LM and myometrium as well as endometrium. They disclosed that a panel of h-caldesmon, CD10, and CD44v3 might be useful in distinguishing ESS from cellular LM in most cases. Additionally, they postulated a need for further investigation and interpretation of mast cells count as a part of a multivariate approach to the differential diagnosis [140].

Moreover, IHC expression of different markers may have prognostic value. Park et al. (2018) evaluated the expression of hormone receptors, i.e., ER, PR, and AR. Their increased expression was associated with significantly better overall survival. When the patients were categorized according to ER, PR, and AR immunoreactivity, triple-positive ESS had the best overall survival, and triple-negative ESS had the worst overall survival. The expression of hormone receptors was associated with favorable survival outcome in ESS. Their predictive value needs further investigation [141].

3.1.8. Steroid Receptors

In light of the fact that the tumors discussed in this article originated from hormonally-active tissue, a short description of steroid receptors seems in order. Specialists who evaluate tumors are obligated to verify the receptor status, but this is not used for differentiation but only for prognostic evaluation (Figure 1). The last decade marked the rise of targeted therapies for various types of neoplastic diseases. The effectiveness of targeted therapies is the result of drug action and patient selection. It is necessary to identify the factor that would supply information about patient response to treatment [142]. As far as uterine smooth muscle cell lesions are concerned, steroid receptors might play such a role. The literature offers numerous publications about studies on ERs and PRs expressions in tissues derived from uterine smooth muscle cells [22,69,72,74,85,94,143]. ER and PR are confirmed in most cases of LMs, and various authors have confirmed different expression of ER and PR receptors in other mesenchymal tumors of the uterine body [13,17,18,85]. The number of publications on the presence of the aforementioned markers in LMS remains limited but, so far, the results presented have been consistent and confirmed decreased ER and PR receptor expression in LMS cells. No differences in the expression of the androgen receptor (AR) in LM and LMS were found (32% vs. 40%, $p = 0.75$) [144]. Azimpouran et al. (2016), reported lack of ER and PR expression in 20 out of 24 and 24 out of 24 patients with LMS, whereas such expression was detected in every case of LM. At the same time, high steroid receptor expression, as was the case with LM, was also confirmed in STUMP changes [81]. These results are consistent with the findings of other authors, who observed a reverse correlation between the ER and PR expression levels and tumor malignancy. Mittal and Demopoulos found a significantly lower number (%) of positive reactions to ER and PR in LMS cells as compared to LM [69]. Similar results were reported by Zhai et al. (1999), who confirmed ER and PR receptor expression in 7% and 36% of LMS cases, respectively [145]. According to other authors, e.g., Lusby et al. (2013), uterine LMS samples exhibited loss of ER and PR expression (with the overexpression of Ki-67 and altered p53, p16 and others) [74]. Liang et al. (2015), demonstrated that high p16 expression and low PR expression suggest the diagnosis of LMS [94]. Additionally, as far as steroid receptors are concerned, ERs can be used to support the gynecologic origin of LMS (3% non-gynecologic vs. 50% gynecologic LMS; $p < 0.001$) [71].

Little is known about the presence of AR in myometrial lesions [27]. Leitao et al. (2004), described positive IHC reactions for AR in 32% of leiomyomas [144]. In a recent study by Baek et al., AR expression was found to be an independent factor for disease-free survival in patients with LMS. No deaths were noted in the AR expression group, and the 5-year overall survival in the AR-negative expression group was 54.8% ($p = 0.014$). Moreover, the co-expression of different steroid receptors (ER and/or PR) with AR was associated with significantly better 5-year disease-free survival and overall survival [146].

In light of the fact that the relationship between these receptors and the development of LMs has been well-documented, it is possible to apply a therapy using selective ER modulators [147,148], GnRH analogs, e.g., leuprolide acetate [99,149], as well as the well-documented therapy using selective PR modulators, e.g., UPA [150–152]. It seems prudent to investigate the effects of these drugs on the aforementioned proliferation markers, and in consequence on the histopathological diagnosis. Extensive research is necessary, especially since reports in the literature are scarce.

3.1.9. Future Directions

A vast majority of myometrial lesions leave no room for doubt as far as histopathology diagnosis is concerned. However, there is a need for very thorough differential diagnostics of smooth muscle cell tumors in some patients due to significant clinical and morphological similarities between benign lesions of unknown potential and malignant tumors [22,27,29].

Surgical intervention remains the most common method of treating uterine tumors, especially using minimally invasive techniques [153,154]. The histopathological diagnosis is usually made post-operatively, which might be a cause for concern in case of STUMP, atypical LMs, and LMS. Unfortunately, pre-operative detection rates for these lesions are still very low. Naturally, some changes are suspicious and, in many cases, a seemingly benign lesion might turn out to be lethal. In 2016, Cui and Wright reviewed papers on the prevalence of all uterine cancers in patients who underwent morcellation, and found that the rates of uterine sarcomas in presumed uterine fibroids ranged from 0.00% to 0.49%, and that LMS was the most commonly reported malignancy [155]. Accurate pre-operative diagnosis of LMS was reported in 65% [156] to 84.1% [157] of patients. The topic of undetected LMS was publicized after the 2014 Food and Drug Administration (FDA) warning. The FDA discouraged the use of laparoscopic power morcellation during hysterectomy or myomectomy for LMs due to the risk of dissemination of previously undetected malignancies [158]. The 2017 FDA statement on power morcellation is less radical but still presents the method as potentially risky for the patient and advises caution [159]. Therefore, an accurate and unambiguous diagnosis is important not only due to pre-operative patient eligibility for laparoscopic or classical interventions, but also the subsequent monitoring, therapy and prognosis. Studies on LMS-specific markers which might be used for pre-operative monitoring are practically non-existent, although cytokines, e.g., TNF-α, seem to be a hopeful direction [60].

In doubtful cases, especially differentiating between LM with high proliferative index and LMS, when data obtained from the morphological studies supported by IHC seem to be insufficient or inconclusive, we may also consider molecular analysis. Shikeeva AA et al., investigated the loss of heterozygosity and microsatellite instability to find out genetic differences between the aforementioned types of uterine lesions. In comparative analysis, they disclosed that patients with LMS had much higher frequencies of genetic changes than those with benign tumors. Specificity and sensitivity of the loss of heterozygosity and/or microsatellite instability markers were 92% and 95%, respectively [160].

3.1.10. Summary

Smooth muscle cell tumors belong to the most common lesions of the genital tract [1,2]. Due to their high prevalence, but also a multitude of histopathological subtypes, it is necessary to use molecular techniques in the diagnostic process of suspicious cases. Markers of cell proliferation such as Ki-67 [31], p53 protein [76], and ER and PR expression [144] have been successfully used in differential diagnosis for years. LMs are characterized by low expression of Ki-67 [68–70,74,145], p53 [31,74,80,81], and high expression of ER and PR receptors [69,74,145]. It is also a well-known fact that the risk for malignant transformation increases with higher expression of proliferation markers [145,161].

The available markers are undeniably good but some data are missing, and more evidence is necessary to better differentiate smooth muscle cell tumors. In light of that, it is reasonable, if not necessary, to search for new IHC markers of higher sensitivity and specificity [22,27,39,74,82]. Identification of new proliferation markers might herald a breakthrough in tumor differentiation and allow for a more individually tailored approach to therapy planning.

4. Conclusions

IHC is easily applicable due to the methodology of obtaining the material for examination (which is similar to routine HE staining). However, it still has its weak sides. Nowadays, in many doubtful cases it is replaced, or better, supported by molecular methods (in first line by: Western Blot, polymerase chain reaction (PCR), enzyme-linked immunosorbent assay (ELISA) or others) as well genetic methods (e.g., evaluation of heterozygosity or microsatellite instability analysis). These are in some cases techniques irreplaceable in differential diagnostics of LMS and proliferative LM of the uterus.

Based on our present knowledge about the applicability of IHC markers in differential diagnosis of lesions originating from the myometrium, we propose focusing on the aforementioned markers, especially gicen that so far, we have not been able to discover unambiguous markers. The use of panels of antibodies is strongly recommended, especially if we bear in mind tumor heterogeneity and semi-quantitative character of this laboratory method. As was clearly demonstrated, markers with strongest differential value are those responsible for proliferation and apoptosis balance. In accordance with global trends, the use of other, new techniques might offer a chance to increase sensitivity and specificity. However, they would also need to be easily applicable, which might present a challenge.

Author Contributions: P.R., M.C., L.H., M.Z., T.Ł., P.D. and C.K. analyzed the data and wrote the paper. M.C. and M.Z. draw the tables. P.R., C.K., P.D. prepared the figures. M.C., L.H., P.D. and C.K. supervised the work. P.R., M.C., L.H., M.Z., T.Ł., P.D. and C.K. accepted the final version of the paper.

Funding: This research received no external funding.

Conflicts of Interest: The authors declare no conflict of interest.

Abbreviations

AR	androgen receptor
Bcl-2	B-cell lymphoma 2
COMT	catechol-O-methyltransferase
EGCG	epigallocatechin gallate
EGFR	epidermal growth factor receptor
ER	estrogen receptor
ESN	endometrial stromal nodule
ESS	endometrial stromal sarcoma
GnRH	gonadotropin-releasing hormone
HE	hematoxylin and eosin staining
HG-ESS	high-grade endometrial stromal sarcoma
IGF	insulin-like growth factor
IHC	immunohistochemistry
LDH	lactate dehydrogenase
LG-ESS	low-grade endometrial stromal sarcoma
LM	leiomyoma
LMS	leiomyosarcoma
MCM	minichromosome maintenance protein
MITF	microphthalmia associated transcription factor
MMP	matrix metalloproteinase
MVP	major vault protein

PCNA	proliferating cell nuclear antigen
PEComa	perivascular epithelioid cell tumor
PR	progesterone receptor
SMA	smooth muscle actin
SPRM	selective progesterone receptor modulator
STUMP	smooth muscle tumor of uncertain malignant potential
TGF	transforming growth factor
TNF-α	tumor necrosis factor alpha
TOP-2A	DNA topoisomerase 2-alpha
TSP1	thrombospondin 1
UPA	ulipristal acetate
WHO	World Health Organization
VEGF	vascular endothelial growth factor

References

1. Stewart, E.A.; Cookson, C.L.; Gandolfo, R.A.; Schulze-Rath, R. Epidemiology of uterine fibroids: A systematic review. *BJOG* **2017**, *124*, 1501–1512. [CrossRef] [PubMed]
2. Stewart, E.A.; Laughlin-Tommaso, S.K.; Catherino, W.H.; Lalitkumar, S.; Gupta, D.; Vollenhoven, B. Uterine fibroids. *Nat. Rev. Dis. Primers* **2016**, *2*, 16043. [CrossRef] [PubMed]
3. Baird, D.D.; Dunson, D.B.; Hill, M.C.; Cousins, D.; Schectman, J.M. High cumulative incidence of uterine leiomyoma in black and white women: Ultrasound evidence. *Am. J. Obstet. Gynecol.* **2003**, *188*, 100–107. [CrossRef] [PubMed]
4. Wise, L.A.; Palmer, J.R.; Stewart, E.A.; Rosenberg, L. Age-specific incidence rates for self-reported uterine leiomyomata in the Black Women's Health Study. *Obstet. Gynecol.* **2005**, *105*, 563–568. [CrossRef] [PubMed]
5. Cramer, S.F.; Patel, A. The frequency of uterine leiomyomas. *Am. J. Clin. Pathol.* **1990**, *94*, 435–438. [CrossRef] [PubMed]
6. Oi, Y.; Katayama, K.; Hirata, G.; Ishidera, Y.; Yoshida, H.; Shigeta, H. Significance of postmenopausal uterine leiomyomas: Focus on variants. *J. Obstet. Gynaecol. Res.* **2018**, *44*, 1445–1450. [CrossRef] [PubMed]
7. Ciebiera, M.; Wlodarczyk, M.; Wrzosek, M.; Meczekalski, B.; Nowicka, G.; Lukaszuk, K.; Ciebiera, M.; Slabuszewska-Jozwiak, A.; Jakiel, G. Role of transforming growth factor beta in uterine fibroid biology. *Int. J. Mol. Sci.* **2017**, *18*, 2435. [CrossRef] [PubMed]
8. Islam, M.S.; Greco, S.; Janjusevic, M.; Ciavattini, A.; Giannubilo, S.R.; D'Adderio, A.; Biagini, A.; Fiorini, R.; Castellucci, M.; Ciarmela, P. Growth factors and pathogenesis. *Best Pract. Res. Clin. Obstet. Gynaecol.* **2016**, *34*, 25–36. [CrossRef] [PubMed]
9. Protic, O.; Toti, P.; Islam, M.S.; Occhini, R.; Giannubilo, S.R.; Catherino, W.H.; Cinti, S.; Petraglia, F.; Ciavattini, A.; Castellucci, M.; et al. Possible involvement of inflammatory/reparative processes in the development of uterine fibroids. *Cell Tissue Res.* **2016**, *364*, 415–427. [CrossRef] [PubMed]
10. Al-Hendy, A.; Myers, E.R.; Stewart, E. Uterine fibroids: Burden and unmet medical need. *Semin. Reprod. Med.* **2017**, *35*, 473–480. [CrossRef] [PubMed]
11. Parazzini, F.; Di Martino, M.; Candiani, M.; Vigano, P. Dietary components and uterine leiomyomas: A review of published data. *Nutr. Cancer* **2015**, *67*, 569–579. [CrossRef] [PubMed]
12. Ciebiera, M.; Lukaszuk, K.; Meczekalski, B.; Ciebiera, M.; Wojtyla, C.; Slabuszewska-Jozwiak, A.; Jakiel, G. Alternative oral agents in prophylaxis and therapy of uterine fibroids-an up-to-date review. *Int. J. Mol. Sci.* **2017**, *18*, 2586. [CrossRef] [PubMed]
13. Bulun, S.E. Uterine fibroids. *N. Engl. J. Med.* **2013**, *369*, 1344–1355. [CrossRef] [PubMed]
14. Makinen, N.; Mehine, M.; Tolvanen, J.; Kaasinen, E.; Li, Y.; Lehtonen, H.J.; Gentile, M.; Yan, J.; Enge, M.; Taipale, M.; et al. MED12, the mediator complex subunit 12 gene, is mutated at high frequency in uterine leiomyomas. *Science* **2011**, *334*, 252–255. [CrossRef] [PubMed]
15. Moroni, R.M.; Vieira, C.S.; Ferriani, R.A.; Reis, R.M.; Nogueira, A.A.; Brito, L.G. Presentation and treatment of uterine leiomyoma in adolescence: A systematic review. *BMC Womens Health* **2015**, *15*, 4. [CrossRef] [PubMed]

16. Tolvanen, J.; Uimari, O.; Ryynanen, M.; Aaltonen, L.A.; Vahteristo, P. Strong family history of uterine leiomyomatosis warrants fumarate hydratase mutation screening. *Hum. Reprod.* **2012**, *27*, 1865–1869. [CrossRef] [PubMed]

17. Ishikawa, H.; Ishi, K.; Serna, V.A.; Kakazu, R.; Bulun, S.E.; Kurita, T. Progesterone is essential for maintenance and growth of uterine leiomyoma. *Endocrinology* **2010**, *151*, 2433–2442. [CrossRef] [PubMed]

18. Maruo, T.; Ohara, N.; Wang, J.; Matsuo, H. Sex steroidal regulation of uterine leiomyoma growth and apoptosis. *Hum. Reprod. Update* **2004**, *10*, 207–220. [CrossRef] [PubMed]

19. Chill, H.H.; Safrai, M.; Reuveni Salzman, A.; Shushan, A. The rising phoenix-progesterone as the main target of the medical therapy for leiomyoma. *Biomed. Res. Int.* **2017**, *2017*, 4705164. [CrossRef] [PubMed]

20. Kim, J.J.; Sefton, E.C. The role of progesterone signaling in the pathogenesis of uterine leiomyoma. *Mol. Cell. Endocrinol.* **2012**, *358*, 223–231. [CrossRef] [PubMed]

21. Ciarmela, P.; Islam, M.S.; Reis, F.M.; Gray, P.C.; Bloise, E.; Petraglia, F.; Vale, W.; Castellucci, M. Growth factors and myometrium: Biological effects in uterine fibroid and possible clinical implications. *Hum. Reprod. Update* **2011**, *17*, 772–790. [CrossRef] [PubMed]

22. Hanley, K.Z.; Birdsong, G.G.; Mosunjac, M.B. Recent developments in surgical pathology of the uterine corpus. *Arch. Pathol. Lab. Med.* **2017**, *141*, 528–541. [CrossRef] [PubMed]

23. Ducimetiere, F.; Lurkin, A.; Ranchere-Vince, D.; Decouvelaere, A.V.; Peoc'h, M.; Istier, L.; Chalabreysse, P.; Muller, C.; Alberti, L.; Bringuier, P.P.; et al. Incidence of sarcoma histotypes and molecular subtypes in a prospective epidemiological study with central pathology review and molecular testing. *PLoS ONE* **2011**, *6*, e20294. [CrossRef] [PubMed]

24. Kho, K.A.; Lin, K.; Hechanova, M.; Richardson, D.L. Risk of occult uterine sarcoma in women undergoing hysterectomy for benign indications. *Obstet. Gynecol.* **2016**, *127*, 468–473. [CrossRef] [PubMed]

25. Graebe, K.; Garcia-Soto, A.; Aziz, M.; Valarezo, V.; Heller, P.B.; Tchabo, N.; Tobias, D.H.; Salamon, C.; Ramieri, J.; Dise, C.; et al. Incidental power morcellation of malignancy: A retrospective cohort study. *Gynecol. Oncol.* **2015**, *136*, 274–277. [CrossRef] [PubMed]

26. Zhao, W.C.; Bi, F.F.; Li, D.; Yang, Q. Incidence and clinical characteristics of unexpected uterine sarcoma after hysterectomy and myomectomy for uterine fibroids: A retrospective study of 10,248 cases. *Oncol. Targets Ther.* **2015**, *8*, 2943–2948. [CrossRef]

27. Gantzer, J.; Ray-Coquard, I. Gynecological sarcomas: What's new in 2018, a brief review of published literature. *Curr. Opin. Oncol.* **2018**, *30*, 246–251. [CrossRef] [PubMed]

28. Bonafede, M.M.; Pohlman, S.K.; Miller, J.D.; Thiel, E.; Troeger, K.A.; Miller, C.E. Women with newly diagnosed uterine fibroids: Treatment patterns and cost comparison for select treatment options. *Popul. Health Manag.* **2018**, *21*, S13–S20. [CrossRef] [PubMed]

29. Seagle, B.L.; Alexander, A.L.; Strohl, A.E.; Shahabi, S. Discussing sarcoma risks during informed consent for nonhysterectomy management of fibroids: An unmet need. *Am. J. Obstet. Gynecol.* **2018**, *218*, 103.e101–103.e105. [CrossRef] [PubMed]

30. Tanos, V.; Berry, K.E. Benign and malignant pathology of the uterus. *Best Pract. Res. Clin. Obstet. Gynaecol.* **2018**, *46*, 12–30. [CrossRef] [PubMed]

31. Chen, L.; Yang, B. Immunohistochemical analysis of p16, p53, and ki-67 expression in uterine smooth muscle tumors. *Int. J. Gynecol. Pathol.* **2008**, *27*, 326–332. [CrossRef] [PubMed]

32. Demura, T.A.; Revazova, Z.V.; Kogan, E.A.; Adamyan, L.V. [The molecular mechanisms and morphological manifestations of leiomyoma reduction induced by selective progesterone receptor modulators]. *Arkh. Patol.* **2017**, *79*, 19–26. [CrossRef] [PubMed]

33. Gannon, B.R.; Manduch, M.; Childs, T.J. Differential immunoreactivity of p16 in leiomyosarcomas and leiomyoma variants. *Int. J. Gynecol. Pathol.* **2008**, *27*, 68–73. [CrossRef] [PubMed]

34. Schaefer, I.M.; Hornick, J.L.; Sholl, L.M.; Quade, B.J.; Nucci, M.R.; Parra-Herran, C. Abnormal p53 and p16 staining patterns distinguish uterine leiomyosarcoma from inflammatory myofibroblastic tumour. *Histopathology* **2017**, *70*, 1138–1146. [CrossRef] [PubMed]

35. Lee, C.H.; Ali, R.H.; Rouzbahman, M.; Marino-Enriquez, A.; Zhu, M.; Guo, X.; Brunner, A.L.; Chiang, S.; Leung, S.; Nelnyk, N.; et al. Cyclin D1 as a diagnostic immunomarker for endometrial stromal sarcoma with ywhae-fam22 rearrangement. *Am. J. Surg. Pathol.* **2012**, *36*, 1562–1570. [CrossRef] [PubMed]

36. Buza, N.; Hui, P. Immunohistochemistry in gynecologic pathology: An example-based practical update. *Arch. Pathol. Lab. Med.* **2017**, *141*, 1052–1071. [CrossRef] [PubMed]

37. Courtoy, G.E.; Donnez, J.; Marbaix, E.; Barreira, M.; Luyckx, M.; Dolmans, M.M. Progesterone receptor isoforms, nuclear corepressor-1 and steroid receptor coactivator-1 and B-cell lymphoma 2 and AKT and AKT phosphorylation status in uterine myomas after ulipristal acetate treatment: A systematic immunohistochemical evaluation. *Gynecol. Obstet. Investig.* **2018**, *83*, 443–454. [CrossRef] [PubMed]

38. Guan, R.; Zheng, W.; Xu, M. A retrospective analysis of the clinicopathologic characteristics of uterine cellular leiomyomas in China. *Int. J. Gynaecol. Obstet.* **2012**, *118*, 52–55. [CrossRef] [PubMed]

39. Soltan, M.M.; Albasry, A.M.; Eldosouky, M.K.; Abdelhamid, H.S. Immunoexpression of progesterone receptor, epithelial growth factor receptor and galectin-3 in uterine smooth muscle tumors. *Cell. Mol. Biol.* **2018**, *64*, 7–12. [CrossRef] [PubMed]

40. Tal, R.; Segars, J.H. The role of angiogenic factors in fibroid pathogenesis: Potential implications for future therapy. *Hum. Reprod. Update* **2014**, *20*, 194–216. [CrossRef] [PubMed]

41. Dixon, D.; He, H.; Haseman, J.K. Immunohistochemical localization of growth factors and their receptors in uterine leiomyomas and matched myometrium. *Environ. Health Perspect.* **2000**, *108* (Suppl. 5), 795–802. [CrossRef] [PubMed]

42. Weissenbacher, T.; Kuhn, C.; Mayr, D.; Pavlik, R.; Friese, K.; Scholz, C.; Jeschke, U.; Ditsch, N.; Dian, D. Expression of mucin-1, galectin-1 and galectin-3 in human leiomyosarcoma in comparison to leiomyoma and myometrium. *Anticancer Res.* **2011**, *31*, 451–457. [PubMed]

43. Garg, G.; Mohanty, S.K. Uterine angioleiomyoma: A rare variant of uterine leiomyoma. *Arch. Pathol. Lab. Med.* **2014**, *138*, 1115–1118. [CrossRef] [PubMed]

44. Oliva, E.; Young, R.H.; Amin, M.B.; Clement, P.B. An immunohistochemical analysis of endometrial stromal and smooth muscle tumors of the uterus: A study of 54 cases emphasizing the importance of using a panel because of overlap in immunoreactivity for individual antibodies. *Am. J. Surg. Pathol.* **2002**, *26*, 403–412. [CrossRef] [PubMed]

45. Zhu, X.Q.; Shi, Y.F.; Cheng, X.D.; Wu, Y.Z. [The differential diagnosis between uterine leiomyosarcoma and the special subtypes of leiomyoma]. *Zhonghua Yi Xue Za Zhi* **2003**, *83*, 1419–1421. [PubMed]

46. Sarlomo-Rikala, M.; Tsujimura, T.; Lendahl, U.; Miettinen, M. Patterns of nestin and other intermediate filament expression distinguish between gastrointestinal stromal tumors, leiomyomas and schwannomas. *APMIS* **2002**, *110*, 499–507. [CrossRef] [PubMed]

47. Abeler, V.M.; Nenodovic, M. Diagnostic immunohistochemistry in uterine sarcomas: A study of 397 cases. *Int. J. Gynecol. Pathol.* **2011**, *30*, 236–243. [CrossRef] [PubMed]

48. Gupta, M.; Laury, A.L.; Nucci, M.R.; Quade, B.J. Predictors of adverse outcome in uterine smooth muscle tumours of uncertain malignant potential (STUMP): A clinicopathological analysis of 22 cases with a proposal for the inclusion of additional histological parameters. *Histopathology* **2018**, *73*, 284–298. [CrossRef] [PubMed]

49. Kefeli, M.; Caliskan, S.; Kurtoglu, E.; Yildiz, L.; Kokcu, A. Leiomyoma with bizarre nuclei: Clinical and pathologic features of 30 patients. *Int. J. Gynecol. Pathol.* **2018**, *37*, 379–387. [CrossRef] [PubMed]

50. Juhasz-Boss, I.; Jungmann, P.; Radosa, J.; von Heesen, A.; Stroder, R.; Juhasz-Boss, S.; Meyberg-Solomayer, G.; Solomayer, E. Two novel classification systems for uterine fibroids and subsequent uterine reconstruction after myomectomy. *Arch. Gynecol. Obstet.* **2017**, *295*, 675–680. [CrossRef] [PubMed]

51. Kurman, R. WHO classification of tumours of female reproductive organs. In *World Health Organization Classification of Tumours*, 6th ed.; Kurman, R., Ed.; WHO Press: Geneva, Switzerland, 2014.

52. Flake, G.P.; Moore, A.B.; Sutton, D.; Kissling, G.E.; Horton, J.; Wicker, B.; Walmer, D.; Robboy, S.J.; Dixon, D. The natural history of uterine leiomyomas: Light and electron microscopic studies of fibroid phases, interstitial ischemia, inanosis, and reclamation. *Obstet. Gynecol. Int.* **2013**, *2013*, 528376. [CrossRef] [PubMed]

53. Quade, B.J. Uterine smooth muscle tumors. In *Pathology of the Female Reproductive Tract*, 2nd ed.; Robboy, S., Ed.; Churchill Livingstone: London, UK, 2009; p. 474.

54. Bell, S.W.; Kempson, R.L.; Hendrickson, M.R. Problematic uterine smooth muscle neoplasms. A clinicopathologic study of 213 cases. *Am. J. Surg. Pathol.* **1994**, *18*, 535–558. [CrossRef] [PubMed]

55. Stanford University. Leiomyosarcoma of Deep Soft Tissue, Retroperitoneum, Mesentery and Omentum. Available online: http://surgpathcriteria.stanford.edu/softsmoothmuscle/soft_tissue_leiomyosarcoma/differentialdiagnosis.html (accessed on 8 October 2018).

56. Harlow, B.L.; Weiss, N.S.; Lofton, S. The epidemiology of sarcomas of the uterus. *J. Natl. Cancer Inst.* **1986**, *76*, 399–402. [PubMed]

57. Berchuck, A.; Rubin, S.C.; Hoskins, W.J.; Saigo, P.E.; Pierce, V.K.; Lewis, J.L., Jr. Treatment of uterine leiomyosarcoma. *Obstet. Gynecol.* **1988**, *71*, 845–850. [PubMed]

58. Duffaud, F.; Ray-Coquard, I.; Salas, S.; Pautier, P. Recent advances in understanding and managing leiomyosarcomas. *F1000Prime Rep.* **2015**, *7*, 55. [CrossRef] [PubMed]

59. Bi, Q.; Xiao, Z.; Lv, F.; Liu, Y.; Zou, C.; Shen, Y. Utility of clinical parameters and multiparametric mri as predictive factors for differentiating uterine sarcoma from atypical leiomyoma. *Acad. Radiol.* **2018**, *25*, 993–1002. [CrossRef] [PubMed]

60. Ciebiera, M.; Wlodarczyk, M.; Wrzosek, M.; Wojtyla, C.; Meczekalski, B.; Nowicka, G.; Lukaszuk, K.; Jakiel, G. TNF-alpha serum levels are elevated in women with clinically symptomatic uterine fibroids. *Int. J. Immunopathol. Pharmacol.* **2018**, *32*, 2058738418779461. [CrossRef] [PubMed]

61. Croce, S.; Ducoulombier, A.; Ribeiro, A.; Lesluyes, T.; Noel, J.C.; Amant, F.; Guillou, L.; Stoeckle, E.; Devouassoux-Shisheboran, M.; Penel, N.; et al. Genome profiling is an efficient tool to avoid the STUMP classification of uterine smooth muscle lesions: A comprehensive array-genomic hybridization analysis of 77 tumors. *Mod. Pathol.* **2018**, *31*, 816–828. [CrossRef] [PubMed]

62. Dall'Asta, A.; Gizzo, S.; Musaro, A.; Quaranta, M.; Noventa, M.; Migliavacca, C.; Sozzi, G.; Monica, M.; Mautone, D.; Berretta, R. Uterine smooth muscle tumors of uncertain malignant potential (STUMP): Pathology, follow-up and recurrence. *Int. J. Clin. Exp. Pathol.* **2014**, *7*, 8136–8142. [PubMed]

63. Ip, P.P.; Cheung, A.N.; Clement, P.B. Uterine smooth muscle tumors of uncertain malignant potential (STUMP): A clinicopathologic analysis of 16 cases. *Am. J. Surg. Pathol.* **2009**, *33*, 992–1005. [CrossRef] [PubMed]

64. D'Angelo, E.; Prat, J. Uterine sarcomas: A review. *Gynecol. Oncol.* **2010**, *116*, 131–139. [CrossRef] [PubMed]

65. Kuhn, E.; Ayhan, A. Diagnostic immunohistochemistry in gynaecological neoplasia: A brief survey of the most common scenarios. *J. Clin. Pathol.* **2018**, *71*, 98–109. [CrossRef] [PubMed]

66. Gerdes, J.; Schwab, U.; Lemke, H.; Stein, H. Production of a mouse monoclonal antibody reactive with a human nuclear antigen associated with cell proliferation. *Int. J. Cancer* **1983**, *31*, 13–20. [CrossRef] [PubMed]

67. Key, G.; Becker, M.H.; Baron, B.; Duchrow, M.; Schluter, C.; Flad, H.D.; Gerdes, J. New Ki-67-equivalent murine monoclonal antibodies (MIB 1-3) generated against bacterially expressed parts of the Ki-67 cDNA containing three 62 base pair repetitive elements encoding for the Ki-67 epitope. *Lab. Investig.* **1993**, *68*, 629–636. [PubMed]

68. Mayerhofer, K.; Lozanov, P.; Bodner, K.; Bodner-Adler, B.; Kimberger, O.; Czerwenka, K. Ki-67 expression in patients with uterine leiomyomas, uterine smooth muscle tumors of uncertain malignant potential (STUMP) and uterine leiomyosarcomas (LMS). *Acta Obstet. Gynecol. Scand.* **2004**, *83*, 1085–1088. [CrossRef] [PubMed]

69. Mittal, K.; Demopoulos, R.I. MIB-1 (Ki-67), p53, estrogen receptor, and progesterone receptor expression in uterine smooth muscle tumors. *Hum. Pathol.* **2001**, *32*, 984–987. [CrossRef] [PubMed]

70. Petrovic, D.; Babic, D.; Forko, J.I.; Martinac, I. Expression of Ki-67, p53 and progesterone receptors in uterine smooth muscle tumors. Diagnostic value. *Coll. Antropol.* **2010**, *34*, 93–97. [PubMed]

71. Lee, C.H.; Turbin, D.A.; Sung, Y.C.; Espinosa, I.; Montgomery, K.; van de Rijn, M.; Gilks, C.B. A panel of antibodies to determine site of origin and malignancy in smooth muscle tumors. *Mod. Pathol.* **2009**, *22*, 1519–1531. [CrossRef] [PubMed]

72. Akhan, S.E.; Yavuz, E.; Tecer, A.; Iyibozkurt, C.A.; Topuz, S.; Tuzlali, S.; Bengisu, E.; Berkman, S. The expression of Ki-67, p53, estrogen and progesterone receptors affecting survival in uterine leiomyosarcomas. A clinicopathologic study. *Gynecol. Oncol.* **2005**, *99*, 36–42. [CrossRef] [PubMed]

73. Mayerhofer, K.; Lozanov, P.; Bodner, K.; Bodner-Adler, B.; Obermair, A.; Kimberger, O.; Czerwenka, K. Ki-67 and vascular endothelial growth factor expression in uterine leiomyosarcoma. *Gynecol. Oncol.* **2004**, *92*, 175–179. [CrossRef] [PubMed]

74. Lusby, K.; Savannah, K.B.; Demicco, E.G.; Zhang, Y.; Ghadimi, M.P.; Young, E.D.; Colombo, C.; Lam, R.; Dogan, T.E.; Hornick, J.L.; et al. Uterine leiomyosarcoma management, outcome, and associated molecular biomarkers: A single institution's experience. *Ann. Surg. Oncol.* **2013**, *20*, 2364–2372. [CrossRef] [PubMed]

75. D'Angelo, E.; Espinosa, I.; Ali, R.; Gilks, C.B.; Rijn, M.; Lee, C.H.; Prat, J. Uterine leiomyosarcomas: Tumor size, mitotic index, and biomarkers Ki-67, and Bcl-2 identify two groups with different prognosis. *Gynecol. Oncol.* **2011**, *121*, 328–333. [CrossRef] [PubMed]

76. Surget, S.; Khoury, M.P.; Bourdon, J.C. Uncovering the role of p53 splice variants in human malignancy: A clinical perspective. *Oncol. Targets Ther.* **2013**, *7*, 57–68. [CrossRef]

77. Ashcroft, M.; Kubbutat, M.H.; Vousden, K.H. Regulation of p53 function and stability by phosphorylation. *Mol. Cell. Biol.* **1999**, *19*, 1751–1758. [CrossRef] [PubMed]

78. Levine, A.J.; Oren, M. The first 30 years of p53: Growing ever more complex. *Nat. Rev. Cancer* **2009**, *9*, 749–758. [CrossRef] [PubMed]

79. O'Neill, C.J.; McBride, H.A.; Connolly, L.E.; McCluggage, W.G. Uterine leiomyosarcomas are characterized by high p16, p53 and MIB-1 expression in comparison with usual leiomyomas, leiomyoma variants and smooth muscle tumours of uncertain malignant potential. *Histopathology* **2007**, *50*, 851–858. [CrossRef] [PubMed]

80. Dastranj Tabrizi, A.; Ghojazadeh, M.; Thagizadeh Anvar, H.; Vahedi, A.; Naji, S.; Mostafidi, E.; Berenjian, S. Immunohistochemical profile of uterine leiomyoma with bizarre nuclei; comparison with conventional leiomyoma, smooth muscle tumors of uncertain malignant potential and leiomyosarcoma. *Adv. Pharm. Bull.* **2015**, *5*, 683–687. [CrossRef] [PubMed]

81. Azimpouran, M.; Vazifekhah, S.; Moslemi, F.; Piri, R.; Naghavi-Behzad, M. Immunohistochemical profile of uterine leiomyomas; a comparison between different subtypes. *Niger Med. J.* **2016**, *57*, 54–58. [CrossRef] [PubMed]

82. Zhou, Y.; Huang, H.; Yuan, L.J.; Xiong, Y.; Huang, X.; Lin, J.X.; Zheng, M. CD146 as an adverse prognostic factor in uterine sarcoma. *Eur. J. Med. Res.* **2015**, *20*, 67. [CrossRef] [PubMed]

83. Layfield, L.J.; Liu, K.; Dodge, R.; Barsky, S.H. Uterine smooth muscle tumors: Utility of classification by proliferation, ploidy, and prognostic markers versus traditional histopathology. *Arch. Pathol. Lab. Med.* **2000**, *124*, 221–227. [CrossRef] [PubMed]

84. Blom, R.; Guerrieri, C.; Stal, O.; Malmstrom, H.; Simonsen, E. Leiomyosarcoma of the uterus: A clinicopathologic, DNA flow cytometric, p53, and mdm-2 analysis of 49 cases. *Gynecol. Oncol.* **1998**, *68*, 54–61. [CrossRef] [PubMed]

85. Maltese, G.; Fontanella, C.; Lepori, S.; Scaffa, C.; Fuca, G.; Bogani, G.; Provenzano, S.; Carcangiu, M.L.; Raspagliesi, F.; Lorusso, D. Atypical uterine smooth muscle tumors: A retrospective evaluation of clinical and pathologic features. *Oncology* **2018**, *94*, 1–6. [CrossRef] [PubMed]

86. Stanescu, A.D.; Nistor, E.; Sajin, M.; Stepan, A.E. Immunohistochemical analysis in the diagnosis of uterine myometrial smooth muscle tumors. *Rom. J. Morphol. Embryol.* **2014**, *55*, 1129–1136. [PubMed]

87. Nobori, T.; Miura, K.; Wu, D.J.; Lois, A.; Takabayashi, K.; Carson, D.A. Deletions of the cyclin-dependent kinase-4 inhibitor gene in multiple human cancers. *Nature* **1994**, *368*, 753–756. [CrossRef] [PubMed]

88. Ohtani, N.; Yamakoshi, K.; Takahashi, A.; Hara, E. The p16ink4a-rb pathway: Molecular link between cellular senescence and tumor suppression. *J. Med. Investig.* **2004**, *51*, 146–153. [CrossRef]

89. Serra, S.; Chetty, R. P16. *J. Clin. Pathol.* **2018**, *71*, 853–858. [CrossRef] [PubMed]

90. Skubitz, K.M.; Skubitz, A.P. Differential gene expression in leiomyosarcoma. *Cancer* **2003**, *98*, 1029–1038. [CrossRef] [PubMed]

91. Missaoui, N.; Mestiri, S.; Bdioui, A.; Zahmoul, T.; Hamchi, H.; Mokni, M.; Hmissa, S. Hpv infection and p16(ink4a) and tp53 expression in rare cancers of the uterine cervix. *Pathol. Res. Pract.* **2018**, *214*, 498–506. [CrossRef] [PubMed]

92. Bodner-Adler, B.; Bodner, K.; Czerwenka, K.; Kimberger, O.; Leodolter, S.; Mayerhofer, K. Expression of p16 protein in patients with uterine smooth muscle tumors: An immunohistochemical analysis. *Gynecol. Oncol.* **2005**, *96*, 62–66. [CrossRef] [PubMed]

93. Hakverdi, S.; Gungoren, A.; Yaldiz, M.; Hakverdi, A.U.; Toprak, S. Immunohistochemical analysis of p16 expression in uterine smooth muscle tumors. *Eur. J. Gynaecol. Oncol.* **2011**, *32*, 513–515. [PubMed]

94. Liang, Y.; Zhang, X.; Chen, X.; Lu, W. Diagnostic value of progesterone receptor, p16, p53 and PHH3 expression in uterine atypical leiomyoma. *Int. J. Clin. Exp. Pathol.* **2015**, *8*, 7196–7202. [PubMed]

95. Moldovan, G.L.; Pfander, B.; Jentsch, S. Pcna, the maestro of the replication fork. *Cell* **2007**, *129*, 665–679. [CrossRef] [PubMed]

96. Mailand, N.; Gibbs-Seymour, I.; Bekker-Jensen, S. Regulation of pcna-protein interactions for genome stability. *Nat. Rev. Mol. Cell Biol.* **2013**, *14*, 269–282. [CrossRef] [PubMed]

97. Vu, K.; Greenspan, D.L.; Wu, T.C.; Zacur, H.A.; Kurman, R.J. Cellular proliferation, estrogen receptor, progesterone receptor, and Bcl-2 expression in gnrh agonist-treated uterine leiomyomas. *Hum. Pathol.* **1998**, *29*, 359–363. [CrossRef]

98. Garnock-Jones, K.P.; Duggan, S.T. Ulipristal acetate: A review in symptomatic uterine fibroids. *Drugs* **2017**, *77*, 1665–1675. [CrossRef] [PubMed]

99. Yun, B.S.; Seong, S.J.; Cha, D.H.; Kim, J.Y.; Kim, M.L.; Shim, J.Y.; Park, J.E. Changes in proliferating and apoptotic markers of leiomyoma following treatment with a selective progesterone receptor modulator or gonadotropin-releasing hormone agonist. *Eur. J. Obstet. Gynecol. Reprod. Biol.* **2015**, *191*, 62–67. [CrossRef] [PubMed]

100. Luo, X.; Yin, P.; Coon, V.J.; Cheng, Y.H.; Wiehle, R.D.; Bulun, S.E. The selective progesterone receptor modulator cdb4124 inhibits proliferation and induces apoptosis in uterine leiomyoma cells. *Fertil. Steril.* **2010**, *93*, 2668–2673. [CrossRef] [PubMed]

101. Lorenz, M.; Urban, J.; Engelhardt, U.; Baumann, G.; Stangl, K.; Stangl, V. Green and black tea are equally potent stimuli of no production and vasodilation: New insights into tea ingredients involved. *Basic Res. Cardiol.* **2009**, *104*, 100–110. [CrossRef] [PubMed]

102. Zhang, D.; Al-Hendy, M.; Richard-Davis, G.; Montgomery-Rice, V.; Rajaratnam, V.; Al-Hendy, A. Antiproliferative and proapoptotic effects of epigallocatechin gallate on human leiomyoma cells. *Fertil. Steril.* **2010**, *94*, 1887–1893. [CrossRef] [PubMed]

103. Hardwick, J.M.; Soane, L. Multiple functions of bcl-2 family proteins. *Cold Spring Harb. Perspect. Biol.* **2013**, *5*. [CrossRef] [PubMed]

104. Otake, Y.; Soundararajan, S.; Sengupta, T.K.; Kio, E.A.; Smith, J.C.; Pineda-Roman, M.; Stuart, R.K.; Spicer, E.K.; Fernandes, D.J. Overexpression of nucleolin in chronic lymphocytic leukemia cells induces stabilization of bcl2 mrna. *Blood* **2007**, *109*, 3069–3075. [CrossRef] [PubMed]

105. Wu, X.; Blanck, A.; Olovsson, M.; Henriksen, R.; Lindblom, B. Expression of Bcl-2, Bcl-x, mcl-1, Bax and Bak in human uterine leiomyomas and myometrium during the menstrual cycle and after menopause. *J. Steroid Biochem. Mol. Biol.* **2002**, *80*, 77–83. [CrossRef]

106. Reed, J.C.; Talwar, H.S.; Cuddy, M.; Baffy, G.; Williamson, J.; Rapp, U.R.; Fisher, G.J. Mitochondrial protein p26 Bcl2 reduces growth factor requirements of nih3t3 fibroblasts. *Exp. Cell Res.* **1991**, *195*, 277–283. [CrossRef]

107. Matsuo, H.; Maruo, T.; Samoto, T. Increased expression of bcl-2 protein in human uterine leiomyoma and its up-regulation by progesterone. *J. Clin. Endocrinol. Metab.* **1997**, *82*, 293–299. [CrossRef] [PubMed]

108. Khurana, K.K.; Singh, S.B.; Tatum, A.H.; Schulz, V.; Badawy, S.Z. Maintenance of increased Bcl-2 expression in uterine leiomyomas after gnrh agonist therapy. *J. Reprod. Med.* **1999**, *44*, 487–492. [PubMed]

109. Bodner-Adler, B.; Bodner, K.; Kimberger, O.; Czerwenka, K.; Leodolter, S.; Mayerhofer, K. Expression of matrix metalloproteinases in patients with uterine smooth muscle tumors: An immunohistochemical analysis of mmp-1 and mmp-2 protein expression in leiomyoma, uterine smooth muscle tumor of uncertain malignant potential, and leiomyosarcoma. *J. Soc. Gynecol. Investig.* **2004**, *11*, 182–186. [CrossRef] [PubMed]

110. Zhai, Y.L.; Nikaido, T.; Toki, T.; Shiozawa, A.; Orii, A.; Fujii, S. Prognostic significance of Bcl-2 expression in leiomyosarcoma of the uterus. *Br. J. Cancer* **1999**, *80*, 1658–1664. [CrossRef] [PubMed]

111. Banas, T.; Pitynski, K.; Okon, K.; Czerw, A. DNA fragmentation factors 40 and 45 (dff40/dff45) and B-cell lymphoma 2 (Bcl-2) protein are underexpressed in uterine leiomyosarcomas and may predict survival. *Oncol. Targets Ther.* **2017**, *10*, 4579–4589. [CrossRef] [PubMed]

112. Conconi, D.; Chiappa, V.; Perego, P.; Redaelli, S.; Bovo, G.; Lavitrano, M.; Milani, R.; Dalpra, L.; Lissoni, A.A. Potential role of Bcl2 in the recurrence of uterine smooth muscle tumors of uncertain malignant potential. *Oncol. Rep.* **2017**, *37*, 41–47. [CrossRef] [PubMed]

113. Bodner, K.; Bodner-Adler, B.; Kimberger, O.; Czerwenka, K.; Mayerhofer, K. Bcl-2 receptor expression in patients with uterine smooth muscle tumors: An immunohistochemical analysis comparing leiomyoma, uterine smooth muscle tumor of uncertain malignant potential, and leiomyosarcoma. *J. Soc. Gynecol. Investig.* **2004**, *11*, 187–191. [CrossRef] [PubMed]

114. de Graaff, M.A.; de Rooij, M.A.; van den Akker, B.E.; Gelderblom, H.; Chibon, F.; Coindre, J.M.; Marino-Enriquez, A.; Fletcher, J.A.; Cleton-Jansen, A.M.; Bovee, J.V. Inhibition of Bcl-2 family members sensitises soft tissue leiomyosarcomas to chemotherapy. *Br. J. Cancer* **2016**, *114*, 1219–1226. [CrossRef] [PubMed]

115. Bodner-Adler, B.; Bodner, K.; Kimberger, O.; Czerwenka, K.; Leodolter, S.; Mayerhofer, K. MMP-1 and MMP-2 expression in uterine leiomyosarcoma and correlation with different clinicopathologic parameters. *J. Soc. Gynecol. Investig.* **2003**, *10*, 443–446. [CrossRef] [PubMed]

116. Bodner-Adler, B.; Nather, A.; Bodner, K.; Czerwenka, K.; Kimberger, O.; Leodolter, S.; Mayerhofer, K. Expression of thrombospondin 1 (TSP 1) in patients with uterine smooth muscle tumors: An immunohistochemical study. *Gynecol. Oncol.* **2006**, *103*, 186–189. [CrossRef] [PubMed]

117. Marin, F.; Luquet, G.; Marie, B.; Medakovic, D. Molluscan shell proteins: Primary structure, origin, and evolution. *Curr. Top. Dev. Biol.* **2008**, *80*, 209–276. [CrossRef] [PubMed]

118. Liu, F.T.; Rabinovich, G.A. Galectins: Regulators of acute and chronic inflammation. *Ann. N. Y. Acad. Sci.* **2010**, *1183*, 158–182. [CrossRef] [PubMed]

119. Oda, K.; Matsuoka, Y.; Funahashi, A.; Kitano, H. A comprehensive pathway map of epidermal growth factor receptor signaling. *Mol. Syst. Biol.* **2005**, *1*, 2005.0010. [CrossRef] [PubMed]

120. Chuang, T.D.; Luo, X.; Panda, H.; Chegini, N. Mir-93/106b and their host gene, MCM7, are differentially expressed in leiomyomas and functionally target f3 and il-8. *Mol. Endocrinol.* **2012**, *26*, 1028–1042. [CrossRef] [PubMed]

121. Nowinska, K.; Dziegiel, P. [the role of mcm proteins in cell proliferation and tumorigenesis]. *Postepy Hig. Med. Dosw. (Online)* **2010**, *64*, 627–635. [PubMed]

122. Lintel, N.J.; Luebker, S.A.; Lele, S.M.; Koepsell, S.A. MVP immunohistochemistry is a useful adjunct in distinguishing leiomyosarcoma from leiomyoma and leiomyoma with bizarre nuclei. *Hum. Pathol.* **2018**, *73*, 122–127. [CrossRef] [PubMed]

123. Sharan, C.; Halder, S.K.; Thota, C.; Jaleel, T.; Nair, S.; Al-Hendy, A. Vitamin D inhibits proliferation of human uterine leiomyoma cells via catechol-o-methyltransferase. *Fertil. Steril.* **2011**, *95*, 247–253. [CrossRef] [PubMed]

124. Zhang, D.; Rajaratnam, V.; Al-Hendy, O.; Halder, S.; Al-Hendy, A. Green tea extract inhibition of human leiomyoma cell proliferation is mediated via catechol-O-methyltransferase. *Gynecol. Obstet. Investig.* **2014**, *78*, 109–118. [CrossRef] [PubMed]

125. Napoli, J.L. Functions of intracellular retinoid binding-proteins. *Subcell. Biochem.* **2016**, *81*, 21–76. [CrossRef] [PubMed]

126. Orlandi, A.; Ferlosio, A.; Ciucci, A.; Sesti, F.; Lifschitz-Mercer, B.; Gabbiani, G.; Spagnoli, L.G.; Czernobilsky, B. Cellular retinol-binding protein-1 expression in endometrial stromal cells: Physiopathological and diagnostic implications. *Histopathology* **2004**, *45*, 511–517. [CrossRef] [PubMed]

127. Orlandi, A.; Francesconi, A.; Clement, S.; Ropraz, P.; Spagnoli, L.G.; Gabbiani, G. High levels of cellular retinol binding protein-1 expression in leiomyosarcoma: Possible implications for diagnostic evaluation. *Virchows Arch.* **2002**, *441*, 31–40. [CrossRef] [PubMed]

128. Zaitseva, M.; Vollenhoven, B.J.; Rogers, P.A. Retinoic acid pathway genes show significantly altered expression in uterine fibroids when compared with normal myometrium. *Mol. Hum. Reprod.* **2007**, *13*, 577–585. [CrossRef] [PubMed]

129. Warburg, O. On the origin of cancer cells. *Science* **1956**, *123*, 309–314. [CrossRef] [PubMed]

130. Song, K.J.; Yu, X.N.; Lv, T.; Chen, Y.L.; Diao, Y.C.; Liu, S.L.; Wang, Y.K.; Yao, Q. Expression and prognostic value of lactate dehydrogenase-A and -A subunits in human uterine myoma and uterine sarcoma. *Medicine (Baltimore)* **2018**, *97*, e0268. [CrossRef] [PubMed]

131. Baiocchi, G.; Poliseli, F.L.; De Brot, L.; Mantoan, H.; Schiavon, B.N.; Faloppa, C.C.; Vassallo, J.; Soares, F.A.; Cunha, I.W. TOP2A copy number and TOP2A expression in uterine benign smooth muscle tumours and leiomyosarcoma. *J. Clin. Pathol.* **2016**, *69*, 884–889. [CrossRef] [PubMed]

132. Cossu, A.; Paliogiannis, P.; Tanda, F.; Dessole, S.; Palmieri, G.; Capobianco, G. Uterine perivascular epithelioid cell neoplasms (PEComas): Report of two cases and literature review. *Eur. J. Gynaecol. Oncol.* **2014**, *35*, 309–312. [PubMed]

133. Bennett, J.A.; Braga, A.C.; Pinto, A.; Van de Vijver, K.; Cornejo, K.; Pesci, A.; Zhang, L.; Morales-Oyarvide, V.; Kiyokawa, T.; Zannoni, G.F.; et al. Uterine PEComas: A morphologic, immunohistochemical, and molecular analysis of 32 tumors. *Am. J. Surg. Pathol.* **2018**, *42*, 1370–1383. [CrossRef] [PubMed]

134. Vang, R.; Kempson, R.L. Perivascular epithelioid cell tumor ('PEComa') of the uterus: A subset of hmb-45-positive epithelioid mesenchymal neoplasms with an uncertain relationship to pure smooth muscle tumors. *Am. J. Surg. Pathol.* **2002**, *26*, 1–13. [CrossRef] [PubMed]

135. Musella, A.; De Felice, F.; Kyriacou, A.K.; Barletta, F.; Di Matteo, F.M.; Marchetti, C.; Izzo, L.; Monti, M.; Benedetti Panici, P.; Redler, A.; et al. Perivascular epithelioid cell neoplasm (PEComa) of the uterus: A systematic review. *Int. J. Surg.* **2015**, *19*, 1–5. [CrossRef] [PubMed]

136. Conklin, C.M.; Longacre, T.A. Endometrial stromal tumors: The new WHO classification. *Adv. Anat. Pathol.* **2014**, *21*, 383–393. [CrossRef] [PubMed]

137. Nucci, M.R.; O'Connell, J.T.; Huettner, P.C.; Cviko, A.; Sun, D.; Quade, B.J. H-caldesmon expression effectively distinguishes endometrial stromal tumors from uterine smooth muscle tumors. *Am. J. Surg. Pathol.* **2001**, *25*, 455–463. [CrossRef] [PubMed]

138. Rush, D.S.; Tan, J.; Baergen, R.N.; Soslow, R.A. H-caldesmon, a novel smooth muscle-specific antibody, distinguishes between cellular leiomyoma and endometrial stromal sarcoma. *Am. J. Surg. Pathol.* **2001**, *25*, 253–258. [CrossRef] [PubMed]

139. Chu, P.G.; Arber, D.A.; Weiss, L.M.; Chang, K.L. Utility of cd10 in distinguishing between endometrial stromal sarcoma and uterine smooth muscle tumors: An immunohistochemical comparison of 34 cases. *Mod. Pathol.* **2001**, *14*, 465–471. [CrossRef] [PubMed]

140. Zhu, X.Q.; Shi, Y.F.; Cheng, X.D.; Zhao, C.L.; Wu, Y.Z. Immunohistochemical markers in differential diagnosis of endometrial stromal sarcoma and cellular leiomyoma. *Gynecol. Oncol.* **2004**, *92*, 71–79. [CrossRef] [PubMed]

141. Park, J.Y.; Baek, M.H.; Park, Y.; Kim, Y.T.; Nam, J.H. Investigation of hormone receptor expression and its prognostic value in endometrial stromal sarcoma. *Virchows Arch.* **2018**, *473*, 61–69. [CrossRef] [PubMed]

142. Olszewski, W.P.; Olszewski, W.T. The role of pathologist in cancer patients selection for EGFR-targeted therapy. *Onkol. Prak. Klin.* **2010**, *6*, 228–235.

143. Hewedi, I.H.; Radwan, N.A.; Shash, L.S. Diagnostic value of progesterone receptor and p53 expression in uterine smooth muscle tumors. *Diagn. Pathol.* **2012**, *7*, 1. [CrossRef] [PubMed]

144. Leitao, M.M.; Soslow, R.A.; Nonaka, D.; Olshen, A.B.; Aghajanian, C.; Sabbatini, P.; Dupont, J.; Hensley, M.; Sonoda, Y.; Barakat, R.R.; et al. Tissue microarray immunohistochemical expression of estrogen, progesterone, and androgen receptors in uterine leiomyomata and leiomyosarcoma. *Cancer* **2004**, *101*, 1455–1462. [CrossRef] [PubMed]

145. Zhai, Y.L.; Kobayashi, Y.; Mori, A.; Orii, A.; Nikaido, T.; Konishi, I.; Fujii, S. Expression of steroid receptors, Ki-67, and p53 in uterine leiomyosarcomas. *Int. J. Gynecol. Pathol.* **1999**, *18*, 20–28. [CrossRef] [PubMed]

146. Baek, M.H.; Park, J.Y.; Park, Y.; Kim, K.R.; Kim, D.Y.; Suh, D.S.; Kim, J.H.; Kim, Y.M.; Kim, Y.T.; Nam, J.H. Androgen receptor as a prognostic biomarker and therapeutic target in uterine leiomyosarcoma. *J. Gynecol. Oncol.* **2018**, *29*, e30. [CrossRef] [PubMed]

147. Borahay, M.A.; Asoglu, M.R.; Mas, A.; Adam, S.; Kilic, G.S.; Al-Hendy, A. Estrogen receptors and signaling in fibroids: Role in pathobiology and therapeutic implications. *Reprod. Sci.* **2017**, *24*, 1235–1244. [CrossRef] [PubMed]

148. Deng, L.; Wu, T.; Chen, X.Y.; Xie, L.; Yang, J. Selective estrogen receptor modulators (SERMs) for uterine leiomyomas. *Cochrane Database Syst. Rev.* **2012**, *10*, CD005287. [CrossRef] [PubMed]

149. Donnez, J.; Tomaszewski, J.; Vazquez, F.; Bouchard, P.; Lemieszczuk, B.; Baro, F.; Nouri, K.; Selvaggi, L.; Sodowski, K.; Bestel, E.; et al. Ulipristal acetate versus leuprolide acetate for uterine fibroids. *N. Engl. J. Med.* **2012**, *366*, 421–432. [CrossRef] [PubMed]

150. Ali, M.; Chaudhry, Z.T.; Al-Hendy, A. Successes and failures of uterine leiomyoma drug discovery. *Expert Opin. Drug Discov.* **2018**, *13*, 169–177. [CrossRef] [PubMed]

151. Powell, M.; Dutta, D. Esmya((r)) and the pearl studies: A review. *Womens Health (Lond.)* **2016**, *12*, 544–548. [CrossRef] [PubMed]

152. Murji, A.; Whitaker, L.; Chow, T.L.; Sobel, M.L. Selective progesterone receptor modulators (sprms) for uterine fibroids. *Cochrane Database Syst. Rev.* **2017**, *4*, CD010770. [CrossRef] [PubMed]

153. Flyckt, R.; Coyne, K.; Falcone, T. Minimally invasive myomectomy. *Clin. Obstet. Gynecol.* **2017**, *60*, 252–272. [CrossRef] [PubMed]

154. Bretschneider, C.E.; Jallad, K.; Paraiso, M.F.R. Minimally invasive hysterectomy for benign indications: An update. *Minerva Ginecol.* **2017**, *69*, 295–303. [CrossRef] [PubMed]

155. Cui, R.R.; Wright, J.D. Risk of occult uterine sarcoma in presumed uterine fibroids. *Clin. Obstet. Gynecol.* **2016**, *59*, 103–118. [CrossRef] [PubMed]

156. Sagae, S.; Yamashita, K.; Ishioka, S.; Nishioka, Y.; Terasawa, K.; Mori, M.; Yamashiro, K.; Kanemoto, T.; Kudo, R. Preoperative diagnosis and treatment results in 106 patients with uterine sarcoma in hokkaido, japan. *Oncology* **2004**, *67*, 33–39. [CrossRef] [PubMed]

157. Nagai, T.; Takai, Y.; Akahori, T.; Ishida, H.; Hanaoka, T.; Uotani, T.; Sato, S.; Matsunaga, S.; Baba, K.; Seki, H. Novel uterine sarcoma preoperative diagnosis score predicts the need for surgery in patients presenting with a uterine mass. *Springerplus* **2014**, *3*, 678. [CrossRef] [PubMed]

158. US Food and Drug Administration. Laparoscopic Uterine Power Morcellation in Hysterectomy and Myomectomy: FDA Safety Communication. Available online: https://wayback.archive-it.org/7993/20170406071822/https:/www.fda.gov/MedicalDevices/Safety/AlertsandNotices/ucm393576.htm (accessed on 8 October 2018).

159. US Food and Drug Administration. FDA Updated Assessment of the Use of Laparoscopic Power Morcellators to Treat Uterine Fibroids. Available online: https://www.fda.gov/downloads/MedicalDevices/ProductsandMedicalProcedures/SurgeryandLifeSupport/UCM584539 (accessed on 8 October 2018).

160. Shikeeva, A.A.; Kekeeva, T.V.; Zavalishina, L.E.; Andreeva, I.; Frank, G.A. [The loss of heterozygosity and microsatellite instability analysis in differential diagnostics of leiomyosarcoma and proliferative leiomyoma of the uterus]. *Arkh. Patol.* **2011**, *73*, 47–50. [PubMed]

161. Amada, S.; Nakano, H.; Tsuneyoshi, M. Leiomyosarcoma versus bizarre and cellular leiomyomas of the uterus: A comparative study based on the MIB-1 and proliferating cell nuclear antigen indices, p53 expression, DNA flow cytometry, and muscle specific actins. *Int. J. Gynecol. Pathol.* **1995**, *14*, 134–142. [CrossRef] [PubMed]

International Journal of
Molecular Sciences

MDPI

Communication

Renal Injury during Long-Term Crizotinib Therapy

Taro Yasuma [1,2], Tetsu Kobayashi [3], Corina N. D'Alessandro-Gabazza [1,*], Hajime Fujimoto [3], Kentaro Ito [4], Yoichi Nishii [4], Kota Nishihama [2], Prince Baffour Tonto [1], Atsuro Takeshita [1,2], Masaaki Toda [1], Esteban C. Gabazza [1], Osamu Taguchi [4], Shigenori Yonemura [5] and Osamu Hataji [4]

[1] Department of Immunology, Faculty and Graduate School of Medicine, Mie University, Edobashi 2-174, Tsu, Mie 514-8507, Japan; t-yasuma0630@clin.medic.mie-u.ac.jp (T.Y.); 316MS02@m.mie-u.ac.jp (P.B.T.); johnpaul0114@yahoo.co.jp (A.T.); t-masa@doc.medic.mie-u.ac.jp (M.T.); gabazza@doc.medic.mie-u.ac.jp (E.C.G.)
[2] Department of Diabetes and Endocrinology, Faculty and Graduate School of Medicine, Mie University, Edobashi 2-174, Tsu, Mie 514-8507, Japan; kn2480@gmail.com
[3] Department of Pulmonary and Critical Care Medicine, Faculty and Graduate School of Medicine, Mie University, Edobashi 2-174, Tsu, Mie 514-8507, Japan; kobayashitetsu@hotmail.com (T.K.); genfujimoto1974@yahoo.co.jp (H.F.)
[4] Respiratory Center, Matsusaka Municipal Hospital, Tonomachi 1550, Matsusaka, Mie 515-8544, Japan; kentarou_i_0214@yahoo.co.jp (K.I.); mchnishii@city-hosp.matsusaka.mie.jp (Y.N.); taguchio@clin.medic.mie-u.ac.jp (O.T.); mch1031@city-hosp.matsusaka.mie.jp (O.H.)
[5] Department of Nephrology, Matsusaka Municipal Hospital, Tonomachi 1550, Matsusaka, Mie 515-8544, Japan; amikurumika@yahoo.co.jp
*   Correspondence: immunol@doc.medic.mie-u.ac.jp; Tel.: +81-59-231-5037; Fax: +81-59-231-5225

Received: 20 August 2018; Accepted: 11 September 2018; Published: 25 September 2018

Abstract: Crizotinib is highly effective against anaplastic lymphoma kinase-positive and c-ros oncogen1-positive non-small cell lung cancer. Renal dysfunction is associated with crizotinib therapy but the mechanism is unknown. Here, we report a case of anaplastic lymphoma kinase positive non-small cell lung cancer showing multiple cysts and dysfunction of the kidneys during crizotinib administration. We also present results demonstrating that long-term crizotinib treatment induces fibrosis and dysfunction of the kidneys by activating the tumor necrosis factor-α/nuclear factor-κB signaling pathway. In conclusion, this study shows the renal detrimental effects of crizotinib, suggesting the need of careful monitoring of renal function during crizotinib therapy.

Keywords: lung cancer; renal injury; fibrosis; crizotinib; anaplastic lymphoma kinase; cystic formation

1. Introduction

Crizotinib is a selective inhibitor of several receptor tyrosine kinases including anaplastic lymphoma kinase (ALK), hepatocyte growth factor receptor (HGF receptor, proto-oncogene c-Met) and c-ros oncogene 1 (ROS1) [1,2]. *ALK* rearrangements are found in approximately 5% of patients with non–small cell lung cancer (NSCLC) [3–5]. Crizotinib is highly effective against *ALK*-positive and *ROS1*-positive NSCLC and its clinical use has been approved in many countries [6–8]. The Food and Drug Administration of United States approved the clinical use of crizotinib in 2011 for *ALK*-positive NSCLC and in 2016 for *ROS1*-positive NSCLC. However, several adverse effects such as gastrointestinal complaints, visual disturbances and interstitial lung disease have been reported during clinical trials of crizotinib [8,9]. Renal cysts and functional impairment of the kidneys have also been reported in patients treated with crizotinib [10–14]. However, the underlying mechanism of renal complication associated with crizotinib therapy is unknown.

In this study, we reported a case of *ALK*-positive NSCLC with multiple renal cysts and renal dysfunction during crizotinib therapy and described the functional and pathological changes observed after long-term administration of crizotinib in an experimental mouse model.

2. Results

2.1. Case Report

A 71-year-old woman consulted the Respiratory Center of Matsusaka Municipal Hospital. The patient was being treated with amlodipine because of arterial hypertension. Lung adenocarcinoma with *ALK* arrangement was diagnosed based on clinical and pathological findings. Therapy with crizotinib (500 mg/day) was associated with marked tumor shrinkage and clinical improvement (Figure 1A–C). Parameters of kidney function were normal before the initiation of crizotinib. Three weeks following crizotinib administration, the blood level of creatinine increased from 0.73 mg/dL (pre-treatment value) to 1.21 mg/dL and remained at similar levels thereafter, but there were no abnormal findings in the kidneys upon computed tomography CT (Figure 1D). Eleven months after starting crizotinib treatment, the blood level of creatinine increased further (1.68 mg/dL) and multiple (>3) renal cysts were detected by CT examination (Figure 1E). Multiseptated renal cysts were detected by CT thirteen months after initiation of crizotinib (Figure 1F). Ultrasound study showed cystic formations, normal renal size and normal blood flow in the kidneys. Laboratory analysis of the cream-colored liquid obtained by ultrasound-guided cyst aspiration showed no cancer cells and microbial culture was negative. Urine analysis showed a mild proteinuria. Crizotinib was stopped and alectinib was started instead for the control of lung tumor. The blood level of creatinine decreased to 0.86 mg/dL after three weeks and the renal cysts regressed after three months of crizotinib withdrawal (Figure 1G).

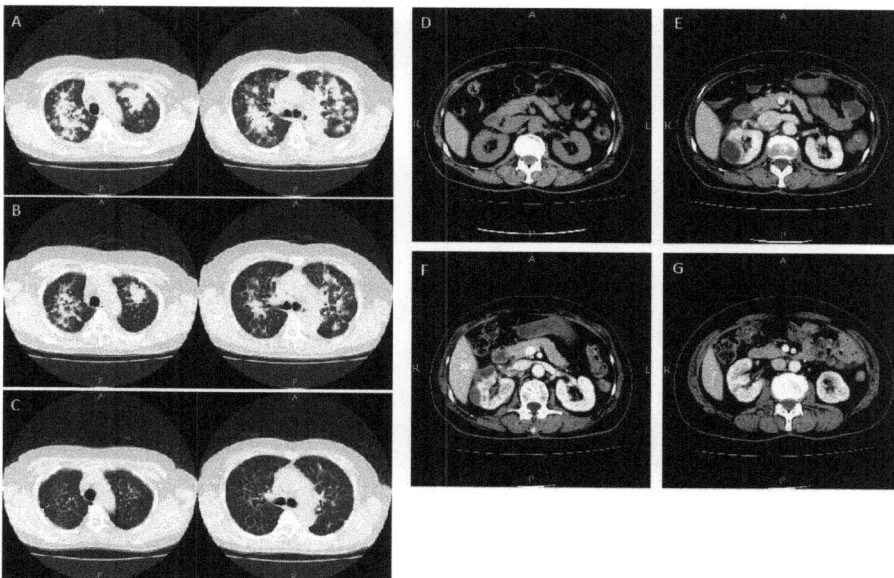

Figure 1. Chest and abdominal computed tomography (CT) in the present case. Chest CT of the patient with *ALK*-positive NSCLC at diagnosis (**A**), after 1 month (**B**), and after 11 months (**C**) of crizotinib therapy. Abdominal CT of the patient before therapy with crizotinib (**D**), 11 months after crizotinib therapy (**E**), 13 months (**F**) after crizotinib therapy, and after stopping the drug (**G**). NSCLC: non–small cell lung cancer; *ALK*: anaplastic lymphoma kinase.

2.2. Experimental Animal Model

Pre-existing renal cysts enlarged during crizotinib administration in mice.

We performed an in vivo experiment to evaluate the long-term effect of crizotinib on renal function and pathology. Mice were allocated to a control group and a crizotinib-treated group. Mice of the crizotinib-treated group received long-term crizotinib administration. To assess the renal morphological changes, we performed micro-CT of kidneys before and after crizotinib treatment. Upon micro-CT scanning, one mouse showed a preexisting cystic lesion that enlarged during crizotinib administration (Figure 2A). The volume of the cyst as evaluated by contrast computed tomography increased from 0.51 cm^3 before treatment to 0.72 cm^3 after treatment. Periodic acid–Schiff staining of the cyst showed empty cysts with compressed renal parenchymal structures (Figure 2B). Apart from this mouse, no other mouse showed cystic formation in the kidneys before or after crizotinib treatment (Figure 2C).

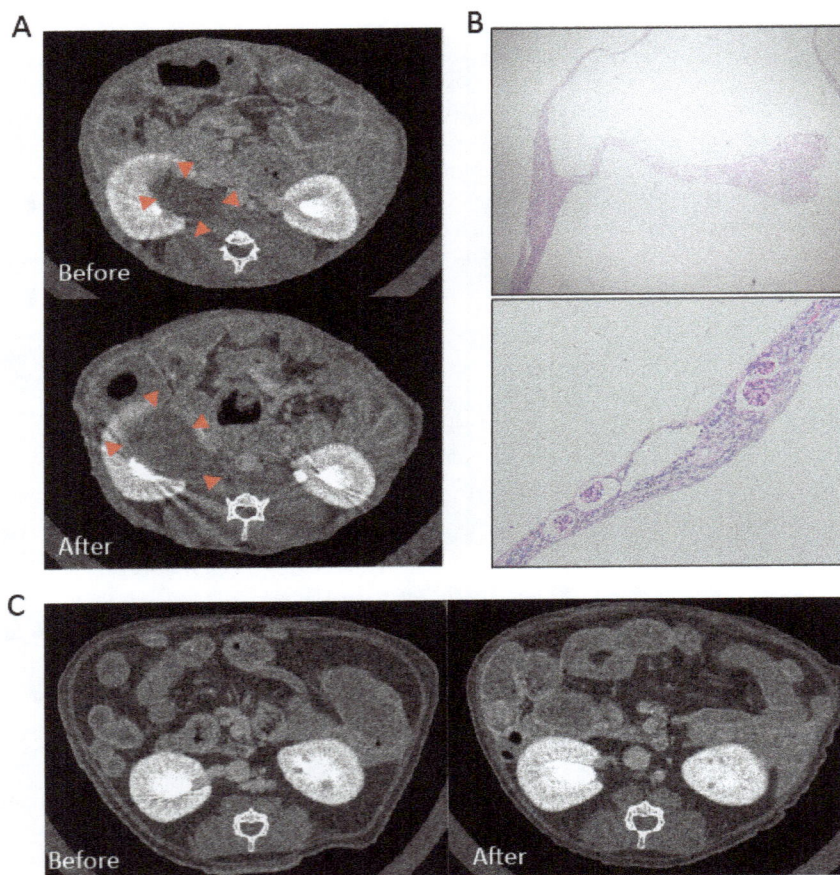

Figure 2. Enlargement of pre-existing cyst after crizotinib administration. Micro-CT images of a mouse show enlargement of kidney cyst after crizotinib treatment (**A**) red arrowheads. Periodic acid–Schiff staining showed compressed glomeruli and tubules (**B**) upper panel at ×40 and lower panel at ×100 magnification. The micro-CT of other mice with no pre-existing cyst show no cystic formation after crizotinib (**C**).

2.2.1. Mesangial Expansion in Crizotinib-Treated Mouse

Enhanced deposition of periodic acid-Schiff (PAS) positive substances was observed in glomeruli from the crizotinib-treated mice compared to untreated control mice (Figure 3A).

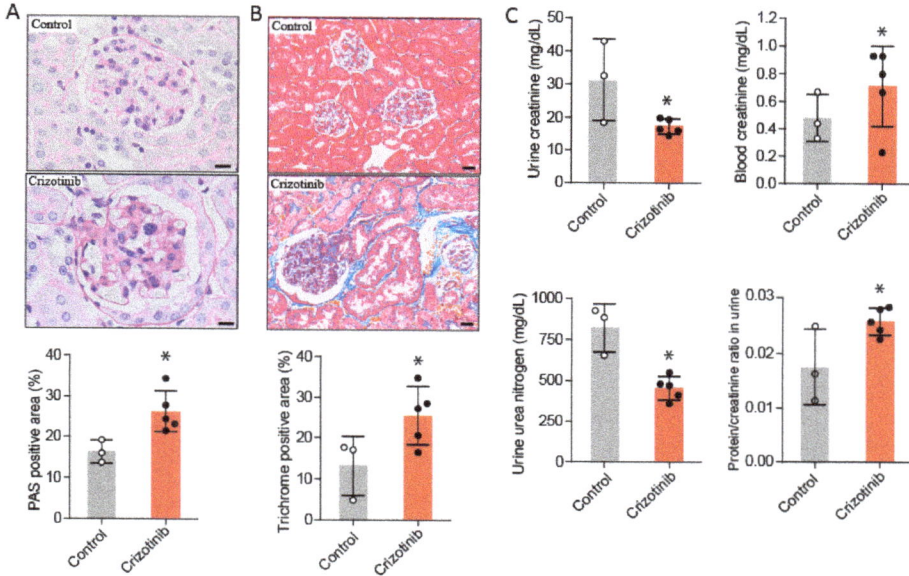

Figure 3. Renal fibrosis and increased markers of renal failure after crizotinib administration. Periodic acid-Schiff staining shows glomerular mesangial expansion in mice treated with crizotinib compared to untreated mice (**A**), scale bar indicates 10 μm. Masson's trichrome staining shows increased glomerular and interstitial fibrosis in the kidneys from mice treated with crizotinib compared to those from untreated mice (**B**), scale bar indicates 20 μm. The blood level of creatinine, urine levels of urea nitrogen and creatinine and the ratio of total protein to creatinine in urine were significantly different between control and crizotinib groups (**C**). Data are mean ± SD. Control group $n = 3$; crizotinib group $n = 5$. * $p < 0.05$ versus control group.

2.2.2. Crizotinib Caused Renal Histopathological Changes

Increased staining for collagen in glomerular and renal interstitial areas was observed in mice treated with crizotinib compared to untreated mice (Figure 3B).

2.2.3. Crizotinib Impaired Renal Function

Compared to the control mice, the plasma concentration of creatinine and the ratio of urine total protein to creatinine were significantly increased in the crizotinib-treated mice. Furthermore, urine concentration of creatinine and urea nitrogen were significantly decreased in the crizotinib-treated mice compared to control mice (Figure 3C).

2.3. Crizotinib Associated with Enhanced Inflammatory Markers in the Kidneys

The relative mRNA expressions of IL-6, TNFα, TGFβ1, MMP2, and collagen I were increased in mice treated with crizotinib compared to untreated mice (Figure 4A). The plasma concentrations of IL-6 and HGF, and the kidney tissue levels of TNFα were also significantly increased in mice treated with crizotinib compared to control mice (Figure 4A).

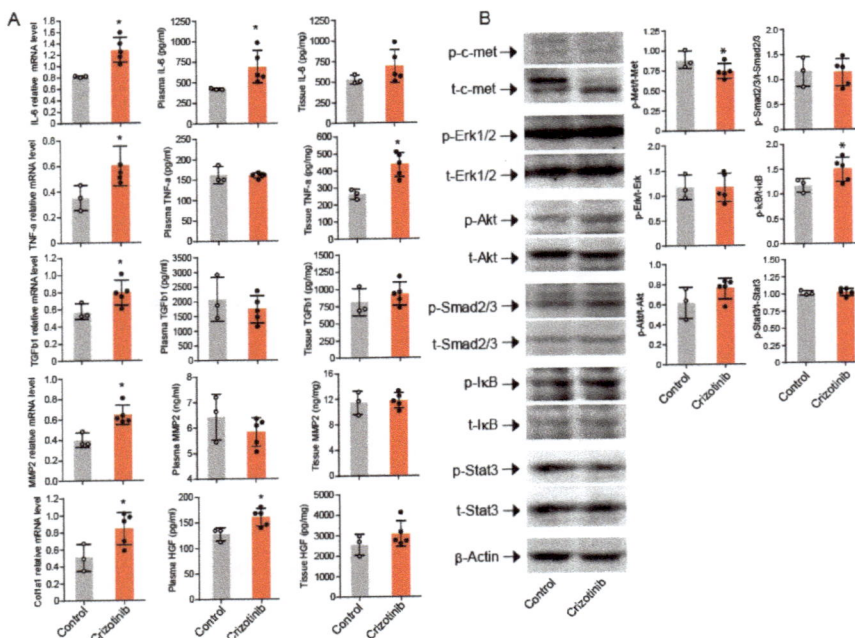

Figure 4. Cytokines, proteases and signal pathways after crizotinib administration. Increased mRNA expression of Col1a1, TGFβ1, IL-6, TNFα, and MMP2 in mice treated with crizotinib compared to untreated mice (**A**). Significant difference in phosphorylation level of c-Met and IκB between mice treated with and without crizotinib (**B**). Data are mean ± SD. Control group n = 3; crizotinib group n = 5. * p < 0.05 versus control group.

2.4. Activation of NF-κB in the Kidneys after Crizotinib Therapy

As expected, c-Met activation was significantly decreased in the kidneys from mice treated with crizotinib compared to untreated mice (Figure 4B). Phosphorylated IκB was significantly increased in mice treated with crizotinib compared to untreated counterparts, but there was no significant difference in phosphorylation of Erk, Akt, Smad2/3, or Stat3 between treated and untreated mice (Figure 4B).

3. Discussion

The development of complex renal cysts associated with crizotinib treatment has been previously documented [10–14]. In a retrospective study among thirty-two Taiwanese patients with ALK-positive NSCLC treated with crizotinib, seven patients presented renal cysts that regressed after drug withdrawal [12]. In another retrospective analysis among seventeen patients with renal cysts associated with crizotinib treatment, seven patients showed compression of adjacent structures by cystic growth although the majority of patients were asymptomatic [13]. The evolution pattern of renal cysts during crizotinib treatment is variable but most renal cysts are asymptomatic, enlarge or spontaneously regress without crizotinib withdrawal [12,15–17]. In some instances the cysts regress after drug discontinuation [18]. Here, we also showed a case of ALK-positive non-small cell lung cancer with multiple renal cysts that developed during crizotinib administration. Although this case report is not the first, it is presented here to further illustrate the relevance of this treatment-related adverse effect in clinical practice and to emphasize the urgent need to clarify the mechanistic pathway.

The mechanistic pathways leading to cystic formation and renal dysfunction during crizotinib therapy remain unknown. A previous study showed that hepatocyte growth factor (HGF) and its receptor c-Met promote cystogenesis [19]. HGF-mediated activation of Mapk/Erk and/or Stat3 appears to be an important mediator of cystic formation [20–22]. Crizotinib inhibits c-Met and thus the involvement of c-Met in drug action would be paradoxical. Here, we confirmed inhibition of c-Met by crizotinib, but found no significant activation of Mapk/Erk or Stat3 signal pathway in mice treated with crizotinib. This suggests that alternative mechanisms may be involved in crizotinib-mediated cystogenesis in the kidneys. It is worth noting that, in the current report, the renal cyst of the patient worsened during crizotinib therapy and that the pre-existing renal cyst of a mouse enlarged after long-term administration of the drug. Based on these findings, it is reasonable to speculate that renal cysts develop only in subjects with pre-existing renal cysts that subsequently enlarge and become radiologically detectable during crizotinib administration. In our study, mice receiving crizotinib therapy showed no new development of cysts. The lack of cystic formation in our present experimental mouse model may be explained by too little exposure to the drug or the absence of a species-dependent propensity for developing the disease.

Patients with NSCLC may also have lower estimated glomerular filtration rates or dysfunction of the kidneys during crizotinib therapy that improves after drug withdrawal [23–25]. Patients with NSCLC and renal dysfunction before crizotinib administration have impaired renal function if they are treated with crizotinib [26]. The cause of the renal dysfunction is unknown. Renal biopsy performed in one case during the acute phase of the kidney dysfunction disclosed histopathological findings of mesangiolysis and acute tubular necrosis, but there is no report of biopsy findings in the chronic phase of the disease [25]. Our results are consistent with these observations. Here, we showed that mice receiving crizotinib over the long-term have renal dysfunction as demonstrated by high blood levels of creatinine, elevated urine total protein to creatinine ratio, as well as low levels of urine creatinine and urine urea nitrogen. Interestingly, PCR analysis showed high mRNA expression of collagen I and the histopathological study disclosed abnormal glomerular mesangial expansion and increased interstitial collagen deposition in the kidneys of mice treated with crizotinib compared to untreated counterparts, suggesting a pro-fibrotic activity of crizotinib in the kidneys.

Fibrosis and impaired dysfunction of the kidneys during crizotinib therapy may be explained by blockade of the protective and anti-fibrotic activity of the HGF/c-Met pathway in the kidneys. The HGF/c-Met signaling pathway is known to promote: (1) inhibition of renal interstitial myofibroblasts by intercepting Smad2/3 signal transduction; (2) reduction of TGFβ1-mediated proliferation; (3) differentiation and secretory activity of fibroblasts; and (4) amelioration of podocyte injury and proteinuria [27–33]. Here we found no changes in Smad2/3 phosphorylation in mice treated with crizotinib compared to untreated mice. In addition, Akt phosphorylation, which may promote fibrosis by inhibiting apoptosis of myofibroblasts, remained unaffected in the kidneys after crizotinib administration [34]. However, we found increased activation of the NF-κB pathway as demonstrated by the increased p-IκB/t-IκB ratio in association with decreased c-Met phosphorylation, as well as increased levels of TNFα and IL-6 in crizotinib-treated mice compared to control counterparts. TNF family cytokines can activate the NF-κB signaling pathway and NF-κB activation can induce collagen expression and cause tissue fibrosis in a TGFβ-independent fashion [35–37]. The NF-κB pathway may also cause tissue fibrosis by promoting differentiation of epithelial cells to fibroblasts [38,39]. The HGF/c-Met axis has been reported to decrease activation of NF-κB [40–42]. Therefore, it is conceivable that inhibition of c-Met by crizotinib causes renal injury and subsequent fibrosis by triggering activation of the NF-κB pathway [43]. However, the potential role of other TGFβ/Smad-independent pathways in crizotinib-associated renal fibrosis should also be evaluated in future studies [44].

The report of only one case, the small number of mice used in the experimental study and the fact that de novo cystic formation was not observed after crizotinib administration are limitations of the present study.

In brief, this study provides new evidence on possible mechanistic pathways causing morphological abnormalities and dysfunction of the kidneys in patients with lung cancer treated with crizotinib.

4. Materials and Methods

4.1. Experimental Animal Model

Wild-type C57BL/6 male mice (8 to 10 weeks old) weighing 19 to 22 g were used in the experiment. Mice were maintained in a specific pathogen-free environment under a 12 h light/dark cycle in the animal house of Mie University. Mice were allocated to a control group ($n = 3$) and a crizotinib-treated group ($n = 5$). The dose of crizotinib prescribed to patients with cancer is usually 400 to 500 mg (6–7 mg/kg) per day and therapy is generally continued for several months or years as long as the drug is beneficial to the patient [7]. In experimental animals, crizotinib has shown effective anti-tumor activity at doses of 10, 25 or 100 mg/kg/day after 4 or more weeks of treatment [2,45,46]. In the present study, to ensure chronic exposure to the drug, we treated mice with crizotinib by oral administration at a dose of 25 mg/kg/day for a period of 50 days. Mice were sacrificed on day 51 after the initial treatment.

4.2. Micro CT of Kidneys

Contrast-enhanced micro-CT of kidneys was performed with an X-ray CT system (Latheta LCT-200, Hitachi Aloka Medical Ltd., Tokyo, Japan) before crizotinib treatment started and 47 days after crizotinib treatment began. Under anesthesia with isoflurane, mice received an intravenous infusion of Iohexol (Daiichi-Sankyou, Tokyo, Japan), an iodine contrast medium, at a dose of 10 mL/kg before CT scanning. CT scanning was performed under conditions previously described [47]. Quantitative assessment of the lesion area was performed using the La Theta software version 3.30 (Hitachi-Aloka Medical Ltd., Tokyo, Japan).

4.3. Mouse Sacrifice and Sampling

Euthanasia of the experimental animals was performed using an overdose of intraperitoneal pentobarbital. Samples for biochemical and histological examinations were subsequently taken. Blood sampling was carried out by closed-chest heart puncture and samples were collected in tubes containing 10 U/mL heparin. Urine spot collection was also done for biochemical analysis.

4.4. Biochemical Analysis

Plasma and urine creatinine levels were measured by Jaffe's reaction (Creatinine Companion Kit; Exocell, Philadelphia, PA, USA) and the concentration of total protein was measured using a dye-binding assay (BCA^{TM} protein assay kit; Pierce, Rockford, IL, USA). Urea nitrogen was measured by colorimetric method (NCal^{TM} NIST-Calibrated Kit; Arbor Assays, Ann Arbor, MI, USA) according to the manufacturer's instructions. The concentrations of interleukin (IL)-6 and tumor necrosis factor (TNF)-α were measured using enzyme immunoassay kits from BD Biosciences (Tokyo, Japan). The concentrations of transforming growth factor (TGF)-β1, metalloproteinase (MMP)-2 and hepatocyte growth factor (HGF) were measured using a commercial enzyme immunoassay kit from R&D System (Minneapolis, MN, USA).

4.5. Tissue Preparation and Staining

Kidneys were dissected, dehydrated, embedded in paraffin, cut into 3-μm-thick sections and prepared for periodic acid-Schiff (PAS) and Masson's trichrome staining. An investigator blinded to the treatment group calculated the areas of glomeruli (>30 per mouse) stained positive for PAS or trichrome using an Olympus BX50 microscope with a plan objective, combined with an Olympus DP70 digital camera (Tokyo, Japan) and WinROOF image processing software (Mitani Corp., Fukui, Japan).

4.6. Western Blotting

Kidney tissues were homogenized in a radioimmunoprecipitation assay buffer with protease inhibitors and then centrifuged at 14,000 rpm for 30 min at 4 °C to remove debris. Protein concentration was measured by the bicinchoninic acid method. Protein extract (10 µg) was resolved using sodium dodecyl sulfate polyacrylamide gel electrophoresis, transferred to a polyvinylidene difluoride membrane and blocked using 5% non-fat milk in Tris-buffered saline with 0.1% Tween-20. After blocking, the membranes were washed and then incubated overnight with the primary antibody at 4 °C. After three washes, the membranes were incubated with horseradish peroxidase-conjugated secondary antibody for 2 h, washed again and then incubated with enhanced chemiluminescence solution. The fluorescent intensity of signals was quantified using ImageJ software (National Institutes of Health, Bethesda, MD, USA). Supplemental Table S1 describes antibodies used in the study.

4.7. Reverse Transcription Polymerase Chain Reaction

Total RNA was extracted from kidneys using Sepasol RNA I super G (Nacalai). All RNA samples had a 260/280 nm ratio between 1.8 and 2.0. Reverse transcription was performed with oligo-dT primers, and the DNA was then amplified by PCR. Supplemental Table S2 describes the sequences of the primers. The PCR products were separated on a 1.5% agarose gel containing 0.01% ethidium bromide, and the intensity of the stained bands was quantitated with ImageJ software (National Institutes of Health, Bethesda, MD, USA). The amount of mRNA was normalized to the expression of glyceraldehyde-3-phosphate dehydrogenase.

4.8. Ethical Statement

The Mie University Committee for Animal Investigation approved the protocol of the study (Approval number 29–23; date: 15 January 2018) and the experimental procedures were performed following the institutional guidelines and internationally approved principles of laboratory animal care (NIH publication no. 85–23, revised 1985; http://grants1.nih.gov/grants/olaw/references/phspol. htm). Written informed consent was obtained from the patient.

4.9. Statistical Analysis

Data are expressed as the mean ± standard deviation (SD). The statistical difference between variables was calculated by Student t-test. Statistical analyses were done using the GraphPad Prism package software for Windows (GraphPad Software Inc., La Jolla, CA, USA). Statistical significance was considered as $p < 0.05$.

Supplementary Materials: Supplementary materials can be found at http://www.mdpi.com/1422-0067/19/10/2902/s1.

Author Contributions: Conceptualization, T.K., S.Y., E.C.G.; formal analysis, O.T., K.N.; investigation, T.Y., C.N.D.-G., P.B.T., Y.N., M.T.; methodology, A.T.; resources, H.F., K.I., O.H.

Funding: This work was financially supported in part by a grant from the Ministry of Education, Culture, Sports, Science, and Technology of Japan (Kakenhi No. 15K09170).

Conflicts of Interest: The authors report no declarations of interest regarding data reported in this manuscript.

References

1. Christensen, J.G.; Zou, H.Y.; Arango, M.E.; Li, Q.; Lee, J.H.; McDonnell, S.R.; Yamazaki, S.; Alton, G.R.; Mroczkowski, B.; Los, G. Cytoreductive antitumor activity of PF-2341066, a novel inhibitor of anaplastic lymphoma kinase and c-Met, in experimental models of anaplastic large-cell lymphoma. *Mol. Cancer Ther.* **2007**, *6*, 3314–3322. [CrossRef] [PubMed]
2. Zou, H.Y.; Li, Q.; Lee, J.H.; Arango, M.E.; McDonnell, S.R.; Yamazaki, S.; Koudriakova, T.B.; Alton, G.; Cui, J.J.; Kung, P.P.; et al. An orally available small-molecule inhibitor of c-Met, PF-2341066, exhibits cytoreductive antitumor efficacy through antiproliferative and antiangiogenic mechanisms. *Cancer Res.* **2007**, *67*, 4408–4417. [CrossRef] [PubMed]

3. Blackhall, F.H.; Peters, S.; Bubendorf, L.; Dafni, U.; Kerr, K.M.; Hager, H.; Soltermann, A.; O'Byrne, K.J.; Dooms, C.; Sejda, A.; et al. Prevalence and clinical outcomes for patients with ALK-positive resected stage I to III adenocarcinoma: Results from the European Thoracic Oncology Platform Lungscape Project. *J. Clin. Oncol.* **2014**, *32*, 2780–2787. [CrossRef] [PubMed]

4. Rikova, K.; Guo, A.; Zeng, Q.; Possemato, A.; Yu, J.; Haack, H.; Nardone, J.; Lee, K.; Reeves, C.; Li, Y.; et al. Global survey of phosphotyrosine signaling identifies oncogenic kinases in lung cancer. *Cell* **2007**, *131*, 1190–1203. [CrossRef] [PubMed]

5. Soda, M.; Choi, Y.L.; Enomoto, M.; Takada, S.; Yamashita, Y.; Ishikawa, S.; Fujiwara, S.; Watanabe, H.; Kurashina, K.; Hatanaka, H.; et al. Identification of the transforming EML4-ALK fusion gene in non-small-cell lung cancer. *Nature* **2007**, *448*, 561–566. [CrossRef] [PubMed]

6. Facchinetti, F.; Rossi, G.; Bria, E.; Soria, J.C.; Besse, B.; Minari, R.; Friboulet, L.; Tiseo, M. Oncogene addiction in non-small cell lung cancer: Focus on ROS1 inhibition. *Cancer Treat. Rev.* **2017**, *55*, 83–95. [CrossRef] [PubMed]

7. Hanna, N.; Johnson, D.; Temin, S.; Baker, S., Jr.; Brahmer, J.; Ellis, P.M.; Giaccone, G.; Hesketh, P.J.; Jaiyesimi, I.; Leighl, N.B.; et al. Systemic Therapy for Stage IV Non-Small-Cell Lung Cancer: American Society of Clinical Oncology Clinical Practice Guideline Update. *J. Clin. Oncol.* **2017**, *35*, 3484–3515. [CrossRef] [PubMed]

8. Shaw, A.T.; Kim, D.W.; Nakagawa, K.; Seto, T.; Crino, L.; Ahn, M.J.; de Pas, T.; Besse, B.; Solomon, B.J.; Blackhall, F.; et al. Crizotinib versus chemotherapy in advanced ALK-positive lung cancer. *N. Engl. J. Med.* **2013**, *368*, 2385–2394. [CrossRef] [PubMed]

9. Camidge, D.R.; Bang, Y.J.; Kwak, E.L.; Iafrate, A.J.; Varella-Garcia, M.; Fox, S.B.; Riely, G.J.; Solomon, B.; Ou, S.H.; Kim, D.W.; et al. Activity and safety of crizotinib in patients with ALK-positive non-small-cell lung cancer: Updated results from a phase 1 study. *Lancet Oncol.* **2012**, *13*, 1011–1019. [CrossRef]

10. Di Girolamo, M.; Paris, I.; Carbonetti, F.; Onesti, E.C.; Socciarelli, F.; Marchetti, P. Widespread renal polycystosis induced by crizotinib. *Tumori* **2015**, *101*, e128–e131. [CrossRef] [PubMed]

11. Heigener, D.F.; Reck, M. Crizotinib. *Recent Results Cancer Res.* **2014**, *201*, 197–205. [PubMed]

12. Lin, Y.T.; Wang, Y.F.; Yang, J.C.; Yu, C.J.; Wu, S.G.; Shih, J.Y.; Yang, P.C. Development of renal cysts after crizotinib treatment in advanced ALK-positive non-small-cell lung cancer. *J. Thorac. Oncol.* **2014**, *9*, 1720–1725. [CrossRef] [PubMed]

13. Schnell, P.; Bartlett, C.H.; Solomon, B.J.; Tassell, V.; Shaw, A.T.; de Pas, T.; Lee, S.H.; Lee, G.K.; Tanaka, K.; Tan, W.; et al. Complex renal cysts associated with crizotinib treatment. *Cancer Med.* **2015**, *4*, 887–896. [CrossRef] [PubMed]

14. Souteyrand, P.; Burtey, S.; Barlesi, F. Multicystic kidney disease: A complication of crizotinib. *Diagn. Interv. Imaging* **2015**, *96*, 393–395. [CrossRef] [PubMed]

15. Cameron, L.B.; Jiang, D.H.; Moodie, K.; Mitchell, C.; Solomon, B.; Parameswaran, B.K. Crizotinib Associated Renal Cysts [CARCs]: Incidence and patterns of evolution. *Cancer Imaging* **2017**, *17*, 7. [CrossRef] [PubMed]

16. Halpenny, D.F.; McEvoy, S.; Li, A.; Hayan, S.; Capanu, M.; Zheng, J.; Riely, G.; Ginsberg, M.S. Renal cyst formation in patients treated with crizotinib for non-small cell lung cancer-Incidence, radiological features and clinical characteristics. *Lung Cancer* **2017**, *106*, 33–36. [CrossRef] [PubMed]

17. Klempner, S.J.; Aubin, G.; Dash, A.; Ou, S.H. Spontaneous regression of crizotinib-associated complex renal cysts during continuous crizotinib treatment. *Oncologist* **2014**, *19*, 1008–1010. [CrossRef] [PubMed]

18. Taima, K.; Tanaka, H.; Tanaka, Y.; Itoga, M.; Takanashi, S.; Tasaka, S. Regression of Crizotinib-Associated Complex Cystic Lesions after Switching to Alectinib. *Intern. Med.* **2017**, *56*, 2321–2324. [CrossRef] [PubMed]

19. Horie, S.; Higashihara, E.; Nutahara, K.; Mikami, Y.; Okubo, A.; Kano, M.; Kawabe, K. Mediation of renal cyst formation by hepatocyte growth factor. *Lancet* **1994**, *344*, 789–791. [CrossRef]

20. Maeshima, A.; Zhang, Y.Q.; Furukawa, M.; Naruse, T.; Kojima, I. Hepatocyte growth factor induces branching tubulogenesis in MDCK cells by modulating the activin-follistatin system. *Kidney Int.* **2000**, *58*, 1511–1522. [CrossRef] [PubMed]

21. Weimbs, T.; Olsan, E.E.; Talbot, J.J. Regulation of STATs by polycystin-1 and their role in polycystic kidney disease. *JAKSTAT* **2013**, *2*, e23650. [CrossRef] [PubMed]

22. Weimbs, T.; Talbot, J.J. STAT3 Signaling in Polycystic Kidney Disease. *Drug Discov. Today Dis. Mech.* **2013**, *10*, e113–e118. [CrossRef] [PubMed]

23. Brosnan, E.M.; Weickhardt, A.J.; Lu, X.; Maxon, D.A.; Baron, A.E.; Chonchol, M.; Camidge, D.R. Drug-induced reduction in estimated glomerular filtration rate in patients with ALK-positive non-small cell lung cancer treated with the ALK inhibitor crizotinib. *Cancer* **2014**, *120*, 664–674. [CrossRef] [PubMed]

24. Camidge, D.R.; Brosnan, E.M.; DeSilva, C.; Koo, P.J.; Chonchol, M. Crizotinib effects on creatinine and non-creatinine-based measures of glomerular filtration rate. *J. Thorac. Oncol.* **2014**, *9*, 1634–1637. [CrossRef] [PubMed]

25. Gastaud, L.; Ambrosetti, D.; Otto, J.; Marquette, C.H.; Coutts, M.; Hofman, P.; Esnault, V.; Favre, G. Acute kidney injury following crizotinib administration for non-small-cell lung carcinoma. *Lung Cancer* **2013**, *82*, 362–364. [CrossRef] [PubMed]

26. Martin Martorell, P.; Huerta Alvaro, M.; Solis Salguero, M.A.; Insa Molla, A. Crizotinib and renal insufficiency: A case report and review of the literature. *Lung Cancer* **2014**, *84*, 310–313. [CrossRef] [PubMed]

27. Dai, C.; Saleem, M.A.; Holzman, L.B.; Mathieson, P.; Liu, Y. Hepatocyte growth factor signaling ameliorates podocyte injury and proteinuria. *Kidney Int.* **2010**, *77*, 962–973. [CrossRef] [PubMed]

28. Iekushi, K.; Taniyama, Y.; Azuma, J.; Sanada, F.; Kusunoki, H.; Yokoi, T.; Koibuchi, N.; Okayama, K.; Rakugi, H.; Morishita, R. Hepatocyte growth factor attenuates renal fibrosis through TGF-β1 suppression by apoptosis of myofibroblasts. *J. Hypertens.* **2010**, *28*, 2454–2461. [CrossRef] [PubMed]

29. Kwiecinski, M.; Noetel, A.; Elfimova, N.; Trebicka, J.; Schievenbusch, S.; Strack, I.; Molnar, L.; von Brandenstein, M.; Tox, U.; Nischt, R.; et al. Hepatocyte growth factor (HGF) inhibits collagen I and IV synthesis in hepatic stellate cells by miRNA-29 induction. *PLoS ONE* **2011**, *6*, e24568. [CrossRef] [PubMed]

30. Li, L.; He, D.; Yang, J.; Wang, X. Cordycepin inhibits renal interstitial myofibroblast activation probably by inducing hepatocyte growth factor expression. *J. Pharmacol. Sci.* **2011**, *117*, 286–294. [CrossRef] [PubMed]

31. Shukla, M.N.; Rose, J.L.; Ray, R.; Lathrop, K.L.; Ray, A.; Ray, P. Hepatocyte growth factor inhibits epithelial to myofibroblast transition in lung cells via Smad7. *Am. J. Respir. Cell Mol. Biol.* **2009**, *40*, 643–653. [CrossRef] [PubMed]

32. Yang, J.; Dai, C.; Liu, Y. Hepatocyte growth factor suppresses renal interstitial myofibroblast activation and intercepts Smad signal transduction. *Am. J. Pathol.* **2003**, *163*, 621–632. [CrossRef]

33. Yi, X.; Li, X.; Zhou, Y.; Ren, S.; Wan, W.; Feng, G.; Jiang, X. Hepatocyte growth factor regulates the TGF-β1-induced proliferation, differentiation and secretory function of cardiac fibroblasts. *Int. J. Mol. Med.* **2014**, *34*, 381–390. [CrossRef] [PubMed]

34. Kulasekaran, P.; Scavone, C.A.; Rogers, D.S.; Arenberg, D.A.; Thannickal, V.J.; Horowitz, J.C. Endothelin-1 and transforming growth factor-β1 independently induce fibroblast resistance to apoptosis via AKT activation. *Am. J. Respir. Cell Mol. Biol.* **2009**, *41*, 484–493. [CrossRef] [PubMed]

35. Hayden, M.S.; Ghosh, S. Regulation of NF-κB by TNF family cytokines. *Semin. Immunol.* **2014**, *26*, 253–266. [CrossRef] [PubMed]

36. Peng, Y.; Kim, J.M.; Park, H.S.; Yang, A.; Islam, C.; Lakatta, E.G.; Lin, L. AGE-RAGE signal generates a specific NF-κB RelA "barcode" that directs collagen I expression. *Sci. Rep.* **2016**, *6*, 18822. [CrossRef] [PubMed]

37. Urtasun, R.; Lopategi, A.; George, J.; Leung, T.M.; Lu, Y.; Wang, X.; Ge, X.; Fiel, M.I.; Nieto, N. Osteopontin, an oxidant stress sensitive cytokine, up-regulates collagen-I via integrin $\alpha_V \beta_3$ engagement and PI3K/pAkt/NFκB signaling. *Hepatology* **2012**, *55*, 594–608. [CrossRef] [PubMed]

38. Julien, S.; Puig, I.; Caretti, E.; Bonaventure, J.; Nelles, L.; van Roy, F.; Dargemont, C.; de Herreros, A.G.; Bellacosa, A.; Larue, L. Activation of NF-κB by Akt upregulates Snail expression and induces epithelium mesenchyme transition. *Oncogene* **2007**, *26*, 7445–7456. [CrossRef] [PubMed]

39. Liu, M.; Ning, X.; Li, R.; Yang, Z.; Yang, X.; Sun, S.; Qian, Q. Signalling pathways involved in hypoxia-induced renal fibrosis. *J. Cell. Mol. Med.* **2017**, *21*, 1248–1259. [CrossRef] [PubMed]

40. Bendinelli, P.; Matteucci, E.; Dogliotti, G.; Corsi, M.M.; Banfi, G.; Maroni, P.; Desiderio, M.A. Molecular basis of anti-inflammatory action of platelet-rich plasma on human chondrocytes: Mechanisms of NF-κB inhibition via HGF. *J. Cell. Physiol.* **2010**, *225*, 757–766. [CrossRef] [PubMed]

41. Romero-Vasquez, F.; Chavez, M.; Perez, M.; Arcaya, J.L.; Garcia, A.J.; Rincon, J.; Rodriguez-Iturbe, B. Overexpression of HGF transgene attenuates renal inflammatory mediators, Na⁺-ATPase activity and hypertension in spontaneously hypertensive rats. *Biochim. Biophys. Acta* **2012**, *1822*, 1590–1599. [CrossRef] [PubMed]

42. Tamada, S.; Asai, T.; Kuwabara, N.; Iwai, T.; Uchida, J.; Teramoto, K.; Kaneda, N.; Yukimura, T.; Komiya, T.; Nakatani, T.; et al. Molecular mechanisms and therapeutic strategies of chronic renal injury: The role of nuclear factor κB activation in the development of renal fibrosis. *J. Pharmacol. Sci.* **2006**, *100*, 17–21. [CrossRef] [PubMed]

43. Sattler, M.; Salgia, R. c-Met and hepatocyte growth factor: Potential as novel targets in cancer therapy. *Curr. Oncol. Rep.* **2007**, *9*, 102–108. [CrossRef] [PubMed]

44. Oga, T.; Matsuoka, T.; Yao, C.; Nonomura, K.; Kitaoka, S.; Sakata, D.; Kita, Y.; Tanizawa, K.; Taguchi, Y.; Chin, K.; et al. Prostaglandin F2α receptor signaling facilitates bleomycin-induced pulmonary fibrosis independently of transforming growth factor-β. *Nat. Med.* **2009**, *15*, 1426–1430. [CrossRef] [PubMed]

45. Gumusay, O.; Esendagli-Yilmaz, G.; Uner, A.; Cetin, B.; Buyukberber, S.; Benekli, M.; Ilhan, M.N.; Coskun, U.; Gulbahar, O.; Ozet, A. Crizotinib-induced toxicity in an experimental rat model. *Wien. Klin. Wochenschr.* **2016**, *128*, 435–441. [CrossRef] [PubMed]

46. Smith, M.A.; Licata, T.; Lakhani, A.; Garcia, M.V.; Schildhaus, H.U.; Vuaroqueaux, V.; Halmos, B.; Borczuk, A.C.; Chen, Y.A.; Creelan, B.C.; et al. MET-GRB2 Signaling-Associated Complexes Correlate with Oncogenic MET Signaling and Sensitivity to MET Kinase Inhibitors. *Clin. Cancer Res.* **2017**, *23*, 7084–7096. [CrossRef] [PubMed]

47. Urawa, M.; Kobayashi, T.; D'Alessandro-Gabazza, C.N.; Fujimoto, H.; Toda, M.; Roeen, Z.; Hinneh, J.A.; Yasuma, T.; Takei, Y.; Taguchi, O.; et al. Protein S is protective in pulmonary fibrosis. *J. Thromb. Haemost.* **2016**, *14*, 1588–1599. [CrossRef] [PubMed]

MDPI

St. Alban-Anlage 66

4052 Basel

Switzerland

Tel. +41 61 683 77 34

Fax +41 61 302 89 18

www.mdpi.com

International Journal of Molecular Sciences Editorial Office

E-mail: ijms@mdpi.com

www.mdpi.com/journal/ijms

www.ingramcontent.com/pod-product-compliance
Lightning Source LLC
Chambersburg PA
CBHW051711210326
41597CB00032B/5446